INTERVENTIONS, CONTROLS, AND APPLICATIONS IN OCCUPATIONAL ERGONOMICS

T0179191

THE OCCUPATIONAL ERGONOMICS HANDBOOK
SECOND EDITION

FUNDAMENTALS AND ASSESSMENT TOOLS FOR OCCUPATIONAL ERGONOMICS

INTERVENTIONS, CONTROLS, AND APPLICATIONS IN OCCUPATIONAL ERGONOMICS

THE OCCUPATIONAL ERGONOMICS HANDBOOK
SECOND EDITION

INTERVENTIONS, CONTROLS, AND APPLICATIONS IN OCCUPATIONAL ERGONOMICS

Edited by

William S. Marras
The Ohio State University
Columbus, Ohio, U.S.A.

Waldemar Karwowski
University of Louisville
Louisville, Kentucky, U.S.A.

CRC Press
Taylor & Francis Group
Boca Raton London New York

CRC Press is an imprint of the
Taylor & Francis Group, an **informa** business
A TAYLOR & FRANCIS BOOK

CRC Press
Taylor & Francis Group
6000 Broken Sound Parkway NW, Suite 300
Boca Raton, FL 33487-2742

First issued in paperback 2019

ISBN-13: 978-0-8493-1938-9 (hbk)
ISBN-13: 978-0-367-86418-7 (pbk)

Library of Congress Cataloging-in-Publication Data

Interventions, controls, and applications in occupational ergonomics / edited by William S. Marras and Waldemar Karwowski.
    p. cm.
  Includes bibliographical references and index.
  ISBN 0-8493-1938-2 (alk. paper)
  1. Human engineering--Handbooks, manuals, etc. 2. Industrial hygiene--Handbooks, manuals, etc. I. Marras, William S. (William Steven), date. II. Karwowski, Waldemar, date.

TA166.I5685 2005
620.8'2--dc22
                                           2005052860

Visit the Taylor & Francis Web site at
http://www.taylorandfrancis.com

and the CRC Press Web site at
http://www.crcpress.com

Preface

Development of the 2nd edition of the *Occupational Ergonomics Handbook* was motivated by our desire to facilitate a wide application of ergonomics knowledge to work systems design, testing and evaluation in order to improve the quality of life for millions of workers around the world. Ergonomics (or human factors) is defined by the International Ergonomics Association (www.iea.cc) as the scientific discipline concerned with the understanding of interactions among humans and other elements of a system, and the profession that applies theory, principles, data, and methods to design in order to optimize human well-being and overall system performance. Ergonomists contribute to the design and evaluation of tasks, jobs, products, environments, and systems in order to make them compatible with the needs, abilities, and limitations of people.

The ergonomics discipline promotes a holistic approach to the design of work systems with due consideration of the physical, cognitive, social, organizational, environmental, and other relevant factors. The application of ergonomics knowledge should help to improve work system effectiveness and reliability, increase productivity, reduce employee health care costs, and improve the quality of production processes, services, products, and working life for all employees. In this context, professional ergonomists, practitioners, and students should have a broad understanding of the full scope and breadth of knowledge of this demanding and challenging discipline.

Interventions, Controls, and Applications in Occupational Ergonomics contains a total of 52 chapters divided into eight parts.

Part I, Ergonomics Processes, describes the elements of the ergonomics processes, including success factors for implementation of industrial ergonomics programs, practical interventions in industry that utilize participatory approaches, surveillance, Occupational Safety and Health Administration (OSHA) recordkeeping requirements, and psychosocial work factors. Part II, Surveillance, discusses the injury surveillance database systems and corporate health management for the design and evaluation of health in industrial organizations. Part III, Industrial Process Applications, is devoted to ergonomics processes from the small and large industry perspectives. The main focus of Part IV, Upstream Ergonomics, is on human digital modeling, as well as facilities planning and organizational design and macroergonomics.

Part V and Part VI deal with the engineering and administrative controls of work-related musculoskeletal disorders. Part V, Engineering Controls, presents the methods and techniques for engineering control, including the knowledge of what works and what does not, general solutions for the control of upper extremity and low back disorders, application of lift assist rail systems and hand tools, as well such engineering controls as gloves, wrist supports, and lower extremity supports. Part VI, Administrative Controls, discusses the theory and practice of ergonomics interventions in the workplace, worker selection for physically demanding jobs, physical ability testing, training, secondary assessment of worker functional capacities, interventions for low back pain, human resource management and selection, work day length, shiftwork, and job rotation issues. This part also provides an update on the use of back belts in industry, and discusses best practices for the prevention of musculoskeletal disorders at work. Part VII, Medical Management, addresses the issues of medical management of work-related musculoskeletal disorders, systems approach to rehabilitation, wrist splints, and application of the clinical lumbar motion monitor.

Finally, Part VIII, Ergonomics Industrial Interventions, provides an overview of best practices for ergonomics interventions with respect to the design and use of chairs and office furniture, as well as design of computer keyboards and notebook computers. This part also discusses ergonomics interventions in a variety of industries, including meat and poultry processing, agriculture, distribution centers (case picking), healthcare (patient handling), as well as service systems, that is, grocery stores/bakery, furniture manufacturing, and construction industry. This section also addresses the problem of medical errors, discusses the challenges and rewards of applying ergonomics in developing areas, and looks into the future of human work.

The knowledge presented in this book should help the readers to improve their understanding of the complex interactions between the people at work and other systems, which must be considered in the context of rapidly changing Occupational Ergonomics technology and evolving social needs. We hope that this book will be useful to the professionals, students, and practitioners who aim to optimize the design of systems, products, and processes, manage the workers' health and safety, and improve the overall quality and productivity of contemporary businesses.

William S. Marras
The Ohio State University

Waldemar Karwowski
University of Louisville

About the Editors

William S. Marras, Ph.D., D.Sc. (Hon), C.P.E., holds the Honda endowed chair in transportation in the department of industrial, welding, and systems engineering at The Ohio State University. He is the director of the biodynamics laboratory and holds joint appointments in the departments of orthopedic surgery, physical medicine, and biomedical engineering. He is also the co-director of The Ohio State University Institute for Ergonomics. Dr. Marras received his Ph.D. in bioengineering and ergonomics from Wayne State University in Detroit, Michigan. He is also a certified professional ergonomist (CPE).

His research is centered around occupational biomechanics. Specifically, his research includes workplace biomechanical epidemiologic studies, laboratory biomechanic studies, mathematical modeling, and clinical studies of the back and wrist. His findings have been published in over 170 refereed journal articles, 7 books, and over 25 book chapters. He also holds several patents, including one for the Lumbar Motion Monitor (LMM). Professor Marras has been selected by the National Academy of Sciences to serve on several committees investigating causality and musculoskeletal disorders. He also serves as the chair of the Human Factors Committee for the National Research Council within the National Academy of Sciences.

His work has attracted national and international recognition. He has been twice winner (1993 and 2002) of the prestigious Swedish Volvo Award for low back pain research as well as Austria's Vienna Award for Physical Medicine. He recently won the Liberty Mutual Prize for injury prevention research. Recently, he was awarded an honorary doctor of science degree from the University of Waterloo for his work on the biomechanics of low back disorders.

In his spare moments, Dr. Marras trains in Shotkan karate (a black belt), enjoys playing and listening to music, sailing, and fishing.

Waldemar Karwowski, Sc.D., Ph.D., P.E., C.P.E., is professor of industrial engineering and director of the center for industrial ergonomics at the University of Louisville, Louisville, Kentucky. He holds an M.S. (1978) in production engineering and management from the Technical University of Wroclaw, Poland, and a Ph.D. (1982) in industrial engineering from Texas Tech University. He was awarded the Sc.D. (dr hab.) degree in management science by the Institute for Organization and Management in Industry (ORGMASZ), Warsaw, Poland (June 2004). He is also a board certified professional ergonomist (BCPE). He also received doctor of science honoris causa from the South Ukrainian State K.D. Ushynsky Pedagogical University of Odessa, Ukraine (May 2004). His research, teaching, and consulting activities focus on human system integration and safety aspects of advanced manufacturing enterprises, human–computer interaction, prevention of work-related musculoskeletal disorders, workplace and equipment design, and theoretical aspects of ergonomics science.

Dr. Karwowski is the author or co-author of more than 300 scientific publications (including more than 100 peer-reviewed archival journal papers) in the areas of work systems design, organization, and management; macroergonomics; human–system integration and safety of advanced manufacturing; industrial ergonomics; neuro-fuzzy modeling in human factors; fuzzy systems; and forensics. He has edited or co-edited 35 books, including the *International Encyclopedia of Ergonomics and Human Factors*, Taylor & Francis, London (2001).

Dr. Karwowski served as a secretary-general (1997–2000) and president (2000–2003) of the International Ergonomics Association (IEA). He was elected as an honorary academician of the *International Academy of Human Problems in Aviation and Astronautics* (Moscow, Russia, 2003), and was named the alumni scholar for research (2004–2006) by the J. B. Speed School of Engineering of the University of Louisville. He has received the Jack A. Kraft Innovator Award from the Human Factors and Ergonomics Society, USA (2004), and serves as a corresponding member of the European Academy of Arts, Sciences and Humanities.

Contributors

David C. Alexander
Auburn Engineers, Inc.
Auburn, Alabama

W. Gary Allread
Institute for Ergonomics
The Ohio State University
Columbus, Ohio

Charles K. Anderson
Advanced Ergonomics, Inc.
Dallas, Texas

A. Asmus
Industrial, Welding and Systems
    Engineering
The Ohio State University
Columbus, Ohio

Ann E. Barr
Physical Therapy Department
College of Health Professions
Temple University
Philadelphia, Pennsylvania

Patricia Bertsche
Ross Laboratories
Chicago, Illinois

R.R. Bishu
Department of Industrial and
    Management Systems
    Engineering
University of Nebraska
Lincoln, Nebraska

Marilyn Sue Bogner
Institute for the Study of
    Human Error, LLC
Bethesda, Maryland

Gene Buer
Crane Equipment and
    Service, Inc.
Subsidiary of Columbus
    McKinnon Corp.
Eureka, Illinois

David Caple
David Caple & Associates Pty Ltd
Melbourne, Victoria, Australia

Pascale Carayon
Ecole des Mines de
Nancy, France

Ernesto Carcamo
WISHA Services
Washington State Department of
    Labor and Industries
Olympia, Washington

Don B. Chaffin
University of Michigan
Ann Arbor, Michigan

A.-M. Chany
Industrial, Welding and Systems
    Engineering
The Ohio State University
Columbus, Ohio

Larry J. Chapman
Department of Biological
    Systems Engineering
University of Wisconsin
Madison, Wisconsin

David J. Cochran
Department of Industrial
    Engineering
University of Nebraska
Lincoln, Nebraska

Marvin J. Dainoff
Psychology Department
Miami University
Oxford, Ohio

Joseph M. Deeb
ExxonMobil Biomedical
    Sciences, Inc.
Annandale, New Jersey

Jack Dennerlein
Harvard School of Public Health
Boston, Massachusetts

C.G. Drury
Department of Industrial
    Engineering
University of Buffalo
Buffalo, New York

Bradley Evanoff
Washington University
    School of Medicine
St. Louis, Missouri

Susan Evans
Sue Evans & Associates, Inc.
Fairfax, Virginia

Fadi A. Fathallah
Biological and Agricultural
    Engineering Department
University of California
Davis, California

Sue A. Ferguson
Industrial, Welding and Systems
    Engineering
The Ohio State University
Columbus, Ohio

Kaori Fujishiro
Environmental and
 Occupational Health Science
University of Illinois at Chicago
Chicago, Illinois

Sean Gallagher
National Institute for
 Occupational Safety
 and Health
Pittsburgh, Pennsylvania

Robert J. Gatchel
University of Texas at
 Arlington and University
 of Texas Southwestern
 Medical Center at Dallas
Arlington, Texas

V. Gnaneswaran
Department of Industrial and
 Management Systems
 Engineering
University of Nebraska
Lincoln, Nebraska

Thomas Hales
National Institute for
 Occupational Safety and
 Health
Cincinnati, Ohio

Christopher Hamrick
Ohio Bureau of Workers'
 Compensation
Pickerington, Ohio

Catherine A. Heaney
Psychology and Human Biology
Stanford University
Stanford, California

Steven F. Hecker
Labor Education and
 Research Center
University of Oregon
Eugene, Oregon

Jennifer Hess
Labor Education and
 Research Center
University of Oregon
Eugene, Oregon

Steven L. Johnson
Department of Industrial
 Engineering
University of Arkansas
Fayetteville, Arkansas

Michael J. Jorgensen
Department of Industrial and
 Manufacture Engineering
Wichita State University
Wichita, Kansas

Bradley S. Joseph
Ford Motor Company
Dearborn, Michigan

Ben-Tzion Karsh
Department of Industrial and
 Systems Engineering
University of Wisconsin
Madison, Wisconsin

Waldemar Karwowski
University of Louisville
Louisville, Kentucky

Glenda L. Key
Key Method Assessments, Inc.
Minneapolis, Minnesota

Åsa Kilbom
National Institute for
 Working Life
Stockholm, Sweden

Laurel Kincl
Labor Education and
 Research Center
University of Oregon
Eugene, Oregon

Brian M. Kleiner
Virginia Polytechnic Institute
 and State University
Blacksburg, Virginia

Peter Knauth
Department of Ergonomics
Institute of Industrial Production
University of Karlsruhe
Hertztrasse, Germany

Stephan Konz
Department of Industrial
 Engineering
Kansas State University
Manhattan, Kansas

Steven A. Lavender
Industrial, Welding and Systems
 Engineering
The Ohio State University
Columbus, Ohio

Karen E.K. Lewis
Honda of America
 Manufacturing, Inc.
Marysville, Ohio

Soo-Yee Lim
Ecole des Mines de
Nancy, France

Veikko Louhevaara
University of Kuopio and
 Kuopio Regional Institute
 of Occupational Health
Kuopio, Finland

Richard W. Marklin
Department of Mechanical and
 Industrial Engineering
Marquette University
Milwaukee, Wisconsin

William S. Marras
Institute for Ergonomics
The Ohio State University
Columbus, Ohio

Tom G. Mayer
University of Texas
Southwestern Medical Center at
 Dallas and PRIDE Research
 Foundation
Dallas, Texas

Stuart M. McGill
Department of Kinesiology
University of Waterloo
Faculty of Applied Health
 Sciences
Waterloo, Ontario, Canada

James M. Meyers
Agricultural Ergonomics
 Research Center
University of California
Berkeley, California

Gary Mirka
Department of Industrial
Engineering
North Carolina State University
Raleigh, North Carolina

A. Muralidhar
Auburn Engineers, Inc.
Auburn, Alabama

Audrey Nelson
James A. Haley VAMC
Tampa, Florida

Gary B. Orr
ORR Consulting
Alexandria, Virginia

J. Parakkat
Industrial, Welding and Systems
Engineering
The Ohio State University
Columbus, Ohio

Nils F. Petersson
National Institute for Working Life
Stockholm, Sweden

Skye Porter
PRIDE Research Foundation
Dallas, Texas

Peter M. Quesada
University of Louisville
Louisville, Kentucky

Robert G. Radwin
Industrial Engineering Department
University of Wisconsin
Madison, Wisconsin

David Rempel
University of California, Berkeley
Richmond, California

David Rodrick
Florida State University
Tallahassee, Florida

Annina Ropponen
Department of Physiology
University of Kuopio
Kuopio, Finland

Scott P. Schneider
Laborers' Health and Safety
Fund of North America
Washington, District of
Columbia

Barbara Silverstein
SHARP, Washington State
Department of Labor
and Industries
Olympia, Washington

Thomas J. Slavin
Occupational Safety and Health
International Truck and Engine
Corporation
Warrenville, Illinois

Juhani Smolander
ORTON Research Institute and
ORTON Orthopaedic Hospital
Helsinki, Finland

Stover H. Snook
Harvard School of Public Health
Boston, Massachusetts

Carolyn M. Sommerich
Industrial, Welding and
Systems Engineering
The Ohio State University
Columbus, Ohio

Peregrin Spielholz
SHARP, Washington State
Department of Labor and
Industries
Olympia, Washington

Carol Stuart-Buttle
Stuart-Buttle Ergonomics
Philadelphia, Pennsylvania

Brian R. Theodore
PRIDE Research Foundation
Dallas, Texas

Martin J. Thul
Institute of Technology
and Work
University of Kaiserslautern
Kaiserslautern, Germany

G. Yang
Industrial, Welding and
Systems Engineering
The Ohio State University
Columbus, Ohio

Xudong Zhang
University of Illinois
Urbana-Champaign, Illinois

Klaus J. Zink
University of Kaiserslautern
Kaiserslautern, Germany

Contents

V   Engineering Controls

VI   Administrative Controls

VII Medical Management

VIII Ergonomic Industrial Interventions

I

Ergonomics Processes

1

Elements of the Ergonomic Process

Åsa Kilbom
Nils F. Petersson
*National Institute for
Working Life*

1.1 Introduction

During the ergonomic process, problems, potential or already apparent, are gradually worked out and brought to a solution. Problems may be similar but the contexts in which they appear are almost unique. Thus, the ergonomic process will hardly ever be the same, and experiences gained in one case cannot be applied mechanically to another place.[15] Furthermore, there is very seldom only one possible solution, but many and probably quite different ones, depending on the culture and awareness at the workplace, its size and level of technology, and the human and financial resources available. Moreover, the chosen solution must comply with the aims of the organization which again influences the choice of solution. The ergonomic process in practice will consequently take different ways and differ considerably from time to time. However, some important steps or phases in the process can be traced in most cases[14,19] and need to be handled for a successful outcome. These phases are: organization of the process, identifying the problem, analyzing the problem, developing a solution, implementing the solution, and evaluating the result (Figure 1.1).

All of these phases normally have to be present to achieve a good result. In most cases it is important to treat the different phases separately from each other and not to begin one phase until the preceding is completely finished. Too often solutions are presented before a thorough analysis is carried out, maybe resulting in a suboptimization.

Schneider[24] suggests three basic elements to implement an effective ergonomics program:

1. Establish ergonomics as a business function
2. Establish a predefined return on investment profile for workplace improvements

3. Establish goals and measure performance

He also argues for not using the word *ergonomics* as it often seems to be synonymous with costs; instead it should be emphasized that healthy people perform and work better. Improvements based on ergonomic investments as well as other investments should be evaluated on the basis of value. Other, more complex elements of the ergonomic process have been reviewed by Wilson, for example, the ergonomic design process.[29]

Apart from the phases included in the process (Figure 1.1), a routine to take advantage of the experiences for the next project is also desired.

Ergonomic programs need time, and an underestimation may lead to failure to fulfil the program. One problem is modern management's focus on short-term goals and profit which can be hard to handle.[3]

Finally, an ergonomic process is characterized by its comprehensive view and multidisciplinary approach, taking both productivity and human

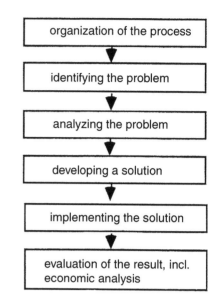

FIGURE 1.1 Important steps in the ergonomic process.

aspects into consideration. The multidisciplinary approach must be considered in forming the team of the process, taking the participants' background into account.[23] Actually the next most important requirement, after involvement of the employees and management commitment, for a successful implementation of ergonomics, is the multidisciplinary approach.[2]

1.2 Organization of the Process

Although the organization of the project or the process* many times does not start until the problem is identified or during this step, it is advisable to treat it separately from the problem analysis and problem-solving phases. It is common to have some sort of a fixed organization taking care of ergonomics. Big companies often have their ergonomics department or a health care service with ergonomic expertise. For example, Faville recently presented one approach for a large manufacturing company faced with a high rate of musculoskeletal problems.[8] Another example is the nationwide occupational health service for the construction industry "Bygghälsan" which has been in operation in Sweden since the late 1960s.[7] Through medical check-ups every 2 to 3 years a high prevalence of musculoskeletal disorders were identified. An ergonomic program was linked to the health surveys and gradually introduced improved work practices and technical improvements at the building sites.

Obviously, the preexistence of an organization dedicated to ergonomics will facilitate and speed up the identification, analysis, and implementation phases. Ideally, such an organization will already have developed methods for surveillance of risk factors and health, a system for collaboration with production engineering and personnel departments, and pathways for feedback to management and employees.

In many countries, for example, the Scandinavian nations, formal joint union–management committees, taking care of ergonomics, are prescribed and in other countries voluntary union–management groups appear in companies.[18,22]

---

*Throughout this text we use the term *project* for time-limited activities and *process* for those activities going on without any predetermined termination.

The process of continuous improvement, kaizen, has been introduced in many western companies and ergonomic projects may be integrated in these processes. When a preexisting organization is not in place, the emergence of an ergonomic problem can be an important factor triggering its development.

Lacking a preexisting organization, many different ways to organize an ergonomics project or process exist. One extreme is to hire an external consultant expert of ergonomics, and the other is to involve all concerned employees in the process without an expert. Although both systems appear and may be adequate in some cases the most effective organization, in general, lies somewhere in between, taking the advantages of both. It is hardly possible for employees to master all different fields of ergonomics, and therefore not using an expert may result in nonoptimal solutions. Especially for small companies with limited technical expertise, the implementation of high-tech solutions without external know-how may result in costly mistakes. On the other hand, not using the knowledge of the employees can imply that important basic factors are not taken into account. A process where employees take an active part in all phases of the problem is usually called "participatory ergonomics" and is used successfully all over the world.[13,20] Involving the employees most often facilitates the implementation through a greater acceptance of change, and adds problem-solving capabilities as the person doing the job often has the best insights in how to improve the work.[2,11] By teaching workers fundamental ergonomic principles they can become responsible partners in the ergonomic process.[3] The involvement of workers in the process leads to empowered workers,[3,26] who can interpret accurately ergonomic needs and who become increasingly responsible for pushing the processes. The ergonomic training has to be "just-in-time-training" adapted to the real acute situation. Starting the training from one's own situation, for example, an evaluation of the worker's own workstation, seems to be an appropriate method.[9] Apart from training in ergonomics, training in team work and communication is also important.[11]

Another prerequisite for success is management commitment.[3] An ergonomic program is doomed to failure without visible support of the management and financial backing.[23] Schneider[24] too puts a heavy emphasis on management commitment to ergonomic improvements and argues that the onus for managing the ergonomic agenda is with line management. Active participation of all involved including management and supervisors has been found to be the most important factor in implementing ergonomics.[2] Chavalitsakulchai et al.[2] also argue for regarding government officers as vital for ergonomics intervention programs.

An organization pleaded for by some authors is the "system group" approach.[1] Representatives of all those concerned with the problem, that is, ergonomics or health department, management, employees, production engineering, personnel, sales representatives, and customers (of the product or the service) form a group that meets to analyze the problem and develop a solution.

Important issues for team building and participative approaches have recently been summarized in a report from NIOSH[11] and are:

- *Management commitment*, from top management to supervisors
- *Training* in ergonomics, communication, feedback skills, technology, all tailored to the participants' skills
- *Composition* of teams, tailored to the problem at hand
- *Information sharing*, within teams, to and from management and employees
- *Activities and motivation*, includes meetings, data gathering and analysis, and planning of remedies. Motivation is achieved by goal setting and feedback, commitment from management and rewards
- *Evaluation*, for example, of team efforts and outcome of activities

In most cases a change agent, for example, a production engineer or a safety professional, is recommended. Unfortunately, ergonomists often occupy low organizational positions which makes their negotiating position weak.[14] Apart from the change agent, dedicated persons are of great value and an early identification of them — if they are present — is recommended.

1.3   Identifying the Problem

The title of this phase is not entirely appropriate as many ergonomic processes do not start with a real "problem" but are part of a development strategy. In fact the reasons to start a process or project can vary widely and be divided either regarding their abode — a productivity or a health issue — or in how they occur. For example, the problem can appear as an acute incidence, for example, an accident which has to be remedied. It can also emerge after a survey to identify critical factors, for example, a bottleneck in the production or a work task creating musculoskeletal disorders. Third, the problem can be identified as the result of a conscious and continuous improvement procedure, for example, activities in quality circles. Depending on what type of problems and by whom they are identified, the ensuing process will differ regarding both organization and solutions identified.

Both for a temporary survey and a long-lasting improvement procedure some supporting devices exist to facilitate the problem identification. Group discussions based upon photos or videos are often efficient. Brainstorming is another method both for the problem identification and also later in the solution phase. Short courses in ergonomics, including a workplace-based project, can make problems evident. An ergonomic training program including an evaluation of workstations can be effective. The evaluation of the worksite should include not only physical aspects of the workplace but also analysis of work methods, product flow, and maintenance of tools.[23]

Other sources for problem identification may be internal statistics, for example, sick-leave or numbers of health care visits. Benchmarking and other quality methods[5] are still other ways to be aware of problems.

It is important to have systems which encourage workers who have early symptoms to report these.[3] The problem phase may end up with frame settings regarding the costs and time accepted for the project.

1.4   Analyzing the Problem

The analysis phase includes, apart from a thorough analysis of all the components of the problem, also the analysis of the consequences if the problem remains unsolved and the obstacles remain for a solution.

Both in the step of identifying the problems as well as in the analyzing phase one way to start is to ask (1) What is the purpose of the work performed? (2) How are the functions in the work process allocated between humans and technology?[26] For work-related problems a task analysis is a good base[27] and in product development a function analysis is recommended.[29]

When analyzing the components of the problem, it is important not to limit the scope only to the imminent problem. Sometimes the optimal solution is not confined to the work process where the problem was identified, but to work processes or technology used in a preceding process.

The analysis should also contain the goals and the criteria for the solution. Goals are preferably expressed in measurable quantitative terms, for example, a certain increase in productivity or reduction of sick leave. Fuzzy goals have to be operationalized, that is, translated and expressed into concrete ones.[17] Skill in defining goals is essential for measuring ergonomic progress.

1.5   Developing a Solution

As stated in the introduction there is seldom only one solution. If the analysis is carried out thoroughly, the solutions are normally easy to find.

Solutions are traditionally subdivided according to their approach into engineering, administrative, and behavioral.[12] Engineering approaches can be redesign of a machine, a workplace or a tool; administrative approaches are changes of work processes, that is, job rotation, job enlargement, or reallocation of tasks between machines and humans; and behavioral approaches attempt to influence attitudes or behaviors toward risks and changes at work. Training is sometimes classified as an administrative, and sometimes as a behavioral approach depending on the content.

The problem analysis phase should ideally identify the most promising and feasible approach in the individual case. A cost benefit analysis of the chosen solution should be undertaken, and in the case of several feasible approaches such an analysis can assist in choosing the best. Although one approach usually dominates, the solution commonly contains (and should contain) elements of all three approaches. Most administrative approaches focus on work organization which is one part of the macro-ergonomic concept. Commonly engineering and behavioral solutions are the first ones considered, especially in organizations with limited experience in ergonomic problem solving. However, in a process of continuous changes it will soon be evident that administrative approaches are also necessary. These changes sometimes meet more resistance in an organization and are more far-reaching, and they often require more experience.

The analysis phase also includes a thorough time planning and an allocation of tasks to those concerned. Information about the stages of the analysis, from concepts to detailed plans, is important for acceptance by all those directly or indirectly involved.

Many of the methods used in the problem identification phase are also appropriate in the solving phase, for example, group discussions and brainstorming. Other means are sketches, models, full-scale mock ups. Literature review, net search, and visits to other sites should not be neglected.

1.6   Implementing the Solution

In many cases the implementation phase seems to be the most critical one, calling for special care and time. Many projects have turned out unsuccessful due to an underestimation of the problems in the implementation phase. This is common in projects carried out by an external expert with no or little involvement by the employees/users during the preceding phases.

A change process, generative or innovative,[6] is very seldom a straightforward action which can be planned in detail. Instead it is a movement with many loops and to's and fro's. Those responsible must be adaptable and ready to change the plans, keeping the main purpose in mind.

As mentioned above, all projects include organizational changes, even those regarded as purely technical projects. Neglecting this fact may be disastrous. Organizational changes are a threat to most people. According to Gardell, resistance to change originates in a threat to the following circumstances: job security, material standard, social status, social relations and freedom of movement.[10] Szilagyi and Wallace, using a somewhat different classification, distinguish the following reasons for resistance: fear of economic loss, potential social disruption, inconvenience, fear of uncertainty, and resistance from groups.[28]

The resistance to change is a serious threat to the implementation of a program, and the best way to handle it is by continuous information. Access to necessary information is one of the most important factors promoting a program.[2] Misunderstandings are often a source of resistance. The main goals must be stated clearly and very early to all concerned.

A participative approach or a system group approach with representatives among all those involved in the change is very useful to forestall the resistance. The following approaches will facilitate the implementation:

- Provide possibilities to influence the solutions
- Offer longer time to comprehend
- Reconcile differences between different personnel categories
- Set up effective channels for information and communication

Resistance may also occur due to a fear of new technical equipment if those involved are not given enough and appropriate training. Training should be planned for and incorporated in all projects of change. The training has to be adjusted to the special situation and occur at the proper occasion, and should preferably be conducted at the worksite rather than in classrooms.[16]

Another source of resistance may be former changes which have not been accomplished in a proper way or have resulted in deteriorations.

1.7   Evaluating the Result

Commonly, the evaluation of an ergonomic change process is based only on perceptions and random observations, without quantitative data support. Such an evaluation implies a risk that unspecific, short-lasting effects are recorded as specific consequences of the change (Hawthorne effects). As a consequence, the effectiveness of the change can be seriously misjudged.

Although ergonomic changes at the workplace are usually not research projects, some lessons can be learned from intervention research.[25] One important experience is to plan for the evaluation as soon as after the analysis stage, when the goals of the process have been identified. Moreover, data describing the situation before the change process are needed for the evaluation. Since the implementation of changes is a dynamic process, the evaluation should ideally be a continuous process where short- as well as long-lasting effects are monitored. Efforts to push the change process may result in a temporary dip of productivity during the changes. Such dips must not be interpreted as a failure. In line with the multidisciplinary character of ergonomics the evaluation should include productivity and economic, as well as health aspects. Obviously, this evaluation is vastly facilitated if a change agent, or an ergonomics department, is in charge of the program and if a system for monitoring is already in place.

The outcome of the ergonomic process should also be evaluated in economic terms.[23,24] The easiest way for this analysis is to balance the costs of implementing the changes, including the investments, against the likely savings, that is, reduced incidence of injuries, increases in productivity and quality, reduced staff turnover (including training new staff), etc. In a similar way, alternative approaches to improvements can be compared to select the most cost-effective solution. As demonstrated in many case studies, the payback period for ergonomic improvements is frequently only a few months.[21]

1.8   Using the Results and Experiences for the Next Process

The ergonomic process creates a vast amount of experience and knowledge among all concerned. This experience must not be discarded but should be used for future processes. The evaluation of the program should focus not only on the outcomes in quantitative terms. The process of development and implementation of the program can also be assessed and expressed in qualitative rather than quantitative terms.[4] A protocol where the process is described and where the experience obtained is documented is therefore a valuable tool for future work. It enables those responsible to analyze the reasons for successes and failures, and in this way the ergonomic process can achieve results far beyond those of a single project.

References

1. Andersson E. A systems approach to product design and development; an ergonomic perspective. *Int J Indust Ergon* 1990; 6: 1–8.
2. Chavalitsakulchai P, Okubo T, Shahnavaz H. A model of ergonomics intervention in industry: case study in Japan. *J Human Ergol* 1994; 23: 7–26.
3. Dawkins S. Does ergonomics work? *Manag Office Tech* 1995; 40(3): 12–14.
4. Dehar M-A, Casswell S, Duignan P. Formative and process evaluation of health promotion and disease prevention programs. *Evaluation Review* 1993; 17(2): 204–220.
5. Drury C. Ergonomics and the quality movement. *Ergonomics* 1997; 40(3): 249–264.
6. Ellegård K. Förändringsarbete och förändring — perspektiv på Volvo Uddevallaverken i efterhand. (The process of changes and its result — perspectives of the Volvo Uddevalla plant. In Swedish). Nordiska Ergonomisällskapets Årskonferens NES '94, 1994: 170–173.
7. Engholm G, Englund A. Morbidity and mortality patterns in Sweden. *Occ Med State of the Art Reviews.* Construction safety and health. 1995; 10: 261–268.
8. Faville B. One approach for an ergonomics program in a large manufacturing environment. *Int J Indust Ergon* 1996; 18: 373–380.

9. Fragala G. Get more from your ergonomics training. *Manag Office Tech* 1995; 40(11): 45–46.
10. Gardell B. Arbetsanpassning och teknologisk miljö (Work adaptation and technological environment. In Swedish). In Luthman G, Åberg U, Lundgren N, Eds. *Handbok i ergonomi.* 2nd ed. Stockholm: Almqwist och Wiksell, 1969: 546–573.
11. Gjessing C, Schoenborn T, Cohen A. *Participatory Ergonomics Interventions in Meatpacking Plants.* NIOSH, Cincinnati, 1994.
12. Goldenhar L, Schulte P. Methodological issues for intervention research in occupational health and safety. *Am J Indust Med* 1996; 29: 289–294.
13. Imada A. The rationale and tools of participatory ergonomics, in Noro K, Imada A, Eds. *Participatory Ergonomics.* London: Taylor & Francis, 1991: 30–49.
14. Kuorinka I. Tools and means of implementing participatory ergonomics. *Int J Indust Ergon* 1997; 19: 267–270.
15. Kvarnström S. Organizational approaches to reducing stress and health problems in an industrial setting in Sweden. *Conditions of Work Digest* 1992; 11(2): 227–232.
16. Luopajärvi T. Workers' education. *Ergonomics* 1987; 30: 305–311.
17. McCoy T. Getting to the seat of the matter: ergonomics can improve worker safety and productivity. *Rehab Management* 1994; 2: 116–119.
18. Moir S, Buchholz B. Emerging participatory approaches to ergonomic interventions in the construction industry. *Am J Indust Med* 1996; 29: 425–430.
19. Moore J, Garg A. Use of participatory ergonomics teams to address musculoskeletal hazards in the red meat packing industry. *Am J Indust Med* 1996; 29: 402–408.
20. Nagamachi M. Requisites and practices of participatory ergonomics. *Int J Indust Ergon* 1995; 15: 371–377.
21. Oxenburgh M. *Increasing Productivity and Profit Through Health and Safety.* Sydney: CCH International, 1991.
22. Reid P. *Well Made in America: Lessons from Harley-Davidson on Being the Best.* New York: McGraw-Hill, 1990.
23. Ross P. Ergonomic hazards in the workplace; assessment and prevention. *AAOHN J.* 1994; 42(4): 171–176.
24. Schneider F. Targeting ergonomics in your business plan. *Manag Office Tech* 1995; 40(9): 28–30.
25. Skov T, Kristensen T. Etiologic and prevention effectiveness intervention studies in occupational health. *Am J Indust Med* 1996; 29: 378–381.
26. Springer T. Managing effective ergonomics. *Manag Office Tech* 1994; 39(3): 19–24.
27. Stammers R, Shepherd A. Task analysis, in Wilson J, Corlett E, Eds, *Evaluation of Human Work.* London: Taylor & Francis, 1995: 144–168.
28. Szilagyi A, Wallace M. *Organizational Behavior and Performance.* (5th ed.) Glenview: Foresman, 1990.
29. Wilson J. A framework and a context for ergonomics methodology, in Wilson J, Corlett E, Eds, *Evaluation of Human Work.* London: Taylor & Francis, 1995: 1–29.

2

Success Factors for Industrial Ergonomics Programs

David C. Alexander
Auburn Engineers, Inc.

Garry B. Orr
ORR Consulting

2.1 Introduction

Many ergonomics programs are not successful. A survey performed by Auburn Engineers (Auburn, Alabama) found that only 25% of the ergonomics programs they surveyed were successful. The data, shown in Figure 2.1, separates the organizations into small, medium, and large sizes. Four different outcomes are possible:

- Successful
- Too new to call
- Floundering due to management issues
- Floundering due to technical issues

The term floundering was chosen to indicate that more effort is going into the ergonomics program than is appropriate for the results being achieved. While this is not technically a failure, it is clearly headed in that direction. Floundering due to management issues was the result of lack of vision or program direction, inadequate resources, lack of coordination, and other management issues.

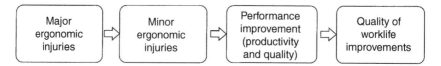

FIGURE 2.1   Progression of the results of ergonomics projects.

Floundering due to technical issues included programs where there was a fundamental lack of technical skills such as job analysis or ability to generate appropriate solutions.

This chapter lists a number of success factors which will make an ergonomics program effective. It is not easy to develop a list such as this. After reviewing hundreds of programs, success factors and common flaws begin to appear. Many of these are found in this chapter. At the same time, a successful ergonomics program does not have to include all of these factors. Many, but not all, of these factors are found in successful programs.

The success factors can be roughly divided into four groups:

- Meet business needs
  1. Emphasize business objectives
  2. Avoid too many low-value/high-cost solutions
  3. Ensure that ergonomics projects are evaluated quantitatively
  4. Maintain a tabulation of the cost of projects
  5. Use resources efficiently (the self-help/skilled-help/expert-help strategy)
- Avoid common traps
  6. Identify and overcome barriers
  7. Training should be supported by suitable infrastructure
  8. Avoid using "ergo-babble"
- Create a strong purpose
  9. Clearly define the purpose of your ergonomics program
  10. Plan the stages of the ergonomics culture change
  11. Create a strategic plan
- Maintain the program
  12. Understand the difference between an ergonomics program and the practice of ergonomics
  13. Create a tactical plan
  14. Ensure that there are regular quantitative evaluations of the overall ergonomics program
  15. Do not wait for top management to push the program down
  16. Maintain political support

2.2   Discussion of Success Factors

2.2.1   Emphasize Business Objectives

For an ergonomics program to sustain itself over the long term, it must be anchored to business objectives. The best way to do this is to ensure that its results improve the business objectives of the organization. There is a progression in the types of results achieved, as shown in Figure 2.1. Most ergonomics applications are initially targeted toward the elimination of major injuries, and then minor injuries become important. Once injuries are under control, the emphasis should shift to improving performance in the areas of productivity and quality, and eventually, improvements in the quality of worklife should occur. A successful ergonomics program will, as soon as practical, ensure that business objectives are improved, documented, and shared with management.

This may be difficult to do because many ergonomists only see ergonomics as a technology and not as business enhancement tools. In business, however, technology is a tool to achieve business objectives.

Unfortunately, when ergonomics is applied primarily as a technology without firm business goals, the results rarely amount to more than training and scattered job analyses.

Fortunately, though, when ergonomics is viewed as a tool which helps drive important business measures, it continues to be important over an extended time period, and will be retained. However, when ergonomics is seen either as an "add-on" with little business value or as a "must do" from a compliance standpoint, an ergonomics program will only be mildly supported and eventually discontinued.

2.2.2 Avoid Too Many Low-Value/High-Cost Solutions

Cost is or soon becomes an issue for most ergonomics programs. When costs are too high relative to the value received, the ergonomics program is regarded as a money pit and is stopped or slowed down (see Figure 2.2).

Expensive solutions usually result from a misunderstanding of the role of people and equipment such as an "automation mentality" which requires that automation be used to remove the person totally from the job. In the office area, many office ergonomists go on a "chair buying binge" and spend too much on new seating with correcting the hand/wrist problems common in office areas.

Typically, these problems result from less experienced ergonomists who have difficulty creating lowcost solutions that address the root cause. Many ergonomists have developed skills to identify and analyze ergonomics problems. Unfortunately, if they have not developed skills for the efficient resolution of those problems, the result can easily be overly expensive solutions.

When the common solution for ergonomics problems is to "automate the job," a single solution can be very costly, often in the range of $100,000 to $1,000,000. It is difficult to offset this high cost with the benefits gained, and this gives management the perception that ergonomics is (and always will be) prohibitively expensive. It does not take management long to tire of these types of "low-value/high-cost solutions." And once that occurs, the ergonomics program usually has a short future.

An example can illustrate this concept. Suppose that a material handling problem has been uncovered, and the question is how to resolve it. A less experienced ergonomist might use automation, when other alternatives such as scissors-lifts, spring-loaded levelers, and turntables are available. Typically, an administrative control is also considered, and one might even add the consideration of back belts to the decision. For many lifting situations, simply getting the lift to correct height and close to the body will resolve the problem. The two-dimension matrix in Figure 2.3 easily shows where the value/cost benefit lies. Figure 2.4 provides a generic view of the value cost matrix and how to assess different solutions for their own value/cost relationship.

Management's View of Ergonomics

FIGURE 2.2 The money pit.

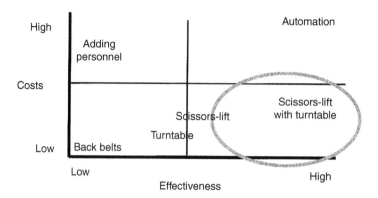

FIGURE 2.3   Possible solutions for a material handling problem.

2.2.3   Ensure That Ergonomics Projects Are Evaluated Quantitatively

Only the most successful ergonomics programs have instituted a systematic method for quantitatively evaluating individual ergonomic projects. Ergonomic projects should be evaluated for both ergonomic improvement and for cost/benefits. The degree of ergonomic improvement can be measured by changes in such lagging indicators as incident rate, severity rate, or losses for workers' compensation. It is also possible to use leading indicators such as a symptoms survey or pain/discomfort body parts survey. It is best to use these surveys more than once, and a suitable timetable for surveys is based on "Dave's Rule of Twos" — perform a survey after the changes have been implemented for two days, then again after two weeks, and finally, after two months.

In addition, each ergonomics project should also be evaluated financially. The costs and benefits can be measured for each project. These dollar figures for costs and benefits can be translated into net overall improvement for all the ergonomics projects, or a cost/benefit ratio for each project can be maintained. Each project does not have to pay off, but, overall, the program should be able to pay for itself.

These basic items can become part of composite measures such as those listed below:

- Cumulative stress reduction (CSR) index — Index of stress reduction × number of people affected
- Cumulative stress reduction per $1,000 — Ratio of CSR index to costs

One additional project evaluation measure which has created a lot of interest is the time to complete each project. Usually this is just the elapsed time from the day the ergonomic project is initiated until the recommended solution is implemented. If the ergonomic problem-solving skills are increasing, then the time to complete individual projects should be decreasing.

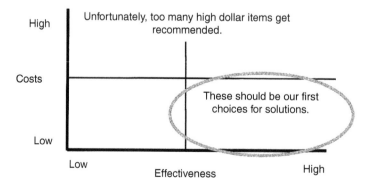

FIGURE 2.4   Avoid too many low-value/high-cost solutions.

2.2.4   Maintain a Tabulation of the Cost of Projects

Cost is an important issue for many ergonomics programs and one additional method of cost measurement is valuable. Since so many projects are incorrectly assumed to be overly expensive, a simple tabulation of the costs of a number of projects can help dispel the notion that all projects must be costly. Table 2.1 shows the costs of 29 ergonomics projects completed by Auburn Engineers in late 1994 and early 1995. The interested reader will note that about half of these project solutions cost less than $500 per project. Most of these projects (98%) cost less than $5,000.

Tracking costs like this places an emphasis on the low-cost (yet effective) solutions, and it sends a clear and simple message to ergonomic problem solvers. The message is "Solve the problem, but spend as little money as possible. If we save money on one project, then we have more left for additional projects."

2.2.5   Use Resources Efficiently (The Self-Help/Skilled-Help/Expert-Help Strategy)

A successful ergonomics program will ensure that it uses resources as efficiently as possible. The major costs are personnel and hardware. Hardware costs were outlined above. Personnel costs can be controlled by delegating the ergonomics problems to the correct skill level. Too many ergonomists get involved in projects that do not require their level of skill, and which they should delegate to others.

Efficient problem solving uses a stratification based on the difficulty of problems using three levels called self-help problems, skilled-help problems, and expert-help problems. For most organizations, the number of self-help problems is the largest. Following is a list of the three types of problem-solving groups, a brief description of each type, and the training required:

- *Self Help.* Self-help is the lowest cost method to resolve an ergonomic problem. Self help also requires the lowest level of problem-solving skill. Self help solutions are usually generated by a worker along with his or her supervisor. Awareness training provides the workers with the necessary skills to determine the self help solutions by familiarizing them with symptoms of musculoskeletal disorders, workplace risk factors, and ways to reduce risk factors.
- *Skilled Help.* This is the second lowest cost method of problem-solving help. Skilled help typically involves a problem-solving team comprised of workers who have had ergonomics problem-solving training. Awareness training is inadequate to perform at this level.
- *Expert Help.* Expert help is provided by an expert ergonomist, typically a corporate ergonomist or an outside ergonomics consultant. This is the most expensive level of help and should be utilized for problems that are too difficult to solve or go beyond the knowledge and skill of the other two help levels.

A summary is shown in Table 2.2.

2.2.6   Identify and Overcome Barriers

Successful ergonomics programs identify and overcome barriers to their success. There are many barriers which occur with any new initiative, and to be successful, the barriers must be identified, resolved, and

TABLE 2.1   Cost of Ergonomics Solutions

Cost per Project	Number of Projects	Cumulative % of All Projects
Less than $100	11	22%
$100 to $500	12	47%
$500 to $1,000	11	69%
$1,000 to $2,000	7	84%
$2,000 to $5,000	7	98%
Over $5,000	1	100%

TABLE 2.2   Efficiently Using Problem-Solving Skills

Skill Level	Self-Help Problems	Skilled-Help Problems	Expert-Help Problems
Occurrence	50–70% of problems	20–40% of problems	5–15% of problems
Typical problem	Simple and "Quick Fix" problems	Multiple workplace changes or complex diagnosis	Most complex problems, unique problem, complex multi-part solution, expensive solutions
Typical solution	Adjust workplace, proper pace, use proper tools, awareness of early warning symptoms	Workstation redesign, modification of production process, new assembly tools	Unique tool, design of entire factory, redesign of production process
Expertise	With minor training, can be resolved by many people	Requires training, practice and possible guidance	Require special expertise or professional judgment
Typical training	2–4 hours of ergonomics awareness training	2–3 days of team-based ergonomics problem solving	Graduate degree plus professional experience

eliminated. Unsuccessful ergonomics programs will either be stopped by these barriers, or will be overcome by them. There are four typical barriers within an ergonomics program as shown in Table 2.3. In addition, some typical methods to overcome these barriers are also listed.

It is important to address barriers to the success of the ergonomics program early and often. The most successful programs will address barriers during the initial strategic planning, and will discuss barriers during each planning session. Once the barriers are identified, corrective actions are planned and implemented, and follow-up is done to ensure that the barriers are not impeding progress with installation.

2.2.7   Training Should Be Supported by Suitable Infrastructure

Training is a valuable part of an ergonomics program, but it can be done before the organization is ready for it. One important objective of an ergonomics program is to ensure that people are aware of the signs and symptoms of musculoskeletal disorders. General awareness training of these disorders is an appropriate way to meet that need. But what happens too often is that this awareness training occurs much too early in the program. Once this training takes place, many ergonomic concerns quickly come to the surface — some very important, but many of less significance. The dilemma is that each situation should be evaluated reasonably soon or the ergonomics program loses credibility. When an operator

TABLE 2.3   Four Typical Barriers

Barriers	Methods to Overcome Barrier
Not enough time	• Determine "Top Five" or "Dirty Dozen" problem areas • Avoid "paralysis by analysis" • Enable others and get them involved • Buy additional time
Too little money	• Use low-cost/high-value solutions • Use nickel and dime solutions • Avoid cost/benefit justification • Cluster projects • Get "refillable pot" of funds • Use two-step solutions
Gaps in skills	• Provide specific training (and only as needed) • Look for existing solutions (remember worker modified solutions) • Use teams for simple problems and experts for difficult problems
Management concerns	• Propose a specific plan • Answer the 5 questions managers ask • Develop an ergonomics culture • Understand change management

recognizes an ergonomics problem and asks for help, there is a limited amount of time before the worker suspects that ergonomics is just another management fad.

When numerous situations surface at once, particularly if the ergonomics program is new, there are insufficient resources to deal with everything in a timely manner. This creates a big problem for the ergonomics program — too many requests, not enough time, and lots of frustration, discontent, and loss of credibility.

This is clearly a situation which a successful ergonomics program should avoid, and the recommendation to wait for the ergonomics program to mature a bit and develop a suitable infrastructure of people with some basic skills prior to conducting widespread training. An organization which has performed passive or active surveillance is well aware of the more serious problems anyway, and is probably already dealing with them. Therefore, little is lost by waiting to perform training, and the ergonomics program will avoid generating substantial negative publicity and discontent.

2.2.8 Avoid Using "Ergo-Babble"

One common contributor to the lack of success of an ergonomics program success is the extensive use of technical jargon. This "ergo-babble" is rarely understood and results in a loss of support by managers and the line workforce. Often, after a long (and difficult-to-understand) medical term has been used to describe a workplace illness, the operator is left in the dark and is only too willing to leave the situation to the "doctors and engineers" to fix. Unfortunately, many ergonomic problems which could be corrected by operators are left untouched once this scenario begins to play out.

It often takes more effort to describe a musculoskeletal illness in terms a lay person can understand, but the effort is worthwhile in gaining support for the ergonomics program.

2.2.9 Clearly Define the Purpose of Your Ergonomics Program

The successful ergonomics program has a clearly defined objective. Programs which flounder and fail have unclear objectives. Programs which have been put in place to "do ergonomics" will likely fail relatively quickly. Defining the objectives may take the form of a mission statement or vision statement. Typically, vision statements are longer than mission statements, and "paint a picture of what life will be like at some point in the future." Some examples are provided in Table 2.4 and Table 2.5.

2.2.10 Plan the Stages of the Ergonomics Culture Change

Successful ergonomics programs are guided by a knowledge of organizational culture change models. There is a body of knowledge generally titled "change management" which deals with change in organizations. One especially helpful part of this technology is that it outlines the steps which must be followed before commitment to successful organizational change occurs. As the ergonomics program matures, it goes through six distinct stages, each with separate concerns and issues. The six stages, using layman's language, are:

1. *Awareness* that a change is necessary (e.g., injuries are excessive)
2. *Acceptance* of ergonomics as a tool that can help
3. *Trial* using ergonomics to see if it works
4. *Regular use* of ergonomics because it does work
5. *Procedures* written to include ergonomics
6. And finally, a *culture* that is totally supportive of the use of ergonomics

These stages are outlined in Table 2.6, along with brief comments about some key issues.

Unfortunately, few people are familiar with the stages of change or with the pitfalls that occur when the stages are not followed. The most common problem results from the tendency to skip the first two stages of creating awareness and acceptance. The next most common problem is to attempt to jump over stages,

TABLE 2.4 Examples of Mission Statements

Mission Statement	Comments/Observations
Example #1 The Ergonomics Committee will develop systems for the multidisciplinary study of the problems that exist between people, the tools and machinery they use and their work environment. These systems will initially focus on reducing injuries/illnesses related to cumulative trauma disorders of the upper extremities and backs by hazard prevention and control, medical management, and training/education. This will lead to an increased level of comfort at work, improved quality of product, and greater productivity. This mission will be obtained with management support and associate involvement.	Positive Points 1. Takes ergonomics beyond injury/illness into quality and productivity 2. Will not do all the work itself, but will develop systems for ... Areas of Concern 1. Over emphasis on CTDs 2. Somewhat long 3. Mentions how this will be achieved which may be unnecessary
Example #2 The Ergonomics Committee will develop and manage systems for the improvement of the conditions between people, tools, machinery, and their work environment.	Positive Points 1. Short 2. Will develop systems for... Areas of Concern 1. When is the mission completed?
Example #3 The ergonomics program provides education, analysis, and guidance to prevent and alleviate ergonomically related stress and illness in order to protect the health, and further, the productivity of the plant.	Positive Points 1. Mentions prevention 2. Relatively short Areas of Concern 1. The program should set up systems to provide... because when the program ends ergonomics will be part of the culture 2. Ergonomics can go beyond stress and illness (it's not just CTDs)
Example #4 To put into place the will and skill to eliminate and/or prevent ergonomics problems (pain, illness, injuries), and to capture quality and efficiency benefits so that ergonomics becomes institutionalized.	Positive Points 1. Will put into place ... 2. Relatively short 3. ... becomes institutionalized Areas of Concern 1. A little wordy 2. Stilted language

for example by going from "Stage 1 — Awareness" to "Stage 5 — Procedure." While jumping any stage causes resistance to the change, jumping multiple stages creates even more resistance.

Another way to look at the program stages is to think in terms of child development, from infant, toddler, child, adolescent, young adult, and finally mature adult. As the child grows, its needs change and it no longer responds to things the way it once did. An ergonomics program is very similar.

2.2.11 Create a Strategic Plan

Successful ergonomics programs have a strategic plan. A strategic plan is necessary to guide the ergonomics program. The strategic plan defines what the ergonomics program intends to accomplish over the long term. Some organizations use vision statements or mission statements to describe program objectives, and these are very helpful concepts for the program.

However, a mission or vision statement, by itself, is not sufficient to fully describe the strategic plan. To develop the strategic plan, the following questions need to be discussed and answered:

1. *What do we want the ergonomics program to do?* This usually becomes the mission, vision and scope of the ergonomics program
2. *How do we monitor results?* What data do we measure to demonstrate progress with ergonomics?

TABLE 2.5   Examples of Vision Statements

Example	Vision Statement
Vision Statement #1	The vision was addressed from two different time frames — long term and over the next 12 months. Both are important because they provide the framework necessary to build the appropriate program. Long Term • Ergonomics is part of our culture. We don't think about it separately any more. • Ergonomics improves safety and health, improves plant performance (worker productivity, product quality, cost control) and improves the quality of worklife (QWL) of workers. • Ergonomics is used before as well as after injuries/illnesses occur. Prevention is common. • All aspects of human performance are considered part of ergonomics, including such issues as heat stress and human error. Within One Year • Ergonomics will be more commonplace with a great deal more awareness. There will be successful projects completed, ergonomics reviews of new designs, supervisors evaluation of jobs, and illness investigation procedures. • Ergonomics will still be primarily a safety and health issue. • The plant will be addressing all known problems. • The major areas of emphasis will be cumulative trauma disorders along with manual material handling type injuries.
Vision Statement #2	The program will undergo a change in focus and activities. The focus will go from pain reduction to maximizing effectiveness on the job. Maximizing effectiveness on the job includes all aspects of performance, such as a safe and healthy workplace, the ability to produce high quality goods, highly productive workplaces that don't waste time and energy of the workers, and high quality of worklife. Ergonomics is one of several tools used to maximize effectiveness on the job. During and after the transition, there will be technical ergonomic resources available to the plants for projects and for auditing assistance. Auditing for ergonomic concerns will become part of normal auditing procedures used for other safety and health audits.
Vision Statement #3	The guiding principles involved with this organization are: • To push problem solving down to the working level in the organization • To spread ergonomics throughout the organization by heavily involving others • To avoid making ergonomics an "overlay" that just adds work • To ensure that ergonomics solutions dovetail with other changes being made These principles dictate that the ergonomics task force be more of a facilitator, technical resource and trainer than a problem solving group. Where appropriate, the Ergonomics Committee will either work with a group who is already studying a job or task, or will request a specific problem solving team to consider the ergonomics issues. If a team is not available, then the Ergonomics Committee may work on the project itself.

3. *What are the barriers?* And how can they be overcome?
4. *What policy issues* are likely to be affected?
5. *Who is/should be involved, and what are their roles?* This includes both the ergonomics committee and ergonomics problem-solving groups
6. *What is the priority?* How long should it take to finish the job? What resources are available to us?
7. *When and how should we review our plans* and progress with management? Are there others who need to hear our story?

2.2.12   Understand the Difference between an Ergonomics Program and the Practice of Ergonomics

Many ergonomists have had training in the identification, analysis, and resolution of ergonomics problems. They have not typically had training in the management of complex programs such as the ergonomics program. Without this background, they have difficulty visualizing the roll-out of the program, planning for resources, estimating degrees of success, and conceptualizing how the ergonomics program will mature and what it will be like once it is completed.

TABLE 2.6   Stages of an Ergonomics Program

Area	Stage 1 Awareness	Stage 2 Acceptance	Stage 3 Trial	Stage 4 Regular Use	Stage 5 Procedure	Stage 6 Culture
Brief description of stage	Learning about ergonomics	Positive image of ergonomics	Willingness to give it a try	Multiple ergonomics projects	Ergonomics in operating and design	Inconceivable not to use ergonomics
Ergonomics	"Ergo-What?"	Oh, yes, ergonomics. That sounds interesting	Ergonomics — it should reduce injuries	Ergonomics is more than injury prevention	Ergonomics is human performance	Ergonomics helps with every aspect of our business.
Results	None	None (But wants to hear about success stories from others in this industry)	Very limited (Results only with specific projects — still going on faith)	Paying off (Still used mainly for injury prevention)	Solid benefits (Results in safety and health, performance, cost reduction)	Solid benefits (But little need to measure benefits any longer)
Management feelings	Skeptical	Acceptance (grudging to willing)	Prove it to me! (On our site)	Yes, it works, but can you do it again? And lets show some payoff	This stuff really works. I'll have everyone use it	"And why didn't you think of ergonomics? We always use it"
Ergonomics Committee feelings	Why us?	Learning	OK, I hope it works.	I hope these other people understand it like we do	All we do is training. Will we ever get done?	That was a great committee. I'm glad I was on the team!
Role of ergonomist	Advocate	Assurance of others	Leading the effort	Facilitator and training	Builder of others; ensures systems in place	Maintenance

It is important to clearly distinguish the two areas and the skills necessary to be successful at each. If it is necessary, provide both program management support and ergonomics technical support for the ergonomics committee. Ergonomics technical support may be provided by the ergonomist while program management support is provided by a mid-level manager. Figure 2.5 provides a distinction between the two areas.

2.2.13 Create a Tactical Plan

The strategic plan is necessary but not sufficient for ergonomics program success. There are two types of plans which are necessary for a successful ergonomics program:

- Strategic plans determine what are we trying to accomplish
- Tactical plans determine specifically how we accomplish those goals

A tactical plan is the month-to-month and week-to-week plan which outlines the jobs to be analyzed, the procedures to be written or reviewed, the training to be accomplished, and the solutions to be implemented. Many ergonomics programs fail to develop tactical plans. Some people develop initial plans but then fail to maintain them.

High-quality tactical plans should be developed initially, then monitored to ensure that the planned activities are accomplished. Experience with the most successful ergonomics programs indicates that tactical plans must be reviewed monthly, revised quarterly, and fully reviewed and updated semiannually.

Sound tactical plans are the key to obtaining funding and personnel to accomplish the ergonomics program goals. Once the tactical plans are developed, management can easily see the costs (financial resources and personnel) as well as the benefits (expected improvements from proposed projects). This makes approval of the ergonomics program plans easier, and allows management to clearly understand what to expect from the ergonomics program.

In developing the tactical plan, the following questions must be answered many times:

- What activities should be done?
- When should they be done?
- Who should do them?
- What are the quality standards?

An example of some activities found in a tactical plan are shown in Table 2.7.

2.2.14 Ensure That There Is Regular Quantitative Evaluations of the Overall Ergonomics Program

Successful ergonomics programs measure themselves. And they do it regularly and often. The adage "You can't manage what you don't measure" is certainly true when it comes to ergonomics programs. Yet, few

(a) (Technical skills)
Job analysis
Solving Problems
Preventing problems

(b) (Managerial skills)
Planning
Coordination
Evaluation

FIGURE 2.5 (a) Ergonomics practice and (b) Ergonomics programs.

TABLE 2.7   Abbreviated Example of a Tactical Plan

What?	Will be Done by Whom?	By When?	Check When Complete
1. Plant strategy developed	Committee	May 5	
2. Ergonomics policy written	Committee	May 5	
3. Roles/responsibilities developed	Committee	May 5	
4. First problem solving training	Don	June 3	
5. Initial 2 projects started	Todd, Mary and teams	June 6	
6. Baseline data collected for measurement systems	Alice	June 12	
7. Initial project recommendations implemented	Todd, Mary	July 15	
8. Next 2 projects initiated	Maria, Jude and teams	July 6	
9. Illness investigation procedure tested	Bruce	July 20	
10. Project recommendations implemented	Maria, Jude	August 20	
11. Next 2 projects initiated	Cindy, Ron and teams	August 4	
12. Detailed surveillance completed	Bruce	August 16	
13. Policy issues identified	Mike	August 24	
14. Project recommendations implemented	Cindy, Ron	September 24	
15. Next 2 projects initiated	Todd, Maria and teams	September 7	
16. Plans for proactive ergonomic program developed	Ivan	September 9	
17. Project recommendations implemented	Todd, Maria	October 12	
18. Next 2 projects initiated	Mike, Randy and teams	October 5	
19. Engineering training started	Ivan	October 14	
Continued			

ergonomics programs are quantitatively measured on any regular schedule. With beginning programs, rigorous evaluations are seldom required. However, as an ergonomics program begins to mature and the "honeymoon period" ends, management often asks about progress and results. Without ongoing evaluations, it is difficult to respond with any specificity to these questions. There are a number of audit and assessment tools available to evaluate ergonomics programs, and an example is provided using the assessment tool *Assessing Your ERGONOMICS Program* developed by Auburn Engineers. This audit tool is widely used, in part because of its simplicity and ease of use. It has 50 multiple choice statements, and requires less than one hour to complete.

An example of the scoring provided by *Assessing Your ERGONOMICS Program* is shown in Figure 2.6.

The interpretation of this assessment is equally easy. From the bars shown in Figure 2.6, these conclusions can be drawn:

1. The areas of organization, medical management, and correction are adequate
2. There is a need for more balance in this program because there is a large gap between the best and worst areas assessed

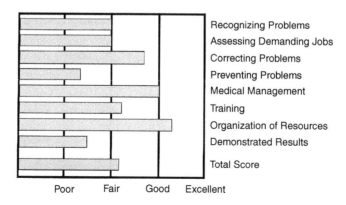

FIGURE 2.6   Scoring the success of an ergonomics program.

3. There is a need for more work on prevention and on demonstrated results
4. This is average progress for an ergonomics program which is 6 to 12 months old

2.2.15 Do Not Wait for Top Management to Push the Program Down

Unfortunately, for many programs, a litmus test of acceptance is the degree to which top management supports the program. Top management support is too often seen as a requirement for ergonomics program success, and without an endorsement, nothing happens.

Successful ergonomics programs do not require top management support. A caveat is important: this is not to suggest that one simply take on ergonomics projects with no management support or against management's directions. There are usually several layers of management within an organization, and one does not need to wait for top management endorsement to begin. Even with the support of lower level management, much can be accomplished, thus laying the foundation for a larger program later on. This initial success with positive activities and projects now permits the easy endorsement of the program by management.

There are several distinct advantages to getting things going, at least on a small scale, with some important projects. Clearly, this will begin the process of eliminating some difficult tasks, thus making the workplace a little safer. By starting with some projects, one can determine how to go about the process of solving ergonomic problems, thus gaining valuable experience. Many of these projects become examples of the types of changes which can take place with ergonomics, and therefore serve as good illustrations to use once the program begins to grow.

Finally, it is important to note that good ideas "catch on" on their own, while poorly conceived ideas are stopped or die from lack of interest. If ergonomics is seen to work well, it will take hold and people will soon be asking for help with more ergonomic projects. However, if the ergonomics program seems to go nowhere without the push from top management, then perhaps it is a poorly planned and illmanaged program.

2.2.16 Maintain Political Support

Successful ergonomics programs have internal political support when they need it. It is necessary to develop and maintain strong political support with the organization's safety and health committee, with senior management, and with key staff groups like engineering and health services.

This is done with frequent contact, seeking and using input, and by openly sharing what is happening. Even small successes should be shared with others, and credit should passed liberally around the organization. Publicity plans should be developed which permit everyone who contributes to share in the limelight. If these things are done as the program grows, then the political support will be there when it is needed.

2.3 Conclusion

There is enough information regarding ergonomics programs to know what factors contribute to their success. An ergonomics program manager should review the sixteen factors, assess their presence in the program, and if missing, seek to implement them as soon as practical.

For Further Information

The Practice and Management of Industrial Ergonomics, David C. Alexander, Prentice-Hall, 1986.
The Top Ten Reasons Ergonomics Programs Fail, Auburn Engineers, Inc., Auburn, AL, 1994.
Assessing Your ERGONOMICS Program, Auburn Engineers, Inc., Auburn, AL.
The Economics of Ergonomics, Auburn Engineers, Inc., Auburn, AL, 1994.

The Economics of Ergonomics, Part II, Auburn Engineers, Inc., Auburn, AL, 1995.

Advanced Techniques for Managing Your Ergonomics Program, a short course sponsored by Auburn Engineers, Inc. Some specific items used for this paper which are included in that course are:

- Selling Ergonomics to Management
- A Model Ergonomics Program
- Stages of an Ergonomics Program
- Defining the Ergonomics Culture
- Building Commitment to Ergonomics

3

Practical Interventions in Industry Using Participatory Approaches

Barbara Silverstein
Peregrin Spielholz
*SHARP, Washington State
Department of Labor and
Industries*

Ernesto Carcamo
*WISHA Services, Washington
State Department of Labor
and Industries*

Producing change in any organization is a challenge, whether it is a small business or a multinational corporation. Different approaches may be used to reach the same goal. However, existing frameworks are available to develop positive solutions addressing ergonomics issues in any environment. Elements consistent with most effective programs include: worker participation, management commitment, employee training, methods for identifying problems and controls, and medical management. Previous study has shown that successful programs most often identified practical, low-cost solutions that led to significant benefits.[8] These benefits may include not only reduced injuries and workers compensation costs, but also reduced absenteeism, as well as increased morale, productivity, and product quality. Achieving and sustaining these positive results can be attained through cooperative work toward a common goal.

Musculoskeletal risk factors and subsequent solutions may go unrecognized until a valuable worker is disabled unless workers participate at some level of the process. Participatory ergonomics is a way of harnessing the knowledge and ingenuity of workers into improving the work environment and work product. Participation can occur at many levels and in many forms. The result is usually the implementation of very practical interventions once participants have been educated as to the program goals, methods, and processes.

This chapter focuses on practical interventions that have been implemented in a variety of work environments in Washington State. Although some were developed prior to the promulgation of the Washington State Ergonomics Rule (http://www.lni.wa.gov/WISHA/Rules/generaloccupationalhealth/ PDFs/ErgoRulewithAppendices.pdf), most projects identified interventions that would be in compliance with the Rule, as it was phased in over a period of 6 years. The strategies and successes presented were developed through either individual company or industry group initiatives using different types of participatory approaches.

A "Hazard Zone" checklist was included in the Ergonomics Rule to assist those who needed help with identifying high-risk activities (see Appendix). This structured approach for those with limited knowledge is thought to be more effective than an open-ended approach for identifying risk factors and interventions.[12] Use of multiple information sources will increase intervention opportunities.[13] Most projects went well beyond the rule requirements in implementing good and sustainable ergonomics processes and practices that included:

- Strong management commitment to and employee involvement in the ergonomics process
- Systematic review of existing jobs for musculoskeletal risk factors
- Providing ergonomic awareness education for employees so they could participate in the process
- Detailed assessment of jobs that might put employees at risk of developing or exacerbating musculoskeletal disorders
- Developing and testing potential control measures
- Implementing controls
- Training employees in the use of new control measures
- Evaluating the effectiveness of the controls in reducing exposures to acceptable levels

Most of the companies and industry groups also reviewed injury statistics to identify additional jobs that required intervention even if they did not exceed regulated levels. As companies and workplaces became more comfortable with identifying problems, they began to move more upstream in the process. More information about these specific projects, as well as the Washington State Department of Labor and Industries Ergonomics *Ideas Bank* can be found at: http://www.lni.wa.gov/Safety/KeepSafe/ReduceHazards/ErgoBank/. An *ergonomics program checklist* (see following box item) can also serve as a reminder of the need to assess sustainability of a new or existing intervention focusing on ergonomics improvements.

Ergonomics Program Checklist	Yes	No
1. Program Goals	☐	☐
2. Management Commitment	☐	☐
3. Communication Plan	☐	☐
4. Program Resources Identified and Allocated	☐	☐
5. Formal or Informal Employee Involvement	☐	☐
6. Employee Training	☐	☐
7. Supervisor and Management Training	☐	☐
8. Risk Prioritization Protocol	☐	☐
9. Risk Assessment Plan	☐	☐
10. Solution Development Plan	☐	☐
11. Employee Participation in Solution Ideas	☐	☐
12. Follow-through on Identified Solutions	☐	☐
13. Program Evaluation Plan	☐	☐

Practical ergonomic interventions for industrial use can be developed, implemented, and evaluated with different levels of participation by potential users of the intervention. This can happen at the societal level

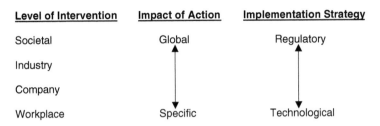

Level of Intervention	Impact of Action	Implementation Strategy
Societal	Global	Regulatory
Industry		
Company		
Workplace	Specific	Technological

FIGURE 3.1 The impact and strategies of different levels of ergonomics interventions.

(regulations and guidelines), an industry-wide level (best practices), company level, or individual workplace level (Figure 3.1).

- *At the societal level,* industrial partners (business and labor representatives, health/safety/ ergonomics experts, professional societies, and government officials) participate in identifying at least minimally acceptable criteria for regulation and codes of practice.[2,3,22–24] Similar partners are engaged in the draft Machine Directives of the European Union (pCEN 159). The history of the recently defunct ANSI z365 committee that was developing a voluntary consensus standard on the control of work-related musculoskeletal disorders suggests that some social partners do not share the same objectives. Societal regulatory and potentially regulatory (ISO; ANSI) interventions usually take many years to develop and implement. There has been considerable debate on how detailed guidance should be included in standards, guidelines, and regulations, with some suggesting that small employers need more specific information than larger employers.[7] With a few exceptions, systematic evaluation of their impact on implementing ergonomics in regulated communities has not been addressed.
- *Industry-specific guidelines* (e.g., American Furniture Manufacturing Association[1]) usually involve industry, vendors, government, and consulting experts but often exclude labor unions or hourly workers. When government is the initiator of industry-wide guidelines, there tends to be labor participation as well as concern by business partners that the guidelines will be used for enforcement purposes. Systematic evaluation of the impact of voluntary guidelines has not been well reported. Participatory ergonomics activities with workers in the same industry but working alone, such as in home health care, can also lead to significant improvement in both psychosocial aspects of work and work ability.[19] The participatory ergonomics approach was very successful in the Dutch bricklaying industry where more than half had implemented the interventions because they were feasible, and improved comfort and production.[4]
- *At the multisite company level,* there have been a number of successful ergonomics programs developed which have resulted in ergonomic improvements in existing processes and increasingly upstream in the design of production processes and products.[11,21,26] The nature of employee participation in ergonomics activities can vary depending on a variety of factors related to organizational complexity and culture. A primary consideration is the number of different groups affected by the decision-making. A larger company may involve labor unions, different levels of management, as well as different affected groups of employees. Monitoring and maintaining supervisor "buy-in" is necessary for the viability of implementation of ergonomic interventions[20] although a "sense of empowerment" may be the initial view of worker teams without supervisory involvement.[16]
- *At the small single location workplace,* the process of implementing ergonomics improvements is often less structured and is dependent on the complexity of the change.

The more that ergonomics is integrated into existing organizational structures and involves workers throughout the process, the more sustainable and accepted continuous ergonomics improvement will become.[8,15,17] Workers are usually the most knowledgeable about the existing jobs and often have ideas for improvement. They also provide invaluable insight into new prototype design and "build ability." However, unless knowledge of ergonomics is an integral part of their training, full utilization of their knowledge and skills may not be realized.[10] Different models of participatory ergonomics have been used with success by a wide range of organizations.[5,6,11] Worker participation in ergonomics activities is germane to both industrially developed and developing countries.[9,25] As with all workplace improvement efforts, without strong management support, these efforts are not sustainable. There are different participation models (Figure 3.2) including quality improvement work teams, labor–management committees, direct worker involvement, industry-wide focus groups, and worker consultation.[18]

3.1   Work Teams

Many companies employ teams of workers and supervisors as part of on-going quality improvement programs in their work areas. These types of formal programs have taken different forms and names over the years, from quality circles and lean manufacturing to Motorola's Six Sigma™ process. The primary focus of these groups is identifying inefficiencies and putting improvements in place. This type of regular evaluation process offers a readymade framework for implementing ergonomics. Generally, improvements in a production environment will include facilities modifications that may be performed by a maintenance or engineering department. Inclusion of these workers can increase buy-in and aid in the timely implementation of the identified improvements.

The following are examples of the use of work teams in diverse industries in developing and implementing ergonomic improvements:

Company: Aladdin Hearth Products, Colville, WA, medium-size employer
Products: Manufactures wood, gas, and pellet stoves and fireplace inserts

The metal components of a stove are heavy and cumbersome to handle. The stove assembly provides for parts to be moved from workstation to workstation on rollers. The rollers are mounted on height-adjustable scissors lifts (Figure 3.3 and Figure 3.4), so that employees can work on the top and sides, without lifting or in awkward postures. The recent move to a new facility created an opportunity to design workstations and methods incorporating ergonomics principles. The company has closely integrated safety, ergonomics, and quality functions to make changes, which enables a threefold impact. Ergonomics Rapid Continuous Improvement (ERCI) is the Kaizen model followed to introduce ergonomics principles. This involves dedicating 1 week to have problem-solving teams (employees heavily involved) and focused on just one workstation, identifying problems and brainstorming solutions.

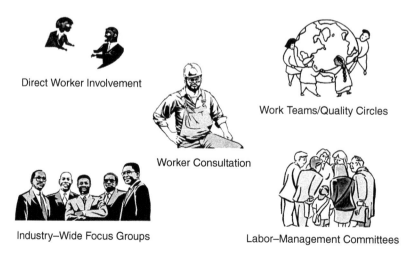

FIGURE 3.2   Models of participation in the ergonomics process.

FIGURE 3.3 Scissors lift tables with rotating tops allow height adjustability for the worker, and prevention of neck and back awkward postures.

Key features of the model at Aladdin Hearth Products:
New facility design brings an opportunity for ergonomics interventions to workstations and methods
Production employees, office employees, engineers, and outsider are the problem-solving team
Kaizen-type model: Ergonomics Rapid Continuous Improvement (ERCI)
Focus on one workstation per week

Key outcomes:
Several solutions developed in-house at low cost; some investments in new equipment
Improved working postures and reduced lifting heavy parts
Lower injury rate
Reduced costs
Remain competitive and in business

Company: Precor Inc., Bothell, WA, medium-size employer
Products: Treadmills, cross-trainers, stair climbers, exercise cycles, stretch training and strength training equipment

This company has taken the lead in applying lean* ergonomics principles to provide safer and simpler work processes to their employees. Their model uses "cells" that include the participation of line operators, design engineers, sales representatives, and management. Teams were formed throughout the assembly areas. The primary objective of each cell team was to solicit ideas from assembly level

---

*Lean ergonomics is elimination of inefficiencies to achieve the most production for the least cost. There are further opportunities to incorporate efficiency into the existing lean systems in the form of ergonomics or ergonomic solutions. Lean is fundamentally a relentless commitment to eliminate waste in all forms — starting with the cost of injuring employees on the job. The ordered priorities of a lean system are, in fact: safety, quality, delivery, and cost. This has strong intuitive appeal.

FIGURE 3.4   Roller conveyors on the assembly line eliminate lifting heavy parts.

workers and track/implement them on a large scale. What happened quickly was a system of idea generation and quick follow-up that snowballed into a process that drove down assembly times and resulted in an unprecedented willingness to contribute by all people.

The cell meetings were conducted in a way that everyone felt at ease with sharing. Employees feel rewarded when they perform the change themselves or they see the change happen at someone else's hand, because it was their idea. After hundreds of these small to large ideas are driven to completion, more ideas get generated. Precor believes that as long as they support cell meetings, fund the changes, and continue "every idea counts" in the cell meetings, that the "idea mill" will continue to produce positive outcomes.

The results of the Precor program show that recordable injuries at the company have decreased in the same time period, from a rate of 11.81 in 1999 to fewer than 3.4 in 2002. Workers compensation costs fell from $0.17/paid hour to less than $0.05 per hour in the same time, resulting in a direct cost savings of $92,000.

Key features of the model at Precor Inc.:
     Work cell meetings and employee contributions
     On-going implementation of changes by the engineers and facility maintenance
     Continuous improvement and idea generation

Key outcomes:
     Reducing heavy lifting in manufacturing and assembly operations.
     Involving its employees in continously finding and testing solutions
     Implementing a systematic approach to improving materials handling
     Investing in lift-assist equipment and hoists
     Reduction in recordable injuries and compensation costs
     Increase in productivity

Company: Bare Root Tree Nursery, Western Washington, medium-size employer
Products: Bareroot tree seedlings

The work team process (Figure 3.5) was also implemented successfully at a bareroot tree nursery in Washington State. The program was initiated by identification of a team with representatives that included management, supervisors, and workers from different areas of the nursery. The team was given a 2-day training on ergonomics. The team then identified risk factors present in the nursery

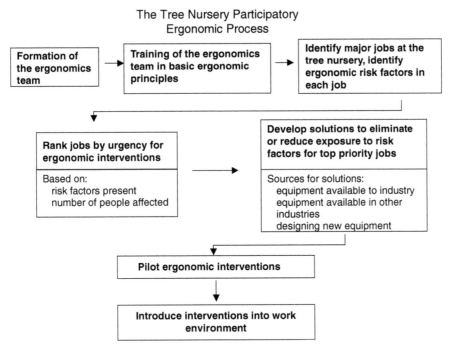

FIGURE 3.5 Participatory process diagram used in implementation of an ergonomics program in bareroot tree nurseries.

jobs. These risk factors and jobs were priority ranked by each of the team members to achieve an overall priority list. Solutions were brainstormed for each of the highest-priority tasks. Ideas that appeared feasible were further investigated by consulting ergonomists and implemented in the nursery (Figure 3.6 through Figure 3.9). Several low-cost tool modifications and process improvements were identified, which reduced injury risk factors and helped improve productivity.

Key features of the model at the Bare Root Tree Nursery:
 Training with formal evaluation and solution identification program
 Team approach to risk identification and prioritization
 Implementation of low-cost solutions for prioritized jobs
 Ergonomics consultant discussing solutions

Key outcomes:
 Development of solutions for high-risk activities
 Implementation of engineering solutions in packing shed and greenhouse
 Process improvements reduced risk factor levels and increased productivity

---

Company: Ace Hardware Distribution Center, Yakima, WA, medium-size employer
Products: Goods distribution and transportation

A project at an Ace Hardware distribution center in Washington State demonstrated how safety team members could quickly identify hazardous risk factors and implement low-cost, effective solutions. Jobs evaluated at the site included stockers, order fillers, truck loaders, truck drivers, janitors, and administrative support. Company commitment to increasing productivity and decreasing injuries was critical to program success. Process improvements included palletizing more goods to make forklifts and

FIGURE 3.6 *Before*: Hand washing of blocks required static, prolonged gripping, prolonged standing in one position, and exposure to a cold, wet environment.

pallet jacks the primary means of truck loading instead of manually handing each item. Order fill totes are limited to 50 lb by company policy to better match physical capabilities. New containers are transparent, which allow the workers to visually inspect the contents and make informed handling decisions while reducing unexpected lifting loads.

Several specific engineering solutions were implemented at the Ace Hardware distribution center. Figure 3.10 and Figure 3.11 show how storage bins were raised from floor level in the racks to reduce awkward lifting postures. An inexpensive hook, shown in Figure 3.12, was designed and fabricated in-house to allow the case order filler to pull items in the back of the bin forward, reducing awkward reaching postures.

FIGURE 3.7 *After*: An automated block washing system to reduce physical risk factors and increase productivity.

FIGURE 3.8 *Before*: Carrying individual blocks in and out of the greenhouse requiring repetitive handling.

Key features of the model at the Ace Hardware Distribution Center:
 Formal evaluation by team members of most jobs
 Identification of high-level risk factors and solutions
 Follow-through on development and implementation of interventions
Key outcomes:
 Process improvements that included more use of forklifts
 Implementation of a 50-lb lifting limit policy
 Introduction of new containers and improvements in material storage
 Development of low-cost interventions such as an order-picking hook

FIGURE 3.9 *After*: Pallets with blocks are wheeled in and out of the greenhouse.

FIGURE 3.10   *Before*: Boxes of products used to be stored on pallets on the ground, requiring back bending to lift and handle the boxes.

Company: Bessie Burton Nursing Home, Seattle, WA, medium-size employer
Products: Nursing and geriatric care

Another ergonomics demonstration project took place in a 139-bed skilled nursing facility. The primary objective of the project was to show how a nursing home could comply with the Washington State Ergonomics Rule. The project found that, while there were a number of risk factors for work-related musculoskeletal disorders (WMSDs) present in this nursing home facility, only manual handling of residents

FIGURE 3.11   *After*: The stored boxes were raised to a more comfortable lifting height to reduce awkward lifting postures.

FIGURE 3.12   An order filler uses a hook designed to reach items in the back of the bins to reduce awkward reaching postures.

would reach the hazard level under the ergonomics rule. The staff at the facility had already implemented controls to address this hazard, and had implemented voluntary improvements to address other risk factors not covered by the rule. A physical therapy assistant, who chaired the joint labor–management health and safety committee, also led the work team implementing ergonomics (Figure 3.13 through Figure 3.15).

Key features of the model at Bessie Burton Nursing Facility:
   Identify jobs in the facility that fell into the "caution zone" job category and would therefore be
      covered by the ergonomics rule
   Determine which jobs had risk factors that would reach a hazard level under the rule, and
      therefore would require controls to reduce risk factor exposure
   Implementation of modifications to decrease exposure to below hazard levels

Key outcomes:
   Patient-handling intervention implementation
   Medication-dispensing solution development
   Patient-bathing solution identification and implementation
   Interventions implemented for dietary, laundry, and housekeeping

3.2   Labor–Management Committees

In many jurisdictions, joint labor–management health and safety committees are required in larger workplaces. In Washington State, more than 50% of larger workplaces address ergonomics issues through the joint health and safety committee compared to 18% of small employers. In some very large organizations, the ergonomics committees may function separately from the joint health and safety committees. Active management representation on a working team or committee is critical to implementation effectiveness. Some participatory ergonomics programs have functioned with outside experts acting as trainers, guides, and consultants during the process, primarily larger employers. In Washington State, only 4% of employers reported turning to private consulting firms for ergonomics

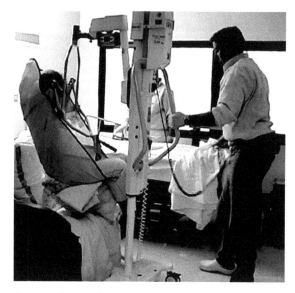

FIGURE 3.13   A total lift for dependent residents to control lifting hazards while handling residents.

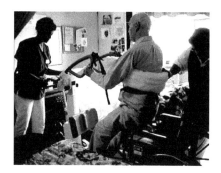

FIGURE 3.14   A sit-to-stand lift for partially weight-bearing residents to control lifting hazards.

FIGURE 3.15   A tilting soup pot allows for easier soup dispensing in dietary services.

information. If the program is internal, having a champion of the program with both technical knowledge and power within the organization can greatly increase the benefit derived from the work. Whoever is responsible for the implementation process should have good organizational navigation skills for working in complex environments.

The following are examples of labor–management participation in ergonomics programs:

Company: Boeing Co., Auburn, WA, large-size employer
Products: Airplane wing skins and spars

Boeing has implemented a successful ergonomics program at the Skin and Spar Manufacturing Business Unit in Auburn, Washington.[26] Management commitment and a participatory process were key factors in the initiative. Team members analyzed jobs following the Washington State Ergonomics Rule and developed solutions for jobs identified as having "Hazard Zone" risk factors and previous injuries. Simple and relatively low-cost solutions were found, which greatly reduced the physical risk factors.

Many ideas at Boeing were off-the-shelf products such as overhead air-balancing systems, low-vibration tools, lighter-weight tools, and material-handling carts. Heavier items were also placed at waist level, and job rotation strategies were implemented where needed. Several creative solutions developed by the team reduced risk factors in relatively simple ways. Nylon handles were fabricated for ultrasonic thickness gauges used frequently to check the diameter of parts. The new handles were three times wider resulting in lower pinch grip forces (Figure 3.16 and Figure 3.17). Chord forming blocks weighing 35 lbs were drilled with holes to reduce the weight by 45% and decreasing the lifting load (Figure 3.18 and Figure 3.19).

Key features of the model at the Boeing Company:
    Communicating the importance of the program to all employees
    Assigning and allowing employees the time to participate
    Resolving difficult issues for the team
    Reviewing the team's recommendations
    Developing plans to implement the recommendations
    Make resources available for ergonomic solutions
Key outcomes:
    Identification of simple, low-cost, and off-the-shelf solutions
    Significant reductions in injury risk-factors through engineering controls
    Administrative and process controls also produced good solutions

---

FIGURE 3.16 *Before*: Quality inspectors frequently used small-diameter gauges requiring forceful pinch grips. (From Boeing Commercial Airplane Group Seattle, WA 2003 and Zawitz Ben, *Ergonomics at Boeing*. Ergosolutions magazine July/August 2003. With permission.)

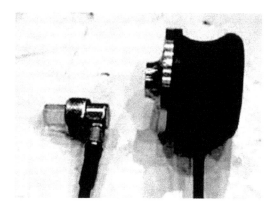

FIGURE 3.17   *After*: Working with in-house resources, newly designed nylon handles were fabricated. These handles are three times larger, thereby reducing the pinch force. (From Boeing Commercial Airplane Group Seattle, WA 2003 and Zawitz Ben, *Ergonomics at Boeing*. Ergosolutions magazine July/August 2003. With permission.)

FIGURE 3.18   *Before*: Mechanics had to repetitively lift 35-lb forming blocks. (From Boeing Commercial Airplane Group Seattle, WA 2003 and Zawitz Ben, *Ergonomics at Boeing*. Ergosolutions magazine July/August 2003. With permission.)

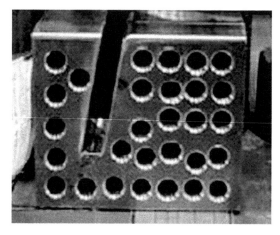

FIGURE 3.19   *After*: A 45% reduction in the weight of the forming blocks was achieved by carefully adding lightening holes to the blocks. (From Boeing Commercial Airplane Group Seattle, WA 2003 and Zawitz Ben, *Ergonomics at Boeing*. Ergosolutions magazine July/August 2003. With permission.)

3.3　Direct Worker Involvement

Inclusion of the users of any product or environment in the evaluation and design phases is crucial. The people that interact with the products have learned through trial and error what the pitfalls may be in a given design. Including internal workers or external customers in the evaluation and design assist with acceptance of change when it is introduced. A common example of direct worker participation is the design of new or remodeled facilities such as the following example:

Company: Metropolitan Market, Seattle, WA, large-size employer
Products: Retail grocery stores chain

Worker involvement in design decisions at a Washington grocery store chain working on a new check stand design for future store openings avoided costly mistakes. Workers were involved in the initial planning stage via interviews with checkers in each store to compile comments on important design elements. These comments were presented at planning and design meetings by a representative checker. A prototype workstation was developed and evaluated with the representative checker and the engineers. Having an experienced person as part of the team with the added input of a cross section of workers proved invaluable. Designers had focused largely on maximizing customer interaction and hand motion. The actual checkers stated that the back and legs were more of a problem that motivated the product designers to change the workstation height and include areas to raise and rest the feet while working.

Key features of the model at Metropolitan Market:
　Interviewing and involving line checkers at the grocery stores
　Inclusion of a worker representative in design meetings
　Prototyping and design-review with workers

Key outcomes:
　Key problem areas were identified by the workers
　Issues in the prototype were corrected before production
　The resulting check stands included dimensions and features preferred by the actual checkers

---

3.4　Industry-Wide Focus Groups

The participatory approach can extend beyond a company to an entire industry. Changes often occur throughout an industry either due to changes in codes and regulations or through changes in technology. These changes can affect people both upstream and downstream. In the case of a product change, the manufacturers and workers will likely be affected as well as the producers. Many issues can be resolved up-front by negotiation and problem solving in industry-wide focus groups.

The perspectives of different groups represented in any committee are often very different. These points of view are likely best communicated face to face. In the case of extremely divergent opinions, the outcome can heavily rely on a mediator and the ability to compromise. Managing change across an industry can be a time-consuming process but can result in standardization and agreement that makes everyone's job easier and leads to better cooperation between partners. Concerns about the trade offs between productivity and musculoskeletal load come from both workers and employers.[14]

The following examples present industry-wide projects in Washington State that produced identification of risk factors and solutions across employers:

Company: Washington State Drywall Industry, Seattle, WA, small-, medium-, and large-size employers
Products: Home and commercial drywall installation

A focus group was formed that included employer, labor, and state representatives as well as manufacturers and industry-wide representatives. Regular meetings were held for over a year and often resulted in somewhat heated arguments between sides. An agreement was reached that allows the work to continue not only by using the same material but also by reducing the heaviest lifting. The lengths of drywall that

TABLE 3.1 Acceptable Lengths of 48 in.-Wide Drywall Boards for *One-Worker Lifts* as Determined by Industry Focus Group

Wallboard Type	Maximum Length
$\frac{1}{2}$ in. regular panels	12 ft
$\frac{1}{2}$ in. all other types	10 ft
$\frac{5}{8}$ in. regular panels	10 ft
Firecode X	10 ft
Firecode C	9 ft
$\frac{5}{8}$ in. W/R Regular Panels	9 ft 6 in
W/R Firecodes X, C	9 ft
Exterior panels	8 ft
Dense glass	8 ft
Coreboard (shaftliner)	10 ft
Abuse resistant ($\frac{5}{8}$ in.)	—
($\frac{1}{2}$ in. only)	9 ft 6 in.

Notes: All abuse-resistant ($\frac{5}{8}$ in.) panels must be lifted by two (or more) workers. All 54 in. width panels must be lifted by two (or more) workers.
Acceptable for the emeritus Washington state ergonomics rule.

one person was allowed to lift were limited to 8 or 9 ft depending on the type of material (Table 3.1). Ultimately, what appeared to be an extremely difficult process produced a relatively simple agreement that could improve working conditions and productivity without negatively affecting the bottom-line of industry member. With time and experience with this method, it may be possible to identify and implement more protective solutions.

Key features of the model in the Washington State Drywall Industry:
  Labor groups wanted protection from what they felt were excessive lifting demands
  Contractor and manufacturer representatives stated that business could not be conducted efficiently by reducing material size
  Government ergonomist acted as mediator in the discussion process
Key outcomes:
  Companies agreed that drywall-handling size requirements were reasonable given the heavy material weights
  Labor representatives were satisfied that the extreme lifts were eliminated while still allowing workers to achieve current production levels
  Manufacturers did not have to retool equipment to reduce material widths

Company: Washington State Sawmills, Seattle, WA, small-, medium-, and large-size employers
Products: Lumber and wood products

Another industry-wide example from Washington State involved the sawmill industry with participants from six mills (general manager, health and safety manager, union representative) and government ergonomists. These companies came together because the industry as a whole had high workers' compensation claims rates. Each member went through the complete process as outlined in the following text. Committee members learned from each other and implemented solutions adapted from learning about how other mills solved similar problems.

Key features of the model in the Washington State Sawmill Industry:
  Review industry-wide workers compensation data
  Identify all types of lumber-handling jobs
  Risk analysis done by ergonomists in the laboratory and presented to industry committee
  Solutions were developed through brainstorming methods

Key outcomes:
  Majority of musculoskeletal disorders were among "lumber handlers"
  High-risk tasks and processes were identified
  Different interventions were identified and implemented based on the type and size of operation
  Industry-specific ergonomics workshops and training materials were developed and provided to
    the entire industry

3.5 Worker Consultation

Working together in regular dialogue with all workers is a technique more often used in very small companies or at temporary or satellite locations such as construction sites. In these cases, either a formal group or a select group of representative workers may not appear feasible or appropriate. However, the principles of participation remain the same. Regular discussion of issues and ideas for change can be part of a formal or informal process. In smaller group situations of less than five or ten workers, an informal process usually makes more sense.

Construction work sites often use worker consultation. Most work sites have regular safety meetings that present a good forum for the inclusion of ergonomics. Typically a very short presentation on a specific topic is given, which could be on an ergonomics issue such as back bending or stairways lighting. The work group then identifies issues or unsafe conditions they have observed. A group discussion can often lead to a speedy resolution.

The following are examples of worker consultation in previous ergonomics projects:

Company: GLY Construction, Bellevue, WA, medium-size employer
Product: Commercial building construction

The worker consultation process was used at a large commercial building project in Seattle. Meetings occurred weekly, where ergonomics issues for tasks in the coming week were identified. These tasks were observed by ergonomists and supervisors and the following week, discussion focused on brainstorming ideas to correct any hazards that were found. Back bending during repetitive tasks and occasional heavy lifting were the most frequent risk factors identified. Each of the hazards was corrected within a week by introducing an intervention, using on-site equipment, rotating worker tasks, or training on work methods. Table 3.2 shows an example of risk factors and solutions developed using this method for commercial carpentry tasks.

TABLE 3.2 Practical Interventions Developed for Hazard Zone Tasks in Commercial Carpentry

Task	Potential Hazard Zone Risk Factors	Identified Controls
Moving equipment	Heavy, frequent, awkward lifting	Use mechanical equipment such as cranes or forklifts Get help from another worker if equipment is not available

Moving material	Heavy, frequent, awkward lifting	Use mechanical equipment or get help if >90 lb
		  Limit loads of multiple pieces of material of 70 lb maximum
		  Lift up 90 lb occasionally using the walk-up technique
		  Place load close to work location, use pallet jacks/slide if frequent

Install deck form sheeting	Back bending >45° more than 2 h	Use a nail-gun with handle extension
		  Rotate between deck support and deck sheet installation
		  Perform cuts on saw horses or plywood stack

Gang form construction	Back bending >45° more than 2 h	Use a screw-gun with handle extension Perform cuts on saw horses or plywood stack

Key features of the model at GLY Construction:
  Weekly discussion with workers about upcoming high-risk activities
  Observation of tasks identified by the workers and superintendent
  Weekly consultation with workers to identify and implement solutions
Key outcomes:
  Identification of hazards through phases of construction
  Real-time intervention and follow-up weekly
  Low-cost, simple solutions were available for all Hazard Zone activities

Company: PW Pipe, Tacoma, WA, medium-size employer
Product: Extruded plastic pipe

Palletizing manufactured plastic (PVC, polyvinyl chloride) pipes required back bending several hundred times during a work shift (Figure 3.20). The ergonomics committee analyzed the operation and proposed a simple cart at knee height to reduce the repetitive back bending (Figure 3.21). Ergonomics had been in use before the Washington State Ergonomics Rule required it. In 2001, the company started an ergonomics committee with employees and supervisors participating, and ergonomics issues are on the agenda in monthly safety meetings. At all times, employees may bring in ideas and deposit them at the box located at the Ergonomics Awareness Board (Figure 3.22). Jobs are analyzed using videotape taken by employees and members of the committee. When the data analysis shows that the job needs to be fixed, solutions are identified and implemented.

Key features of the model at PW Pipe:
  Management commitment
  Employee awareness education and participation
  Joint ergonomics committee
  Ergonomics Awareness Board to communicate to employees about the progress of the projects
Key outcomes:
  Several solutions developed in-house, at low cost
  Improved working postures
  Reduced repetitive lifting
  Management commitment together with employee involvement became part of the continuous
    improvement process

Company: Empire Bolt and Screw, Spokane, WA, small-size employer
Product: Fastener distribution

Program-level initiatives are not possible and do not make sense for many smaller companies. These employers most often rely on worker consultation to identify risk factors and develop solutions. A demonstration project at a bolt and screw distribution company showed what a smaller employer in this type of operation might target to reduce the risk of work-related musculoskeletal injuries. The major operations of this company were similar to most U.S. fastener distribution companies that import fasteners from other countries, repacking them, sell, and distribute in the domestic market.
  The goals of this demonstration project were to:

- Identify caution zone and hazard zone jobs as defined in the Washington State Ergonomics Rule
- Identify engineering and administrative controls to eliminate hazards
- Identify additional controls to improve comfort and productivity even though the Washington State Ergonomics Rule did not require this

FIGURE 3.20   Palletizing PVC pipes required repetitive bending of the back several hundred times during the work shift.

FIGURE 3.21   The ergonomics committee designed this knee-height cart to reduce bending of the back.

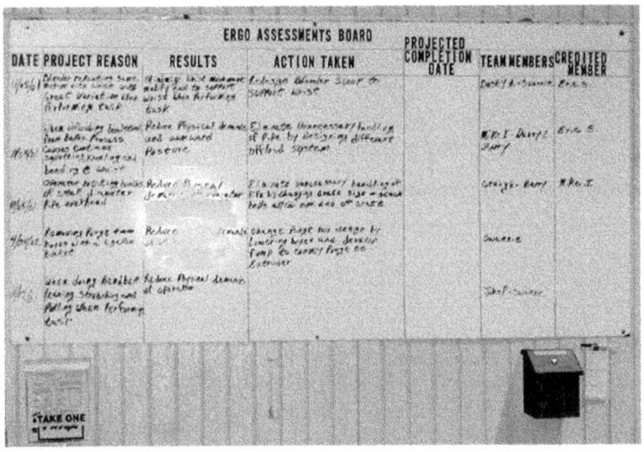

FIGURE 3.22   The Ergonomics Awareness Board is the communication instrument utilized at PW Pipe to receive employee ideas and provide updates on projects being implemented.

FIGURE 3.23   Vertical stacking of material presented a safety hazard and required manual lifting.

FIGURE 3.24   Vertical stacking of material was replaced with horizontal stacking that allows forklifts to do more material handling.

The primary areas of the facility evaluated included: warehouse operations, packaging tasks, customer service duties, and office work. Many of the risk factors listed in the Washington State Ergonomics Rule are present in the different activities of this industry. For instance, handling heavy boxes in warehouse activities, repetitive movements and high hand force in the packaging tasks, and awkward postures in customer service and office activities. With the emphasis on safe work practices from management and active involvement of employees, this company was able to reduce the hazards related to exposure to these risk factors. Hazardous (heavy, frequent, and awkward) lifting was one of the highest priority issues identified (Figure 3.23 and Figrue 3.24).

Key features of the model at Empire Bolt and Screw:
  Identification caution zone and hazard zone jobs as defined in the Washington State
     Ergonomics Rule
  Engineering and administrative control development to eliminate hazards
  Identification of additional controls that could be used to improve comfort and productivity
     beyond requirements of the Washington State Ergonomics Rule

Key outcomes:
   Interventions to address heavy lifting issues
   Worker awareness training development and implementation
   Lifting assist devices and sharing of heavy lifting tasks
   Work methods improvement

3.6 Summary

Ergonomics interventions in industry cross three major structural layers in the society: the societal level, the industry level, and the company level. At the societal level, compromise between various social partners is required. This often results in regulations or guidelines that define the minimally acceptable levels of tolerable risk or activity. Employee involvement at this level is usually present through organized labor. At the industry level, most of the energy is invested by the private sector, where the motivation is driven more by financial and productivity goals than human well-being. The employee involvement at this point appears to be scarce compared to societal- and company-level approaches. Ergonomics guidelines are the primary means of intervention. However, occasionally, product development may be an outcome (e.g., better knives for the meat packing industry). The practitioner of ergonomics is most familiar with the company level where the day-to-day worker–machine–organization interaction is occurring.

   This chapter has presented several models of ergonomics interventions in industries in Washington State. It is likely that some of the models described are applicable elsewhere. The ergonomics rule provided an impetus for participation in ways to define compliance and in virtually all cases, solutions went beyond the requirements of the rule. The likelihood of actually implementing voluntary guidelines at the industry and company level is less clear.

   The practitioner may easily conclude that the common element to all the models described here is the participation of the workers with their expertise and ingenuity on the jobs they do and the active commitment of management to supporting the process through time and resources. Regardless of the model, there are key elements that an intervention in ergonomics should consider. These elements are summarized in the 13-point checklist at the beginning of the chapter.

Acknowledgments

The Washington State Department of Labor and Industries ergonomists developed the demonstration project reports from which most of the examples in this chapter were derived. We are thankful for their hard work and determination. They include Rick Goggins, Gary Davis, Diane Lee, Ninica Howard, Stephen Bao, Kathleen Rockefeller, and Bruce Coulter. They were supported by numerous field staff to whom also we are grateful.

References

1. American Furniture Manufacturers Association. *AFMA Voluntary Ergonomics Guideline for the Furniture Manufacturing Industry.* 2002. High Point, NC, American Furniture Manufacturers Association.
2. Brazil. Regulation No. 3.435 of June 1990 Modifying Standard No. 17 on Ergonomics. Fundacentro-Atualidades em Prevencao de Acidentes. 1990.
3. California Code of Regulations. Title 8, Section 5110. Repetititve Motion Injuries. 1997.
4. de Jong AM, Vink P, de Kroons JCA. Reasons for adopting technological innovations reducing physical workload in bricklaying. *Ergonomics* 2003; 46(11):1091–1108.
5. de Looze MP, Urlings IJ, Vink P et al. Towards successful physical stress reducing products: an evaluation of seven cases. *Applied Ergonomics* 2001; 32(5):525–534.

6. Evanoff BA, Bohr PC, Wolf LD. Effects of a participatory ergonomics team among hospital orderlies. *American Journal of Industrial Medicine* 1999; 35(4):358–365.

7. Fallentin N, National Institute of Occupational Health. Regulatory actions to prevent work-related musculoskeletal disorders — the use of research-based exposure limits. *Scandanavian Journal of Work Environment and Health* 2003; 29(4):247–250.

8. General Accounting Office. Worker protection: private sector ergonomics programs yield positive results. Report to Congressional Requesters. 1997.

9. Hagg GM. Editorial. *Applied Ergonomics* 2003; 34:1.

10. Hagg GM. Corporate initiatives in ergonomics — an introduction. *Applied Ergonomics* 2003; 34(1):3–15.

11. Joseph BS. Corporate ergonomics programme at Ford Motor Company. *Applied Ergonomics* 2003; 34:23–28.

12. Keyserling WM, Hankins SE. Effectiveness of plant-based commitees in recognizing and controlling ergonomic risk factors associated with musculoskeletal problems in the automotive industry. *Rehabilitation-IEA* 1994; 3:346–348.

13. Keyserling WM, Ulin SS, Lincoln AE, Baker SP. Using multiple information sources to identify opportunities for ergonomic interventions in automotive parts distribution: a case study. *American Industrial Hygiene Association Journal* 2003; 64:690–698.

14. Mirka GA, Monroe M, Nay T, Lipscomb H, Kelaher D. Ergonomic interventions for the reduction of low back stress in framing carpenters in the home building industry. *International Journal of Industrial Ergonomics* 2003; 31:397–409.

15. Moore JS. Office ergonomics programs. *Journal of Occupational and Environmental Medicine* 1997; 39(12):1203–1211.

16. Morken T, Moen B, Riise T et al. Effects of a training program to improve musculoskeletal health among industrial workers-effects of supervisors role in the intervention. *International Journal of Industrial Ergonomics* 2002; 30:115–127.

17. Munck-Ulfsfalt U, Falck A, Forsberg A, Dahlin C, Eriksson A. Corporate ergonomics programme at Volvo Car Corporation. *Applied Ergonomics* 2003; 34:17–22.

18. National Institute for Occupational Safety and Health (NIOSH). Elements of ergonomics programs. 1997. Cincinnati, Ohio, U.S. Department of Health and Human Services, Public Health Service, Centers for Disease Control, National Institute for Occupational Safety and Health.

19. Pohjonen T, Punakallio A, Louhevaara V. Participatory ergonomics for reducing load and strain in home care. *International Journal of Industrial Ergonomics* 1998; 21:345–352.

20. Rice VJB, Pekarek D, Connolly V, King I, Mickelson S. Participatory ergonomics: determining injury control "buy-in" of US Army cadre. *Work* 2002; 18(2):191–203.

21. St-Vincent M, Chicoine D, Beaugrand S. Validation of a participatory ergonomic process in two plants in the electrical sector. *International Journal of Industrial Ergonomics* 1998; 21:11–21.

22. State of Victoria. Occupational health and safety (manual handling) regulations 1999. Statutory Rules 1999. 1999. Victoria, Australia, Anstat Pty Ltd.

23. Swedish National Board of Occupational Safety and Health. Ergonomics for the prevention of musculoskeletal disorders. Elanders Gototab, Stockholm, Swedish National Board of Occupational Safety and Health on Ergonomics for the Prevention of Musculoskeletal Disorders, 1998.

24. Workers Compensation Act. Part 4 General Conditions. Ergonomics (MSI) Requirements. 2001.

25. Zalk DM. Grassroots ergonomics: initiating an ergonomics program utilizing participatory techniques. *Annals of Occupational Hygiene* 2001; 45(4):283–289.

26. Zavitz B. Ergonomics at Boeing. Ergo Solutions 1 [July/August], 26–31. 2003.

Appendix: Hazard Zone Risk Factor Checklist

Criteria for analyzing and reducing WMSD hazards for employers who choose the Specific Performance Approach

For each "caution zone job" find any physical risk factors that apply. Reading across the page, determine if all of the conditions are present in the work activities. If they are, a WMSD hazard exists and must be reduced below the hazard level or to the degree technologically and economically feasible (see WAC 296-62-05130(4), specific performance approach).

Awkward Posture				Check (✓) here if this is a WMSD hazard
Body Part	**Physical Risk Factor**	**Duration**	**Visual Aid**	
Shoulders	Working with the hand(s) above the head or the elbow(s) above the shoulder(s)	More than 4 hours total per day		☐
	Repetitively raising the hand(s) above the head or the elbow(s) above the shoulder(s) more than once per minute	More than 4 hours total per day		☐
Neck	Working with the neck bent more than 45° (without support or the ability to vary posture)	More than 4 hours total per day		☐
Back	Working with the back bent forward more than 30°(without support, or the ability to vary posture)	More than 4 hours total per day		☐
	Working with the back bent forward more than 45°(without support or the ability to vary posture)	More than 2 hours total per day		☐
Knees	Squatting	More than 4 hours total per day		☐
	Kneeling	More than 4 hours total per day		☐

(continued)

APPENDIX *Continued*

High Hand Force					Check (✓) here if this is a WMSD hazard
Body Part	**Physical Risk Factor**	**Combined with**	**Duration**	**Visual Aid**	
Arms, wrists, hands	Pinching an unsupported object(s) weighing 2 or more pounds per hand, or pinching with a force of 4 or more pounds per hand (comparable to pinching half a ream of paper)	Highly repetitive motion	More than 3 hours total per day		☐
		Wrists bent in flexion 30° or more, or in extension 45° or more, or in ulnar deviation 30° or more	More than 3 hours total per day		☐
		No other risk factors	More than 4 hours total per day		☐
Arms, wrists, hands	Gripping an unsupported object(s) weighing 10 or more pounds per hand, or gripping with a force of 10 pounds or more per hand (comparable to clamping light duty automotive jumper cables onto a battery)	Highly repetitive motion	More than 3 hours total per day		☐
		Wrists bent in flexion 30° or more, or in extension 45° or more, or in ulnar deviation 30° or more	More than 3 hours total per day		☐
		No other risk factors	More than 4 hours total per day		☐

(*continued*)

APPENDIX *Continued*

Highly Repetitive Motion				
Body Part	**Physical Risk Factor**	**Combined with**	**Duration**	Check (✓) here if this is a WMSD hazard
Neck, shoulders, elbows, wrists, hands	Using the same motion with little or no variation every few seconds (excluding keying activities)	No other risk factors	More than 6 hours total per day	☐
	Using the same motion with little or no variation every few seconds (excluding keying activities)	Wrists bent in flexion 30° or more, or in extension 45° or more, or in ulnar deviation 30° or more **AND** High, forceful exertions with the hand(s)	More than 2 hours total per day	☐
	Intensive keying	Awkward posture, including wrists bent in flexion 30° or more, or in extension 45° or more, or in ulnar deviation 30° or more	More than 4 hours total per day	☐
		No other risk factors	More than 7 hours total per day	☐

Repeated Impact				
Body Part	**Physical Risk Factor**	**Duration**	**Visual Aid**	Check (✓) here if this is a WMSD hazard
Hands	Using the hand (heel/base of palm) as a hammer more than once per minute	More than 2 hours total per day		☐
Knees	Using the knee as a hammer more than once per minute	More than 2 hours total per day		☐

(continued)

APPENDIX *Continued*

This analysis only pertains if you have "caution zone jobs" where employees lift 10 lbs. or more (see WAC 296-62-05105, Heavy, Frequent, or Awkward Lifting) and you have chosen the specific performance approach.

Step 1 Find out the actual weight of objects that the employee lifts.

Actual Weight = _____ lbs.

Step 2 Determine the Unadjusted Weight Limit. Where are the employee's hands when they begin to lift or lower the object? Mark that spot on the diagram below. The number in that box is the Unadjusted Weight Limit in pounds.

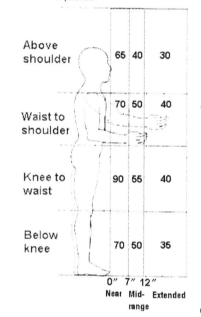

Above shoulder	65	40	30
Waist to shoulder	70	50	40
Knee to waist	90	55	40
Below knee	70	50	35

0" 7" 12"
Near Mid- Extended
range

Unadjusted Weight Limit: _____ lbs.

Step 3 Find the Limit Reduction Modifier. Find out how many times the employee lifts per minute and the total number of hours per day spent lifting. Use this information to look up the Limit Reduction Modifier in the table below.

How many lifts per minute?	For how many hours per day?		
	1 h or less	1 h to 2 h	2 h or more
1 lift every 2–5 min	1.0	0.95	0.85
1 lift every min	0.95	0.9	0.75
2–3 lifts every min	0.9	0.85	0.65
4–5 lifts every min	0.85	0.7	0.45
6–7 lifts every min	0.75	0.5	0.25
8–9 lifts every min	0.6	0.35	0.15
10+ lifts every min	0.3	0.2	0.0

Note: For lifting done less than once every five minutes, use 1.0

Limit Reduction Modifier: _____._____

Step 4 Calculate the Weight Limit. Start by copying the Unadjusted Weight Limit from Step 2.

Unadjusted Weight Limit: = _____ lbs.

If the employee twists more than 45 degrees while lifting, reduce the Unadjusted Weight Limit by multiplying by 0.85. Otherwise, use the Unadjusted Weight Limit

Twisting Adjustment: = _____._____

Adjusted Weight Limit: = _____ lbs.

Multiply the Adjusted Weight Limit by the Limit Reduction Modifier from Step 3 to get the Weight Limit.   X

Limit Reduction Modifier: _____.___

Weight Limit: = _____ lbs.

Step 5 Is this a hazard? Compare the Weight Limit calculated in Step 4 with the Actual Weight lifted from Step 1. If the Actual Weight lifted is greater than the Weight Limit calculated, then the lifting is a WMSD hazard and must be reduced below then hazard level or to the degree technologically and economically feasible.

Note: If the job involves lifts of objects with a number of different weights and/or from a number of different locations, use Steps 1 through 5 above to:
1. Analyze the two worst case lifts — the heaviest object lifted and the lift done in the most awkward posture.
2. Analyze the most commonly performed lift. In Step 3, use the frequency and duration for *all* of the lifting done in a typical workday.

(continued)

APPENDIX *Continued*

Hand-Arm Vibration

Use the instructions below to determine if a hand–arm vibration hazard exists.

Step 1. Find the vibration value for the tool. (Get it from the manufacturer, look it up at this website: http://umetech.niwl.se/vibration/HAVHome.html, or you may measure the vibration yourself). The vibration value will be in units of meters per second squared (m/s^2). On the graph below find the point on the left side that is equal to the vibration value.

Note: You can also link to this website through the L&I WISHA Services Ergonomics web site http://www.lni.wa.gov/wisha/ergo

Step 2. Find out how many total hours per day the employee is using the tool and find that point on the bottom of the graph.

Step 3. Trace a line in from each of these two points until they cross.

Step 4. If that point lies in the crosshatched "Hazard" area above the upper curve, then the vibration hazard must be reduced below the hazard level or to the degree technologically and economically feasible. If the point lies between the two curves in the "Caution" area, then the job remains as a "Caution Zone Job." If it falls in the "OK" area below the bottom curve, then no further steps are required.

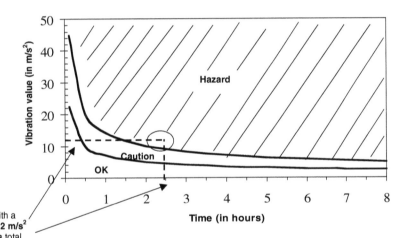

Example:
An impact wrench with a **vibration value of 12 m/s^2** is used for **2½ hours** total per day. The exposure level is in the Hazard area. The vibration must be reduced below the hazard level or to the degree technologically and economically feasible.

Note: The caution limit curve (bottom) is based on an 8-hour energy-equivalent frequency-weighted acceleration value of 2.5 m/s^2. The hazard limit curve (top) is based on an 8-hour energy-equivalent frequency-weighted acceleration value of 5 m/s^2.

4

OSHA Recordkeeping

David J. Cochran
University of Nebraska

Injury and illness recordkeeping is a requirement for most places of employment in the United States. It is also a good idea from a moral and a business standpoint. Without good and consistent records of injuries and illnesses, it is difficult to impossible to determine the status of the safety and health of a job site, facility, or organization. It is difficult to track the progress of methods and procedures implemented to improve the safety and health of an organization. If injury and health data are not tracked, problems that could have been corrected may go undetected sometimes resulting in serious injuries and illnesses and unnecessary costs that could have been avoided. Finally, injury and illness records allow the tracking of the bigger picture for facilities, companies, industries, and the whole country so that needs can be identified and resources can be allocated appropriately.

This chapter is not meant to be a substitute for 29 CFR Part 1904 — Recordkeeping and Reporting Occupational Injuries and Illnesses (currently available at http://www.osha.gov/recordkeeping). It is meant to familiarize the reader with the most pertinent and most commonly encountered requirements of that regulation. This document very closely follows, and in many cases copies, the contents of the Power Point presentation "Rkcomprehensive_rev4" revised 1/1/04 that was created by the Occupational Safety and Health Administration (OSHA) and is currently available at http://www.osha.gov/recordkeeping/RKpresentations.

Special attention needs to be paid to the terms "recordable" and "reportable" as they are often confused. A "recordable" is an injury or illness that must be recorded on the OSHA Form 300. A "reportable" is an injury or illness (death or hospitalization of three or more employees) that must be reported to OSHA orally within 8 h.

This rule is presented in the following four subparts:

A — Purpose
B — Scope

C — Forms and Recording Criteria
D — Other Requirements

4.1 Subpart A — Purpose

The purpose of this rule is to require employers to record and report work-related fatalities, injuries, and illnesses. Recording or reporting an injury or illness does not mean that the employer was at fault and it is independent of Workers' Compensation.

4.2 Subpart B — Scope

All employers covered by the Occupational Safety and Health Act are covered by this regulation. This includes all industries in agriculture, construction, manufacturing, transportation, utilities, and wholesale trades. Companies that had ten or fewer employees at all times during the last calendar year are partially exempt. Business establishments that are in one of the specific low hazard retail, service, finance, insurance, or real estate industries listed in Appendix A of Subpart B do not need to keep OSHA injury and illness records unless the government specifically asks that business to keep records. Even those who are exempt must report any fatality or hospitalization of three or more employees.

4.3 Subpart C — Forms and Recording Criteria

This subpart is broken into the following nine parts:

1904.4	Recording Criteria
1904.5	Work-Relatedness
1904.6	New Case
1904.7	General Recording Criteria
1904.8	Needlesticks and Sharps
1904.9	Medical Removal
1904.10	Hearing Loss
1904.11	Tuberculosis
1904.29	Forms

4.3.1   1904.4 Recording Criteria

Covered employers must record each fatality, injury, or illness that:

Is work-related
Is a new case
Meets one or more of the criteria contained in Sections 1904.7–1904.11

4.3.2   1904.5 Work-Relatedness

In most cases, an injury or illness is presumed to be work-related if it results from events or exposures occurring in the work environment. In reality, it is presumed work-related if, and only if, an event or exposure in the work environment is a discernable cause of the injury or illness or of a significant aggravation of a preexisting condition. The work event or exposure needs only to be one of the discernable causes; it does not need to be the only or even the predominant cause. The work environment includes the establishment and other locations where employees are as a condition of employment. A preexisting injury or illness is significantly aggravated when an event or exposure in the work environment results in death, loss of consciousness, days away from work, days on restricted work, job transfer, or medical treatment.

There are exceptions to the previous paragraph. The general public; those participating in wellness, fitness, recreational programs; eating, drinking, or preparing food for personal use; personal grooming; common cold or flu; and travel when not actually engaged in work activities are some of the exemptions. Injuries that occur while working at home are work-related if they occur while engaged in work activities and they are directly related to the performance of work.

4.3.3   1904.6 New Case

An injury or illness is a new case if the employee has not experienced a recordable injury or illness of the same type that affects the same part of the body. It is also a new case if the employee has experienced a recordable injury or illness of the same type that affects the same part of the body but had completely recovered and a new event or exposure caused the reappearance. The absence of an event or exposure indicates it is not a new case.

4.3.4   1904.7 General Recording Criteria

An injury or illness is recordable if it results in one or more of the following:

Death
Days away from work
Restricted work activity
Transfer to another job
Medical treatment beyond first aid
Loss of consciousness
Significant injury or illness diagnosed by a properly licensed health care professional

4.3.4.1   Days Away from Work and Restricted Work Activity

Days away from work cases include any cases where the employee missed one or more days from work because of injury or illness. Restricted work activity cases involve situations where the employee is unable to work the full workday or is unable to perform one or more of the routine job functions. If an employer assigns a work restriction for the purpose of preventing a more serious condition from developing, it is not a recordable case. This exemption is narrow. It applies only to musculoskeletal disorders (MSDS) and applies only if an employee experiences minor musculoskeletal discomfort, a health care professional determines that the employee is fully able to perform all of his or her routine job functions, and the employer assigns a work restriction to that employee for the purpose of preventing a more serious condition.

The number of days, for either days away or restricted, does not include the day of occurrence of the injury or illness. The number of days includes all of the calendar days the employee is unable to work. This number is capped at 180 days or when the employee leaves the company.

4.3.4.2   Transfer to Another Job

If an injured or ill employee is assigned to a job other than their regular one for all or part of a day because of that injury or illness, it is recordable. If the employee is permanently assigned to a job that has been permanently changed such that that employee can perform all of the routine functions of the job, the day count can be stopped. Also, if the employee is permanently assigned to a different job that the employee can perform all of the routine functions of the job, the day count can be stopped.

4.3.4.3 Medical Treatment beyond First Aid

If an employee receives medical treatment because of a work related injury or illness it is considered a recordable case. First aid is not considered medical treatment but using nonprescription drugs at prescription strength is considered medical treatment. The rule contains a specific list of first aid treatments and all other treatment is considered medical treatment.

4.3.4.4 Loss of Consciousness

All work-related cases involving loss of consciousness must be recorded.

4.3.4.5 Significant Injury or Illness

Additionally, any case of significant, diagnosed injury or illness such as cancer or chronic irreversible disease, or fractured or cracked bone or tooth must be recorded.

4.3.5 1904.8 Needlesticks and Sharps

Record all work-related needlesticks and cuts from sharp objects that are contaminated with another persons blood or other potentially infectious bodily fluids. Record splashes or other exposures to blood or other potentially infectious material if it results in a bloodborne disease or meets the general recording criteria.

4.3.6 1904.9 Medical Removal

If an employee is medically removed under the medical surveillance requirements of an OSHA standard, the case must be recorded as either one involving days away from work or days of restricted work activity. If a case involves voluntary removal below the removal levels required by the standard, the case need not be recorded.

4.3.7 1904.10 Hearing Loss

Hearing loss cases must be recorded where the employee has experienced a Standard Threshold Shift and that employee's hearing level is 25 dB or more above the audiometric zero. See Section 1904.10 for more detail on recording hearing loss cases.

4.3.8 1904.11 Tuberculosis

Record a case where an employee is exposed at work to someone with a known case of active tuberculosis (TB) and subsequently develops TB.

4.3.9 1904.29 Forms

The following three forms are used for recording injuries and illnesses:

OSHA Form 300 — Log of Work-Related Injuries and Illnesses — Figure 4.1
OSHA Form 300A — Summary of Work-Related Injuries and Illnesses — Figure 4.2
OSHA Form 301 — Injury and Illness Incident Report — Figure 4.3

These forms are available on OSHA's web site (www.osha.gov) in PDF or Microsoft Excel XLS formats. The forms must be retained for 5 yr. Form 300 should be updated during that period but Forms 300A and 301 need not be updated. Equivalent forms may be used as long as they have the same information and are readable and understandable and they use the same instructions as the OSHA forms they replace. Additionally, forms can be kept on a computer as long as they can be produced as needed.

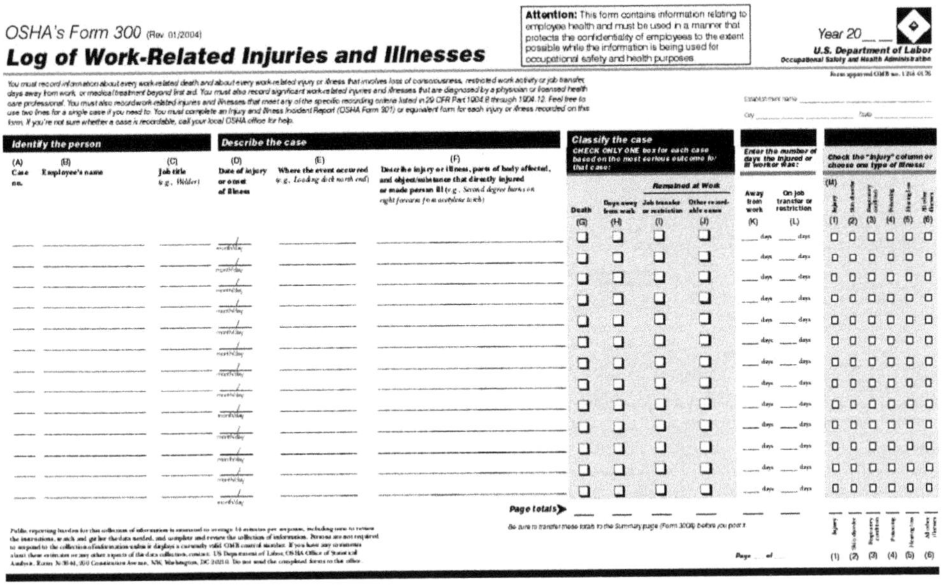

FIGURE 4.1   OSHA Form 300.

4.3.9.1   OSHA Form 300 — Log of Work-Related Injuries and Illnesses

Columns A–F are self-explanatory. Check only one of the columns G–J (the one that is most serious) and update it if the injury or illness progresses and the outcome is more serious than originally recorded. Columns K and L are used to record the number of days away from work or days on job transfer or restriction. Some cases will involve both days away from work and days on job transfer or restriction. Column M requires that the employer classify the injury or illness into one of six types. This form contains information referring to an employee's health and must be used in a manner that protects the confidentiality of the information to the extent possible.

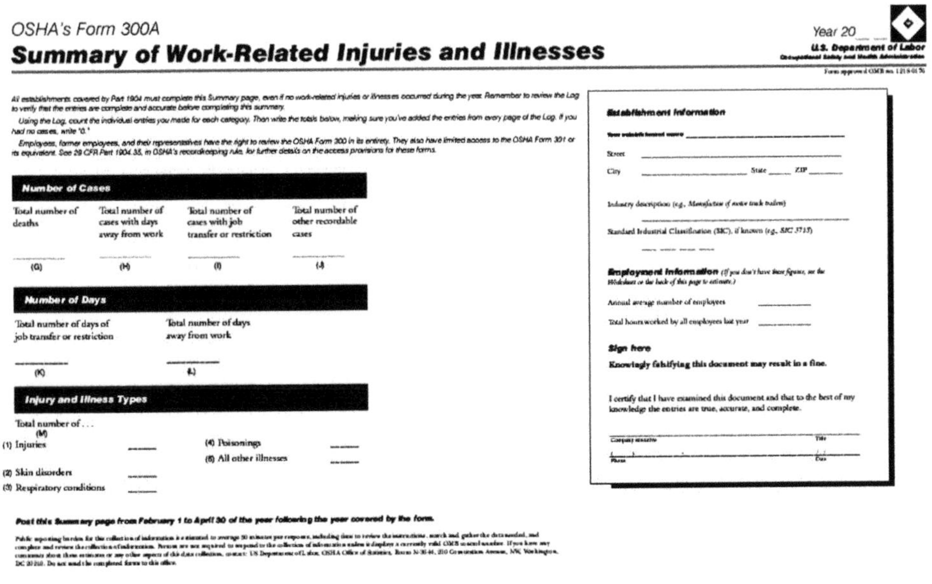

FIGURE 4.2   OSHA Form 300A.

OSHA's Form 301
Injury and Illness Incident Report

Attention: This form contains information relating to employee health and must be used in a manner that protects the confidentiality of employees to the extent possible while the information is being used for occupational safety and health purposes.

U.S. Department of Labor
Occupational Safety and Health Administration

Form approved OMB no. 1218-0176

This *Injury and Illness Incident Report* is one of the first forms you must fill out when a recordable work-related injury or illness has occurred. Together with the *Log of Work-Related Injuries and Illnesses* and the accompanying *Summary*, these forms help the employer and OSHA develop a picture of the extent and severity of work-related incidents.

Within 7 calendar days after you receive information that a recordable work-related injury or illness has occurred, you must fill out this form or an equivalent. Some state workers' compensation, insurance, or other reports may be acceptable substitutes. To be considered an equivalent form, any substitute must contain all the information asked for on this form.

According to Public Law 91-596 and 29 CFR 1904, OSHA's recordkeeping rule, you must keep this form on file for 5 years following the year to which it pertains.

If you need additional copies of this form, you may photocopy and use as many as you need.

Completed by ______
Title ______
Phone (___) ___-____ Date ___/___/___

Information about the employee

1) Full name ______
2) Street ______
City ______ State ______ ZIP ______
3) Date of birth ___/___/___
4) Date hired ___/___/___
5) ☐ Male
☐ Female

Information about the physician or other health care professional

6) Name of physician or other health care professional ______
7) If treatment was given away from the worksite, where was it given?
Facility ______
Street ______
City ______ State ______ ZIP ______
8) Was employee treated in an emergency room?
☐ Yes
☐ No
9) Was employee hospitalized overnight as an in-patient?
☐ Yes
☐ No

Information about the case

10) Case number from the Log ______ (Transfer the case number from the Log after you record the case.)
11) Date of injury or illness ___/___/___
12) Time employee began work ______ AM / PM
13) Time of event ______ AM / PM ☐ Check if time cannot be determined
14) **What was the employee doing just before the incident occurred?** Describe the activity, as well as the tools, equipment, or material the employee was using. Be specific. *Examples:* "climbing a ladder while carrying roofing materials"; "spraying chlorine from hand sprayer"; "daily computer key-entry."
15) **What happened?** Tell us how the injury occurred. *Examples:* "When ladder slipped on wet floor, worker fell 20 feet"; "Worker was sprayed with chlorine when gasket broke during replacement"; "Worker developed soreness in wrist over time."
16) **What was the injury or illness?** Tell us the part of the body that was affected and how it was affected; be more specific than "hurt," "pain," or sore." *Examples:* "strained back"; "chemical burn, hand"; "carpal tunnel syndrome."
17) **What object or substance directly harmed the employee?** *Examples:* "concrete floor"; "chlorine"; "radial arm saw." *If this question does not apply to the incident, leave it blank.*
18) **If the employee died, when did death occur?** Date of death ___/___/___

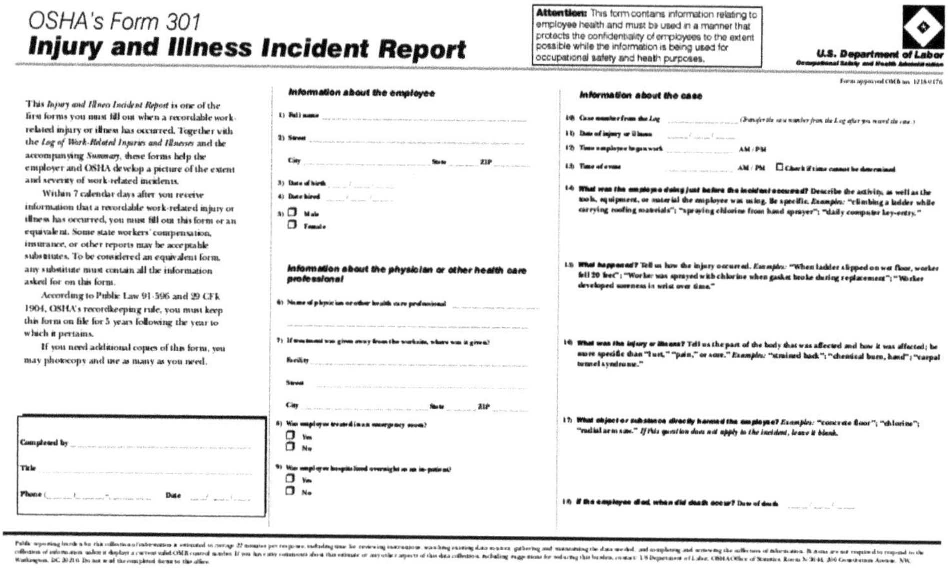

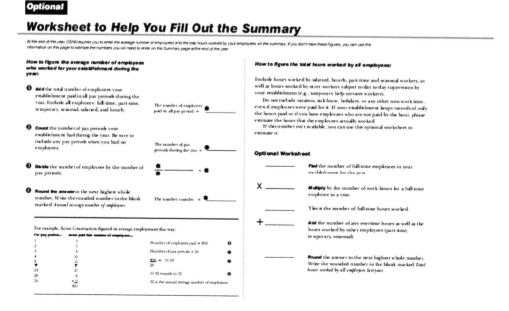

FIGURE 4.3 OSHA Form 301.

4.3.9.2 OSHA Form 300A — Summary of Work-Related Injuries and Illnesses

This form must be posted from February 1 to April 30 for the preceding year. OSHA provides a worksheet with instructions to help employers correctly fill out this form (Figure 4.4).

4.3.9.3 OSHA Form 301 — Injury and Illness Incident Report

There must be an OSHA Form 301 for every entry on the OSHA Form 300 and it must be filled in within seven calendar days after information that a recordable work-related injury or illness has occurred.

Optional

Worksheet to Help You Fill Out the Summary

At the end of the year, OSHA requires you to enter the average number of employees and the total hours worked by your employees on the summary. If you don't have these figures, you can use the information on this page to estimate the numbers you will need to enter on the Summary page at the end of the year.

How to figure the average number of employees who worked for your establishment during the year:

❶ **Add** the total number of employees your establishment paid in all pay periods during the year. Include all employees: full-time, part-time, temporary, seasonal, salaried, and hourly. — The number of employees paid in all pay periods = ❶ ______

❷ **Count** the number of pay periods your establishment had during the year. Be sure to include any pay periods when you had no employees. — The number of pay periods during the year = ❷ ______

❸ **Divide** the number of employees by the number of pay periods. — ❶/❷ ______ = ❸

❹ **Round the answer** to the next highest whole number. Write the rounded number in the blank marked *Annual average number of employees*. — The number rounded = ❹ ______

For example, Acme Construction figured its average employment this way:

For pay period...	Acme paid this number of employees...
1	0
2	0
3	5
4	30
5	40
▼	▼
24	20
25	5
26	+0
	830

Number of employees paid = 830 ❶
Number of pay periods = 26 ❷
830/26 = 31.92 ❸
31.92 rounds to 32 ❹
32 is the annual average number of employees

How to figure the total hours worked by all employees:

Include hours worked by salaried, hourly, part-time and seasonal workers, as well as hours worked by other workers subject to day to day supervision by your establishment (e.g., temporary help services workers).

Do not include vacation, sick leave, holidays, or any other non-work time, even if employees were paid for it. If your establishment keeps records of only the hours paid or if you have employees who are not paid by the hour, please estimate the hours that the employees actually worked.

If this number isn't available, you can use this optional worksheet to estimate it.

Optional Worksheet

______ **Find** the number of full-time employees in your establishment for the year.

X ______ **Multiply** by the number of work hours for a full-time employee in a year.

______ This is the number of full-time hours worked.

\+ ______ **Add** the number of any overtime hours as well as the hours worked by other employees (part-time, temporary, seasonal)

______ **Round** the answer to the next highest whole number. Write the rounded number in the blank marked *Total hours worked by all employees last year*.

FIGURE 4.4 Worksheet for OSHA Form 300A.

This form contains information referring to an employee's health and must be used in a manner that protects the confidentiality of the information to the extent possible.

4.4   Injury and Illness Incidence Rates

Injury and illness incidence rates are usually calculated relative to 100 full time workers over a year. These rates can be used to compare one facility with another or with industry rates (see www.bls.gov/ iif). The two most common are the Recordable case rate and the days away from work and days of restricted work (DART). Figure 4.5 shows the calculation of these rates. OSHA gives further guidance on the calculation of these rates on its website.

4.5   Privacy Protection

Where a case involves an injury or illness to an intimate body part; results from a sexual assault; involves mental illness, HIV infection, hepatitis, tuberculosis, needlesticks or sharps; or the employee requests anonymity, enter "privacy case" in the name column on the OSHA Form 300 and keep a list of case numbers and corresponding employee names. Employee requested privacy cases are limited to illnesses.

4.6   Multiple Business Establishments

Where multiple business establishments are involved, the employer must keep a separate OSHA Form 300 for each establishment that is expected to be in operation for more than a year. The employer may keep one OSHA Form 300 for all short-term establishments. Each employee must be linked to one establishment.

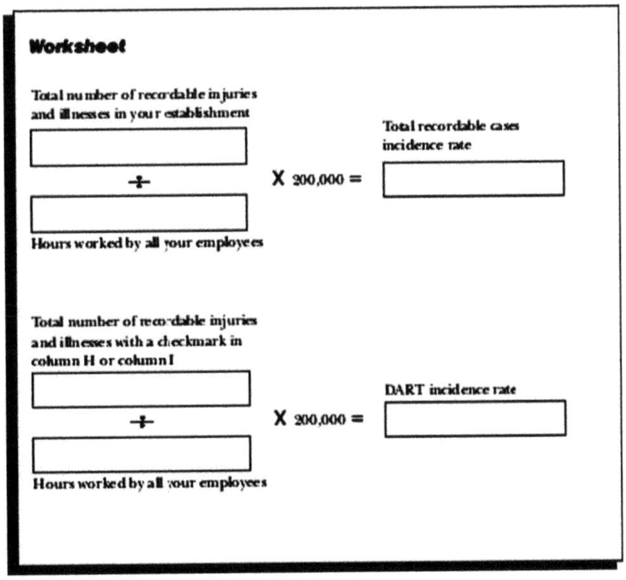

FIGURE 4.5   Calculation guidance for injury and illness rates.

4.7 Covered Employees

Covered employees are those who are on the payroll and those not on the payroll but supervised on a day-to-day basis. Self-employed persons and partners are excluded. Temporary help agencies should not record the cases experienced by temporary workers who are supervised by the using firm.

4.8 Reporting a Fatality or Catastrophe

Report orally within 8 h any work-related fatality or incident involving three or more inpatient hospitalizations. Motor vehicle accidents that occur on a public street or highway or accidents that involve a commercial airplane, train, subway, or bus need not be reported.

4.9 Musculoskeletal Disorders

MSDs are the most likely occupational injury or illness related to the field of ergonomics. MSDs are recorded just like any other work-related injury or illness. It is highly unlikely that an MSD will result in death or loss of consciousness. Those MSD cases involving days away from work, restricted work activity, transfer to another job, or medical treatment beyond first aid are recordable cases. They must meet the criteria of work-relatedness and new case to be considered recordable.

OSHA entered into a settlement agreement with the National Association of Manufacturers (NAM) to resolve NAM's legal challenge to OSHA's revised recordkeeping regulation. The following is taken verbatim from that settlement agreement.

Section 1904.5(a) states that "[the employer] must consider an injury or illness to be work-related if an event of exposure in the work environment either caused or contributed to resulting condition or significantly aggravated a pre-existing condition. Work-relatedness is presumed for injuries and illnesses resulting from events or exposures occurring in the work environment ...." Under this language, a case is presumed work-related if, and only if, an event or exposure in the work environment is a discernable cause of the injury or illness or of a significant aggravation to pre-existing condition. The work event or exposure need only be one of the discernable causes; it need not be the sole or predominant cause.

Section 1904(b)(2) states that a case is not recordable if it "involves signs of symptoms that surface at work but result from a non-work-related event or exposure that occurs outside the work environment." This language is intended as a restatement of the principle expressed in 1904.5(a) described above. Regardless of where signs or symptoms surface, a case is recordable only if a work event or exposure is a discernable cause of the injury or illness or a significant cause of the injury or illness or of a significant aggravation to a pre-existing condition.

Section 1904(b)(3) states that if it is not obvious whether the precipitating event or exposure occurred in the work environment or elsewhere, the employer "must evaluate the employee's work duties and environment to decide whether or not one or more work events or exposures in the work environment caused or contributed to the resulting condition or significantly aggravated a pre-existing condition." If the employer decides the case is not work-related, and OSHA subsequently issues a citation for failure to record, the Government would have the burden of proving that the injury or illness was work-related.

4.10 Assistance

OSHA, at all levels, offers assistance in recordkeeping. There is a recordkeeping page on the website — www.osha.gov/recordkeeping. There are regional recordkeeping coordinators. State Plan States offer assistance and the OSHA Training Institute Education Centers offer courses in recordkeeping.

Acknowledgments

Thanks to the people at OSHA for their help.

References

Recording and Reporting Occupational Injuries and Illnesses, OSHA, 19 CFR 1904, currently available at http://www.osha.gov/recordkeeping under Regulatory Text.

Power Point presentation, "Rkcomprehensive_rev4" revised 1/1/04, developed by OSHA, currently available at http://www.osha.gov/recordkeeping/RKpresentations under "Comprehensive Presentation."

5

Psychosocial Work Factors

Pascale Carayon

Soo-Yee Lim

Ecole des Mines de

5.1 Introduction

This chapter examines the concept of psychosocial work factors and its relationship to occupational ergonomics. First, we provide a brief historical perspective of the development of theories and models of work organization and psychosocial work factors. Definitions and examples are then presented. Several explanations are given for the importance of psychosocial work factors in occupational ergonomics. Finally, measurement issues and methods for controlling and managing psychosocial work factors are discussed.

The role of "psychosocial work factors" in influencing individual and organizational health can be traced back to the early days of work mechanization and specialization, and the emergence of the concept of division of labor. Taylor (1911) expanded the principle of division of labor by designing efficient work systems accounting for proper job design, providing the right tools, motivating the individuals, and sharing of responsibilities between management and labor, and sharing of profits. This is known as the era of scientific management in which scientific methods are used to objectively measure work with the aim of improving its efficiency. These scientific methods involved breaking the tasks into small components or units, thus making work requirements and performance evaluations easy to define and monitor. Under these methods, work is simplified and standardized, therefore having a great impact on job and work processes. An analysis of psychosocial work factors in a job in this system would reveal that skill variety is minimal, workers have no control of the work processes, and the job is highly repetitive and monotonous. Such work system design can still be found in numerous workplaces.

As the workforce became more educated, individuals became more aware of their working conditions and environment, and began to seek avenues for improving their quality of working life. This is when the human relations movement emerged (Mayo, 1945), which raised the issue of the potential influence of the work environment on an individual's motivation, productivity, and well-being. Individual needs and wants were emphasized (Maslow, 1970). Thus, job design theorists incorporated worker behavior and work factors in their theories. The two theories of job enlargement and job enrichment formed the basis for many job design theories thereafter. These theories conceptualize the role of worker behavior and perception of the work environment in influencing personal and organizational outcomes. Job enlargement theory emphasized giving a larger variety of tasks or activities to the worker. While this was an improvement from the era of scientific management, the additional tasks or activities could be of a similar skill level and content: workers were performing multiple tasks of the same "kind." This has been called "horizontal loading" of the job, and is the opposite of job enrichment, which focused on the "vertical loading" of the job. Job enrichment aims at expanding the skills used by workers, while at the same time increasing their responsibility. Herzberg (1966), the father of the job enrichment theory, defined intrinsic and extrinsic factors (or motivation versus hygiene factors) that are important to worker motivation, thus leading to satisfaction or dissatisfaction, and psychological well-being. Intrinsic factors are related to the work (or job) conditions, such as having additional control over work schedules or resources, feedback, client relationships, skill use and development, better work content, direct communications, and personal accountability (Herzberg, 1974). Extrinsic factors are related to aspects of financial rewards and benefits and also to the physical environment. Herzberg indicated that extrinsic factors could lead to dissatisfaction with work, but not to satisfaction, while intrinsic factors could increase satisfaction with work. Herzberg's work demonstrated the complex relationships of job conditions, the individual's motivation, satisfaction, dissatisfaction, and psychological well-being. In a way similar to Herzberg's job enrichment theory, the Job Characteristics Theory (Hackman and Oldham, 1976) focused on the idea that specific characteristics of the job (i.e., skill variety, task identity, task significance, autonomy, and feedback) in combination with individual characteristics (growth need strength) would determine personal and work outcomes.

The Sociotechnical Systems Theory recognized two inter-related systems in an organization: the social system and the technical system. The main principle of the Sociotechnical Systems Theory is that the social and technical systems interact with each other, and that the joint optimization of both systems can lead to increased satisfaction and performance. The social system focused on the workers' perception of the work environment (i.e., job design factors) and the technical system emphasized the technology and the work processes used in the work (e.g., automation, paced systems, and monitoring systems). In a study of coal mining (Trist and Bamforth, 1951), it was demonstrated that the technical system could impact the social system. In this study where semi-autonomous work groups were set up, workers were given opportunities to make decisions related to their work, and experienced better interactions with workers in their group, as well as task significance and completeness (see also Trist, 1981). Work by Trist and his colleagues showed that technological factors could influence both organizational and job factors. However, it was Davis (1980) who provided a conceptual framework and a set of principles that formulated the Sociotechnical theory. His framework called for a flattened management structure that would promote participation, interaction between and across groups of workers, enriched jobs, and most important, meeting individual needs. The Sociotechnical Systems Theory laid down the groundwork for the current understanding of how psychosocial work factors can be related to ergonomic factors by examining the interplay between the social and technical systems in organizations. Other recent theories and models of psychosocial work factors will be discussed later.

This rapid overview of the development of job design theories in the 20th century demonstrates the increasing role of psychological, social, and organizational factors in the design of work.

5.2 Definitions

Within the last decade, the role of psychosocial work factors on worker health has gained much popularity. However, the term of "psychosocial work factors" has been used loosely to define and represent many factors that are a part of, attached to or associated with the individuals. Some would consider what has been traditionally termed socioeconomic factors such as income, education level, and demographic or individual factors (e.g., age and marital status) as part of the psychosocial factors (Hogstedt and Vingard et al., 1995; Ong et al., 1995). In order to understand psychosocial factors in the workplace, one needs to take into account the ability of an individual to make a psychological connection to his or her job, thus formulating the relationship between the person and the job. For instance, the International Labour Office (ILO, 1986) defines psychosocial work factors as "interactions between and among work environment, job content, organizational conditions and workers' capacities, needs, culture, personal extra-job considerations that may, through perceptions and experience, influence health, work performance, and job satisfaction." Thus, the underlying premise in defining psychosocial work factors is the inclusion of the behavioral and psychological components of job factors. In the rest of the chapter, we will use the definitions proposed by Hagberg and his colleagues (Hagberg et al., 1995) because they are most highly relevant for occupational ergonomics.

Work organization is defined as the way work is structured, distributed, processed, and supervised (Hagberg et al., 1995). It is an "objective" characteristic of the work environment, and depends on many factors, including management style, type of product or service, characteristics of the workforce, level and type of technology, and market conditions. Psychosocial work factors are "perceived" characteristics of the work environment that have an emotional connotation for workers and managers, and that can result in stress and strain (Hagberg et al., 1995). Examples of psychosocial work factors include overload, lack of control, social support, and job future ambiguity. Other examples are described in the following section.

The concept of psychosocial work factors raises the issue of objectivity–subjectivity. Objectivity has multiple meanings and levels in the literature. According to Kasl (1987), objective data is not supplied by the self-same respondent who is also describing his distress, strain, or discomfort. At another level, Kasl (1987) feels that "psychosocial factor perception" can be less subjective when the main source of information is the employee but that this self-reported exposure is devoid of evaluation and reaction. Similarly, Frese and Zapf (1988) conceptualize and operationalize "objective stressors" (i.e., work organization) as not being influenced by an individual's cognitive and emotional processing. Based on this, it is more appropriate to conceptualize a continuum of objectivity and subjectivity. Work organization can be placed at one extreme of the continuum (that is the objective nature of work) whereas psychosocial work factors have some degree of subjectivity (see definitions above).

Psychosocial work factors result from the interplay between the work organization and the individual. Given our definitions, psychosocial work factors have a *subjective*, perceptual dimension, which is related to the *objective* dimension of work organization. Different work organizations will 'produce' different psychosocial work factors. The work organization determines to a large extent the type and degree of psychosocial work factors experienced by workers. For instance, electronic performance monitoring, or the on-line, continuous computer recording of employee performance-related activities, is a type of work organization that has been related to a range of negative psychosocial work factors, including lack of control, high work pressure, and low social support (Smith et al., 1992). In a study of office workers, information on psychosocial work factors was related to objective information on job title (Sainfort, 1990). Therefore, psychosocial work factors are very much anchored in the objective work situation, and are related to the work organization.

5.3   Examples of Psychosocial Work Factors

Psychosocial work factors are multiple and various, and are produced by different, interacting aspects of work. The Balance Theory of Job Design (Smith and Carayon-Sainfort, 1989) proposed a conceptualization of the work system with five elements interacting to produce a "stress load." The five elements of the work system are: (1) the individual, (2) tasks, (3) technology and tools, (4) environment, and (5) organizational factors. The interplay and interactions between these different factors can produce various stressors on the individual which then produce a "stress load" which has both physical and psychological components. The stress load, if sustained over time and depending on the individual resources, can produce adverse effects, such as health problems and lack of performance. The models and theories of job design reviewed at the beginning of the chapter tended to emphasize a small set of psychosocial work factors. For instance, the human relations movement (Mayo, 1945) focused on the social aspects

TABLE 5.1   Selected Psychosocial Work Factors and their Facets

1. Job demands	Quantitative workload
	Variance in workload
	Work pressure
	Cognitive demands
2. Job content	Repetitiveness
	Challenge
	Utilization and development of skills
3. Job control	Task/instrumental control
	Decision/organizational control
	Control over physical environment
	Resource control
	Control over work pace: machine-pacing
4. Social interactions	Social support from supervisor and colleagues
	Supervisor complaint, praise, monitoring
	Dealing with (difficult) clients/customers
5. Role factors	Role ambiguity
	Role conflict
6. Job future and career issues	Job future ambiguity
	Fear of job loss
7. Technology issues	Computer-related problems
	Electronic performance monitoring
8. Organizational and management issues	Participation
	Management style

of work, whereas the job characteristics theory (Hackman and Oldham, 1976) lists five job characteristics, i.e., skill variety, task identity, task significance, autonomy, and feedback. However, research and practice in the field of work organization has demonstrated that considering only a small number of work factors can be misleading and inefficient in solving job design problems. The balance theory proposes a systematic, global approach to the diagnosis and design or redesign of work systems that does not emphasize any one aspect of work. According to the balance theory, psychosocial work factors are multiple and of diverse nature.

Table 5.1 lists eight categories of psychosocial work factors and specific facets in each category. This list cannot be considered as exhaustive, but is representative of the most often studied psychosocial work factors.

The study of psychosocial work factors needs to be tuned in to the changes in society. Changes in the economic, social, technological, legal, and physical environment can produce new psychosocial work factors. For instance, in the context of office automation, four emerging issues are appearing (Carayon and Lim, 1994): (1) electronic monitoring of worker performance, (2) computer-supported work groups, (3) links between the physical and psychosocial aspects of work in automated offices, and (4) technological changes. The issue of technological changes applies nowadays to a large segment of the work population. Employees are asked to learn new technologies on a frequent, sometimes continuous, basis. Other trends in work organization include the development of teamwork and other work arrangements, such as telecommuting. These new trends may produce new psychosocial work factors, such as high dependency on technology, lack of socialization on the job and identity with the organization, and pressures from teamwork. Two APA publications review psychosocial stress issues related to changes in the workforce in terms of gender, diversity, and family issues (Keita and Hurrell, 1994), and some of the emergent psychosocial risk factors and selected occupations at risk of psychosocial stress (Sauter and Murphy, 1995).

5.4   Occupational Ergonomics and Psychosocial Work Factors

The emergence of macroergonomics has strongly contributed to the increasing interest in psychosocial work factors in the occupational ergonomics field (Hendrick, 1991; Hendrick, 1996). As shown above, the work factors can be categorized into the individual, task, tools and technologies, physical environ-ment, and the organization (Smith and Carayon-Sainfort, 1989). They can also be described as either physical or psychosocial (Cox and Ferguson, 1994). Cox and Ferguson (1994) developed a model of the effects of physical and psychosocial factors on health. According to this model, the effects of work factors on health are mediated by two pathways: (1) a direct physicochemical pathway and (2) an indirect psychophysiological pathway. These pathways are present at the same time, and interact in different ways to affect health. Physical work factors can have direct effects on health via the physicochemical pathway, and indirect effects on health via the psychophysiological pathway, but can also moderate the effect of psychosocial work factors on health via the psychophysiological pathway. This model demonstrates the close relationship between physical and psychosocial work factors in their influence on health and well-being.

The importance of psychosocial work factors in the field of occupational ergonomics emerges from several considerations.

1. Physical and psychosocial ergonomics are interested in the same job factors
2. Physical and psychosocial work factors are related to each other
3. Psychosocial work factors play an important role in physical ergonomics interventions
4. Physical and psychosocial work factors are related to the same outcome, for instance, work-related musculoskeletal disorders

First, some of the concepts examined in the physical ergonomics literature are similar to concepts examined in the psychosocial ergonomics literature. For instance, the degree of repetitiveness of a task is very important from both physical and psychosocial points of view. Physical ergonomists are more interested in the effect of the task repetitiveness on motions and force exerted on certain body parts, such as hands; whereas psychosocial ergonomists are concerned about the effect of task repetitiveness on monotony, boredom, and dissatisfaction with one's work (Cox, 1985). In the physical ergonomics literature, an important job redesign strategy for dealing with repetitiveness is job rotation: workers are rotated between tasks which require effort from different body parts and muscles, therefore reducing the negative effects of repetitiveness of motions in a single task. From a psychosocial point of view, job rotation is one form of job enlargement (see above for a discussion of job enlargement). However, as discussed earlier, the psychosocial benefits of job rotation are limited because workers may be simply per-forming a range of similar, nonchallenging tasks. From a physical ergonomics point of view, job rotation is effective only if the physical variety of the tasks is increased; whereas from a psychosocial ergonomics point of view, job rotation is effective only to the extent of the content and meaningfulness of the tasks.

Second, physical and psychosocial work factors can be related to each other. For instance, the model proposed by Lim (1994) states that the psychosocial factor of work pressure can influence the physical factors of force and speed of motions. According to this model, workers may change their behaviors under the influence of work pressure, and, therefore, tend to exert more force or to speed up their work. Empirical evidence tends to confirm this relationship between work pressure (i.e., a psychosocial work factor) and physical work factors (Lim, 1994). Another form of relationship between physical and psychosocial work factors is evident in the literature on control over one's physical environment. In this case, the psychosocial work factor of control is applied to one particular facet of the work, that is the physical environment. Control over one's physical environment can, therefore, have benefits from a phys-ical point of view (i.e., being able to adapt one's physical environment to one's physical characteristics and task requirements), but also from a psychosocial point of view (i.e., having control is known to have many psychosocial benefits [Sauter et al., 1989]).

Third, psychosocial work factors are a crucial component of physical ergonomics interventions. In particular, the concept of participatory ergonomics uses the benefits of one psychosocial work factor, that is participation, in the process of implementing physical ergonomics changes (Noro and Imada, 1991). From a psychosocial point of view, using participation is important to improve the process and outcomes of ergonomic interventions. In addition, any type of organizational interventions, includ-ing ergonomic interventions, can be stressful because of the emergence of negative psychosocial work factors, such as uncertainty and increased workload (i.e., having more work during the intervention or the transitory period). Therefore, in any physical ergonomics intervention, attention should be paid to psychosocial work factors in order to improve the effectiveness of the intervention and to reduce or minimize its negative effects on workers.

Fourth, physical and psychosocial work factors can be related to the same outcome. One of these outcomes is work-related musculoskeletal disorders (WMSDs). There is increasing theoretical and empirical evidence that both physical and psychosocial work factors play a role in the experience and development of WMSDs (Hagberg et al., 1995; Moon and Sauter, 1996). Several mechanisms for the joint influence of physical and psychosocial work factors on WMSDs have been presented (Smith and Carayon, 1996). Therefore, in order to fully prevent or reduce WMSDs, both physical and psychosocial work factors need to be considered.

5.5 Measurement of Psychosocial Work Factors

From the occupational ergonomics point of view, the purpose of examining psychosocial work factors is to investigate their influence on and role in worker health and well-being. Thus, psychosocial work factors can be considered as predictors (i.e., independent variables), while worker health and well-being serve as the dependent variables or outcomes. The measurement or assessment of well-being can be classified into two levels of measures in terms of "context-free" (that is, life in general or general satisfaction) and "context-specific" (for example, job-related well-being) (Warr, 1994). It is the latter level of measure, "context-specific" that is relevant to the assessment of psychosocial work factors in the workplace. Table 5.1 shows a selected sample of the many different dimensions of jobs (for example, job demands, control, social support) that have been studied extensively. Furthermore, each dimension is made up of different facets that define and operationalize that particular dimension. For example, as shown in Table 5.1, the dimension of job demands consists of various facets, such as quantitative workload, variance in workload, work pressure, and cognitive demands; the dimension of job content includes repetitiveness, challenge on the job, and utilization of skills. It should be noted that Table 5.1 is not an exhaustive list of psychosocial work factors.

The most often used method for measuring psychosocial work factors in applied settings is the questionnaire survey. Difficulties with questionnaire data on psychosocial work factors are often due to the lack of clarity of the definitions of the measured factors or poorly designed questionnaire items that measure "overlapping" conceptual dimensions of the psychosocial work factor of interest. Measures of any one facet typically include several items that can be grouped in a "scale." Reliability of the scale is often being assessed by the Cronbach-alpha score method in which the intercorrelations among the scale items are examined for internal consistency. In general, it is recommended that existing, well-established scales be used in order to ensure the "quality" of the data (i.e., reliability and validity) and to be able to compare the newly collected data with other groups for which data has been collected with the same instrument (benchmarking).

The level of objectivity/subjectivity of the measures of psychosocial work factors will depend on the degree of influence of cognitive and emotional processing. For example, ratings of work factors by an observer cannot be considered as purely objective because of the potential influence of the observer's cognitive and emotional processing. However, ratings of work factors by an outside observer can be considered as more objective than an evaluative question answered by an employee about his/her work environment (e.g., "How stressful is your work environment?"). However, self-reported measures of psychosocial work factors can be more objective when devoid of evaluation and reaction (Kasl and Cooper, 1987). As discussed earlier, any kind of data can be placed somewhere on this objectivity/subjectivity continuum from "low in dependency on cognitive and emotional processing" (e.g., objective) to "high in dependency on cognitive and emotional processing" (e.g., subjective).

We discuss three different questionnaires which include numerous scales of psychosocial work factors. In addition, validity and reliability analyses have been performed on all three questionnaires. Two of these questionnaires have been developed and used to measure psychosocial work factors in various groups of workers or large samples of workers: (1) the NIOSH Job Stress questionnaire (Hurrell and McLaney, 1988) and (2) the Job Content Questionnaire (JCQ) (Karasek, 1979). The NIOSH Job Stress questionnaire is often used in the Health Hazard Evaluations performed by NIOSH. Translations of Karasek's JCQ exist in many different languages, including Dutch and French. The University of Wisconsin Office Worker Survey (OWS) is a questionnaire developed to measure psychosocial work factors in office/computer work (Carayon, 1991). This questionnaire covers a wide range of psychosocial work factors of importance in office and computer work. In addition to many of the psychosocial work factors measured by the NIOSH Job Stress Questionnaire or Karasek's JCQ, the OWS measures psychosocial work factors related to computer technology, such as computer-related problems (Carayon-Sainfort, 1992). The OWS questionnaire has been translated into Finnish, Swedish, and German. For all three questionnaires, data exist for various groups of workers in numerous organizations of multiple countries. This data can serve as a comparison to newly collected data and for benchmarking. Numerous other questionnaires for measuring psychosocial work factors exist, such as the Occupational Stress Questionnaire in Finland (Elo et al., 1994) and the Occupational Stress Indicator in England (Cooper et al., 1988). Other questionnaires are listed in Cook et al. (1981).

5.6 Managing and Controlling Psychosocial Work Factors

It is clear from the job design and occupational stress literature that jobs with negative psychosocial work factors, such as repetitiveness, no opportunity to develop skills, and low control, can have adverse effects on job performance and mental and physical health. Various approaches have been proposed to improve the design of jobs, such as job rotation and other forms of job enlargement, and job enrichment (see above). These strategies can be efficient to increase the variety in a job, to reduce the dependence on a particular technology or tool, and to increase worker control and responsibility. In particular, lack of job control is seen as a critical psychosocial work factor (Sauter et al., 1989). Providing a greater amount of control can be achieved by, for instance, allowing workers to determine their work schedules in accordance with organizational policies and production requirements, by allowing workers to give input into decisions that affect their jobs, by letting workers choose the best work procedures and task order, and by increasing worker participation in the production process. An experimental field study of a participation program showed the positive effects of participation on emotional distress and turnover (Jackson, 1983). According to the Sociotechnical Systems theory, autonomous work groups can be an effective strategy for increasing worker control and enriching jobs. Beyond increased control and improved job content, some forms of teamwork can have other positive psychosocial benefits, such as increased opportunity for socialization and learning.

Achieving the perfect job without any negative psychosocial work factors may not be feasible or realistic, given individual, organizational, or technological constraints and requirements. The balance theory (Smith and Carayon-Sainfort, 1989) proposes a job redesign strategy that aims to achieve an optimal job design. In this process, negative psychosocial work factors need to be eliminated or reduced as much as possible. However, when this is not possible, positive psychosocial work factors can be used to reduce the impact of negative psychosocial work factors. This balancing, or compensating, effect is based on the concept of the work system of the balance theory. The five elements of the work system (the individual, tasks, technology and tools, environment, and organizational factors) are interrelated: they can influence each other, and they can also influence the impact or effect of each other or their interactions. In this systems approach, negative psychosocial work factors can be balanced out or compensated by positive work factors.

Some trends in the field of organizational design and management may have positive characteristics from a psychosocial point of view. For instance, under certain conditions, the use of quality engineering and management methods can positively affect the psychosocial work environment, such as increased opportunity for participation, and learning and development of quality-related skills (Smith et al.,

1989). However, other trends in the business world can have negative effects on the psychosocial work environment. For instance, downsizing and other organizational restructuring and reengineering may create highly stressful situations of uncertainty and loss of control (DOL, 1995).

5.7 Conclusion

This chapter has demonstrated the importance of psychosocial work factors in the research and practice of occupational ergonomics. In order to clarify the issue at hand, we presented definitions of work organization and psychosocial work factors. It is important to understand the long research tradition on psychosocial work factors that has produced numerous models and theories, but also valid and reliable methods for measuring psychosocial work factors. At the end of the chapter, we presented examples of methods for managing and controlling psychosocial work factors.

Psychosocial work factors need to be taken into account in the research on and practice of occupational ergonomics. We have discussed the important role of psychosocial work factors with regard to physical ergonomics. In addition, given the constantly changing world of work and organizations, we need to pay even more attention to the multiple aspects of people at work, including psychosocial work factors.

References

Carayon, P. (1991). *The Office Worker Survey.* Madison, WI, Department of Industrial Engineering, University of Wisconsin-Madison.

Carayon, P. and Lim, S.-Y. (1994). Stress in automated offices. *The Encyclopedia of Library and Information Science.* Kent, A., New York, Marcel Dekker. Vol. 53, Supplement 16: 314–354.

Carayon-Sainfort, P. (1992). The use of computers in offices: impact on task characteristics and worker stress. *International Journal of Human Computer Interaction* 4(3): 245–261.

Cook, J. D., Hepworth, S. J. et al. (1981). *The Experience of Work.* London, Academic Press.

Cooper, C. L., Sloan, S. J. et al. (1988). *Occupational Stress Indicator.* Windsor, England, NFER-Nelson.

Cox, T. (1985). Repetitive work: Occupational stress and health. *Job Stress and Blue-Collar Work.* Cooper, C. L. and Smith, M. J., New York, John Wiley & Sons: 85–112.

Cox, T. and Ferguson, E. (1994). Measurement of the subjective work environment. *Work and Stress* 8(2): 98–109.

Davis, L. E. (1980). Individuals and the organization. *California Management Review* 22(2): 5–14.

DOL (1995). *Guide to Responsible Restructuring.* Washington, D.C. 20210, U.S. Department of Labor, Office of the American Workplace.

Elo, A.-L., Leppanen, A. et al. (1994). The occupational stress questionnaire. *Occupational Medicine.* Zenz, C., Dickerson, O. B., and Horvarth, E. P., St. Louis, Mosby: 1234–1237.

Frese, M. and Zapf, D. (1988). Methodological issues in the study of work stress. *Causes, Coping and Consequences of Stress at Work.* Cooper, C. L. and Payne, R. Chichester, John Wiley & Sons.

Hackman, J. R. and Oldham, G. R. (1976). Motivation through the design of work: test of a theory. *Organizational Behavior and Human Performance* 16: 250–279.

Hagberg, M., Silverstein, B. et al. (1995). *Work-Related Musculoskeletal Disorders (WMSDs): A Reference Book for Prevention.* London, Taylor & Francis.

Hendrick, H. W. (1991). Human Factors in organizational design and management. *Ergonomics* 34: 743–756.

Hendrick, H. W. (1996). Human factors in ODAM: an historical perspective, in *Human Factors in Organizational Design and Management*, Vol. V, Brown, O. J. and Hendrick, H. W., Eds., Amsterdam, The Netherlands, Elsevier Science Publishers: 429–434.

Herzberg, F. (1966). *Work and the Nature of Man.* New York, Thomas, Y. Crowell Company.

Herzberg, F. (1974). The wise old turk. *Harvard Business Review* (September/October): 70–80.

Hogstedt, C., Vingard, E. et al. (1995). *The Norrtalje-MUSIC Study — An Ongoing Epidemiological Study on Risk and Health Factors for Low Back and Neck-Shoulder Disorders.* PREMUS'95-Second International Scientific Conference and Prevention of Work-Related Musculoskeletal Disorders, Montreal, Canada.

Hurrell, J. J. J. and M. A. McLaney (1988). Exposure to job stress — A new psychometric instrument. *Scandinavian Journal of Work Environment and Health* 14(suppl. 1): 27–28.

ILO (1986). *Psychosocial Factors at Work: Recognition and Control.* Geneva, Switzerland, International Labour Office.

Jackson, W. E. (1983). Participation in decision-making as a strategy for reducing job-related strain. *Journal of Applied Psychology* 68: 3–19.

Karasek, R. A. (1979). Job demands, job decision latitude, and mental strain: implications for job redesign. *Administrative Science Quarterly* 24: 285–308.

Kasl, S. V. (1987). Methodologies in stress and health: past difficulties, present dilemmas, future directions, in *Stress and Health: Issues in Research and Methodology,* Kasl, S. V. and Cooper, C. L., Eds., Chichester, John Wiley & Sons: 307–318.

Kasl, S. V. and Cooper, C. L., Eds. (1987). *Stress and Health: Issues in Research and Methodology.* Chichester, John Wiley & Sons.

Keita, G. P. and Hurrell, J. J. J. (1994). *Job Stress in a Changing Workforce — Investigating Gender, Diversity, and Family Issues.* Washington, D.C., APA.

Lim, S. (1994). An integrated approach to cumulative trauma disorders in computerized offices: the role of psychosocial work factors, psychological stress and ergonomic risk factors. *IE.* Madison, WI, University of Wisconsin-Madison.

Maslow, A. H. (1970). *Motivation and Personality.* New York, Harper and Row.

Mayo, E. (1945). *The Social Problems of an Industrial Civilization.* Andover, MA, The Andover Press.

Moon, S. D. and Sauter, S. L., Eds. (1996). *Beyond Biomechanics — Psychosocial Aspects of Musculoskeletal Disorders in Office Work.* London, Taylor & Francis.

Noro, K. and Imada, A. (1991). *Participatory Ergonomics.* London, Taylor & Francis.

Ong, C. N., Jeyaratnam, J. et al. (1995). Musculoskeletal disorders among operators of video display terminals. *Scandinavian Journal of Work Environment and Health* 21(1): 60–64.

Sainfort, P. C. (1990). *Perceptions of Work Environment and Psychological Strain Across Categories of Office Jobs.* The Human Factors Society 34th Annual Meeting.

Sauter, S. L., Hurrell, J. J. Jr. et al., Eds. (1989). *Job Control and Worker Health.* Chichester, John Wiley & Sons.

Sauter, S. L. and Murphy, L. R. (1995). *Organizational Risk Factors for Job Stress.* Washington, D.C., APA.

Smith, M. J. and Carayon, P. (1996). Work organization, stress, and cumulative trauma disorders. *Beyond Biomechanics — Psychosocial Aspects of Musculoskeletal Disorders in Office Work.* Moon, S. D. and Sauter, S. L. London, Taylor & Francis: 23–41.

Smith, M. J., Carayon, P. et al. (1992). Employee stress and health complaints in jobs with and without electronic performance monitoring. *Applied Ergonomics* 23(1): 17–27.

Smith, M. J. and Carayon-Sainfort, P. (1989). A balance theory of job design for stress reduction. *International Journal of Industrial Ergonomics* 4: 67–79.

Smith, M. J., Sainfort, F. et al., Eds. (1989). *Efforts to Solve Quality Problems,* Secretary's Commission on Workforce Quality and Labor Market Efficiency, U.S. Department of Labor, Washington, D.C.

Taylor, F. (1911). *The Principles of Scientific Management.* New York, Norton and Company.

Trist, E. (1981). *The Evaluation of Sociotechnical Systems.* Toronto, Quality of Working Life Center.

Trist, E. L. and Bamforth, K. (1951). Some social and psychological consequences of the long-wall method of coal getting. *Human Relations* 4: 3–39.

II

Surveillance

6

Injury Surveillance Database Systems

Carol Stuart-Buttle
Stuart-Buttle Ergonomics

6.1   Introduction

A database is a collection of organized, related data typically in electronic form that can be accessed and manipulated by computer software. When the interface between the data and user is developed beyond just managing the data, but with a particular function in mind, a database system evolves. Features of the interface may include data dictionaries, data security, statistical and graphical capabilities, and report writing.

Injury surveillance database systems focus on the collection of data that are pertinent to injuries and illnesses, the management of that data, and data interpretation. The goals of interpretation may include determining trends, promoting prompt treatment, and directing prevention strategies. The general use of the term injury refers to all types of injuries and illnesses. Conducting epidemiological studies or other research that attempt to find causal relationships of injuries may be practical with databases of large organizations. However, most industries have small databases and do not have the resources for more sophisticated study or analyses. This may change in the future as a study by the National Research Council (NRC) and Institute of Medicine, of the National Academy of Sciences, reviewed musculoskeletal disorders and the workplace and recommended that the National Institute for Occupational Safety and Health (NIOSH) develop a model active surveillance program in which companies could participate (NRC, 2001). This would provide statistical power for better surveillance. However, presently, many companies remain independent with both active and passive surveillance so the use of databases for research is not discussed in this chapter. Some large, global corporations have developed their own sophisticated database systems (Evans et al., 2000; Joseph, 2000; Shulenberger and Gladu, 2004). Injury surveillance databases and the relationship of surveillance to ergonomics are addressed from the perspective of the practicing industrial safety and health professional in a moderate to small company.

Surveillance of the workplace for injuries and potential injuries is an important aspect of the ergonomics process as injuries are an indicator of poor design. Often, the effectiveness of ergonomic processes are measured by the rates of injury and lost workdays. Most employers are required to record work-related injuries and illnesses in accordance with the Occupational Safety and Health (OSH) Act (OSHA, 1970). As personal computers (PCs) have become commonplace in industry, the most basic information pertaining to the OSH Act can be easily computerized, in other words, a company database can be established. Considerable time and effort can be saved in the update and analysis of the data with the right software interface.

The rate of change and growth in the computer software market is so great that for many industries the task of exploring, evaluating, and choosing a database system for the company is formidable. This chapter neither attempts to predict the future of information systems, nor does it try to summarize the existing safety and health software since such information rapidly changes. However, some safety and health software directories are provided. The main points and discussion of this chapter focus on the fundamental components of injury surveillance as they relate to ergonomics and tips on how to assess commercial injury surveillance database systems. A section touches on the development of an injury surveillance database only to provide an opportunity to understand injury surveillance in the context of system development.

6.2 Injury Surveillance

6.2.1 History

"Surveillance is the ongoing and systematic collection, analysis and interpretation of data related to health." (Baker et al., 1989). Surveillance historically began in the public health setting, as a method of watching out for certain serious diseases or illnesses such as syphilis and smallpox, so that isolation could be instituted at the first signs of symptoms. Since 1950, the term "surveillance" has been formally applied to the systematic collection of relevant disease data with the purpose to improve the control of specific diseases. During the 1960s, surveillance in the United States expanded such that it applied not only to communicable diseases but also to noninfectious diseases including environmental and occupational hazards (Langmuir, 1976).

During the 1960s, some states developed reporting requirements for selected occupational diseases or used data from the workers' compensation system for surveillance. Not until the federal OSH Act of 1970 was there a standardized scheme to monitor injuries and illnesses (Wegman and Froimes, 1985). It is not surprising that the chemical industry was a pioneering group in developing occupational surveillance systems, since there was much potential for disease. Hazard surveillance, as a means of predicting work-related health problems in the chemical industry, was developed to complement the basic surveillance of recording injuries and illnesses required by the OSH Act. Many companies then started hazard profiles that included information such as industry demographics (employment size in the geographic area), use of chemicals in the workplace, levels of the chemicals to which workers were exposed, and data on the dose–response relationship for each chemical (Froines et al., 1986).

The hazard surveillance model has been applied to ergonomics-related issues in the workplace. One problem in applying such a model to ergonomics is that suspected risk factors for musculo-skeletal disorders (MSDs) are treated discretely without taking into account the interactions of the risk factors. The effects of the interactions and the dose–response relationships are still being discovered.

"Monitoring" is another term that refers to the methods to anticipate a disease or medical disorder before it occurs (Yodaiken, 1986). Monitoring complements medical surveillance in a way similar to that of hazard surveillance. The difference lies in semantics and the use of the word "hazard," which implies that the dose–response is known. As stated earlier, it is difficult to determine when one or more risk factors are hazardous for MSDs.

Monitoring may be considered a more general term, one that refers to looking at the workplace in order to anticipate problems and prevent their occurrence. An illustration of the practical difference between medical surveillance and monitoring is the distinction between a symptom survey and a discomfort survey (Stuart-Buttle, 1994). A symptom survey identifies the employees with physical problems through the collection of detailed information about symptoms. On the other hand, a dis-comfort survey attempts to find out the discomfort that employees are experiencing while performing their jobs. Discomfort indicates a physical stress at performing the job, not necessarily an active medical problem. The implication of a discomfort survey is that discomfort could be a precursor to injury, hence a discomfort survey is a monitoring method for determining where to focus prevention efforts.

Another common distinction of the term surveillance is whether it is "passive" or "active" (BNA, 1993a). Passive surveillance is the analysis of data from existing databases, such as the occupational safety and health administration (OSHA) 300 logs that record work-related injuries and illnesses, or workers' compensation records. Active surveillance is collecting data to "actively" seek information, such as using surveys to determine the physical conditions that individuals have not reported, or defining aspects of the jobs that may be risk factors for injury. The distinction between passive and active surveil-lance is almost the same as that between surveillance and monitoring, or between medical surveillance and hazard surveillance.

6.2.2 Goals and Objectives

Injury surveillance should entail both active and passive surveillance. Existing injury data as well as additional information collected about the workers and their jobs must be analyzed. The goals of a surveillance system include "the prevention of occupational disease [*injuries*] through control of the causative agents" (Sundin et al., 1986). Surveillance helps to direct the prevention programs to control or eliminate preventable disorders (Baker, 1989).

The objectives to fulfill the goals of surveillance are to locate and monitor groups of workers who are exposed to risk, to determine the risk factors and take corrective action or reduce exposure to those factors. Discovering the relationships between exposure and the injury is a research domain that may be undertaken by some corporations with substantial databases or as state, national or university projects. The interaction and dosage of risk factors for MSDs are not yet fully known, but industry can take action based on the best knowledge to date and refine the decisions as research progresses. Gerard Scannell, former president of the National Safety Council asks "How do you direct your safety and health program if you don't know what is happening in your workplace?" (Smith, 1995). In addition, surveillance is a primary method of assessing the efficacy of prevention measures that are instituted to control the identified problems (Hanrahan and Moll, 1989). Therefore, a company benefits from collecting the data that are useful in reducing the risk of injuries and improving job performance.

6.3 Sources of Primary Injury Surveillance Data

6.3.1 Injury Data

There are several primary sources of injury data that companies either keep or to which they have access without having to survey the workforce.

- Injury logs that are mandated by OSHA
- Workers' compensation records
- Medical records
- Accident records

6.3.1.1 OSHA Recordkeeping

Since the OSH Act of 1970 companies have had to keep a record of their injuries. In 1981, OSHA began to look at the log data as indicators for an inspection. Industry responded to this new focus in one of two ways. They either improved the safety and health in the facility or became more selective in what was recorded on the logs. OSHA began to discover significant under-reporting on companies' logs and stronger enforcement action was instituted in 1986 (Tyson, 1991). Total penalties have become so substantial that companies now pay much more attention to keeping accurate records (BNA, 1993b, 1994). However, some of the difficulties of accurate recordkeeping have come to light and influence injury surveillance.

The increase in cumulative trauma disorders in the workplace highlights the difficulty of keeping the records up-to-date. As a medical condition progresses the diagnosis sometimes changes, but the OSHA records may not reflect the new diagnosis. For example, if an injury originally recorded as a strained wrist eventually led to carpal tunnel syndrome, the diagnosis is typically not changed in the logs. This may be a common oversight if nonmedical personnel keep the records or if there is poor communication with the health care professional (HCP). Computerized databases help a company, even if there is no medical department, as the system could be designed to prompt for confirmation of diagnosis while a case is open.

Despite guidelines on recordkeeping (OSHA, 2003), there remains considerable confusion among employers about when and how to record occupational injuries and illnesses. Historically, the ambiguity of some of the requirements has been a point of litigation, and one company issued a digest of official interpretations of earlier recordkeeping guidelines primarily based on legal decisions in enforcement cases or in the courts (Duvall, 1996). In 1996, OSHA proposed revised recordkeeping requirements and since January 2004 new OSHA logs and forms, along with recordkeeping guidelines are used (OSHA, 2003).

In addition to recording injuries, OSHA and other government and state agencies mandate that several types of records are to be kept by the person or people responsible for safety and health. Records pertaining to hearing conservation, tracking of and training in the use of personal protective equipment (PPE), and compliance with confined space are just a few examples of the requirements. These records may be easily tracked and integrated into a safety and health database system.

6.3.1.2 Workers' Compensation Data

Small industries commonly rely on workers' compensation data supplied by the company's insurance carrier as an injury surveillance system. The benefit of this practice is the convenience of the data, especially if the insurance carriers summarize the information. There are several downsides to using only these data. Not all injuries become workers' compensation cases, and therefore the injury profile is only partial. By the time a claim is processed and appears on a summary, a considerable length of time has passed during which preventative action could have been initiated. However, the claims on the workers' compensation insurance originate in the company, so that the company can have immediate access to the information if good in-house records are kept. If the company can track their own workers' compensation data effectively, then the company can also track the accident data that might not become workers' compensation. The company would have richer data on which to base prevention and improvements.

The cost of injuries is one type of information that workers' compensation data do provide that is difficult to obtain any other way. Cost data are useful for calculating cost–benefit equations for improvements. Since workers' compensation databases are often large they can be used effectively for major epidemiological studies or to complement large-scale surveillance (Seligman et al., 1986; Liberty Mutual, 2004).

6.3.1.3 Medical Records

Medical records can be the most reliable source of injury data since there is no interpretation required for recording as there is for the OSHA logs and workers' compensation filing. As discussed later in this chapter, a difficulty that can arise is access to the medical records. Very few workplaces have onsite medical services, and medical records can reflect both work- and non-work-related injuries, which can confuse data collection for injury surveillance.

First-aid treatment is often recorded by a company but may not be entered in an employee's medical record. Although by definition the incidents are minor, there may be indication of a design problem. For example, it is not uncommon to find frequent treatment for grazed knuckles. If this is happening to the same employee or to a group of employees under the same circumstances, it could be a clearance or access issue.

6.3.1.4 Accident Records

Accident reports are filled out by many companies and are the earliest records of an injury within a company's safety and health process. Not all injuries from accidents are recorded on the OSHA forms or claimed on workers' compensation, so that accident reports potentially provide the most comprehensive record of injuries, although maybe without medical detail. Some companies treat accident reports as full accident investigations and may also include investigations of near misses. If good root cause analyses are conducted and recorded there may be valuable information indicating job redesign.

6.3.2 Job Descriptions

Job information is useful from many standpoints. The causes of injuries need to be identified in order to prevent them. Although the causes for musculoskeletal injuries are not well defined, identifying risk factors of the job can be helpful. For example, risk factors for cumulative trauma disorders include extreme postures, high forces, high repetitions, inadequate recovery times, and excessive duration. Some of the reasons to consider building a database of job descriptions are to:

- Have a list of risk factors present in each job
- Track employees' exposure
- Aid in the initial placement of employees
- Determine a job rotation system to create a balance of the job demands
- Provide information to the HCPs to help them with decisions regarding the work-relatedness of conditions
- Provide the context for functional capacity evaluations or to tailor workhardening or conditioning programs
- Guide appropriate placement of employees returning to work and to determine temporary alternative jobs or jobs that could be modified
- Help determine jobs that may be modified for accommodating a person under the Americans with Disabilities Act (ADA) (EEOC, 1990)

The downside of developing a job description database is the expense of establishing it and keeping it up-to-date. Such costs will vary considerably according to the type and size of industry. Some industries change their processes frequently, perhaps seasonally. To keep job descriptions current in a rapidly changing environment may be difficult. In recent years, there has been transition by industry from having discrete jobs to cross-training employees for a cluster of jobs or cell production. In which case, each employee undertakes several different skills and tasks to make up their job. This complicates employee exposure tracking but does not make it more difficult to develop a job description database. Some companies update job descriptions each time an accident investigation is conducted or production change is made. The remainder of the descriptions are put on a schedule for audit and updating.

Computerized injury surveillance database systems can provide a tremendous benefit for a small company that is communicating with offsite services. If an electronic network is established, the records can be efficiently updated. Information about the jobs could be online for the medical service to access, including formats such as computer-captured video. This level of electronic communication is possible with today's security systems but medical record confidentiality remains important.

Industrial hygiene information and sampling results can also be included with job descriptions. If the facility or company is large, the database may be separate. Sophisticated systems have been adopted on global information systems (GIS) where there may be a mapping of occupational exposure data that is accessible to any site (Shulenberger and Gladu, 2004).

6.3.3 Worker Tracking

Keeping up with job descriptions may be a challenge for some industries, as is tracking the worker. Many of the companies that developed computerized databases early on stated that tracking the workers was the most complex and difficult task (Hagstrom et al., 1982; Kuritz, 1982; Smith et al., 1982; Sugano, 1982). These pioneer companies reported that effective tracking required frequent updating of employee assignments including the length of time at the job in the specific area. In addition, each job in every area required a profile so that there was a history of exposure. The job profiles or descriptions also needed to be kept up-to-date by periodic job analyses or audits. The pioneers also found employee tracking costly to implement.

Employee tracking remains worthwhile and in some circumstances critical if there are hazardous materials in the industry. Each company must assess the importance of worker tracking, and determine the type and extent of the hazards of the industry and to develop surveillance that is the most reasonable, practical, and cost effective. There are trade-offs to consider when developing a tracking system. The companies that pioneered employee tracking systems observed these trade-offs but they are still pertinent today. The trade-offs to consider with the four basic qualities of an employee tracking system follow (Sugano, 1982):

- *Uniformity within the exposure group.* Are the exposures the same across many jobs, and if so, may some of the descriptions be combined?
- *Accuracy of tracking the worker.* What are the means for communicating when an employee moves area or job? How does job rotation affect exposure? What differences are there between plants?
- *Adequacy of job and location description.* How often does the job change, and how frequently should the database be updated? Is environmental monitoring conducted regularly? How difficult is it to keep the physical, environmental and material descriptions up to date?
- *Cost effectiveness.* How much money and manpower is needed to develop and maintain the database? Is too much data being collected and too often?

At present, worker tracking for exposure to risk factors for musculoskeletal injuries remains an area of research by large corporations or universities. However, companies without the volume of data for statistical analyses still need to collect job information to help with risk assessments. The NRC study recommended that much more information be collected than is currently required by law so that greater deduction can be drawn from the data. The study suggests employee demographics such as age, gender, race, time on the job, and occupation are collected. In addition, for an injury, the event, source, nature, body part involved, time on the job, and rotation schedule should be routinely collected (NRC, 2001).

6.4 Benefits of an Injury Surveillance Database System

A complete and standardized reporting system for occupational injuries does not exist in the United States. OSHA mandates almost all employers with more than 11 employees to keep a current record of injuries and illnesses and details of how the injury occurred. The OSHA forms 300 and 301 are available electronically but substitution by equivalent ones is allowed. Exposure information, such as, the demands of the job, a history of the different jobs the employee has held, or the job characteristics such as the state of the environment or the presence of chemicals, is typically not included. However, even without the enhanced job information, the two mandated OSHA forms contain considerable amounts of data that are not easily analyzed without a system common to both forms and other pertinent information. What is important to remember is that the quality of the data depends on the detail of the information that is entered into the system. The OSHA forms filled thoroughly and thoughtfully can provide a company database. A basic computerized injury surveillance database system can have the following benefits:

- Standardized and accurate recordkeeping, improving the quality of health records and compliance with OSHA
- Efficient handling of data, reducing paper work and data reentry
- Easy and effective reporting system
- Quick statistical analyses for the evaluation of potential problems and the measurement of effectiveness of interventions
- Enhanced tracking ability, particularly of employees' job exposures, and such tracking also facilitates epidemiological studies
- Quick verification in disability and compensation cases
- Potential for integration with other data, for example, occupational sampling data
- Ease of cost tracking for cost–benefit analyses
- Efficient corporate response to litigation, community complaints, and public relations requirements

There are many commercial software packages available for computerizing the OSHA forms combined with basic surveillance functions. The sophistication of the available programs varies considerably, from large database systems with an injury surveillance module combined with advanced statistical analyses, to simple OSHA form recording and tracking. A few software directories that are regularly updated are listed in the appendix.

6.4.1 Integration of Database Systems

The data collection mandated by OSHA is useful but a broader collection of data would be more beneficial (NRC, 2001). Many industries also go further and see benefit from greater integration with other company functions (Finucane and McDonagh, 1982; OMPC, 1992; Holzner et al., 1993; Dieterly, 1995). Early on, during the development of surveillance within the chemical industry it was realized that exposure measures were needed to interpret the medical findings, thus integrating medical surveillance programs with industrial hygiene surveillance (Parkinson and Grennan, 1986). A survey conducted by OSHA, found medical surveillance programs were most effective when they were integrated into a comprehensive and systematic approach for identifying and addressing workplace hazards (BNA, 1991, 1993c). A comprehensive approach requires the involvement of different expertise within a company and hence involves more than one discipline.

Injury surveillance can be incorporated into comprehensive safety and health programs. The Occupational Medical Practice Committee (OMPC) of the American College of Occupational and

Environmental Medicine (ACOEM) provided guidelines for the scope of occupational and environmental health programs. The committee identified 14 essential components of a program (OMPC, 1992).

1. Health and evaluation of employees
2. Diagnosis and treatment of occupational and environmental injuries or illnesses, including rehabilitation
3. Emergency treatment of nonoccupational injury or illness
4. Education of employees in jobs where potential occupational hazards exist, including job specific instruction, instruction on methods of prevention, and recognition of possible adverse health effects
5. Implementation of programs for the use of indicated personal protective devices — ear protection (plugs, muffs), safety spectacles, respirators, etc.
6. Evaluation, inspection, and abatement of workplace hazards
7. Toxicological assessments, including advice on chemical substances that have not had adequate toxicological testing
8. Biostatistics and epidemiology assessments
9. Maintenance of occupational medical records
10. Immunization against possible occupational infections
11. Medical interpretation and participation in development of governmental health and safety regulations
12. Periodic evaluation of the occupational or environmental health program
13. Disaster preparedness planning for the workplace and the community
14. Assistance in rehabilitation of alcohol- and drug-dependent employees or those with emotional disorders

Other professionals such as those in industrial hygiene, safety, or ergonomics, can address many of these items. The OMPC stressed that the practitioner in occupational medicine should understand how to enlist and collaborate with the skills of colleagues from the many other professions involved with industry. The aforementioned list is a useful one for a company to consider. Some elements will be more important than others, depending on the industry. If many services are contracted offsite, the overall program may not be managed by a HCP but perhaps by another professional, for example, in safety or industrial hygiene.

6.5　Ergonomics and Injury Surveillance

Several professions in industry, such as safety, industrial hygiene, medicine, risk management as well as ergonomics have an interest injury surveillance. The different disciplines may have a part in contributing to the database or one department may be responsible for it overall, while others access it. Different professions may develop their own interface or module with the same database so as to manipulate and interpret the data as they wish. If a person wears the hats of several job functions, for example, all aspects of safety and health, then they might work with a number of discrete modules, such as safety, industrial hygiene, and ergonomics. However, a small company may choose to integrate the modules to a basic safety and health one. The scale of module for each discipline depends on the level of detail desired or degree of expertise of the person using the system. If expertise is outsourced, then an in-house module in that field may not be needed.

Many injuries are related to poor design of jobs or the workplace and this is the domain of ergonomics. In particular, musculoskeletal injuries such as low back disorders and cumulative trauma disorders have become informally termed as ergonomics-related disorders. This is a limited view as other injuries can occur due to poor design, for example, accidents can arise because of inadequate accessibility or because of an error provoked by poor layout and labeling of controls and displays. However, there is considerable focus on MSDs as they account for one-third of all lost workday cases in the United States and such injuries are the leading cause of disability of people in their working years (BLS, 2003a). The cost of lost workdays of MSDs, based on lost earnings and workers' compensation, has been estimated at $13–20 billion annually and as high as $45–54 billion annually if indirect costs such as lost productivity are included (NRC, 2001). The most common source of injury and illness for lost workday cases is "sprains and strains" and overexertion is the primary event causing injury and illness across all the divisions of industry (BLS, 2003a). Injury surveillance, therefore, is an important method by which to determine indicators of poor job design and the need for ergonomics intervention. Injury surveillance is also an important component in an ergonomics process and the data are used to measure effectiveness of interventions.

6.5.1 Ergonomics Process

An ergonomics process in a company commonly looks at health and medical indicators of MSDs, which includes injury surveillance; worksite analysis which is active surveillance for risk factors; prevention and control which also entails measures of effectiveness of the controls; and training and education (NIOSH, 1997). Good injury surveillance promotes prompt treatment and directs the prevention programs. The greater the understanding of the causes of the injuries, the more proactive can be the prevention. If prevention measures are not taken, then the system consists only of treatment and remains reactive to the occurrences of the workplace. Surveillance systems that are set up from only a medical standpoint can be weak on the prevention aspect, even if there are good data. The "who, what, when and where" (Garrett, 1982) of injuries can often be captured on the accident reports and on the OSHA forms or equivalent if they are thoroughly completed. Typically, the "how and why" of an injury are not well reported and are dependent upon the quality of the accident investigation and whether it includes an ergonomics perspective.

Some caution is needed in relying upon a medical focus for improving the workplace. The workplace system may suffer if workstation changes are in the form of accommodation or excessive tailoring on a case by case basis, as a response to medical injuries. A balance must be kept between appropriate design of workstations and the overall function of the system, including organizational design aspects. Involvement of an interdisciplinary team helps to ensure the work system is addressed. Disciplines such as, professionals in safety, engineering, management, health care, industrial hygiene, and ergonomics, as well as the operators should participate in the teams.

Responsibility has to be assigned to an individual or a group within the company to apply the results of the surveillance data by looking at the workplace, identifying the problems and root causes, and developing potential solutions. The process does not stop there, as interventions have to be implemented and then followed to ensure effectiveness. It is more likely that preventive measures will not be taken when the medical facilities are offsite, which is the case for the majority of workplaces. It is the company's responsibility to bridge the communication gap between the company and offsite services. Assigning a company contact person helps promote communication.

6.5.2 Ergonomics Module of a Database System

An ergonomics module that interfaces with an injury surveillance database system has its own list of needs to fulfill the ergonomics function. The main tasks are:

- Collecting data, apart from injury data, that indicate ergonomics intervention or else help with the interpretation of injury surveillance data
- Interpretation of the data, for example, prioritization methods and graphical trends or statistical analysis of surveys
- Tracking of information such as, project status, timelines, workplace changes, effectiveness and training

To accomplish data collection and interpretation in the context of task analyses, an ergonomics module may also interface with technical software such as the NIOSH Lifting Index. Job profiles may be generated from job analyses or analyses results could be added to profiles that may have been developed by another department, such as human resources. Data that can be gathered apart from injuries include:

- Turnover rates
- Absenteeism rates
- Productivity rates and quotas
- Quality-related figures
- Workers' compensation costs
- Job profiles or descriptions
- Employees' jobs record or work history
- Discomfort survey results
- Ergonomics audit results
- Employee suggestions or reports of problems
- Ergonomic analyses results
- Ergonomic training records

Although an old concept, a simple way to categorize most of this information is by using three groups: People, Places and Things (Hillman, 1982). An ergonomics module would add ergonomics software tools of choice, statistical analysis programs, analyses and survey results, and tracking systems. Specific customization will be required to interface the desired software of an ergonomics module to an injury surveillance database, whether that database is company developed or a commercial package. The degree of software tailoring and the extent of the company's capabilities and resources will determine how much would be in an ergonomics module. It may be practical to keep some components independent such as the ergonomics tools.

6.6   Early Databases

The earliest computerized surveillance databases were purchased or developed during the 1970s primarily by the chemicals and petroleum industry. Computer-based information systems became a necessity to manage the data growth more effectively. Manual data systems in large corporations were hand-written and paper-based, lacked consistency and standardization and could not adequately capture the data on important monitoring aspects such as job history and workplace exposures (Finucane and McDonagh, 1982). The models that were conceived were comprehensive and sophisticated and many currently continue their development (Holzner et al., 1993). The focus of the systems were occupational health and at a minimum incorporated (Finucane and McDonagh, 1982):

- Detailed worker and job histories and demographic data
- An inventory of potential exposures and associated adverse health effects related to specific workplace location
- Worksite exposure data
- Employee medical information collected throughout the worker's career

These extensive models depended on effective interaction between the departments and the readily available retrieval and analysis of the information. Wolkonsky (1982) described the following information as necessary for a health data system at Amoco:

- *Employee* — job history, location codes, sickness and disability
- *Claims* — workers' compensation payments, death certificates, internal cost reports
- *Medical* — personal and family history, physical examination, laboratory results, immunizations
- *Industrial Hygiene* — personal samples, area samples, potential exposures by location, exposure levels
- *Toxicology* — animal research, materials safety data sheets
- *Safety* — accident data, physical conditions, supervisor data, DOT data

These early medical surveillance database systems remain but are continually upgraded and developed as computer hardware and software change, regulations alter and knowledge about exposures and risk assessment models progress through research.

6.6.1 Challenges That Were Met

Descriptions of the early computerized systems highlighted many difficulties in establishing comprehensive databases systems. Of the large and sophisticated models being developed by the major corporations in the country, none were the same. Those who attempted to purchase another's system ended up making significant modifications and customizations to fit the needs of the company (Joiner, 1982). However, despite the individuality of the systems there were some common factors that made them successful. The common factors of the systems were: flexibility, interaction, user friendliness, modular design, valid databases, economy, innovation, key staff, commitment, and phased approach.

One of the first difficult decisions for the developers was in deciding whether to develop a system in-house or buy a commercial system and modify it to the meet the company's needs. Building the system gradually by modules was a common approach, but the modules of the database systems varied considerably between companies. On a basic level the core data modules can be summarized as People, Places, and Things (Hillman, 1982). Deciding how much information about each area was needed, what was cost effective, and what hardware to accomplish the goal varied according to company philosophy and existing computer experience or investments.

Early developers reported some modules as a challenge to design. In particular, tracking workers was the most complex and difficult task (Hagstrom et al., 1982; Kuritz, 1982; Smith et al., 1982; Sugano, 1982). Centralization of information, such as personnel data, was another issue for a few companies. Despite the large corporate size, not all systems, such as the personnel system, were consistent or centralized across facilities (Hagstrom et al., 1982). Privacy and confidentiality of medical information were of general concern and were resolved in different ways. Some companies encrypted information in case it was routed to the wrong terminal, while others limited access and used passwords. Many of the issues and difficulties that were met by those who developed the early medical surveillance systems are pertinent today. Some of these issues are further addressed in the context of developing or choosing an injury surveillance database system.

6.7 Defining the Database Model

A team effort is necessary for a company to successfully define the model of an injury surveillance database system (Wrench, 1990). The computer software professionals need to work closely with the safety and health professionals and any other users of the system. Whether a system is to be developed or purchased the goals and objectives of the system should be clear.

An assessment of the company itself is necessary to ensure that the goals of the surveillance system integrate with the company's needs, vision, and resources. There are common phases of development for a computerized health surveillance program. Designing the system and defining the components are just one stage (Joiner, 1982). If commercial software packages are being deliberated, take particular care to consider them in context of the overall system because the result can be a different program for each departmental function. If a later decision is made to combine the programs, it may be technically too difficult to do so and efficiency is lost in the long run. The following program development phases are also pertinent for defining programs to purchase.

- Feasibility study
- Planning and identifying potential users
- Designing the system and defining components (system development and modification)
- Implementing
- Operating

- Maintaining
- Upgrading the system

6.7.1 Designing the System and Defining Components

The step of designing the system and defining the components requires particular comment because it is important for determining the scope of an injury surveillance system. Deciding whether injury surveillance should be part of a larger system or if it should be a small discrete system may not be an easy task for a company. The following are some steps to help determine the components and functions that you either wish to have developed or to purchase.

1. Define the present injury surveillance system in the company, however manual that may be, and critique for aspects that you wish to maintain and determine areas of improvements.
2. Define the basic components of the injury surveillance database system that you would like to have. Keep in mind the primary goal of an injury surveillance database system, that is, determining the who, what, where, when of employee exposure (Garrett, 1982) and using that information to guide prevention. As a database, this may be organized conceptually as three subsets: People, Places, Things (Hillman, 1982).
3. Determine the functions or modules of the groups that could utilize the database and compare to the needs of the company, e.g., an industrial hygiene function is possible but may not be needed as a full module by the company.
4. Compare the present surveillance system in the company with the defined database model. Prioritize the development of the modules.
5. Assess the status of computerization in the company. Allocate which modules warrant computerization now versus in the future. Allow for growth in both software and hardware.

6.7.2 Company Size

The size of a company influences the models and decisions about injury surveillance database systems. In the past, database systems utilized mainframes and analyzed volumes of data that was at a scale that was cost prohibitive to smaller industries. Now computerization is readily available. Despite the availability, many companies have not devoted resources to developing injury surveillance database systems despite the time-saving advantage. One possible reason is that the system is not developed in any form in those companies. Basic recordkeeping that entails filling out the OSHA forms does not constitute even a simple injury surveillance system, unless there is some analysis of the data and preventive action is taken. Another possibility that resources have not be devoted is that a company experiences no recordable injuries (or so rarely) such that attention to injury surveillance is deemed unwarranted. A third reason may be that improving the efficiency of surveillance is not a top priority relative to other company improvements.

6.7.2.1 Large Companies

As noted earlier, large corporations were the first to develop injury surveillance database systems. This is not surprising for several reasons. There was a large volume of data due to the size of the company. The companies had a number of potential exposures from chemicals and the environment, and they had the resources for epidemiological studies.

6.7.2.2 Small and Medium Companies

Establishment sizes of 50–249 employees incur the highest injury incidence rates (per 100 full-time workers) (Stuart-Buttle, 1998; BLS, 2003b). The Small Business Administration (SBA) defines small as typically less than or equal to 500 employees (SBA, 2003). Regardless of definition, companies of about 50–500 employees in size should be particularly vigilant of their injury statistics. Despite national statistics, there will be some small companies with such low injury prevalence that a formal injury surveillance system is not necessary (other than mandated recordkeeping).

Some of the characteristics of a small industry indicate that conducting injury surveillance may be difficult (Stuart-Buttle, 1998). The main common characteristics are:

- Less formality
- Responsibility for several positions
- Greater responsiveness
- Less specific knowledge
- More management involvement
- Less data-oriented approach

However, it is precisely some of these characteristics, for example, the responsibility for several positions that would make an investment in an injury surveillance database system such an aid. The system could provide the benefits listed earlier and enhance an individual's and the company's efficiency.

6.7.3 Onsite or Offsite Medical and Surveillance Services

Only 6.3% of U.S. workplaces provide onsite medical surveillance that includes treatment and recording of occupational injuries (BNA, 1995). These facilities are large corporations and medium size companies with a sufficient number of employees or injuries or both for an onsite medical department to be cost effective. Other workplaces use offsite treatment services that may or may not track data for the company. Usually a company employee — typically a person without medical knowledge — fulfills the basic clerical duty of recording the OSHA logs and filling out workers' compensation forms.

If the OSHA logs and accident records are the minimum data for surveillance tracking, the quality of the information is very important if it is to be useful. The questionable quality of offsite surveillance has been raised by NIOSH. Murthy of NIOSH suggested that physicians away from the worksite were less likely to probe a patient about work history and often had very little occupational medicine training to assess work-related conditions effectively (BNA, 1995). Although the company is dependent upon the health care professional for accuracy and detail of diagnosis, the designated contact person for the company can exert some control by communicating the requirements of the company to the medical service (Stuart-Buttle, 1998). Close communication between the medical service and company is needed for all injury cases to keep the information accurate and up-to-date.

Brewer et al. (1990) described a clinic-based occupational injury surveillance system that served many industries. The system connected several clinics that treated employees for even minor injuries. All injuries, regardless of whether they were compensable or recordable on the OSHA log, were recorded. Statistical analyses were routinely conducted and results provided to the companies. The companies learned the type of injuries that were occurring, the frequency, part of the body, severity, and how the injury was caused. Of all the cases analyzed within the system, 78.4% of the patients received care for minor trauma, which were classified as cuts or lacerations, sprains, strains, or contusions. Such analyses showed that collecting data of first-aid treatment is beneficial and that a company could reduce costs by preventing the occurrence of minor trauma. The study by Brewer et al. also demonstrated the feasibility of clinic-based occupational injury surveillance as a means of assisting employers with the control of work-related conditions. Since this study, there have been many occupational medicine groups, with web-based data exchange systems, offering such services to industry.

Medical screening is often used to complement environmental control measures. The purpose of screening is to detect disorders early and implement appropriate intervention. Screening also serves as a monitor of the effectiveness of prevention measures already instituted. Many medical screening functions, such as auditory testing and pulmonary function screening, can be conducted and supported by an offsite contract service. Tracking and timetabling the routine screening of employees is easily performed by a database.

6.7.4 Types of Database Models

An injury surveillance database system can be developed anywhere along the continuum of simple to sophisticated. It may be a system that is customized, commercially purchased, or contracted as an offsite service. Whichever type of system is chosen it is most likely to be modular. The most basic injury surveillance database model fulfills the recordkeeping function and provides some descriptive statistics of the data. Many PC packets provide the interface to enter the information required for the OSHA forms, analyze the data, and produce summary graphs by department, for the month, quarter, or year, and sometimes across years. Printouts of the filled forms are also commonly available.

Beyond a recordkeeping function, the fundamental database model can be represented by the three interacting modules of Hillman (1982): People, Places, Things. Although there is more regulation since Hillman wrote of this concept, simplicity is still relevant. Figure 6.1 illustrates a modification of Hillman's categories to People, Places, Plant. People includes demographics (gender, age, date of hire, job), health data, worker evaluations and incidents; Places includes a basic coding of the areas or jobs, a description and other specific job information including history of specific interventions or modifications; and Plant includes information files of plant or department wide data such as chemicals or materials that are in the company. Some small companies may have little information in these files. A large corporation could expand this model with considerable databases of information under all three modules.

With this concept, potentially there may be many users of the database, depending what data is entered in the first place. Human resources, safety, medical, industrial hygiene, ergonomics, and risk management are some of the possible parties interested in an injury surveillance database system. Each group interfaces with the system from their own perspective, using their own programs to extract appropriate data, perform statistical functions and risk assessments as required, and present data in the format

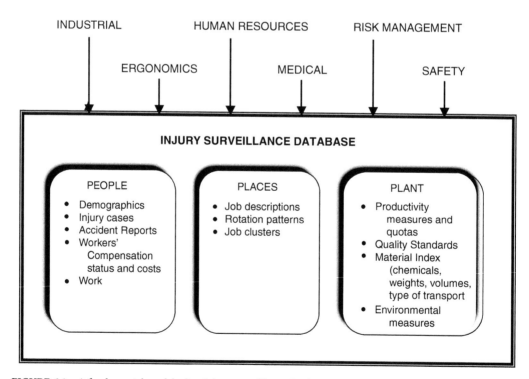

FIGURE 6.1 A fundamental model of an injury surveillance database system and possible user groups. (Based on Hillman, 1982.) Listings under People, Places, and Plant are examples of database information of common interest to the users.

desired and appropriate for the user's job function. Injury surveillance is dependent upon someone being responsible for looking at the data of the injury reports that has been entered into the system. The responsibility may be almost any one of the users indicated, as the job function depends on the company size and structure. Each interface of a user group is a module that could be developed as the system evolves according to the company's needs.

6.7.5 Confidentiality

A database that is shared between many users raises concern about confidentiality and the need to limit access to certain information. Systems with distinct modules for each user group typically make each module a limited access to a defined group of people, usually by use of a password. A network can be designed to monitor passwords and terminal identification. Medical data are particularly sensitive information. Joyner and Pack (1982) described the Shell Oil Company's system that encrypts the data and stores it centrally. Decryption programs were installed in the medical department to decode output to the terminal, therefore precluding disclosure of confidential information in the event the output inadvertently arrived at another terminal. Another technique that has been adopted to ensure confidentiality is the scrambling of social security numbers (Hillman, 1982; Wolkonsky, 1982).

6.7.6 Data Entry

Many of the early large and sophisticated systems were invested with efficient methods of data entry. For example, the Amoco occupational health system recorded a patient's medical history and examination on mark-sense forms that were computer input sheets that were entered by optical scanner. This eliminated any possible error from transcription (Wolkonsky, 1982). Now, there are many personal digital assistants (PDAs) and commercial software for field data collection that can be downloaded into the database (Shulenberger and Gladu, 2004).

6.7.7 Commercial Software

The compatibility of commercial software to an existing system is influenced by the age of the system. Companies with relatively new computer systems are more likely to be able to take advantage of the large commercial market, tailor the packages with less technical problems, and support the systems with in-house personnel. Considering the available choices of programs, purchase of a system can save considerable design and development costs (see Appendix for a list of software directories). The choice of system also depends upon the level of computer skill of the users or resources of the company.

Most companies have a network but some may choose to have an independent safety and health system. Many commercial software programs are available that provide a recordkeeping function with OSHA log entry and analysis. In addition, there is software for specific functions such as, maintaining Material Safety Data Sheets (MSDS), tracking Personal Protective Equipment (PPE) and tracking employee training.

Web based company systems are becoming commonplace. The intranet is a convenient method for employee access to a central database (Evans et al., 2000). For large corporations a global information system (GIS) is the only practical method to communicate globally (Joseph, 2000; Shulenberger and Gladu, 2004).

6.8 Statistical Treatment of Data

A computer can perform sophisticated data analyses only if specifically instructed to do so. A good understanding of statistical methods is required to develop a data analysis program, and if necessary, a statistician should be consulted. There are commercial statistical packages that may be interfaced with a database system, and some of the packages guide a user in interpreting data. There is potential for a vast array of analyses of the data in an injury surveillance database, depending on the extent and sophistication of the overall system. For example, if there is a focus on industrial hygiene issues, numerous calculations may be required to interpret data from workplace monitoring. The safety and health professional needs some knowledge of statistics to interpret data beyond the OSHA logs. The commercial software market has responded to demands by specific professions, for example, by producing programs for industrial hygienists, such as those that assist with industrial hygiene formulae (Cameron, 1996). There are other programs on the market designed to meet the technical needs of the ergonomist and safety professional.

The majority of workplaces are small businesses in which those responsible for safety and health often do not have a background in statistics. In addition, they may not know what to look for in the basic data collected from OSHA logs and accident records. Some fundamental information is, what volume and type of injuries are occurring, to which parts of the body while performing which jobs or tasks. Most of these questions can be answered by descriptive statistics. The commercial market offers many programs that focus on OSHA recordkeeping and provide basic data analysis with graphic output. After careful assessment of these programs there is likely to be a software package that meets the company's needs. However, even more useful information can be obtained if data collection and analysis extend beyond the minimum requirements for compliance. The following are a few points that will help to make the data more useful. The points will also help when assessing software programs.

- Convert the number of injuries into an incidence rate as described in OSHA's recordkeeping guideline (OSHA, 2003). This provides a figure that has a common denominator with other companies and national statistics. The national injury incidence rate calculated by the Bureau of Labor Statistics (BLS) or the rate for the company's Standard Industrial Classification (SIC) code [or North American Industrial Classification System (NAICS) code (OMB, 1997)] can serve as benchmarks for the company. Incidence rate can be calculated for groups of types of injuries and also for departments. Direct comparison between departments can be made when the numbers are expressed as incidence rates.
- The number of lost work days is usually considered an indication of severity of the injury. An incident rate can also be calculated for the lost workdays. At times a given incidence may be so severe that it accounts for most of the recorded lost workdays. In such situations the incidence rate should be calculated with and without that particular case. The restricted days rate should also be calculated. When injuries begin to be detected early and undergo conservative treatment, then lost workdays are less likely to be experienced than restricted days.
- Record the side of the body that was injured. This is not often recorded yet it can be useful to know. If the information is related to a record of the employee's dominant hand a useful picture of the stresses and injuries at the job begin to build, which helps with job analyses and prevention strategies.
- Look at the incidence rate by age, gender, length of time on the job, and whether the employees are full-time or part-time. All of these analyses will help determine if a certain group is susceptible. Those who are very new on the job are typically suffering from lack of conditioning. If this is found to be the case, strategies can be implemented to gradually introduce the worker to the job. For example, orientation sessions could be scheduled throughout the first week rather than in the first couple of days.
- Look at the injury rate by supervisor or line leader if these are different groupings from the department or area. This is useful to determine if a certain team has a problem and suggests there may be some administrative issues to investigate. Similarly, data should be compared between shifts. Often differences are related to management.
- Incorporate workers' compensation costs into the database and determine the mean for the type of injury and cause and for each area or job. This helps with cost benefit analyses and prioritizing areas to be improved. Sometimes the incidence rate in a plant has increased, but the workers' compensation costs decreased. This may indicate that more people are reporting before the condition is severe and more expensive to treat.

- Track the work status of all those injured, whether they are off work on workers' compensation or on work restrictions. Develop a list on a regular basis, such as once a week, to ensure that each case is kept up to date and progressed toward full return to work, preferably at the previous job. It is not unusual for an employee to "fall between the cracks" and be on workers' compensation for a period of time without any functional change or decision about the case.
- Identify the causes and mechanisms of injuries based on the accident record, and in particular, relate the findings to the areas of high incidence rates.
- Look at the data to see if there are patterns according to time of day, days of the week, the month, quarter or year. Compare across the years as well.
- If possible, collect first aid data, as well as the data mandated by OSHA. Treating minor cuts and bruises takes time and there can also be an associated loss of productivity. If this information is collected and analyzed similar to the logs, there may be some useful emerging patterns that can help with identifying the causes and lead to prevention measures. Examples of common problems are blunt tools that cause the user to slip, or inadequate clearance around equipment that provokes bruising.

More sophisticated analyses can be conducted on data from OSHA forms. However, some understanding of statistics is needed to make use of such features in a software program, even if the interface is user friendly. Goldberg et al. (1993) describes an injury surveillance database with some useful, advanced analysis features. A description of some of these features follows.

- Actual exposure hours were collected from payroll to calculate the incidence rate. This allowed for a better picture of incidence rate related to overtime hours, especially since overtime was inconsistent and potentially a reason for increased injuries.
- Regression analysis helped determine the factors leading to lost and restricted workdays. The software interface displayed a list of independent variables and interactions from which to choose. For example, the number of restricted workdays could be looked at by gender, age, and length of employment. Regression analysis could also be used to predict the number of expected lost or restricted workdays for an injured employee, and this helped management plan for the situation.
- Time series modeling provided insight as to whether there were certain periods of higher incidence rates. Accidents are sometimes based on cyclical seasonal weather change, employment relationships, or perhaps rapid expansion.
- A forecasting feature smoothed the data to remove known cyclical fluctuations and provided indication of the data trend, whether the incidence rate was increasing or decreasing.
- Inclusion of cost and billing information was used to generate cost–benefit analyses.

6.8.1 Report Generation

Report generation of a surveillance system should be in electronic formats to include in internal electronic reports as well as good hardcopy formats. They should also be able to be saved. The OSHA form has to be posted annually for employees but otherwise other reports are based on what the company needs. Reports from a database have two main purposes: further analyses and communication. These can also be thought of as two levels of reports, one for the professional to explore and interpret the data, and another that summarizes the findings for presentation to a team or management.

The maxim "pictures speak a thousand words" is true when it comes to concise presentation. Good graphic presentation straight from the software program saves considerable time, especially when the alternative is transporting the data to another application program. However, it is possible to set up standard links to a program that produces charts and graphs, so that graphs can be routinely produced.

6.9   Choosing a Commercial Injury Surveillance Database

"One size does not fit all" when it comes to setting up an injury surveillance database system (Wrench, 1990). If a commercial program is purchased, it will require professional tailoring to fit the needs of the company. For example, along with social security numbers, company identification numbers are commonly used for personnel tracking, yet every company has a different numbering system that requires modification of that data field. Likewise, the job coding system is different in every company.

Involve the employees as much as possible in decisions about a database. The employees know the information needed for their jobs and can contribute significantly to the needs assessment and system design and to evaluation of existing programs. In addition, the success of a system is dependent upon their acceptance of it.

The following are some aspects to consider when choosing a commercial software program (Menzel, 1994).

Evaluating needs and resources

- Identify data management goals and objectives.
- Define requirements. (This may be accomplished by mapping out the existing system.) Include anticipated growth and potential new functions.
- Define company constraints, such as type of hardware system, technical equipment, and the characteristics of other company databases.
- Assess company resources (personnel, budget, hardware, and software).
- Develop a list of criteria, based on the needs and resources evaluation, by which to assess the software programs. There may not be a program that satisfies each criterion so developing a spreadsheet can be useful for deciding the trade-offs for a final selection. A scoring system could also be used to determine several final contenders.

Finding the software

- Use directories that may be available through professional safety and health associations, see Appendix
- Research articles on software in safety and health magazines
- Respond to software advertisements in occupational safety and health magazines
- Ask colleagues about the software they use and their experience with it
- Request demonstration disks or web demonstrations and onsite demonstrations

Evaluating the software

- Assess the basic features of the programs against the criteria. This will narrow the selection for more critical view
- Evaluate for qualities such as:
  - Ease of use
  - Provision of high quality, onsite training
  - Variety and quality of reports
  - Ability to transfer data to and from other programs
  - Potential degree of customization
  - Presence of security systems to safeguard confidentiality and data integrity
  - Reasonable initial and continuing cost
- Select a few final contenders based on the extent to which criteria are satisfied
- With program specifications attached to the proposal, request price proposals with details of features, service, and renewal fees
- Request names of current users who can be contacted
- Assess the responsiveness and quality of the support available
- Review the length of time each company has been in operation

- Assess their financial stability and long-term viability. (This is not easy but the number of clients being supported may give an indication.)

Selecting the vendor

- Negotiate the price
- Negotiate the extent of customization and support
- Negotiate for additional features such as training

Evaluating the effectiveness

- Ensure that the program is thoroughly reviewed before the first anniversary (when maintenance fees usually begin)
- If necessary, negotiate for modifications before paying maintenance fees

When implementing a new system, ensure that parallel systems are run for at least a month after full installation. The accuracy of the system must be checked before abandoning the old methods. If there are software modifications or connections made to an existing network, it is better to assure that the system is running smoothly before relying upon it. New hardware should be tried also before being relied upon.

6.10 Summary

The cost benefits of injury surveillance database systems are becoming more apparent since there has been greater focus on accurate recordkeeping, and as software programs have become more available. There are several benefits including standardized and accurate recordkeeping, time saving and efficient data entry and analysis, enhanced tracking abilities, and effective guidance to prevention strategies.

A company should give considerable and careful thought to the design of the injury surveillance database system. The design is dependent upon the company's needs, size, constraints, potential growth, and resources. An assessment of the company's needs prevents a quick purchase of a commercial system that might not have the growth potential, adequate support and customization services to make the program a successful addition to the company.

References

Baker, E. L., Honchar, P. A., and Fine, L. J. 1989. I. Surveillance in occupational illness and injury: concepts and content. In *Surveillance in Occupational Safety and Health*, ed. E. L. Baker, American Journal of Public Health. Supplement. 79:9–11.

Baker, E. L. 1989. IV. Sentinel event notification system for occupational risks (SENSOR): the concept. In *Surveillance in Occupational Safety and Health*, ed. E. L. Baker, *American Journal of Public Health.* Supplement. 79:18–20.

Brewer, R. D., Oleske, D. M., Hahn, J., and Leibold, M. 1990. A model for occupational injury surveillance by occupational health centers. *Journal of Occupational Medicine.* 32(8):698–702.

Bureau of Labor Statistics (BLS). 2003a. Lost-worktime injuries and illnesses: characteristics and resulting days away from work, 2001. News, United States Department of Labor, Bureau of Labor Statistics, USDL 03-138. Washington, D.C.

Bureau of Labor Statistics (BLS). 2003b. Charts from annual BLS survey: 2002 OSH summary estimates chart package. http://www.osha.gov/oshstats/work.html.

Bureau of National Affairs (BNA). 1991. Preliminary results from OSHA survey said to show benefit of 'integrated' plan. *Occupational Safety and Health Reporter.* 5-22-91:1712–1713.

Bureau of National Affairs (BNA). 1993a. All work sites need passive surveillance job checklist for risk factors, ANSI group told. *Occupational Safety and Health Reporter.* 1-20-93:1444–1445.

Bureau of National Affairs (BNA). 1993b. OSHA's egregious penalty policy for recordkeeping violations upheld. *Occupational Safety and Health Reporter*. 2-10-93:1601–1602.

Bureau of National Affairs (BNA). 1993c. Comprehensive risk management programs effective at finding illness, draft report says. *Occupational Safety and Health Reporter*. 3-31-93:1899–1900.

Bureau of National Affairs (BNA). 1994. Recordkeeping violations affirmed against general dynamics as non-serious. *Occupational Safety and Health Reporter*. 3-16-94:1353–1354.

Bureau of National Affairs (BNA). 1995. NIOSH staffer says onsite surveillance plays important role in treating workers. *Occupational Safety and Health Reporter*. 11-8-95:817.

Cameron, M. 1996. IH calculator. *The Synergist*. April:17–18.

Dieterly, D. L. 1995. Industrial injury cost analysis by occupation in an electric utility. *Human Factors*. 37(3):591–595.

Duvall, M. N. 1996. Digest of official interpretations of the bureau of labor statistics recordkeeping guidelines for occupational injuries and illnesses (second revised edition). *Occupational Safety and Health Reporter*. 1-31-96:1144–1171.

Equal Employment Opportunity Commission (EEOC), 1990. *Americans with Disabilities Act of 1990*. Public Law 101-336 101st Congress, July, 26. US Government Printing Office, Washington, D.C.

Evans, S. M. et al. 2000. Ergonomics@Ford.com: using the intranet to enhance program effectiveness. *Proceedings of the IEA 2000/HFES 200 Congress*. Paper 1–467.

Finucane, R. D. and McDonagh, T. J. 1982. Foreword. In *Medical Information Systems Roundtable. Journal of Occupational Medicine*. Supplement. 24(10):781–782.

Froines, J. R., Dellenbaugh, C. A., and Wegman, D. H. 1986. Occupational health surveillance: A means to identify work-related risks. *American Journal of Public Health*. 76(9):1089–1096.

Garrett, R. W. 1982. Environmental tracking at Eli Lilly and Company. In *Medical Information Systems Roundtable. Journal of Occupational Medicine*. Supplement. 24(10):836–839.

Goldberg, J. H., Leader, B. K. and Stuart-Buttle, C. 1993. Medical logging and injury surveillance database system. *International Journal of Industrial Ergonomics*. 11:107–123.

Hagstrom, R. M., Dougherty, W. E., English, N. B., Lochhead, T. J., and Schriver, R. C. 1982. SmithKline Environmental Health Surveillance System. In *Medical Information Systems Roundtable. Journal of Occupational and Medicine*. Supplement. 24(10):799–803.

Hanrahan, L. P. and Moll, M. B. 1989. VIII. Injury surveillance. In *Surveillance in Occupational Safety and Health*, ed. E. L. Baker. *American Journal of Public Health*. Supplement. 79:38–45.

Hillman, G. 1982. ECHOES: IBM's environmental, chemical and occupational evaluation system. In *Medical Information Systems Roundtable. Journal of Occupational Medicine*. Supplement. 24(10):827–835.

Holzner, C. L., Hirsh, R. B., and Perper, J. B. 1993. Managing workplace exposure information. *American Industrial Hygiene Association Journal*. 54(1):15–21.

Joiner, R. L. 1982. Occupational health and environmental information systems: basic considerations. In *Medical Information Systems Roundtable. Journal of Occupational Medicine*. Supplement. 24(10):863–866.

Joseph, B. S. 2000. Ford Motor Company global ergonomics process. *Proceedings of the IEA 2000/HFES 2000 Congress*. Paper 2–454.

Joyner, R. E. and Pack, P. H. 1982. The Shell Oil Company's computerized health surveillance system. In *Medical Information Systems Roundtable. Journal of Occupational Medicine*. Supplement. 24(10):812–814.

Kuritz, S. J. 1982. The Ford Motor Company environmental health surveillance system. In *Medical Information Systems Roundtable. Journal of Occupational Medicine*. Supplement. 24(10):844–847.

Langmuir, A. D. 1976. William Farr: founder of modern concepts of surveillance. *International Journal of Epidemiology*. 5(1):13–18.

Liberty Mutual Research Institute for Safety. 2004. Numerous research reports of workers' compensation data, for example: Dempsey, P. G. and Hashemi, L. 1999. Analysis of workers' compensation claims associated with manual materials handling. *Ergonomics*. 42(1):183–195. www.libertymutual.com.

Menzel, N. N. 1994. Occupational health software: selecting the right program. *AAOHN Journal.* 42(2):76–81.

National Institute of Occupational Safety and Health (NIOSH). 1997. *Elements of Ergonomics Programs: A Primer Based on Workplace Evaluations of Musculoskeletal Disorders.* DHHS (NIOSH) Publication No. 97–117. NIOSH, Cinncinnati, OH

National Research Council (NRC) and the Institute of Medicine. 2001. *Musculoskeletal Disorders and the Workplace: Low Back and Upper Extremities.* Panel on Musculoskeletal Disorders and the Workplace. Commission on Behavioral and Social Sciences and Education. National Academy Press, Washington, D.C.

Occupational Safety and Health Administration (OSHA). 1970. *Occupational Safety and Health Act.* Public Law 91-596 91st Congress S. 2193 Dec. 29. U.S. Government Printing Office, Washington, D.C.

Occupational Safety and Health Administration (OSHA). 2003. *Record Keeping Compliance Directive (CPL 2-0.131).* Department of Labor/OSHA, Washington, D.C. http://www.osha.gov/record keeping/.

Office of Management and Budget's (OMB) Economic Classification Policy Committee. 1997. *North American Industry Classification System — United States.* PB98-127293. National Technical Information Service (NTIS), Springfield, VA. http://www.ntis.gov/products/bestsellers/naics.asp? loc=4-2-0

OMPC. 1992. Scope of occupational and environmental health programs and practice. Committee Report of the Occupational Medical Practice Committee. *Journal of Occupational Medicine.* 34(4):436–440.

Parkinson, D. K. and Grennan, M. J. Jr. 1986. Establishment of medical surveillance in industry: problems and procedures. *Journal of Occupational Medicine.* 28(8):773–777.

Seligman, P. J., Halperin, W. A. E., Mullan, R. J., and Frazier, T. M. 1986. Occupational lead poisoning in Ohio: Surveillance using workers' compensation data. *American Journal of Public Health.* 76(11): 1299–1302.

Shulenberger, C. and Gladu, N. 2004. Cost-effective use of computer technologies in EHS management. *Occupational Hazards.* January:25–29.

Small Business Administration (SBA) 2003. *Small Business Size Regulations.* http://www.sba.gov/size.

Smith, F. R., Gutierrez, R. R., and McDonagh, T. J. 1982. Exxon's health information system. In *Medical Information Systems Roundtable. Journal of Occupational Medicine.* Supplement. 24(10):824–826.

Smith, R. B. 1995. Recordkeeping rule aims for accuracy, wiser use of injury and illness data. *Occupational Health and Safety.* January:37–68.

Stuart-Buttle, C. 1994. A discomfort survey in a poultry-processing plant. *Applied Ergonomics.* 25(1):47–52.

Stuart-Buttle, C. 1998. How to set-up ergonomic processes: a small industry perspective. In *Handbook of Occupational Ergonomics,* W. Karwowski and W. S. Marras, Eds, CRC Press, Boca Raton, FL.

Sugano, D. S. 1982. Worker Tracking — a complex but essential element in health surveillance systems. In *Medical Information Systems Roundtable. Journal of Occupational Medicine.* Supplement. 24(10): 783–784.

Sundin, D. S., Pedersen, D. H., and Frazier, T. M. 1986. Occupational hazard and health surveillance. Editorial. *American Journal of Public Health.* 76(9):1083–1084.

Tyson, P. R. 1991. Record-high OSHA penalties. *Safety and Health.* March:17–20.

Wegman, D. H. and Froines, J. R. 1985. Surveillance needs for occupational health. Editorial. *American Journal of Public Health.* 75(11):1259–1261.

Wolkonsky, P. 1982. Computerized recordkeeping in an occupational health system: the Amoco system. In *Medical Information Systems Roundtable. Journal of Occupational Medicine.* Supplement. 24(10):791–793.

Wrench, C. P. 1990. *Data Management for Occupational Health and Safety: A User's guide to Integrating Software*, Van Nostrand Reinhold, New York.

Yodaiken, R. E. 1986. Surveillance, monitoring, and regulatory concerns. *Journal of Occupational Medicine.* 28(8):569–571.

Appendix

American Society of Safety Engineers (ASSE). 2004, *Virtual Exposition.* Search for exhibitors at the annual national ASSE conference at the ASSE website under the menu item annual conference and exposition. There are many software vendors. www.asse.org

Canadian Centre for Occupational Health and Safety (CCOHS). 2004. Web Information Service (WIS) of Databases of MSDS and other useful information. www.ccohs.com

Occupational Hazards. 2004. Website has past articles about software and vendor advertisements. www.occupationalhazards.com

Occupational Health and Safety, 2004. The magazine has an annual product directory available online and includes software. www.ohsonline.com

Peterson K. W., Ed. 1997. *Handbook of Occupational Health and Safety Software*, Version 10.0.

American College of Occupational and Environmental Medicine (ACOEM), Arlington Heights, IL. This directory provides descriptions, uses, hardware requirements, training, user support, costs, and other details for software programs (450 pp., spiral-bound). www.acoem.com

7

Corporate Health Management: Designing and Evaluating Health in Organizations

Klaus J. Zink
Martin J. Thul
University of Kaiserslautern

7.1 Health in Organizations

In many countries, the topic of health is integrated in occupational health and safety approaches, which are described in worker protection laws and respective rules. Most companies in Western, industrialized

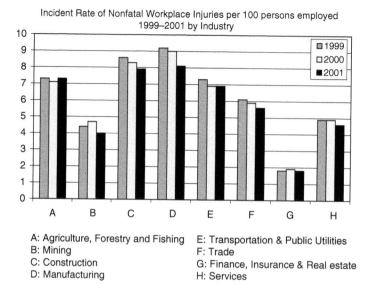

FIGURE 7.1 Workplace injuries in the United States of America. (Taken from Bureau of Labor Statistics, U.S. Department of Labor, 2002. With permission.)

countries have some sort of an occupational health and safety system. Because of the legalities involved, such activities are not popular, which are perceived as being primarily cost intensive. But this is not true at all, if we look at the yearly work place injuries and the costs that employee's illness incur each year (Figure 7.1).

Table 7.1 shows the estimation of costs of work-related diseases in Germany for the year 2001.

In this sense, management should have an interest in improving employees' health for increasing competitiveness.

Overall, the topic of health is becoming increasingly important — the sole cause is not restricted to absenteeism or physical illness. There is a slowly changing understanding of health, and an increasing acceptance of the necessity of well-being and work-related satisfaction.

Therefore, the preservation and improvement of health is a necessity that should not only focus on humanization of the work place or the fulfillment of legal demands, but also on improving the sustainability of human resources and the competitiveness of an organization.

- The commitment of employees regarding the goals of an organization and their motivation and capabilities are growing success factors for competitiveness. Dissatisfied and unhealthy employees will not be able to fulfill such demands — and at the least, productivity will suffer.

TABLE 7.1   Estimation of Costs of Work-Related Diseases in Germany 2001

34.810 million employees × 14.6 days absence due to illness = 508.57 million days absence due to illness =	1.39 million years loss of productivity
Estimated costs caused by loss of productivity based on labour costs (loss of output) 1.39 million years loss of productivity × €32.200 average remuneration = loss of output due to unfitness for work	€44.76 billion
Estimated loss of output (loss of gross value added) 1.39 million years loss of productivity × €50.900 average gross value added = loss of gross value added	€70.75 billion

Source: From the Federal Institute for Occupational Safety and Health, Germany, 2001. With permission.

- Globalization and a growing competition has led to a series of cost-reduction strategies, for example, lean production or lean management, which are focused on a reduction of workforce. Therefore the same amount of work — with growing customer demand — has to be done by less people.
- The growing interest in corporate social responsibility (CSR) (see, e.g., The European Commission, 2002a; also see Business for Social Responsibility, 2003 and others) and the ongoing discussion about the redefinition of corporations (Post et al., 2002) will gain more and more importance for companies. It could be a realistic scenario where customers will buy goods only from companies that demonstrate their engagement in CSR. Within CSR, occupational health and safety is an important issue (The European Commission, 2002b), in addition to the responsibility for the health of customers or community members.
- In this context, socially responsible investment is another topic of growing importance. Reflecting these developments, different new share indices have been created — among others the Dow Jones Sustainability Group Index. Sustainability is seen as precondition for long-term shareholder value. The sustainability criteria are applied to assess opportunities and risks deriving from economic, environmental, and social dimensions. Here, not only occupational health and safety, but also employee participation and satisfaction are playing an important role (SAM Group, 2003).
- The demographical development, especially in European countries, will reduce the work force— and therefore also reduce the number of qualified people for professional and managerial positions. The quality of working conditions will be a predominant criterion in the competition for companies recruiting these employees.
- The demographical development in Europe will also lead to an older workforce. The preservation of health has to contribute to the preservation of human resources to avoid increasing costs for work-related diseases and to maintain the know-how.

These few examples show the growing importance of health (preservation and promotion) for the executives of an organization.

7.1.1 The Changing Understanding of Health

Whether individuals are or will become healthy or ill depends on a number of different determinants: their behavior, their physiological and mental preconditions, or their life and work environment. As people normally spend most of their time at work, working conditions are of particular importance: work contents, social relations at work, or corporate sources of endangerment can be causes for impairment of health. However, if designed properly, they also provide chances for the promotion of health. To this extent, organizations bear a far reaching responsibility for the health of their employees. Therefore working contents and working conditions should be designed in a way that the health of employees will not be damaged but improved.

In this context, we also have to consider the fact that task demands and working conditions have changed over the last years. In some cases job demands have changed dramatically: the reduction of physiological or physical strain has been substituted by mental or psychological exposure. Though traditional health problems like muscular–skeletal disorders still exist within a great number of working places, the sole concentration on physiological and safety problems would not be adequate. The growing importance of psychosocial determinants of the health of employees (Bödecker et al., 2002) brings new challenges to management. The European Union carries out a survey on changing working conditions every 5 years — the last one was in 2000. The main findings of the last survey are given in Table 7.2.

The changes in working conditions followed by a redefined health definition have led to a change of paradigm regarding the definition of health. In the past, there was a dominance of a biomedical health understanding, which was based on a model of pathogenesis. Health and illness have been understood as complementary states regarding the organic functions of an individual. Mental or social factors have been neglected nearly completely. Today health is understood in a broader sense: "Health is a state of complete physical, mental and social well-being and not merely the absence of disease and infirmity"

TABLE 7.2   Working Conditions in the European Union

Main Findings

The most common work-related health problems are:
    Backache (reported by 33% of respondents)
    Stress (28%)
    Muscular pains in the neck and shoulders (23%)
    Overall fatigue (23%)
There is a direct relationship between poor health outcomes and adverse working conditions, arising in particular from a
    high level of work intensity and repetitive work
Exposure to physical risk factors (noise, vibrations, dangerous substances, heat, cold, etc.) and to poor design (carrying heavy
    loads and painful positions) remains prevalent
Work is getting more and more intensive: over 50% of workers work at high speed or to tight deadlines for at least a quarter of
    their working time
Control over work has not increased significantly: one third of workers say they have little or no control over their work while
    only three out of five workers are able to decide when to take holidays
The nature of work is changing: it is less dependent on machinery and production targets and more driven by customer
    demand
The number of people working with computers has increased: from 39% in 1995 to 41% in 2000
Flexibility is widespread in all aspects of work: working time ("round-the-clock" and part-time work); work organization
    (multiskilling, teamwork and empowerment); and employment status (18% of all employees work under nonpermanent
    contracts)
Temporary workers (employees with fixed-term contracts and temporary agency workers) continue to report more exposure
    to risk factors than permanent employees
Gender segregation and gender discrimination—both highly disadvantageous to women—are prevalent.
Violence, harassment, and intimidation remain a feature of the workplace: from 4% to 15% of workers in different countries
    report that they have been subjected to intimidation

Source: From the European Foundation for the Improvement of Living and Working Conditions, Dublin, 2000. With
permission.

(WHO, 1948). Furthermore, health is not only to be understood in a result-oriented way but also as being
process-oriented (Kickbusch, 1995) Health as a process includes the personal development of a person and
his or her ability to compensate environmental influences aiming to physical and mental well-being
(Badura et al., 1997). This redefinition of health also indicates a change of paradigm from pathogenesis
to salutogenesis (Antonovsky, 1979, 1987). In this respect the improvement of health has to include:

- Mental aspects (like design of work (content))
- Decrease of stress or possibilities for qualification and personal development)
- Social aspects (like relationship to superiors and coworkers)
- Physical aspects (like industrial safety, handling and reduction of dangerous substances, elimination of accident risks)

In the field of industrial safety and physical workload, concepts for (re)designing the task have been in
existence since many years. The influence of work on mental health and its consequences is a younger
branch of research. Nevertheless, there are scientific results explaining the origination of psychosocial
stress and its influence on physical health. Here two empirically proofed models of explanation shall
be referred to. According to the demand control model, mental stress mainly results from of the interdependence of individual resources and the conditions of the working environment (Karasek and
Theorell, 1990). The model of the gratification crisis focuses on the importance of recognition for the
mental and social health of employees (Siegrist, 1996). Both models show the importance of the work
situation and the potential consequences for physical health.

    Looking at the dimension with a specific importance for the psychosocial health of employees, one can
find the following (Karasek, 1979):

- Changes in the working conditions (e.g., frequency of changes, job security, opportunity for
advancement)

TABLE 7.3 European Union Strategy on Health and Safety at Work 2002–2006: Novel Features

It adopts a global approach to well-being at work, taking account of changes in the world of work and the emergence of new risks, especially of a psychosocial nature. As such, it is geared to enhancing the quality of work, and regards a safe and healthy working environment as one of the essential components

It is based on consolidating a culture of risk prevention, on combining a variety of political instruments — legislation, the social dialogue, progressive measures and best practices, corporate social responsibility, and economic incentives — and on building partnerships between all the players on the safety and health scene

It points up the fact that an ambitious social policy is a factor in the competitiveness equation and that, on the other side of the coin, having a "nonpolicy" engenders costs which weigh heavily on economies and societies

Source: From the European Commission, 2002. With permission.

- Work content (e.g., possibilities to participate, clarity of demands, broad job demands)
- Social working conditions (e.g., social support by colleagues and superiors, feedback, team cohesion)
- Working time (e.g., shift systems, flexible working hours)
- Financial aspects (e.g., remuneration system, incentives)
- Social aspects (e.g., status and prestige of a profession)

This listing and the fact that psychosocial health is a result of the interdependence of individual resources and working conditions show that measures for improvement need both an individual approach (e.g., personal development) and a situational approach (e.g., working conditions). Regarding working conditions and health promotion one has to differentiate between situations with negative (mental stress) and positive (mental demands) potentials (Leitner et al., 1993). Harmful conditions are to be eliminated, whilst positive influences have to be enhanced.

Even though the responsibility for the improvement of employees' health lies with the company, political support by the respective state, federal, or regional government can promote these activities. The novel features of the strategy on Health and Safety at Work 2002–2006 published by the European Union (Table 7.3) provide a good example.

Similar thoughts can be found in the U.S. Department of Labor's 2003–2008 Strategic Management Plan of the Occupational Safety and Health Administration (OSHA) (U.S. Department of Labor, 2002).

7.1.2 Approaches for Improving Health in Organizations

As described earlier, health includes physical and mental dimensions. Though health improvement activities should take into account both fields of action, this has not always been the case in the past. One of the reasons for this not-so-satisfactory situation can be found in the legal responsibilities for health and safety in some countries. This responsibility was delegated to specialists, departments, or external organizations with special qualification — who therefore had a special view on the problem. One consequence, which is quite often found in countries having a broad legal basis for health care (e.g., Germany), is related to different responsibilities for health promotion and occupational health and safety (OHS).

If there is only one department for OHS, one has to state that these specialists concentrated mostly on harmful working conditions (Schröer, 1996). For the protection of health, a multitude of rules and regulations, and mainly technical measures have been used. If behavioral aspects played a role, extrinsic motivational approaches stood in the foreground (Zink, 1980). Mental and social factors have been mainly neglected (Gaertner, 1998). These factors and the fact that regulations focused on minimal standards led to limited success. The effectiveness has also been reduced because these activities have been mostly reactive and concentrated on the fulfillment of regulations and on their control. The protection of health has been the task of specialists, who had a specific point of view. Their work has been considered disruptive to the daily routine and their image within the organization has remained so (Kerkau, 1997; Pischon and Liesegang, 1997).

Corporate health promotion — if it exists — uses a more behavioral oriented approach. Healthy behavior of individuals at work and at home has been the main issue (Asvall, 1995). The aim of health promotion programs has been the empowerment of employees to improve their health resources (Greiner, 1998). In this regard, personal engagement and organizational deployment strategies were the main concepts in improving health. The strength of such approaches can be seen in a broader understanding of health and in the participation of the employees, which generally happened voluntarily. As weaknesses one can state the following: very loose connection to the respective working conditions, a more or less unsystematic approach, and the lack of an adequate legal basis (Breucker, 1998). Therefore, the implementation of corporate health promotion remains voluntary for organizations.

The situation in North America (e.g., the United States) is quite similar. There are two organizations working on OHS — the Occupational Safety and Health Administration (OSHA) as part of the U.S. Department of Labor (see http://www.osha.gov) and the National Institute for Occupational Safety and Health (NIOSH) related to the Department of Health and Human Services (see http://www.cdc.gov/niosh/homepage.html). Looking at practical issues of workplace health promotion, one can find a more general understanding quite often supported by external organizations. In this sense, OHS and voluntary health practices, which concentrate on nutrition, physical activities, tobacco, and so on, are offered from one consulting company. On the one hand, specialized knowledge in both fields is easily available to the organization. On the other hand, it bears the risk that the responsibility for employees' health is delegated to external consultants. In Europe, OHS and workplace health promotion are promoted by different actors. Mainly small, and medium-sized companies not having their own department for OHS, are using external organizations to fulfill their legal obligations (e.g., work place analysis). Preventive measures in health promotion are mainly organized by health insurance companies, of which one has to be a member by law (compulsory insurance). The prevention in the workplace is legally fixed (Employer's Liability Insurance Association) — also because of a compulsory insurance. Therefore, there has sometimes been a split off between "individual health" and "occupational health."

Both approaches, industrial safety or OHS and corporate health promotion, have led to successes in the past. But there are limits under changed conditions, which are to be overcome. Isolated concepts without integration of all relevant target groups — especially management — that is based on a narrow understanding of health will not be able to meet the challenges of the present and the future.

7.1.3　Necessity for Integration

Taking together the situation described so far, it becomes obvious that there is a need for integration in a multiple sense: not only integration of contents, but also integration of systems — and within systems, people.

Regarding the integration of contents we have to bring together OHS and corporate health promotion, where it has been a divided action in the past. The negative image of OHS has to be changed into a positive one of health or healthy and motivated employees as an important precondition for competitiveness. This approach is to be included in a broader one, focusing on corporate social responsibility as a part of sustainability with growing importance in a globalized society.

To reduce the economic efforts, health has to become a part of an integrative management system, which may also include topics like quality or environmental protection.

But these concepts of integration are not sufficient. The main effort of integration is concerned with (top) management. Corporate health promotion has to be understood as an important way to do ones' business — and therefore has to be a part of the corporate policy and strategy. This includes managers understanding the importance of health and acting as role models in health promotion. Therefore health promotion cannot be outsourced, as has happened quite often in the past.

The following section will present different integration strategies and evaluate their results by referring to the demands formulated earlier in the text.

7.2   Strategies of Integration — Necessity for a Corporate Health Management System

To overcome the deficits discussed earlier, OHS and corporate health promotion have to be integrated in a corporate health management (CHM) system. This CHM has to be based on a broad health definition formulated by the World Health Organization (WHO, 1948) many, many years ago (1948!) and has to involve management and all other relevant target groups. It has to be a multidimensional approach, which has to simultaneously focus on working conditions in a broader sense and employee behavior. The aim is to realize sustainable successes. Therefore, CHM has to be a specific management system using the idea of continuous improvement and be integrated into daily work. The term "integrated" is to be understood in a way that a new management system need not be created without the systems, that have been in existence. Furthermore, health must be understood as an integral part of the management system and be related to the organization's specific frame conditions (Thul, 2003a).

7.2.1   Goals of a Corporate Health Management System

Goal of a CHM is the sustainable promotion of the employees' health by creating the respective frame conditions. The resulting activities are based on a broad understanding of health, not only focusing on physical aspects, but also including mental and social well-being. Approaches for respective design activities can be found within the behavior of the members of the organization (e.g., leadership behavior, observance of security regulations, or cooperation and communication) and within the working conditions (e.g., climate, ergonomic design, or use of technique). In contrast to the traditional OHS approaches, there is no one-sided concentration on hazards (e.g., elimination of safety problems), but rather a proactive focus on working conditions that have a positive influence on health (like work contents).

The broad understanding of health leads to broad sets of strategies in realizing improvement measures. But there should not be the misunderstanding that there is a necessity for a huge amount of activities. Further, it is important to secure the efficiency and effectiveness of these measures. In other words, the expenditures should be in an acceptable relation to the returns. New processes and structures should only be implemented if this is unavoidable. It is much more suitable to use and — if necessary — adapt existing approaches to optimize the corporate health situation. Therefore, the integration of the topic health in an existing management system is an important precondition. It also reveals ways of proofing the economic effects of corporate health promotion. This is an important precondition to make health within management acceptable and guarantee the durability of this approach.

To understand the approach of CHM discussed here, one has to first analyze the term "management."

7.2.2   Management in the Context of Health

Referring to management literature (Staehle, 1999) one can differentiate an institutional and a functional management term. Institutional management means health as a task of an institution: the management or the managers. From a functional understanding of management it can be concluded that managing corporate health includes a sum of management principles one has to use. Based on this differentiation of the term "management" one has to discuss the respective approaches necessary and the frame conditions, which have to be built up.

7.2.2.1   Management in a Functional Understanding

The work on specific topics using management principles requires, in many cases, the use of adequate management systems. Management systems are result oriented and include all organizational measures, which are suitable to support the achievement of a defined goal (Schwaninger, 1994). Related to a corporate health management the following points merit discussion (Thul, 2003a):

- *Open systems approach.* A management approach to improve the corporate health situation has to be based on an open systems approach (von Bertalanffy, 1981). This means inclusion of the relations to the socioeconomic environment of the company. To optimize an organization's health situation, it is not sufficient to reduce the focus on its inner social structure. The connections with the economy and society also have to be regarded. This means, at the end, that a corporate health management has to include the interests of all relevant internal and external stakeholders (stakeholder-orientation). In this understanding the realization of a CHM is a possibility to take care of the social responsibility of a company and to meet the growing importance of corporate social responsibility (Zink, 2003a, 2005).

- *Integration in a holistic approach.* Activities to improve health in an organization have to be embedded in a holistic concept, which ensures the systematic coordination of all single measures. In addition, the respective activities should not be developed and implemented isolated from the (other) daily business. Therefore, health has to be integrated in the general management processes to prevent or minimize conflicts of targets. The necessity of such an approach is especially given at a time when corporate change processes have to be realized. If health-relevant questions have been included in early planning phases, there is a chance for multiple possibilities to prevent health stress in advance. Where such a prevention is not realized, in general, corrective actions are needed, which causes additional costs. As a consequence of these necessities of integration, one can conclude that approaches focusing on the improvement of health in organizations need a strategic orientation. A strategic health management in this sense has not only a long-term orientation, but also anticipates future developments based on a systematic analysis of relevant factors within the organization and its environment. This is a good basis for a broader setting with regard to health in organizations.

- *Establishment of adequate feedback control systems.* A regular review of the effectiveness and efficiency of all realized measures is an essential characteristic of a management-oriented approach. Therefore, adequate feedback control and indicator systems have to be established (Bleicher, 2004). This is the same for health projects. Only if there is a regular measurement, whether realized measures are still effective (also under changed frame conditions), the necessities to act can be identified and (limited) resources be allocated in the correct way. But this can only be achieved if the relevant indicators are available. Furthermore, these indicators should not only show the respective results, but also the preconditions ("enablers") for these results. Only such a comprehensive measurement enables an organization to judge, whether results are a consequence of realized preconditions — and not a consequence of external frame conditions, which cannot be influenced. In addition, one gets information on whether inefficient and ineffective systems should be given up or developed further.

- *Establishing a continuous improvement process.* From the necessity of adequate feedback control systems, it may be concluded, that the improvement of the corporate health situation has to be understood as a continuous improvement process (CIP). As mentioned previously, the quality of all health-relevant processes has to be measured on a broad basis of indicators, not only showing results but also preconditions. The extension of a result-oriented measurement to a potential-oriented evaluation leads to the advancement of including a future-oriented view of the corporate health situation. First signs may appear in relation to how the health situation will develop in the future and which measures will be necessary to preserve and promote corporate health. A CIP based on such a broad database allows for a systematic analysis of the effectiveness of existing processes and structures of corporate health management and to draw respective conclusions purposefully. In addition, a continuous measurement and analysis of this set of data also allows the valuation of realized improvement activities.

- *Giving up a pure expert orientation.* In the past, the improvement of an organization's health situation was governed by dominating expert views. Another characteristic of a CHM approach is the inclusion of additional interest groups. As mentioned earlier, an expert orientation leads to the danger that health problems are dealt with under a specific point of view created by a specific professional discipline. In many cases, solutions coming up under such conditions are only

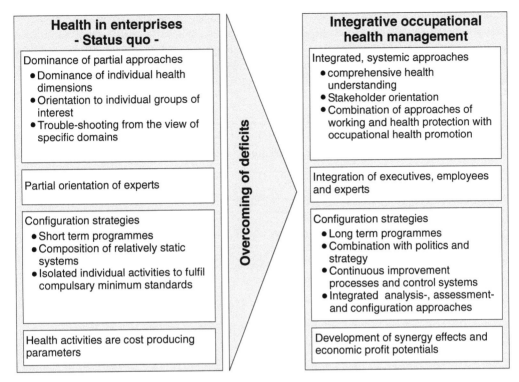

FIGURE 7.2 Relevant characteristics of an integrative corporate health management. (Taken from Thul, M. J., Benz, D., and Zink, K. J. in: Landau, K. Ed. Good Practice — Ergonomie und Arbeitsgestaltung, 2003, p. 502. With permission.)

suboptimal, because single aspects have been optimized at the expense of neglecting others. To prevent such problems, CHM includes all relevant target groups: professionals, managers, and employees. The participation of employees aims to include their experiences and their creativity. This results in measures to improve the corporate health situation, which are developed and implemented to meet their requirements. In addition, the healthy behavior of employees will be influenced positively and sustainably.

The most relevant aspects of a CHM, which is based on the demands described in the preceding list are summarized in Figure 7.2.

7.2.2.2 Institutional Management: The Role of Leadership

There are different reasons to include managers in a CHM concept: First and premost, managers are by law (see, e.g., the European legislation for occupational health and safety) obliged to take care of health and safety. But much more important for a successful implementation of a CHM are two further aspects:

- Only senior management has the power to successfully install a new approach in health preservation and promotion. Their decisions influence the working conditions directly.
- Furthermore, they have to act as role models concerning a health promoting behavior. If top management does not develop and act as a role model in following respective values and ethics, the whole approach will soon lose its authenticity and not survive in the long term.

Regarding the competencies and possibilities of action of managers referring to the improvement of the corporate health situation, the following points are to be emphasized. The professional qualification of managers is normally focused on technical or economical aspects of management and less, or not at all, on health-oriented topics. As a consequence, there are specific deficits in recognizing and solving corporate health problems. Even if respective qualifications exist, strategies, targets, and plans of an organization may potentially restrict the possibilities to act. In this sense, management normally acts in a field of conflict between economic interests and those of the demands for decent working conditions.

In view of this, the following requirements for a promotion of corporate health by management can be stated (Thul et al., 2003):

- Creation of awareness concerning the meaning, basics, and effects of corporate health. This is necessary to have a corresponding sensibility and therefore have a disposition to support measures for improvement.
- Qualified professional support (e.g., consulting) when solving problems in daily work and specific health problems. This is a precondition for effective and efficient problem-solving approaches.
- Review of strategies, plans, and objectives to enable managers to contribute to the avoidance and the reduction of health-related problems.
- Inclusion of remuneration systems is another critical success factor, because some individual objectives, which influence the salary of a manager, may be counterproductive for a comprehensive engagement in the field of reduction of health-related strain.

7.2.3   Fields of Action within Corporate Health Management

Following the institutional and functional perspectives of CHM, it is obvious that corporate health preservation and promotion is influenced on the one side by specific measures of a health management and on the other side by the impacts of business in general. To realize a systematic harmonization between daily business and health management, a strategy-oriented approach is necessary. This delivers a respective frame of action for managers and allows the use of management principles in including all relevant aspects of health within general management.

Table 7.4 shows, how and by which measures health can be established as a management task. In differentiating between institutional and functional management and between effects on daily business and specific approaches for health promotion, one needs a two-dimensional perspective. As a consequence, the measures are allocated to four different fields of action.

TABLE 7.4   Fields of Action within Corporate Health Management

Referring to the person of the manager (institutional)	Integration of goals related to workplace corporate health into the balanced scorecard Aspects of corporate health as part of the training and development of managers	Active involvement in improvement projects for promotion of corporate health Development of scopes through appropriate strategic guidelines (e.g., financial budgets, empowerment, objectives)
Referring to management-oriented action (functional)	Integration of aspects of corporate health into the ratio system for the supervision of an organization (business excellence model) Consideration of aspects of workplace partnership and corporate health in the planning processes (e.g., change of work contents, cooperation and communication structures within change processes)	Identification and supervision of core processes of corporate health (e.g., handling of conflicts recognition systems) Regular review of the effectiveness of measures based on suitable ratios (e.g., results of employee surveys)
	Aspects of corporate health within day-to-day business	**Specific aspects in promoting corporate health**

7.3 Integrating Health Management in Management Systems

Whereas till now the tasks of a CHM have been described in general, it is now pertinent to discuss those possibilities of integration, especially regarding management systems, which do exist.

7.3.1 Integrating OHS and Health Promotion into a CHM

As discussed several times before, the first step has to be "content integration." Depending on different legal preconditions in different parts of the world, and depending on the economic development of a country, we may find different approaches to improve the health and safety situation in an organization (Zink, 2003a, 2005).

If there are different approaches for OHS and health promotion, we have to bring together contents — and organizational structures. As mentioned earlier, CHM approaches refer to the work situation and the behavior of employees regarding health and safety. So the attachment of contents should not be a big problem. The greatest problem may exist where procedures and organizational structures are given by law, which have to be met within a common approach. More progressive organizations are also regarding health and safety problems outside the working sphere like traffic safety or safety and health problems regarding sports or leisure, because absenteeism can also be caused in these fields. Therefore the structures and processes within a CHM should also be open to health topics located in an organization's socio-economic environment.

7.3.2 Integrating OHS and Health Promotion in Respective Management Systems

As mentioned before, it is reasonable to integrate health and safety topics into existing management systems. The reduction of additional effort, a higher level of acceptance, and the possibility to gain synergy effects (Eklund, 1999) are good reasons to do so. For this type of integration, different roadmaps are possible:

- Integrating health and safety topics into the structure and processes of another management system, like quality management
- Creating a new integrated management system, which can be a basis for different topics like quality, environment, and health
- Integrating health topics in holistic management systems based on an existing excellence model like that of the European Foundation for Quality Management (EFQM, 1999)

These possibilities of integration will now be described in a more or less extensive way, before discussing their pros and cons — and coming to a final proposition.

7.3.2.1 Integrating the Health and Safety Topics into Another Management System

The preconditions for such an attempt improved just "recently"; when the ISO 9000 based quality management system was reviewed and restructured as ISO 9000-2000. Among others the reason for this revision was to increase the possibilities for integrating other topics like environment management based on ISO 14001. As written down in ISO 9004-2000, Point 0.4 "Compatibility with other Management Systems": "This International Standard does not include guidance specific to other management systems, such as those particular to environment management, occupational health and safety management, financial management or risk management. However, this International Standard enables an organization to align or integrate its own quality management system with related management systems. It is possible for an organization to adapt its existing management system(s) in order to establish a quality management system that follows the guidelines of this International Standard.

Figure 7.3 shows the ISO 9000-2000 structure. As there is no internationally accepted OHS or CHM management system — besides the usually mentioned BS 18001 (BSI, 1999) — ISO 9000-2000 could be the basis for such an integration. As the principal problems and approaches of quality, environment and

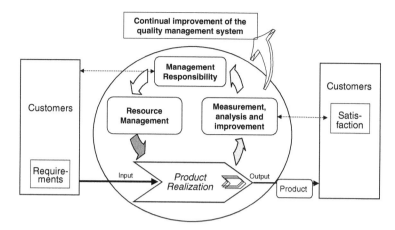

FIGURE 7.3 Model of a process-based quality management system. (Taken from EN ISO 9001:2000, p. 13. With permission.)

health are very similar (Zink, 1999), there are several arguments supporting the use of such synergies — based on a structure, which proves the fulfillment of all respective legal demands.

The problem of these management systems in the past has been their poor image, like for OHS. The whole process is mainly performed because of external demands and the structure is seen as bureaucratic and management is not really engaged (Conti, 1999). As of today, there are no research results that reported of a change in this scenario with the revised ISO 9000 structure.

7.3.2.2 Creating a New Integrated Management System

Before the review of ISO 9000 there have been several other approaches to create a management system (like in New Zealand, the AS/NZ 3 4581:1999), which could be used for all possible topics. Figure 7.4 shows the structure of such a management system. As known so far, these management systems have not received a high acceptance.

7.3.2.3 Integrating Health Topics into Holistic Management Systems Based on So-Called Excellence Models

In contrast to the aforementioned integrated management systems, so-called excellence models (or Organizational Excellence, Business Excellence or Performance Excellence Models) have been developed

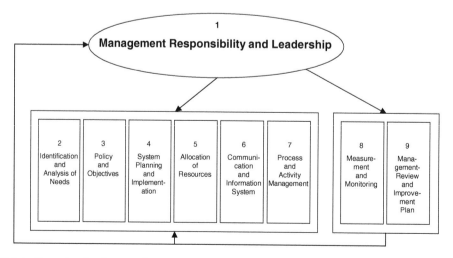

FIGURE 7.4 Example of an integrated management system.

and used all over the world (e.g., Japan, Australia, United States, Europe). Primarily based on a broader understanding of quality (like the Japanese Deming Prize, which was developed in the early 1950s), these models — all awarded as regional or national quality awards to create awareness for a broader understanding of "enterprise quality" — have the huge advantage of being accepted by management.

Although excellence models are primarily used for assessment purpose, they are also very useful for building holistic management systems. Their underlying philosophy, their content, and their assessment systematically deliver detailed information on how to design, implement, and deploy a holistic management system. As mentioned before, excellence models have their origin in the field of quality management. Based on a comprehensive definition of quality these models showed ways how "quality" could be measured and which prerequisites have to be fulfilled to reach a high performance level. As time passed by, the term "quality" was substituted by the word "excellence." This was mainly done because in earlier versions of excellence models the term "quality" was often misinterpreted, which led to a limited scope of application. The implementation of the model contents was frequently delegated to quality departments or quality specialists who focused on technical aspects. In many cases, this resulted in a lack of management support and responsibility. Necessary changes in an organization's structure and processes were omitted, followed by low effectiveness and efficiency. The use of the term "excellence" in recent versions of excellence models was especially aligned with two objectives. On the one hand, the (top-)management's responsibility for the achievement of excellence should be pointed out, and on the other hand, the claim for a holistic approach was to be strengthened. The letter point was of particular importance. Excellence models are not limited to special areas of application; they show ways how to handle widely different topics by using management principles. This can be validated in taking a closer look at, for example, the European Model for Excellence, promoted by the European Foundation for Quality Management (EFQM). The prerequisites for excellence within this model might be used — after minor adaptations — to handle topics like quality, environment, or health. According to this, a special model for corporate health has been developed by the Institute for Technology and Work at the University of Kaiserslautern in Germany (Zink and Thul, 1998). Based on the Business Excellence Model of the EFQM, which is responsible for the European Quality Award, a health-oriented self-assessment concept has been developed.

The principal idea behind this approach is again the use of synergies and acceptance by senior management. Synergies exist under one of the following circumstances: if a company is already working with this excellence model and can now use it for health reasons as well, or if a company is starting with this health evaluation concept, and then has no problem to enlarge the approach to measure business excellence.

Figure 7.5 through Figure 7.7 show, respectively, the EFQM Business Excellence Model used, the "transformation" from business excellence to health excellence, and the model for the evaluation of CHM.

As the CHM Assessment Model is based on the European Model for Total Quality Management, until 1999 known as EFQM Model for Business Excellence, one needs to first describe the specific approach of the EFQM Model (Zink, 2003a, 2005). As shown in Figure 7.5, one has to distinguish between two groups of criteria: the so-called "enablers" and "results."

The model fulfills the requirements of a comprehensive concept that integrates all relevant aspects for securing the future of an organization.

The "enablers" ensure:

- The involvement of all managers
- The integration of a culture of excellence in the organization's policy and strategy
- That the organization brings out the potential of its people to improve its business continuously
- The management, utilization, and preservation of resources
- The continuous optimization of all processes

The aim of investing in "enablers" or potentials is to improve business results through:

- Increased customer satisfaction and retention
- People motivation and satisfaction
- Regarding environmental aspects

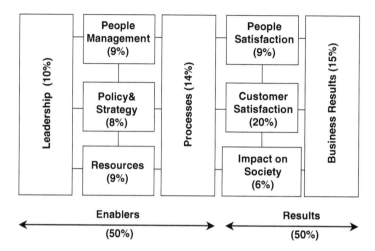

FIGURE 7.5   The EFQM model for business excellence 1999. (Taken from EFQM, Self-Assessment based on the European Model for Business Excellence 1999 — Guidelines for Companies, 1999. With permission.)

Looking at the results criteria we can find a stakeholder approach — as asked for in the context of sustainability.

Apart from this broad set of criteria, the evaluation is based on the ideas of:

- Prevention and systematic approaches
- Continuous improvement
- Integration in daily business and in planning processes
- Comprehensive deployment
- String relationship to policy, strategy, and respective goals
- Orientation toward the "best in class" (benchmarking)

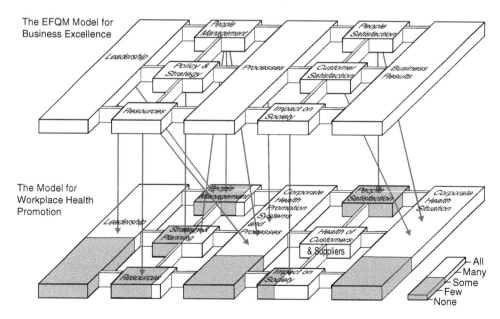

FIGURE 7.6   Transformation from business excellence to health excellence.

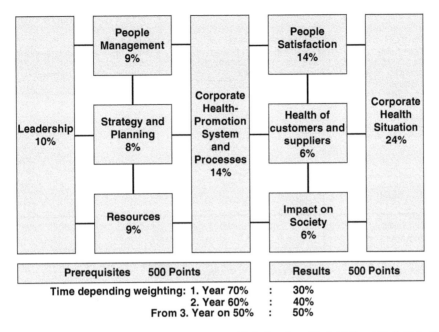

FIGURE 7.7 Corporate health management assessment model. (Taken from Thul, M. J. and Zink, K. J., *Zentralblatt für Arbeitsmedizin, Arbeitsschutz und Ergonomie*, Vol. 49, Issue 8, 1999, p. 282. With permission.)

As this evaluation model fulfills all the general requirements formulated for a health management system, it has been slightly modified to bring health topics within the purview of top management (see Figure 7.6 and Figure 7.7).

Like the EFQM model, the CHM assessment model is not a design model but an evaluation model. Nevertheless, the criteria and subcriteria give hints on how to realize a comprehensive management system.

Again the results criteria show a broad spectrum of health results, which are created by respective enablers. So far the model represents a feedback control system, whereas the results show defined goals or objectives and on the one hand, indicators to evaluate the effectiveness of the underlying processes on the other. The model is not restrictive; there are only hints, which preconditions are to be created fundamentally, and which results should be looked at to meet the requirements of a comprehensive understanding of health. Each organization has to decide, how specific preconditions have to be deployed and which results should be measured in which dimension. This has to be based on respective frame conditions and objectives (Thul and Zink, 1999).

The contents of the criteria are roughly the following:

- *Leadership.* The leadership criterion refers to how leaders implement a CHMS via appropriate actions and behaviors, personal involvement, role modeling, and support.
- *Strategy and planning.* Via strategy and planning, the objectives of a corporate health policy or the respective employee orientation are prescribed and frame conditions defined, which are the basis for all other activities. In addition to the inclusion of health-relevant aspects within the general strategy and planning, it also forms the basis for how strategies and plans of a corporate health management are developed, communicated, and implemented.
- *People orientation.* The employee orientation criterion focuses primarily on operative approaches of CHM (regarding employees). It deals systematically how working conditions, work contents, competencies, and needs are harmonized. In addition, questions of qualification, employee involvement regarding processes of a corporate health management, and respective possibilities of recognition have to be answered.

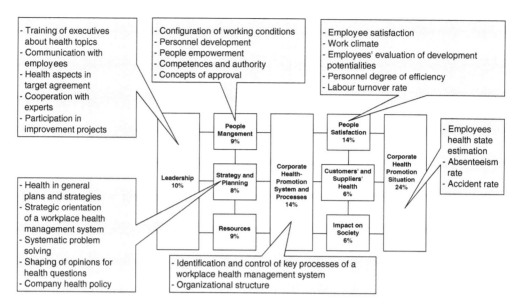

FIGURE 7.8   Examples of how to deploy the criteria of a corporate health management model.

- *Resources.* Resources refer to the processes related to management of financial support, health information and knowledge, buildings, equipment, materials, and technology with regards to health aspects.
- *Corporate health system and processes.* An effective CHM depends on an organization with respective structures for all relevant tasks, clarifying responsibilities, and competences. Furthermore, processes of analysis, evaluation, design, and implementation have to be defined.
- *Health of customers and suppliers.* Within the open systems approach, CHM has to integrate external target groups as well. Whereas the preconditions for such activities have to be allocated to the "resources" criterion, here the effectiveness of those measures is to be demonstrated.
- *People satisfaction.* There is a strong linkage between people orientation and people satisfaction. Measures with a focus on people orientation may have a strong influence on people satisfaction and their mental and social well-being. Respective indicators have to be delivered within this context.
- *Impact on society.* A company is always involved in social relationships, for example, as a corporate citizen. So far, CHM has been the part of a company's social responsibility. Here, it is of interest whether a company is promoting the idea of corporate health outside the organization and which economic results regarding society can be shown.
- *Corporate health situation.* This most important results criterion includes all relevant health indicators describing the health situation of a company. This includes "performance indicators" like absenteeism rates, number of injuries, and others as well as "perception indicators" like the image of an organization regarding corporate health or the well-being of employees.

Figure 7.7 shows only the first level of criteria. Most of the criteria are described by subcriteria focusing on specific aspects. On a third level, the evaluation model includes areas to address, for example, how other organizations realized the specific model aspect. Figure 7.8 gives some examples for contents of respective enabler-criteria and how to measure the resulting effects.

7.3.3   Summary of Different Approaches to Integrate Health and Safety or Health Promotion

The aforementioned examples show that there are different ways of integrating health topics into management systems. As we know from quality management, integration strategies can have a narrower or a comprehensive approach. Integrating health and safety in the structure of another management system — may be based on ISO 9000 — is the first concept discussed.

Integrated management systems evolved out of the need for synergies, and health management could be one of the management systems that needs integration. As all these management systems are mainly the task of professionals and their department (e.g., OHS), they have the same limitations as all management systems that focus on these target groups.

Therefore, the need for such a management approach had always, but they appeared when the so-called National Quality Awards — later called Business or Performance Excellence Models — came into being. The German health excellence model using the Business Excellence Model of the EFQM is based on the broad definition of health given by the WHO (1948) — and therefore has a broad focus. It includes several indicators for measuring health in a comprehensive way and prerequisites which are of particular importance for a sustainable progress in improving an organization's health situation. The special advantages of this model can be seen in an active management involvement and the usage of management principles to handle health topics. This ensures that there is no longer an insurmountable gap between daily business and health-improving activities. Rather, there are possibilities to integrate health into an organization's management system.

7.4 Evaluation of Corporate Health Management Systems

A CHMS aims at a long-term and sustainable improvement of an organization's health situation. Therefore, relevant activities should be based on a comprehensive understanding of health and managed in such a manner that the employee's health and satisfaction is preserved and promoted by suitable measures. However, effectiveness and efficiency aspects must also be given due consideration. In effect this means that health and satisfaction improvement measures are also subject to an effort and benefit consideration. Whether the implemented measures are effective or if other concepts and strategies are required due to, for example, changed levels of aspiration of employees must be constantly reviewed.

The comprehensive evaluation of a CHMS plays a decisive role in its development and improvement. In both cases, compliance with specific quality criteria has to be reviewed regularly to certify positive performances and to systematically identify improvement areas. Using this feedback control systems that are essential for the continuous development of a CHM can be established. This results in extensive evaluation requirements and raises the question of which evaluation tools can fulfill these.

7.4.1 The Evaluation Problem in the Corporate Health Management System

An evaluation which meets demands mentioned in the previous paragraph should provide a conclusion pertaining to the "quality" of a CHMS. Especially in the field "health," quality cannot be defined by single indicators like absenteeism or accident rates alone. Numerous factors that are not controllable by the CHMS have a significant impact on these health results. A mere consideration of results cannot show if low absenteeism rates are due to pertinent efforts of the CHMS or must be attributed to the fear of becoming redundant. Such result-oriented assessment approaches are focused on the past. All measured health results are the consequence of certain activities and events that have occurred in the past. Therefore, it is nearly impossible to determine if a company has created the necessary prerequisites for the preservation of the health and satisfaction of its employees in the future. Therefore, the result-focused assessment perspective has to be extended by one which also includes the preconditions for results. This precisely means reviewing of structures that the corporation has established and measures and processes that have been implemented to ensure the health and satisfaction of its employees in the long term. This linkage of results and their preconditions facilitates the development of substantiated feedback control systems, which allow an effective control and an adaptation to changed frame conditions.

7.4.2   The Need for an Adequate Evaluation Approach

Against the background of the requirement drafted earlier, the need arises for a suitable evaluation instrument. Thereby the eligibility of such an instrument has to be judged in two dimensions: the way the evaluation data are gained and the type of data gathered.

Concerning the type of data gathered, it can be derived from the earlier discussion that a data acquisition focused solely on "health-results," is not sufficient. Health indicators extracted from surveys, absenteeism, or accident statistics may indicate areas for improvement in general. But in many cases it is impossible to derive ascertained improvement measures. Therefore any evaluation instrument that is used to deploy a CHMS has to take health results as well as their prerequisites into account. Another point that has to be mentioned is related to the underlying philosophy of an evaluation instrument. Health results as well as their prerequisites can be measured in a broad or a narrow sense. An evaluation that, for example, focuses solely on hazardous working environments and thus includes only absenteeism, injury, or accident rates cannot meet the demands of a comprehensive health definition. Therefore important areas to address will remain unconsidered, which leads to reduced effectiveness and efficiency of a CHMS.

The way data are gathered particularly deals with the question of whether the evaluation is accomplished by external specialists or by the company itself. Analogous to a quality management system based on ISO 9000 it would be conceivable that a CHMS is evaluated by an external "health-auditor." In addition, during a quality audit, the health-auditor would have to examine health-related documents and he or she is compelled to make a site-visit. But in addition to an ISO 9000 audit, he or she would also have to review different health indicators to appraise the effectiveness and efficiency of an CHMS. As long as the "health-auditors" qualifications are sufficient, such an evaluation approach may deliver reliable information. But it has to deal with a fundamental problem: Whenever an external person or institution reviews the structures, processes or results, an organization aspires to present itself in the most positive way. This means that it will demonstrate the strength of its CHMS and will hide the areas for improvements. Similar experiences, resulting from the certification of quality management systems based on ISO 9000, demonstrate that such an evaluation approach can be counterproductive for a continuous improvement process (Kamiske et al., 1994).

An external assessment always has to deal with the aforementioned disadvantages. However, there is the advantage that it will, in many cases, deliver more objective data. A well-trained external assessor is normally more neutral than an internal one and often he or she also has knowledge about the CHMS of other organizations. Therefore he or she has a better overview of possible solutions, making his or her evaluations more reliable.

One of the most suitable evaluation instruments for a continuous improvement process is a self-assessment approach. Self-assessment in the context of a CHMS can be defined as a comprehensive, regular, and systematic evaluation of activities and results of a CHM based on a respective model. This points out the importance of a suitable assessment model. Depending on the model's content and underlying philosophy, a self-assessment may have a broad or a narrow view on an organization's health situation and the assessment results are, to a greater or lesser extent, useful for the deployment of a CHMS. Therefore, the requirements concerning the evaluation data discussed earlier are of special importance for self-assessment (Zink, 1998).

The crucial point of a self-assessment is that the organization applies this assessment tool itself and is not evaluated by an external auditor. This increases the likelihood that critical improvement areas are really identified. There is no need to justify the CHMS externally and therefore to present only best practices (like in quality management audits). A self-assessment rather creates the basis to reflect one's own approaches critically and to improve them continuously. In addition, the internal execution of the assessment is of significant psychological importance. It points out that an organization actively looks into the subjects of health or illness, implements the required changes, and accepts the responsibility for the health of its employees. While it can be reasonable to use the support of external specialists for solving specific problems, it would be wrong to delegate the responsibility for the CHM to external agents. This applies, in particular, to the critical analysis of the corporate health situation and the required frame conditions. The attempt to refer the evaluation solely to, external hands leads to the fact that management is released from its responsibility toward the health of the employees. Thus, the credibility of the CHM will be forfeited and the lasting implementation of improvement activities will be complicated. Especially the active participation in the self-assessment process provides, for managers, the opportunity to deal with the CHM intensively and shows their commitment.

Normally a self-assessment is more suitable to support a continuous improvement process than an externally conducted evaluation (Conti, 1999). But this assumes that the neutrality of the assessors can be taken for granted. So an adequate assessment system and notably trained assessors are essential preconditions for reliable assessment results.

7.4.3 Assessment Approaches to Evaluate CHMS

In the following, different assessment approaches will be introduced. They differ with respect to the underlying assessment model, the way the assessment data are gathered, and the extent of experiences resulting from practical applications.

7.4.3.1 Assessment Based on ISO 9004:2000

In Section 7.3 the possibilities of integrating health topics into quality management systems based on ISO 900x:2000 were discussed. As shown earlier, an externally conducted "health-audit," which may be an integral part of a quality management audit might not be very useful for the continuous improvement of a CHMS. But a closer look at the standards show that there are also possibilities to build up a self-assessment approach. ISO 9004:2000 offers the possibility to evaluate the maturity of a quality management system by answering questions pertaining to 27 crucial QM-elements (ISO 9004:2000). Table 7.5 shows the performance maturity levels used to guide the self-assessment.

Although health is not an explicit object of the ISO 9004:2000 it can be found, that at least implicit health-relevant aspects are attended to. Furthermore there are possibilities to "translate" the crucial quality management elements in a way that they are suitable for evaluation of health topics.

7.4.3.2 Self-Assessment Based on the CHAA Award

The Corporate Health Achievement Award (CHAA) is awarded by the American College of Occupational and Environmental Medicine (ACOEM) and sponsored by a couple of health care companies.

The purpose of the CHAA is (ACOEM, 2004):

- To foster awareness of quality of occupational and environmental medical programs
- To identify model programs and outstanding practices with measurable results
- To encourage organizational self-assessment and continuous improvement

There are four categories of assessment:

1. Healthy People
2. Healthy Environment

TABLE 7.5 Performance Maturity Levels

Maturity Level	Performance Level	Guidance
1	No formal approach	No systematic approach evident, no results, poor results, or unpredictable results
2	Reactive approach	Problem — or corrective-based systematic approach; minimum data on improvement results available
3	Stable formal system approach	Systematic process-based approach, early stage of systematic improvements; data available on conformance to objectives and existence of improvement trends
4	Continual improvement emphasized	Improvement process in use; good results and sustained improvement trends
5	Best-in-class performance	Strongly integrated improvement process; best-in-class benchmarked results demonstrated

Source: Based on ISO 9004:2000, p. 83.

TABLE 7.6 Corporate Health Achievement Award — Criteria and Point Weighting Overview

	Categories	Point Value
1.0	*Healthy People*	**300**
	1.1 Health evaluation of employees	60
	1.2 Diagnosis and treatment of occupational and environmental injuries or illnesses, including rehabilitation	60
	1.3 Assistance in rehabilitation of alcohol, and drug-dependent employees and of those with emotional disorders	50
	1.4 Emergency treatment of nonoccupational injury or illness, and palliative treatment of disorders to allow completion of work shift or for conditions for which an employee may not ordinarily consult a physician	50
	1.5 Immunization against possible occupational infections and other infectious diseases	40
	1.6 Collaborative treatment of nonoccupational conditions	40
2.0	*Healthy Environment*	**300**
	2.1 Evaluation, inspection, and abatement of workplace hazards	70
	2.2 Education of employees in jobs where potential occupational hazards exist which may be specific to the job; instruction on methods of prevention and on recognition of possible adverse health effects	70
	2.3 Implementation of programs for the use of indicated personal protective devices	60
	2.4 Toxicologic assessments, including advice on chemical substances that have not had adequate toxicological testing	40
	2.5 Environmental protection programs	30
	2.6 Disaster preparedness planning for the workspace and the community	30
3.0	*Healthy Company*	**300**
	3.1 Health education and counseling	80
	3.2 Assistance in control of illness-related absence from job	50
	3.3 Participation in planning, providing and assessing the quality of employee health benefits	50
	3.4 Assistance in evaluation of personal health care	50
	3.5 Medical interpretation/participation in development of government health and safety regulations	50
	3.6 Termination and retirement administration	20
4.0	*Management and Leadership*	**100**
	4.1 Administration, organization, innovation, and values	30
	4.2 Periodic evaluation of the occupational and environmental health program	20
	4.3 Biostatics and epidemiology assessments	20
	4.4 Participation in systematic research	20
	4.5 Maintenance of occupational medical records	10
Total Points		**1000**

Source: ACOEM, 2004.

3. Healthy Company
4. Management and Leadership

This award is given to North American companies with over 1000 employees in service and manufacturing sectors. Like in other award models, the programs of the companies are reviewed by professionals and a feedback report can be used for improvement activities. Table 7.6 shows the criteria and the weights of the respective subcriteria.

In addition to the external evaluation of an organization's CHMS there is also an easy-to-use Corporate Health Excellence Checklist to support the self-assessment. This assessment approach shows — like in ISO 9004:2000 — that internal and external assessments may be based on the same fundamental model.

7.4.3.3 An Assessment System for Occupational Health and Safety Performance Based on the Malcolm Baldrige Self-Assessment Model

As mentioned in Section 7.3, the so-called excellence models may form the basis for the implementation of comprehensive management systems in which health topics may be integrated. Therefore excellence models are also suitable for the development of holistic assessment approaches to evaluate CHMS. One approach using this advantage was introduced by Rantanen et al. (1999). They used the Malcolm Baldrige Self-Assessment Model to improve OHS performance. As the Center for Excellence Finland (CEF) model is based on the Malcolm Baldrige Criteria for Performance Excellence (NIST, 2003), Rantanen et al. developed a model that is related to the CEF and the Malcolm Baldrige models which can easily be used by companies that use the CEF model. Figure 7.9 illustrates the CEF model for business excellence.

The objectives of the study of Rantanen et al. (1999) were:

- To assess the state of OHS management in companies using the CEF model according to the application of the quality award criteria
- To draw up and carry out a plan for development of safety activities
- To devise a model adapting the set of quality award criteria to develop safety activities and make further improvements to OHS management

To arrive at a systematic approach for improvements, the initial status of OHS management has to be reviewed. Therefore, a series of questions based on the assessment criteria of the CEF Model/Malcolm Baldrige Award (also including the European Business Excellence Model and the BS 8800 (BSI, 1996) Guide to Occupational Health and Safety Management Systems standard and OHS audit checklists) have been formulated.

The criteria used include 85 questions related to the following Malcolm Baldrige Award topics (Rantanen et al., 1999):

1. *Leadership* (14 questions). The purpose of this category is to examine OHS leadership development, and maintenance as a part of the management system. In addition to this, the commitment of senior management to safety issues, their participation, and continuous development of safety activities are examined.
2. *Human resources development* (15 questions). The purpose of this category is to examine the competence of the personnel for their work and their motivation and welfare. In addition to this, improvements in working environment and atmosphere and their influence on performance and commitment of personnel are examined.
3. *Policy and strategy* (8 questions). The purpose of this category is to examine how safety issues are organized as a part of strategic planning and main action plans. In addition to this, an examination of how safety plans are converted into practical actions is undertaken.
4. *Information and analysis* (11 questions). The purpose of this category is to examine, choose, control, and utilize information in design, process control, development of activities, and a company's competitive position.

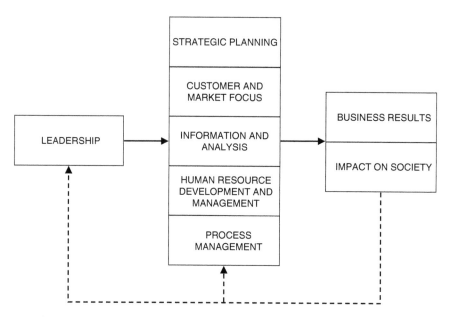

FIGURE 7.9   The Center for Excellence Finland (CEF) model for business excellence. (Taken from Rantanen, S.; Levä, K., Liuhamo, M., Mattila, M., Sulameri, R., and Uusitalo, T. in: Axelson, J., Bergmann, B., and Eklund, J. Eds., *Proceedings of the International Conference on TQM and Human Factors — Toward Successful Integration*, 1999, p. 452. With permission.)

5. *Processes* (14 questions). The purpose of this category is to examine process control, that is, process safety, development of safe products and services, supportive actions, and cooperation with different interest groups.

6. *Customer and market focus* (11 questions). The purpose of this category is to examine how customers' and other interest groups' safety demands for a product, services, manufacturing process are taken into consideration. In addition to this, how safety is utilized with clients and marketing is also examined.

7. *Impact on society* (8 questions). The purpose of this category is to examine the social effects of a company's activities and products on society, interest groups, welfare, living environment, and natural resources.

8. *Results* (4 questions and 19 examples of different indicators). The purpose of this category is to examine safety measures and to compare and monitor the development of these measures.

Prior to the assessment some information was given to the assessor team. Apart from an explanation of the model, the kind of material and information that must be collected for the self-assessment was also explained. During the assessment workshop the chairman introduced each assessed category and each participant wrote down three strengths and three areas for improvement individually. Each strength and area for improvement was then discussed until a consensus had been found. All consensus topics were then included in a feedback report. It was found (Rantanen et al., 1999) that the questionnaire used was suitable for reviewing the safety status and offers initiatives for continuous improvement of occupational health and safety management.

7.4.3.4   Self-Assessment in Corporate Health Management Based on the EFQM Model for Business Excellence

The last evaluation concept to be outlined in this section, is a self-assessment approach based on the CHM model derived from the EFQM model for Business Excellence. As shown in Section 7.3, the

CHM model is primarily an assessment model which is based on a comprehensive health definition that embraces physiological health dimensions as well as psychological ones. One of the main advantages of this model can be seen in its content. On the one hand, there is a broad variety of different health indicators to measure health in an adequate way. On the other hand, the model also includes the prerequisites, that are crucial for the sustainable improvement of an organization's health situation.

When carrying out a self-assessment, an organization orients itself on the criteria and subcriteria of the CHM model and documents the structures and processes that have been implemented up to a certain date and the results that have been achieved. This generates a comprehensive picture ("snap shot"), which encompasses, in addition to the multidimensional result parameters, the prerequisites which should ensure a permanent stabilization of the CHM. A qualified assessment team of the organization's employees evaluates the descriptions. Its primary task is to identify strengths and areas for improvement and there upon to appraise the efficiency and effectiveness of the CHM approach. Therefore, self-assessment is an approach which not only answers the question if an organization "does things right," but also delivers information whether the "right thing is done."

The models contents have already been described in Section 7.3. In the following the evaluation systematic, the assessment-process, and the derivation of improvement measures will be introduced in detail.

7.4.3.5 The Evaluation Systematic

The self-assessment tool presented here includes content dimensions outlined in the assessment model, as well as the methodical procedure. The method refers to the question on how to evaluate the prerequisites and results. Similar to the primary assessment systematic of the EFQM, results as well as their prerequisites are evaluated in two dimensions. The prerequisites with regard to the quality of the approach and their degree of deployment, and the results with regard to their excellence as well as the scope of their achievement. Table 7.7 presents the corresponding criteria assigned to the four assessment dimensions. The concepts that an organization has realized in context with the prerequisite criteria and the classifications of the result criteria are evaluated by means of these evaluation dimensions.

TABLE 7.7   Evaluation Dimensions

Prerequisite Criteria (Enablers)	
Approach	Deployment
Appropriateness of the methods, tools, and techniques used	Vertically through all relevant levels
Degree to which the approach is systematic and prevention based	Horizontally through all relevant areas and activities
The use of review cycles in regard to the effectiveness of the implemented concepts	In all relevant processes
The implementation of improvements resulting from review cycles	To all relevant products and services
The degree to which the approach has been integrated into the normal operations	Related to all health dimensions
Result Criteria	
Quality	Scope
Indications that negative trends are understood and addressed	The extent to which the results cover all relevant areas and activities of the company
Existence of positive trends	The extent to which a full range of results, relevant to the criterion are presented
Comparisons with own targets	The extent to which the results presented are suitable for the control of health relevant processes
Comparison with external organizations, including "best-in-class" organizations	
Signs that the results can be put down to the approach	
The company's ability to sustain its performance	

TABLE 7.8   Extract from the Evaluation Scheme for Assessment of Prerequisites

The prerequisites
The prerequisites are evaluated in regard to the degree of which
1. the corporate concept is suitable
2. the concept has been actually deployed

Approach	Score	Deployment
The described approaches are not systematic and not suitable for the implementation of the corporate health management	0%	Little effective usage
Some evidence of soundly based approaches of the corporate health management and prevention based systems. The organization uses suitable methods, tools, and techniques	25 %	The corporate approach is applied to about one quarter of the potential in regard to all relevant levels, areas, processes, products, services, and health dimensions
The described approaches are occasionally reviewed in regard to their effectiveness		
Some areas of integration of the corporate health management concepts into normal operations and planning		
Evidence of soundly based approaches of CHM and prevention-based systems. The organization uses suitable methods, tools and techniques	50%	The corporate approach is applied to about half of the potential in regard to all relevant levels, areas, processes, products, services, and health dimensions
The described approaches are reviewed in regard to their effectiveness regularly		
Well-established integration of the corporate health management concepts into normal operations and planning		

The evaluation systematic presented in the previous paragraph should allow a quantitative appraisal of concepts and results. Suitable scales, which assist in assigning the defined percentage values, were developed for this purpose. Table 7.8 shows an extract of such an evaluation scheme.

The evaluation of the self-assessment documentation is a consensus process that can only be performed by a qualified team. Therefore, the members of such an assessment team, who are called "assessors," have to fulfill special qualification requirements. In addition to particular knowledge in the field of self-assessment, they also need adequate knowledge in the management and health fields. As the self-assessment competencies can be mediated in training sessions, the second group of qualification requirements involves specific education and training as well as professional experience. This again illustrates that organizations with prior experience in the implementation of an EFQM model or other excellence models have an advantage. If experiences with the instrument of self-assessment exist already, the additional qualification effort is reduced and a positive effect on the quality of self-assessment results is given.

The demand for an internal execution of the self-assessment has to be corrected in view of the high competences necessary for the assessment procedures. External support for the assessors' qualification is needed. It can also be helpful to integrate an experienced external assessor to lead the consensus workshop. His or her neutrality can be an essential contribution to a goal-oriented and constructive discussion, and as objective an assessment as possible.

7.4.3.6   The Self-Assessment Process

In general, the execution of a self-assessment process is subdivided into three phases. First, the required data have to be collected and documented in an applicable format. Then the assessment of the documentation, generated by the assessors, takes place and finally the improvement activities have to be derived from the assessment results. These activities have to be prioritized and adequate improvement measures deployed. The success and the deployment of the improvement measures is reviewed in the next self-assessment cycle. Figure 7.10 shows this basic coherence.

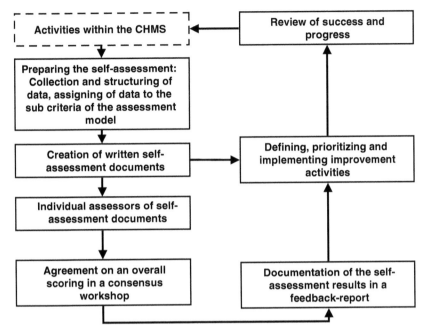

FIGURE 7.10 Self-assessment process. (Taken from Thul, M. J. in: Franke, D. and Boden, M. Eds., *Das Personal Jahrbuch 2004, Wegweiser für Zeitgenäße Personalarbeit*, Neuwied, 2003b, p. 702. With permission.)

7.4.3.7 Self-Assessment Documents

The documents for the self-assessment are based on the criteria and subcriteria of the evaluation model outlined earlier. The documentation includes the structures, processes, and measures of the CHM that have been implemented up to a certain time and the results achieved.

In accordance with the differentiation between prerequisites and results, the documents of the self-assessment show two focus points. The prerequisites should clearly outline what (systematic and preventive) measures the organization has taken for the design and development of the CHM. Their actual deployment has to be proven by suitable examples. Planned or just intended activities are not considered in the assessment of the documentation. What specific measures and concepts have to be described per criterion can be derived from the areas to address of the assessment model. These are examples of possibilities on how to put the requirements of the particular model criteria into practice. The description of the prerequisites also has to be orientated on the evaluation systematic outlined earlier. The self-assessment documentation does not only have to illustrate the development of the respective approaches, but also has to provide statements regarding the deployment concerning areas, processes, products, and services. In addition, the health parameters by which the success of the realized measures can be evaluated have to be denominated.

As the descriptions for the prerequisites are mainly given in words, the comments regarding the result criteria are limited to numbers, data, and facts. In this respect, graphics and tables are the preferred means of presentation. The listed results have to be edited in such a way that their quality (e.g., by comparison with industry average or "best in class" companies) and the scope (e.g., define in which part of the organization the results were achieved) is identifiable. The respective requirements of the assessment systematic also need consideration.

Though the requirements and results are always described in different parts of the documentation, comments should illustrate as to what extent the results can be attributed to the efforts.

7.4.3.8   Analysis of Self-Assessment Documentation

The analysis of the self-assessment documentation is a two-step process. First, each assessor receives a complete version of the self-assessment documentation which he or she evaluates individually. He or she then identifies and documents strengths and areas for improvement for each description of a sub-criterion and criterion.

A strength in realizing the demands of the assessment model can be defined by a deployed approach, a review cycle leading to activities for improvement or a positively assessed result of the CHMS (like employee satisfaction with positive trends or a favorable comparison with own targets). Areas for improvement are given if demands of the CHM assessment model or its assessment systematic are not, or are only partially, fulfilled. Therefore these areas for improvement are not to be understood as criticism but as possibilities for improvement. Areas for improvement could also focus on the presentation of results (e.g., non existence of benchmarking or comparisons with one's own targets) — or the missing use of results for improvement activities.

Given these strengths and areas for improvement, the assessor is in a position to perform a quantitative evaluation. For each criterion and subcriterion, a percentage of fulfillment (score) can be described regarding "approach" and "deployment" of prerequisites, and "excellence" and "scope" of results.

In a second step, all assessors meet in a consensus workshop to reach consensus on strength and areas for improvement and the quantitative evaluation of each subcriterion and criterion. This is a critical step for the quality of the whole assessment process. Only the bringing together and discussion of separate perspectives guarantees that all relevant aspects are assessed correctly, no relevant information has been neglected, the documentation is showing the overriding connection of single criteria, and the qualitative statements go well with the quantitative evaluation.

The results of the consensus workshop are documented and form the basis for the so-called feedback report. The content of this report can be used in different ways. First, the score (of the quantitative evaluation) may show the distance from "best in class" organizations. Using this approach for an internal assessment, the score is of minor interest. Much more important are the strengths and areas for improvement. They give hints on which strengths of a CHM can be further developed and which areas for improvement could be used. These areas for improvement form the basis for improvement activities.

7.4.3.9   Deduction of Improvement Activities

In general, a self-assessment leads to a large number of hints for improvements. Therefore it would not be possible to work on this simultaneously. Priorities have to be set and improvement projects be brought in an order. The evaluation of a single criterion can give some first hints. Improvement projects differ not only with respect to their content and their target of fulfillment, but also with respect to factors such as the time needed to realize them, the focus on different stakeholders, the sustainability of the result, and so on. Therefore, when fixing the priorities, these aspects also have to be taken into consideration. If the priorities are fixed, improvement measures are to be developed and a suitable deployment has to be chosen. The deployment needs a systematic planning and a review regarding its effectiveness.

The development of measures regarding the corporate health situation has to include all other activities in the company. Especially, the integration in daily operations is a must. All improvement activities must be coordinated with other corporate activities.

The process of deployment and realization of improvement activities based on a self-assessment is shown in Figure 7.11.

Independent of which measures are taken, some general hints can be given:

- Realize measures, that come to an early success, showing that something is happening
- Show the relationship with self-assessment results
- If measures cannot be successfully deployed (especially if measures related to the interests of employees are concerned) explain, why it does not work
- Avoid the delay of deployment processes
- Do not apply improvement activities only to one stakeholder group

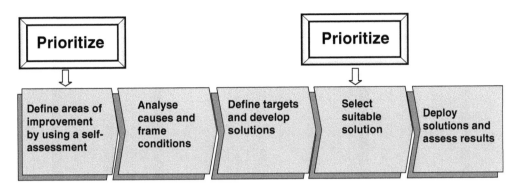

FIGURE 7.11   Deployment of improvement activities.

- Do not start with a very difficult improvement project at the cost of blocking all other projects
- Clarify in time which resources are needed and provide them
- Recognize and reward the engagement for improvement activities
- Define targets for improvement measures and review their fulfillment

7.5   Some Empirical Results

As mentioned before, the implementation of a CHMS is a promising way of obtaining long-lasting improvements of an organization's health situation. But to show such results there is a need for an adequate evaluation, to judge by facts whether the achieved benefits exceed the efforts. To evaluate a CHMS in the appropriate way, it is not adequate to focus on isolated activities and their (short-term) effects on specific indicators for employees' health, like absenteeism or accident rates. Instead, an adequate evaluation must consider a wide variety of health indicators derived from the requirements of the relevant stakeholders. Although the use of a broad spectrum of health indicators is an essential fact, it is not sufficient: The evaluation also has to take the complex framework of a workplace health management system into account.

As part of three research cooperations with local health insurance companies (AOK) in Germany and the WHO, the Institute for Technology and Work, University of Kaiserslautern (ITA) developed and tested a holistic evaluation concept that meets the aforesaid requirements. The main aim of these projects has been to verify whether a financial incentive, may stimulate the implementation of CHMS to improve the corporate health situation. The precondition to receive such a bonus is an application showing the results of a CHMS. This document is evaluated via a self-assessment based on the modification of the European Excellence Model discussed earlier. But in contrast to an internal self-assessment approach, the applications are assessed by external assessors.

The evaluation concept to judge the effectiveness of CHMS reflects the needs of interest groups representing three different levels: the society, the organization, and the individual. In a first step, important indicators — based on a broad health definition — have been selected for each stakeholder to measure resulting effects. In a second step, when and how the data have to be collected has been proposed. To prove the relations between indicators and organizational activities, self-assessments are used. The correlation of assessment results concerning prerequisites and specific health indicators shows whether there is a linkage between changes in employees' health and organizational activities.

In a project located in Lower Saxony, the CHMS of 16 companies was assessed over a time period of at least 4 years. Therefore, evaluation at two different levels was made possible. First, the effectiveness and efficiency of a CHMS can be assessed on the company level. Second, there is a chance to reflect on all companies.

Figure 7.12 shows the assessment results. These data clearly substantiate the implementation and deployment of profound CHMS. The available data also indicate that many organizations gained success in improving people's health and reducing the absenteeism rates.

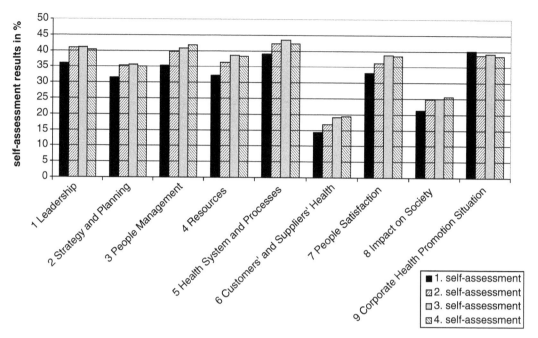

FIGURE 7.12 Self-assessment results.

To indicate the correlation between health-related results and their preconditions, it is reasonable to differentiate three types of companies. The basis for this classification is the ratio between the assessment of preconditions and results in total during the first self-assessment compared with those achieved during the fourth assessment. The following circumstances arise:

Cluster 1

- Dominance of enablers at the first assessment ($n = 6$)
- On average results of the fourth application have been evaluated about 8.5% higher than the first application

Cluster2

- Balanced assessment results between enablers and results with the first application ($n = 5$)
- Results evaluated were worse (with 0.8%) than the first application

Cluster 3

- Dominance of results at the first application ($n = 5$)
- Results evaluated were 3.8% worse than the first application

These results reveal that the companies investing more in the implementation of required processes and structures were also considerably more successful as to the results. However, those organizations with positive results, but comparatively worse preconditions in the first assessment, have clearly deteriorated in the course of time in terms of results.

But the significance of the assessment results in the characterization of health effects is limited. This is mainly due to the fact that only the absolute score of health indicators are considered during the self-assessment process. For example, the fact that a company has defined its own targets for health indicators or delivers benchmarking data also has an influence on the final assessment results. Because of this it is reasonable to have a closer look at selected single measures. But this part of the evaluation is also associated with problems. Many health indicators are also affected by external

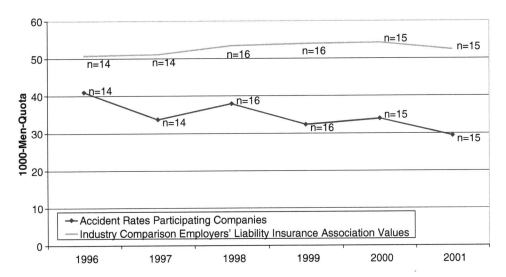

FIGURE 7.13   Accident rate.

factors, which are beyond the control of the company. Insofar, conclusions on the effectiveness of the CHMS can be made if the data of an enterprise are set in relation to comparable industry values. Figure 7.13 through Figure 7.16 show the developments of some selected indicators over a period of several years.

Figure 7.13, shows the development of the accident rate in relation to comparable values originated from the German Employer's Liability Insurance Association. The figure clearly shows that the implementation of a CHMS resulted in a distinct reduction in the number of accidents. Figure 7.14 outlines some economical effects of a successful implementation of CHMS. This chart shows the progression of paid benefits concerning the health insurance company's members. This indicator includes hospital expenditures, cost of ambulant treatment, substitutes for wages, and cost of pharmaceuticals.

Although the complete data set is only available for the time period 1999–2001, the development outlined is another proof of the effectiveness of a CHM.

Apart from the overall view, the analysis of individual enterprises provides additional information to confirm the aforesaid statements. Figure 7.15 and Figure 7.16 show the developments of self-assessment results and several health indicators in the case of a business which has systematically developed its

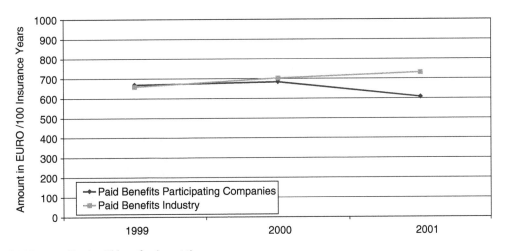

FIGURE 7.14   Total paid benefits ($n = 16$).

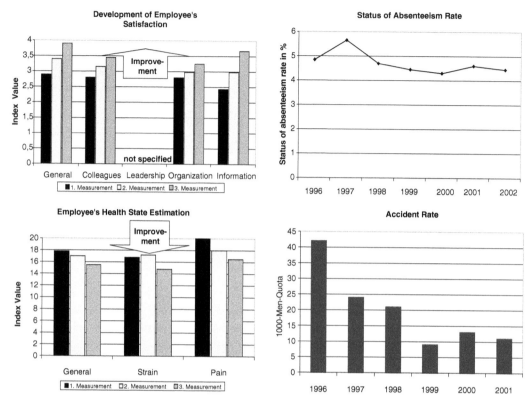

FIGURE 7.15 Results of a successful company (active since 1997).

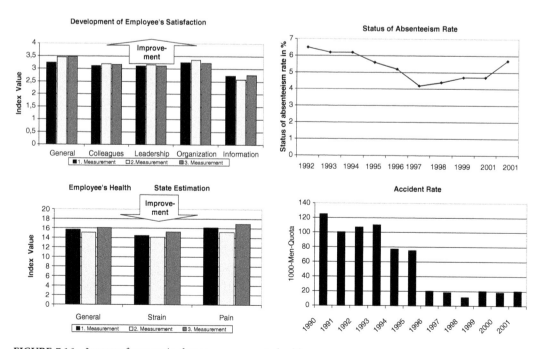

FIGURE 7.16 Impact of economic changes on corporate health-management.

CHMS approach. The results clearly reveal that in this case a comprehensive approach has been realized, which has very positive effects on the most different health dimensions. Furthermore, the data show that the effects have not been short-term, but rather long-term and sustainable.

As mentioned earlier, CHM must not be considered separate from other activities within an organization. Figure 7.15 explains why. It concerns the results of an organization which implemented very profound and systematic concepts in the beginning of the project in 1997. At first, improved results have been achieved. But the chart also shows that the third satisfaction survey in 2001 and the absenteeism rate of the same year achieved clearly worse results. This was caused by a dramatic deterioration of the organization's economic situation during 2001. The uncertainty as to the future of the organization and to the preservation of jobs causes effects that can be compensated by CHM only up to a certain limit. This again shows the necessity to understand CHM as part of a sustainability concept.

7.6 Implementation of a Corporate Health Management System

The development and implementation of a CHMS can partially lead to very challenging tasks for senior management. Therefore, it cannot be excluded that — especially because of decreasing resources of time and personnel — the senior management refrains from the implementation of such an approach. Thus a two-step approach is reasonable. Initially the basic structures have to be established and in a second step the strategic integration has to be promoted. The specific procedure of implementation is outlined next.

7.6.1 Basic Organizational Structures

The development and implementation of a CHMS has to consider existing structures and processes dealing with health topics. Legal guidelines regarding OHS as well as a company's initiatives in the past will almost always lead to substructures that could be used for the implementation of a CHMS. In view of the already emphasized efficiency and effectiveness orientation these existing systems have to be implicitly considered in the development of a CHMS and — if reasonable — have to be integrated into the new systems. This assures that established concepts — after a potentially required adaptation — are still utilized and reduce the development efforts for a CHMS. Furthermore, resistance against the new approach, which otherwise is to be expected of individuals responsible for the concepts that have to be given up, can be prevented.

To install a CHMS permanently and to implement the initially outlined contents effectively, an adequate organizational structure must be established. Therefore, its organizational structure is to be developed in such a way that specialized staff and executives, as well as the employees of the shop floor level, are involved in a suitable way. In the following, an example of the possibility for the implementation of the required organizational structure is briefly outlined (see Figure 7.17).

- *Steering Committee.* The steering committee is of central importance within the organizational structure. Its tasks are of a rather strategic nature. Superior decisions are made, which, for example, appear in the context of the adaptation of the business objectives or the implementation of organizational and personnel development measures. The composition of human resources for this committee results from the tasks to be fulfilled. In addition to a responsible project manager, the works council, external consultants, and members of the senior management are represented. The necessary management support is ensured by their direct involvement.
- *Project group.* The project group is responsible for the functional implementation of measures. This team has to deal with tasks like planning, implementing and controlling activities concerning an integrated CHMS. In addition, the project group coordinates the continuous improvement process with regard to the corporate health situation after completion of the implementation phase. Members of this committee are (apart from the project manager) mainly specialists who can contribute to the problem solution, like the company medical officer and the company physician, the specialist for OHS, or external consultants.

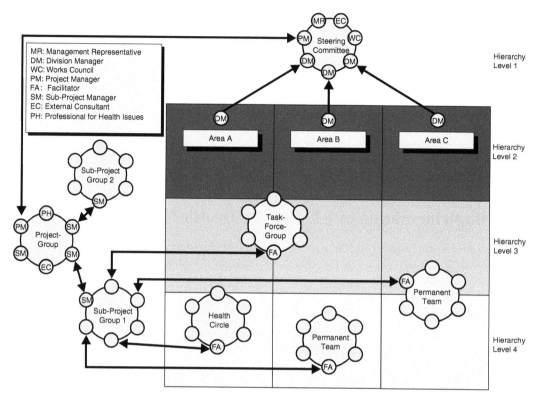

FIGURE 7.17 Organizational structure of a corporate health management system (Taken from Thul, M. J. and Zink, K. J., *Zentralblatt für Arbeitsmedizin, Arbeitsschutz and Ergonomie*, Vol. 49, Issue 8, 1999, p. 277. With permission.)

- The differentiation between project group and steering committee is not always necessary. Especially in smaller organizations the project group can also adopt the functions of the steering committee. The processing of professional issues will then be — at least partially — shifted to subordinate boards and its composition will be modified accordingly. This means that representatives of the senior management become members of the project groups, while specialists commit themselves to subordinate working groups.

- Due to the specific tasks of the project group, the cooperation with the committees for OHS (which, e.g., in Germany are governed by law) is a critical factor for success. A suitable agreement could be, for example, realized if individuals contribute to both committees at the same time. An advanced approach is to develop the occupational health and safety committee and to include the functions of the project group.

- *Site or sub project groups.* For corporations with different sites and/those of large size, it can be reasonable to establish additional project groups. They report to the project group and adopt the planning and coordination functions on site or they handle defined projects, which, for instance, arise in the implementation of an integrated CHMS.

- *Employee involvement.* The committees ensure an appropriate involvement of senior management and specialists. An adequate involvement of the shop floor employees is equally important. The organizational concept presented here provides different possibilities: moderated health circles establish a lasting forum in which employees of the same area of operation meet to discuss area-specific health problems. Task force groups, which can consist of overlapping divisions and hierarchies, have a limited life time. They are created to process given problems that require specific specialized knowledge. These teams are suspended after the problem-solving step. Already existing problem-solving teams can also be integrated meaningfully, for instance, quality circles or CIP teams can process problems with regard to health.

- The organizational concept presented in the earlier text was successfully tested in the context of participatory approaches for the improvement of OHS and the implementation of new technologies (Zink and Ritter, 1993; Zink et al., 1993; Thul, 1995; Zink and Thul, 1995). Although the original application context was different, it could be confirmed that these organizational models are optimally suitable, after minor modifications, in realizing an integrated CHMS (Drupp and Osterholz 1998).

7.6.2   Design of the Implementation Process

A systematic procedure during the implementation process is another important success factor. The following outlined measures are to be emphasized.

- *Information of the involved.* At the beginning of the project, the support of the management and the works council has to be ensured. This requires suitable information events in which information that meets the specific demands of the respective target group has to be presented. Furthermore, appropriate information strategies have to be established in the early stage to inform the employees on results and procedures in a timely manner.
- *Kick-off meeting.* After the decision to realize the project has been taken, the members of the steering committee and the project group have to be selected in a second step. In a kick-off meeting the fundamental goals have to be defined, the competencies and areas of responsibility of the individual committees and persons involved have to be clarified, and the further procedure has to be determined.
- *State-of-the-Art analysis of the corporate health situation.* A comprehensive state-of-the-art analysis of the initial situation is the basis for the derivation of measures. As different data are required, there is a need for detailed analysis. "Health-related results" as well as facts causing health or illness are topics of this data collection. Causes and effects should consist of subjective data (e.g., employee surveys) as well as objective data (e.g., accident and illness rates analysis).
- *Self-assessment.* The aims, the procedure, and the results of a self-assessment, for example, on the basis of the integrated CHM models have been described earlier. Based on these data, the following can be evaluated in detail: (i) which existing health promotion concepts will be pursued, (ii) in which areas adaptations are required, and (iii) in which areas new measures are to be implemented.
- *Transition into a continuous improvement process.* Since a CHM is not a short-term program but a long-term approach, the implementation project has to be transitioned into a continuous improvement process. The required feedback control system refers, on the one hand, on single measures and on the other hand, on the entire health management system. As regards the health management system, regular self-assessments according to the aforementioned model create the required data.

Figure 7.18 shows, in summary, the fundamental procedure for the implementation of a CHMS.

7.6.3   Strategic Alignment

The "Mission Statement Health" creates a significant base for the determination of goals and strategies in the CHM. Its development particularly serves the purpose of generating and establishing a common understanding of the term health. In addition, this document should characterize the individual features of the CHM and illustrate the significance the topic "health" has within the company, which health state of the employees the company strives for, or which essential influences and design factors have to be considered. In this respect, the mission statement provides an orientation which ensures that the efforts for the improvement of health are aligned towards a joint goal. Furthermore, it includes a self-commitment of the corporation, which can be used to assess the actually implemented measures and the results achieved. With regard to the development of the mission statement it has to be considered that health

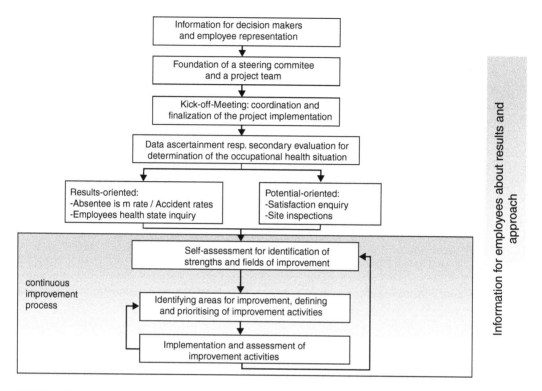

FIGURE 7.18   Principal procedures for implementation of corporate health-management system.

is not restricted to physical aspects alone. It has to be ensured that psychological and social dimensions are also properly taken into consideration. Especially, the managers' participation in the development of the "Mission Statement Health" provides a good possibility of making the organization sensitive toward the significance and the determinants of health. The discussion of these topics can be an important contribution to consider health topics in their managerial decisions and to increasingly support the implementation of corresponding improvement measures — especially in the course of their function as promoters (Thul, 2003a).

Comprehensive goals for a CHM can then be derived from the "Mission Statement Health" directly. Nevertheless, the risk exists that these goals are unrealistic or cannot be achieved (due to goal conflicts) if they are developed solely based on the mission statement. This possibly restricted view may lead to the fact that relevant frame conditions within the corporation and its environment are not considered sufficiently. Therefore, it is essential to analyze the corporate situation as well as the corporations environment systematically with regard to the influences on the CHM. A methodical tool for this purpose is the SWOT (strength, weaknesses, opportunities and threats) analysis, one of the most common tools in strategic planning. With its assistance, the strengths and weaknesses of the initial situation are determined as well as future chances and risks that are part of the implementation of a CHM. The results of this analysis open further perspectives for the completion of the goal system. The goal system itself represents, on the one hand the fundamentals for plans and strategies, and on the other, it plays an important role in the assessment of the implementation success. But to bring these functions to bear it has to be guaranteed that the achievement of the goals can be verified. The description of the result criteria of the CHM model provides assistance for this (Thul, 2003a).

As Figure 7.19 shows, strategies can be partially derived directly from the goal system. The systematic comparison of the CHM goals and the corresponding measures for securing the goal achievement constitute another fundamental of the strategy development. Their evaluation with regard to the

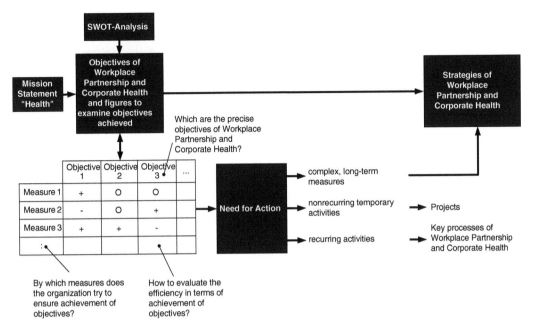

FIGURE 7.19   Use of a goal system. (Taken from Thul, M. J., in: Knauth, P. and Wollert, A. Eds., *Human Resources Management — Kapitel 2, Personalmanagement A–Z*, 45. Erg-LFg., Cologne 2003a, p. 25. With permission.)

effectiveness of the matrix or the lack of appropriate approaches provides indications for activities. Amongst the determination of strategies, which are rather of long-term and complex nature, this procedure delivers information to where the implementation of temporary health projects is required and where recurring activities (processes) have to be optimized under consideration of health aspects. A peculiarity of the outlined procedure is to consider not only measures that the corporation has initiated especially for the achievement of individual health goals, but also such activities that can be related to the general day-to-day business operation. By this the strict separation between day-to-day business and the CHM is withdrawn and starting points arise where health issues can be subject of discussion within the overall corporate management (Thul, 2003a).

7.7   Summary and Outlook

After several years of experience, one is able to describe promoting and hindering factors in introducing a CHMS (Zink, 2003b). The following points can be stated as critical success factors:

- Health has to be a part of the company's policy and strategy
- There must be a strong support by (top) management
- We need a harmonization with existing industrial safety committees
- The existing corporate approaches need continuous improvement
- A sufficient broad basis of health dimensions and health indicators as control systems are a relevant precondition
- CHM has to be integrated in existing management systems — or at least should be coordinated with them
- Long-term approaches should overcome short-term programs

Convincing companies to invest in health promotion is also related to the possibility of improvements in the corporate health situation in connection with improving economic results. Therefore, these connections have to be shown in respective assessment approaches. In the best case CHM activities should lead to win-win-win situations. One example in this context is the restructuring of the work organization. If we organize work in a more process-oriented way, we can create advantages for customers, employees, and the company. Process-orientation gives the opportunity to focus on the customer at first. But process-oriented work organization can also be used for a different division of work leading to a more holistic work content for each single worker — or at least for teams. This again leads — as experience shows — to more satisfaction and motivation. In sum, the result is normally an increase in productivity (even generated by less days of absence) and a higher customer and employee satisfaction leading to better economic results. This way of thinking must be increasingly included in respective assessment procedures.

Apart from the economic way of thinking, the story about competitiveness through sustainability, with health as an important element, has to be told. This should be demonstrated in a CHM policy.

Since health has to become a management topic, "management approaches" as demonstrated by integration into so-called excellence models could be a successful instrument. Though the potential of the approach described in the preceding text is by far not exhausted, positive experiences with holistic concepts can be shown. In discussing the future relevance of the topic one can take a last look at the so-called Kondratieff Cycles describing the most relevant influences in the development of society. The last publication (Nefiodow, 2001) in discussing the next Kondratieff Cycle shows that "Psychological and Social Health" is at the top of the agenda. So it is really worthwhile using all the possibilities to promote this topic — in promoting Ergonomics.

References

American College of Occupational and Environmental Medicine (ACOEM) (2004) (Ed.): *Corporate Health Achievement Award*, http://www.chaa.org.

Antonovsky, A. (1979): *Health, Stress, and Coping: New Perspectives on Mental and Physical Well-being*, San Francisco, CA.

Antonovsky, A. (1987): *Unraveling the Mystery of Health. How People Manage Stress and Stay Well*, San Francisco, CA.

AS/NZ 3 4581:1999, Council of Standards Australia/Council of Standards New Zealand (1999) (Eds.): *Management System Integration — Guidance to Business, Government and Community Organizations*, Homebush, Wellington.

Asvall, J.E. (1995): Introduction: a new area of health development, in: Kaplun, A. and Wenzel, E. Eds., *Health Promotion in the Working World*, Papers presented at an International Conference of the Federal Centre for Health Education in collaboration with World Health Organization, Regional Office for Europe, Berlin, Heidelberg, New York, pp. IX–XIII.

Badura, B., Münch, E., and Ritter, W. (1997): *Partnerschaftliche Unternehmenskultur und betriebliche Gesundheitspolitik. Fehlzeiten durch Motivationsverluste?* Gütersloh.

Bleicher, K. (2004): *Das Konzept integriertes Management*, 7th ed., Frankfurt.

Bödecker, W., Friedel, H., Röttger, Chr., and Schröer, A. (2002): Kosten arbeitsbedingter Erkrankungen, Schriftenreihe der Bundesanstalt für Arbeitsschutz und Arbeitsmedizin, Fb 946, 2nd ed., Dortmund, Berlin.

Breucker, G. (1998): Entwicklungen im Gesundheits- und Arbeitsschutz im europäischen Vergleich, in: Müller, R. and Rosenbrock, R. Eds., *Betriebliches Gesundheitsmanagement, Arbeitsschutz und Gesundheitsförderung–Bilanz und Perspektiven*, St. Augustin, pp. 247–264.

BSI 8800:1996, British Standards Institution (BSI) (Ed.) (1996): *Guide to Occupational Health and Safety Management Systems*, London.

BSI 18001:1999, British Standards Institution (BSI) (Ed.) (1999): *Occupational Health And Safety Management Systems–Specifications*, London.

Bureau of Labor Statistics, U.S. Department of Labor (2002): *Workplace Injuries in the United States of America*, http://www.bls.gov.

Business for Social Responsibility (2003): http://www.bsr.org.

Conti, T. (1999): Vision 2000: positioning the new ISO 9000 standards with respect to total quality management models, in: Total Quality Management, Special Issue – *Proceedings of the 4th World Congress for Total Quality Management*, Vol. 10, Issue 4 and 5, pp. 454–464.

DIN EN ISO 14001, Deutsches Institut für Normung, (DIN) (Ed.) (1996): DIN EN ISO 14001:1996, *Umweltmanagementsysteme – Spezifikation mit Anleitung zur Anwendung*, Berlin.

DIN EN ISO 9000, Deutsches Institut für Normung (DIN) (Ed.) (2000): DIN EN ISO 9000:2000, *Qualitätsmanagement – Grundlagen und Begriffe*, Berlin.

DIN EN ISO 9004 Deutsches Institut für Normung (DIN) (Ed.): DIN EN ISO 9004:2000, *Qualitätsmanagementsysteme – Leitfaden zur Leistungsverbesserung*, Berlin.

Drupp, M. and Osterholz, U. (1998): Prospektiver Beitragsbonus – Ein Projekt der AOK Niedersachsen zur Förderung von integrativen Gesundheitsmaßnahmen in der Arbeitswelt, in: Müller, R. and Rosenbrock, R. Eds., *Betriebliches Gesundheitsmanagement, Arbeitsschutz und Gesundheitsförderung – Bilanz und Perspektiven*, St. Augustin, pp. 349–371.

Eklund, J.A.E. (1999): Ergonomics and quality management – humans in interaction with technology, work environment, and organizations, *International Journal of Occupational Safety and Ergonomics*, 5, 2, pp. 143–160.

Europäische Kommission (2002): *Die Soziale Verantwortung der Unternehmen*, Brüssel (Generaldirektion Beschäftigung und Soziales, Referat D1).

European Foundation for Quality Management (EFQM) (Ed.) (1999): *Self Assessment Based on the European Model for Business Excellence 1999 – Guidelines for Companies*, Brussels.

European Foundation for the Improvement of Living and Working Conditions (Ed.) (2000): *Third European Survey on Working Conditions 2000*, Dublin, http://www.eurofound.eu.int/publications.

Federal Institute for Occupational Safety and Health, Germany (Ed.) (2001): *Volkswirtschaftliche Kosten durch Arbeitsunfähigkeit 2001* [Estimation of costs of work related diseases in Germany 2001], http://www.baua.de/info/statistik/stat_201/cost01.htm.

Gaertner, I. (1998): Mensch und Arbeit, *Social Management, Magazin für Organisation und Innovation*, 8, 1, pp. 15–16.

Greiner, B.A. (1998): Der Gesundheitsbegriff, in: Bamberg, E.; Ducki, A. and Metz, A.-M. Eds., *Handbuch der betrieblichen Gesundheitsförderung*, Göttingen, pp. 42–43.

International Organization for Standardization (ISO) (Ed.): ISO 14000, Geneva, http://www.iso.ch.

Japanese Deming Prize, http://www.deming.org/demingprize.

Kamiske, G., Malorny, C., and Michael, H. (1994): Zertifiziert – die Meinung danach, QZ, 39, 11, pp. 1215–1224.

Karasek, R. (1979): Job demand, job decision latitude and mental strain: implications for job redesign, in: *Administrative Science Quarterly*, Vol. 24, pp. 285–308.

Karasek, R. and Theorell, T. (1990): *Healthy Work*, New York.

Kerkau, K. (1997): Betriebliche Gesundheitsförderung, *Faktoren für die erfolgreiche Umsetzung des Gesundheitsförderungskonzeptes in Unternehmen*, Hamburg.

Kickbusch, I. (1995): Health promotion: a new approach at the workplace, in: Kaplun, A. and Wenzel, E. Eds., *Health Promotion in the Working World*, Papers presented at an International Conference of the Federal Centre for Health Education in collaboration with World Health Organization, Regional Office for Europe, Berlin, Heidelberg, New York, pp. 5–7.

Leitner, K., Lüders, E., Greiner, B., Ducki, A., Niedermeier, R. and Volpert, W. (1993): Analyse psychischer Anforderungen und Belastungen in der Büroarbeit. Das RHIA/VERA-Büro-Verfahren, Handbuch, Göttingen.

Nefiodow, L.A. (2001): *Der sechste Kondratieff – Wege zur Produktivität und Vollbeschäftigung im Zeitalter der Informationen* [The Sixth Kondratieff – Ways to Productivity and Full Employment in the Information Age], 5th ed., St. Augustin.

NIST — National Institute of Standards and Technology (Ed.): Criteria for Performance Excellence, Gaithersburg 2003.

Pischon, A. and Liesegang, D.G. (1997): *Arbeitssicherheit als Bestandteil eines umfassenden Managementsystems, Bestandsaufnahme, Modellbildung, Lösungsansätze*, Heidelberg.

Post, J.E., Preston, L.E., and Sachs, S. (2002): *Redefining the Corporation—Stakeholder Management and Organizational Wealth*, Stanford.

Rantanen, S., Levä, K., Liuhamo, M., Mattila, M., Sulameri, R., and Uusitalo, T. (1999): Improving occupational health and safety performance with an application of the Malcolm Baldrige Self-Assessment Model, in: Axelson, J., Bergmann, B., and Eklund, J. Eds. *Proceedings of the International Conference on TQM and Human Factors — towards Successful Integration*, Linköping, Vol. 1, pp. 451–457.

SAM Corporate Sustainability Assessment Group (2003) (Ed.): *SAM Corporate Sustainability Assessment Questionnaire*, http://www.sam-group.com.

Schröer, A. (1996): Workplace health promotion—conception and definition, experience and prospects, in: Breucker, G.; Schröer, A. (Eds.): *International Experiences in Workplace Health Promotion, European Health Promotion Series No. 6*, WHO/Europe, Copenhagen/BKK Bundesverband, Essen, pp. 9–13.

Schwaninger, M. (1994): *Managementsysteme*, Frankfurt.

Siegrist, J. (1996): Soziale Krisen und Gesundheit, *Gesundheitspsychologie*, Vol. 5, Göttingen, Bern.

Staehle, W.H. (1999): *Management*, 8th ed., Munich.

The European Commission (2002a), Employment and Social Affairs, Corporate Social Responsibility: New Commission Strategy to Promote Business Contribution to Sustainable Development, Brussels.

The European Commission (2002b): Adapting the change in work and society: a new community strategy on health and safety at work 2002–2006, Communication from the Commission, Brussels.

Thul M.J. (1995): Effiziente Prozeßinnovationen durch partizipatives Projektmanagement, in: Zink, K.J., Ed., *Erfolgreiche Konzepte zur Gruppenarbeit—aus Erfahrungen lernen. Human Resource Management in Theorie und Praxis*, Neuwied, pp. 237–251.

Thul, M.J. (2003a): Betriebliches Gesundheitsmanagement, in: Knauth, P. and Wollert, A. *Human Resource Management—Kapitel 2 Personalmanagement A–Z*, 45. Erg.-Lfg., Cologne, pp. 1–30.

Thul, M.J. (2003b): Betriebliches Gesundheitsmanagement, in: Franke, D. Boden, M. *Das Personal Jahrbuch 2004, Wegweiser für Zeitgemäße Personalarbeit*, Neuwied, pp. 696–704.

Thul, M.J. and Zink, K.J. (1999): Konzepte und Instrumente eines integrativen Gesundheitsmanagements, in: *Zentralblatt für Arbeitsmedizin, Arbeitsschutz und Ergonomie*, 49, 8; pp. 274–284.

Thul, M.J., Benz, D., and Zink, K.J. (2003): Betriebliches Gesundheitsmanagement — Ziele, Kennzeichen und praktische Umsetzung im Global Logistics Center, Germersheim, in: Landau, K. Ed. *Good Practice—Ergonomie und Arbeitsgestaltung*, Stuttgart, p. 499–517.

United States Department of Labor (2002): *OSHA 2003–2008 Strategic Management Plan*, U.S. Department of Labor Occupational Health and Safety Administration, Washington, DC, http://www.osha.gov/StratPlanPublic/strategicmanagementplan-final.html.

von Bertalanffy, L. (1981): The theory of open systems in physics and biology, in: Emery, F.E. Ed. *Systems Thinking*, Vol. 1, 7th ed., Bunday, pp. 83–99.

World Health Organization (WHO) (1948): Preamble to the Constitution of the World Health Organization as adopted by the International Health Conference, New York, 19–22 June 1946; signed on 22 July 1946 by the representatives of 61 states (Official Records of the World Health Organization, no. 2, p. 100) and entered into force on 7 April 1948.

Zink, K.J. (1980): Arbeitssicherheit als Akzeptanzproblem aus motivationstheoretischer Sicht, *Zentralblatt für Arbeitsmedizin, Arbeitsschutz und Prophylaxe*, 30, 2, pp. 33–48, see also *Ergonomia, Zeitschrift der Polish Ergonomics Society*, 3, 2, pp. 205–219.

Zink, K.J. (1998): Self assessment—a holistic approach for the evaluation of health promotion concepts, in: Scott, P. A., Bridger, R. S. and Charteris, J. Eds. *Global Ergonomics, Proceedings of the Ergonomics Conference*, Cape Town, Amsterdam, Lausanne, New York, pp. 233–238.

Zink, K.J. (1999): Safety and quality issues as part of a holistic (i.e. sociotechnological) approach, *International Journal of Occupational Safety and Ergonomics*, 5, 2, pp. 179–290.

Zink, K.J. (2003a): From industrial safety to health management, in: The Ergonomics Society of Korea (Ed.): Ergonomics in the Digital Age, *Proceedings of the XV Triennal Congress of the International Ergonomics Association*, Seoul, Korea, Vol. 1, Plenary Session III.B (pdf-document 00008.pdf).

Zink, K.J. (2003b): Corporate social responsibility promoting ergonomics, in: Luczak, H. and Zink, K.J. Eds. *Human Factors in Organizational Design and Management–VII: Re-Designing Work and Macroergonomics–Future Perspectives and Challenges*, IEA Press Santa Monica, CA, pp. 63–72.

Zink, K.J. (2005): From industrial safety to corporate health management, Ergonomics, 48, 5, pp. 534–546.

Zink, K.J. and Ritter, A. (1993): Sicherheitsgruppen einführen und verankern, *Ein Leitfaden für Fachkräfte für Arbeitssicherheit*, Wiesbaden.

Zink K.J. and Thul M.J. (1995): Kleingruppenunterstütztes Projektmanagement–ein systematischer Ansatz zur Mitarbeiterbeteiligung bei der Einführung neuer Technologien, *zfo*, 64, 4, pp. 221–226.

Zink, K.J. and Thul, M.J. (1998): Gesundheitsassessment–ein methodischer Ansatz zur Bewertung von Gesundheitsförderungsmaßnahmen, in: Müller, R.; Rosenbrock, R. (Eds.): *Betriebliches Gesundheitsmanagement, Arbeitsschutz und Gesundheitsförderung–Bilanz und Perspektiven*, St. Augustin, pp. 327–348.

Zink, K.J., Ritter, A. and Thul, M. (1993): *Kleingruppenunterstützte Prozeßinnovationen–Leitfaden für Projektleiter und Projektkoordinatoren*, Bremerhaven.

Industrial Process Applications

8

Ergonomics Process in Small Industry

Carol Stuart-Buttle
Stuart-Buttle Ergonomics

8.1 Introduction

The ergonomics process for a small business differs from that of a large corporation. The main components of the ergonomics process are similar but the characteristics and limitations related to the size of a business or industry need to be taken into account. Each of the primary elements of the ergonomics process is discussed in this chapter in terms of the associated difficulties and benefits related to industry size.

Many of the issues raised may pertain to a small plant that is part of a large corporation. The main difference between a small industry and a small facility of a larger entity is that the latter usually has a safety and health process defined or mandated at the corporate level. The small facility may turn to the corporation for guidance and information and perhaps assistance and expertise. This chapter focuses on the issues faced by small businesses but does not attempt to contrast these issues with those of large industries.

8.2 Definition of Small Business

Small businesses account for more than 99% of all employers in the United States of America and employ 51% of the private sector workforce (SBA, 2003a). About 5.4% of all businesses (14% of the workforce) are in manufacturing, 41% of which have less than 500 employees (6.0% of the workforce) (U.S. Census Bureau, 2000). The U.S. Small Business Administration (SBA) defines small businesses under the North American Industry Classification System (NAICS), assigning each code in the industry division a maximum number of employees or maximum of average annual receipts, depending on the type of industry (SBA, 2003b). NAICS was formed in 1997 and the system defines and codes each industry (OMB, 1997). In October 2000, the SBA replaced the Standard Industrial Classification (SIC) code system of 1987 with the NAICS (OMB, 1987; SBA, 2003b).

Typically a small business is defined as having less than or equal to 500 employees (although a maximum of 1500 employees is defined as small business for a few NAICS codes) (SBA, 2003b). The contribution of small businesses to the overall economy is important. Small businesses generate 52% of the gross domestic product and are the source for two thirds of net new jobs (SBA, 2003a). Small to medium companies are viewed as vitally important to the economic health of other industrialized nations as well, and there is specific outreach by some governments toward small companies to improve their work environments (Wortham, 1994; Sundström, 2000).

8.3   Characteristics of Small Industry

Implementing an ergonomics program and evolving a process are challenging for any company. If the program is sound, then it will mature to become an integral part of the processes of the company (Hägg, 2003). There are constraints that all companies face in establishing and managing a process and small industries have unique characteristics that carry associated constraints. These constraints impact the decisions that are required to develop the best ergonomics program model to fit the culture and business needs of the company. Small industries often have the following characteristics (Stuart-Buttle, 1993, 1998; Headd, 2000):

- *Less formality.* Procedures and communication methods are informal and often have minimal documentation. Teams or task forces are often used but are loosely structured and formed ad hoc.
- *Responsibility for several positions.* Personnel perform several job functions, and therefore positions ergonomics is often included at the same time as addressing other issues. There is reduced team input because there are fewer people in different job positions, although the range of perspectives remains.
- *Greater responsiveness.* The company tends to be project based. Therefore, it focuses on issues and brings them to conclusion.
- *Less specific knowledge.* Having personnel with several job positions means that less time can be spent focused on one specific area such as ergonomics. Therefore, the degree of in-house expertise is usually limited.
- *More management involvement.* Management tends to be involved with the details of plant activities and show their responsiveness to projects by making decisions and coordinating project efforts. However, the culture of the company also affects the extent of management commitment and employee involvement.
- *Less data-oriented approach.* Once the company has decided on a course of action and is convinced that the approach is a good one, quantification is typically downplayed. This reduces the data available for prioritizing problem areas, cost–benefit decisions, and determining project effectiveness.

8.4   Reasons to Implement an Ergonomics Process

A productive, competitive business is important for all sizes of business and small businesses are "a dynamic force in the economy, bringing new ideas, processes, and vigor to the marketplace" (Headd, 2000). In addition, small firms have a larger share in manufacturing occupations, that is, production-type jobs, than large companies (500 employees or more) (Headd, 2000). Workers' compensation costs can be a primary drain on any company's productivity. High workplace injuries and illnesses were the reason for the passing of the Occupational Safety and Health Act of 1970 in the United States (OSHA, 1970). The Occupational Safety and Health Administration (OSHA) recognized that there were several elements in the management practice of safety and health that affected prevention of injuries and illnesses. As a result, in 1989, OSHA issued safety and health program management guidelines (OSHA, 1989). A federal ergonomics standard and some state ergonomics standards have been issued, although all the main U.S. standards for general industry have been rescinded except for the California ergonomics standard. Generally, small businesses have been resistant to any ergonomics legislation due to the perceived cost and inflexibility of such standards. Groups on their behalf have spoken out against the short-lived ergonomics standard as well as the potential of another standard (Keating, 2000; NSBA, 2003).

Some industries, especially very small versus medium sized ones, may have lower costs associated with workers' compensation cases, yet their competitiveness may remain suboptimal because the production process is inefficient and potentially unsafe due to poor design. Ergonomics offers more to industry than reduced injuries and illnesses. It is a science that applies knowledge about people to the design of the workplace. When the workplace is designed so that people can perform at their best and machines at their most effective, then productivity is optimized.

8.4.1 Competitiveness

The greatest return from and the prime reason for addressing ergonomics is the increase in the bottom line through production and quality. This may be partly realized by the reduction in workers' compensation and related costs or by the increase in work by those no longer in discomfort. However, making the job easier for the worker by improving the work layout or reducing rehandling also improves production. According to the government report in 1993 on the state of small business: "The drive for manufacturing competitiveness in both domestic and overseas markets means that there must be fewer person hours per unit of production" (SBA, 1993, p. 45). The drive remains the same today. Too often operating with fewer workers and automating production are seen as the primary means to increase competitiveness. Automation has certainly given increased productivity but as Sheble stated: "The humanless factory and plant will not materialize yet for many years. Highly automated plants are more flexible than totally automated plants. Moreover, the former has higher availability as well as lower maintenance and investment costs" (Sheble, 2004). A study by the National Academy of Sciences (NAS) also states that though automation will characterize work in the future, manual labor will remain important. Therefore, the study contends, evaluation of work tasks will be necessary to design effectively for an aging workforce and as more women enter the workforce (NRC, 2001, p. 10).

Quality is part of the productivity equation. If products are reworked or wasted then the unit cost is increased. Quality is influenced by human performance, and that performance is directly impacted by the design of the job and workplace. Insufficient time to perform the task, unsuitable lighting and environment, and ineffective training are examples of aspects that can affect the quality of someone's work. Excessive reaching and awkward postures increase the time it takes to perform the task, and quality standards may not be met. If a worker begins to fatigue at the task, then he or she will slow down or make mistakes, hence affecting quality and production. Some companies are becoming aware of the great competitive gains that can come from many small improvements in how people do their jobs and these companies are focusing on ergonomics, even if the term ergonomics is not used (Newman, 2003).

A survey found small enterprises (100 or fewer employees) find time to be scarce and also lack a perceived need to implement occupational safety and health programs. In addition, the majority of small business owners showed disdain toward OSHA and the government (Dyjack et al., 2003). "Money signs and hassles are what most small business owners see when you talk about health and safety" (*The Synergist*, 1994). This is a common view, especially when the full cost benefits are not understood or considered when assessing a safety and health problem and the potential solutions. When the only measure of benefit is the return from reduced workers' compensation costs the solution may sometimes appear expensive. This may be particularly noticed in a small company with just one or two low-cost medical cases. It is important to consider the full cost benefits of workplace improvement. If the solution or improvement is not cost effective with a return on investment in a reasonable time frame, then the approach may not be the right one, and an alternative needs to be sought.

8.4.2 Workers' Compensation Costs

The cost of workers' compensation was found to be a primary concern amongst companies of less than 100 employees (Dyjack et al., 2003). For a number of years there has been a trend across most industry divisions that shows the highest injury incidence rates (per 100 full-time workers) are incurred by establishment sizes of 50 to 249 employees (BLS, 2003a). Figure 8.1 illustrates the injury and illness incidence rates for private industries by employment size.

Days away from work are particularly expensive to a company. Sprains and strains are the most common injury in manufacturing. The most common source for injury or illness for lost work day cases in manufacturing was reported to be the "worker motion or position." Overexertion was the primary event causing injury and illness across all the divisions of industry in 2001 (BLS, 2003b). Such data strongly suggest the presence of design issues and the job not matching worker capabilities, that is, poor ergonomics.

Although the national statistics shows higher incidence rates for small- and medium-sized companies (Figure 8.1), small companies often have only a few cases. This precludes identifying trends and makes it difficult to decide if there is a larger problem. A proactive approach, that is not dependent upon the medical or workers' compensation data, helps to prevent the company from becoming entangled in such trend debates. Workers' compensation and the safety and health of the workforce are important, and the jobs associated with the problems should be addressed first.

Lost workdays are expensive for any company, but the burden may be greater on a small industry because there are fewer employees. If an employee serves several job functions the loss may be even greater. The hidden or indirect costs that go hand in hand with a workplace injury should also be remembered in the cost–benefit equation. Examples of cost include training of a temporary replacement, the associated lost production, productive time lost while attending to medical needs, and the time for accident investigation.

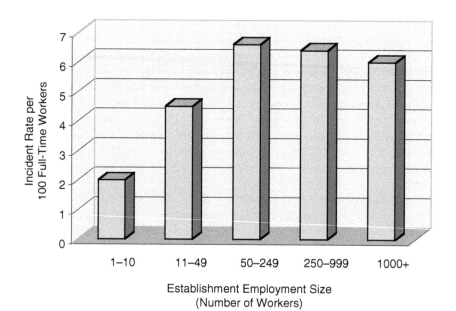

FIGURE 8.1 2002 Injury and illness incidence rates by industry employment size for private industry. (Based on data from Bureau of Labor Statistics (BLS). 2003. *Charts from Annual BLS Survey: 2002 OSH Summary Estimates Chart Package.* December 18, p. 7. http://www.osha.gov/oshstats/work.html.)

8.4.3 Compliance

Dyjack et al. (2003), at the National Institute of Occupational Safety and Health (NIOSH), who were conducting a study on small business and health and safety programs, found small business, especially with less than 20 employees, unaware that their states may require formal written occupational safety and health programs. The researchers also found that the small businesses generally would not use the free OSHA consultation.

However, compliance to government regulations should not be the sole reason for implementing an ergonomics program. If the workplace is designed well for most people to perform their jobs effectively on a long-term basis, then the workplace will be safe and healthy. However, there are laws that oblige employers to provide a safe and healthy workplace. In the United States, the OSH Act of 1970 has a general duty clause (section 5 (a) (1)) that requires employers to furnish each employee a place of "employment which is free from recognized hazards that are causing or are likely to cause death or serious physical harm" (OSHA, 1970). Also, OSHA is generating industry guidelines, although not mandated standards, that reflect a consensus of best practice. If an ergonomics program is implemented with compliance as the only goal, redesigns are likely to fall short of the best solutions to the problems and limit the efficiency improvements.

Although, at present, in the United States there are no national ergonomics or safety and health standards, there are some guidelines. Many companies have incorporated ergonomics into general safety and health programs following the Safety and Health Management Guidelines issued by OSHA in 1989 (OSHA, 1989). In 1990, OSHA published similar ergonomics program guidelines for meatpacking plants, which have since been adopted by other industries (OSHA, 1990). The same basic program components appear in the *OSHA Handbook for Small Businesses* (OSHA, 1992). A more recent document was produced by NIOSH called "Elements of Ergonomics Programs" that guides companies just starting to establish an ergonomics process, and the same core elements of previous documents are reflected (NIOSH, 1997). The Voluntary Protection Program (VPP) is also based on the Safety and Health Management Guidelines. A survey of 650 companies indicated a 26% decline in workers' compensation costs, due in part to safety and health programs (BNA, 1996). Companies that have participated in VPP have on average a 50% drop in workers' compensation premiums since 1989 (Esposito, 1996). Total quality improvement or management programs incorporate the primary elements encouraged by OSHA in their guidelines (Robinson, 1991). The NAS study on musculoskeletal disorders (MSDs) and the workplace also concluded that the application of ergonomics principles can be effective to reduce MSDs, but to be effective intervention should include "employee involvement, employer commitment, and the development of integrated programs that address equipment design, work procedures and organizational characteristics" (NRC, 2001, p. 10). However, the effectiveness of an ergonomics program is dependent also on how well it is implemented.

Other government regulations that require compliance and are related to ergonomics are the laws of Equal Employment Opportunity (EEO) (EEOC, 1964) and the Americans with Disabilities Act (ADA), which falls under the EEO Commission (EEOC, 1990). Many manufacturing tasks are performed by men and are difficult for most women to undertake. However, if a job is improved so that more of the population, including women, can perform it, then costs associated with finding the right people are lowered and compliance with the EEO is not in question. Likewise, if the workplace is designed to accommodate as much of the general population as possible, then specific accommodation for the disabled is easier because the workplace is more flexible. The ADA requires employers to make "reasonable accommodation" for a worker with a disability. During the first 18 months in which the ADA had been in effect, about 12,000 charges were filed. Nearly 20% dealt with back ailments (BNA, 1993). When a company develops a culture that attempts to accommodate those with disabilities, it also makes it easier for people to return to work. The company becomes more creative and flexible, so that jobs can be modified, allowing injured workers to be re-introduced gradually back to a fully productive job (Olsheski and Brelin, 1996; Eastman Kodak, 2003, p. 29–30). Therefore, there are many benefits to designing well for a wide sector of the population, so compliance in itself need not be a burden to industry.

8.5   Elements of the Ergonomics Process

To establish an ergonomics process a program has to be initiated. The process sustains the program using methods such as systematic evaluation and revision, based on effectiveness. In the long term, a program without a process is not successful. The model of the program and the degree of integration into other company processes is dependent upon several factors such as company size, responsibilities of the personnel, available resources, and the company culture. For success, it is important for a company to find an approach that is most suitable and effective, rather than to apply a theoretical model that may conflict with the organization's needs and culture. Whatever the model the ergonomics program goal remains the same: "design the workplace so that it is healthy, safe, and optimally productive."

8.5.1   Choosing the Ergonomics Approach

The primary four elements in the Safety and Health Management Guidelines are: management commitment and employee involvement; worksite analysis; hazard prevention and control; and safety and health training (OSHA, 1989). These components are generally considered essential in any approach to ergonomics in industry (GAO, 1997; NRC, 2001). Medical management and a written program are elements in the meatpacking guidelines and often included in companies' programs (OSHA, 1990; GAO, 1997; NIOSH, 1997). The success of ergonomics in a company remains dependent on how well the program is structured and carried out. A "one size fits all" method does not work when implementing an ergonomics process (GAO, 1997; Hägg, 2003). A program and process must be tailored to fit the company culture. The following factors influence the development of a program and should be taken into consideration:

- Size
- Culture
- Resources
- Type of industry
- Types of ergonomics controls implemented
- Compatibility with other programs and processes

A program does not have to reach full integration with other company processes to be successful. The factors listed and particularly the characteristics of small industry determine the appropriate extent of integration. However, if a program is minimally integrated into the company, it is more likely that the program remains driven by reacting to problems and precludes the development of a proactive approach. In other words, job improvements tend to be made in response to problems and not before problems occur. Incorporating ergonomics proactively requires shared responsibility, involvement, and commitment throughout departments (Stuart-Buttle, 1998; Eastman Kodak, 2003). The more ergonomics becomes part of the company culture the greater its effectiveness. But it takes time to evolve and integrate a process into a system. An ergonomics process is dynamic and regular evaluation and revision is necessary to maintain or improve effectiveness.

8.5.2   Process Elements

The elements of an ergonomics process are discussed from the perspective of small businesses, highlighting the problems and advantages due to size. The order in which the elements are discussed is the recommended order that small industry could follow when establishing their ergonomics programs. However, there may be several elements already in place in the company, so there cannot be a cook book approach. To begin with, management commitment and employee involvement, or participatory approach, are essential for any degree of success with an ergonomics process (GAO, 1997; Hägg, 2003). If there are many work-related medical cases, establish or enhance medical management because addressing this component early in program development can have a dramatic effect on

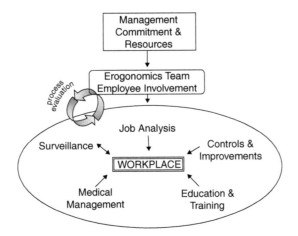

FIGURE 8.2 Diagram of the elements of an ergonomics process.

employees' well-being and workers' compensation costs. Next, acquire some education about ergonomics and initiate a plan for training. Then, having learnt more about what to monitor in the workplace, establish a surveillance system. Conduct job analyses as prioritized by the surveillance system and proactively assess new designs. Finally, implement improvements to the process for a safe, healthy, and productive business. Figure 8.2 illustrates the process components, which are discussed next.

1. **Management Commitment and Employee Involvement**
   a. **Management commitment**
   Management commitment is fundamental to the success of an ergonomics program. Several aspects of commitment are discussed.
   - *Is essential*
   Lack of management commitment undermines any ergonomics program, however well that program may have been conceived and initiated. Small companies have an advantage, as top management is more involved in the details of the production or service and closer alliances with the line workers are forged. Demonstrate commitment to the ergonomics process by presenting a positive attitude. Management attitude indicates the true priorities of the company to the employees. Commitment also affects whether a program matures to become a way of doing business. If a culture change is required for the success of the program, then the vision, expertise, and example of management is essential.
   - *Must be communicated*
   To enhance communication consider the following points:
     - Verbally communicate the importance of ergonomics as the means to a safe, healthy, and efficient workplace.
     - Express interest to give a message that the program is important and to instill energy, pride, and quality work in employees. Reinforce the commitment by an ergonomics policy or including the statement in the overall company policy and mission statement.
     - State the program objectives clearly to ensure that employees understand the reasons for changes.
   - *Is set by example*
   The behavior of an owner and the managers will convey more than a written statement or formally expressed policy. Set an example by displaying an active interest in how employees perform their work and whether the job demands are reasonable. Closely monitor the injury records and follow safety requirements when in the work areas, such as donning personal protective equipment. Convey the importance of ergonomics in productivity by showing

concern for the best environment and methods when looking to improve the process and production.

- *Entails employee involvement*
Employees are a valuable source of information and ideas. Involve employees in the process to accomplish effective ergonomics. This also demonstrates the respect and expectations management has for employees.

- *Requires clear responsibilities*
Involve the employees in program development to initiate active participation. As the program develops, clearly delineate the roles, assignments, and responsibilities of those involved. A common downfall of an ergonomics process is the lack of project follow-through for implementation and confirmation of effectiveness. Establish responsibility for project follow-through.

- *Entails the provision of resources*
Provide employees with the training and the time to contribute to the program. Provide authority for decisions to be made and money allocated to implement the projects to meet the program objectives.

b. **Employee involvement**
Management should support and facilitate employee involvement. The employees know the jobs best. They are an excellent source of ideas for improvements and are the ones required to work with any implemented changes.

- *Set up a core task force*
Certain knowledge and perspectives are needed to contribute to an ergonomics program. How the individuals are included depends on the structure of the program and the formality of a task force or team. To establish a core task force, designate a person to monitor the program and individuals to be responsible for specific aspects of the program. Clearly define the roles and responsibilities. The identification of the primary problems, their root causes, and potential solutions come from the contribution of many perspectives during problem-solving. The following are the departments or areas that have personnel who provide a particular perspective and should be considered on an ergonomics team. In a small business, some of these departments are combined:
    - *Operators.* Those who perform the job usually know the most about it. How the job is done is essential information and may be contrary to the way it was theoretically designed. The operators know the difficulties in performing a job and often have ideas for improvement. The operators will continue doing their jobs after improvements are made so their acceptance of any changes is important.
    - *Human resources.* In small companies information about turnover, absenteeism, accident records, OSHA logs, and workers' compensation records are often kept in the human resources office. Turnover is typically low in small companies and less likely to be an indicator of a problem job. Absenteeism and sick days, however, should be monitored, as they may reflect an employee's attempt to counteract the high demands of a job.
    - *Safety and health/medical.* Accident information and related records may reside in the safety and health department rather than in human resources. Apart from medical treatment, medical personnel generally handle return-to-work issues and job modifications to accommodate medical restrictions.
    - *Engineering/maintenance/facilities.* Knowledge about the mechanical aspects of the production system and the environment helps when determining the workplace redesign potentials and the related costs. The involvement of engineering personnel is also necessary to ensure that ergonomics is incorporated.
    - *Production engineers and quality control.* Personnel overseeing production and quality contribute information that helps to identify problem areas. Their perspective is useful in problem-solving and assessing potential solutions.

– *Management/supervisors.* In small companies management is likely to be involved, therefore providing decision-making authority. Management support is necessary for effective implementation of administrative controls.

– *Purchasing.* Involvement of the purchasing department ensures that the best products are purchased based on their function and quality as well as cost. For example, an inexpensive drill bit may seem a good bargain at first, but uneven wear of the bit may cause the worker to use excess force and may increase the risk of injury from drill slipping.

- *Set clear responsibilities*
An ergonomics process that is fully integrated into the company processes is likely to be the responsibility of everyone. However, one person or group usually becomes the resident "expert" or ergonomics resource. That person or group may be responsible for establishing and keeping ergonomics "alive" within the company.

- *Use existing task forces or teams if possible*
Combining committees or teams helps to prevent redundancy of efforts and reduces the time employees are away from production. However, this reduces the opportunity for employee involvement. A rotation onto a task force and ad hoc involvement helps to include more people.

c. **Written program**
A written program is often emphasized as essential for a successful program. However, the typical reaction of a small business to this suggestion is "more paper work." There are many advantages to putting down in writing the basics such as the objectives and who is involved. Writing the program basics helps to:
- Get the program started more efficiently
- Organize thoughts and the best plan of action
- Clearly communicate the process
- Make it easier to introduce the process to a newcomer
- Establish the goals and achievements by which the program can be assessed for success and improvement

The written program need not be a lengthy document but rather a clear statement conveying the objectives, goals, and processes to establish and continue ergonomics within the company. The schematic in Figure 8.2 helps provide an overall picture. Consider writing the document so that it includes other functions such as training. For example, a section that outlines the medical management procedure for a work injury could be used during orientation training of new employees. A written program need not be an independent document but may be incorporated into other processes within the company, such as the safety and health program or quality management process. Consider including the following in the document:
- An overall schematic of the components and the process
- Program objectives
- A list of those involved (or of job positions) and their responsibilities
- The program process (e.g., who gets training, to what extent, and by whom)
- A section for the project action list, analyses, and record of changes and their effectiveness (these could be the minutes from project meetings)

In addition, it is important to record the goals and action items to establish the program, so that there is accountability. The methods of measuring program effectiveness should also be included.

2. **Medical Management**
A primary responsibility of a company is to respond to work-related medical problems reported by employees. The medical conditions require treatment and correction of the cause to prevent reoccurrence. Medical management per se is not in the realm of ergonomics; however, early detection, prompt treatment, and quick return to work have direct bearing on the recorded injuries and illnesses. If poor design at work caused or contributed to the medical condition, then improvement of the workplace is

essential to prevent recurrence. Good communication between the medical community and the company is important to successfully improve the incidence of work-related injuries.

An industry with a large number of medical cases should start their ergonomics process by addressing medical management. Many companies have dramatically reduced their workers' compensation costs in as short a time as 1 yr by improving the medical management of existing and emerging cases (Stuart-Buttle, 1993). Small industries do not usually have in-house medical departments, thus it can be difficult to control medical management. For a company with 600 employees or more, the employment of an in-house health care professional (HCP) may be beneficial, particularly if there are a large number of employees requiring treatment. Contracting with a medical group to provide on-site services or using an occupational medicine group in the community are possible alternatives. A small company may have difficulty in finding the services, or having access to the service, especially if the company is located in a rural area (which is characteristic of small businesses; Headd, 2000). However, it may be possible to develop a relationship with a local medical group that is prepared to acquire the expertise the company needs. Trade groups, professional groups, or the company's workers' compensation carrier may be able to provide guidance and referral to occupational medical groups.

The company has some control over the quality of the services rendered. The following suggestions may help to establish a good working relationship:

- *Establish close communication*
  Assign a primary contact person in the company and have him or her establish a rapport with the medical group or HCP.
- *Communicate the expertise expected of the medical group*
  The company should make clear the type of medical expertise that is required. For example, if vibration exposure is inherent in the job, then the medical practice needs to have some experience or willingness to acquire training in screening for vibration white finger.
- *Introduce the medical group to the plant's culture and processes*
  When the HCP understands the jobs, he or she may be able to give specific job restrictions and helpful guidance on job modifications or alternative jobs. This promotes return to work more quickly and successfully. The HCP may also assist in progressing the worker through various modified jobs until the worker is able to return to work at the previous job or an equivalent.
- *Consider the medical community as a training resource*
  The local medical or rehabilitation group may also be a source for education of employees in health-related topics, such as back care and exercise programs.

3. **Education and Training**

   First, some general education in ergonomics is needed by someone in the company so as to plan an ergonomics program. In-depth training is recommended for a specified team, and general awareness training is recommended for managers, supervisors, and line workers. The extent of investment in training may depend upon the financial resources and the turnover rate. A smaller team may be trained to save costs, but everyone should receive some awareness training. If a company suspects there are many problems requiring improvement it may be helpful to train gradually. The team should be trained to respond effectively to the increase in reports that may occur after the general awareness training of the employees. Lack of response from the team may invite cynicism and reduce participation of the employees.

   Some companies feel it is cost effective to use external expertise on an as needed basis, rather than invest in in-depth training for one or a few individuals. Whatever the choice that is made, ergonomics expertise has been found crucial for a successful program (Hägg, 2003).

   a. **In-depth training**

      At the very least (unless outside expertise is called on, as needed), in-depth training is recommended for the person or team primarily responsible and involved with ergonomics. This does not mean that the recipients have to become ergonomists but rather the training objectives are to:

- Understand the overall program objectives, goals, and process
- Understand the injury and illness system for treatment, return to work, and job modifications
- Be able to correctly record and interpret medical records and OSHA logs for surveillance purposes
- Know how to conduct basic problem-solving job analysis for ergonomics issues
- Recognize risk factors for injury and illness in workplace design
- Be able to develop, implement, and affirm effectiveness of solutions to basic problems
- Understand basic ergonomics principles to apply to solutions and new designs
- Be familiar with outside resources and methods for finding resources

b. **Awareness training**

An awareness level of ergonomics training prepares employees to participate in the ergonomics process. A program can become less reactive and more prevention oriented when there is greater understanding and contribution of all the employees. The objectives of an awareness level of training are for the employees to:

- Generally understand the ergonomics program
- Appreciate their role and responsibilities in the program
- Recognize the early indicators of physical problems
- Understand the medical management system of the company
- Understand basic risk factors for injuries and illnesses
- Know basic ergonomics principles
- Understand their participation in job analyses

To train only those employees who work in jobs with identified physical risk factors may cause the company to lose the opportunity to improve other areas from a production or quality standpoint. An awareness level of training is recommended for all employees. New employees should be trained to maintain the knowledge base in the workforce. There are many resources for basic training. Commercial videos and training programs are available, as are companies that conduct training programs. Those who have received more extensive ergonomics training may also be able to provide the awareness training to the general workforce. It is not unusual to give separate programs for supervisors, management, and production workers as there may be differences in educational level and perspectives of each group.

c. **Refresher training**

Refresher sessions help to maintain the employees' interest in ergonomics. As with the ergonomics program or process itself, the refresher sessions can be part of other processes in the company, such as continuous improvement or safety and health.

Additional training may be beneficial for either the task force or the in-house "expert." The extent of extra training can vary according to the degree of investment the company wishes to make. Some companies have set up self-assessment approaches and found that when the employees receive training and are empowered to contribute, the employees address many problems themselves (Hägg, 2003).

4. **Surveillance**

An initial step to improving the workplace is to identify the areas with problems or the potential for improvement. This entails looking at data that have already been collected by the company. Such data include information on injuries and accidents, production and quality measures, and personnel records.

a. **Methods**

- *Medical information*

  The first priority is for jobs at which injuries or illnesses have been reported. The OSHA 300 logs, workers' compensation records, first aid logs, and accident records are all sources of data from which to determine problem areas. For small companies, the numbers can be small and hard to interpret. For example, only one person in a department may have a physical problem, which may or may not mean other employees will also develop a problem. Statistical analyses on small samples are less feasible, making trends difficult to determine. Although it may not be statistically feasible to detect trends, it may be useful to informally determine if there are patterns in data or performance.

Government statistics for given SIC or NAIC codes can be used as a benchmark against which company performance can be compared. If the number of cases on the OSHA 300 logs is at least five for the year, and this figure is converted to an incidence rate, it can be compared to the annually published figures of the Bureau of Labor Statistics (BLS). The BLS can also be contacted to find the rate for a given NAIC code if it has not been published. Calculation of incidence rates is described in the record-keeping guidelines for the OSHA 300 logs (OSHA, 2003).

- *Discomfort surveys*

Occupational medical cases often go unrecorded, because employees tend to report to their personal physicians rather than to the company. This can occur especially in smaller companies in which there is no medical department. Furthermore, the collaborative nature of a small company often deters absence from the job and hence the reporting of a problem. Finding out who is working with discomfort attributable to the workplace and treating those conditions promptly are important for two main reasons:

- The medical outcome is more successful and occurs more quickly, and it prevents the person from experiencing unnecessary hardship.
- The treatment and lost time costs are lower when the condition is less severe.

The best way to encourage early reporting is through education of the employees so that they understand the benefit in the long term. In addition, supervisors and managers need to understand the benefit of early treatment as well as encourage the employees to report. Detailed symptom surveys can be used as a formal mechanism for finding those with work-related medical problems. However, such surveys may be of limited value to the small company due to statistical constraints and the need for medical knowledge for interpretation.

Improving the jobs that provoke discomfort and fatigue can reduce the likelihood of an illness or injury occurrence. These jobs provoke a performance decrement, for example, a reduction in productivity, a high number of rejects, or an increase in absenteeism or sick days. Data demonstrating reduced performance and subjective discomfort can be collected either informally through interviews and focus groups or formally through questionnaires and discomfort surveys (Stuart-Buttle, 1994). A discomfort survey typically contains a simple body diagram that is shaded by the respondent to indicate areas of discomfort. Ratings of the intensity of the discomfort is also marked for each shaded area. A valuable addition to a survey is questions about the cause of the problem and requests for suggestions for improvement (Stuart-Buttle, 1994).

To attempt to uncover physical issues before they become medical problems does not necessarily mean "opening a can of worms" necessitating responses to a flood of reports. If there is likelihood of too many reports to handle at one time, the departments or areas can be surveyed sequentially. Soliciting information consecutively by area, or specifically by task, helps to gather information pertinent to improvements. Surveying the whole plant provides a useful baseline and helps to prioritize the departments or jobs to be assessed. If the number of jobs is large, the circumstances of the job may change before the area is addressed, thereby making the survey less useful. Job rotation can make the interpretation of surveys difficult as the connection of discomfort to a particular task is less distinct. There are also inherent problems with discomfort surveys, for example, discomfort would not be reported by those who are "survivors" at the job.

- *Absenteeism and turnover*

Both of these can be indicators of difficult or stressful jobs that may warrant redesign. There may be insufficient data due to company size upon which to draw formal conclusions, but exit interviews and informal investigation may indicate the roots of a problem.

- *Production and quality data*

  Data related to worker performance can be indicators of a mismatch of the workplace to the employee. Production may be lower than necessary if the worker takes longer to perform the job because of awkward postures or an inefficient layout. A change in production rate can also occur if the worker slows due to fatigue or discomfort. Rework, rejects, and mistakes are measurable aspects of performance that job redesign can improve. Errors may be due to a variety of causes including physical fatigue, cognitive or perceptual limitations, environmental distractions, or psychological and social stresses. So finding the primary cause is important to improve the job.

- *Accident investigation*

  Accident investigations can be a powerful tool to identify ergonomics-based problems. Less rigorous investigations can guide a company to accept the workplace design and look for behavioral shortcomings or system and equipment failures. However, root cause analysis can begin to question existing designs that may need improvement. Near misses should also be investigated with ergonomics in mind because inadequate design may be a contributing factor.

- *Audits*

  Conducting an audit to determine problem areas or areas of potential improvement is proactive. The scale of a small business lends itself to auditing ergonomics at the same time as conducting other system checks. Consider including ergonomics audits with safety and production improvement reviews. A checklist approach is common, and there are many checklists available on the market. However, consider tailoring one to capture your industry-specific issues. Checklists should be used only as a reminder of what to look for since they do not adequately address the interactions of risk factors or of the worker with the workplace.

- *Interview and employee reports*

  Open-ended interviews, focus groups, and employee suggestions are other ways to identify problem areas.

b. **Prioritizing**

Problem areas need to be prioritized to develop an effective action plan. A scoring system, which assigns a number for the presence of an indicator of a problem, is a questionable method by which to prioritize, because different data are gathered as indicators of a problem area. Another approach that can be useful is to develop a spreadsheet with types of indicators or data by departments, areas, or jobs. The amount and extent of the indicators can be used to qualitatively rank the problem areas in conference with other team members or employees. Consider factors such as the anticipated scope and difficulty of the project or whether the area or job is about to be changed for production reasons, because these factors also influence the priority of a project. At the beginning of a program undertake projects that appear to be relatively easy and inexpensive so that there are some early successes.

5. **Job Analyses**

Job analyses can be conducted from a reactive or proactive stance. A reactive analysis is evaluating a job known to have problems. A program usually starts from a reactive standpoint in response to injuries and illnesses that have occurred. A proactive analysis is looking at a new design, recent installation, or redesign to anticipate problems and ensure that the workstation incorporates ergonomic principles. A proactive approach should be implemented as early as possible in an ergonomics program not only to prevent new problems arising, but also to avoid remaining in a reactive position that is more expensive in the long run compared to designing well initially.

Start the program with small and simple projects and complete many of them. This is an effective way to involve employees from many areas and to build interest and full support for ergonomics. Learn by doing, getting more training, and taking on more challenging projects as experience builds. However, remember that using an expert may be necessary for complex problems and may be a time- and cost-saving measure.

a. **Responsibility for conducting analyses**

One of the decisions in designing the ergonomics process centers is who will conduct the actual ergonomics analyses. A team approach can be used in which the group collects the data, looks at the problem, brainstorms for the root causes, and generates potential solutions. Alternatively, one person may actually go and collect the data using interviews or video-taping, for example, and present to the team for a group brainstorming. It is not uncommon, particularly in a small company, for a process to develop in which one person collects data, analyses it, and generates the solutions with informal input from others. Caution should be used when adopting an isolationist approach, as over time, any joint responsibility for ergonomics might lessen because of a lack of involvement, employees may stop applying ergonomics principles, and the program might be perceived as only one person's responsibility. If this occurs ergonomics has not been sufficiently integrated into the company processes.

There are many sources of data that indicate ergonomics issues. Who collects the data depends on company resources and the complexity of the problems. In small companies an individual may be responsible, for example, for human resource issues, safety and health as well as fulfill the function of a sales manager. Owing to the multiple positions typically held, the ergonomics team would be small with less input, although with multiple perspectives.

Some companies choose not to invest in in-depth education of an employee but rather work with an external expert who gets to know the company process and culture. The decision about whether to have an internal or external "expert" depends upon the size and resources of the company and the other responsibilities of the "expert" within the company. The approach chosen also depends on the amount and complexity of the issues to be addressed. There may be benefit in having more in-house knowledge if the production process changes frequently. The company needs sufficient understanding of ergonomics to know when assistance is needed, how to find it, and where to get further information to address the issues.

b. **Problem-solving**

The root causes of poor design are identified through careful analyses. A good problem list helps generate the best possible solutions. For example, after detailed analyses it may be determined that a material is re-handled unnecessarily and that the handling can be eliminated, whereas a more casual look at the job may have focused on improving the handling, incurring higher costs to make improvements, and less productivity savings by not eliminating the step. Ineffective or expensive improvements may also be decided upon unnecessarily if a casual approach to analyses is adopted. The traditionally informal structure of small industry does not preclude good problem-solving, nor does careful, detailed problem-solving always require extensive quantification. A key to ergonomics problem-solving is to repeatedly question if a job can be performed a better way. If the issue does not appear straightforward, assistance should be sought.

c. **Quantification**

Quantification assures that what is perceived as a problem is in fact a problem, and to what extent. Quantification also helps to assess improvements to determine the best alternative, and it provides a measure of effectiveness and cost benefit. However, quantification is unusual in small businesses, and is especially discouraged when the decision makers are already convinced of the benefit. Measurement should be selective, that is, collecting necessary data rather than what is possible to collect. Taking unnecessary measurements increases the cost of analysis. A simple problem with easy improvements may be more informally approached although thorough problem-solving is always essential. When the situation is more complex, selective quantification assures appropriate design decisions are made and helps prevent the generation of new problems.

As Brough (1996) pointed out, the overall program objectives need to be kept in mind during analyses; that is, to design a safer, more productive workplace. The budget can easily be consumed by using elaborate materials and computer programs that produce more data than needed and that which require more time for analyses than correcting the workplace problems.

d. **Analysis**

Discussion of the many analysis methods and tools is outside the scope of this chapter. However, some discussion about checklists is appropriate since they are widely used. An overview of the basic steps in an applied analysis is provided as a guide.

- *Checklist*

  A checklist is a popular applied method for identifying and prioritizing the problem areas. However, caution should be taken when using a checklist as an analysis tool, because it does not address the interactions in a job. Checklists may serve as a reminder of the issues that indicate a poor design. A checklist can be useful for assessing a new design.

- *Workstation analysis*

  The following steps provide a general overview of an approach in the analysis of a workstation:

  – Clearly define the job function and the tasks, so that they are understood in context with the overall system
  – Collect pertinent job information such as performance rates and quality expectations
  – Interview employees
  – Describe the component actions of each task (possibly video-tape them if the task is complex or fast)
  – Identify the risk factors and job components that place excessive demand on the worker or that make the job awkward or inefficient
  – Assess the risk factors and job demands quantitatively and qualitatively
  – Determine the root causes of the risk factors, job demands and awkward methods
  – Develop a primary problem list with possible causes
  – Brainstorm for several short- and long-term solutions to the problems, especially ones that are inexpensive
  – Assess the cost benefits of alternative solutions
  – Develop an implementation plan including a trial stage if necessary
  – Reassess the solution after implementation to determine its effectiveness
  – Record the project

- *New design*

  Incorporate ergonomics into the process of purchasing new equipment, designing new layouts, or redesigning an existing area or workplace in the plant. If ergonomics is addressed at this stage there will be minimal subsequent concerns and better productivity from the beginning of equipment startup. A checklist may help to consider all the ergonomic aspects, but the interaction of the operators with the equipment or layout should be especially anticipated and critiqued. Company engineers should have at least basic knowledge of ergonomics and work closely with other personnel to ensure that all safety and quality standards are met.

  Maintenance requirements are commonly forgotten. Evaluate new designs for access and ease of maintenance. The time lost from awkward access, for example, the need to fetch and use a ladder to reach a control, can be considerable. The physical toll on the employee from difficult maintenance is also a cost to the company. Set up a preventative maintenance schedule for the equipment and facility. Equipment that is not properly maintained can increase the forces or control required of the operator, making the job harder to perform.

- At times corporate memory can be very short. Design mistakes can be repeated, often when there are new people who are unaware of the history of changes at a workplace. A small industry does not experience such forgetfulness as much as a larger company because turn-over or movement within the company is less. However, a project record may still be helpful to prevent repeating earlier trials and errors.

6. **Controls and Improvements**

Effective problem-solving usually generates more than one solution, typically long- and short-term ones with various costs. The solutions may incorporate administrative and engineering approaches. Although engineering solutions are perceived as more permanent and less dependent upon human behavior, it is not uncommon to need an administrative measure to accompany an engineering change (Hägg, 2003). An advantage in a smaller company is that employees are more likely to be cross-trained so that administrative controls, such as job rotation, are easier to implement.

a. **Engineering changes**

Engineering solutions vary considerably by type of industry. The company has knowledge of equipment that is specific to the industry but often finds it difficult to locate any other equipment that might be a design solution. A small company has less time to search the market due to the multiple job functions held by personnel. Therefore, a company may find it worthwhile to develop outside resources that can reduce search time.

b. **Administrative changes**

Administrative changes can entail organizational and policy changes. Quota systems, break patterns, rotation methods, and overtime, amongst others, are important aspects to address to improve the balance of job demands. Unfortunately, these approaches are often over-looked (Hägg, 2003). Sometimes changes in purchasing policies may be indicated. The common practice of buying the cheapest is not always cost effective, particularly if the item is of poor quality and adds stress and time to do the job.

c. **Follow-through**

A small company has the advantage of closer communication and more management involvement than a larger company, so that there is focus when the decision is made to address a project. Therefore, follow-through to ensure implementation and effectiveness is easier than in a large corporation, where often many jobs are analyzed but few changes are made. Even so, roles must be defined clearly for accountability, particularly if there is a team approach. After ergonomics awareness training, many small improvements are made by employees and supervisors at the plant floor. These changes typically miss being recorded and can occur outside of the process, unless there is a focus to capture them as part of the measure of implementations.

d. **Cost benefits**

In a small industry the cost benefit of an improvement cannot be based on injury and illness alone because there are fewer incidents and they may be scattered throughout the facility. Therefore, gains in productivity and quality become important in the cost–benefit equation. Of course, productivity and quality are important components for all businesses but companies with more employees often have more incidents (but not necessarily higher incidence rates) and potentially more people who may be affected, so that injury and illness costs may be more of the basis for justification.

8.6   Summary

When a small industry sets up an ergonomics process there are many factors that influence the development of the process. The main factors relate to the unique characteristics associated with small company size. Characteristics commonly found are less formality, responsibility for several positions, greater responsiveness, less specific knowledge, more management involvement, and less data-oriented approach. By being aware of these influences a small business may successfully incorporate ergonomics, especially if it focuses on the productivity and quality benefits.

References

Brough, W.R. 1996. Make your ergonomics program a powerful tool. *Safety and Health*, February:109.

Bureau of Labor Statistics (BLS). 2003a. *Charts from Annual BLS Survey: 2002 OSH Summary Estimates Chart Package.* December 18, p. 7. http://www.osha.gov/oshstats/work.html.

Bureau of Labor Statistics (BLS). 2003b. Lost-worktime injuries and illnesses: characteristics and resulting days away from work, 2001. *News.* Bureau of Labor Statistics, Washington, DC. March:27.

Bureau of National Affairs (BNA). 1993. Effective programs reduce liability from OSHA- and ADA-related charges, group told. *Occupational Safety and Health Reporter.* 10-13-93:527.

Bureau of National Affairs (BNA). 1996. Safety programs cited as factor in 18 percent drop in employer costs. *Occupational Safety and Health Reporter.* 1-3-96:1053.

Dyjack, D., Redinger, C., and Palassis J. 2003. Describing occupational safety and health programs in small businesses. Abstract in *Proceedings of American Industrial Hygiene Conference and Exposition*, AIHA, Dallas, TX. May 10–15:12–13.

Eastman Kodak Company. 2003. *Kodak's Ergonomic Design For People at Work.* S. Chengalur, S. Rodgers, and T. Bernard (Eds). 2nd Edition. John Wiley, Hoboken, NJ.

EEOC. 1964. *Title VII, Civil Rights Act of 1964.* U.S. Government Printing Office, Washington, DC.

EEOC. 1990. *Americans with Disabilities Act of 1990.* Public Law 101-336 101st Congress, July, 26. U.S. Government Printing Office, Washington, DC.

Esposito, P. 1996. Program management guidelines: a safety and health management system. *The Synergist*, August:27–29.

General Accounting Office (GAO). 1997. *Worker Protection: Private Sector Ergonomics Programs Yield Positive Results.* GAO/HEHS 97-163. General Accounting Office.

Hägg, G. 2003. Corporate initiatives in ergonomics — an introduction. *Applied Ergonomics*, 34:3–15.

Headd, B. 2000. The characteristics of small-business employees. *Monthly Labor Review*, April:13–18.

Keating, R.J. 2000. The pain of new ergonomics rules and the regulatory process. *SBSC'sWeekly Cybercolumn.* Small Business Survival Committee. The Entrepreneural View #130.

National Institute of Occupational Safety and Health (NIOSH). 1997. *Elements of Ergonomics Programs: A Primer Based on Workplace Evaluations of Musculoskeletal Disorders.* DHHS (NIOSH) Publication No. 97-117. NIOSH, Cinncinnati, OH.

National Research Council (NRC) and the Institute of Medicine. 2001. *Musculoskeletal Disorders and the Workplace: Low Back and Upper Extremities.* Panel on Musculoskeletal Disorders and the Workplace. Commission on Behavioral and Social Sciences and Education. National Academy Press, Washington, DC.

National Small Business Association (NSBA). 2003. *Ergonomics Standards.* Press Release. National Small Business Association, Washington, DC. Feburary 1.

Newman, R. 2003. Productivity Payoff. *US News and World Report.* February 24/March 3.

Occupational Safety and Health Administration (OSHA). 1970. *Occupational Safety and Health Act.* Public Law 91-596 91st Congress S. 2193 December 29. U.S. Government Printing Office, Washington, DC.

Occupational Safety and Health Administration (OSHA). 1989. Safety and health program management guidelines. *Federal Register*, 54(16):3904–3916.

Occupational Safety and Health Administration (OSHA). 1990. *Ergonomics Program Management Guidelines for Meatpacking Plants*. OSHA 3123. Department of Labor/OSHA, Washington, DC.

Occupational Safety and Health Administration (OSHA). 1992. *OSHA Handbook for Small Businesses*. OSHA 2209. Department of Labor/OSHA, Washington, DC.

Occupational Safety and Health Administration (OSHA). 2003. *Record Keeping Compliance Directive (CPL 2-0.131)*. Department of Labor/OSHA, Washington, DC. http://www.osha.gov/record keeping/.

Office of Management and Budget (OMB). 1987. *Standard Industrial Classification Manual*. PB 87-100012. National Technical Information Service, Springfield, VA.

Office of Management and Budget's (OMB) Economic Classification Policy Committee. 1997. *North American Industry Classification System — United States*. PB98-127293. National Technical Information Service (NTIS), Springfield, VA. http://www.ntis.gov/products/bestsellers/naics.asp?loc = 4-2-0.

Olsheski, J.A. and Breslin, R.E. 1996. The Americans with Disabilities Act: implications for the use of Ergonomics in rehabilitation. In *Occupational Ergonomics: Theory and Applications*. A. Bhattacharya and J.D. McGlothlin (Eds). pp. 669–683. Marcel Dekker, Inc., New York, NY.

Robinson, A. (Ed.). 1991. *Continuous Improvement in Operations: A Systematic Approach to Waste Reduction*. Productivity Press, Cambridge, MA.

Sheble, N. 2004. Difficult Growth: Manufacturing is following the same path as agriculture. As productivity rises, employment falls. *InTech*, January:22–24.

Small Business Administration (SBA). 1993. *The State of Small Business: A Report of the President*. U.S. Government Printing Office, Washington, DC.

Small Business Administration (SBA). 2003a. *Research Publications 2002*. Office of Economic Research, Office of Advocacy, Washington, DC. http://www.sba.gov/library/reportsroom.html.

Small Business Administration (SBA). 2003b. *Small Business Size Regulations*. http://www.sba.gov/size.

Stuart-Buttle, C. 1993. Small industry programs. Paper presented at *"Low Back Injury Symposium"*, Ohio State University, Columbus, OH.

Stuart-Buttle, C. 1994. A discomfort survey in a poultry-processing plant. *Applied Ergonomics*, 25(1):47–52.

Stuart-Buttle, C. 1998. How to set-up ergonomic processes: a small industry perspective. In *Handbook of Occupational Ergonomics*. W. Karwowski and W. S. Marras (Eds). CRC Press, Bora Ratoo, FL.

Sundström, L. 2000. Better working environment for small companies in Sweden. *Proceedings of the IEA 2000/HFES 2000 Congress*. Vol. 6. pp. 175–177.

The Synergist. 1994. Reach out to small business owners to improve health and safety, IH professionals told. *The Synergist*. June/July:12.

U.S. Census Bureau. 2000. *Statistics of U.S. Businesses: 2000: All Industries, United States*. http://www.census.gov/epcd/susb/2000/us/US–.HTM.

Wortham, S. 1994. Britain's small businesses face big safety issues. *Safety and Health*, February:66-67.

9

An Ergonomics Process: A Large Industry Perspective

Bradley S. Joseph
Ford Motor Company

Susan Evans
Sue Evans & Associates, Inc.

9.1 Introduction

9.1.1 Background

Ergonomics in the manufacturing arena is a fairly new science and not well known. Even though it has been taught in various forms through traditional academic institutions, it has just recently been viewed as a useful approach to "fitting jobs to people" in order to reduce work-related injuries and illnesses, and to produce a host of other operational benefits. Although almost every facet of a manufacturing company stands to benefit from improved job design through ergonomics, certain organizational and knowledgebased barriers to change make the adoption and wide application of ergonomic principles a difficult process.

Many recommendations in this chapter are based on research into traditional and more modern theories of organizational politics and of organizational change. In addition, these recommendations are based on a set of basic assumptions and beliefs that are reinforced by experiences in the auto industry.

- The practice of ergonomics works most effectively through the participative, rather than the expert, approach
- Adequate training of participants at all levels is essential. General introductory training should be provided for all participants, while special topic training should be available for designated individuals
- The structuring of ergonomics committees, teams, task forces or other groups must suit the organization within which they function
- The selection of committee members is critical. Representatives from workers, supervision, management, and union should be included; all areas affected by ergonomics should be represented
- Securing top management commitment is essential for success

9.1.2   Chapter Organization

This chapter is divided into five main sections:

- The introduction is devoted to background information on ergonomics
- Section two discusses the theory and practice of participative management in large organizations
- The third section, Recommended Methods for Implementing a Plant Ergonomics Program, outlines a protocol for the establishment and maintenance of an effective participative ergonomics program
- The fourth section, Recommended Methods of Data Collection, details existing and proposed data collection methods, for use in identifying ergonomics problems, justifying their correction, and evaluating their effectiveness
- The fifth and final section discusses computer, internet, and communication issues in an ergonomics program

9.1.3   Statement of Need

Why is a chapter like this needed in an ergonomics handbook for practitioners? Neglecting ergonomic considerations in the workplace contributes heavily to the incidence of musculoskeletal disorders, e.g., low back pain and upper extremity cumulative trauma disorders. In addition, poor job design has been blamed for reduced productivity, poor quality, and increased absenteeism. In some industries, these disorders can occur in epidemic proportions, costing billions of dollars in health costs annually. Research findings suggest that changes in work practices and equipment design can substantially reduce the number of musculoskeletal injuries and substantially increase productivity and improve quality.

This concept has led to great advances in ergonomics research and knowledge. The output of this research is often in the form of models and guidelines to aid engineers in designing machines. However, despite increasing knowledge about ergonomics, changes to the workplace, whether to new or existing machines, still largely fail to incorporate ergonomic principles.

Therefore, even as more research is completed and more is known about the limits of the human body in the workplace, without an understanding of the organizational process that is now widely used, ergonomic factors will continue to be inadequately used in the workplace.

9.1.4   Definition of Ergonomics

Ergonomics is defined as the study of work. Chaffin and Andersson (1984) further define ergonomics as "fitting the work to the person." Ergonomics is concerned with the problems and processes involved in designing things for effective human use, and in creating environments that are suitable for human living and work. It recognizes that work methods, equipment, facilities, and tool design all influence the worker's motivation, fatigue, likelihood of sustaining an occupational injury or illness, and productivity.

Properly designed workplaces, equipment, facilities, and tools can:

1. Reduce occupational injury and illness
2. Reduce workers' compensation and sickness and accident costs
3. Reduce medical visits
4. Reduce absenteeism
5. Improve productivity
6. Improve quality and reduce scrap
7. Improve worker comfort on the job

The primary goal of ergonomics is "improving worker performance and safety through the study and development of general principles that govern the interaction of humans and their working environment" (Chaffin and Andersson, 1984). Rohmert (1985) states that ergonomics "deals with the analysis of problems of people in their real-life situations." Further, he urges that ergonomists "design these relations, conditions, and real-life situations with the aim of harmonizing people's demands and capacities, claims and actualities, longings and constraints."

Ergonomics should not be associated with the measurement of work in the traditional sense. The ergonomist does not measure work only to set standards of time and productivity. That is a task best left to the industrial engineer, who performs motion and time studies using scientific measuring systems. Instead, the ergonomist identifies elements of the job that reduce the quality of the interface between the human operator and the workstation. A poor interface can cause unnecessary stress to the operator, leading to an increased risk of injuries and errors (which in turn may lead to an accident, poor quality, or a loss in productivity).

9.2 Organizational Issues in Developing a Comprehensive Ergonomics Process

Why has the workplace been deprived of something so useful as ergonomics? One explanation is that there is something internal to industrial organizations, such as communication breakdowns between designers and operators of workstations, or that there is something lacking in the knowledge base of the key actors, such as engineers who do not understand the principles of anthropometry, that erects barriers to proper ergonomic design. The operators, who often know the problems associated with workstations, do not communicate with the designers of the workstations early enough, and the designers of the workstation may not have the ergonomic expertise to design the workplace properly. This leads to design flaws that can contribute to ergonomic stress.

It is important to understand the complexity of industrial organizations when attempting to implement a health and safety (or any) program in the workplace. These organizations are usually very large and resist change. In order to be effective, the ergonomist or the organizer of the ergonomics program must understand how the organizations work and what can be done to overcome the barriers to change.

Below is a summary of basic organizational theory. This summary will help to explain and justify the need for developing specific types of ergonomics programs in plants.*

9.2.1 Organizational Models — Traditional and New

Traditional organizational theory is largely based on the assumption that organizations are rational entities (Shafritz and Whitbeck, 1978). Allison (1971) calls this the Rational Actor Model. That is, an

---

*Note that parts of this section can be found in "A Participative Ergonomic Control Program in a U.S. Automotive Plant: *Evaluation and Implications*" by Bradley S. Joseph.

organization is a "system within which individuals and groups will act in internally consistent ways to reach explicit objectives" (Tushman and Nadler, 1980). Therefore, organizations and their structures are "planned and coordinated for the most efficient realization of explicit objectives" (Tushman and Nadler, 1980). The theory blames behavior that violates these assumptions on ignorance, miscalculation, or managerial error within the organization, that is, error independent of either the organizational structure or the management approach. The theory assumes that organizational directions are explicitly and rationally planned. That is, organizations rationally choose goals that optimize an objective function. Managers' roles, as defined by traditional organizational theory, are rationally to plan, organize, coordinate, and control the organization's objectives (Koontz, 1964).

Traditionally, organizations are hierarchical systems involving top-down decision making. Top management decides on an objective for the organization (i.e., what it should be pursuing over the next several months) and instructs subordinates to follow a particular plan designed to reach that objective. These organizations are characterized by an extensive division of labor, including detailed job descriptions, tightly controlled departmental budgets, and narrow spans of supervisory control. This pattern effectively limits the opportunity of people to interact with one another either vertically or horizontally within the organization.

This model has done a satisfactory job in explaining why and how an organization, as a whole, reacts to a particular stimulus (e.g., a change in raw material prices). However, it often fails to explain or predict important aspects of organizational life (Tushman and Nadler, 1980). For example, while the organization as a whole may appear to be reacting consistently to crises or other stimulation from the environment, components within the organization often do not react in consistent ways. Too often, these inconsistencies cannot be explained solely by ignorance, miscalculations, or management error. Instead, the inconsistencies happen so often and are so overwhelming that they cause changes in the organization. Therefore, a new organizational model must be developed to explain these disparities.

One model that explains many of the disparities in industrial settings is called the Organizational Politics Model (see Figure 9.1). Allison (1971) describes this model (based on the work of Cyert and March, 1963, and March and Simon, 1958) in a book which examines the decisions leading up to and during the 1962 Cuban Missile Crisis. He argues that in this case study the rational actor model does not adequately explain many of the critical events or answer critical questions.

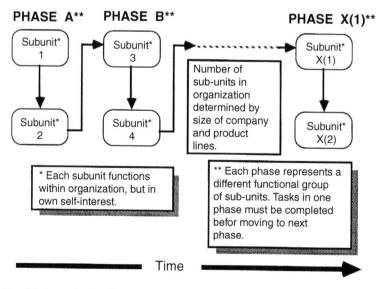

FIGURE 9.1 Simplified organizational politics model.

The basic principle of the organizational politics model is the concept of organizational units. Within the bureaucracy, for example, a corporation, these organizational units will act in their own self-interest to achieve their desired goals. This behavior will lend to conflicts between units with different goals, differing perceptions of how to reach a common goal, or joint dependence on scarce resources (March and Simon, 1958; Pfeffer, 1977; Schmidt and Kochan, 1972).

The organizational politics model fits well in current U.S. industrial settings, particularly now that budget cutting and increased workloads are placing increasing pressures on individual subunits. All these pressures have occurred without subsequent change in the basic organizational structure. Units are still expected to operate with the same or even greater productivity as before the budget cuts. Because they fear for their very survival, subunits consult their self-interest to a point where it is detrimental to the effective operation of the whole organization. Often these subunits may be organized into a matrix that attempts to break down individual unit self-interest. However, this only works if subunits work with each other as a team.

An example of the organizational politics model is shown in Figure 9.2. The organizational process of a division and plant that produces chassis and suspension components and rear axles was analyzed. Figure 9.2 shows the complexity of the organizational process required for the installation of new and the maintenance of existing equipment within this division.

In this example, a series of units are grouped together into phases. The phases represent periods of time during which groups of units have to complete a task before passing it on to the next phase. First, new processes are studied (study phase) and designed (design phase) at division engineering. Little plant input is solicited in this phase.

Next, division and plant engineering together install and debug machinery (implementation phase). This procedure involves a complex series of actions whereby process plans are sent to selected vendors, interpreted and built to specifications, delivered, and installed in the plant, using resources from the plant, the vendor, and the division. Unless the plant is willing to bear costly delays and excessive

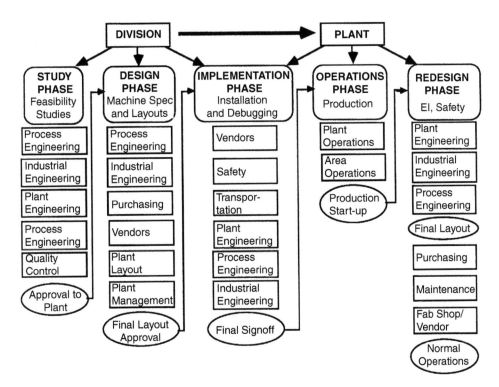

FIGURE 9.2 Organizational process for the installation of new and the maintenance of existing equipment.

expenditures, few changes can be made on machinery, for ergonomic or any other reason, between the time the vendor builds and the time he delivers it, because the vendor has signed off on machine specifications and is under contract to build it to standards agreed upon by the plant and division. Consequently, the plant must wait until the machines are delivered and operating under those specifications (known as final sign-off) before changes can be made.

After debugging, normal operation and maintenance proceeds (operations phase). However, there is often a need for process improvements or other redesigns to update equipment (redesign phase). Depending on the cost, the plant usually controls these activities. However, due to limitations on cash and manpower resources, lost production from shutting down the machines, and other plant priorities, this activity is often limited in scope and takes a considerable time to complete.

According to the organizational politics model (Allison, 1971), this complex process, involving the interdependence of so many parties, offers a certain prospect of conflict. Moreover, the fact that those who make key decisions on manufacturing processes are geographically and organizationally separated creates a high probability of communications breakdown (Allen, 1977).

What are the implications of all this for ergonomics and other health and safety programs? In order to implement sound ergonomic design, we have to overcome two kinds of barriers:

- *The knowledge-based barriers.*
  1. A lack of general ergonomics knowledge (knowledge of ergonomic principles) — Ergonomics is a technical science. Persons involved in the designing, operating, and maintaining of machinery who lack the technical knowledge of ergonomics will be more likely to design workstations poorly.
  2. A lack of specific job knowledge by workplace designers — People who operate jobs are most familiar with them. People who design jobs often do not know the specific information that pertains to daily operation (e.g., the process sheets and industrial engineering studies may vary from the designer to actual operation). This information is important when trying to determine job stresses, etc.
- *The organizational-based barriers.*
  1. A lack of communication between personnel involved in workplace designs — Players in each organizational unit must be able to interact with adjacent units to ensure that ergonomics is properly transferred along the organizational pathway. If for some reason this does not occur, then ergonomics, and many other considerations, may not be incorporated into the new job design.
  2. A conflict between subunit interests — Each subunit has its unique set of goals. Therefore, things like budget, manpower, and time can all be important aspects of subunit performance and may reduce the cooperation between competing and/or successive units along the job installation pathway.

In order to correct this situation and eliminate the barriers, ergonomic or other changes to the workplace will best be accomplished by organizational interventions in combination with technical ergonomics training. Several mechanisms for managing technical change have been researched. They can be grouped into two categories: expert methods and participative methods. These will be discussed below.

9.2.2 Ways to Effect Change in Industry

9.2.2.1 Traditional Ways to Effect Change — The Expert Approach

Traditionally, major operational changes in large organizations are effected with the help of professionally trained experts. In the case of ergonomics, most manufacturing plants have to import this expertise from the outside. The experts bring their special knowledge to the plant, collect data, return to their labs to analyze it, and make recommendations for change based on their investigations. Once the experts have done their work, there is likely to be no one in the plant with sufficient initiative and interest to follow through with improvements.

Lack of involvement is more often responsible than lack of knowledge for this failure to follow up. In fact, many people in the plant possess potentially beneficial knowledge and skills, but they are rarely asked to play a part in implementing ergonomic changes. For example, workers who do a job every day know it better than anyone else yet they are usually excluded from the job design process. As a result, workers may resist job changes and workplace designs that make sense technically, but in which they have had no stake.

9.2.2.2 New Ways to Effect Change — The Participative Approach

Worker–management participation itself is not new. However, industry in the United States has only recently started using worker–management participation on a large scale. In fact, there are ongoing debates if participation should be used in all industries. These debates are discussing a variety of issues including — who owns the problem (labor or management) and do these committees undermine the collective bargaining process. This trend has primarily resulted in growing realization that American productivity, labor–management communication, and the overall competitiveness of American goods worldwide have not been keeping up with the world pace (Peters and Waterman Jr., 1982). By looking at companies in other countries, notably Japan, American managers have learned to make increasing use of their employees as a source of information for all areas of plant operations. Truly enlightened managers perceive people as their most important resource.

Experience suggests that the participative approach can ensure the effective continuation of a program long after the expert or consultant is gone. However, when contemplating the use of participation, management and labor must consider how effective it will be in accomplishing *specific* goals of the workplace.

9.2.2.3 Participation in Health and Safety Programs

Recently, health and safety issues have become of greater concern in industry, partly because of the creation of the Occupational Safety and Health Administration in 1970 and partly because of the realization that health and safety is a core process that can affect the bottom line as much or more than other traditional programs. OSHA has increased both management's and labor's awareness of employee health and safety rights under the law, and legislation has inspired workers to take a more aggressive stance against observed violations. In fact, once a labor contract has been negotiated and ratified, most United Auto Worker shops throughout the country can strike for only one of two reasons: health and safety issues and productivity issues. This emphasis on health and safety has given managers who were reluctant to act an incentive to solve health and safety problems.

Several studies on the effectiveness of worker participation have been conducted. The W. E. Upjohn Institute funded a study (Kochan, Dyer, and Lipshy, 1977) to survey plants with union/management health and safety committees. Its intent was to determine how these committees function and to make a preliminary judgment regarding their effectiveness. General findings indicate that the committees with a high degree of continuity or high levels of interaction exist where OSHA pressure is strong, the local union itself is strong, rank and file involvement in health and safety is substantial, or management approaches health and safety in a problem-solving manner. This indicates that the most important attribute predicting a successful program was management and union commitment to solving health and safety problems rather than objective attributes such as frequency and length of meeting, number of members, existence of an agenda, and whether the committee was mandated by the collectivebargaining agreement.

9.2.2.4 Obstacles to Effective Use of Participative Problem Solving

Participative management and participative problem solving are not universal answers. Participation should be used for the kinds of programs in which it is known to be effective and for problems which are best addressed by a group process.

The quality circle is a good example of how participation can both succeed and fail (Lawler and Mohrman, 1985). In a manufacturing plant, a quality circle is a group which concentrates on solving workplace problems, usually those affecting the quality of the product, the quality of worklife, and working conditions. The quality circle works well with such problems, especially in early stages, and especially with easy problems. However, when the problems become more difficult or if the quality circle program is expanded too rapidly, the confidence and the effectiveness developed early on quickly becomes eroded. Frustration and the increasing cost of the expanded program usually spell disaster.

An additional threat to quality circles, and one which is relevant to the participative approach in ergonomics, comes in the form of supervisor resistance. Supervisors, like many middle managers, often feel that quality circles (and other participative programs) undermine their authority and control. Their unwillingness to support the quality circle program greatly diminishes its possible effectiveness.

Another barrier to success in participative programs arises in the differing perceptions of what constitutes participation, and how it should be administered. Workers often have unrealistic expectations about what they can do. Managers often treat participation as a special program or campaign rather than a viable technique; or else they abruptly embrace participation, something which can throw the workforce into confusion. Finally, managers may not have the patience to wait for the long-term benefits of participation to appear before they scrap a program when they fail to secure early success.

In her book *The Change Masters: Dilemmas of Managing Participation*, Rosabeth Moss Kanter provides guidelines for appropriate and inappropriate uses of participation. These include:

Appropriate Use of Participation:

1. To assemble sources of expertise and experience among the workforce
2. To tackle a problem that no one "owns" by organizational assignment
3. To address conflicting approaches or views
4. To develop and educate people through their participation (i.e., to develop new skills, acquire new information, and make new contacts)

Inappropriate Use of Participation:

1. When there is a "hip pocket solution" (i.e., the manager already knows the solution)
2. When nobody really cares much about the issues
3. When there is insufficient time for discussion and the group process

Kanter also draws attention to five critical challenges, or what she calls dilemmas, which any participatory group must face and which must be overcome if the participative program is to succeed. These are the situations which have no easy resolution. They are as follows:

1. The beginning or setting up the program
2. The organization of the program, in terms of structure and management
3. The prioritizing of issues to be addressed
4. The linking of teams with their environment so as to make them compatible with the existing organization
5. The evaluation process, that is, determining whether the program is working

9.2.3 Concluding Statement on Needs of a Participative Program

9.2.3.1 Role of Training in Participation

One of the most important requirements for success of any participative program is adequate training. In general, effective workplace education should consist of a process of instruction, reinforcement, and establishment of norms of behavior for workers (Vojtecky, 1985; Klein, 1984). It should provide guidelines on problem-solving skills and techniques in running a meeting. In essence, it gives participants the tools to perform their required functions in such programs. Many types of participative training programs are available. For example, Ford Motor Company trains all its employees before they are involved in participative problem-solving groups, emphasizing the basic skills outlined above.

For effective application of ergonomics, Shackel (1980) suggested that six factors must be addressed:

1. Ergonomics should be considered a science and a technology
2. Ergonomists should be researchers and practitioners
3. Ergonomics training and its content need constant updates and review
4. Presentation of data must be in a usable form for engineers, designers, and producers
5. The status of ergonomics must be high enough in the organization to make an impact
6. The ergonomists must have the necessary social skills to use ergonomics in the organization

Training addresses several of these factors. For example, training should teach theory and practice in order to give participants the skills to conduct research and to implement practical changes.

Even though ergonomics is a complex science, often requiring specialization in one area for the development of expertise, the complexity should not discourage the training of persons with average to lower levels of education. Knowledge is not necessarily a function of education level — a lot of knowledge comes from practical experience. The worker does not need to understand ergonomic models that explain the biomechanical cause of injury. Rather, the training needs to emphasize workplace configurations that lead to health problems and to provide understanding of how to reduce the risk associated with poor ergonomic design. Technical experts should be available to aid participants if they request more information or need more knowledge (Allen, 1977); they should play a resource, not an expert, role.

9.3 Recommended Methods for Implementing a Plant Ergonomics Program

This section outlines a recommended methodology for implementing a participative in-plant ergonomics program. The description of the three major steps is followed by a discussion of training needs. Note that although the steps are presented sequentially, some activities are best carried out concurrently in order to increase efficiency and reduce project overhead.

It is the opinion of the author that a participative approach to ergonomics is most effective. If your plant decides against the participative approach, only Step One (Securing Top Management Support) and Part Three of Step Two (Training) are relevant.

9.3.1 Setting Up and Implementing a Plant Ergonomics Program

All plants possess an operating organization which directs daily procedures. This organization requires that the proper authority be secured to begin implementing a program. In particular, health and safety programs require top management and labor support. Figure 9.3 outlines a process by which a program should develop at a plant.

Step 1: Secure top management and labor support — In order for a health and safety, in particular an ergonomics, program to be successful, management and labor must commit to the following:

1. Both management and labor must agree that the problem exists
2. Management and labor must agree that the problems can be corrected
3. Management and labor must agree that they will work together on solving the problems
4. Top management must commit to the program by giving a high priority to the implementation of job changes recommended by the program. This includes, but is not limited to, a commitment from the maintenance department to fabricate and install the changes, a commitment of plant funds to pay for the changes, and a commitment from management to install the changes in a timely manner
5. Top labor officials must commit and give a high priority to implementing and to using the changes

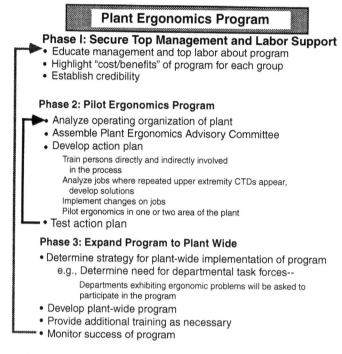

FIGURE 9.3   Ergonomics process flow chart.

Therefore, top plant management and labor must be in agreement that this program is important to the overall operation of the plant. In order to get this buy-in, it is often necessary to educate them. This education usually involves a presentation that defines the program, describes how it fits into existing plant programs and plans, and outlines its benefits and risks. Often this educational process can be supplemented by showing the audience case studies from their own and other facilities. These case studies demonstrate that the problems are real and widespread. The use of an in-house expert or outside consultant is helpful in conducting these case studies.

The presentation to plant management and labor representatives must be designed to demonstrate the "costs and benefits" of the programs and to highlight areas that positively affect each group's selfinterest. This task is relatively easy for an ergonomics program because all parties stand to benefit.

Dollar costs are those that affect the bottom line of the plant. They can be assessed through the use of traditional accounting techniques and cost/benefit analysis. For example, successful implementation of proper ergonomic design may reduce the numbers of injuries and costs associated with them, while increasing quality and productivity. Simple rate of return charts can be used to demonstrate these costs. An example of a rate of return chart is shown in Figure 9.4. With this chart, an experienced person, knowing the cost of poor job design and the associated plant profitability margin, can show the sales necessary to offset the costs.

Emphasis must also be placed on "people benefits." These include reductions in injury prevalence and increases in employee job satisfaction. Accurate records for employee injuries and worker satisfaction are often difficult to obtain, but are important assets to the success of the program.

Step 2: Pilot ergonomics program — Only after top management and labor support have been granted should you proceed to the Step 2. Often this support is conditional — they will want to review the results of the program after a trial period. In order to perform this step, several activities must be initiated.

First, a thorough study of the plant's organizational structure must be done. Typically, a manufacturing plant has a bipartite structure: production and support. Production's purpose is to produce whatever the plant sells. Support's purpose is to provide the expertise and facilities to ensure a smooth production

Accident Costs (Dollars)	Company Profit Margin				
	2%	4%	6%	8%	10%
$50,000	2,500,000	1,250,000	833,000	625,000	500,000
$100,000	5,000,000	2,500,000	1,667,000	1,250,000	1,000,000
$250,000	12,500,000	6,250,000	4,167,000	3,125,000	2,500,000
$500,000	25,000,000	12,500,000	8,333,000	6,250,000	5,000,000
$1 Million	50,000,000	25,000,000	16,667,000	12,500,000	10,000,000
$10 Million	500,000,000	250,000,000	166,667,000	125,000,000	100,000,000
$20 Million	1,000,000,000	500,000,000	333,333,000	250,000,000	200,000,000
	Sales Necessary to Offset the Cost of injuries at Different Profit Margins				

FIGURE 9.4   Rate of return calculation for health and safety table.

process. Some plants are highly automated, and have few production personnel but a large support function. Figure 9.5 shows the operational organization of a large automotive assembly plant.

Second, a team must be assembled to form the plant ergonomics committee. The members should be selected by high-level management and union representatives. These leaders should also be represented on the team or be closely associated with it. This support helps to ensure the timely implementation of ergonomic improvements. A letter, signed by these leaders and sent to the appointed individuals and their supervisors, helps to reinforce support for the program.

One criterion for membership should be a familiarity with the plant's culture, since this will ease the team's way in getting things done. Because of the interdisciplinary nature of ergonomics, the committee should consist of people responsible for:

- Identifying problem jobs (a representative from the medical department)
- Determining job stresses (labor representative from the jobs, engineering representative, health and safety engineer, and union health and safety representative)

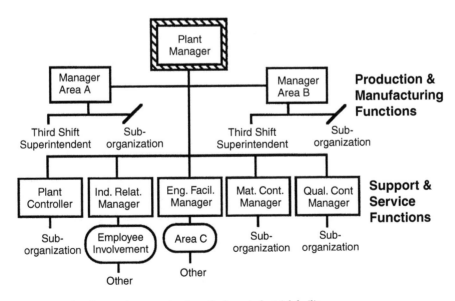

FIGURE 9.5   Functional operating organization of a large industrial facility.

- Developing solutions (process engineers, manufacturing engineers, maintenance)
- Implementing change (maintenance, industrial engineering, labor representative)
- Follow-up (medical department, etc.)
- Facilitating meetings (a facilitator or group leader to organize meetings and document changes)

A possible list of candidates for a plant ergonomics committee are:

- Industrial Relations Manager or Plant Manager
- Union chairperson
- Plant safety engineer
- Union health and safety representative
- Process engineering supervisor, industrial engineering supervisor, or manufacturing engineering supervisor
- Maintenance supervisor
- Union committee representative from each of the plant area units or departments
- Hospital representative

Third, participants should be trained in the basics of ergonomics. Different levels of training will be required by different people associated with the program. See section on *Training Needs*, below.

Fourth, the program concept must be introduced and piloted in one or two representative areas of the plant. This start-up phase is extremely important for the long-term success of the program. In this phase, management and union support for the program is confirmed. Although the initial step was to gain verbal support for the program, this phase actually secures their long-term commitment, which is necessary for the expansion of the program throughout the plant. Action plans are developed stipulating the "rules and regulations" under which the program will run. These rules may be changed several times before they are agreed upon by all interested parties. If this phase is not properly nurtured, the program may never mature.

Figure 9.6 shows an action plan used by a successful ergonomics program in a manufacturing facility. This action plan includes global issues that specifically affect the operating procedures of the program. Note that the action plan addresses issues for *existing* and *new* projects, as well as for the *people* of the plant.

Step 3: Expand the program plant-wide — It is important that before implementing the program plant-wide, the start-up and pilot phases be given adequate time and resources for the various components to mature. Often, false starts during the start-up phase will necessitate the revision of the rules and regulations. Management and labor support will have to be reinforced. Implementing a plant-wide program during one of these false starts can cause a severe and possibly fatal set-back.

The right time and method for expanding the program to the entire plant will vary, depending on the plant organization, culture, union/management relationship, etc. However, the expansion may involve the development of area ergonomics committees assigned to particular areas of the plant. These area committees may be organized around departments that manufacture and assemble unique products (e.g., building a line of gauges for a particular vehicle) or around areas in the plant that perform a specific set of operations as a subset of the entire assembly and manufacturing process (e.g., the body shop, paint shop, and the trim department of a large auto assembly plant). Because each plant area involves different ergonomic stresses, these teams can concentrate their efforts on identifying their unique problems and developing solutions.

It is often necessary to develop separate guidelines under which each team operates. These guidelines help define the scope of the each team's responsibilities as well as outlining the way the program operates. These guidelines, together with the plant action plan, will help resolve questionable issues. The guidelines must outline when an ergonomics problem will be reviewed by the committee, methods used to evaluate the job, and actions taken (including time limits) for problem resolution.

Figure 9.7 and Figure 9.8 display two ergonomics programs in large industrial plants. Note that both plants decided to expand their programs to the area levels. In the assembly plant, the expansion involved

Action Plan

Organization
The Ergonomics Organization will consist of
an Area Ergonomics Committee.

Training
Provide education and training to increase
ergonomics awareness.

New Projects
All new projects will utilize ergonomic principles
and considerations in the design of products, manufacturing
and assembly processes, and equipment.

Existing Projects
Based on medical data, safety considerations, or employe
responses, existing processes identified as ergonomically
stressful will be reviewed for ergonomic problems and improved

People
Employes exhibiting the effects of ergonomic problems in
their work will be encouraged to identify those problems to
the ergonomics task force and participate in their solution.

FIGURE 9.6   Ergonomics action plan from manufacturing facility.

forming departmental ergonomics committees; in the manufacturing facility, where different processes are needed to make a single product, ergonomics committees were formed in each of the seven manufacturing areas.

The composition of area committees depends on the size of the respective areas. Typically, a member of supervision, the union committee person, a maintenance person associated with the job, and several hourly employees are associated with such committees.

9.3.2   Training Needs for a Plant Ergonomics Program

One of the significant barriers to the successful implementation of plant ergonomics programs is a lack of ergonomic knowledge. Studies (Joseph, 1986) indicate that training can increase the ergonomics knowledge of all participants, regardless of education level or job status. The nature of the training program depends on the needs of the plant. In general, it is suggested that at least two levels of training be used — awareness training and expert training.

9.3.2.1   Awareness Training

The basic purpose of the awareness training is to make plant persons familiar with the program and to get buy-in from key members of the plant staff. This buy-in is essential in launching and sustaining ergonomics programs. Training should be short, less than two hours, and should be designed to give participants a basic understanding of the principles of ergonomics, an outline of methodologies for identifying job stresses, and an overview of alternative solutions. In addition, this training should address the procedure by which the ergonomics program and its organization will function within the plant organization. Training should be made available to everyone in the plant to make them aware of the efforts being undertaken. In particular, management personnel responsible for implementing and paying for projects, and union representatives responsible for securing worker buy-in to the program should receive awareness training.

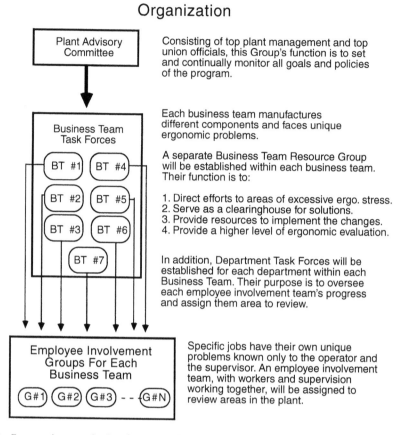

Organization

Plant Advisory Committee

Consisting of top plant management and top union officials, this Group's function is to set and continually monitor all goals and policies of the program.

Business Team Task Forces

BT #1 BT #4
BT #2 BT #5
BT #3 BT #6
BT #7

Each business team manufactures different components and faces unique ergonomic problems.

A separate Business Team Resource Group will be established within each business team. Their function is to:

1. Direct efforts to areas of excessive ergo. stress.
2. Serve as a clearinghouse for solutions.
3. Provide resources to implement the changes.
4. Provide a higher level of ergonomic evaluation.

In addition, Department Task Forces will be established for each department within each Business Team. Their purpose is to oversee each employee involvement team's progress and assign them area to review.

Employee Involvement Groups For Each Business Team

G#1 G#2 G#3 - - G#N

Specific jobs have their own unique problems known only to the operator and the supervisor. An employee involvement team, with workers and supervision working together, will be assigned to review areas in the plant.

FIGURE 9.7   Ergonomics organization for a manufacturing plant.

9.3.2.2   Expert Training

The expert training should be designed to give participants the necessary skills to understand the theory behind an ergonomic problem, the procedures for job analysis, the problem-solving skills to correct job stresses, and methodologies to evaluate the solution. All members of the ergonomics committee should receive this training.

This training may be further divided into two categories:

- *General introductory training* — which covers a broad range of topics
- *Special topic training* — which focuses on one or two special topics in ergonomics that are necessary for risk factor analysis and correction of job stresses

The length of the *general introductory* training can vary between a day and a week, depending on the number of topics covered. Regardless of length, the training should combine both lecture-based and hands-on sessions. The contents should include at least the following components:

1. Introduction and General Principles — A short session on general ergonomic principles, including an explanation of the theoretical man–machine interface. Often these principles are demonstrated by defining the man–machine interface, the components that make up the interface, and the consequences of poor design.
2. Risk Factors in Poor Job Design — A session on specific problems associated with poor ergonomic design. Generally, four types of problems are highlighted, along with their associated risk factors:

Action Plan

Organization

The Ergonomics Organization will consist of
an Area Ergonomics Committee.

Training

Provide education and training to increase
ergonomics awareness.

New Projects

All new projects will utilize ergonomic principles
and considerations in the design of products, manufacturing
and assembly processes, and equipment.

Existing Projects

Based on medical data, safety considerations, or employe
responses, existing processes identified as ergonomically
stressful will be reviewed for ergonomic problems and improved

People

Employes exhibiting the effects of ergonomic problems in
their work will be encouraged to identify those problems to
the ergonomics task force and participate in their solution.

FIGURE 9.6   Ergonomics action plan from manufacturing facility.

forming departmental ergonomics committees; in the manufacturing facility, where different processes are needed to make a single product, ergonomics committees were formed in each of the seven manufacturing areas.

The composition of area committees depends on the size of the respective areas. Typically, a member of supervision, the union committee person, a maintenance person associated with the job, and several hourly employees are associated with such committees.

9.3.2   Training Needs for a Plant Ergonomics Program

One of the significant barriers to the successful implementation of plant ergonomics programs is a lack of ergonomic knowledge. Studies (Joseph, 1986) indicate that training can increase the ergonomics knowledge of all participants, regardless of education level or job status. The nature of the training program depends on the needs of the plant. In general, it is suggested that at least two levels of training be used — awareness training and expert training.

9.3.2.1   Awareness Training

The basic purpose of the awareness training is to make plant persons familiar with the program and to get buy-in from key members of the plant staff. This buy-in is essential in launching and sustaining ergonomics programs. Training should be short, less than two hours, and should be designed to give participants a basic understanding of the principles of ergonomics, an outline of methodologies for identifying job stresses, and an overview of alternative solutions. In addition, this training should address the procedure by which the ergonomics program and its organization will function within the plant organization. Training should be made available to everyone in the plant to make them aware of the efforts being undertaken. In particular, management personnel responsible for implementing and paying for projects, and union representatives responsible for securing worker buy-in to the program should receive awareness training.

Organization

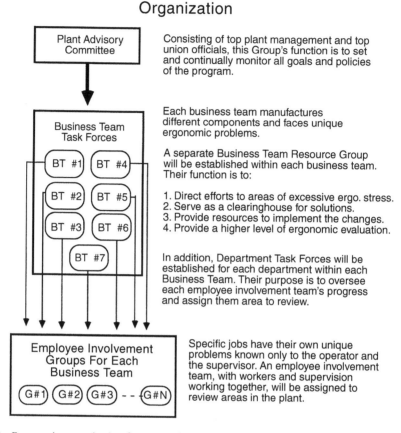

FIGURE 9.7   Ergonomics organization for a manufacturing plant.

9.3.2.2   Expert Training

The expert training should be designed to give participants the necessary skills to understand the theory behind an ergonomic problem, the procedures for job analysis, the problem-solving skills to correct job stresses, and methodologies to evaluate the solution. All members of the ergonomics committee should receive this training.

This training may be further divided into two categories:

- *General introductory training* — which covers a broad range of topics
- *Special topic training* — which focuses on one or two special topics in ergonomics that are necessary for risk factor analysis and correction of job stresses

The length of the *general introductory* training can vary between a day and a week, depending on the number of topics covered. Regardless of length, the training should combine both lecture-based and hands-on sessions. The contents should include at least the following components:

1. Introduction and General Principles — A short session on general ergonomic principles, including an explanation of the theoretical man–machine interface. Often these principles are demonstrated by defining the man–machine interface, the components that make up the interface, and the consequences of poor design.
2. Risk Factors in Poor Job Design — A session on specific problems associated with poor ergonomic design. Generally, four types of problems are highlighted, along with their associated risk factors:

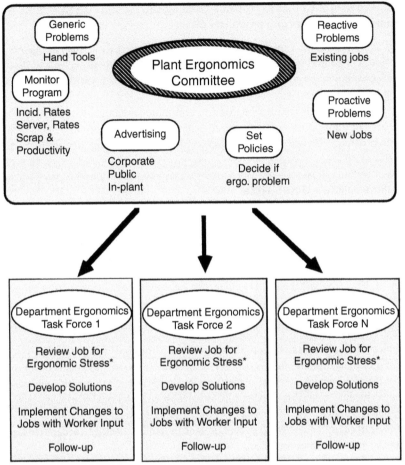

FIGURE 9.8   Ergonomics organization for a large auto assembly plant.

  1. Upper extremity cumulative trauma disorders, including localized fatigue, shoulder, hand, and wrist disorders, etc
  2. Manual material handling, including problems with lifting, whole-body fatigue, heat stress, etc
  3. Controls, displays and information processing problems, including problems with machine repair, quality control, etc
  4. General considerations in workplace design and layout, including problems with lighting, seating, etc
3. Job Analysis and Solutions — A session on methods to use in assessing ergonomic stresses, and on developing solutions. Obviously, the introductory level of training will not teach sophisticated job analysis techniques. Instead, it should equip committee members to analyze jobs for basic ergonomic stress. The results of this analysis can often be used to reduce the exposures. However, the training should also emphasize that in some cases, more sophisticated job analysis techniques will be necessary, requiring the assistance of in-house or outside experts.
4. Hands-on — The final session should be devoted to hands-on exercises. Participants should have practice analyzing jobs, determining exposures, and developing solutions. These solutions should

be discussed with the other participants in the classroom. This portion of the training can be supplemented with videotapes or job-site visits.

The *special topics* portion of expert training should be designed to teach participants advanced job analysis techniques. It should include workshops for individuals in the program who have very specialized tasks — for example, the medical persons in charge of medical surveillance may need training in recordkeeping techniques.

It is recommended that this training be given to committee members who require special skills and who are in charge of implementing and fitting the changes onto the shop floor. This includes the practitioners in the program who will have direct responsibility for analyzing jobs and developing solutions (e.g., engineers, etc.).

In all levels of training, it should be emphasized that ergonomics is a broad-based science requiring a team approach to identify and fix problem jobs.

9.4   Recommended Methods of Data Collection

Whenever a health and safety program (or any program) is implemented in the plant, constant feedback is essential. This feedback helps participants to determine the success of the program in accomplishing its goals. Data collection should begin as early as possible during the start-up phase because the results can help management decide how and when to begin plant-wide implementation of the program.

Data collection methods need to be compatible with the plant's operating procedures in order to minimize the resistance to change so often encountered. In some cases, existing data collection methods can be adapted for the requirements of the project, thus obviating the need to invent new ones. However, any method used in the data collection process has to be reviewed, and possibly changed, to ensure that it suits both the plant's system and the needs of ergonomic recordkeeping.

There are two important areas of data collection that should be studied to accurately assess a program. They are Process and Outcomes measures.

9.4.1   Proposed Methods of Data Collection

Accurate and complete data collection is essential to the success of an ergonomics program, both to ensure that efforts to reduce injuries and illnesses are working, and to demonstrate this success to a facility's decision makers. This is especially true in large organizations where "corporate management" is often removed from the daily operations.

Data should be collected in two areas: Process measures and outcome measures. Process measures determine if the system is performing properly. It evaluates systems and organizations to determine if they are doing what they are suppose to be doing. Outcome measures determine if the process is delivering the correct product to the customer. In the case of ergonomics programs, at least two different measures should be used — medical and intervention information.

Medical data: Medical reporting and recording systems in most industrial plants are inadequate for the purposes of effective ergonomics data collection. For the purpose of identifying problem jobs, the chief weakness of the system is the difficulty of tracing a particular injury to a particular job. Another problem is that the nature of ergonomics-related injuries and illnesses makes "early detection" difficult. It is almost impossible, in the present state of affairs, to assess the real numbers of work-related injuries and illnesses.

It is recommended that existing medical data collection be supplemented by active surveillance (diagnostic examinations and self-administered questionnaires are discussed) and by the use of other data sources, such as job process sheets. It is further recommended that medical costs be more accurately tracked through the use of a relational database system for the recording of medical visits.

The report reviews the sources currently available which reflect operational changes due to ergonomic improvements. Suggestions are made for recommendations in procedures for collecting the following data:

- Absenteeism
- Scrap/quality measurables
- Productivity
- Worker satisfaction
- Reductions in known risk factors for ergonomics-related injuries

Intervention costs: The report discusses the need for and the difficulties in securing accurate cost/ benefit information for ergonomic changes. Recommendations are made for the development of a form which details the various costs involved in implementing job changes. Such costs should be weighed against the documented reductions in medical and other costs associated with work-related injuries in particular jobs, departments, or areas of the plant.

9.4.2   Data Collection for Medical

One of the primary benefits of health and safety programs is the reduction in job-related injuries. Therefore, the proof of any program's effectiveness depends on a demonstration that injuries and the associated costs have declined. Evaluating the effectiveness of an ergonomics program means showing that injuries and medical costs have gone down for jobs or areas where ergonomic improvements have been made. The present medical recording system makes this task very difficult. In many plants, after an employee develops a medical condition, the procedure works as follows:

Step 1.   Employee reports a medical problem in four different ways:
  a. Reports to foreman and gets permission to go to medical.
  b. Goes straight to Emergency Room of affiliated hospital.
  c. Reports to main hospital next day before shift.
  d. Report problem to family doctor — If the doctor judges it to be work-related, then the employee will usually report the problem to plant medical within a few working days.

Step 2.   Plant hospital case history or equivalent — Once the employee reports the problem to the medical department, a case history report or equivalent is filled out. Typically, this form has two sides or has multiple pages. On one of the additional pages or the back side of the form is information about the injury itself and any follow-up notes concerning the progress of the injury and the employee's recovery. On the front side of the form is information that identifies the employee by his/her name, social security/employee number, home address, telephone number, sex, age, shift, plant department number, date/time of injury, date/time injury reported, cause of accident/injury, the job classification at time of injury, and injury/illness code. In some cases, these reports will include a statement from the employee as to the cause of the incident, including a description of how it occurred.

Step 3.   Depending on the severity of the case an accident investigation may be done. This investigation often has two parts — one done by the supervisor and another by the safety engineer. The purpose of this report is to determine exactly how the incident occurred and what corrective actions are being taken to prevent further occurrences. If the case results in lost or restricted workdays, then the number of days will be recorded.

Step 4.   Application for compensation and lost time — If the injury is sufficiently severe to require medical attention beyond initial treatment at the medical office, or if the injury prevents the employee from performing normal work duties, then the employee can apply for workers' compensation. Depending on the state and the rules, compensation for specific types of injuries may be available. Usually every claim must be reviewed by the plant compensation office before being accepted or rejected by workers' compensation. Often workers' compensation records only the most severe cases and does not provide an accurate picture of the true incidence of illness or injury in a plant.

This medical recording system is often called passive surveillance. It has a number of weaknesses for use as a basis for evaluating ergonomics programs.

First, the system is not designed to associate a specific injury with a specific job. Instead, it provides a picture of "global" trends of injury for large areas of the plant. This picture makes it possible to evaluate the overall effectiveness of health and safety efforts, and to pinpoint large "hot spots" where further attention is needed. But because one cannot use the data to relate the incidence of injuries to specific jobs, its value for use in cost analysis of ergonomic improvements is limited.

Second, passive surveillance systems alone, or analysis of existing medical records, may not indicate the extent of injury. Cumulative trauma disorders and related musculoskeletal disorders often have nonspecific symptoms that occur after hours and on weekends. Employees often do not relate these symptoms to the job and do not seek medical attention until after the symptoms have progressed enough to hamper their work efforts. Often these late-stage cases indicate only the tip of the iceberg of job effects.

For every compensable case, there are many more cases of employees with subclinical complaints. Consequently, OSHA logs and workers' compensation data often identify late-stage (tip of iceberg) disorders and complaints, reflecting only a subset of the population afflicted by these disorders. However, the plant medical case reports may be useful as an early indicator of medical incidents. Because these reports are filled out for all cases other than simple first aid, they can give a more accurate indication of the number of work-related injuries and illnesses in the plant. Still, the accuracy of these records depends on many uncontrollable factors.

Third, the period elapsing from the time when the operator notices the initial symptoms of the injury and when he seeks medical attention often is lengthy, and the employee may have moved to a new department or job. Consequently, it is hard to pinpoint the job that caused the injury or to correlate the job type with specific injury type.

Fourth, passive surveillance systems depend on the employee to report the injury to the plant medical department. If employees are not knowledgeable about the symptoms and do not associate them with their work activity, or if employees do not have a good relationship with the plant medical department, they may neglect to report them. This can result in an under-estimation of the numbers of work-related injuries in the plant.

Fifth, the use of medical records in passive surveillance systems may hamper the participative approach since only certain personnel have access to them.

An alternative method for determining the extent of work-related injuries is active surveillance. Active surveillance can be done in a number of ways, of which two are discussed here.

One method involves noninvasive diagnostic examinations by trained medical professionals. The examinations are designed to detect symptoms of cumulative trauma disorders and other ergonomics-related injuries. Although accurate, this method is expensive and inconvenient. The employee must be off the job for the period of the examination. Furthermore, it requires the services of trained medical specialists.

The other method uses self-administered questionnaires. The main advantage of this method is that it is inexpensive. The employee can fill out the questionnaire on his or her own time. No special personnel are required to administer or interpret the results. However, the questionnaires are probably not as reliable as the examinations in producing data about subclinical cases. Furthermore, the success of the program depends on the timeliness and accuracy with which the questionnaires are completed and returned.

The best strategy is to combine these two methods. For example, one approach is to administer the questionnaire to all employees, and then, based on their responses, to select a subset of employees who have a high probability of disease. These employees are then examined by medical specialists in order to confirm the diagnosis.

As one might suspect, even though active surveillance systems are more accurate than passive surveillance, they can be costly to administer. In addition, they are time consuming and may disrupt normal plant operations. Therefore, even if the resources are available to conduct active surveillance properly (e.g., a well-trained staff, arrangements for workers to be off the job for up to an hour), it should be used with discretion.

Because of the complications involved in establishing an active surveillance method for ergonomics, it is often desirable, at least initially, to adapt the plant's existing system. Typically, the information in the existing plant medical system is sufficient for the current needs of the medical department and any reports they have to generate. However, for evaluation purposes, there needs to be a way to identify the individual job. One cannot simply develop a new medical recordkeeping system without adequate support from the medical department, plant management, and the corporation. This support can take months or years to gain, and once there, the system itself can take longer to implement. In the meantime, it is proposed that modifications should be made to the current system.

Because of the availability of the plant case history report and the frequency with which it is used, most modifications should be tied to this report. These modifications should help link reported injuries to specific jobs, and they should include at least the following information. Below is a summary of proposed modifications.

1. Job identification system — Most plants have a structure for dividing up responsibilities on the production floor. Typically, this structure has at least three levels: superintendent, general supervisor, and supervisor. The superintendent has responsibility for all the general supervisors in a particular area, which is defined by various factors such as similarity of products or operations. A general supervisor controls several supervisors, each of whom oversees from 20 to 30 operators. Therefore, a method should be developed to match a particular employee to a specific supervisor so that a medical incident can be tied to 1 of 20 jobs.

It should be noted that a mechanism already exists in most plants to help identify specific jobs — namely, the job process sheets completed by the plant engineers. These sheets are typically developed and maintained by industrial and process engineering. They outline the specific tasks the operator must go through to complete the job. All jobs have a unique sheet.

2. Medical cost system — There are several costs associated with a medical incident, whether it involves a single medical visit or lost-time compensation. The cost of a medical visit includes the time away from the job, the time for the doctor or nurse to make an examination, treatment time, and materials associated with treatment. (In fact, some studies estimate that medical visit costs may average $50.) These costs should be recorded, on existing medical record forms, along with the major costs associated with workers' compensation, days restricted, or days lost from work.

Because of the nature of these injuries, it is important to use "relational" database systems. Many of the larger database systems used in corporations (payroll, workers' compensation) are relational. A relational database is very powerful and allows one to record multiple injuries for a single individual. These injuries may be differentiated by time of occurrence or by body location. Whichever is the case, accurate records of the data can only be done efficiently if the data collection system relates each incident of an indicator to a common denominator. These denominators can either be the person or the job the person is working on. (In many cases, it is both.) A relational database system can record such data properly.

Figure 9.9 depicts the logic of a simple relational database that can be used for recording medical records. There are two components — employee information and injury data. For each employee, demographic information identifying the employee and worker number are recorded to identify exactly who the injured is and where he/she was working at the time of injury. For each injury (there may be more than one injury for each employee), the type of injury, the location of the job at the time of injury, the date of the injury, and the cost of immediate medical attention can be recorded. If the employee receives lost time or restriction, then the employee number can be used to access an existing medical records management system.

Success in using the modified medical records system depends on certain characteristics of the plant. First, it is necessary that the project be implemented in a plant where the medical department and the employees have a good relationship, and where they are aware of and sensitive to the symptoms and causes of CTDs and low back pain.

Second, only plants located where the workers' compensation system recognizes that CTDs and low back pain can be work-related can employ this modified system. For example, in the state of Ohio, because of a recent state Supreme Court ruling in favor of allowing workers' compensation for CTDs,

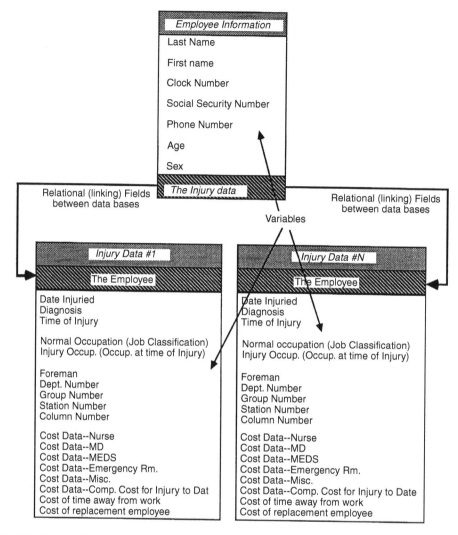

FIGURE 9.9 Database logic for personal computer-based medical records analysis system.

employers and employees alike are keenly aware of the problem. Consequently, they are reporting the problems earlier to the medical department.

9.4.3 Data Collection for Outcomes of Ergonomics Projects that Benefit the Plant

In order to determine whether ergonomic improvements have the desired effect, it is essential to record before and after measurements of certain operational variables of the job. Desired effects include:

- Lower absenteeism rates
- Reduction in errors and scrap (quality measurables)
- Improved productivity
- Greater job satisfaction
- Reduction in known risk factors leading to musculoskeletal disorders

Analysis of all these variables should be done as a part of the job analysis, and again later as a part of project evaluation.

The current sources for this information in most manufacturing facilities are as follows:

1. *Absenteeism.* Absenteeism records are typically kept for each department by the foreman. These records are sent to the personnel office and recorded.
2. *Scrap/Quality.* Scrap rates are kept for each operation. A scrap budget is set, by operation, for each department. This scrap rate is standardized by comparing it to the number of vehicles assembled.
3. *Productivity.* Productivity rates are determined by traditional time study procedures. The plant operates at a specific rate, expressed in the number of units per hour. These rates are changed only after engineering has made a process change that warrants a reevaluation of the time standard.
4. *Worker Satisfaction.* In many plants, no data are *routinely* collected.
5. Reduction in the Risk Factors Contributing to CTDs and Low-Back Pain. No data are routinely collected.

Data for scrap rates, quality rates, and productivity are currently collected in most industrial plants. However, the data are usually collected by department and sent to a central processing area to be summarized. This makes it difficult to associate the data with a particular job change. In addition, data on worker satisfaction and ergonomic risk factors are not currently collected in most plants. These data need to be collected before and after job changes to determine if the projects are reducing exposures to known risk factors.

In all cases, current operating procedures should be maintained. Any new collection systems should be developed to work within the existing systems. With this in mind, it is proposed that the following changes be made to the existing databases or other data collecting mechanism in order to facilitate data collection. A form should be developed to record and compare the data before and after projects have been implemented on the floor. Every entry onto the form should record the department and specific job location.

1. *Absenteeism.* Current operating procedures should be maintained. A copy of the data should be intercepted and entered into the project database for analysis by department and, if possible, by job.
2. *Scrap/Quality.* The ergonomics task force should devise a new form to keep track of scrap and quality measurements, by department and by job, and to determine improvements on jobs changed by the ergonomics committee. Since plants measure quality in different ways, a QC inspector should be assigned to the team to help with the data collection.
3. *Productivity.* Current operating procedures should be maintained. The industrial engineering process studies and process sheets should be used. A copy of this data should be intercepted and entered into the project database, by department and, if possible, by job before and after job changes.
4. *Worker Satisfaction.* Worker satisfaction may give an indication of the success of the job change since ergonomics is a tool to improve the quality of worklife. The ergonomics task force should devise a form that assesses participant satisfaction in the program. This form may be modeled after one used by Bradley Joseph in his study of an ergonomics program in an automotive plant.
5. *Reduction in the Risk Factors Contributing to CTDs and Low-Back Pain.* This information will be reflected in the data collection for medical because job improvements should show a related reduction in the medical incidence rates. However, because of the nature of these injuries and the time it takes for them to develop, this information may take months or years to collect. Therefore, in the meantime, it is important to keep accurate records on changes, by department and by job.

A standardized job analysis system should be developed using existing technology. For example, the NIOSH Work Practices Guide and the University of Michigan Static Strength Model can assess many of the risk factors associated with low back pain. These instruments can be used together to analyze stresses to the low back and determine the static strength requirements of the job.

Analysis of the risk factors for upper extremity cumulative trauma disorders can be done in a variety of ways. However, an economical and convenient way uses a checklist approach that assesses the repetitiveness and postural requirements of the job. The forcefulness of the job is more difficult to assess without the aid of electronic equipment. In addition, ongoing research is being conducted that may lend insight into estimating force through the measurement of several simple variables.

If this research is not available at the time of the study, an estimation of force can still be made by comparing the EMG values to several known forces and stating if it is above or below the known level.

Finally, it is proposed that several characteristics of the job before and after the changes are made will be recorded on paper and videotaped for archival purposes, for further analysis when new risk information becomes available, and for use in the estimation of other risk factors, including force.

9.4.4 Data Collection for Intervention Costs

Interventions or job changes represent a major cost of any ergonomics program. Often, before management will allow the installation of any new project, it has to be cost-justified. Therefore, since cost justification of projects can often decide their fate, it is important that these costs be monitored during the program. In most plants, the current procedure for implementing a project is as follows:

Step 1. Write up proposal — At this stage, the project specifications are developed and summarized through engineering on a project form. A cost justification (cost analysis) statement is attached that allows the decision-makers to determine if the costs of the project will justify the benefits — projects often have to be justified on the basis of traditional cost/benefit analysis and computed in terms of productivity and completed pieces per hour rather than in terms of health and safety costs (see discussion below).

Step 2. Approval of the project — There are two stages in this step.

Stage 1: Corporate Approval — Depending on the estimated cost of the project, engineering and other approval, have to be obtained either at the corporate level or at the plant level. For example, some plants specify a dollar value cutoff above which the project must be approved by the corporation, and below which the project must be approved through normal plant channels. If a project requires corporate approval, then the corporation pays for the project. However, if a project does not require corporate approval, then the project must be paid through existing plant resources. Projects requiring corporate approval must also go through plant approval.

Stage 2: Plant Approval — If necessary, and once approval has been obtained from the corporate engineering functions, the project must also be approved through various plant functions. The director of manufacturing at the plant, industrial engineering or equivalent, and the controller's office must all approve the project.

Regardless of the stage, the project form is sent through proper channels for approval. If approved, the responsible supervisor signs the form and retains a copy.

Step 3. Purchasing — The purchasing department sends the completed project out for bids.

Step 4. Vendors bid on project — Usually the lowest bid with the highest quality wins the bid and builds the project. It should be noted that this step may be omitted if the work is done in-house.

Step 5. Installation of project — The project may be installed with existing plant resources or, in special cases, with contracted professional services.

Step 6. Payment for project — Depending on the costs, either the corporation or the plant pays for the completed project.

Intervention costs can be assessed through normal operating procedures. If the project is approved, the project form should be copied by a member of the plant ergonomics team (and the departmental task force, if there is one). This information will be useful in determining the direct costs (materials, design time, etc.) to implement the job changes.

Problems with Cost/Benefit Analysis for Ergonomics Projects

Currently, as in most manufacturing facilities and depending on the costs, all business projects must go through normal purchasing channels to be approved for funding (see above discussion). Unless costs are nominal, these projects must be reviewed for cost/benefits. Funding is awarded based on traditional cost/benefit analysis calculations and expected savings due to work standards, work practices, or quality.

Below is a list of some of the costs involved in installing new equipment. All these costs should be considered in order to determine accurately the costs of implementing ergonomics projects and changes on the plant floor. It is recommended that a form be developed that records these costs for later analysis.

1. *Design time* — The time and resources involved in designing projects.
2. *Engineering time* — The time and resources involved in engineering the project.
3. *Tool change* — The fabrication costs and time necessary to fabricate a set of tools for the project.
4. *Skilled trades time* — Manpower needs for installing, testing, and maintaining the projects.
5. *Materials* — Cost of materials for the new project.
6. *Machine down time* — If the project is going to directly affect an existing line, that line may have to schedule down time to properly install the project. Therefore, down time and lost production must be budgeted into the installation costs.
7. *Training* — When new equipment and/or processes are implemented on the plant floor, operators responsible for running and maintaining the equipment must receive training.

It may be difficult to use traditional cost systems to justify an ergonomics project. This is because ergonomics projects often do not show significant savings, in the traditional sense, immediately after installation. Instead, the type of savings often seen in ergonomic projects are reductions in health care costs. These are often difficult to justify when the relationship between injuries and the responsible jobs is not well established (see above section).

This lack of an obvious link between an injury and a job yields two results: First, medical costs associated with worker accidents and chronic musculoskeletal disorders are usually not charged directly to the production department responsible for causing the injury. Instead, they are charged to a separate central account in the plant's Industrial Relations Department (or equivalent), thereby partitioning the true costs over the entire plant. This makes it difficult to justify a job change because the benefits are hidden. Consequently, projects often have to be justified on the basis of traditional cost/benefit analysis and computed in terms of plant-wide and area productivity (e.g., or completed pieces per hour). Projects that cannot show a cost/benefit advantage based on these measures often have little chance of implementation.

The second result of the absence of a known relationship between injuries and the production department deals with the poor recording of data. Often, CTDs are recorded only on sickness and absence reports with little or no follow-up. Consequently, employers have little data to go on in establishing a relationship between a job and injury.

Figure 9.10 depicts the relationship between the cost and benefits of ergonomics. Because of the problems of using traditional cost/benefit analysis, it becomes more important to document all the costs associated with poor job design and all the benefits after ergonomic intervention. Therefore, it is often best to make simple, inexpensive changes first. As poorly designed jobs are identified, the data (as outlined above) should be collected and analyzed before and after the proposed job changes. As more data are collected and the cost/benefit equation becomes better defined, it should become less difficult to justify job changes.

9.5   Computers, Internets, and Ergonomic Communication

9.5.1   Introduction

Communication breakdowns between workstation designers and operators has been stated as a barrier to proper ergonomic design. Effective ergonomic programs rely on communication to train the operators

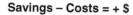

Savings – Costs = + $

1. Injury Changes
2. Absenteeism
3. Scrap/Quality
4. Productivity
5. Worker Satisfaction
6. Risk Factors

1. Design Time
2. Engineering Time
3. Tool Change Time
4. Skille Trades Time
5. Materials

FIGURE 9.10   Cost/benefit summary.

to recognize ergonomics problems, provide appropriate measurement systems for detection and remediation, and ensure timely feedback to engineers and designers to ensure that existing problems are not replicated and new designs fully consider ergonomic factors.

This section explores the use of computer technology to overcome communications breakdowns within a plant ergonomics committee and across the entire corporation. Characteristics of participative ergonomics programs which impact communication and the flow of information are presented. Challenges to effectively manage communication throughout corporate ergonomics programs are presented, followed by an approach which uses computer technology to overcome some of these challenges.

9.5.2   Communication and Information Flows

The characteristics of the in-plant ergonomics program discussed in this chapter have several direct implications on the role of communication to ensure its effectiveness. These characteristics include:

- *Monitoring and feedback systems* to track and record injuries, costs, and benefits of job changes
- *Participative teams*, involving representatives from all sectors of the manufacturing and support areas, which are tailored to the existing organizational structures
- *Available training* to attain general and specialized levels of ergonomics expertise among team members

The implications deal with what, who, or how ergonomics program information is being communicated. A considerable amount of data must be collected, analyzed, and reviewed to ensure the program is working. Having consistent procedures in place across the organization helps to ensure that plants can talk to other plants and share information, lessons learned, and discuss problems from a common base of understanding. Table 9.1 suggests several implications for effective communication in a corporationwide ergonomic program.

Gathering and tracking the myriad data needed to support an effective ergonomics program are no small feat. Examples of the types of information flowing through these programs are presented in Table 9.2. All the information is useful within the plant committee to manage the program. Much of these data are also of interest to a broader corporation-wide ergonomics community. This utility is indicated in the table.

The benefits for sharing information across the corporation are significant. Developing effective mechanisms for achieving this is but one challenge to those developing corporate ergonomics programs.

TABLE 9.1 Ergonomics Program Communication Characteristics and Implications

Characteristic	Implication
Monitoring and feedback systems	Process and procedures must be in place to document incidents, causes, solutions, and effectiveness, and report to appropriate communities of interest.
	Data collection methods must be identified, tested, and implemented within the plant to gather medical, cost intervention, and outcome data.
	Paper and pencil or automated tracking systems are needed to log incidents, assign responsibility, and track progress.
	Mechanisms for tracing design issues back to the designer, along with any in-plant solutions and operator feedback must be established.
Participative teams	Mechanisms must be established for selecting team members with an appropriate cross-functional background.
	Team-building exercises and management commitment documents are encouraged to ensure common goals and objectives are adopted among members from divergent experience and background.
	Team member rosters must be kept current and include active and adjunct members, and their position on the team as well as contact data.
Available training	Training classes must be made available to all team members to ensure informed participation and common bases for problem solving.
	Training must be consistent across teams to achieve common understanding.

9.5.3 Challenges to Effective Communication

The task of keeping ergonomics programs running smoothly are full of communications challenges. Challenges relevant to the plant-level programs, and those which occur as a result of adopting the program corporation-wide are discussed.

TABLE 9.2 Ergonomic Process Information and the User Community

Information used by →	In-plant Committee	Corporate Community
Best Practices	1	1
Case Studies	1	1
Committee Meeting Minutes	1	
Cost data (injury, design solutions)	1	1
Engineering Drawings	1	1
Ergonomic Incidence Logs	1	
Ergonomic Indicators	1	1
Ergonomic Risk factors	1	1
Ergonomic Standards	1	1
Incidence Action Tracking Log	1	
Job/operation codes	1	1
Lessons Learned	1	1
Management commitment documents	1	1
Medical data (visits, case reports)	1	
OSHA Logs	1	1
Part code, characteristics	1	1
Team roster/phone directory	1	1
Training classes, schedules	1	1
Vendor/Supplier data	1	1
Workforce population data	1	1
Workers Compensation data	1	1
Workstation layouts	1	1
Workstation/Ergonomic checklists	1	1

Setting up in-plant programs clearly must accommodate plant-specific cultures, management styles, and work patterns. Maintaining effective plant programs faces the challenges of:

1. Applying appropriate ergonomics expertise to a broader community of cross-functional team members, with safeguards to prevent misapplication (through training)
2. Effectively supporting an administrative documentation process without burdening committee members
3. Allowing plant-control of work in progress but provide sufficient corporate oversight to ensure accurate data gathering and consistent process documentation

All of these challenges pose a manpower resource dilemma for ergonomics programs. Unless the information is captured and documented, the process cannot document its effectiveness. Nor can it provide insights and lessons to use later. The documentation becomes the database. Supporting the documentation process through labor-saving means should become a number-one priority.

As the process is adopted for corporation-wide use, an additional set of challenges surfaces. These include:

1. Providing access to corporate repositories for common data (e.g., manufacturing part information, sanitized workers' compensation costs, approved ergonomic risk factors, etc.)
2. Applying consistency to the ergonomics process to ensure an acceptable level of corporation-wide legal compliance
3. Providing for informal, risk-free but timely means of communication across committees
4. Providing means and media to share information on ergonomic best practices, lessons learned, costs, and recurring problems

The common threads among these challenges seem to be

- Access to data
- Access to people
- Minimize the impact on resources

An effective solution to these challenges must weave these issues together.

9.5.4 Using Computers to Overcome Challenges

The traditional mode of communication employed in participative teams has been face-to-face. The time constraints and geographic dispersion of today's corporate work environment (even for in-plant programs) demand alternative modes of collaboration to achieve effective communication. Just as counterparts in the manufacturing, production, and back-office components of the corporation have adopted computer technology as the means to do more with less, so must the ergonomist employ the computer as a tool to leverage resources effectively.

Experience in developing computer tools for ergonomics processes has taught us several things:

1. The applications must effectively reduce the administrative burden at the plant level first (i.e., the tools must reduce the documentation workload and free the committee for solving ergonomic problems)
2. Users from several local committees must be involved in the design, pilot testing, and fielding to ensure broadest acceptance
3. The life-cycle costs of fielding, training, distributing, and supporting traditional software applications can far exceed the cost of initial development
4. Internet technology presents significant advantages for addressing communication challenges and managing application life cycle costs

Figure 9.11 presents the components for a proven computer-based solution. The approach is based around a centralized ergonomics evidence database application and uses corporate internet technology to support plant access and data exchange.

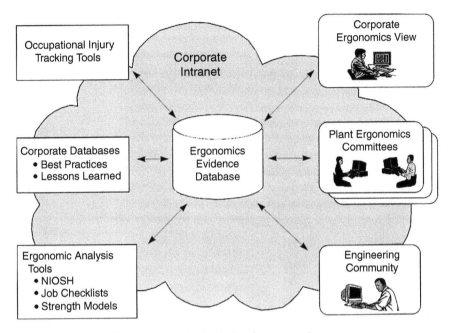

FIGURE 9.11 Components of a computer-assisted solution for ergonomic processes.

The central database allows all plants to read and write to the database as they access and manage their ergonomic documentation data. Data sharing across plants is permitted only at the discretion of the source plant. Once the documentation process is completed for an ergonomics incident, the plant can release the data for corporation-wide viewing.

The Automated Ergonomics Evidence Book (AEEB) uses an advanced internet browser front end to guide the user through the features and functions of the application. It links all plants to the central database, but maintains plant-level control over documentation in process. Builds an electronic ergonomics incident evidence book and easily supports the administrative tasks of maintaining the book, documenting the progress, and generating required and *ad hoc* reports. It also reduces the time spent documenting by sharing common data across reports.

By using a centralized database, data generated by or of use to a corporate audience can be maintained centrally but accessed globally. Corporate commitment documents are updated centrally. Schedules for ergonomic training classes, as well as corporation- or division-wide ergonomics news releases are maintained in one place, but are accessible for all local committees.

The database application also lets the plants track and maintain their roster of plant committee members. These rosters are viewable by other plants and provide an effective means of connecting one with another. Other electronic communication aids provided by the system are bulletin boards for posting news releases, and discussion databases to share ergonomics problems and solutions over an informal but timely communication channel.

The centralized database also permits corporate oversight through read-only access to plant data. Corporate and division ergonomics personnel can conduct audits of each plant's electronic evidence book remotely. For global corporations, this can reduce travel costs and improve process effectiveness significantly. Spot-audits conducted remotely can permit remedial coaching and process improvement on a more timely basis.

This approach has provided a flexible, accessible, and maintainable solution with key features. These include:

1. Centralized data and software location reduces maintenance and distribution costs as well as simplifying corporate oversight and process audits
2. Leverage of existing corporate intranet technology for data access, distribution, and sharing

3. Enhanced support for communication among diverse corporate groups, including the ergonomics, engineering, and health and safety communities
4. Customized computer support for all the documentation requirements of the established ergonomics process

As the AEEB internet approach is fielded, new requirements are being added to expand access to a broader range of data sources and user communities. Web-based databases for engineering data, material safety, process descriptions, and personnel rosters are all potential sources which can be accessed through corporate internets.

In addition to corporate web sites, an increasing number of ergonomic sites exist on the world wide web. Commercial, government, and professional organizations are employing the WWW to promote data sharing and communication to the widest audience. Jumping on the web has appeal as a near-term solution to solving the communication and information dilemma. Some degree of corporate ergonomic oversight and control is still advisable to ensure that the information derived from public sites is consistent with the corporate vision or intent.

In order for the ergonomics documentation process to be effective, the data must be accessible and usable by the broadest community. However, the documentation process itself must not be so burdened by recordkeeping that it fails under its own weight. Well-designed computer applications have been proven to be an effective means to the end.

While they may sound obvious, the following lessons bear repeating here:

1. Develop solutions for nontechnical users
2. Understand the user's requirements thoroughly — this is an ergonomics solution
3. Realize that local plant committees will not have the high-tech hardware of engineers and designers at their disposal
4. Develop a single solution with sufficient flexibility to be applicable across the organization
5. Try to share data from existing corporate databases wherever possible
6. In the days of global corporations, consider foreign language issues in the interface
7. Factor the distribution, training, and ongoing support costs into the budget
8. Dealing with corporate gear-heads presents new challenges!

By employing the power of the web and corporate internet resources in developing computer solutions for ergonomics, the challenges of data access, communication, and cost constraints can be effectively managed.

9.6 Lessons Learned

First, it takes a considerable amount of time to start up, pilot, and expand an ergonomics (or any health and safety) program in a large industrial facility. All parties must exercise patience.

Second, the start-up and pilot phase are important steps. Do not attempt to expand the program before these steps have matured.

Third, developing plant monitoring systems is crucial to the cost justification process. Because traditional methods of cost justification do not readily fit ergonomics projects, "cost justification" must be based on other indicators. Proper monitoring can help show how these indicators help in the overall operating efficiency of the plant — people benefits and cost benefits.

Fourth, a plant action plan is very important to develop. Without it, the program will lose focus and direction. This should be jointly developed with management and labor.

Fifth, and most important, top management and top labor support are essential to launch the program. Without this support, the program will have too low a priority to justify participants' time and resources.

Because of competitive pressures and quality specification and because employees are no longer willing to accept the pain and suffering resulting from poorly designed jobs, these and other safety programs will become the norm rather than the exception. It is hoped that with this report, other plants will develop successful ergonomics programs.

Acknowledgments

The authors would like to thank the American Automobile Manufacturers Association (AAMA) for their support in this effort. The research behind this chapter would have not been possible without the commitment from the plant where the research was done and the support of the AAMA. A list of those who contributed to this effort would be too long to include here, but a special thanks must go to the Plant Safety Supervisor, the Engineering Staff, the workers, and all those individuals and groups that are part of a joint effort. The views, opinions, and conclusions expressed herein are, of course, those of the author and are not necessarily those of the supporting institutions.

References

Allen, T. *Managing the Flow of Technology.* Cambridge, MA: The MIT Press, 1977.

Allison, G. *Essence of Decision.* Boston: Little, Brown and Co., 1971.

Cyert, R. and March, J. *The Behavioral Theory of the Firm.* Englewood Cliffs, NJ: Prentice-Hall, 1963.

Chaffin, D. and Andersson, G. *Occupational Biomechanics.* New York: John Wiley & Sons, 1984.

Kanter, R.M. *The Change Masters.* New York: Simon and Schuster, 1983.

Klein, J.A. Why supervisors resist employee involvement. *Harvard Business Review*, Sept.–Oct. 1984.

Kochan, T., Dyer, L., and Lipski, D. *The Effectiveness of Union Management Safety and Health Committees.* Michigan: W.E. Upjohn Institute for Employment Research, 1977.

Koontz, H. and O'Donnell, C. *Principles of Management* (3rd Ed.). New York: McGraw-Hill Book Co., 1964.

Lawler, E., III, and Mohrman, S.A. Quality circles after the fad. *Harvard Business Review*, Jan.–Feb. 65–71, 1985.

March, J. and Simon, H. *Organizations.* New York: John Wiley & Sons, 1958.

Peters, T. and Waterman, R. in: *Search of Excellence: Lessons from America's Best-Run Companies.* New York: Harper and Row, 1982.

Pfeffer, J. Power and resource allocation in organizations, in *New Directions in Organizational Behavior* (M.B. Straw and G. Salancik, Eds.). Chicago: St. Clair Press, 1977.

Rohmert, W. Ergonomics and manufacturing industry. *Ergonomics*, 28(8), 1115–1134, 1985.

Schmidt, S. and Kochan, T. Conflict: toward conceptual clarity. *Administrative Science Quarterly*, 17, 359–370, 1972.

Shafritz, J.M. and Whitbeck, P.H. *Classics of Organization Theory.* Oak Park, Ill.: Moore Publishing Co., 1–67, 1978.

Shackel, B. Factors influencing the application of ergonomics in practice. *Ergonomics*, 23(8), 817–820, 1980.

Tushman, M. and Nadler, D. Implications of political models of organization. Reprinted from *Resource Book in Macro-Organizational Behavior* (R.H. Miles, Ed.). Santa Monica, CA: Goodyear Publishing, 170–190, 1980.

Vojtecky, M.A. Workplace health education: principles in practice. *Journal of Occupational Medicine*, 27(1), 1985.

IV

Upstream Ergonomics

10

Digital Human Modeling for Computer-Aided Ergonomics

Xudong Zhang
*University of Illinois at Urbana-
Champaign*

Don B. Chaffin
University of Michigan

10.1 Introduction

While human modeling has always been at the forefront of ergonomics research, it is being propelled at an unprecedented tempo in the digital age by the advancement of computer technology. Digital human modeling is rapidly emerging as an enabling technology and a unique line of research, with the promise to profoundly change how products or systems are designed, how ergonomics analyses are performed, how disorders or impairments are assessed, and how therapies or surgeries are conducted. Digital human representations in various forms are increasingly being incorporated in the computer-aided design of human–machine systems, such as a driver–vehicle system or a manufacturing workstation. With the computing power and computational methods available today, we are able to render digital human models that are an order of magnitude more sophisticated and realistic than the ones produced a decade ago. There is, however, still a long way to go for achieving the "ultimate" digital human surrogates — ones that look, act, and even think like we do. Some of the limitations associated with the current digital human models are unlikely to disappear as computer performance further advances.

This chapter discusses digital human modeling both as a technology and as a fundamental research area, in the context of computer-aided ergonomics or human-centric design. As a technology, digital human modeling is a means to create, manipulate, and control human representations and human–machine system scenes on computers for interactive ergonomics and solving of design problems. As a fundamental research area, digital human modeling refers to the development of mathematical models that can predict human behavior in response to minimal command input and allow real-time computer graphic visualization. The discussion of the latter will emphasize on the modeling of human movements particularly those produced during physical acts under industrial settings. Such an emphasis heeds the fact that not only is movement the prime form of human physical work actions and interactions with machines, tools, and products, but its modeling is also where most research challenges arise.

10.2 Digital Human Modeling Technology

As we expect new products to be designed and manufactured in a short time frame, and demand a high level of convenience, comfort, and safety, the capabilities to create a digital human with specific population attributes and merge it with three-dimensional (3D) graphic renderings of proposed work environments is much desired. Figure 10.1 is an illustration of such a capability being utilized to assess a proposed manufacturing workplace layout design. The implementation of digital human modeling technology allows easier and earlier identification of ergonomics problems, and reduces or sometimes even eliminates the need for physical mock-ups and real human subject testing (Badler et al., 1993; Morrissey, 1998; Zhang and Chaffin, 2000). While there is an additional implementation and training cost in the initial stage, the use of digital human modeling, digital prototyping, and virtual testing in a computer-aided design or ergonomics process can quickly lead to reduced cost and shorten time. This is illustrated by a conceptual model of cost profile comparison between a conventional ergonomics design process and a computerized process (Figure 10.2).

The conceptual model shown in Figure 10.2 appears to be consistent with the fact that the past decade has witnessed significant growth in both the software systems offering digital human modeling capabilities and the companies adopting the technology. Features of various digital human modeling software systems and how they can be utilized in ergonomics and design problem solving have been described in the literature (e.g., McDaniel, 1990; Porter et al., 1995; Chaffin, 2001; Delleman et al., 2004).

The practical need for digital human modeling technology may be better argued using the success stories of companies that are utilizing this technology. A collection of seven case studies demonstrating successful applications of various digital human models in proactive ergonomics analysis and digital human-centric design was recently put together by Chaffin (2001). These cases studies were performed

FIGURE 10.1 A digital human figure model in a virtual manufacturing workstation rendered by Jack for static reach, fit, and visibility assessment. (Courtesy of Ulrich Raschke, UGC-PLM Solutions. With permission.)

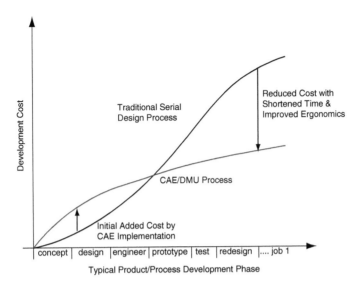

FIGURE 10.2 A conceptual model comparing cost profiles between a traditional ergonomics design process and a computer-aided ergonomics design process with digital mock-up (DMU) capability. (Modified from Chaffin, D.B., 2001. *Digital Human Modeling for Vehicle and Workplace Design*, SAE, Warrendalte, PA.)

at different organizations to address a wide variety of ergonomics and design issues, ranging from manufacturing work cell design to space station maintenance task design. Collectively, these studies suggested that the most prevalent use of digital human modeling in industry is to simulate people of extreme sizes (i.e., to perform 3D anthropometric analyses) for the purpose of providing designs that will accommodate a large variety of people to allow them to reach, see, or fit in a given workspace. In a few cases, a digital human model was needed for predicting a population's reach and clearance capability, including the mitigating effects of different clothing or personal protective equipment such as heavy gloves or helmets. In some other cases the issue was how much human strength or endurance was required to perform a manual exertion, with a special concern of design compliance with guidelines or policies (e.g., National Institute of Occupational Safety and Health guidelines or Department of Transportation policies). Furthermore, in a few cases it was identified that one of the most important features of a digital human model was that it would allow product or process designers to understand better the potential problems and associated risks a particular population subgroup could face when operating or servicing a proposed design.

What also was demonstrated by the case studies is the real possibility that designers or engineers with limited or no ergonomics expertise are enabled to examine ergonomics issues comprehensively and expeditiously. Digital human modeling software systems are meant to be integrative expert systems that capture experts' knowledge in areas including anthropometry, mechanical work capacity (e.g., range of motion and muscular strength), human kinematics and kinetics, work physiology, tissue stress and tolerance, visual process, motor control, and so forth. Corporations without an ergonomic staff line can therefore be empowered via the software systems that are readily accessible and user-friendly. In that regard, digital human models also serve as an effective vehicle for advocating ergonomics and disseminating its principles and knowledge.

Nevertheless, a review of existing digital human modeling tools and the case studies discussed earlier seemed to converge on one limited aspect that warrants future development and research more than anything else: that is the difficulty in predicting complex human postures and movements in a timely and realistic manner. Modern digital human modeling and ergonomics design software systems are just beginning to incorporate empirically validated, perceptual-motor, and biomechanical models that predict how humans move, despite the fact that physical actions are largely manifested as movements. A general practical problem revealed in the case studies is that the designers were incapable of specifying how a person of certain demographic and anthropometric characteristics should be positioned in the virtual workplace, especially when dynamic activities or motions are involved.

10.3 Human Motion Modeling Research

At the very core of digital human modeling and simulation is a model — a biomechanical representation of the human body along with the computational algorithms that configure or drive the representation to produce postures or motions. Two model attributes are essential: first, the biomechanical construct should have sufficient level of sophistication (e.g., number of degrees of freedom, DOFs) to realistically represent the human body and the range of movement capability by most people, and the model-reproduced postures or motions should resemble those of real human beings; second, the computational algorithms should be time-efficient, allowing real-time or near real-time rendition. Here, the argument for physical realism is not so much based on naturalistic and appealing look of the computer-synthetic humans, but on the validity of ensuing biomechanical or ergonomic analyses. It has been shown that small errors in the posture or motion specification can lead to significant errors in joint loading and muscle force estimation (Chaffin and Erig, 1991; Desjardins et al., 1998; Chaffin et al., 2000; Riemer et al., 2004). The computational efficiency is not only the overarching goal of making human modeling digital but also a basic software usability requirement that is in fact heightened by computer-aided design or ergonomics applications. Digital human models that act significantly (e.g., several orders of magnitude) slower than real human beings simply would not serve the purpose. These two attributes will be used as criteria later as the current human motion or posture (as a special form of motion) prediction models are evaluated.

10.3.1 A Review, a Taxonomy, and the Challenges

Before a review of human motion prediction models is given, it is necessary to briefly describe four basic computational procedures used and two principal computational problems encountered in human movement modeling. The four basic computational procedures are: forward or direct kinematics, inverse kinematics, forward or direct dynamics, and inverse dynamics. Forward or direct kinematics refers to the procedure of computing joint and end-point (e.g., fingertip) coordinates from known joint or segmental angles. Inverse kinematics is the procedure of determining the joint or segmental angles from known joint coordinates, or most often end-point coordinates. Forward or direct dynamics refers to the procedure that starts from muscle activation or neural excitation and derives the body motions. This procedure, also called a forward solution, arguably represents the true sequence of events in movement production (Winter, 1990). In contrast, inverse dynamics or an inverse solution estimates the joint reaction forces and moments from known or measured body motions. In biomechanical models of human movement, normally the number of joint angles (i.e., DOFs) is greater than the dimension of end-point position, and the number of muscles is greater than the number of DOFs specifying the movement. Therefore, two types of redundancies can occur: kinematic redundancy in inverse kinematics and muscle redundanci, when muscle forces need to be determined in either forward or inverse dynamics. Indeed, redundancy is what gives rise to a very fundamental problem in the modeling of human motion — the so-called Bernstein's problem (Bernstein, 1967). The most viable and widely used means for solving redundancy is optimization, in which various objective functions as performance criteria or cost functions are formulated to mathematically represent an optimal strategy in determining joint motions or muscle forces. The nature of the formulated optimization problems will largely influence the computational complexity associated with modeling.

Four classes of representative existing human motion simulation or prediction models are reviewed here. These four classes are jointly defined by two dimensions: with or without musculature; static (time independent) or dynamic (time dependent). The intent of using such a taxonomy is to offer a more fundamental perspective and to unveil more explicitly where the modeling challenges lie.

1. *Static models without musculature.* Models in this class use static optimization and inverse kinematics to solve a discrete posture determination problem. The earliest documented human simulation model "BOEMAN" was a 3D static model developed for cockpit design by Boeing Company in a Joint Army–Navy Aircraft Instrument Research (JANAIR) project (Ryan, 1970). This model consisted of a large-scale linkage system representing the torso, neck, and upper extremity of a seated person. The optimization routine in the model minimized an objective function of total postural deviation from a "neutral" reference posture. Given a specification of end-point (the hand) trajectory, the model could generate a sequence of individually solved static postures to emulate a motion. Similar approaches but with different objective functions including minimum joint loading, discomfort, and energy expenditure have since been proposed and evaluated (Byun, 1991; Dysart and Woldstad, 1996; Park, 1973). Some of the optimization-based human-figure positioning algorithms for computer animation (e.g., Witkins et al., 1987; Zhao and Badler, 1994) also belong to this class. There are several limitations with the use of static models in a sequential- or quasi-static manner for motion simulation. First, it is computationally highly intensive: the determination of each single static posture corresponds to a sizable, often nonlinear optimization problem that has to be repeatedly solved as many as times as the number of time frames composed in a simulated motion. Second, some key characteristics possessed by real human motions, such as smooth velocity and acceleration profiles, are usually absent in the sequential-static compositions or at least not guaranteed by the models.

2. *Static models with musculature.* The static models with musculature were built mainly for the purpose of estimating muscle forces and joint loading (Seireg and Arvikar, 1975; Hardt, 1978; Crowninfields and Brand, 1981; Dul et al., 1984; Bean et al., 1988). Such models apply static optimization and inverse dynamics to a static posture or a series of discrete postures in a movement. The optimization routines in these models typically minimize an objective function quantifying muscle stress to effectively solve the muscle redundancy problem. The same model structure can be utilized conversely to, for instance, determine at each time step which postural variation results in minimum total muscle stress or forces in a particular group of muscles, and then produce sequential- or quasi-static simulation of a motion. Such a modeling approach has not been attempted, but imaginably would bear the same limitations as the models in the first class and additional computational burden. The difficulty in accurately representing muscle physiology such as the length–tension or velocity–tension relationship would also pose a barrier to achieving model realism.

3. *Dynamic models with musculature.* The static models with musculature evolved into dynamic models, incorporating neuromusculoskeletal forward dynamics and dynamic optimization, for rendering simulated human movement. Models have been developed to simulate human gait (Pandy and Berme, 1987; Yamaguchi, 1990; Anderson and Pandy, 2001), jumping (Hatze, 1981; Pandy et al., 1990, 1992; Anderson and Pandy, 1999), and arm motions (Yamaguchi et al., 1995; Lemay and Crago, 1996). Elaborate representations of musculature, including muscle paths, musculo-tendon actuation, and excitation–contraction coupling, are usually featured in these models (Pandy, 2001). While these dynamic models seem to be able to generate physically more plausible synthetic motions, so far they have not allowed movement simulation of large-scale systems (with a large number of DOFs) in a reasonable time frame. The primary reason is that dynamic optimization or dynamic programming suffers from the "curse of dimensionality" (Bellman, 1957) — that the computational complexity increases exponentially with the number of variables included. An in-depth analysis of how the computational complexity can quickly become prohibitive in forward dynamics modeling was presented in Yamaguchi (1990). A computationally efficient forward dynamic model of arm motion developed by Yamaguchi et al. (1995), consisting of only 5 DOFs and 30 muscles, needed 3.5 h of CPU time to complete a

3-sec motion simulation. With perhaps the most sophisticated human walking model which possessed 23 DOFs and 54 muscles (Anderson and Pandy, 2001), simulation of a single gait cycle took about 10,000 h of CPU time to converge a solution. It is important to note that, since the number of muscles far exceeds the number of DOFs in a musculoskeletal system representation, a tremendous "overhead" computational cost comes with the attempt to increase the level of sophistication in simulated motion typically measured by the number of DOFs.

4. *Dynamic models without musculature.* It would seem sensible to exclude the musculature in order to enhance computational tractability and efficiency, particularly when quantitative insights into the neuromuscular control or musculoskeletal dynamics are not of primary interest. This is the rationale behind this fourth class of models sought to render dynamic motion simulation without involving the neuromuscular or musculoskeletal interactions. These models employ estimates of "high-level" dynamic or kinematic parameters to drive forward dynamics or forward kinematics procedures expeditiously. In that regard, three distinct approaches have been proposed to explore this idea. One approach popular in the computer animation field relies on heuristics or estimates of joint torques to simplify forward dynamics, thus effectively avoiding the muscle redundancy problem (Bruderlin and Calvert, 1989; Zordan and Hodgins, 1999). The validity of these heuristics or estimates nevertheless has not been empirically confirmed by systematic biomechanical studies. When human motion data are combined with simplified forward dynamics to add behavioral veracity, case-by-case "fine-tuning" is often a necessity. Another approach proposes the use of statistical or curve-fitting techniques to directly model measured body segment or joint angles. Such a seemingly straightforward approach, however, did not begin to flourish until the ability to acquire, process, and organize large, dense volume of 3D human motion data became mature. For example, a series of movement-prediction methods blending curve-fitting and optimization have been developed and refined by Ayoub and his associates (Ayoub et al., 1974; Hsiang and Ayoub, 1994; Lin et al., 1999). While the methods are general or readily expandable, the model validation and applicability have so far been restricted to planar sagitally symmetric lifting movements, due mainly to a lack of databases of more complex movements. The recent success of University of Michigan Human Motion Simulation Laboratory in establishing large databases of reaching (both seated and standing) and materials handling movements has allowed the development of a new functional regression model to statistically predict such motions (Faraway, 1997) and also to reveal which predictive factors are more dominant than others (Chaffin et al., 2000). It is, however, well recognized that in general statistical models have bounded predictive power, especially when extrapolated to novel, untested conditions. A third approach, named optimization-based differential inverse kinematics (ODIK) approach (Zhang et al., 1998), resolves the kinematic redundancy in the velocity domain efficiently through a parameterized pseudoinverse of the Jacobian matrix that maps the end-point position change to the joint angle changes. The parameters quantify how a motion is apportioned among various DOFs. The integration of the velocity–domain solution results in smooth joint angle profiles, which is not achievable using ordinary inverse kinematics described earlier. All three approaches have one property in common; that is, once the heuristics, coefficients, or parameters are known, movement profiles are delivered via efficient mathematical operations.

It should become clear from the listed items that human motion modeling research is presented with two major challenges. The first challenge is that the human motion modelers have to cope with two somewhat conflicting goals: one is to improve the level of physical realism in mathematical representations of the human musculoskeletal system; the other is to enhance the computational tractability and speed toward real-time simulation of complex motions of the system. As the level of realism increases, the number of variables involved in the modeling increases, and so does the level of difficulty in the quest for real-time simulation. In order words, there is an inherent trade-off between the biomechanical realism and computational efficiency (Zhang, 2001). The second challenge pertains to the acquisition of high-quality experimental data: human motion measurements for model development as well as model-prediction verification, and estimates of the geometry and properties of human neuromusculoskeletal system components for model construction. Ideally, data acquisition should be *in vivo*, noninvasive, and subject to minimal artifacts and interferences. It is, however, extremely difficult, with the current methodologies, to achieve these goals simultaneously.

10.3.2 Human Motion Data Acquisition for Model Development and Validation

There is a rich collection of reviews and tutorials on human motion data acquisition in the literature. For general principles and different hardware systems, the readers are referred to Allard et al. (1995). A summary of various human motion analysis methods in the context of occupational biomechanics is available in Chaffin et al. (1999). Winter's classic text (1990) also contains an excellent treatment on kinematic data processing. Here, a succinct critique of three human motion measurement systems is presented, with an emphasis on their potential in acquiring data for development as well as validation of high-realism digital human models.

Optoelectronic systems are the most commonly used systems for measuring human motions *in vivo*, and require the placement of markers on consistently identifiable and palpable body surface landmarks. Systems are called passive if the markers are light reflecting, active if light emitting. Optoelectronic systems have enjoyed great improvement in their efficacy in recent years (see Richard, 1999, for a survey), and now possess the needed accuracy and resolution for tracking the movement of markers on individual finger segments or spinal processes. Figure 10.3 shows the marker placement schemes adopted in a study of vertebral kinematics during manual materials handling (Figure 10.3a), and a study of finger segmental motion during manipulative tasks (Figure 10.3b). Figure 10.4 displays the respective motions recorded and visualized as "stick-figure" representations by a Vicon motion capture system (Oxford Metrics, U.K.), one of the passive optoelectronic systems. Motion data containing such levels of detail facilitate building biomechanical models with empirically verifiable realism and subtleties.

Two requirements limiting the applicability of optoelectronic systems are that a marker must be seen by at least two cameras simultaneously in order to determine the marker coordinates in a 3D space, and that coordinates of three noncollinear markers attached to a body segment must be known in order to

(a)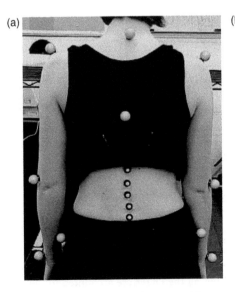
(b)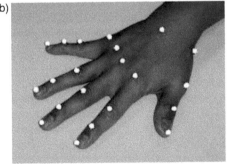

FIGURE 10.3 Marker placement schemes used in measuring (a) manual materials handling motions and (b) hand manipulative motions by an optoelectronic system.

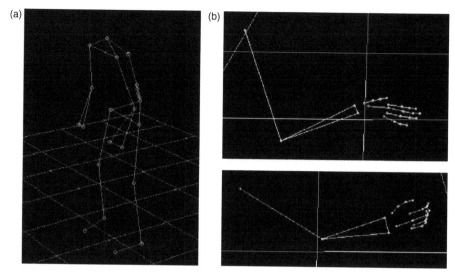

FIGURE 10.4 Motions recorded and visualized as "stick-figures" by an optoelectronic system: (a) manual materials handling motion; (b) hand manipulative motion. (From Zhang et al. *International Journal of Industrial Performance*, Elsevier, 2004. With permission.)

determine the position and orientation of the segment. *Nonoptical systems* including electromagnetic systems and ultrasonic systems are meant to do away with the visibility requirement, which may be desirable, for example, in measuring the lumbar motion in a seat with an obstructing seat back. Electromagnetic systems also have the advantage of requiring significantly fewer sensors because each electromagnetic sensor registers both position and orientation, measuring all 6 DOFs of the body segment to which the sensor is attached. Figure 10.5 illustrates a MotionStar electromagnetic motion capture system (Ascension Technology, Vermont, U.S.A.) being used to measure shoulder kinematics *in vivo*. The sensor size and the vulnerability to inference from metals are two noted limitations of the electromagnetic systems. Ultrasonic systems have only been devised to measure relatively simple motion parameters (Berners et al., 1995; Huitema et al., 2002), and have not enjoyed much success in measuring whole-body more complex human motions with a large number of sensors.

Both optoelectronic and nonoptical systems aforementioned are surface measurement systems in that the markers or sensors are placed on the body surface. Of real interest is nevertheless the underlying skeletal kinematics (Cappozzo et al., 1996; Alexander and Andriacchi, 2001). There are well-established methods for extracting the rigid-body kinematic information from the surface measurements, but it would be much desirable to directly measure the body position and orientation. This is the need that *medical imaging systems*, including radiography, computer tomography, and magnetic resonance imaging, are expected to fulfill. So far, the measures obtained through these medical imaging systems have been restricted to static position variables. Whether kinematic information estimated from static measures can be used to compare or verify dynamic movement quantities remains questionable.

10.3.3 High-Fidelity Biomechanical Linkage Representation as the Basis for Modeling

Digital human modeling usually begins with constructing a biomechanical linkage representation of the human body, using motion measurements directly or existing anthropometric databases. The fidelity of this linkage representation largely determines the physical realism of a digital human model. The issue of fidelity is embodied by several specific questions: whether all the DOFs are accounted for, whether the centers of rotation (CORs) or axes of rotation (AORs) are correctly identified, and whether the link segment dimensions are accurately represented.

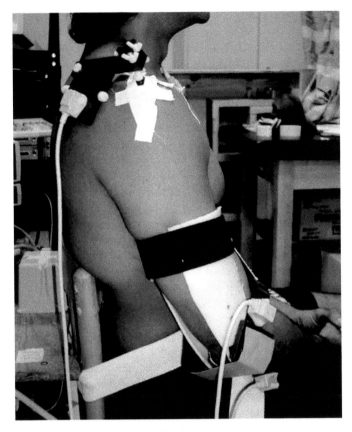

FIGURE 10.5 An electromagnetic system being used to measure shoulder and upper extremity motions. (Courtesy of Richard Hughes, University of Michigan Orthopaedic Surgery Department. With permission.)

It is acknowledged that the human skeletal system is not a perfectly rigid mechanical linkage articulated by idealized spherical or axial joints. But an optimal linkage representation should be such that (1) its joint centers or central axes most closely approximate the centers or axes of the relative rotations between two neighboring body segments, and (2) the link segment lengths (i.e., distances between two adjacent joints) vary minimally over a certain time window or range of motion (Zhang et al., 2004). These two criteria, however, are not easily satisfied. For instance, if a link connects two non-adjacent joint centers or if a certain DOF is omitted, the link length can change considerably beyond what the normal joint elasticity or laxity would allow. The use of the optimal linkage system stated earlier as the basis can alleviate the problems such as nonuniformity in defining and deriving body segment dimensions, or disparity between structural (or static) and functional (or dynamic) data. More importantly, application of computational procedures based on rigid-body mechanics in modeling would be best served by such a linkage representation. Nonrigidity in body link segments is one major source of error leading to computational problems such as end-point offset in forward kinematics (Zhang, 2002), and inverse dynamics failure (Risher et al., 1997) in biomechanical modeling.

Existing human body segment length databases, unfortunately, were not established in accordance with a linkage representation meeting the above two criteria. Nor can they be directly used to form such a representation. Conventional body link length measurements on living subjects are conducted in a static manner with reference to bony landmarks (Roebuck et al., 1975; Kroemer, 1989; Roebuck, 1995). Identification of joint centers or centers of rotation, on the other hand, necessitates recording and interpreting dynamic movement information. In fact, one misconception in developing a linkage model of the human body is treating joint centers, which are kinematic entities, as static anatomical entities. Joint centers and corresponding bony landmarks do not coincide with each other, and thus two kinds of segment length measures which are defined, respectively, as link lengths and bone lengths (Chaffin et al., 1999), are also different from each other. A very common misconstructed linkage is one formed by directly using bone length data or by connecting measured surface markers on bony landmarks (as the stick-figures shown in Figure 10.4).

Therefore, deriving centers of rotation or joint centers from surface marker or sensor movement data is often a necessary and important step in constructing a linkage representation as the basis for subsequent biomechanical modeling or analysis. There has been a persisting interest in inventing and refining computational methods for COR derivation (Spiegelman and Woo, 1987; Veldpaus et al., 1988; Water and Panjabi, 1988; Woltring, 1990; Crisco et al., 1994; Halvorsen et al., 1999; Gamage and Lasenby, 2002). Most of these methods, however, require a minimum number of markers affixed to each body segment — two for planar motions and three (noncollinear) for 3D motions (or alternatively, one 6-DOF electromagnetic sensor per segment). Fulfilling this minimum requirement can be difficult or impossible for many areas of the body, such as the fingers, vertebrae, and even the upper arm, due to spatial or morphological constraints. Yet, it is exactly in these areas that the existing linkage representations need improvement. Recently, facilitated by the contemporary surface-based motion capture systems with much improved resolution, several novel approaches to determining the CORs for inter-finger-segmental and inter-lumbar-vertebral rotations have been proposed (Zhang and Xiong, 2003; Zhang et al., 2003; Cerveri et al., 2004). Zhang et al. created mathematical models to clarify the relationships between external surface marker movements and the internal skeletal movements, and then used the models to guide or better condition the algorithms to locate the CORs (Zhang and Xiong, 2003; Zhang et al., 2003). Figure 10.6 illustrates the difference and relation between the external measurable marker-defined inter-segmental angles (ϕ_i) and the internal vertebral rotation angles (θ_i) for the lumbar spine. Figure 10.7 depicts the relation between the marker-defined surface links (l_k^i) and the internal links (L_k^i) as the surface markers excurse around the finger joints. Cerveri et al. (2004) employed evolutionary optimization to compute the intervertebral CORs for the lower spine, and produced results that were in agreement with those from the earlier work by Zhang and Xiong (2003). These newest methods demonstrate the viability of building linkage representation components that capture perhaps the smallest motions measurable by the current technology. These

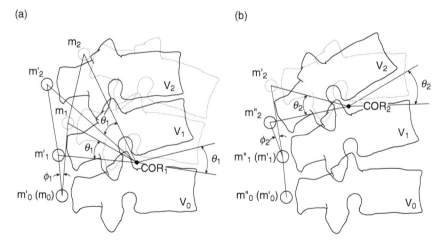

FIGURE 10.6 A model of the relation between the externally estimated intersegmental motion and internal vertebral rotation. The model incorporates three segments (V_0, V_1, and V_2) and three surface markers (m_0, m_1, and m_2) affixed to the segments at respective spinal processes. The three segment-marker pairs undergo two consecutive rotations: (a) a rotation of V_1 and V_2 together by θ_1 with respect to V_0; (b) a rotation of V_2 by θ_2 with respect to V_1. The two rotations change the marker set from $[m_0, m_1, m_2]$ to $[m_0', m_1', m_2']$, and then to $[m_0'', m_1'', m_2'']$. Gray and dark lines illustrate before and after rotation, respectively. (From Zhang et al. *Journal of Biomechanics*, Elsevier, 2001. With permission.)

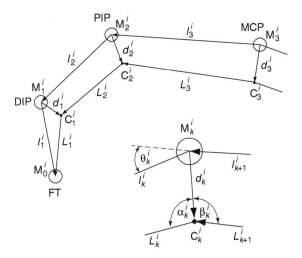

FIGURE 10.7 A model of the geometric relationship between surface markers and internal CORs during the flexion-extension of a digit $i (i = 2-5)$. The model comprises four markers M^i_{0-3} at the finger tip (FT), distal interphalangeal (DIP), proximal interphalangeal (PIP), and metacarpophalangeal (MCP) joints, and three CORs C^i_{1-3} of DIP, PIP, and MCP joints. During the flexion-extension of joint k, l^i_k changes its length and orientation, L^i_k changes its orientation while maintaining a constant length, and $d^i_k(t)$ only rotates around C^i_k, changing its orientation α^i_k relative to L^i_k and β^i_k relative to L^i_{k+1}. (From Zhang et al. *Journal of Biomechanics*, Elsevier, 2001. With permission.)

components constitute the integral parts of a more detailed framework for a realistic digital human model.

10.3.4 Data-Supported Modeling Approaches to Achieving Physical Realism and Computational Efficiency

The latest systematic experimental studies, such as the ones conducted by University of Michigan Human Motion Simulation Laboratory, have yielded a tremendous trove of data along with a body of new knowledge on human movement. This accumulated wealth of data and knowledge can be taken advantage of, not only to develop and validate digital human models, but also to guide, simplify, and accelerate the computational algorithms that drive the models. A collection of modeling approaches have emerged in the past few years and demonstrated great potential for achieving both physical realism and computational efficiency. These approaches belong to the class of dynamic models without musculature according to the above taxonomy, and are supported by robust human motion databases. They either contain parameters that can be empirically estimated from the data, or make optimal modification to known motions (measured or predicted by other models) for prediction in untested conditions.

One such model is the *functional regression model* proposed by Faraway (1997) for predicting target-directed reaching motions. The model takes the following form:

$$\theta(t) = \beta_0(t) + C_\chi \beta_\chi(t) + C_y \beta_y(t) + C_z \beta_z(t)$$
$$+ C_\chi C_y \beta_{\chi y}(t) + C_y C_z \beta_{yz}(t) + C_z C_\chi \beta_{z\chi}(t)$$
$$+ C_\chi^2 \beta_\chi^2(t) + C_y^2 \beta_y^2(t) + C_z^2 \beta_z^2(t) + D$$

where $\theta(t)$ is the predicted joint motion over time, C_χ, C_y, and C_z are target coordinates, $\beta(t)$ are parametric functions to be estimated in the regression, and D represents demographic variables (e.g., age, stature, gender) that modulate the predictions.

This quadratic regression model was found to account for approximately 80% of the joint angle variation. The same methodology was also employed by Chaffin et al. (2000) to discover the dominant role that stature plays in seated reaching motions, compared to gender and age.

The *optimization-based differential inverse kinematics (ODIK) model* (Zhang et al., 1998) hinges upon the following differential kinematic relationship for an *N*-segment *M*-DOF chain-like linkage:

$$\dot{P} = J(\Theta)\dot{\Theta}$$

where $\dot{P}$ is the end-point position change in three dimensions, $\dot{\Theta}$ is a vector of *M*-dimensions representing the joint angle changes, and J is the Jacobian matrix. The system is redundant when the dimension of $\dot{\Theta}$ is greater than that of $\dot{P}$. By taking a weighted pseudoinverse, a unique solution of $\dot{\Theta}$ can be conveniently derived as

$$\dot{\Theta} = W^{-1}[JW^{-1}]^{\dot{P}}$$

where # symbolizes a pseudoinverse and W is an $M \times M$ diagonal weighting matrix: $W = \text{diag}(w_1, w_2, \ldots, w_M)$. Given a specification of end-point (e.g., the hand) trajectory, the weighting parameters w_i can quickly determine the joint motion. They quantify the motion apportionment among the DOFs involved.

This model structure allows empirical estimation of the parameters in a data-prediction fitting process in which a series of hypotheses can be tested to progressively reduce the number of variables thus accelerating the algorithm. For example, a hypothesized time-invariant weighting strategy would mean a reduction of variable number from *MT* (*T* is the number of time frames) to *M*; a hypothesized presence of motion synergy or coordination between certain DOFs (Braido and Zhang, 2004), if tested true with large DOF systems, can further reduce the variables dealt with in the modeling. The ODIK modeling approaches along with the variable reduction scheme have been applied to generate time-efficient simulation of in-vehicle seated reaching motions (Zhang and Chaffin, 2000) and symmetric planar lifting motions (Zhang et al., 2000). The approach, however, has not been applied to modeling larger systems with more than 10 DOFs.

Most recently, a *forward dynamic model incorporating system identification* was proposed for predicting finger flexion–extension motions (Lee et al., 2005). The model drives a 12-link system of digits 2–5 by three torque generators within each digit at the distal interphalangeal (DIP), proximal interphalangeal (PIP), and metacarpophalangeal (MCP) joints, according to the following equation of motion:

$$T^i = -K_p^i(\Theta^i - \Theta_r^i) - K_d^i(\dot{\Theta}^i - \dot{\Theta}_r^i), \quad \text{subject to} \quad T \leq M_{\text{max}}^i$$

where T^i is a 3×1 joint torque vector for digit *i*, Θ^i a joint angle vector for digit *i*, K_p^i a 3×3 diagonal proportional feedback gain matrix, K_d^i a 3×3 diagonal derivative feedback gain matrix, and M_{max}^i a maximum torque vector for digit *i*. The parameter values of the feedback controller, K_p^i and K_d^i, characterize two movement-specific properties of the system, dynamic joint stiffness and damping, respectively. The M_{max}^i poses a strength constraint — a limit to the maximum joint torque that can be produced by the finger muscles. These parameters are empirically estimated in an optimization procedure minimizing the discrepancy between the model-predicted and measured movement profiles.

Once the parameters are given or expressed as functions of anthropometric properties or movement-specific characteristics (e.g., movement speed), the model allows a computationally efficient generation of motion, owing to a simple linear-time-invariant control scheme. What distinguishes this approach from previous torque-based forward dynamic modeling approaches (Bruderlin and Calvert, 1989; Zordan and Hodgins, 1999) is that the model parameters here can be empirically estimated and physically interpreted. They represent system dynamic properties and also control specific kinematic features such as peak velocity and total movement time.

When a large and robust human motion database is available and digitally stored for easy access, prediction may simply mean retrieving the closest match of input condition(s) and if necessarily making some modification to the retried. The *motion modification algorithm* proposed by Park et al. (2004) addresses this particular aspect. The algorithm modifies an original motion according to a minimal dissimilarity principle while satisfying the constraints such as new input conditions and joint range of motion limits. The minimal dissimilarly principle is formulated as an optimization problem:

$$\text{Minimize:} \quad \int_0^T (\dot{\hat{\theta}}_j(t) - \dot{\tilde{\theta}}_j(t))^2 \, dt$$

where $\dot{\hat{\theta}}_j(t)$ and $\dot{\tilde{\theta}}_j(t)$ are the first time-derivatives of the new and original motions, respectively. The minimization is formulated in the velocity domain to best retain the kinematical characteristics of the original motions, and is solved efficiently by simple calculus. The algorithm has been tested against seated reaching motion and whole-body load-transferring motion data, and shown to have prediction accuracy comparable to inherent variability. Conceivably, this modification algorithm can be extended to interpolate or extrapolate motions predicted by other models.

10.4 Future Challenges

Despite the significant strides made in advancing digital human modeling as a technology as well as a basic research area, there are remaining challenges that motivate future investigative and development efforts. First, the prevailing modeling framework is deterministic and cannot reconcile the goal-directed strategy (e.g., optimality) and the richness of motor variability. The models developed to date have done well in explaining "average" movement behavior or responses but not in capturing the intra- or interperson variability. Stochastic modeling may hold the key to unlocking the "mystery" of variability and individual differences. Second, although models that possess large DOFs have been developed in the past few years, many were based on data of limited movement varieties in which not all DOFs were activated. For instance, there are "armless" models of hand manipulative motion models, whereas most reaching motion models have motionless hands or fingers. Seamless integration of these models into more versatile digital human modeling simulation systems warrants further investigative efforts. Third, dynamic measurement of true skeletal movement and musculoskeletal system properties *in vivo* presents an enduring challenge, especially when abnormality or pathology is involved. Finally, modeling the repetition or chronic time effect, for example, muscle fatigue or tissue degeneration, on simulated motions still remains a tantalizing goal.

10.5 Summary

Digital human modeling faces new challenges and opportunities in the digital age. The challenges arise from the fact that human movement modeling for the digital-age ergonomics and system design applications has to meet heightened demand on model realism and computational efficiency. There nevertheless exists an inherent trade-off between biomechanical realism and computational efficiency, which continues to be a major obstacle. The opportunities are afforded by advances in the techniques for measuring and modeling humans in physical acts, computing power, and knowledge about human motor behavior. These advances have enabled more aggressive pursuit of a new generation of digital "human surrogates" that will increasingly populate in the computer-aided ergonomics world.

References

Alexander, E.J. and Andriacchi, T.P., 2001. Correcting for deformation in skin-based marker systems. *Journal of Biomechanics* 34, 355−561.

Allard, P., Stokes, I.A.F., and Blanchi, J. Eds., 1995. *Three-Dimensional Analysis of Human Movement*, Human Kinetics, Champaign, IL.

Anderson, F.C. and Pandy, M.G., 1999. A dynamic optimization solution for vertical jumping in three dimensions. *Computer Methods in Biomechanics and Biomechanical Engineering*, 2, 201–231.

Anderson, F.C. and Pandy, M.G., 2001. Static and dynamic optimization solutions for gait are practically equivalent. *Journal of Biomechanics*, 34, 153–161.

Ayoub, M.A., Ayoub, M.M., and Walvekar, A.G., 1974. A biomechanical model for the upper extremity using optimization techniques. *Human Factors*, 16, 585–594.

Badler, N.I., Phillips, C.B., and Webber, B.L., 1993. *Simulating Humans: Computer Graphics, Animation, and Control*, Oxford University Press, Oxford.

Bean, J.C., Chaffin, D.B., and Schultz, A.B., 1988. Biomechanical model calculation of muscle contraction forces: a double linear programming method. *Journal of Biomechanics*, 21, 59–66.

Bellman, R.E., 1957. *Dynamic Programming*, Princeton University Press, Princeton, NJ.

Berners A.C., Webster, J.G., Worringham, C.J., and Stelmach, G.E., 1995. An ultrasonic time-of-flight system for human movement measurement. *Physiological Measurement*, 16, 203–211.

Bernstein, N., 1967. *The Coordination and Regulation of Movements*, Pergamon Press, Oxford.

Bradio, P. and Zhang, X., 2004. Quantitative analysis of finger motion coordination in hand manipulative and gestic acts. *Human Movement Science*, 22, 661–678.

Bruderlin, A. and Calvert, T.W., 1989. Goal-directed, dynamic animation of human walking. *Computer Graphics*, 23, 233–242.

Byun, S.N., 1991. A Computer Simulation Using a Multivariate Biomechanical Posture Prediction Model for Manual Material Handling Tasks. Unpublished doctoral dissertation, The University of Michigan, Ann Arbor, MI.

Cappozzo, A., Catani, F., Leardini, A., Benedetti, M.G., and Della Croce, U., 1996. Position and orientation in space of bones during movement: experimental artefacts. *Clinical Biomechanics*, 11, 90–100.

Cerveri P., Pedotti, A., and Ferrigno, G., 2004. Evolutionary optimization for robust hierarchical computation of the rotation centres of kinematic chains from reduced ranges of motion the lower spine case. *Journal of Biomechanics*, 37, 1881–1890.

Chaffin, D.B., 2001. *Digital Human Modeling for Vehicle and Workplace Design*, SAE, Warrendale, PA.

Chaffin, D.B., Andersson, G.B.J., and Martin, B.J., 1999. *Occupational Biomechanics*, John Wiley, New York.

Chaffin, D.B. and Erig, M., 1991. Three-dimensional biomechanical static strength prediction model sensitivity to postural and anthropometric inaccuracies. *IIE Transactions*, 23, 125–227.

Chaffin, D.B., Faraway, J.J., Zhang, X., and Woolley, C.B., 2000. Stature, age, and gender effects on reach motion postures. *Human Factors*, 42, 408–420.

Crisco, J.J., Chen, X., Panjabi, M.M., Wolfe, S.W., 1994. Optimal marker placement for calculating the instantaneous center of rotation. *Journal of Biomechanics*, 27, 1183–1187.

Crowninshield, R.D. and Brand, R.A., 1981. A physiologically based criterion of muscle force prediction in locomotion. *Journal of Biomechanics*, 14, 793–801.

Delleman, N.J., Haslegrave, C.M., and Chaffin, D.B. Eds., 2004. *Working Postures and Movements: Tools for Evaluation and Engineering*, CRC Press, Boca Raton, FL.

Desjardins, P., Plamondon, A., and Gagnon, M., 1998. Sensitivity analysis of segment models to estimate the net reaction moments at the L5/S1 joint in lifting. *Medical Engineering & Physics*, 20, 153–158.

Dul, J., Johnson, G.E., Shiavi, R., and Townsend, M.A., 1984. Muscle synergism-II. A minimum-fatigue criterion for load sharing between synergistic muscles. *Journal of Biomechanics*, 17, 675–684.

Dysart, M.J. and Woldstad, J.C., 1996. Posture prediction for static sagittal-plane lifting. *Journal of Biomechanics*, 29, 1393–1397.

Faraway, J.J., 1997. Regression analysis for a functional response. *Technometrics*, 39, 254–261.

Gamage, S.S. and Lasenby, J., 2002. New least squares solutions for estimating the average centre of rotation and the axis of rotation. *Journal of Biomechanics*, 35, 87–93.

Halvorsen, K., Lesser, M., and Lundberg, A., 1999. A new method for estimating the axis of rotation and the center of rotation. *Journal of Biomechanics*, 32, 1221–1227.

Hardt, D.E., 1978. Determining muscle forces in the leg during normal human walking — an application and evaluation of optimization methods. *Journal of Biomechanical Engineering*, 100, 72–78.

Hatze, H., 1981. A comprehensive model for human motion simulation and its application to the take-off phase of the long jump. *Journal of Biomechanics*, 14, 135–142.

Hsiang, M.S. and Ayoub, M.M., 1994. Development of methodology in biomechanical simulation of manual lifting. *International Journal of Industrial Ergonomics*, 13, 271–288.

Huitema, R.B., Hof, A.L., and Postema, K., 2002. Ultrasonic motion analysis system — measurement of temporal and spatial gait parameters. *Journal of Biomechanics*, 35, 837–842.

Kroemer, K.H.E., 1989. Engineering anthropometry. *Ergonomics*, 32, 767–784.

Lemay, M.A. and Crago, P.E., 1996. A dynamic model for simulating movements of the elbow, forearm, and wrist. *Journal of Biomechanics*, 29, 1319–1330.

Lee, S.-W. and Zhang, X., 2005. Dynamic modeling and system identification of finger movements. In *Proceedings of the XXth Congress of the International Society of Biomechanics & 29th American Society of Biomechanics Meeting*.

Lin, C.J., Ayoub, M.M., and Bernard, T.M., 1999. Computer motion simulation for sagittal plane lifting activities. *International Journal of Industrial Ergonomics*, 24, 141–155.

McDaniel, J.W., 1990. Models for ergonomic analysis and design: COMBIMAN and CREWCHIEF. In *Computer-Aided Ergonomics*, W. Karwowski, A.M. Genaidy, and S.S. Asfour, Eds., pp. 138–156, Taylor & Francis, New York.

Morrissey, M., 1998. Human-centric design. *Mechanical Engineering*, 120(7), 60–62.

Pandy, M.G., 2001. Computer modeling and simulation of human movement. *Annual Review of Biomedical Engineering*, 3, 245–273.

Pandy, M.G. and Berme, N., 1987. Synthesis of human walking: a three-dimensional model for single support. Part 2: pathological gait. In *Proceedings of ASME Winter Annual Meeting*, American Society of Mechanical Engineering. New York.

Pandy, M.G., Zajac, F.E., Sim, E., and Levine, W.S., 1990. An optimal control model for maximum-height human jumping. *Journal of Biomechanics*, 23, 1185–1198.

Pandy, M.G., Anderson, F.C., and Hull, D.G., 1992. A parameter optimization approach for the optimization control of large scale musculoskeletal systems. *Journal of Biomechanical Engineering*, 114, 450–460.

Park, K.S., 1973. Computerized Simulation Model of Posture During Manual Material Handling. Unpublished doctoral dissertation, The University of Michigan, Ann Arbor, MI.

Park, W., Chaffin, D.B., and Martin, B.J., 2004. Toward memory-based human motion simulation: development and validation of a motion modification algorithm. *IEEE Transactions on Systems, Man, and Cybernetics — Part A*, 34, 376–386.

Porter, J.M., Freer, M., Case, K., and Bonney, M.C., 1995. Computer aided ergonomics and workspace design. In *Evaluation of Human Work: A Practical Ergonomics Methodology* J.R. Wilson and E.N. Corlett, Eds., pp. 574–620, Taylor & Francis, London.

Richards, J.G., 1999. The measurement of human motion: a comparison of commercially available systems. *Human Movement Science*, 18, 589–602.

Riemer, R., Lee, S.-W., and Zhang, X., 2004. Full body inverse dynamics solutions: an error analysis and a hybrid approach. In *Proceedings of American Society of Biomechanics Annual Meeting*, American Society of Biomechanics.

Risher, D.W., Schutte, L.M., and Runge, C.F., 1997. The use of inverse dynamics solutions in direct dynamics simulations. *Journal of Biomechanical Engineering*, 119, 417–422.

Roebuck, J.A., 1995. *Anthropometric Methods: Designing to Fit the Human Body*, Human Factors Society, Santa Monica, CA.

Roebuck, J.A., Kroemer, K.H.E., and Thomson, W.G., 1975. *Engineering Anthropometry Methods*, Wiley-Interscience, New York.

Ryan, P.W., 1970. *Cockpit Geometry Evaluation* (Joint Army–Navy Aircraft Instrumentation Research Report 700201), The Boeing Company, Seattle, WA.

Seireg, A. and Arvikar, R.J., 1975. The prediction of muscle load sharing and joint forces in the lower extremity during walking. *Journal of Biomechanics*, 8, 89–102.

Spiegelman, J.J. and Woo, S.L.-Y., 1987. A rigid-body method for finding centers of rotation and angular displacements of planar joint motion. *Journal of Biomechanics*, 20, 715–721.

Veldpaus, F.E., Woltring, H.J., and Dortmans, L.J., 1988. A least-squares algorithm for the equiform transformation from spatial marker co-ordinates. *Journal of Biomechanics*, 21, 45–54.

Water, S.D. and Panjabi, M.M., 1988. Experimental errors in the observation of body joint kinematics. *Technometrics*, 30, 71–78.

Winter, D.A., 1990. *Biomechanics and Motor Control of Human Movement*, John Wiley, New York.

Witkins, A., Fleischer, K., and Barr, A., 1987. Energy constraints on parameterized models. *ACM Computer Graphics*, 21, 225–232.

Woltring, H., 1990. Data processing and error analysis. In *Biomechanics of Human Movement* N. Berme and A. Cappozzo, Eds., pp. 203–237, Bertec Corporation, Worthington, OH.

Yamaguchi, G.T. (1990). Performing whole-body simulation of gait with 3-D, dynamic musculoskeletal models. In *Multiple Muscle Systems: Biomechanics and Movement Organization* Eds. J.M. Winters and S. L-Y. Woo, pp. 663–679, Springer-Verlag, New York.

Yamaguchi, G.T., Moran, D.W., and Si, J., 1995. A computationally efficient method for solving the redundant problem in biomechanics. *Journal of Biomechanics*, 8, 999–1005.

Zhang, X., 2001. Biomechanical realism versus algorithmic efficiency: a trade-off in human motion simulation and modeling. *SAE Transaction*, 110 (6), 2184–2191.

Zhang, X., 2002. Deformation of angle profiles in forward kinematics for nullifying end-point offset while preserving movement properties. *ASME Journal of Biomechanical Engineering*, 124, 490–495.

Zhang, X. and Chaffin, D.B., 2000. A three-dimensional dynamic posture prediction model for simulating in-vehicle seated reaching movements: development and validation. *Ergonomics*, 43, 1314–1330.

Zhang, X. and Xiong, J., 2003. Model-guided derivation of lumbar vertebral kinematics in vivo reveals the difference between external marker-defined and internal segmental rotations. *Journal of Biomechanics*, 36, 9–17.

Zhang, X., Kuo, A.D., and Chaffin, D.B., 1998. Optimization-based differential kinematic modeling exhibits a velocity-control strategy for dynamic posture determination in seated reaching movements. *Journal of Biomechanics*, 31, 1035–1042.

Zhang, X., Nussbaum, M.A., and Chaffin, D.B., 2000. Back lift versus leg lift: an index and visualization of dynamic lifting strategies. *Journal of Biomechanics*, 33, 777–782.

Zhang, X., Lee, S.-W., and Braido, P., 2003. Determining finger segmental centers of rotation in flexion-extension based on surface marker measurement. *Journal of Biomechanics*, 36, 1097–1112.

Zhang, X., Lee, S.-W., and Braido, P., 2004. Towards an integrated high-fidelity linkage representation of the human skeletal system based on surface measurement. *International Journal of Industrial Ergonomics* (Special Issue on Anthropometrics and Disability), 33, 215–227.

Zhao, J. and Badler, N.I., 1994. Inverse kinematics positioning using nonlinear programming for highly articulated figures. *ACM Transaction on Graphics*, 13, 313–336.

Zordan, V.B. and Hodgins, J.K., 1999. Tracking and modifying upper-body human motion data with dynamic simulation. In *Proceedings of Eurographic Workshop on Computer Animation and Simulation*.

11

Design for Ergonomics: Facilities Planning*

Stephan Konz
Kansas State University

11.1 Introduction

Figure 11.1 shows a schematic workstation. The total workstation consists of an operator, one or more machines, energy and information input, energy and information output, material input and output, and input and output storage. The ergonomist or engineer needs to decide what tasks are done by people and what tasks are done by machines. Then, while designing the physical workstation, the physical and information flows need to be compared with other workstations. Finally, the environment (lighting/illumination, noise, climate, chemical) needs to be designed.

In manufacturing (both components and assembly), movement of material tends to be important. However, service activities (offices, toilets, eating areas) tend to emphasize movement of information or even convenience/esthetics, rather than movement of physical items.

11.2 Manufacturing Workstations

The first step is to decide the allocation of tasks between the operator and the machine or machines.

In general, people are expensive and machines are cheap. For example, most manufacturing workers will receive a wage of at least $10/h — often $20 or $30/h. But, in addition, there are fringe costs

---

*The material in this chapter is covered in more detail in Hanna and Konz (2004) and Konz and Johnson (2004).

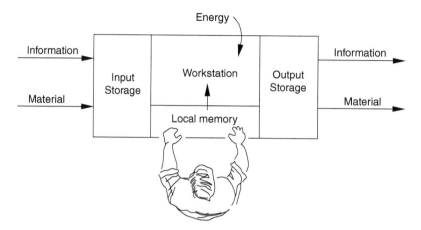

FIGURE 11.1 Schematic workstation showing both material and information flow into and from a workstation. They are stored temporarily in input and output storage. They are transformed in the workstation, using energy and local memory. The workstation is located in a designed environment (lighting, noise, climate, chemicals). (From Hanna, S. and Konz, S. 2004. *Facility Design and Engineering*, 3rd ed., Holcomb Hathaway, Scottsdale, AZ. With permission.)

(vacations, holidays, coffee breaks, medical insurance, etc.), which add 25 to 60% to the wage, making the cost of labor considerably higher.

In general, machines are cheap. Assume a fork truck or machine tool costs $18,000 and has an economic life of 10 yr. Further assume a single shift operation of 1800 h/yr. Thus, the capital cost of $18,000 can be spread over $1800 \times 10 = 18,000$ h or $1/h. A machine with a capital cost of $1800 would only cost $0.10/h. Utility costs are primarily power, of which most is electricity. At $0.08/ kWh, even a large motor (say 1 hp [746 W]) uses only $0.05/h. Maintenance varies but few machines require $1000/yr for maintenance — that is, $0.50/h. Thus, few machines will cost over $1/h while the person is 10 or 20 times higher.

Thus, the design should emphasize maximizing the utilization/productivity of the expensive component, the person, and not be especially concerned about the possible idle time of one or more machines or workbenches.

The following will consider the design of the job at the station, the relation of the station to other stations, and the physical design of the station.

11.2.1 Design of the Job

Table 11.1 gives eight guidelines concerning the design of the job.

TABLE 11.1   Guidelines for Job Design

Use specialization even though it sacrifices versatility
Make several identical items at the same time
Combine operations and functions
Minimize idle capacity
Use filler jobs or filler people
Vary environmental stimulation inversely with task stimulation
Reduce fatigue
Communicate information

Source: From Hanna, S. and Konz, S. 2004. *Facility Design and Engineering*, 3rd ed., Holcomb Hathaway, Scottsdale, AZ. With permission.

1. *Use specialization even though it sacrifices versatility.* Specialization has been the key to progress. It tends to decrease cost per unit and to increase quality. The primary problem is to get sufficient sales to justify the specialization. The fast-food business is an example of specialization. The firms offer only a few standard items and depend on low prices and quick service to attract enough customers.

   Specialization can come from special-purpose equipment, material, labor, and organization. The disadvantages of specialization can be reduced by low setup costs (allowing the firm to quickly shift to another product) and cross-trained workers (allowing shifting people between tasks/jobs).

2. *Make several identical items at the same time.* Tasks can be divided into three states: get ready, do, and put away.

   One alternative is to increase the (one-replication) get ready time so as to decrease the (multiple-replication) do time. For example, the U.S. Post Office forwards mail to people who have moved. When a piece of mail with an incorrect address comes to the carrier, the carrier puts it aside for a special operator. This operator has a disk for each carrier. The operator enters the first three letters of the last name and the last two numbers of the street address onto the disk. The computer then prints out the full forwarding address on a peel-off label that the operator puts on the mail.

   Another alternative is to increase the lot size so that the get ready and put away times are prorated over more units. Consider a person making hamburgers. If he or she makes two or four hamburgers at a time, then the get ready and put away of handtools are prorated over more units. (McDonalds uses specialized equipment that cooks a hamburger on both sides simultaneously.)

3. *Combine operations and functions.* Multifunction can apply to materials, equipment, and information. A multifunction material would be a compound that cleans and waxes at the same time; common applications are automobiles and floors. Multifunction equipment examples are a lathe form tool that forms several surfaces simultaneously and the nut drivers that race car mechanics use to turn all nuts on a wheel simultaneously. Multifunction information would be having your word processor remember the address you key for a letter so that the address does not need to be rekeyed for the envelope.

4. *Minimize idle capacity.* Minimizing idle capacity has two divisions: fixed costs and variable costs.

   Fixed costs. In many situations, the major cost of the facility and personnel varies little with use of the resource. Some examples are capital cost of a factory not varying whether the facility is used 8 or 16 h/day; a cook receiving the same pay whether there are 50 or 100 customers; an airplane flight costing the same whether there are 150 or 200 passengers. If there is better utilization, the incremental cost to the firm is very low. Four solutions (which basically rearrange customer service) are operate more hours per year, use pools, revise schedules, and encourage off-peak use.

   More hours per year. Operating more hours per year may spread a constant number of customers over more hours, thus requiring fewer/smaller facilities. The longer hours also may attract more customers.

   Use pools. Pools of facilities are based on the concept that peaks (or valleys) in one group do not coincide with peaks (or valleys) in another group. Sharing computer support staff between departments is one example. Rental cars are an example between organizations.

   Revise schedules. You may be able to influence the work schedule. A common technique is to divide the month into alphabetic sections. Instead of processing everyone on the first of the month, customers whose last name begins with A are done on day 1, those with B and C are done on day 2, etc. Custodial work can be done during the day instead of the night, making it easier to recruit custodial staff and saving lighting expense.

   Encourage off-peak use. You can encourage customers to be serviced off-peak; that is, in the valleys. Airlines and hotels charge less for use during slack times. Phone companies charge less for phone use during nonbusiness hours.

Variable costs. The general concept is to improve utilization among members of a "team" (where team includes equipment and facilities as well as people). As pointed out above, the cost per hour of people tends to be much higher than the cost of equipment. Three approaches are duplicate components, idle low-cost components, and not one on one.

Duplicate components. If people are cross-trained, a worker can replace a "failed" worker (e.g., one who is absent). If a worker uses a power tool or battery, have a backup tool (or battery) in case the regular one fails.

Idle low-cost components. An example is two Coke dispensers at McDonalds. McDonalds is more concerned with minimizing time for the server (and customer) than increasing utilization of inexpensive dispensers. Duplicate service lines for a single-queue server permit faster customer service time. For example, a bank drive-in teller could serve two lanes. After the first customer is served, the teller can begin serving a customer in the other lane while the first lane is "unloaded" and then "loaded" with a new customer.

Not one on one. Earlier it was assumed that there was one machine for one worker. But with increased computerization, there can be more than one machine per person. In addition, the machines per person need not be an integer. For example, assign two workers to three machines — yielding three machines for every two workers or two or three worker per machine. Machines (and people) also can be shared between departments. Thus, a computer expert or a bookkeeper could be shared between departments.

5. *Use filler jobs or filler people.* The challenge is to match worker time to job requirement time. What does a cook do while the food cooks? Two strategies to minimize idle time are (1) adjust the workload to a fixed workforce (filler jobs) and (2) adjust the workforce to a fixed workload (filler people).

Filler jobs (adjusting the workload). It is difficult to find an additional job when the idle time is in short segments. Thus, try to concentrate the idle time. It is important to have broadly defined job descriptions so that transferring people from task to task is not a problem. Two approaches to filler jobs are (1) short jobs and (2) scheduling. For short jobs, make a list of short, low-priority jobs that can be done in the idle time, rather than in "prime time." Examples are routine maintenance and clerical tasks (e.g., answering e-mail and phone calls). One of McDonald's guidelines is "If there is time to lean, there is time to clean."

Scheduling may increase or decrease idle time. For example, if a meeting starts 15 min after the start of work, most people will waste the time. Meetings also tend to fill the time available. To shorten meetings, have an agenda and schedule the meeting partly on the participant's time rather the organization's time (i.e., schedule them to overlap lunch or quitting time). Supervisors should assign more work to subordinates than the subordinates have time to do. No one wants to come to the boss and say "I am out of work."

Filler people approaches are (1) staggered schedules, (2) temporary workers, and (3) part-time workers. The goal is to make labor a variable cost, not a fixed cost.

Staggered schedules vary starting and stopping times for different workers. This not only spreads the demand on production workstations but also on service facilities (parking lots, food service, etc.). When people do the same task, staggering their work hours permits keeping the business open for more hours at no additional cost. For example, if Mary goes to lunch from 11:30 to 12:00 and Sam goes from 12:00 to 12.30, someone can always be present to answer the phone. If Mary works from 7:30 to 4:30, and Sam works from 8:30 to 5:30, customers can be served from 7:30 to 5:30 instead of 8 to 5.

These flexible schedules imply considerable cross-training of workers since the business must continue when a specific individual is not there. Often firms define core time (say 80 to 90% of people must be present) and flex time (when fewer people are there).

A greater modification is changing the work from the standard 8 h/day, 5 day/week pattern. The many alternatives are called compressed plans, as they compress the number of days worked, through expanding the number of hours per day. One plan is seven shifts of

12 h each during a 14-day period; another is four shifts of 10 h/day during each week. Compressed plans reduce worker commuting.

Temporary workers have been used for many years — especially in agriculture and construction. The big advantage for the firm is that it can have core staff retained even at the bottom of the business cycle. The famous Japanese "lifetime employment policy" only applies to a core staff of males; surges are handled by temporary workers (such as recalled retired employees) and females.

Part-time workers are those hired for less than 30 h/week. Part-time workers are not very common in manufacturing but are common in service and retail firms. In 2000, 13% of the workers in the United Stated worked less than 30 h/week; women made up 70% of part-timers. Some advantages to the firm include: (1) possible better match between work requirements and work availability (e.g., part-time workers for McDonald's at meal times); (2) possibly better quality workers (e.g., students who would work part-time for $8/h); (3) possible lower labor costs (e.g., wages at bottom of the bracket, not the top; low pension costs as many workers quit before retiring; low medical costs as many firms have zero medical benefits for part-time workers); and (4) change in work hours is easier (e.g., it is difficult to increase full-time workers more than 8 h/week and they resist cuts in hours while part-time workers adjust more easily to changes up from 15 h/week to 20 h/week or down from 20 h/week to 15 h/week).

Another alternative is job sharing, where two people, each working part-time, fill one full-time position. It allows an expanded range of skill and experience, both workers can work extra hours during peak load times, noncoverage of the job (due to illness, vacations, absenteeism) becomes minimal, and if one person leaves, the other provides partial coverage until a new person is hired/trained.

Part-time can be (1) part-time within a day, or (2) full-time within a day, but part-time within a week or month (e.g., people can be hired to work just weekends, Wednesdays, or whenever).

6. *Vary environmental stimulation inversely with task stimulation.* Tasks are divided into low and high stimulation.

   Low stimulation. Monitoring and machine-paced tasks can have stimulation added to either the task or the environment. For adding stimulation to the task, the primary technique is adding physical movement. For adding stimulation to the environment, the primary technique is to allow people to talk to each other. Talking is easier when people are face to face, are close, and there are no auditory or visual barriers. Music is another auditory stimulus. Isolated people (guards, truckers) can interact with others through cell phones and radios. Windows permit visual stimulation. It may be possible to improve socialization during time not at the workstation (e.g., on breaks, while getting/disposing of supplies, contacts with supervisors).

   High stimulation. Most high-stimulation tasks are office tasks. The primary problem is the need for auditory privacy. One approach is two workstations: the usual one and one for concentration (typically an enclosed area, scheduled when needed). For visual privacy, obscure the face. Use a barrier from 0.6 to 1.6 m above the floor. Orient chairs so the worker's eyes do not meet the eyes of passersby.

7. *Reduce fatigue.* Solutions divide into fatigue prevention and fatigue reduction.

   Fatigue prevention. Fatigue may be caused by a lack of sleep caused by working too many hours or working hours at the "wrong" time (shiftwork). Fatigue increases exponentially (not linearly) with time. Thus, it is important to have rest breaks. The problem with a conventional break is that there is no productivity during the break.

   Fatigue reduction. Work breaks use a different part of the body to work while resting the fatigued part. On a micro-level, alternate working and resting with the left and right hand or alternate working with the muscles and the brains. If a muscle is used 4 h/day, it has 20 h to recover; a recovery–work ratio of 5. If a specific muscle is used 8 h/day, it has 16 h to recover;

TABLE 11.2   Guidelines for Good Instruction Writing.

Plan

Know the user's ignorance. This requires knowing the user
Decide what will be communicated
Decide upon strategy (medium, text vs. text + pictures)
Decide upon tactics (e.g., organization of message, types of tables, etc.)

Draft

Plan on multiple drafts
Put the information in a logical order. Number steps
Have a specific objective for each instruction (who, what, where, why, when, how)
Use both text and figures
Make the text at the appropriate reading level
Use two short sentences instead of a long sentence with a clause. Instead of, "After cleaning the part with solvent,
   dry in a ventilated location," use
   Clean the part with solvent
   Dry in a ventilated location
Use *active* statements. Avoid *passive* or *negative* statements
   Active: The large lever controls the depth of cut
   Passive: The depth of cut is controlled by the large lever
   Negative: The small lever does not control the depth of cut
For emphasis in sentences, use **bold** or *italics*; avoid underlining or CAPITALIZATION

Revise

Identify the target audience (see *Plan*)
Select a sample audience for testing
Run the test
Revise material
Continue the loop until satisfied
Allow sufficient time for testing and revision

Source: From Konz, S. and Johnson, S. 2004. *Work Design: Occupational Ergonomics*, 6th ed., Holcomb Hathaway, Scottsdale, AZ. With permission.

a recovery–work ratio of 2. But, in a 12-h shift, the work is 12 h and recovery is 12 h, so the recovery–work ratio is 1. (The recovery–work calculation also is valid for exposure to chemicals.)

On a macro-level, job rotation, in which workers switch jobs periodically, not only reduces fatigue but, since everyone shares good and bad jobs, reduces feelings of inequality. Since fatigue recovery is exponential, short breaks often are better than long breaks occasionally.

8. *Communicate information.* Transferring information from the mind of an ergonomist to someone else's mind (operator, technician, boss) is not simple and automatic (see Table 11.2). In general, avoid verbal orders — unless the task is commonly done.

In summary, plan what you will communicate, draft the information, and then check to see how well the information was communicated. Then improve your communication.

11.2.2   Relation among Workstations

Table 11.3 gives three guidelines concerning relations among workstations.

1. *Decouple workstations.* Office workstations usually are decoupled but manufacturing workstations often are arranged in flow lines. The most common type is the assembly line. There are two primary reasons for decoupling (separating workstations with buffers) — (1) line balancing and (2) shocks and disturbances.

TABLE 11.3   Guidelines for Relations Among Workstations

Decouple tasks
Minimize material handling costs
Optimize system availability

Source: From Hanna, S. and Konz, S. 2004. *Facility Design and Engineering*, 3rd ed., Holcomb Hathaway, Scottsdale, AZ. With permission.

Line balancing. Assume that station A took 50 sec, station B took 40 sec, and station C took 60 sec. Then, if there was no buffer, the line would have to index at the speed of the slowest workstation (the bottleneck station), station C. There would be idle time at stations A and B. Actually, it is worse than that as the line speed would be have to be set at the mean time of the slowest operator on the slowest station.

Shocks and disturbances. A second problem is that operator times vary — e.g., talking to the supervisor, sneezing, dropping a part. Thus, if flow line stations are not decoupled, the line must run at the slowest time of the slowest operator at the slowest stations. Thus, designers use buffers to give the line flexibility.

Decoupling techniques. Chapter 6 in Hanna and Konz (2004) discusses decoupling techniques in detail. In summary, you can decouple by changing product flow (subdivided into (1) buffers at or between stations, (2) buffers due to carrier design, and (3) buffers off-line) and decoupling by moving operators (subdivided into (1) utility operator, (2) help your neighbor, (3) *n* operators float among *n* stations, and (4) *n* operators float among more than *n* stations).

2. *Minimize material handling cost*

Material handling cost = capital cost + operating cost
Operating cost = (no. of trips per year)(cost per trip)
Cost per trip = fixed cost per trip + (variable cost per distance)(distance per trip)

The key point of this equation is that distance per trip is only one part of the material handling cost — usually a minor part. Much transportation (and communication) is distance insensitive (i.e., cost varies little with distance). Technology has increased this death of distance. Distance is not completely irrelevant, but has become a minor part of material handling cost.

3. *Optimize system availability*

$$\text{Availability} = \frac{\text{Uptime}}{\text{Total time}}$$
$$= \frac{\text{Uptime}}{\text{Uptime} + \text{downtime}}$$
$$= \frac{\text{MTBF}}{\text{MTBF} + \text{MTR}}$$

where MTBF is the mean time between failures (reliability) and MTR is the mean time to repair (maintainability)

There are three strategies to improve availability: (1) increase uptime (reliability); (2) decrease downtime (increase maintainability); and (3) make loss of availability less costly.

Increase uptime. One approach is to use a "parallel circuit" rather than a "series circuit." In a parallel circuit, the system works if *any* path works; in a series circuit, the system works only if *every* component in the path works. Another possibility is to use a "standby circuit." Standby people are quite common, ranging from the utility operator on a flow line to a relief pitcher in baseball; however, as the relief pitcher example shows, even the backup can fail!

Decrease downtime

Downtime = Fault detection time (time to find the device does not work)
+ Fault location time (time to find the problem)
+ Logistics time (time to get repair parts)
+ Repair time (time to fix unit)

Fault location can be expedited by giving maintenance people cell phones so that they arrive more quickly. Diagnostic advice can be furnished (using an 800 number) by a remote person, "the brain," to a local person, "the hands." Reduce logistics time by stocking the needed spare parts as well as using the nationwide 24-h delivery services. Figure 11.2 gives design features to improve maintenance access during repair time.

Make loss of availability less costly. Preventative maintenance and partial function are the two possibilities.

Preventative maintenance (i.e., scheduled maintenance) has you pick the time when you do not need the availability and then do the maintenance.

Partial function is part of planning. For example, split orders between vendors rather than have

Design/location

Use hinged covers and quick release fasteners. Put hinges at the bottom. Consider removal of the entire panel/cover. Note that the maintainer may use gloves

Consider quick disconnects of utilities

Access ports should not expose the maintainer to hot surfaces, electrical currents, or sharp edges

Locate access apertures on the same side of the machine as related displays, controls, and test points

Access apertures on top surfaces can collect debris and water, which when the aperture is opened, can contaminate the machine

If a fastener is close to the floor, locate it on the horizontal plane; at high heights, locate it on the vertical plane at the rear. Avoid having fasteners on the plane perpendicular to the right side of the opening as they are difficult for right-handed people to manipulate

Size

Access ports provide clearance for tools and allow insertion/removal of machine components

Aperture size depends upon the height off the floor and the depth of reach

Consider visual accessibility and reach accessibility

For tool access, allow straight-line access to fasteners/components

If the tool is inserted "deep" into the machine, allow room for the hand/arm to move. The hand/arm in the aperture may block vision. Avoid aperture widths of less than 25 cm — especially for depths over 30 cm

For inserting/removing a component (two-hand insertion), have aperture width = component width + 8 cm. If the forearms are inserted, have aperture width = component width + 16 cm

FIGURE 11.2 Maintenance access. (From Corlett, E. and Clark, T. 1995. *The Ergonomics of Workspaces and Machines.* Taylor and Francis, London; Helander, M. 1995. *A Guide to the Ergonomics of Manufacturing.* Taylor and Francis, London. With permission.)

one sole-source vendor. A substitute cross-trained worker may not have the capability of the normal operator, but at least the operation is not shut down completely.

11.2.3 Physical Design of the Workstation

Table 11.4 gives 15 guidelines concerning workstation design:

1. *Avoid static loads and fixed work postures.* Ideally, the person can be seated so that most of the body weight is supported. But occasional standing (such as to get supplies) is good. Static standing is fatiguing. Consider a "bar rail" so the legs can alternate positions. Floors should be wood or carpet. If the floor is concrete or tile, use a floor mat (24 × 36 in. is a typical size) that compresses 3 to 7% under adult weight; the underside of the mat should not slip on the floor. Walking aids blood circulation; venous pooling in the legs is eliminated in as few as ten steps.

2. *Reduce musculoskeletal disorders.* From age 18 to 64 yr, more people are disabled from musculoskeletal disorders (also called cumulative trauma) than any other disorder.

3. *Avoid slips and falls.* Falls can occur from slips (unexpected horizontal foot movement, trips (restriction of foot movement), and stepping-on-air (unexpected vertical foot movement).

4. *Do manipulative work in the "strike zone."* The best work zone is similar to the strike zone in baseball — mid-thigh to the chest. Optimum height is the same for sitting and standing. Since elbow height varies with sitting and standing and people vary in size, any workstation that puts the work a fixed distance from the floor is bad design. Note that work height is not table height as most items being worked on have a thickness of 25 to 125 mm. A small departure from the middle of the zone (50 mm below the elbow) for manipulative work is not critical as there no "cliffs" in ergonomics. For nonmanipulative work (i.e., where force is being applied), use 150 to 200 mm below the elbow as the goal.

   There are three design approaches: change worksurface height, change elbow height, and change work height on the worksurface. Changing worksurface height is relatively easy on many conveyors and workstations but, unless there is a motorized adjustment, probably should be reserved for situations in which the same worker uses the workstation for a "long" time (say 1 month). Changing elbow height is simple for seated workers — just adjust chair-seat height. For standing, have adjustments of 25 to 50 mm by adding or

TABLE 11.4   Guidelines for Physical Design of the Workstation

Avoid static loads and fixed work postures
Reduce musculoskeletal disorders
Avoid slips and falls
Do manipulative work in the "strike zone"
Furnish every employee with an adjustable chair
Use both the feet and the hands
Use gravity; do not oppose it
Conserve momentum
Use two-hand motions rather than one-hand motions
Use parallel motions for eye control of two-hand motions
Use rowing motions for two-hand motions
Pivot motions about the elbow
Use the preferred hand
Keep arm motions in the normal work area
Let the small person reach; let the large person fit

Source: From Hanna, S. and Konz, S. 2004. *Facility Design and Engineering*, 3rd ed., Holcomb Hathaway, Scottsdale, AZ. With permission.

subtracting one or more mats. Larger adjustments (say 100 mm) can use wooden platforms. Change the work height on the worksurface by using a spacer under the work.

5. *Furnish every employee with an adjustable chair.* Chair cost is entirely a capital cost since
   there are zero power costs and maintenance costs. Assuming a cost of $180 per chair, a life
   of 10 yr, and 1800 h/yr, the cost is only 1 cent/h.

   When selecting specific chairs for a specific job, a committee (e.g., engineering, purchasing, facilities) should narrow the possible chairs to three to five. Then, however,
   the *users* should try the specific chairs in their specific jobs before the purchase. A 2-h
   trial for each chair should be sufficient. Easy adjustment is more important if multiple
   people use the same chair but even if only one person uses the chair that person may
   wish to make adjustments during the day.

6. *Use both the hands and the feet.* For manipulative movements, the hand/fingers is superior
   to the foot/toes. But, for simple on–off motions, the foot can be used through foot
   switches and pedals. If the pedal is used repeatedly, make it wide so the worker can alternate feet. Knee switches use a lateral movement of the knee so that the leg weight is not
   moved.

7. *Use gravity; do not oppose it.* If you move 25 g of feathers in your hand, you also move
   4500 g of arm. Thus, avoid lifting; make movements horizontal or downward. Gravity
   also can move a product being worked on (paint from a paint brush, welding beads
   from a welding rod). Fixtures should permit the work item to rotate so that the operator
   can use a "downhand" position.

8. *Conserve momentum.* The concept is to avoid unnecessary acceleration/deceleration of an
   arm, leg, or body. Four aspects are circular motions, disposal motions, grasping motions,
   and transport motions.

   Circular motions are better than back-and-forth motions. A leg example is bicycle
   pedals; a hand example is writing versus printing. When mixing products, use a rectangular container (not a circular one) to obtain turbulence and better mixing. A motor, of
   course, is better than hand stirring. Dispose of items without precise placement to minimize arm deceleration. Avoid sharp edges where items are grasped so that less care is
   needed. Minimize the force needed for pushing and pulling so that less force is needed
   to accelerate or decelerate a load.

9. *Use two-hand motions rather than one-hand motions.* For manipulative work, work output
   using two hands is higher than with one hand. If the "work" of one hand is "holding," use
   a mechanical clamp. The limiting factor in hand–arm motions is not the ability of the
   brain to command or the ability of the eyes to supervise; rather, it is the ability of the
   nerves and muscles to carry out the orders. "The spirit is willing but the flesh is weak."

10. *Use parallel motions for eye control of two-hand motions.* When using two hands simul-
    taneously, minimize eye control problems by placing the hands side by side; that is, the
    arms should be parallel to each other, not symmetrical about the body centerline.

11. *Use rowing motions for two-hand motions.* Two-hand motions can alternate (like bike pedal-
    ing with the feet) or have a rowing motion (both hands move out and back together). Alter-
    nation has more shoulder motion and torso twisting than rowing. Thus, rowing is best.

12. *Pivot motions about the elbow.* Pivoting about the elbow (a forearm move) takes less time
    and effort than moving across the body (a move of the entire arm). Since the arm is
    pivoted about the shoulder (not the nose), for best efficiency, put bins ahead of the
    shoulder, not the nose.

13. *Use the preferred hand.* The preferred hand is about 10% faster for reach-type motions, is
    5 to 10% stronger, and is more accurate in movement. If the hand uses a tool, the
    preferred hand time advantage is much greater. About 10% of the population is left-
    handed.

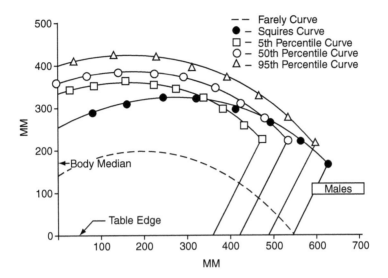

FIGURE 11.3 Male work area (right hand) at elbow height; the left hand is mirrored. (From Hanna, S. and Konz, S. 2004. *Facility Design and Engineering*, 3rd ed., Holcomb Hathaway, Scottsdale, AZ. With permission.)

14. *Keep arm motions in the normal work area.* Figure 11.3 and Figure 11.4 show the normal work area for males and females. The windshield-wiper shape areas are due to the pivoting of the forearm about the elbow.
15. *Let the small person reach; let the large person fit.* Design for *most* of the *user population*. For *most*, social customs and laws against discrimination make it more and more difficult to exclude people. For *user population*, there have been three changes: (1) changing sex/occupation stereotypes (engineers no longer are just males, truck drivers can be male or female, etc.), (2) immigration (resulting in a wider range of dimensions), and (3) multi-person use

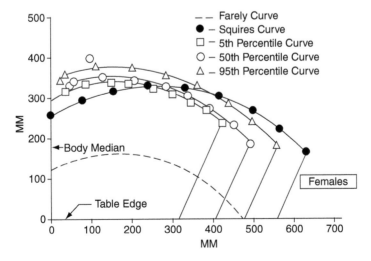

FIGURE 11.4 Female work area (right hand) at elbow height; the left hand is mirrored. (From Hanna, S. and Konz, S. 2004. *Facility Design and Engineering*, 3rd ed., Holcomb Hathaway, Scottsdale, AZ. With permission.)

of equipment and workstations (due to more shiftwork and more part-time work). To include many (exclude few), three alternatives are:

- *One size fits all* (tall door, big bed) has an advantage of simplicity
- *Multiple sizes* (clothing) has manufacturing and inventory problems
- *Adjustability* (adjustable chair) requires a more complex product

The most practical application technique usually is adjustability

11.3 Material Handling

Material handling is divided into at the workstation, between workstations, and aisles.

11.3.1 Material Handling at the Workstation

Material handling at the workstation is divided into four categories: (1) manual, (2) muscle multipliers, (3) robots, and (4) fixed automation:

1. *Manual.* Many components and assemblies are moved manually about the workstation by the operator. Table 11.5 gives some general guidelines.
2. *Muscle multipliers.* Balancers (see Figure 11.5) are often used to suspend handtools; they can drastically reduce the force required to hold handtools. A good design is to suspend them from a small jib crane so that they cover an area and minimize not only vertical force but also horizontal forces. Manipulators (see Figure 11.6) cover a larger area than balancers; they are used to move heavy items (such as castings) to and from pallets, conveyors, and machines. Turntables (see Figure 11.7) reduce reaching distances when loading/unloading; they can be manually powered or motor powered. Scissors lifts (see Figure 11.7) adjust the vertical height of the worksurface.
3. Robots. Robots add a "brain" to the muscle of the manipulator. A human operator is not needed, except for maintenance. The two primary robot types are (1) pick and place and (2) processing. A pick and place robot might pick up an item from a machine and transfer it to a pallet or conveyor or vice versa. A processing robot would have a spray painting or welding "workhead."

   Robots are considered flexible automation since they can be reprogrammed for different tasks.
4. *Fixed automation.* For high-volume production, transport around the workstation (and between workstations) can be dedicated and of special purpose. This equipment positions and moves only one product. A typical name for this alternative is a transfer line. Typically a human is used only for maintenance.

TABLE 11.5 Guidelines for Occasional Lifting

Select individual
   Select strong individuals based on tests
Teach technique
   Bend the knees
   Do not slip or jerk
   Do not twist during the move
Design the job
   Use machines
   Move small weights often
   Put a compact load in a convenient container
   Get a good grip
   Keep the load close to the body
   Work at knuckle height

Source: From Hanna, S. and Konz, S. 2004. *Facility Design and Engineering*, 3rd ed., Holcomb Hathaway, Scottsdale, AZ. With permission.

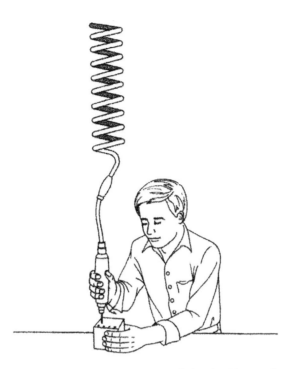

FIGURE 11.5 Balancers permit operators to move loads (typically handtools) around a workstation as if the load had only a small weight. (From Hanna, S. and Konz, S. 2004. *Facility Design and Engineering*, 3rd ed., Holcomb Hathaway, Scottsdale, AZ. With permission.)

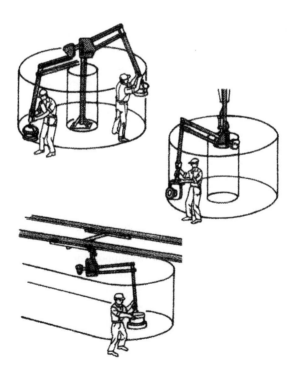

FIGURE 11.6 Manipulators (balancers with "arms") can be used to move a product around the workstation (e.g., to/from pallets, conveyors, machines). (From Hanna, S. and Konz, S. 2004. *Facility Design and Engineering*, 3rd ed., Holcomb Hathaway, Scottsdale, AZ. With permission.)

FIGURE 11.7 Turntables can be rotated to reduce reaches and moves by the operator. One option is to mount the turntable on the floor and place a pallet on it. The turntable (as well as ball casters and air-film tables) also can be mounted at a fixed height (such as even with a conveyor or worksurface). A scissors lift moves up and down to reduce lifting/lowering stress. If it is at the height of the work, the work can be transferred to/from it with a sliding motion instead of lifting/lowering. Of course, a scissors lift can be combined with a turntable. (From Hanna, S. and Konz, S. 2004. *Facility Design and Engineering*, 3rd ed., Holcomb Hathaway, Scottsdale, AZ. With permission.)

11.3.2 Material Handling between Workstations

There are three common alternatives: (1) conveyors, (2) vehicles, and (3) cranes:

1. *Conveyors.* Conveyors not only transport items, but also store items, serve as an assembly or processing worksurface, and even act as a method of work pacing. In general, it is inefficient to machine-pace a person.

   Although a conveyor is thought of as a fixed path (point to point) system, spurs and branches (combined with switches controlled by sensors) allow considerable flexibility in the destination.

   From an ergonomics viewpoint, it is the loading or unloading of conveyors that is important. The key guideline is to *minimize lifting*. As discussed in Section 11.31, lifting by humans can be eliminated by using machines (e.g., robots or fixed automation). The effort of lifting can be reduced by muscle multipliers. Finally, for unassisted manual lifting, the initial object location should be level or higher than the final object location so that the movement can be horizontal (i.e., sliding is a possibility) or downhill (i.e., have a gravity assist, not gravity penalty).

   Figure 11.8 shows a horizontal transfer from a conveyor. Figure 11.9 shows the relative locations of conveyors and pallets for unloading and loading of conveyors.

   Although belt, wheel, and roller conveyors support the load from the bottom, the power and free conveyor supports the load from above and presents the load on hooks. In this case, the challenge is to prevent reaching above the shoulder. Taller workers should obtain/dispose of the item at waist height so that shorter workers can obtain it at shoulder height; very short workers should stand on a platform.

2. *Vehicles.* Vehicles (wide-area equipment) service an unbounded area (in contrast to the fixed areas of conveyors and cranes). Vehicles are divided into operator walks (two-wheel hand trucks, platform trucks, hydraulic hand-pallet trucks, and powered walkie trucks), operator rides (personnel/burden carriers, tow tractors and trailers, and lift trucks), and no driver (automatic guided vehicles).

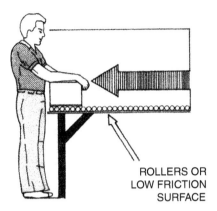

FIGURE 11.8   Slide, do not lift, to reduce back strain. Avoid lips at the end of the conveyor since they require lifting. Allow sufficient toe space (6 in. deep, 6 in. high, 20 in. wide). (From Hanna, S. and Konz, S. 2004. *Facility Design and Engineering*, 3rd ed., Holcomb Hathaway, Scottsdale, AZ. With permission.)

There are many attachments (e.g., forks, clamps, sideshifters, rotators, rams, push-pull, and fork positioners).

3. *Cranes.* As with conveyors and vehicles, there is a wide variety of overhead lifting equipment, sometimes call hoists. Typically cranes/hoists cover a limited area. A common, relatively low-cost crane is a jib crane — a hoist on a rotating horizontal beam. It usually covers a 180° semicircle but, if the beam is mounted on a rotating column, can cover 360°. Air film equipment (air casters) forces compressed air into a chamber under the load (equipment, pallet, die, etc.). A 1-lb push force can move a 1000-lb load so loads up to 80,000 lb can be pushed by one person.

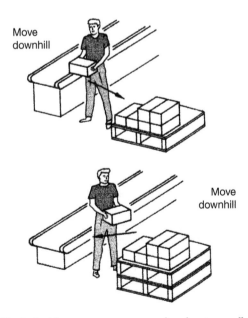

FIGURE 11.9   Move downhill whether from a conveyor or worksurface to a pallet or a pallet to a conveyor or worksurface. When moving to a conveyor/worksurface from a pallet, adjust the conveyor to be low; raise the pallet by having the lift truck driver place it on a platform (such as an empty pallet). When loading a pallet from a conveyor/worksurface, have the conveyor higher and the pallet on the floor. (From Hanna, S. and Konz, S. 2004. *Facility Design and Engineering*, 3rd ed., Holcomb Hathaway, Scottsdale, AZ. With permission.)

TABLE 11.6   Recommended Corridor Widths (Inches) for People-Only Traffic

Number of People	Situation	Minimum	Better
1	Avoid touching equipment or hitting switches	20	24
2	Passing one person standing with back to wall	30	36
3	All three walking abreast in same direction	60	72

11.3.3   Aisles

A key design consideration is aisle width. Aisles in production areas may have a mixture of vehicles and people. A width of 12 ft allows two-way traffic of a sit-down counterbalanced lift truck and people. But this uses a lot of floor space. Thus, there is a pressure to use narrower aisles. One alternative is one-way aisles, but this is difficult to enforce. More commonly a wide variety of narrow-aisle equipment is used. In addition, only vehicles are allowed in these narrow (5 ft) branch aisles.

In offices (i.e., no vehicles), aisles need not be straight. Corridors are aisles with walls. Because people can step outside of aisles to avoid oncoming traffic, aisles can be narrower than corridors. Shoulder width is the key factor. Table 11.6 gives recommended corridor widths. To prevent accidents, doors should open from the corridor or aisle rather than into it.

11.4   Service Areas

Three service (non-production) areas will be discussed: offices, toilets, and food service.

11.4.1   Offices

Although factories process things, offices process information. The movement of information in an office does not require much attention to the arrangement of workstations but the social environment (including color, décor, and esthetics) is important in an office. The open plan with cubicles has become the normal arrangement. Often there is more than one worker/workstation (due to multiple shifts, telecommuting, management by walking around, etc.)

The open-plan office has two key concepts: (1) modular furniture and (2) the needs of each workstation varies. Individual cubicles have panels holding modular furniture; the panels, standing free on feet, have work surfaces, shelves, and storage. Private offices are still used with open plan but only for higher-ranking personnel.

Visual variety is emphasized by using specified colors of partitions and chairs to identify departments (e.g., green for accounting, blue for purchasing). Windows should be shared by putting them at the end of aisles, in areas with low partitions, near elevator banks, and stairwells. Walls without windows can be a different color than the partitions and can have art work with spotlights. Achieve visual privacy with partition placement, interrupted lines of sight, opaque or translucent glass instead of transparent glass, partition heights of 62 to 80 in., and placement of plants.

Acoustic privacy is difficult to achieve. In acoustics, the ratio of signal (or message) to noise (or background) is important. Acoustic treatment of partitions, chairs, and flooring makes them absorb sound and thus makes the message more noticeable. But it may be that the signal (e.g., conversation from an adjacent cubicle) is really a distraction. One technique is to mask the signal by *increasing* the background noise (e.g., by air handling noise or background music or even white noise generators). Another technique is to have "phone booths" and private conference rooms in each department in which people can have conversations not heard by others.

It has been recognized that not everyone is at their cubicle all the time. They may be gone for an hour or two, a day or two, or even weeks. Five alternatives are: shift sharing, hoteling, the virtual office, telecommuting, and management by walking around.

In shift sharing, the same cubicle is shared with people from other shifts. Thus, if there are supervisors for the day, evening, and night shifts, they share one cubicle. In hoteling, the cubicle is considered to be a hotel room. Most information is stored on a central computer. Paper files and personal items (e.g., photos of spouse) are moved from storage to the cubicle on a cart. Example users might be sales and service personnel, part-time workers, and people working at several locations. The virtual office comes from the laptop computer and computer networks. The user (say a sports writer) taps into the computer network for information, processes local information (information on the game), and then sends the message to the central computer. Thus, there would only be occasional use of the cubicle. With telecommuting, most of the work is done at home, and communication is by faxes, e-mail, and phones. Home workers tend to come to the central office occasionally for meetings and other work. In management by walking around, higher management believes first-line managers should be "on the floor," not in the office. Thus, 12 department managers might share a 10×19 ft "manager's office," which has six desks, six chairs, one computer, and eight file cabinets. Area managers each have a mail basket and a three-ring binder, but no desk.

The typical cubicle will have a work surface (desk in old terms), chair, computer, phone, some storage, and wall partitions. There may be a printer for each computer or a centralized printer for multiple computers. There may be task lighting (either under the shelf or a free-standing desk lamp). The occupant chair typically has casters (as there is need for movement around the cubicle) but any guest chair should have a sled base (to avoid tipping over).

Spatial arrangement in the factory emphasizes material handing but in the office where the processing is of information, arrangements may be to support social structures. For example, only higher supervisors get a private office and the top supervisor of the facility gets a corner office with (unique color) chairs with high backs. One design consideration is territorial and personal space. Territory is a visible area in a stationary location; personal space is an invisible boundary that moves with the person. Thus, people might "mark out" their territory with family pictures, art objects, and a calendar; in the cafeteria, they may put their coat on the opposite chair so that no other person sits at the same table. Generally, designers should encourage personalization of long-term space since it costs the organization nothing and satisfies the occupant.

Konz and Johnson (2004 Chapter 14) discuss in detail the design of conference, meeting, and training rooms.

11.4.2 Toilets

Table 11.7 gives the Occupational Safety and Health Administration (OSHA) requirement; see also the National Standard Building Code and local modifications of the Building Code. The Building Code, however, provides equal numbers for men and women (and assumes a male–female ratio of 50:50); it is better to estimate the actual male–female ratio and use a ratio of two water closets for women to one (water closet + urinal) for men. Toilets are relatively permanent and are difficult to expand or move. Therefore, plan ahead for a large number of users.

Architectural barriers for the handicapped are illegal. Accessibility requirements for the disabled are in the *Federal Register* (29 CFR, part 36), July 26, 1991. The requirements cover not just toilet access but drinking fountains, corridor widths, ramp and stair design, and more. The requirements are quite detailed. For example, the grab bar in the toilet stall must be at least 36-in. long, the stall door opening force must be 5 lb or less, the bottom of the mirrors must be no higher than 40 in. above the floor, etc.

Emergency flushing stations (for eyes and body) are required in some work areas where corrosive materials are used (OSHA 1910.94; OSHA 1926.403).

11.4.3 Food Service

Although some larger facilities have either a snack bar or cafeteria, most facilities just have vending machines. The primary advantage of vending machines is low cost as, vending is self-service and the

11-18

Interventions, Controls, and Applications in Occupational Ergonomics

TABLE 11.7 Water Closets Required by OSHA (subpart J, 1910.141) for Employees of Each Sex

Number of Employees	Minimum Number of Water Closets
1–15	1
16–35	2
36–55	3
50–80	4
81–110	5
111–150	6
Over 150	One additional fixture for every 40 employees

Note: When toilet rooms will be occupied by no more than one person at a time, can be locked from the inside, and contain at least one water closet, separated toilet facilities for each sex need not be provided.

Source: From Hanna, S. and Konz, S. 2004. *Facility Design and Engineering*, 3rd ed., Holcomb Hathaway, Scottsdale, AZ. With permission.

only service labor is machine stocking and some equipment repair. The low cost permits 24/7 availability as well as location at multiple locations within the facility, maximizing user convenience.

For serving 100 people, a good mix is:
- One hot drink machine
- Two cold drink machines
- Two sandwich machines
- One ice cream machine
- One candy/snack machine
- One bill changer (changing $1 and $5 bills)

Other features should include microwave ovens, a sink with faucets, a drinking fountain, a refrigerator for "brown baggers," and condiments (napkins, salt and pepper, ketchup, etc.). Storage for the condiments, trash containers, and a serving area are needed. Typically, the machines are grouped along a utility wall to minimize plumbing and electrical costs.

The eating area should have good esthetics. Some possibilities are windows with a view, carpet on the floor, good noise control, artwork on the walls, mural wallpaper, indirect lighting, warm-white instead of cool-white fluorescent lamps, uneven ceiling fixture spacing (giving more unequal illumination), wood instead of plastic chairs and tables, upholstered chairs, artificial flowers on the tables (changed periodically), and glass plates and tumblers (instead of plastic or paper).

The eating area should have a mixture of table sizes and shapes. Circular tables accommodate both even- and odd-number groups and encourage conversation. Rectangular tables do not accommodate odd-number groups well but can be combined with other tables (you can take three tables of two seaters and make seating for six). Avoid long rectangular tables as once the table has more than one or two occupants, newcomers feel they are intruding on the occupants' personal space. Tables should be pedestal style and be moveable.

Provide two bulletin boards — one for official notices and one for personal notices. Pay telephones should be wall mounted and should not be near chairs (to discourage people tying up the phones). A wall-mounted phone book and a roll of paper (toilet paper fashion) for notes are useful.

11.5 Environments

The four environments discussed are (1) illumination, (2) noise, (3) climate, and (4) chemical.

11.5.1 Illumination

In this section, illumination is divided into three categories: (1) lighting basics, (2) sources/fixtures, and (3) lighting strategies.

1. *Lighting basics.* There is a fundamental fact: "Light is cheap; labor is expensive." On a per square foot basis, labor cost per year tends to be over 300 times lighting cost per year. Thus, do not try to save a penny by cutting lighting costs if it costs labor productivity.

   Table 11.8 defines various illumination units. Visibility of a task is affected by quantity of light (illuminance), by quality of light (glare, direction, color), by task factors (size and contrast of the object, time available, speed and accuracy needed), and by people factors (how good are the eyes).

TABLE 11.8   Measures of Illumination

Measure	Unit	Definition and Comments
Luminous intensity, l	Candela	Light intensity within a very small angle, in a specified direction (lumens/steradian). Candela = 4 (pi) lumens
Luminous flux, lm	Lumen	Light flux, irrespective of direction from a source, generally used to: (1) express total light output of a source and (2) express amount incident on a surface
Illuminance, dlm/da	Lux	1 lumen/m^2 = 1 lux = 0.093 footcandle 1 lumen/ft^2 = 1 footcandle = 10.8 lux
Luminance (brightness)	Nit	Luminance is independent of the distance of observation as candelas from the object and area of the object perceived by the eye decrease at the same rate with distance 1 candella/m^2 = 1 nit = 0.29 footLambert 1 candella/ft^2 = 1 footLambert = 3.43 nits
Reflectance	Unitless	Percent of light reflected from a surface

Typical reflectance		Recommended reflectances	
Object	%	Object	%
Mirrored glass	80–90	Ceilings	80–90
White matte paint	75–90	Walls	40–60
Porcelain enamel	60–90	Furniture and eqpt.	25–40
Aluminum paint	60–70	Floor	20–40
Newsprint, concrete	55	Munsell	Reflectance
Dull brass, dull copper	35	Value	%
Cardboard	30	10	100
Cast and galvanized iron	25	9	100
Good quality printer's ink	15	8	78
Black paint	3–5	7	58
		6	40
		5	24
		4	19
		3	6

Measure	Unit	Definition and Comments
Brightness contrast	Unitless	$(L_b - L_t)/l_b$, where L_b is the luminance of the background and L_t is the luminance of the target
Wavelength	Nanometers	The distance between successive waves (a "side view") of light. Wavelength determines the color hue. Saturation is the concentration of the dominant wavelength (the degree to which the dominant wavelength predominates in a stimulus). Of the 60 octaves in the electromagnetic spectrum, the human eye detects radiation in the octave from 380 to 760 nm
Polarization	Degrees	Transverse vibrations of the wave (an "end view") of the light. Most light is a mixture; horizontally polarized light reflected from a surface causes glare.
Coherence	—	Most light is incoherent, analogous to a stadium crowd roaring. Lasers produce coherent light — the crowd sings a song in unison

Source: From Konz, S. and Johnson, S. 2004. *Work Design: Occupational Ergonomics*, 6th ed., Holcomb Hathaway, Scottsdale, AZ. With permission.

Table 11.9 gives the IESNA recommendations for the amount of lighting. Contrast between the target and the background should be over 0.3. Contrast can be improved by modifying the lighting or the task. Environmental contrast, on the other hand, should be minimized to reduce adjustments of the eye. Many people do not have 20-20 corrected vision — especially after age 40.

2. *Sources/fixtures.* The first choice is between artificial illumination and sunlight. The general choice is artificial illumination since it can be controlled for amount, direction, color, etc.

   Fluorescent lamps are the primary choice for industrial use due to their efficient use of electricity and their long life (20,000 h). Sodium lamps have long lives but give a narrow-spectrum yellow light (which makes color recognition difficult) and so tend to be used in parking lots. Fixtures direct the light from the lamp but adsorb some of the light. A variety of designs are available.

3. *Lighting strategies.* The primary strategy is general lighting; the entire area is lit "evenly" by ceiling fixtures. This permits location of equipment and people anywhere in the area as well as permitting easy rearrangement of the equipment and people. However, specific areas (private offices, warehouses) may have equipment and people which will not move. Then the ceiling fixture can be located specifically for the task. For example, in warehouses, put the fixtures over the aisles, not over the racks.

   Another strategy is task lighting (offices, inspection). There is general ceiling lighting but at a lower illumination level with supplementary lighting fixtures for the task itself. Task lighting typically permits the person to change the lighting amount (moving the light closer or farther away) and to change the light/task orientation (by moving the lamp orientation). Lighting for computer screens often uses task lighting. There are specific strategies for special situations such as warehouse aisle lighting, security lighting, safety lighting, sports lighting, highway lighting, etc. (see IESNA, 2000).

TABLE 11.9   Illuminance Categories

Orientation and simple visual tasks. Visual performance is largely unimportant. These tasks are found in public spaces where reading and visual inspection are only occasionally performed. Higher levels are recommended for tasks where visual performance is occasionally important:
Public spaces, 30 lux
Simple orientation for short visits, 50 lux
Working spaces where simple visual tasks are performed, 100 lux
Common visual tasks. Visual performance is important. These tasks are found in commercial, industrial, and residential applications. Recommended illuminance levels differ because of the characteristics of the visual task being illuminated. Higher levels are recommended for visual tasks with critical elements of low contrast or small size:
Performance of visual tasks of high contrast and large visual size, 300 lux
Performance of visual tasks of high contrast and small size, or visual tasks of low contrast and large size 500 lux
Performance of visual tasks of low contrast and small size, 1000 lux
Special visual tasks. Visual performance is of critical importance. These tasks are very specialized, including those with very small or very low contrast critical elements. Recommended illuminance levels should be achieved with supplementary task lighting. Higher recommended levels are often achieved by moving the light source closer to the task.
Performance of visual tasks near threshold, 3000–10,000 lux

Notes: Low contrast, ≤ 0.3 (but not near threshold); high contrast, >0.3.
Small size, $\leq 4 \times 10^{-6}$ solid angle in steradians; large size, $>4 \times 10^{-6}$ sr.
For a reading task from 50 cm: 6-point type $= 1.7 \times 10^{-6}$; 8-point type $= 3.1$; 10-point type $= 4.8$; 12-point $= 6.9$; 14-point $= 9.4$; 24-point $= 28$; 36-point $= 62$.
For a square object viewed from 30 cm, a 7.5×7.5 cm object $= 6.3 \times 10^{-6}$ sr. For a circular hole viewed from 40 cm, a 5.76 mm hole $= 3.1 \times 10^{-6}$ sr.
Source: Adapted from IESNA. 2000. *Lighting Handbook: Reference and Application*, 9th ed., Illuminating Engineering Society of North America, New York. (Tables 10-9 to 10-11). With permission.

11.5.2 Noise

For information on noise, see Chapter 31 in Karwowski (2005) and Konz and Johnson (2004).

11.5.3 Thermal

Interior climate normally is controlled by furnaces and air conditioners, perhaps supplemented by fans. Thermal comfort is defined by the American Society of Heating, Refrigeration and Air Conditioning Engineers (ASHRAE) as from 68 to 75 F (effective temperature) in the winter and from 73 to 79 FET in the summer. The assumption is that people wear warmer clothing in the winter.

Heat stress now typically occurs only in special areas such as steel mills or in outdoor work. Protective strategies include sufficient drinking water and protection from radiant heat (e.g., clothing, fixed shields).

Cold stress usually occurs outdoors but can occur in freezers. The typical protection is from clothing and protection from the wind.

Konz and Johnson (2004) present more detailed information on thermal comfort, heat stress, and cold stress.

11.5.4 Chemicals

This section is divided into (1) toxicology and (2) control of respiratory hazards.

1. *Toxicology.* Toxic compounds are poisons. Almost all the problem compounds used in industry are "slow" poisons — their effect takes place over 20 to 30 yr. The major input route into the body is through the lungs — thus much toxicology work concerns ventilation.

   Threshold limit values (TLVs) give the concentration of a compound that is not safe.

   TLVs issued by the Federal Government are legal requirements. TLVs issued by the American Conference of Governmental Industrial Hygienists are only recommendations, but should be strongly considered. The time weighted average (TWA) is the value for an 8-h exposure; the short-term exposure limit (STEL) is a 15-min TWA that should not be exceeded at any time. For exposures less than 8 h, the TWA is adjusted for the time of exposure. For example, acetone has a TWA for 8 h of 500 ppm. A person is assumed to be able to be exposed in a shift to 8 (500) = 4000. If exposed for 2 h, then the permitted exposure is 4000/2 = 2000 ppm; if exposed for 10 h, then exposure per hour is 400 ppm. This trade-off is obviously simplistic so be careful of high exposures.

2. *Control of respiratory hazards.* There are two basic strategies: (1) engineering controls and (2) administrative controls. See Table 11.10.

TABLE 11.10 Engineering and Respiratory Controls (Engineering Controls are More Desirable than Administrative Controls).

Engineering Controls	Administrative Controls
Substitute a less harmful material	Screen potential employees
Change the machine or process	Periodically examine existing employees (biological monitoring)
Enclose (isolate) the process	Train engineers, supervisors, and workers
Use wet methods	Reduce exposure time
Provide local ventilation	
Provide general ventilation	
Use good housekeeping	
Control waste disposal	

Source: From Hanna, S. and Konz, S. 2004. *Facility Design and Engineering*, 3rd ed., Holcomb Hathaway, Scottsdale, AZ. With permission.

References

Corlett, E. and Clark, T. 1995. *The Ergonomics of Workspaces and Machines*. Taylor and Francis, London.

Hanna, S. and Konz, S. 2004. *Facility Design and Engineering*, 3rd ed., Holcomb Hathaway, Scottsdale, AZ.

Helander, M. 1995. *A Guide to the Ergonomics of Manufacturing*. Taylor and Francis, London.

IESNA, 2000. *Lighting Handbook: Reference and Application*, 9th ed., Illuminating Engineering Society of North America, New York.

Karwowski, W., Ed., *Fundamentals and Assessment Tools for Occupational Ergonomics*, CRC Press, Boca Raton, FL, 2005.

Konz, S. and Johnson, S. 2004. *Work Design: Occupational Ergonomics*, 6th ed., Holcomb Hathaway, Scottsdale, AZ.

Further Reading

Hanna, S. and Konz, S. 2004. *Facility Design and Engineering*, 3rd ed., Holcomb Hathaway, Scottsdale, AZ. This popular textbook on facility design has four major sections: (1) Evaluation of alternative layouts, (2) Building shell and specialized areas, (3) Material handling among workstations, and (4) Services and environment.

12

Organizational Design and Macroergonomics

Brian M. Kleiner

Virginia Polytechnic Institute and State University

12.1 Theoretical Foundations of Organizational Design

Organizations have been around for centuries. Through reverse engineering, we can assume the great pyramids of Egypt required complex formal organizations to mobilize and manage the manpower necessary for design and construction. However, modern organizational design has its foundation in the Classical era, starting around 1900. The classical school of organization and its iconic methods of scientific management, supervision, etc. had its roots in the military and the church. The most noticeable derivation of the classical school is the organization's structure. This structure is the hierarchy or grouping and segmentation of members to help distribute the purpose or mission. In theory, each time strategy changes, a decision should be made whether there should be a respective change in organizational design. According to Pasmore (1988), people come together in an organization in a choiceful manner. That is, their collaboration is represented by informal contracts among people. Sometimes the organization facilitates these contractual arrangements and sometimes, if a mismatch is evidenced, the organizational design inhibits these relationships.

In theory, organizational structures exist to create lines of authority and communication in an organization. Lines of authority and communication are easily envisioned when viewing an organizational chart, the pictorial representation of the organization's structure. In such a portrayal, solid vertical lines represent line of authority or the reporting chain of command. Horizontal, solid lines are the lines of communication exemplified by organizational members coexisting at the same level in the hierarchy. Dotted lines are informal lines of authority. In such cases, the relationship is advisory. The organizational structure is also the manner by which an organization distributes its purpose throughout the organization.

In addition, the organizational structure distributes the organizational purpose throughout the organizational network. The mission or purpose is usually stated in terms of the products and customers in the organization and, ideally, all employees understand the overall purpose and their role in achieving the purpose. If the organizational design is effective, this will occur. Effective designs often result in suboptimization, where employees work to an invalid purpose. For example, in functional organizations, employees often see the function as synonymous with the purpose. This will be discussed further when we describe various types of structures.

12.2   Structural Foundations of Organizational Design

There are three core dimensions of organizational design. These are referred to as complexity, formalization, and centralization (Hendrick and Kleiner, 2001). Complexity has two components — differentiation and integration. Differentiation refers to the segmentation of the organizational design. An organization can be segmented horizontally through, for example, departmentalization. Personal differentiation, a variation of horizontal differentiation, segments personnel, typically on the basis of specialization. Vertically, an organization is segmented into levels. In the past 25 yr, organizations have aggressively been decreasing their levels of vertical differentiation. Organizations can also be differentiated geographically or organized by region. In all cases, an increase in differentiation results in increased complexity, which also is assumed to increase cost.

Integration refers to the coordinating mechanisms in an organization. Coordinating mechanisms serve to link or tie together, the various segments. Coordinating mechanisms can be manual such as meetings and can be automated such as e-mail or other information system-based mechanisms. An increase in integration is also believed to increase complexity and, therefore, cost.

Formalization refers to the degree to which there are standard operating procedures, detailed job descriptions, or other systematic processes or controls. A highly formalized organization would have many of these procedures, processes, or controls.

Centralization refers to the degree to which decision-making is concentrated in a relatively few number of personnel, typically at the top of the organization. Organizations with relatively low levels of centralization are characterized by a more participative decision-making environment than organizations with high levels of centralization. Participatory ergonomics is a process consistent with the goal of decentralization (Hendrick and Kleiner, 2001, 2002). The term "empowerment" is also given to the notion of distributing authority throughout the organization.

12.3   Types of Organization

From the classical school of organization, there is a basic type of organizational design.

12.3.1   Functional Design

The functional organizational design organizes workers by common technical specialization domains as illustrated in Figure 12.1. At the plant level, maintenance, operations, quality assurance, human resources, and shipping would be example functions. Functional organizational design works best in small to moderately sized organizations (up to 250 employees), has standardized practices and products, and has a somewhat stable external environment. This basic design has its roots in the classical movement and, thus, is influenced by the military and the church. Some advantages of the functional organizational design include professional identity, professional development, and the minimization of redundancy. The major weaknesses associated with this structure is suboptimization, a condition that is characterized by competition, coordination, and communication challenges laterally across units at the same level in the hierarchy. The overall purpose of an organization should be stated in terms of its products and customers. The state of suboptimization occurs when a unit begins to behave as if the unit is the purpose of the overall organization rather than the actual purpose of the organization.

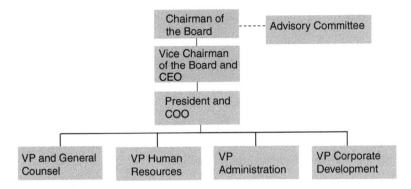

FIGURE 12.1 Functional organization structure.

12.3.2 Product Design

The product or divisional organizational design organizes workers by product cluster as seen in Figure 12.2. In many organizations, divisions characterize the clusters. This variation of the functional design attempts to minimize suboptimization. Instead of focusing on functions, personnel theoretically identify with a product and, therefore, the customer. Another intended advantage with this design is to allow the development and management of profit centers. Each division can be operated as a business within the business. However, within each product cluster or division, functions typically appear. Thus, the functional units still exist albeit at a lower level.

12.3.3 Geographic Design

A variation of the product structure is the geographic structure (see Figure 12.3). Instead of creating divisions or product clusters, this structure is organized by regions. In many retail organizations, customer preference varies by region and, thus, the geographical structure is a useful alternative. Each region can be operated as a profit center in such an organization. Like the product design, the geographic structure typically has functions beneath the geographic level.

12.3.4 Matrix Design

Since all of the previously mentioned alternatives are variations of the functional design, all have major shortcomings. Specifically, suboptimization occurs typically in all designs. That is, communication, competition, and coordination challenges prevail. Thus, designers derived a new alternative, mostly inspired from a combination of the functional and product structures. The function × product matrix (see Figure 12.4) attempts to integrate the best of functional and product structures. Specifically, the benefits associated with professionalism and lack of redundancy from the functional design are retained. From the product structure, focus on a customer reduces the possibility of suboptimization.

The major flaws associated with the matrix structure are the now confused lines of authority and communication. That is, a given employee has a supervisor on both the x and y axes. Having two bosses is not

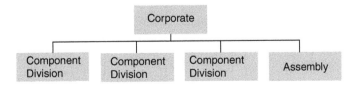

FIGURE 12.2 Product organizational structure.

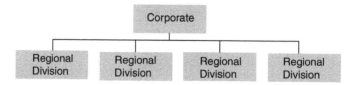

FIGURE 12.3   Geographical organizational structure.

only confusing, but it violates one of the core assumptions of organizational design — to incorporate lines of authority and communication in an organization.

12.4   Challenges in Organization Design

As discussed, none of the major types of organizational design are without flaws. In response to these shortcomings, organizations have attempted to improve conditions. Restructuring is one approach to manage the inherent weaknesses in organization designs. Restructuring refers to the modification of the organizational structure. This could entail adding or eliminating units, although most restructuring has involved combining or eliminating units rather than adding units. Restructuring has typically focused on the performance objective of reducing cost.

Another approach to the need for integration in organizations has been the incorporation of teams. In functional organizations, cross-functional teams have been used to create an informal organizational structure and to foster integration. Other types of teams have been used with other structures.

Some have created virtual organizations in an attempt to bypass the shortcomings of traditional organizations.

12.4.1   Paradox of Restructuring

Organizations typically restructure in order to reduce costs. However, not all organizations in fact reduce cost through this process. The goal of reducing cost coupled with the actual result of having costs increase is referred to as the "paradox of restructuring." In terms of the organizational design structural foundations discussed earlier, in essence what an organization is attempting to do is to reduce complexity via reduced vertical differentiation, which is expected to reduce cost since reduction in complexity

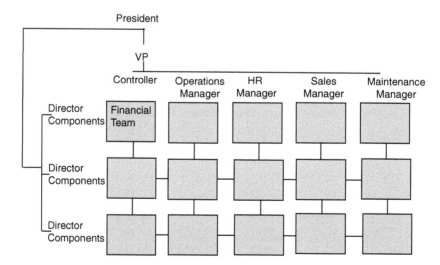

FIGURE 12.4   Matrix organizational structure.

will reduce cost as previously discussed. However, in reducing vertical differentiation, organizations have had to increase horizontal differentiation or integration. For example, with reduced vertical differentiation, many organizations have perceived the need to provide increased integration, which also increases complexity and therefore cost.

12.4.2 Macroergonomics

Macroegonomics is also sometimes referred to as "systems ergonomics" and can deal with systems of systems. Whether the professional is specifically trained in macroergonomics (or one of its allied subdisciplines such as human systems integration) or not an appreciation for the systems context of work can be most beneficial. Three highly interrelated work system design practices that frequently underlie dysfunctional work system development and modification efforts have been identified (Hendrick and Kleiner, 2001). These causes of safety, health, and performance problems are technology-centered design, a "left-over" approach to function and task allocation, and a failure to consider a work system's sociotechnical characteristics and integrate them into its work system design. In addition, it has been recognized that large-scale improvement of performance equires major sociotechnical systems (STS) change. This approach was implemented by Tavistok researchers in post-World War II Europe and was effective in creating an economic development transformation through industrial intervention more recently (Kleiner and Drury, 1999). The STS approach has been tested as an enduring, integrative, productive, and rich framework across a variety of sectors. Drawing inspiration from the socio-technical school as well as theoretical support, macroergonomics has emerged as a subdiscipline of ergonomics. While the term "macroergonomics" was first used in the mid-1980s, informally the approach has been used my some researchers for some time (Hendrick, 1991).

A simplified descriptive model views a work system as being consisting of several important and related subsystems. Figure 12.5 illustrates this work system using the construction occupation as an example. From the literature, five major components of STS systems can be identified. These are the (1) technological subsystem, (2) personnel subsystem, (3) external environment, (4) internal environment, and (5) organizational design. Of special importance are the interfaces among the various subsystems. Inadequate or inappropriate interfaces create a hazardous or under-performing work system. The technological subsystem is concerned with the manner in which work is performed. This includes the technology (e.g., machinery and equipment), tools, methods, and procedures. The personnel subsystem

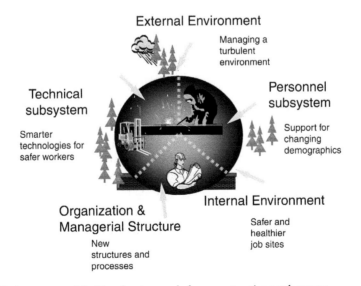

FIGURE 12.5 Work system model with subsystem goals for a construction work system.

concerns the sociocultural and socioeconomic characteristics of the people involved in construction, including their selection and training. The external environment consists of the political, economic, technological, educational, and cultural forces that affect the construction work system. It is recognized that this environment is increasingly turbulent for most industries. The work system must successfully procure resources from its environment and efficiently and effectively produce products back to the environment to be successful. The internal environment comprises the physical and cultural environments. The organizational design is focused on the organization structure and management of the work system. Of special importance are the interfaces among the various subsystems. Inadequate or inappropriate interfaces between a work system's organizational design and external environment can create an internal environment fraught with risk. Safety, health, and performance results are expected through this approach; a dramatic improvement in the culture is an outcome (Hendrick and Kleiner, 2001).

12.5 Organizational Design Transfer to Ergonomics

One of the key attributes of macroergonomics or organizational design and management in ergonomics is the transfer of organizational level attributes to ergonomic level implementation (Hendrick and Kleiner, 2001, 2002). In terms of complexity, formalization, and centralization, optimal prescriptions can be carried through to the group and individual operator level. Specifically, when designing jobs, human–machine interfaces and software interfaces, alignment should be facilitated between the organizational design and these lower-level design considerations. For example, in an organization with low levels of centralization, high degrees of decision-making discretion would be designed into group and individual jobs. Also, degrees and types of decision support can be designed to be consistent with the overall organizational design. Similarly, in an organization with high levels or organizational formalization the designer will have requirements to design standardized jobs and standard operating procedures. Degree of standardization or routine procedures would be affected. Specifications for information system design can also result from understanding the macro-level design requirements. Horizontal differentiation can be transferred through design of group tasks, group interfaces, and the breadth of jobs, including span of control. A lack of compatibility between the organizational design and the lower-level design would typically result in lower productivity and reduced job satisfaction (Argyris, 1971). In conclusion, depending on the extent to which the ergonomics is designed with respect to the overall organizational design, the overall performance can be maximized (Hendrick and Kleiner, 2001). This is what Hendrick and Kleiner (2001) called "optimal ergonomic compatibility."

References

Argyris, C. (1971). *Management and Organizational Development*. New York: McGraw-Hill.
Hendrick, H.W. (1991) Human factors in organizational design and management. *Ergonomics*, 34, 743–756.
Hendrick, H. W. and Kleiner, B.M. (2001). *Macroergonomics: An Introduction to Work System Design*. Santa Monica, CA: The Human Factors and Ergonomics Society Press.
Hendrick, H.W. and Kleiner, B.M. Eds. (2002). *Macroergonomics: Theory, Methods and Applications*. Mahwah, NJ: Lawrence Erlbaum, Associates.
Kleiner, B.M. and Drury, C.G. (1999) Large-scale regional economic development: macroergonomics in theory and practice. *Human Factors and Ergonomics in Manufacturing*, 9(2), 151–163.
Pasmore, W.A. (1988). *Designing Effective Organizations: The Sociotechnical Systems Perspective*. New York: Wiley.

V

Engineering Controls

13

Engineering Controls — What Works and What Does Not

Christopher Hamrick
Ohio Bureau of Workers'
Compensation

13.1 Introduction

In the United States, work-related musculoskeletal disorders (MSDs) are prevalent and expensive to treat. In 2001, there were 216,400 "repeated trauma" cases reported by private industry, which translates to a rate of 23.8 cases per 10,000 worker hours of exposure (United States Department of Labor, 2003). Data from the Ohio Bureau of Workers' Compensation (Hamrick, 2000) showed that MSDs account for 15.5% of all claims in the workplace, and 48.5% of all workers' compensation dollars paid go toward MSDs. Those data also indicated that approximately 40% of these claim costs resulted from back pain claims and 9% resulted from upper extremity claims.

Consequently, given the huge cost of MSDs to society in terms of lost time, money, and human suffering, it is in the interest of employers to control MSDs by reducing the incidence and severity of these disorders. This chapter examines the effect that incorporating engineering controls into the workplace can have upon the musculoskeletal health of the worker. First, a discussion of the physical exposures that increase the likelihood of MSDs along with various control strategies will be presented. Then evidence from intervention studies and selected case studies will be reviewed. Finally, recommendations for appropriate control strategies will be discussed.

TABLE 13.1   Evidence for Causal Relationship between Physical Work Factors and MSDs

Body Part	Risk Factor	Strong Evidence(+++)	Evidence (++)	Insufficient Evidence(+/0)	Evidence of No Effect (−)
Neck and Neck/ shoulder	Repetition	—	++	—	—
	Force	—	++	—	—
	Posture	+++	—	—	—
	Vibration	—	—	+/0	—
Shoulder	Posture	—	++	—	—
	Force	—	—	+/0	—
	Repetition	—	++	—	—
	Vibration	—	—	+/0	—
Elbow	Repetition	—	—	+/0	—
	Force	—	++	—	—
	Posture	—	—	+/0	—
	Combination	+++	—	—	—
Hand/wrist	Carpal tunnel syndrome				
	Repetition	—	++	—	—
	Force	—	++	—	—
	Posture	—	—	+/0	—
	Vibration	—	++	—	—
	Combination	+++	—	—	—
	Tendinitis				
	Repetition	—	++	—	—
	Force	—	++	—	—
	Posture	—	++	—	—
	Combination	+++	—	—	—
	Hand-arm vibration syndrome				
	Vibration	+++	—	—	—
Back	Lifting/forceful movement	+++	—	—	—
	Awkward posture	—	++	—	—
	Heavy physical work	—	++	—	—
	Whole body vibration	+++	—	—	—
	Static work posture	—	—	+/0	—

Source: Reprinted from Bernard (1997). With permission.

13.2   MSD Risk Factors

Controlling MSDs in the workplace requires an understanding of the factors that contribute to their occurrence. Factors that increase the likelihood of the occurrence of MSDs are referred to as "risk factors". Occupational risk factors are those risk factors associated with physical activities at work. A very thorough review of the literature (Bernard, 1997) found epidemiologic evidence to support a causal relationship between physical work factors and MSDs. A summary of these results is given in Table 13.1. The study concluded that a strong body of evidence exists to support an association between MSDs and high levels of risk factor exposure, particularly when multiple risk factors are present.

Further support for the notion that MSDs can be associated with occupational risk factors can be found in a report by the National Academy of Sciences, which concluded that the biology and biomechanics literature provides "evidence of plausible mechanisms for the association between musculoskeletal disorders and workplace physical exposures" (National Research Council, 2001). It stands to reason then that an effective approach to control MSDs must eliminate or substantially reduce worker exposure to these physical risk factors.

13.3 MSD Controls

Various intervention strategies have been employed in an attempt to control MSDs. These intervention strategies can be categorized as follows:

- *Engineering controls* — Design of the physical workplace to reduce exposure to physical hazards by machinery or equipment. Examples of engineering controls include automation, mechanization, hand tool design, and workplace layout changes.
- *Administrative controls* — Changes to organizational factors in the workplace. Such changes could include employee work scheduling, job rotation, and job enlargement.
- *Behavioral controls* — Designed to change an employee's knowledge or behavior. Behavioral controls can include hazard-awareness training, training in work methods, or behavior-based safety programs.
- *Personal protective equipment* (PPE) — Barriers between the individual and the hazard that are worn by the individual. Examples of PPE intended to be used to reduce PPE include knee pads and vibration-absorbing gloves.

Of these categories of controls, only engineering controls address worker exposure to MSD risk factors by controlling the risk factors at the source. With respect to prevention of back pain, Snook (1988) stated that "primary prevention of low back pain ... is best achieved through good ergonomic design of the workplace." It is evident, then, that a comprehensive strategy to control MSDs must include control of worker exposure to risk factors by engineering out the hazards.

Examples of engineering controls include adjustable height work surfaces to allow for work in a neutral posture, automation to reduce forces and repetitive motion, material handling aids to reduce forces associated with lifting, and tools that reduce vibration and forces in the hand.

13.4 Evaluating Effectiveness of Controls

The "gold standard" for evaluating the effectiveness of ergonomic controls, or any other intervention study, involves random selection of subjects to interventions and control groups, blinding the subjects and researchers as to group membership, and a placebo treatment for the control group (Westgaard and Winkel, 1997). These authors go on to state that when performing ergonomic intervention studies, this design is "generally not achievable" in practice because subject membership is given and researchers cannot be blinded to exposure assessment. The difficult challenges facing researchers attempting to meet this standard are evidenced by the fact that none of the papers reviewed in their very comprehensive literature review met all of the above criteria.

Due to the inherent difficulties and practicality of incorporating all of the above experimental controls in a working industrial setting, a common design used in intervention studies is the before-and-after (or pre-and-post) design (Robson et al., 2001). In this design, outcome measures of efficacy or effectiveness are recorded before the intervention has taken place. Then, the same outcome measures are taken after the intervention has been implemented, and the measures are then compared. This design is more useful for demonstrating the immediate impacts of interventions, and less useful for long-term studies because other changes can occur which may affect the results. Controlling these changes, or threats, is critical to the validity of the before-and-after design.

One other factor that must be considered when evaluating the effectiveness of engineering controls is the outcome measure. Many evaluation studies report efficacy measures, which include the effect upon exposure to mechanical stressors, or risk factors. The advantage of these types of measures is that feedback can often be obtained very quickly; once an engineering change is made, one would generally expect an immediate change in worker exposure to the mechanical stressors. However, these stressors may or may not directly relate to improved health for the worker. Consequently, it is desirable to also measure direct health outcomes, that is, workers' compensation claims, Occupational Safety and Health Administration (OSHA)-recordable injuries or illnesses, days away from work, or days of restricted work activity. Collecting these data can provide a more valid assessment of the actual impact of the intervention on health assessments, but it also requires a much longer follow-up measurement period. Longer follow-up periods also increase the chances that an outside influence affects the outcome.

Additional, business measures, such as productivity, quality, and return on investment, provide useful information. Such data can be used by employers when evaluating the economic feasibility of proposed design concepts.

13.5   Results of Intervention Studies

One of the most comprehensive literature reviews of ergonomic intervention studies conducted to date was by Westgaard and Winkel (1997). They examined high-quality studies that looked at the effects of altering mechanical exposures (through engineering controls), changing the production system or organizational culture, and strategies to help the worker deal with existing job demands. Of the 20 intervention studies that addressed mechanical exposure, six measured acute responses, nine measured musculoskeletal health, and five took both types of measures. The authors concluded that "modifier interventions that actively involve the worker often achieve positive results," suggesting that the reduction of mechanical exposures through engineering controls may be a necessary, but not sufficient, strategy to reduce the incidence rate of MSDs. Organizational support and worker involvement should be incorporated into any engineering control strategy. The authors also stated that the more "'passive' measures (health education, relaxation training) do not appear equally successful," indicating that unless the primary risk, mechanical exposure, is addressed, success will probably be limited.

Another literature review studied interventions for the primary prevention of work-related carpal tunnel syndrome (Lincoln et al., 2000). The authors found studies that relay positive results by instituting multiple-component ergonomics programs, alternative keyboard supports, and mouse and tool redesign. However, they also state, "none of the studies conclusively demonstrates that the interventions would result in the primary prevention of carpal tunnel syndrome in a working population."

In an evaluation of the impact of engineering controls placed in 32 repetitive materials handling jobs using a control group of four jobs in which no control was placed, Marras et al. (2000) found that lift tables, which are intended to reduce trunk bending, and lift aids, which are intended to eliminate the external load, significantly reduced the rates of low back disorders. The study also failed to find a significant reduction in low back disorders for nine interventions categorized as "work area redesign" and other "equipment." The study found no reduction in the trunk kinematic variables for these jobs, suggesting that the incidence rates were not reduced because these particular interventions failed to reduce the mechanical stressors.

Due to the high rates of MSDs among nurses and nurses aids, Evanoff et al. (2003) investigated the effects of mechanical patient lifts in four acute-care hospitals and five long-term care facilities. They found significant reduction in injury and lost days rates after patient lifts were implemented. The effect was larger in the long-term care facilities, and the authors attribute the greater effect to a more active encouragement by management.

Another study examining the effects of patient handling equipment was conducted in a long-term care facility by Garg and Owen (1992). The outcome measures included back injury incidence and severity rates, biomechanical demands, and perceived physical stress levels. The study found a 43% reduction in back injury rates and a 50% reduction in back injury severity rates after the controls were put into place. They also documented reduction in compressive forces on the spine in the modified tasks, reinforcing the notion that successful controls must reduce worker exposure to mechanical stressors.

With regards to lessening the severity of existing MSD claims, Arnetz et al. (2003) demonstrated that by incorporating ergonomic improvements together with therapy and training into injured workers' work environment, the number of sick days due to MSDs was decreased. The likelihood of returning to work was also found to be 50% for those in the intervention group. Consequently, at 12 months, the benefit-to-cost ratio was calculated to be 6.8. These results imply that in addition to having a role in primary prevention, engineering controls may also have a role in helping workers return to work by lessening the physical demands of the job.

A review of intervention studies designed to evaluate the effect of interventions on neck and upper limb disorders was conducted by Kilbom (1988). The author stated that "the success or failure of an intervention depends on the effectiveness of the intervention technique — that is, whether the intervention does lead to the expected change in posture, work method, work organization, attitude, knowledge or psychosocial climate, and whether this change in work content is sufficient to influence the development of the disorder." At the time the review was conducted, only one study was found to evaluate both of these steps. It was concluded, then, that intervention studies should use a range of both intermediate and end-point outcome variables.

A National Academy of Sciences panel looked at six different literature reviews to determine the scientific evidence of the impact of ergonomic interventions (National Research Council, 2001). Their report stated, "Ergonomic job redesign, modifying organizational culture, and modifying individual factors can prevent low back pain and/or disability." They concluded, "Interventions must mediate physical stressors, largely through the application of principles of ergonomics."

It is apparent that there are few intervention studies that meet the most rigorous scientific and statistical standards of proof of the effectiveness of ergonomic interventions and, specifically, engineering controls. However, numerous reports of effectiveness are reported through case studies. Despite some methodological shortcomings with these studies, they do provide real-world examples that engineering controls can, under certain circumstances, positively impact worker health outcomes by reducing worker exposure to risk factors and by reducing MSD incidence rates.

For example, one case study documented by Moore (1994) involved an engineering change of a flywheel truing task in an automotive manufacturing facility. The design change was conducted through a committee, which defined the problem, brainstormed solutions, determined the most favored solution, and evaluated the change. The committee had representatives from engineering (including one organized labor representative), supervisors, and machine operators. The change reduced incidence rate for all disorders by 29%, and the severity rate was reduced by 82%. Subjective responses from workers were also solicited by a questionnaire, and those results indicated that the workers' perceptions of the change were also positive.

13.6   Results of Ohio BWC "Safety Grant" Studies

In an effort to demonstrate the effectiveness of engineering interventions on the reduction in MSD incidence rates, the Ohio Bureau of Workers' Compensation (BWC) instituted the "Safety Grants" program in 1999. The BWC Safety Grants program has provided assistance to Ohio employers to help them reduce their risk of MSDs in the workplace by incorporating engineering controls. All controls were reviewed by a panel of trained ergonomists to ensure, as best as possible, that the controls would address worker exposure to the physical stressors present. As part of the program the Ohio BWC collected information about job designs that employers have used to reduce the risk of MSDs in their workplaces.

Participating companies report the effectiveness of the interventions by measuring OSHA-recordable MSD incidence rates, lost days due to MSDs, and restricted days due to MSDs. They also measure the relative risk of injury by completing "risk factor" assessments for affected tasks with an assessment tool. The assessment tool was a slightly modified version of the checklist developed by the OSHA for the 1995 draft standard for prevention of work-related MSDs (Schneider, 1995). These assessments provide a measure of the relative risk of injury for a specific task. Using a "before-and-after" nonexperimental design, the measures were reported for a period prior to implementing the engineering control, and then for a 2-yr period after implementing the engineering control.

Results of these case studies were compiled together in the form of industry "best practices" (Ohio Bureau of Workers' Compensation, 2002a through 2002d). The best practices conglomerate the results of specific types of controls within industry groups. Some of the demonstrated engineering control "best practices" and measures of their effectiveness are presented below.

13.6.1 Best Practices in Manufacturing — Lifting Aids, Lift Assist Devices, and Transport Devices

Lifting aids can be installed to lift, tilt, or turn materials. These devices do not eliminate the need to handle material, but they can aid in locating materials so that they can be handled with minimal trunk flexion and minimal reaching. Hence, the forces on the spine are reduced as is the risk of back injury. Examples of lifting aids are lift tables (Figure 13.1), tilt tables (Figure 13.2), and turn tables.

Lift assist devices use mechanical means to lift materials, thus reducing work exposure to the risk factors commonly associated with manual materials handling. These devices greatly reduce the forces on the body by using mechanical means (usually electric, hydraulic, or pneumatic) to provide the lifting power. Such devices include hoists, cranes, manipulators, vacuum lifters (Figure 13.3), and magnetic lifters (Figure 13.4).

Manually transporting material often involves lifting, carrying, pushing, or pulling. These activities all have the potential to generate large forces in the hands, wrists, elbows, and shoulders. These large forces can, in turn, lead to injuries in the affected body part. Transport devices reduce the risk of injury by providing the force through mechanical means or, as in some cases, eliminating the need to manually handle the material at all. Some examples of transport devices include carts, conveyors, tugs (Figure 13.5), powered dollies, and forklifts.

In 65 manufacturing companies that received lift assists, aids, or transport devices, after an average follow-up period of 214 days, the following results were seen:

- The MSD incident rate (incidents per 200,000 h) changed from 9.8 to 4.9 — a 50% improvement.
- The days lost due to MSDs improved from 110 days per 200,000 h worked to 36.2 — a 70% improvement

FIGURE 13.1   Powered scissor-lift.

FIGURE 13.2 Lift and tilt table reduces bending by providing easy access to parts.

- The restricted days due to MSDs improved from 102 days per 200,000 h worked to 39.5 — a 61% improvement
- The average risk factor score for 120 affected tasks in the 65 companies was 33 before the devices were put into place, and was 19 afterwards — a 42% improvement

13.6.2 Best Practices in Manufacturing — Workstation Design

Many manufacturing facilities have workstations where parts are cleaned, assembled, or packed. The work at these stations can be performed while the worker is sitting, standing, or is able to alternate between the two. It is generally accepted that good ergonomic principles accounting for worker

FIGURE 13.3 Vacuum lifter being used to lift a load.

FIGURE 13.4   Magnetic lift attached to a hoist.

anthropometry should be incorporated into workstation design. Some examples of these principles that have been incorporated through the Safety Grants program are:

- Provide antifatigue floor mats in areas where employees stand for long periods of time
- Where workers must sit, provide chairs with good ergonomic design features (e.g., Chaffin et al., 1999), or allow the worker to alternate between sitting and standing
- Reduce wrist flexion when reaching into boxes by orienting the heights and angles. Furthermore, clips can be used to hold the lids down so that the worker does not have to reach around or over the lids, as shown in Figure 13.6

In nine manufacturing locations that incorporated ergonomic workstation design, after an average follow-up period of 247 days, the following results were seen:

- The MSD incident rate (incidents per 200,000 h) changed from 10.0 to 1.6 — an 84% improvement
- The days lost due to MSDs improved from 24.9 days per 200,000 h worked to 16.0 — a 36% improvement

FIGURE 13.5   Tug being used to pull carts, eliminating the forces associated with manual pushing and pulling.

FIGURE 13.6　Tilted box with clips to reduce wrist bending.

- The restricted days due to MSDs improved from 58.5 days per 200,000 h worked to 6.4 — a 61% improvement
- The average risk factor score for 29 affected tasks in the nine locations was 29.7 before the devices were put into place, and was 17.5 afterwards — a 43% improvement

13.6.3　Best Practices in Manufacturing — Automation

When "automation" is implemented the process is done entirely by machine, and the worker then usually has the role of operating, monitoring, and sometimes loading the machinery. Examples of automation include CNC machines, automatic case packers, and palletizers.

In 24 manufacturing locations that incorporated automation, after an average follow-up period of 214 days, the following results were seen in the affected population:

- The MSD incident rate (incidents per 200,000 h) changed from 10.4 to 7.2 — a 31% improvement
- The days lost due to MSDs improved from 123 days per 200,000 h worked to 23.1 — an 81% improvement
- The restricted days due to MSDs improved from 239 days per 200,000 h worked to 57.4 — a 76% improvement
- The average risk factor score for 46 affected tasks in the 24 locations was 28.5 before the devices were put into place, and was 15.4 afterwards — a 46% improvement

13.6.4　Best Practices in Manufacturing — Semiautomation

When "semiautomation" is implemented, a particularly hazardous part of the job is usually automated, but the intervention still requires some substantial operator involvement, such as operating the machine and providing continuous control. Examples of semiautomation include controlled lathes, saws, grinders, and presses. An example of a semiautomated control that reduces the risk factors associated with a manual torque task described earlier is shown in Figure 13.7.

In 16 manufacturing locations that incorporated semiautomation, after an average follow-up period of 229 days, the following results were seen:

- The MSD incident rate (incidents per 200,000 h) changed from 32.4 to 9.7 — a 70% improvement
- The days lost due to MSDs improved from 215 days per 200,000 h worked to 63.0 — a 71% improvement

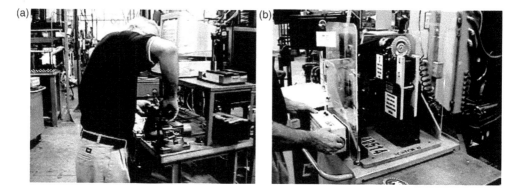

FIGURE 13.7   High force requirements and awkward postures can be seen in the manual torque task (a). The semiautomated torque device reduces worker exposure to these mechanical stressors (b).

- The restricted days due to MSDs improved from 197 days per 200,000 h worked to 0 — a 100% improvement
- The average risk factor score for 17 affected tasks in the 16 locations was 21.4 before the semi-automation was implemented, and was 10.4 afterwards — a 51% improvement

13.6.5   Best Practices in Health Care — Patient Lifting Devices

One of the greatest risks to health-care workers stems from manually lifting and moving patients or residents. Nurses and nurse's aides have among the highest rates of back injuries of any occupational group. Manually moving patients or residents results in very high stresses in the spine; even the safest of manual patient handling tasks under ideal conditions poses significant risk of injury (Marras et al., 1999). These stresses are caused by lifting high weights in awkward postures.

Many powered patient-lifting devices are commercially available that will reduce the forces and awkward postures associated with manually lifting patients. Some devices are on wheels and can be used to mechanically lift and move patients. Devices are available which can provide a total lift or which can aid residents in standing from a seated position (Figure 13.8).

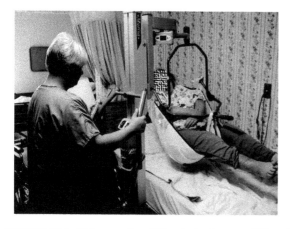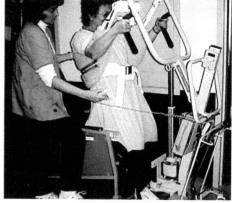

FIGURE 13.8   Using a patient lifting device for a total lift and for a sit-to-stand lift.

The Ohio BWC analyzed the data on injuries from health-care facilities that have received Safety Grants to install patient-lifting devices. For 27 health-care facilities with an average follow-up period of 298 days, the following results were seen:

- The cumulative trauma disorder (CTD) incidence rate has gone from 21.3 CTDs per 200,000 h to 11.9 CTDs per 200,000 h worked — a 44% improvement
- The days lost due to CTDs went from 127.2 per 200,000 h worked to 79.0 per 200,000 h worked — a 38% improvement
- The restricted days due to CTDs changed from 96.6 per 200,000 h worked to 87.0 per 200,000 h worked — a 10% improvement
- The average risk factor score for patient-lifting tasks was 70 before installing the floor-based patient-lifting devices, and was 30.5 after the lifting devices were installed — a 56% improvement

13.6.6 Best Practices in the Office — Computer Workstation Design

One of the most common situations that employers encounter is in the use of computer workstations. Some of the MSD risk factors that can be present in improperly designed or adjusted computer workstations include:

- Repetitive motion from typing
- Awkward postures in the wrists
- Lack of lumbar support in the spine
- Pressure concentrations on the wrists

By reducing worker exposure to these risk factors through workstation design and adjustment, it is theorized that the incidence and severity of MSDs can also be reduced. The Ohio BWC has analyzed data on injuries from organizations that have received Safety Grants to redesign office and computer workstations. The results from 14 facilities, with an average follow-up period of 270 days, are as follows:

- The CTD incidence rate has gone from 12.4 CTDs per 200,000 h to 4.4 CTDs per 200,000 h worked — a 64% improvement
- The days lost due to CTDs went from 45.6 per 200,000 h worked to 8.1 per 200,000 h worked — an 82% improvement
- The restricted days due to CTDs changed from 16.6 per 200,000 h worked to 0 per 200,000 h worked — a 100% improvement
- The average risk factor score office tasks was 26.3 before redesigning the workstations, and was 13.8 after the change — a 48% improvement

13.7 Other Factors Influencing Effectiveness

The collective experience of intervention studies and case studies indicates that engineering controls can play an important role in improving workers' health, at least under some conditions. Engineering controls are designed to eliminate or reduce physical stressors in the workplace. It should be noted that in field studies engineering controls are rarely instituted in a vacuum; other factors are present which undoubtedly have an effect on the outcome of the studies.

Norman and Wells (1998) present a model of workplace outcomes of interest and the influence of other factors. Personal and environmental attributes, including the physical, psychosocial, and organizational variables, can directly influence measured outcomes. Additionally, on a larger scale, the management structure, the industrial processes, and all people involved influence the measured outcomes as well as the personal and environmental systems.

For example, psychosocial factors, from both inside and outside of the workplace, can influence whether an engineering control is accepted and used in the intended task, whether an employee reports an MSD, and whether the MSD results in a compensable workers' compensation claim. A discussion of psychosocial work factors and their relationship to ergonomics can be found in Carayon and Lim (1999).

Additionally, given that workers generally interact with the implemented engineering control, effective implementation should also address the behavioral component of that human–machine interaction. Workers should be given instruction on proper equipment use, including safe operation and maintenance procedures, and follow-up should be performed to ensure that the training was understood and that safe operating procedures are adhered to. So, when implementing and evaluating engineering controls, it is important that administrative and behavioral factors are considered and documented.

13.8   Summary — What Works

Based on available evidence, it appears that reducing mechanical exposure is a necessary condition for reducing work-related MSDs; engineering controls that eliminate or substantially reduce MSD risk factors can be an effective strategy. Furthermore, it is also plausible that the higher the level of reduction in worker exposure to mechanical stressors, the higher the impact the control will have upon MSD rates, although further research may be needed in order to support this assertion.

However, the implementation of engineering controls alone may not be sufficient to eliminate work-related MSDs. Any of the other contributing factors in the workplace can act to undermine even the best engineering controls. Therefore, psychosocial, behavioral, and administrative factors must be addressed when implementing engineering controls.

The National Institute of Occupational Safety and Health (NIOSH) has compiled a document, based upon their experience, of effective ergonomic intervention strategies (Cohen et al., 1997). They concluded that there are seven elements that must be incorporated into an effective ergonomics program. These elements are:

1. Looking for signs of MSDs in the workplace
2. Management commitment and worker involvement
3. Training on evaluating problem areas
4. Gathering data to systematically identify problems
5. Identifying and evaluating controls
6. Establishing health care management
7. Being proactive by incorporating ergonomics into the planning stage

So, in order for engineering controls to have a high likelihood of success, they must reduce mechanical exposure *and* incorporate positive organizational attributes, including committed management and worker involvement. Furthermore, those implementing engineering controls must realize that the control may not initially work as anticipated. Appropriate follow-up efficacy measures should be taken, and worker feedback should be solicited after any control is implemented to ensure that worker exposure to risk factors is reduced and that no new risk factors have been introduced. Based upon the follow-up results, it may be necessary to make adjustments to the intervention to ensure that it is working as intended.

13.9   Need for Additional Intervention Studies

Based on the large number of intervention and case studies reporting the efficacy and effectiveness of ergonomic interventions, it can be inferred that engineering controls can positively influence MSD rates in the workplace, at least in some circumstances. However, despite the relatively large number of studies, very few of them satisfy all elements of scientific rigor. Hence, given the current state of knowledge, it is not possible to specify how effective specific controls will be in all workplace conditions.

Westgaard and Winkel (1997) call for future ergonomic intervention research that puts "considerably more focus on the intervention process" to "improve our understanding of the barriers and facilitators to the implementation of ergonomic measures." Goldenhar and Schulte (1994) concluded from their review of health and safety intervention studies in general that "more occupational health and safety intervention research needs to be conducted, especially as an integral part of a large-scale research agenda ..." More recently, Marras (2000) stated that there is a dire need for high-quality intervention studies in the area of low back disorder research.

Thus, practitioners, researchers, and students can advance our understanding of work-related MSD prevention, using engineering controls and other means, by conducting systematic evaluations of intervention strategies. NIOSH and the Institute for Work and Health have developed a guide for that specific purpose (Robson et al., 2001); the reader is referred to this guide as a resource when embarking upon future intervention studies.

In summary, although the current state of knowledge provides us with direction, and it provides evidence that engineering controls can positively impact worker musculoskeletal health, there remains a need for more research before we can specifically say which ergonomics interventions will work and which will not. Ergonomics practitioners, governmental policy makers, and workers will benefit greatly from the body of knowledge that will be gained by future, scientifically rigorous intervention studies.

References

Arnetz, B.B., Sjogren, B., Rydehn, B., and Meisel, R., 2003. Early workplace intervention for employees with musculoskeletal-related absenteeism: a prospective controlled intervention study, *Journal of Occupational and Environmental Medicine*, 45(5), 499–506.

Bernard, B.B., 1997. Musculoskeletal disorders and workplace risk factors: a critical review of epidemiologic evidence for work-related musculoskeletal disorders of the neck, upper extremity, and low back, DHHS (NIOSH) Publication #97–141.

Carayon, P. and Lim, S., 1999. Psychosocial work factors. In *The Occupational Ergonomics Handbook* (Edited by Karwowski, W. and Marras, W.S.), CRC Press, Boca Raton, pp. 275–284.

Chaffin, D.B., Andersson, G.B.J., and Martin, B.J., 1999. *Occupational Biomechanics*, 3rd ed, John Wiley, New York.

Cohen, A.L., Gjessing, C.C., Fine, L.J., Bernard, B.P., and McGlothlin, J.D., 1997. Elements of ergonomics programs: a primer based on workplace evaluations of musculoskeletal disorders, DHHS (NIOSH) Publication #97–117.

Evanoff, B., Wolf, L., Aton, E., Canos, J., and Collins, J., 2003. Reduction in injury rates in nursing personnel through introduction of mechanical lifts in the workplace, *American Journal of Industrial Medicine*, 44, 451–457.

Garg, A. and Owen, B., 1992. Reducing back stress to nursing personnel: an ergonomic intervention in a nursing home, *Ergonomics*, 35(11), 1353–1375.

Goldenhar, L.M. and Schulte, P.A., 1994. Intervention research in occupational health and safety, *Journal of Occupational Medicine*, 36(7), 763–775.

Hamrick, C.A., 2000. CTDs and ergonomics in Ohio, *Proceedings of the IEA 2000/HFES 2000 Congress*, 5-111–5-114.

Kilbom, A., 1988. Intervention programmes for work-related neck and upper limb disorders: strategies and evaluation, *Ergonomics*, 31(5), 735–747.

Lincoln, A.E., Vernick, J.S., Ogaitis, S., Smith, G.S., Mitchell, C.S., and Agnew, J., 2000. Interventions for the primary prevention of work-related carpal tunnel syndrome, *American Journal of Preventive Medicine*, 18(4S), 37–50.

Marras, W.S., 2000. Occupational low back disorder causation and control, *Ergonomics* 43(7), 880–902.

Marras, W.S., Davis, K.G., Kirking, B.C., and Bertsche P.K., 1999. A comprehensive analysis of low back disorder risk and spinal loading during the transferring and repositioning of patients using different techniques, *Ergonomics*, 42(7), 904–926.

Marras, W.S., Allread, W.G., Burr, D.L., and Fathallah, F.A., 2000. Prospective validation of a low-back disorder risk model and assessment of ergonomic interventions associated with manual materials handling tasks, *Ergonomics*, 43(11), 1866–1886.

Moore, J.S., 1994. Flywheel truing — a case study of an ergonomic intervention, *American Industrial Hygiene Association Journal*, 55(3), 236–244.

National Research Council, 2001. *Musculoskeletal Disorders and the Workplace — Low Back and Upper Extremities*, National Academy Press, Washington, DC.

Norman, R. and Wells, R., 1998. *Ergonomic Interventions for Reducing Musculoskeletal Disorders: An Overview, Related Issues and Future Directions*, Report for the Institute for Work & Health, Royal Commission on Workers' Compensation in British Columbia.

Ohio Bureau of Workers' Compensation, 2002a. *Ergonomics Best Practices for the Construction Industry*, Ohio Bureau of Workers' Compensation, Columbus, OH.

Ohio Bureau of Workers' Compensation, 2002b. *Ergonomics Best Practices for Extended Care Facilities*, Ohio Bureau of Workers' Compensation, Columbus, OH.

Ohio Bureau of Workers' Compensation, 2002c. *Ergonomics Best Practices Manufacturing*, Ohio Bureau of Workers' Compensation, Columbus, OH.

Ohio Bureau of Workers' Compensation, 2002d. *Ergonomics Best Practices for Public Employers*, Ohio Bureau of Workers' Compensation, Columbus, OH.

Robson, L.S., Shannon, H.S., Goldenhar, L.M., and Hale, A.R., 2001. Guide to evaluating the effectiveness of strategies for preventing work injuries: how to show whether a safety intervention really works, DHHS (NIOSH) Publication #2001-119.

Schneider, S. (ed.), 1995. OSHA's draft standard for prevention of work-related musculoskeletal disorders, *Applied Occupational Environmental Hygiene*, 10(8), 665–674.

Snook, S.H., 1988. Comparison of different approaches for the prevention of low back pain, *Applied Industrial Hygiene*, 3(3), 73–78.

United States Department of Labor, Bureau of Labor Statistics (www.bls.com), 2003. Public Data Query, December 16.

Westgaard, R.H. and Winkel, J., 1997. Ergonomic intervention research for improved musculoskeletal health: a critical review, *International Journal of Industrial Ergonomics*, 20, 463–500.

14

General Knowledge Regarding Engineering Controls

Richard W. Marklin
Marquette University

14.1 Introduction

There are three types of controls that ergonomics practitioners can implement to reduce the risk of musculoskeletal disorders (MSDs) and acute trauma in the workplace. These controls are engineering controls, administrative controls, and personal protective equipment (Chengular et al., 2004). Engineering controls, which are usually the most effective long-term controls to reduce risk of MSDs, are modifications to the workplace that fundamentally change workers' exposure through physical modifications to the work or workplace (AFMA, 2003). The purpose of engineering controls is to significantly reduce the physical stressors of a job through fundamental changes, without relying upon the compliance of a worker, because the controls largely do not give the worker the option of performing the work in a different way. Some engineering control are listed below. Refer to Cohen et al. (1997) for an extensive list of engineering controls:

- Changing tools or modifying features of hand or power tools, such as adding more comfortable handles, changing the angle of handles (pistol grip for horizontal exertions and in-line grip for vertical movements), changing the torque, reducing the weight of the tool, moving the tool's center of mass so it is above the center of the hand, etc.
- Changing the manufacturing or production process to reduce risk factors of MSDs, such as new technology that orients parts so that the worker does not have to lean or reach or new processes that reduce exertions and motions that stress the same body parts
- Changing the sequence of assembly or operations to reduce repetitive manual work. Typically, these changes often enhance productivity
- Adjusting the pace of work, either through the speed of the line or quotas, to reduce repetitiveness of the work
- Changing the method by which parts and materials are transported. For example, fully automatic lift systems or mechanical assistive systems can eliminate the physical stress from manual lifting

- Removing physical obstructions or impediments to work flow, which can reduce awkward postures and force exertions
- Adding antifatigue mats to flooring for workers who have to stand in the same place all day
- Purchasing high-quality carts or trolleys that have well-lubricated rolling mechanisms and easy steering to reduce pushing and pulling forces

Administrative controls are changes to the work organization and methods of performing work that do not involve physical changes. Frequently, these controls require formal education and training of workers and management. Administrative controls are not fundamental changes to the workplace, as are engineering controls. Examples of administrative controls are implementing rest breaks, rotating workers, exercise programs, encouraging employees to keep their wrists in a neutral posture while doing manual work, and encouraging workers to perform lifting tasks with good body mechanics — keeping the load as close as possible to the body and using the legs to lift (AFMA, 2003). Administrative controls require the compliance of the employee to implement the control and vigilance of the employer; noncompliance of the worker may not reduce the exposure to risk of MSDs and is the main weakness of administrative controls. For example, a worker who is scheduled to rotate jobs may stay at one job longer than recommended (for a variety of reasons) and thus not benefit from the variable musculoskeletal loading job rotation offers. Although some administrative controls can be very effective, such as worker rotation, administrative controls are generally less effective than engineering controls. In many situations, administrative controls require the employer to vigilantly enforce the controls in order for them to be effective.

The least reliable approach to reducing risk of MSDs is the use of personal protective equipment (PPE). PPEs, in theory, protect the worker from the risks of MSDs from the workplace if properly worn. However, if the PPEs are not worn correctly, they may not provide the necessary protection. Furthermore, PPEs may not be as effective as manufacturers claim.

14.2   Examples of Engineering Controls

Although engineering controls have widely been implemented in industry to "design out" workers' exposures to risk factors of MSDs, many of those controls have not been published and as such are not available to the public. However, several trade groups, research associations, and even governmental organizations have documented engineering controls and made them available to either their own members through handbooks or websites or made them available to the public (either for free or for a fee). The remainder of this chapter will describe engineering controls that various organizations have documented and how these controls can be accessed by the reader.

14.2.1   Graphic Communications Industry

Funded by a grant from the U.S. Occupational Safety and Health Administration (OSHA), the Graphic Arts Technical Foundation (GATF) published an ergonomics guidebook for the printing industry (GATF, 2003), which includes lithography, screen printing, and newspapers. The guidebook shows how common jobs in the printing industry can be improved to reduce risk factors of MSDs. The ergonomics issues and solutions for each job are described in lay language and shown with photos on only one page in an easy-to-read format. Many of the controls in the guidebook are engineering controls, which reduce the risk factors of MSDs such as awkward trunk posture, particularly forward flexion; repetitive twisting of the trunk, especially when lifting and lowering; lifting and lowering heavy loads manually; and extreme ulnar deviation and flexion of the wrist. Examples of engineering controls in the guidebook are a pneumatic pallet lift to raise bundles of paper, a vacuum hoist to lift bundles of paper off the floor, a built-in turntable to rotate large rolls of paper, a trolley to move large rolls from one area to the next, and a pistol grip for a spray gun that rinses mesh. This ergonomics guidebook is available to the public for a nominal fee from the GATF.

14.2.2 OSHA Ergonomics Guidelines

The OSHA published ergonomics guidelines for three work sectors — nursing homes (OSHA, 2003), retail grocery stores (OSHA, 2004a), and poultry processing industry (OSHA, 2004b). These guidelines provide recommendations to employers in the respective sectors to help reduce the number and severity of work-related MSDs, such as low back pain, sciatica, rotator cuff injuries, epicondylitis, and carpal tunnel syndrome. The engineering controls in the guidelines are illustrated with simple line drawings and described in lay language. According to OSHA, the guidelines are advisory in nature, are informational in content, and are not regulatory standards. All the ergonomics guidelines can be downloaded from the OSHA websites listed in the References section.

The nursing home sector was chosen for the guidelines because of the high physical demands placed on nursing home workers and the growing number of workers in nursing homes as the U.S. population ages. These workers often are required to aid patients for walking, bathing, or other normal daily activities. Sometimes, patients must rely upon these workers for their total mobility, and, in particular, manual lifting of these patients expose nursing home workers to MSDs. The large amount of weight lifted or transferred, awkward postures from leaning over a bed or working in a confined area, and shifting of weight all expose nursing home workers to MSDs. The ergonomics guidelines for nursing homes (OSHA, 2003) suggest several engineering controls. Examples of engineering controls are a powered sit-to-stand device for transferring patients, who are partially dependent and can move and place weight on their joints, from bed to chair (or vice versa) or for bathing or using the toilet; portable devices, which can hang from the ceiling or a stand, for lifting totally dependent patients from bed to chair (or vice versa) or to the bathroom; a device for assisting residents, who have weight-bearing capability, to walk safely with a reduced risk of falling; and a board that reduces the friction (and hence pushing or pulling force) from transferring patients laterally between two horizontal surfaces.

OSHA developed ergonomics guidelines for workers in retail grocery stores (OSHA, 2004a), who are a significant portion of the U.S. workforce, because the cost of MSDs was almost 50% of the total costs of serious injuries in 2001 and the cost from MSDs had been rising at a rapid rate, after adjusted for inflation. Recommended engineering controls for front-end workers (checkout, bagging, and carryout) include a powered in-feed conveyor to help cashiers bring the items to the cashiers' optimal work zone (rather than leaning and reaching to get items); a sweeper to move items close to the worker at point of scanning; a height-adjustable keyboard tray to reduce head and trunk bending and reaching; bags with handles because workers can grasp the bags with a power grip, compared to the more stressful pinch grip; and a powered tug vehicle that pushes a line of nested shopping carts from the parking lot to the store. Engineering controls for deli workers include tongs with long handles for grasping meat in the front of the display case; knives with handles bent at various angles to reduce severed wrist deviations; and grinders that drop sliced meat at a height between the waist and the chest. For produce workers, a lightweight, short-handled plastic shovel for ice is recommended for moving ice rather than a heavier shovel or hand scoop because a worker can move more ice more quickly with less physical stress.

The ergonomics guidelines for poultry processing (OSHA, 2004b) were developed because many workers in the poultry processing sector are exposed to risk factors of MSDs and 30% of the lost-days injuries were MSDs in 2002. One salient risk factor to which poultry processing workers are exposed is repetition; some poultry workers make over 25,000 cuts per day processing chickens and turkeys. Examples of recommended engineering controls are cutouts in the work surface that surround the front and sides of the torso so that the worker can get closer to items on the work surface; mechanical tilters and dumpers that unload contents of a container into a machine; embedded scales that are integrated into the production process to eliminate unnecessary handling of poultry parts and waste; chutes under the work surface into which the worker can drop poultry parts or waste for transport, rather than twisting and putting material into a bucket; knives that have handles at various angles to reduce extreme wrist deviations and allow the worker to grasp it with a power grip; and vacuum systems for lifting and transport of materials, rather than manual lifting.

14.2.3 Furniture Manufacturing

In 2003, the American Furniture Manufacturers Association (AFMA) published a voluntary ergonomics guideline for furniture manufacturers to reduce the incidence and severity of MSDs affecting their workers (AFMA, 2003; Mirka, 2005). This guideline was developed with a partnership among the furniture industry, federal and state governments (OSHA and North Carolina Department of Labor, Occupational Safety and Health Division, respectively), the academic community (Department of Industrial Engineering, North Carolina State University), and ergonomics specialists (The Ergonomics Center of North Carolina). This guideline provides practical suggestions for changes in work methods, including engineering controls, to reduce risk factors of MSDs. The guideline is advisory in nature, informational in content, and is not a regulatory standard.

There are multiple engineering controls in the AFMA guidelines, and generally these controls are relatively simple and easy to implement. Examples of controls are lightweight fabric tongs, for picking up fabric pieces, that extend the reach of the worker and reduce the awkward postures of the trunk and shoulder; custom-built spring-loaded carts for raising the work level of components to approximately waist level; inclined tables for sewing operators to reduce neck extension and trunk flexion; a pneumatic stapling tool that staples fabric with a single button press compared to the former tool that required a power grip exertion for every staple; a coiled air hose system that holds the hoses above the worker whereas the previous method had hoses lying on the ground, requiring muscle exertions and awkward postures to lift and untangle the hoses; a dual trigger spray gun that enables the worker to maintain a neutral wrist position while spraying a horizontal surface; and a pneumatic lift for spraying furniture at waist level.

14.2.4 Telecommunications Industry

In 2004, the Ergonomics Subcommittee of the National Telecommunications Safety Panel (NTSP) published Ergonomic Guidelines for Common Job Functions within the Telecommunications Industry (NTSP, 2004). While many of the controls in these guidelines are administrative, such as recommendations to use good body mechanics or use two workers for lifting or handling large objects, there are some engineering controls for typical jobs that telecommunications workers perform. Several engineering controls apply to lifting and manipulating ladders, and examples of these controls are a ladder-tote strap that places some of the ladder's weight on the shoulders rather than the upper extremities bearing all the weight; a wheeled trolley that fits under a ladder and enables a worker to easily push a long ladder over various terrains; and steps that fit over the tire of a telecommunications truck, which increases the height of the worker when removing and handling ladders on the roof of the truck. Other examples of engineering controls include a second-class lever for removing manhole covers, compared to a hook or hook and chain, and a portable seat for workers to sit on while they splice cable in the field.

14.2.5 Washington State Ergonomics Demonstration Projects

In 2001 and 2002, the Washington State Department of Labor and Industries (WSDLI) of the State of Washington published results of demonstration projects in which specific tasks were analyzed for risk of MSDs and controls were suggested to reduce the risk of MSDs. Employers and trade organizations cooperated with WSDLI in the evaluation of the tasks (WSDLI, 2001, 2002). The demonstration projects targeted workers in specific industries: fasteners industry (October 2001); airline customer service agents (October 2001); skilled nursing facility workers (October 2001); carpenters, laborers, rebar, and concrete finishing workers (December 2001); landscaping workers (December 2001); sawmill workers (January 2002); masonry industry (January 2002); trucking industry (January 2002); fruit growing and packing workers (January 2002); wallboard workers (January 2002); hop growing, harvesting, and processing (January 2002); lumber handling in sawmills (February 2002); residential care workers (October 2002); and utilities (December 2002). While many of the recommendations in the demonstration projects are administrative controls, such as carpenters using the walk-up technique to lift boards or using the tilt-up technique for lifting plywood onto sawhorses and rotation of workers in high risk jobs, there are engineering controls recommended for specific tasks. Examples of engineering controls are the following: a powered screed for concrete finishers who regularly screed concrete rather than screeding manually; raised, tilted laundry bin for replacement of bins at floor level for residential care workers; a hydraulic jack to lift a beam to a truck's underbody rather than having two workers lift the beam; a portable dolly with a hand crank for raising a drum dolly to the proper level for truck mechanics, which eliminates manual lifting; a mechanical device for sorting and straightening boards in a sawmill rather than unscrambling them manually; and a mechanical board-turning device with a lug loader for rotating boards in a sawmill. These demonstration projects can be downloaded in .pdf files from the WSDLI website listed in the References section.

14.2.6 Electric Power Utilities

A collaborative effort among a research institution (EPRI — Electric Power Research Institute), a university (Marquette University), and a medium-sized electric utility in the upper Midwest of United States. (We Energies) focused on evaluating common tasks with moderate to high risk for MSDs that are performed by electric power workers who work overhead (i.e., on poles) and underground (i.e., in manholes and vaults). Engineering controls for specific tasks were tested in the field by workers, and those interventions that were efficacious in reducing the magnitude of MSD risk factors were published in three ergonomics handbooks that were distributed to EPRI member electric utilities (EPRI, 2001, 2004, 2005). In all three handbooks, the analysis and recommended controls for each task were described in clear lay language and illustrated with simple line drawings within two to five pages. Most of the interventions were less than $500, and the handbooks were well received by EPRI member utilities. A summary article of the underground handbooks is under review (Stone et al., under review).

For each handbook, two to four of the engineering controls were evaluated quantitatively in a biomechanics laboratory. EMG data from muscles and absolute force exertions (push and pull) were measured, and risk of injury was assessed (Marklin et al., 2004).

In each handbook, a chapter was devoted to a business case for engineering controls that were the most efficacious in reducing the magnitude of MSD risk factors for tasks that were among the most prevalent and most injurious for utility workers in the United States. The business case chapter in each handbook centered on a battery-powered tool as a replacement for manual cutters and crimpers, which utility workers across the United States said were the most injurious tools. The corporate ergonomist for We Energies evaluated injury data dating back 30 yr for injuries attributable to the manual tools that the battery-powered cutters or crimpers replaced. The payback period for each battery-powered tool was 15 months or less (Seeley and Marklin, 2003).

A task from the two most recent EPRI handbooks (EPRI, 2004, 2005) are shown in Appendices A and B. The writing and illustrations of these two tasks are representative of the 32 tasks described in the EPRI handbook for overhead workers (EPRI, 2001), the 16 tasks for manhole and vault workers (EPRI, 2004), and the 17 tasks for direct-buried cable workers (EPRI, 2005).

Appendix A

Copyright by EPRI, Palo, Alto, CA

Removing and Replacing a Manhole Cover Task

Current Work Practice for Some Utilities

A manhole cover is usually a round or square piece of iron that fits inside an iron frame on a manhole's chimney or roof. Workers must remove the manhole cover to enter the manhole and must replace the cover in its frame after exiting the manhole. Round manhole covers typically range in size from 24 to 38 in. in diameter and can weigh between 250 and 500 lbs. A variety of commercially available devices such as a hook and chain, a steel lifting hook, a tripod and a second-class lever can be used to remove

and replace a manhole cover. A common method for utility workers to remove and replace a manhole cover is with a hook and chain or a steel lifting hook, both of which require similar body posture, motion, and force. A typical hook and chain is illustrated in Figure 14.A1. To remove a manhole cover, the worker inserts the hook into a hole in the manhole cover and pulls the T-handle to lift the cover out of its frame; the worker then pulls the T-handle to drag the cover away from the frame so that the worker may enter and exit the manhole safely. To replace the manhole cover, the worker uses the hook and chain to drag the cover until it slides into the frame; the worker may push the cover with his foot so it settles securely in the frame.

FIGURE 14.A1   A worker removing a manhole cover using a hook and chain.

Problems with Using the Hook and Chain

Using the hook and chain device to remove and replace a manhole cover results in high forces in the shoulder and trunk muscles and compression on the lower spine, which are risk factors for MSDs.

Recommended Ergonomic Intervention

While there are a variety of commercially available devices for removing and replacing manhole covers, a laboratory study has shown that the second-class lever substantially reduces forces in the trunk and shoulder muscles and reduces compression on the spine when used on a hard, level surface (such as concrete). A second-class lever has the load (force on the chain) between the base of tool (fulcrum or pivot point) and the applied effort (handle force). The tool shown in Figure 14.A2 is an example of a second-class lever. It has a rod with a chain attached close to its base and a handle on the other end. The chain is attached to a hook that is inserted in a hole in the manhole cover. For a second-class lever, the distance from the fulcrum (base) to the point where effort is applied (handle force) is greater than the distance from the fulcrum (base) to the load (force on chain). Compared to the tension force on the chain, exerting force on the handle at a greater distance to the base provides mechanical leverage and reduces the required handle force.

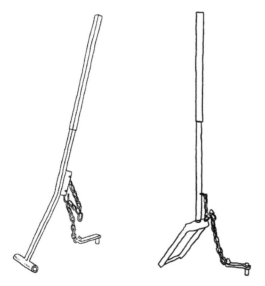

The worker places his foot over the base of the second-class lever and pulls on the handle to lift the edge of the manhole cover out of its frame. The worker repeatedly steps backwards, repositions the lever's base, and pulls the handle to completely remove the cover from its frame.

FIGURE 14.A2   Commercially available second-class lever (left). second-class lever with a modified base (right).

Benefits of Using the Second-Class Lever

The biomechanical benefits of using a second-class lever device, as compared to a hook and chain, are substantial and, in theory, should lower the risk of shoulder and back MSDs. These benefits are summarized as follows:

1. The second-class lever substantially reduces back and shoulder force exertions. As demonstrated in a laboratory study, the second-class lever reduced the peak force of the deltoid muscle required to remove a manhole cover to approximately 30% maximal voluntary contraction (MVC), compared to about 50% MVC for the hook and chain (peak force was indicated by 90th percentile of %MVC EMG data). The second-class lever reduced peak exertion level of the low back muscle that pulls the trunk upward (erector spinae) by almost one third compared to the hook and chain — approximately 70% MVC to remove or replace the manhole cover with the second-class lever compared to about 100% MVC using the hook and chain.
2. The peak forces to pull the T-handle of the hook and chain and the handle of the second-class lever were measured with a Chatillon force gauge to remove a 29 in., 270 lb cover from its frame and drag it across level concrete. The hook and chain required peak forces of 160 lb to remove the cover from its frame and 175 lb to drag it across concrete. The second-class lever required peak handle forces that were at least one third less than the hook and chain — 100 lb to remove most of the cover from its frame and 50 lb to drag it across concrete.

Discussion of Interventions

1. Another commercially available device for removing manhole covers is a lever with a base (fulcrum) between the cover (load) and a long handle (applied force). (Note: this is called a first-class lever in the physics and mechanics literature.) The base of the first-class lever could be wheels or a tripod. While this device reduces the force applied to the handle substantially compared to lifting the cover directly, this device does require more setup time than the hook and chain and second-class lever tested in the EPRI lab study. The first-class lever may be a viable alternative for lifting manhole covers for some utilities.
2. Workers may believe that if they use their legs (instead of their back) and use upright body posture with the hook and chain that the risk for a back injury is minimal. In fact, this is consistent with the training that many workers around the country receive. However, the lab study indicated that there is high EMG activity in the back muscles even when the worker uses their legs and an upright posture. Increased muscle loading in the back muscles increases spinal compression, which can contribute to spine injuries. It is in the utilities' as well as the workers' interests to address the misconception about the hook and chain (and similar devices) and adopt a different device that reduces spinal compression.
3. Compared to the base on the commercially available second-class lever, the modified base reduces twisting of the trunk required to stabilize the base (refer to Figure 14.A2).

Appendix B

Cutting Cable Task

Current Work Practice for Some Utilities

Underground utility workers cut various sizes of primary and secondary copper and aluminum cable. The cable sizes for some utilities range from #6 (conductor diameter 0.41 cm)[20] up to 750 kcmil (conductor diameter 2.54 cm)[20]; other utilities may have a broader range of sizes. The task is performed in work areas such as trenches, manholes, vaults, transformers, splice pits, secondary pedestals, and meter sockets. The frequency of cutting cable varies depending on the nature of the job order, but is usually done at least several times per day. Cutting multiple conductor cable for industrial customers may require 100 or more cuts per day. Traditional cutting tools include a variety of sizes of long-handled cable cutters, ratchet cutters and an AC-powered pump with a remote hydraulic cutter. Depending upon the size of the cable and the tool used, cutting cable with the long-handled cable cutter or ratchet cutter, may take up to 15 sec per cut.

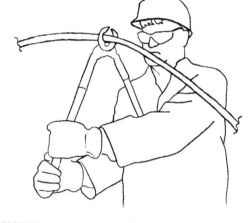

FIGURE 14.B1   Cutting cable, bracing one handle of the manual tool against the chest.

Figure 14.B1 and Figure 14.B2 illustrate the use of a long-handled cable cutter, which is approximately 31 in. long and weighs approximately 8.5 lb. One handle of the tool may be placed against the chest, against the side of the body or under the arm, while pulling on the other handle with both arms.

Problems with Current Work Practices

Current work practices using the long-handled cable cutters have the following risk factors for upper extremity MSDs:

1. Forceful downward exertions with elevated (abducted) shoulders
2. Forceful grip exertions with twisted forearms (supination or pronation) or bent wrists (radial or ulnar deviation)
3. Forceful pulling with arms and shoulders when cutting with the tool braced against the chest or under the arm
4. Awkward bending or twisting of the trunk when cutting with the tool handle on the ground

Recommended Ergonomic Intervention

A battery-powered cutter is recommended for cutting cable. A battery-powered cutter from one manufacturer weighs approximately 10 lb, is 15 in. long, and it can cut cable up to 750 kcmil, as shown in Figure 14.B3. The cable is placed between the tool's jaws, and pressing the trigger closes the jaws. A battery-powered cutter cuts a cable with one stroke.

FIGURE 14.B2   Cutting cable, bracing one handle of the manual tool against the hip.

Benefits of Ergonomic Intervention

Compared to cutting cable with a long-handled cutter, the benefits of using a battery-powered cutter are summarized as follows:

FIGURE 14.B3 Cutting cable using a battery-powered cutting tool.

1. A battery-powered cutter substantially reduces shoulder and arm force exertions. The lab study indicated that the battery-powered cutter reduced the peak force (90th percentile of %MVC EMG data) of the muscle that elevates the shoulder (deltoid — dominant side) to about 20% MVC, compared to approximately 45% MVC or greater for the long-handled cutter. The battery-powered cutter also reduced the peak force in the muscle that pulls the shoulder downward (latissimus dorsi — dominant side) to less than 15% MVC, compared to 50% MVC or more for the long-handled cutters. In addition, peak force of the muscle that pulls the trunk upward (erector spinae) was reduced to less than 50% MVC compared to almost 70% MVC or greater while using the long-handled manual cutter.

2. Using a battery-powered cutter to cut many cables in a day will reduce shoulder and arm muscle fatigue.

3. A battery-powered cutter reduces awkward postures of the arms and trunk.

4. In many situations, a worker is able to keep the forearm and wrists in neutral position when cutting cable with a battery-powered cutter.

5. A battery-powered cutter enables the worker to have one hand free to hold and control the cable during cutting. However, when working with energized cable, another worker may be required to hold the other end of the cable.

6. A battery-powered cutter allows a worker to cut cable in confined space areas, enabling the worker to cut the cable in its original position and handle the cable only once.

Discussion of Intervention

1. The peak pull force to cut 500 kcmil cable (conductor diameter 2.06 cm) was measured with an Instrom Material Testing Systems instrument. When the position of the hands and arms are modeled with the University of Michigan 3D Static Strength Prediction Program (SSPP) to exert the peak force of 115 lb to cut a 500-kcmil cable with a long-handled cutter, less than 3% of females from the general population have the shoulder strength to cut a 500-kcmil cable with a long-handled manual cutter. Less than 50% of males have the maximum shoulder strength to cut a 500-kcmil cable with a long-handled cutter based on 3D SSPP calculations. Virtually everybody from the general population can cut a 500-kcmil cable with the battery-powered cutter.

2. Commercially available battery-powered cutters can cut a wide variety of cable sizes, both copper and aluminum.

3. While a battery-powered cutter weighs more than a long-handled or ratchet cutters (10 lb vs. approximately 3.5 to 8.5 lb), part of the weight of the battery-powered cutters can be held by the cable during cutting of the cable. Therefore, the external forces exerted by the upper extremity that are required to hold the battery-powered cutters during cutting is less than holding an unsupported 10-lb tool. The cable can support part of the battery-powered cutter's weight regardless of its orientation to the cable (horizontal or vertical).

4. A battery-powered cutter initially costs $2000 or more, considerably more than a long-handled cutter. However, the battery-powered cutter significantly reduces the exposure to risk factors of MSDs and may be more cost effective in the long term. Based on an extensive analysis of the cost of workers' compensation, medical expenses, and training of new workers over a 30-yr period at the host utility, the payback period for replacing a long-handled or ratchet cutters with a battery-powered cutter is approximately 1 yr.

References

AFMA (2003). *American Furniture Manufacturers Association Voluntary Ergonomics Guideline for the Furniture Manufacturing Industry.* North Carolina Department of Labor.

Chengalur, S.N., Rodgers, S., and Bernard, T.E. (2004). *Kodak's Ergonomic Design for People at Work*, 2nd edn. John Wiley, New York.

Cohen, A.L., Gjessing, C.C., Fine, L.F., McGlothlin, J.D., and Bernard, B.P. (1997). *Elements of Ergonomics Programs: A Primer Based on Workplace Evaluations of Musculoskeletal Disorders.* DHHS (NIOSH) Publication No. 97–117. U.S. Department of Health and Human Services.

EPRI (2001). *EPRI Ergonomics Handbook for the Electric Power Industry: Overhead Distribution Line Workers Interventions*, EPRI, Palo Alto, CA. 1005199.

EPRI (2004). *EPRI Ergonomics Handbook for the Electric Power Industry: Ergonomic Interventions for Manhole, Vault and Conduit Applications*, EPRI, Palo Alto, CA. 1005430.

EPRI (2005). *EPRI Ergonomics Handbook for the Electric Power Industry: Ergonomic Interventions for Direct-Buried Cable Workers.* EPRI, Palo Alto, CA.

Graphic Arts Technical Foundation (GATF) (2003). *The Ergonomics Guidebook: The How-To Book on Ergonomic Issues and their Solutions in the Graphic Communications Industry.* GATF catalog no. 4725, GATF Press, Sewickley, P.A., www.gain.net.

Marklin, R.W., Lazuardi, L., and Wilzbacher, J. (2004). Measurement of handle forces for crimping connectors and cutting cable in the electric power industry. *International Journal of Industrial Ergonomics*, 34, 497–506.

Mirka, G. (2005). Development of an ergonomics guideline for the furniture manufacturing industry. *Applied Ergonomics*, 36(2), 241–247.

NTSP (2004). *Ergonomic Guidelines for Common Job Functions Within the Telecommunications Industry.* National Telecommunications Safety Panel: Ergonomics Subcommittee, http://www.telsafe.org/ntsp/Publications/FINALversionALL%209-09-04%20.pdf.

OSHA (2003). *Guidelines for Nursing Homes: Ergonomics for the Prevention of Musculoskeletal Disorders.* U.S. Department of Labor, Occupational Safety and Health Administration. Published March 13, http://www.osha.gov/ergonomics/guidelines/nursinghome/final_nh_guidelines.html.

OSHA (2004a). *Guidelines for Retail Grocery Stores: Ergonomics for the Prevention of Musculoskeletal Disorders.* U.S. Department of Labor, Occupational Safety and Health Administration. Published May 28, http://www.osha.gov/ergonomics/guidelines/retailgrocery/retailgrocery.html.

OSHA (2004b). *Guidelines for Poultry Processing: Ergonomics for the Prevention of Musculoskeletal Disorders.* U.S. Department of Labor, Occupational Safety and Health Administration. Document #OSHA 3213-09N, published September 2, http://www.osha.gov/ergonomics/guidelines/poultry processing/poultryprocessing.html.

Seeley, P.A. and Marklin, R.W. (2003). Business case for implementing two ergonomic interventions at an electric power utility. *Applied Ergonomics*, 34: 429–439.

Stone, A. Usher, D. Marklin, R.W., Seeley, P.A., and Yager, J.W. (under review). Case study for underground workers at an electric utility: how a research institution, university and industry collaboration improved occupational health through ergonomics. Submitted to *Journal of Occupational and Environmental Hygiene* for review.

Washington State Department of Labor and Industries (WSDLI) (2001, 2002). *Demonstration Projects* (in.pdf files), http://www.lni.wa.gov/Safety/Topics/Ergonomics/Success/Demo/default.asp.

15

Design and Evaluation of Handtools

Robert G. Radwin
University of Wisconsin–
  Madison

15.1 Introduction

This chapter describes specific hand tool design features that help minimize physical stress and maximize task performance in jobs involving the continuous or repetitive use of hand tools. An important objective of ergonomics in the design, selection, installation, and use of hand tools is the reduction of muscle fatigue onset and the prevention of musculoskeletal disorders of the upper limb. It is not just the tool design, but how a tool is used for a specific task and workstation that imparts physical stress upon the tool operator. Consequently, there is no "ergonomic hand tool" *per se*. What makes sense in one situation can produce unnecessary stress in another.

It is generally agreed that physical stress, fatigue, and musculoskeletal disorders can be reduced and prevented by selecting the proper tool for the task. Tools used so that physical stress factors are minimized, such as reducing stress concentrations in the fingers and hands, producing low force demands on the operator, or minimizing shock, recoil, and vibration are usually the best tools for the job. Control of these factors depends on the tool and the specific tool application. Selection of tools should, therefore, be viewed within the context of the specific job being performed.

Tool selection should be based on (1) process engineering requirements, (2) human operator limitations, and (3) workstation and task factors. Some factors considered for each of these requirements are summarized in Table 15.1. A detailed description of each of these factors is contained in Radwin and Haney (1996). Manufacturing engineers often specify the process requirements with little regard for the operator and the workstation. Hand tool selection should therefore consider how the particular task and workstation relate to the capabilities and limitations of the human operator for a particular tool design. The process is not always simple and often involves an iterative approach, considering individual tool design features and their role in augmenting and mitigating physical stress. This chapter will describe some hand tool design features and the research leading to an understanding of how tool design can help reduce physical stress in hand tool operation.

TABLE 15.1   Requirements for Ergonomic Hand Tool Selection

Process Engineering Requirements	Requirements specified in terms of the production process, such as how fast a drill bit should turn, or how much torque should be applied to a screw being tightened.
	Manufacturing process requirements are often based on the product design and parameters needed for accomplishing the task quickly and reliably at the desired level of quality.
Identify Human Operator Limitations	Consider how process and workstation requirements affect the tool operator's ability to perform the task.
	Human capabilities are limited by strength, fatigue, anthropometry, and manual dexterity.
Workstation and Task Factors	Consider the particular task and workstation where the tool is being used.
	Requirements may include work location, work orientation, tool shape, tool weight, gloves, frequency of operation, tool accessories, work methods and standards.

15.2   Power Tool Triggers and Grip Force

Extended-length triggers (see Figure 15.1) that distribute force among two or more fingers are often suggested for minimizing stress concentrations at the volar aspects of the fingers (Lindquist et al., 1986; Putz-Anderson, 1988). The rationale is that the force for squeezing the trigger and grasping the handle will be distributed among several fingers to reduce the stress in the index finger. Following is a description of a study that investigated how this particular design feature affects the force in the hands.

In order to directly measure finger and hand force exerted during actual tool operation, an apparatus was constructed for simulating a functioning pistol grip pneumatic nutrunner (Oh and Radwin, 1993). Strain gages were installed in two aluminum bars that were used as the handle for measuring force exerted against the fingers and palm. The instrumented bars were constructed so

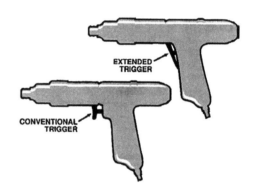

FIGURE 15.1   A conventional trigger and an extended-length trigger on pistol grip power hand tools. (Reprinted from *Human Factors*, 35, 3, 1993. Copyright 1993 by the Human Factors and Ergonomics Society. With permission)

they were insensitive to the point of force application and linearly summed force applied along the length of the handle (Pronk and Niesing, 1981; Radwin et al., 1991). This was accomplished by measuring shearing stress acting in the cross section of the beam. Strain gages were mounted on a thin web that was machined into the central longitudinal plane and aligned at $45°$ with respect to the long axis. The effect of bending stresses were completely removed from the strain gages by selecting a measurement point at the neutral axis of the beam, so that all the strain at the measurement point is strictly due to shear stress. Shear strain is totally independent of the point of application.

The strain gage instrumented handle was mounted on a rigid frame and attached perpendicular to a modified in-line pneumatic nutrunner motor in a configuration resembling a pistol-grip power tool (see Figure 15.2). The two dynamometers were mounted in parallel on a track so the handle span could be continuously adjusted. The apparatus was completely functional. The air motor contained an automatic air shut-off torque control mechanism and was operated at a 6.8 Nm target torque setting.

Plastic caps were formed and attached to each end of the dynamometer so the contours resembled a power tool handle. The handle circumference was 12 cm for a 4 cm span, measured between two points tangent to the handle contact surfaces. The handle circumference increased an additional 2 cm as the handle span was increased by 1 cm. A trigger was mounted on the finger side cap (see Figure 15.2), and a contact switch was installed inside the trigger. A leaf spring was used for controlling trigger tension. When the trigger was squeezed, the switch tripped a relay and a solenoid valve for supplying air to the pneumatic power tool motor.

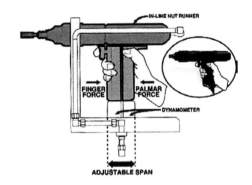

Two different trigger types were tested. One was a conventional power tool trigger, activated using only the index finger. The second was longer than the conventional trigger and was activated using both the index and middle fingers (see Figure 15.1). The conventional trigger was 21 mm long and the extended trigger was 48 mm long. The conventional trigger required 8 N, and the extended trigger required 11 N for activation.

FIGURE 15.2 Dynamometer used for measuring finger and palm forces exerted when operating a completely functional simulation of a pistol-grip pneumatic power hand tool. (Reprinted from *Human Factors*, 35, 3, 1993. Copyright 1993 by the Human Factors and Ergonomics Society. With permission)

Use of the extended trigger was found beneficial for reducing grip force and exertion levels during tool operation. Average peak finger and palm forces were, respectively, 9 and 8% less for the extended trigger than for the conventional trigger. Eleven of eighteen subjects (61) indicated that they preferred using the handle with the extended trigger after just an hour of use in the laboratory. The average finger and palmar holding force was 65 and 48%, respectively, less for the extended trigger, than for the conventional trigger. Since subjects spent 65 to 76% of the operating time holding the tool, using an extended trigger may have an important effect on reducing exposure to forceful exertions in the hand during power hand tool operation.

15.3 Handle Size and Grip Force

Research on handle design has typically focused on finding the optimal handle dimensions. Grip strength is affected through the biomechanics of grip from the relative position of the joints of the hand and by the position and length of the muscles involved. Consequently, grip strength is affected by the handle size. Recommendations for handle size are usually based on the span that maximizes grip strength, or the span that minimizes fatigue.

Hertzberg (1955), in an early Air Force study, reported that a handle span of 6.4 cm maximized power grip strength. Greenberg and Chaffin (1975) recommended that a tool handle span should be in the range between 6.4 and 8.9 cm in order to achieve high grip forces. Ayoub and Lo Presti (1971) found that a 3.8 cm diameter was optimum for a cylindrical handle. This was based on maximizing the ratio between strength and EMG activity, and on the number of work cycles before onset of fatigue. Another study by Petrofsky et al. (1980) showed that the greatest grip strength occurred at a handle span between 5 and 6 cm.

Grip strength is affected by hand size. Fitzhugh (1973) showed that the handle span resulting in maximum grip strength for a 95 percentile male hand length is larger than the handle span for a 50 percentile female. Consequently, a person with a small hand might benefit from using a smaller handle, and a person with a large hand might benefit from using a larger one.

Grip strength data often used for handle design are based on population measurements made using instruments like the Jamar or Smedley dynamometers (Schmidt and Toews, 1970; Young et al., 1989) rather than using handle dimensions representative of an actual tool. In most cases, only one dimension (handle span) has been controlled, while the other handle dimensions were not necessarily similar to a tool handle.

A power hand tool manufacturer considered offering a power hand tool that provided a handle that was adjustable in size. An investigation of grip strength using handle dimensions similar to power hand tool handles was conducted in order to explore the differences against published grip strength data (Oh and Radwin, 1993). Hand length up to 17 cm was classified as small, between 17 and 19 cm as medium, and greater than 19 cm as large. Average grip strength is plotted against handle span and hand size in Figure 15.3. Grip strength increased as hand length increased. Large hand subjects produced their maximum grip strength (mean = 463 N, SD = 128 N) for a handle span of 6 cm, while medium hand (mean = 280 N, SD = 122 N) and small hand (mean = 203 N, SD = 51 N) subjects produced their maximum strength for a handle span of 5 cm.

The span resulting in maximum grip strength agreed with the findings of previous strength studies. Hertzberg (1955) found that subjects exerted more force at a 6.4 cm span than among 3.8, 6.4, 10.2, and 12.7 cm handle spans. Petrofsky et al. (1980) reported that on the average, subjects produced maximum grip force for a handle span between 5 and 6 cm.

Although the span resulting in maximum grip strength and the grip strength function agreed with previous findings, the maximum grip strength for both student and industrial worker subjects was markedly less than what has been previously reported in the literature. Schmidt and Toews (1970) collected grip strength data from 1128 male and 80 female Kaiser Steel Corporation employee applicants, using a Jamar dynamometer. They reported for a handle span of 3.8 cm, an average of 499 N for the dominant male hand and 308 N for the dominant female hand. Swanson et al. (1970) measured the grip strength of 50 females and 50 males using a Jamar dynamometer. Among these subjects, 36 were light manual workers, 16 were sedentary workers, and 48 were manual workers. They reported for a handle span of 6.4 cm, 467 N for the male dominant hand and 241 N for the female dominant hand. These all exceeded the strength levels observed (see Figure 15.3).

A major difference between grip strength measured for tool handles by Oh and Radwin (1993) and previously reported strength data is in the handle dimensions. The Jamar and Smedely dynamometers have smaller circumferences and narrower widths than the tool handle used in this study. The tool handle curvature was also straight while the Jamar dynamometer has a curved surface at the grip center. The handle used in this study closely represented an actual tool handle in circumference and

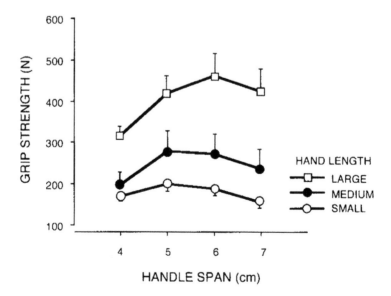

FIGURE 15.3 Average grip strength plotted against handle span for three hand size categories. Error bars represent standard error of the mean. (Reprinted from *Human Factors*, 35, 3, 1993. Copyright 1993 by the Human Factors and Ergonomics Society. With permission.)

width. These size and curvature differences can affect the position of the fingers and grip posture. These dimensional differences must be considered when designing handles based on strength using published grip strength data.

The investigation also found a difference in grip strength between student subjects and industrial workers. Grip strength, averaged over handle span, was 279 N (SD = 133 N) for the students and 327 N (SD = 90 N) for the workers. No significant grip strength differences, however, between the student and worker groups was observed within each hand size.

The underlying assumption in designing handles based on maximum strength is that the actual force exerted is independent of handle size. Exertion level is the ratio of the actual grip force used, to the maximum voluntary force generating capacity. If the grip force used during tool operation is the same for all handle sizes, then the handle span associated with the greatest grip strength should result in the lowest exertion level. If grip force, however is affected by handle size, then the handle span associated with the greatest grip strength may not be the handle span resulting in the minimum exertion level.

A series of experiments were performed using the pistol grip power hand tool with strain gage instrumented handles and an adjustable handle span as described above. Handle span affected peak finger and palmar force. Peak finger force increased 24% for a student subject group, and 30% for an industrial worker group, as handle span increased from 4 to 7 cm. Similarly, peak palmar force increased 21% for the student group and 22% for the worker group, as handle span increased from 4 to 7 cm. Handle span also influenced finger and palmar holding forces. Finger holding force increased 20%, and palmar holding force increased 16%, as handle span increased from 4 to 7 cm for the student subjects.

The study found that hand size was proportional to the handle span operators preferred when offered the opportunity to adjust the handle size to any size they desired. Operators with larger hand sizes reported they preferred using a tool with a larger handle. Preferred handle span is plotted against hand length in Figure 15.4 for both trigger types. There was no difference between the preferred handle span for the conventional trigger and the preferred handle span for the extended trigger. No anthropometric measurements, however, were related to the span resulting in the minimum peak exertion level. Exertion level when holding the tool was less for the large size hands than for the small size

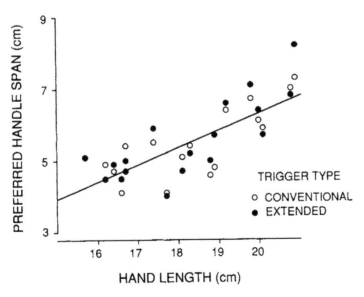

FIGURE 15.4 Preferred handle span plotted against hand length. (Reprinted from *Human Factors*, 35, 3, 1993. Copyright 1993 by the Human Factors and Ergonomics Society. With permission.)

hands. Holding exertion level for the large hands was maximum for the 4 cm handle span, while holding exertion level for the small hands was maximum for the 7 cm handle span. In addition, the tendency for large hand subjects to prefer larger handle spans suggests that selectable size handles may be more desirable than having only a single size handle for power hand tools.

15.4 Static Hand Force

Safe power hand tool operation requires that an operator possess the ability to adequately support the tool in a particular position, apply the necessary forces, while reacting against the forces generated by the tool. Force demands that exceed an operator's strength capabilities can cause loss of control, resulting in an accident or an injury. Design and selection of power hand tools that minimize static grip and hand force will help reduce muscle fatigue and prevent upper limb disorders.

The force necessary for supporting a power hand tool depends on the tool weight, its center of gravity, the length of the tool, and air hose attachments. Power hand tools should be well balanced with all attachments installed. As a general rule, a hand tool center of gravity should be aligned with the center of the grasping hand so the hand does not have to overcome moments that cause the tool to rotate the operator's wrist and arm (Greenberg and Chaffin, 1977).

Psychophysical experiments have provided some insight into the load that power tool operators prefer. When experienced hand tool operators were asked to rate the mass of the power tools they operated, tools weighing 0.9 to 1.75 kg mass were rated "just right" (Armstrong et al. 1989). Other psychophysical experiments showed that perceived exertion for a tool mass of 1 kg was significantly less than for tools with a mass of 2 and 3 kg (Ulin and Armstrong, 1992).

There is a tradeoff between selecting a light tool and the benefit of the added weight for performing operations that require high feed force. The power available for a grinding task increases with increasing mass of the grinder. Reducing the weight of the grinder can increase the feed force the operator must provide and may increase the amount of time necessary for accomplishing the task, consequently subjecting the operator to more stressful work and greater vibration exposure. Heavy grinding tasks should be performed on horizontal surfaces so the weight of the tool does not have to be supported by the operator. Heavy power tools should be suspended using counterbalancing accessories.

In addition to supporting the tool load, power hand tool operators often have to exert push or feed force, or act against reaction forces. Feed force is necessary for starting a threaded fastener, advancing a bit or keeping a bit or socket engaged during the securing cycle. Feed force is affected by the work material and design of the tool, bit, or fastener. Large feed forces are sometimes needed when operating power tools such as drills and screwdrivers. Repetitive or sustained exertions associated with these operations should be minimized. Drill feed force is affected by the drill power and speed, bit type, material, and diameter of the hole drilled. Power screwdriver feed force may be affected by the fastener head and screw tip used. Feed force for a slotted or Phillips head screw generally requires more feed force than for a torx head screw. Self-tapping screws require more force than screws tightened through pre-tapped holes. Material hardness is also a factor for self-tapping screws and drilling. Feed force requirements also increase as torque level increases for cross recess screws.

Power hand tools such as screwdrivers or nutrunners, used for tightening threaded fasteners, are commonly configured as (1) in-line, (2) pistol grip, and (3) right angle. A mechanical model of a nutrunner was developed for static equilibrium (no movement) conditions (Radwin et al., 1995). Hand force, reaction force from the workpiece, tool orientation, weight, and output torque were included in this model. This chapter will describe the model developed for pistol grip nutrunners.

The model uses a Cartesian coordinate system relative to the orientation of the handle grasped in the hand using a power grip. This coordinate system has the x-axis perpendicular to the axial direction of the handle; the y-axis is parallel to the long axis of the handle; and the z-axis is parallel to the tool spindle.

The origin is the end of the tool bit or socket. Hand forces are described in relation to these coordinate axes. To simplify the model, an initial assumption is that orthogonal forces can be applied along the handle without producing coupling moments. This assumption allows force to be considered as having a single point of application. The resultant hand force $\mathbf{F}_H$ at the grip center is the vector sum of the three orthogonal force components

$$\mathbf{F}_H = F_{Hx}\mathbf{i} + F_{Hy}\mathbf{j} + F_{Hz}\mathbf{k} \tag{15.1}$$

where the hand force magnitude is:

$$|\mathbf{F}_H| = \sqrt{F_{Hx}^2 + F_{Hy}^2 + F_{Hz}^2} \tag{15.2}$$

and $\mathbf{i}$, $\mathbf{j}$, $\mathbf{k}$ are the unit vectors. The coordinates and respective force components are illustrated in Figure 15.5.

Consider the free-body diagram for the pistol grip nutrunner in Figure 15.6. The torque, T_s acts in reaction to torque T, applied by the tool to the fastener. The tool operator has to oppose this equal and opposite reaction torque in the counter-clockwise direction by producing a reaction force F_{Hx}. That is not the only force, however, that the operator has to produce. A force acting in the z direction, F_{Hz}, provides feed force and produces an equal and opposite reaction force, F_{Rz}. In addition, the operator has to react against the tool weight in order to support and position the tool by providing a vertical force component, F_{Hy}. The tool weight, W_T and push force, F_{Hz} tend to produce a clockwise moment about the tool spindle in the yz-plane which is countered by this vertical support force.

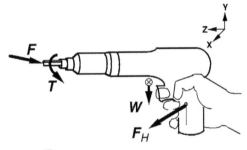

F = Workpiece Reaction Force
T = Spindle Torque
F$_H$ = Hand Force
W = Tool Weight

FIGURE 15.5   Free body diagram and orthogonal force components considered in the pistol grip force model.

When a body is in static equilibrium, the sum of the external forces and the sum of the moments are equal to zero. Using that relationship, the following system of equations was developed for the pistol-grip nutrunner to describe these static forces:

$$
\begin{bmatrix} F_{Hx} \\ F_{Hy} \\ F_{Hz} \end{bmatrix} =
\begin{bmatrix}
-L_{Gz}/L_{Hz} & 0 & 0 & -1/L_{Hz} & 0 \\
0 & -L_{Gz}/L_{Hz} & (-L_{Gy}-L_{Hy})/L_{Hz} & 0 & -L_{Gy}/L_{Hz} \\
0 & 0 & -1 & 0 & -1
\end{bmatrix}
\cdot
\begin{bmatrix} W_{Tx} \\ W_{Ty} \\ W_{Tz} \\ T_{Sz} \\ F_{Rz} \end{bmatrix}
\tag{15.3}
$$

Assuming one-hand operation, resultant hand force magnitude was predicted using the model for the four different tools and plotted as a function of torque in Figure 15.7. Hand force was determined for both low feed force (1 N) and high feed force (50 N) conditions, when operating these tools against a vertical surface. When feed force was small, the resultant hand force was mostly affected by torque reaction force, which increased as torque increased for all four tools. Since the greatest force component in this case was torque reaction force, Tools 3 and 4 had the least resultant hand force since they both had the longest handles. Tool 3, however, had a considerably greater resultant hand force when feed force was high. This effect was not observed for Tool 4, which also had a similar handle, but contained a spindle extension shaft.

15.5   Dynamic Reaction Force

Whereas manual hand tools rely on the human operator for generating forces, power hand tools operate from an external energy source (i.e., electric, pneumatic, and hydraulic) for doing work. The tool operator provides static force for supporting the tool and for producing feed force, and must react against the forces generated by the power hand tool. Power hand tools such as nutrunners produce rapidly building torque reaction forces which the operator must react against in order to maintain full control of the tool.

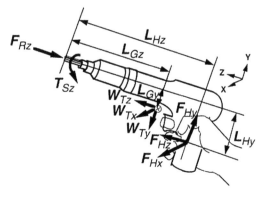

FIGURE 15.6   Power hand tool geometry and variables in the hand tool static force model.

Nutrunner reaction torque is produced by spindle rotation and is affected primarily by the spindle torque output and tool size. Nutrunner spindle torque can range from less than 0.8 Nm to more than 700 Nm. This torque is transmitted to the operator as a reaction force through the moment arm created by the tool and tool handle. A tool operator opposes reaction torque while supporting the tool and preventing it from losing control.

The three major operating modes for nutrunners include (1) mechanical clutch, (2) stall, or (3) automatic shut-off. When a stall tool is used, maximum reaction torque time is directly under operator control by releasing the throttle, which can last as long as several seconds. Stall tools tend to expose an operator to reaction torque the longest. Although clutch tools limit reaction torque exposure, ratcheting clutch tools can expose workers to significant levels of vibration if used frequently (Radwin and Armstrong, 1985). The speed of the shut-off mechanism controls exposure to peak reaction force for automatic shut-off tools. Consequently, automatic shut-off tools have the shortest torque reaction time because these tools cease operating immediately after the desired peak torque is achieved.

As torque is applied to a threaded fastener, it rotates at a relatively low spindle torque until the clamped pieces come into intimate contact. This torque can approach zero with free running nuts or can be rather

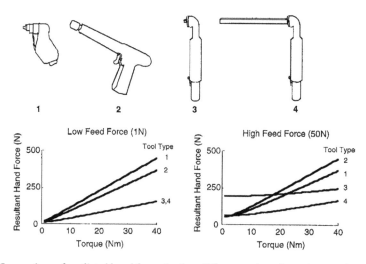

FIGURE 15.7   Comparison of predicted hand forces for four different tool configurations performing the same task for one-hand tool operation.

significant as in the case where locking nuts, thread interference bolts, or thread-forming type fasteners are used. After the fastener brings the clamped members of the joint into initial intimate contact, it continues to draw the parts together until they form a solid joint. When the joint becomes solid, continued turning of the nut results in a proportionally increasing torque. This is the elastic portion of the cycle and is the time when reaction torque forces are produced. Torque build-up, and consequently torque reaction force, continues rising at a fixed rate until peak torque is achieved, which is the clamping force of the joint. Forearm muscle reflex responses when operating automatic air shut-off right angle nutrunners during the torque-reaction phase was more than four times greater than the muscle activity used for holding the tool and two times greater than the run-down phase (Radwin et al., 1989). Flexor EMG activity during the torque–reaction phase increased for tools having increasing peak spindle torque.

Threaded fastener joints are classified as "soft" or "hard" depending on the relationship between torque build-up and spindle angle. The International Standardization Organization (ISO) specifies that a hard joint has an angular displacement less than 30 degrees when torque increases from 50 to 100% of target torque, and a soft joint has an angular displacement greater than 360 degrees (ISO-6544).

Nutrunner torque reaction force is a function of several factors including target torque, spindle speed, joint hardness, and torque build-up time. Some of these factors are interdependent. Faster spindle speed results in shorter torque build-up time, and softer joints are related to longer build-up times. The duration of exertion is directly related to torque build-up time rather than just the speed of the tool or joint hardness.

Studies have shown that torque build-up time as well as the magnitude of torque reaction force has a significant influence on human operators during power nutrunner use. Kihlberg et al. (1995) studied right angle nutrunners having different shut-off mechanisms (fast, slow, and delayed) and found a strong correlation between perceived discomfort, handle displacement, and reaction forces. Radwin et al. (1989) investigated the effects of target torque and torque build-up time using right-angle pneumatic nutrunners and found that average flexor rms electromyography (EMG) activity scaled for grip force increased from 372 N for a low target torque (30 Nm) to 449 N for a high target torque (100 Nm), and that average grip force was 390 N for a long build-up time (2 sec), and increased to 440 N for a medium build-up time (0.5 sec). They also reported that EMG latency between tool torque onset and peak flexor rms EMG for the long torque build-up time (2 sec) was 294 msec and decreased to 161 msec for the short build-up time (0.5 sec). The findings suggested that torque reaction force can affect extrinsic hand muscles in the forearm, and hence grip exertions, by way of a reflex response. Johnson and Childress (1988) showed that low torque was associated with less muscular activity and reduced subjective evaluations of exertion.

Representative torque reaction force, handle kinematics, and EMG muscle activity are illustrated in Figure 15.8. Since torque builds up in a clockwise direction, the reaction torque has a tendency to rotate the tool counterclockwise with respect to the operator. When the operator has sufficient strength to react against the reaction torque, the tool remains stationary or rotates clockwise and the operator exerts concentric muscle contractions against the tool (positive work). However, when the tool overpowers the operator, it tends to move in a counterclockwise direction and the operator exerts eccentric muscle contractions against the tool (negative work). Therefore, measures of handle movement that occur (handle velocity and displacement) and the direction of rotation can indicate relative tool controllability. Handle movement direction was defined as positive when the handle moved in the direction of tool reaction torque (see Figure 15.1). If handle velocity increases after shutoff, it means that the tool and hand are unstable. The work done on the tool–hand system and the power involved in doing work during torque build-up were also assessed. If the operator has the capacity to successfully react against the torque build-up (positive work), then the tool is considered stable. This occurs when handle displacement and velocity were less than zero. If the handle become unstable and the net handle displacement occurs in the direction of torque reaction away from the operator, then work and power are negative.

A computer-controlled right angle nutrunner was used to study power hand tool reaction forces (Oh and Radwin, 1997). A torque transducer and an angle encoder were integrated into the tool spindle head

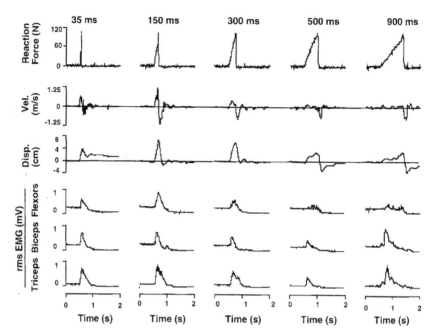

FIGURE 15.8   Representative torque reaction force, handle kinematics and EMG muscle activity for different torque build-up times.

which outputted analog torque and digital angular rotation signals. A threaded fastener joint simulator that could be oriented horizontally or vertically was mounted on a height adjustable platform. The longitudinal axis of the joint head was oriented perpendicular to the ground for the horizontal workstation setting, and oriented parallel to the ground for the vertical workstation setting.

The study showed that workstation orientation and tool dynamics (torque reaction force and torque build-up time) influenced operator muscular exertion and handle stability. In general, handle instability increased when the tool was operated on a vertical workstation (rather than a horizontal workstation), when torque reaction force was high (88.3 and 114.6 N), and for a 150 msec torque build-up time, regardless of torque reaction force.

As torque reaction force increased from 52.1 to 114.6 N, peak hand velocity 89%, and peak hand displacement increased 113%. Peak hand velocity was greatest for a 150 msec build-up time and the least for a 900 msec build-up time. The effect of target torque was consistent with previous studies that showed that target torque was related to muscular exertion, subjective perceived exertion, and handle instability (Johnson and Childress, 1988; Lindqvist, 1993; Oh and Radwin, 1994; Radwin et al., 1989). As torque reaction force increased from 52.1 to 114.6 N, the magnitude of negative work increased by 35%, and the magnitude of average power against the operator increased by 30%. Under these conditions, perceived exertion also increased from 2.7 to 4.3 (as rated on Borg's 10-point scale), and task acceptance rate decreased from 73 to 28%. When the tool was operated on a horizontal workstation, average finger flexor EMG was significantly influenced by torque reaction force. As torque reaction force increased from 52.1 to 114.6 N, the average flexor EMG increased by 14%.

The effect of torque build-up time on power hand tool operators has been studied in terms of perceived exertion, muscular activity, and handle stability (Armstrong et al., 1994; Freivalds and Eklund, 1991; Lindqvist et al., 1986; Oh and Radwin, 1994; Radwin et al., 1989). Torque build-up time is a concern because it is directly related to assembly time and exertion duration. Increased duration may lead to earlier fatigue onset. Although longer build-up time results in longer duration exertions and increases the operation cycle time, it may provide an opportunity for better tool control since it gives the operator a longer time to react.

Peak hand velocity was 46.7% less for horizontal workstations (mean = 0.46 m/sec, SD = 0.26 m/sec) than for vertical workstations (mean = 0.67 m/sec, SD = 0.34 m/sec). A similar trend was observed for peak hand displacement. Peak hand displacement for horizontal workstations (mean = 4.0 cm, SD = 2.2 cm) was 90.2% less than peak hand displacement for vertical workstations (mean = 7.6 cm, SD = 4.6 cm). Previous findings agree that a horizontal workstation is preferable for right angle tool use. Ulin et al. (1992) showed that average subjective ratings of perceived exertion were significantly less when the tool was operated on horizontal workstations rather than vertical workstations. Also, 88% more negative work and 58% more power against the operator were recorded while the tool was operated on a vertical workstation. However, subjective ratings of perceived exertion and task acceptance rates did not differ between horizontal and vertical workstations. This might come from the fact that the torque levels in the current study were much greater than the torque level used for the Ulin et al. study (1992).

Although perceived exertion was less and task acceptance rate was greater for a 35 msec build-up time than for longer build-up times, the operator might not have sufficient time to voluntarily react against torque build-up with the 35 msec build-up time. On the average, the onset of the EMG burst occurred 40 msec after the onset of torque build-up for the 35 msec build-up time. This indicated that the muscles were not activated until a significant amount of torque had built up for the 35 msec build-up time. Lack of muscular contraction during torque build-up might explain why the peak handle velocity was higher for short build-up times. Without muscular contractions, the inertia of the tool and hand had to absorb all of the reaction force. Short exertion duration and lack of muscular contractions due to EMG latencies might contribute to lower subjective ratings of perceived exertion for the 35 msec build-up.

The larger torque variance that occurred for the 35 msec build-up time indicated that even though subjective perceived exertion was less, this condition might result in more target torque error. Also, the probability of increased handle instability after shutoff was significantly greater for the 35 msec build-up time. This suggests that even after shutoff, operators did not have sufficient capacity to control the tool reaction torque. Therefore, the 35 msec build-up time increased handle stability in terms of peak handle displacement and negative work, and reduced subjective perceived exertion, however, the lack of muscular contraction during torque build-up reduced tightening quality.

Methods for limiting reaction force include (1) use of torque reaction bars, (2) installing torque absorbing suspension balancers, (3) providing tool mounted nut holding devices, and (4) using tool support reaction arms. A torque reaction bar sometimes can be used to transfer loads back to the work piece. Tools that can be equipped with a stationary reaction bar adapted to a specific operation so reaction force can be absorbed by a convenient solid object can completely eliminate reaction torque from the operator's hand. These bars can be installed on in-line and pistol-grip tools. Right angle tools can react against a solid object instead of relying on the hand and arm. Reaction devices (1) remove reaction forces from the operator, (2) permit pistol-grip and in-line reaction bar tools to be operated using two hands, (3) free the operator from restricting postures, (4) provide weight improvements over right angle nutrunners, and (5) improve tool fastening performance. The disadvantages are that reaction bars must be custom made for each operation, and the combination of several attachments for one tool can be difficult. Torque reaction bars may also add weight to the tool and can make the tool more cumbersome to handle.

15.6 Vibration

Vibration can be a by-product of power hand tool operation, or it can even be the desired action as is the case with abrasive tools like sanders or grinders. Vibration levels depend on tool size, weight, method of propulsion, and the tool drive mechanism. It is affected by work material properties, disk abrasives, and abrasive surface area. Continuous vibration is inherent in reciprocating and rotary power tools. Impulsive vibration is produced by tools operating by shock and impact action, such as impact wrenches or chippers. The tool power source, such as air power, electricity, or hydraulics can also affect vibration. Vibration is also generated at the tool-material interface by cutting, grinding, drilling, or other actions.

Pneumatic hammer recoil was observed producing a stretch reflex and muscular contractions in the elbow and wrist flexors (Carlsöö and Mayr, 1974). Studies of the short-term neuromuscular effects of hand tool vibration have demonstrated that hand tool vibration can introduce disturbances in neuromuscular force control resulting in excessive grip exertions when holding a vibrating handle (Radwin et al., 1987). The results of these studies demonstrated that grip exertions increased with tool vibration. Average grip force increased for low frequencies (40 Hz) vibration but did not change for higher frequencies (160 Hz) vibration. Since forceful exertions are a commonly cited factor for chronic upper extremity muscle, tendon, and nerve disorders, vibrating hand tool operation may increase the risk of CTDs through increased grip force.

Vibration has also been shown to produce temporary sensory impairments (Streeter, 1970; Radwin et al., 1989). Recovery is exponential and can require more than 20 minutes (Kume et al., 1984). Workers often sand or grind surfaces and periodically inspect their work using tactile inspection to determine if the surface was sanded to the desired level of smoothness. Diminished tactility may result in a surface feeling smoother than it actually is, resulting in a rougher surface than is actually desired.

Vibration has not been shown to be significantly reduced by using resilient mounts on handles. Vibration isolation techniques have been generally unsuccessful for limiting vibration transmission from power tools to the hands and arms. Isolation has been particularly difficult for vibration frequencies less than 100 Hz. This is because attenuation only occurs when the vibration spectrum falls above the resonant frequency of the isolation system or material. When the vibration frequency is less than the resonant frequency of the isolating material, the handle acts as a rigid body and no vibration is attenuated. Grinding tools typically run at speeds near 6000 rpm (100 Hz) making it difficult to have a resilient vibration isolating handle. Furthermore, if the vibration frequency is approximately equivalent to the isolator resonant frequency, the system will actually intensify vibration levels. Weaker suspension systems have lower resonant frequencies, but are often impractical because such a system is usually too flexible for the heavily loaded handles of tools like grinders. Handles loaded with high forces must be very rigid.

References

Armstrong, T.J., Bir, C., Finsen, L., Foulke, J., Martin, B., Sjøgaard, G., and Tseng, K. 1994. Muscle responses to torques of hand held power tools, *Journal of Biomechanics*, 26(6): 711–718.

Ayoub, M.M. and Lo Presti, P. 1971. The determination of an optimum size cylindrical handle by use of electromyography. *Ergonomics*, 14(4): 509–518.

Fitzhugh, F.E. 1973. *Dynamic aspects of grip strength.* (Tech Report). Department of Industrial & Operations Engineering, Ann Arbor: The University of Michigan.

Freivalds, A. and Eklund, J. 1991. Subjective ratings of stress levels while using powered nutrunners, in W. Karwowski and J.W. Yates, Eds., *Advances in Industrial Ergonomics and Safety III*, New York, Taylor & Francis, 379–386.

Greenberg, L. and Chaffin, D.B. 1975. *Workers and Their Tools: A Guide to the Ergonomic Design of Hand Tools and Small Presses.* Midland, MI: Pendell.

Hertzberg, H.T.E. 1955. Some contributions of applied physical anthropology to human engineering. *Annals of NY Academy of Science*, 63(4): 616–629.

International Organization for Standardization 1981. *Hand-held Pneumatic Assembly Tools for Installing Threaded Fasteners — Reaction Torque Reaction Force and Torque Reaction Force Impulse Measurements.* ISO-6544.

Johnson, S.L. and Childress, L.J. 1988. Powered screwdriver design and use: tool, task, and operator effects, *International Journal of Industrial Ergonomics*, 2: 183–191.

Kihlberg, S., Kjellberg, A., and Lindbeck, L. 1995. Discomfort from pneumatic tool torque reaction force reaction: acceptability limits, *International Journal of Industrial Ergonomics*, 15: 417–426.

Lindqvist, B. 1993. Torque reaction force reaction in angled nutrunners, *Applied Ergonomics*, 24(3): 174–180.

Lindquist, B., Ahlberg, E., and Skogsberg, L. 1986. *Ergonomic Tools in Our Time*. Atlas Copco Tools, Stockholm.

Oh, S. and Radwin, R.G. 1993. Pistol grip power tool handle and trigger size effects on grip exertions and operator preference, *Human Factors*, 35(3): 551–569.

Oh, S. and Radwin, R.G., 1994, Dynamics of power hand tools on operator hand and arm stability, in *Proceedings of the Human Factors and Ergonomics Society 38th Annual Meeting*, 602–606, Santa Monica, CA: Human Factors and Ergonomics Society.

Oh, S. and Radwin, R.G. 1998. The influence of target torque and torque build-up time on physical stress in right angle nutrunner operation, *Ergonomics*, 41(2): 188–206.

Petrofsky, J.S., Williams, C., Kamen, G., and Lind, A.R. 1980. The effect of handgrip span on isometric exercise performance, *Ergonomics*, 23(12): 1129–1135.

Pronk, C.N.A. and Niesing, R. 1981. Measuring hand grip force using an application of strain gages, *Medical, Biological Engineering and Computing*, 19:127–128.

Putz-Anderson, V. 1988. *Cumulative Trauma Disorders*. New York: Taylor & Francis.

Radwin, R.G., Masters, G., and Lupton, F.W. 1991. A linear force summing hand dynamometer independent of point of application, *Applied Ergonomics*, 22(5): 339–345, 1991.

Radwin, R.G. and Haney, J.T. 1996. *An Ergonomics Guide to Hand Tools*, Fairfax, VA: American Industrial Hygiene Association.

Radwin, R.G., VanBergeijk, E., and Armstrong, T.J. 1989. Muscle response to pneumatic hand tool torque reaction force reaction forces, *Ergonomics*, 32(6): 655–673.

Radwin, R.G., Oh, S., and Fronczak. 1995. A mechanical model of hand force in power hand tool operation, *Proceedings of the Human Factors and Ergonomics Society 39th Annual Meeting*, Santa Monica: Human Factors and Ergonomics Society: 348–352.

Schmidt, R.T. and Toews, J.V. 1970. Grip strength as measured by the Jamar dynamometer, *Archives of Physical Medicine & Rehabilitation*, 51(6): 321–327.

Swanson, A.B., Matev, I.B., and Groot, G. 1970. The strength of the hand, *Bulletin of Prosthetics Research*, Fall: 145–153.

Ulin, S.S., Snook, S.H., Armstrong, T.J., and Herrin, G.D. 1992. Preferred tool shapes for various horizontal and vertical work locations, *Applied Occupational and Environmental Hygiene*, 7(5): 327–337.

Young, V.L., Pin, P., Kraemer, B.A., Gould, R.B., Nemergut, L., and Pellowski, M. 1989. Fluctuation in grip and pinch strength among normal subjects, *The Journal of Hand Surgery*, 14A(1): 125–129.

16

Low Back Disorders: General Solutions

Carol Stuart-Buttle
Stuart-Buttle Ergonomics

16.1 Introduction

Good design minimizes the risk of low back disorders. However, design entails making trade-off decisions amongst a combination of risks to reduce such exposures. Implementing solutions to reduce the risk of low back disorders within an existing system requires as much caution with trade-off decisions as when designing a new system.

There are many equipment choices on the market and some have features described as "ergonomic." Each purchase should be made with consideration of the context of the use of the equipment. The equipment needs should be clearly identified to meet the operational objectives within the environmental constraints of the facility. Particula.r emphasis should be given to defining the task requirements and assessing the usability of the equipment: the controllability, ease of use, efficiency, and safety (Mack et al., 1995). It should be made sure that employees are trained on the use of the equipment.

Choosing the appropriate intervention is important but so is ensuring the quality of installation. How interventions are integrated into the overall system influences the effectiveness of the control. For example, in one company, the installation of a bank of high scissor lifts was met with resistance by the employees who complained that it was harder to palletize on the lifts than by the traditional method from a conveyor to a pallet on the floor. This was not a case of resistance to change, as the company originally suspected. When analyzed, two installation features were found to make the job harder to perform: a small 2-in. lip along the conveyor edge over which they had to lift a case; and a very high safety railing around the pallet on the scissor lift that caused the case to be lifted up higher. Once these two points were addressed the job was easier and less risky and the employees supported the change (Stuart-Buttle, 1995a).

This chapter discusses general engineering solutions to reduce the exposures to the main risk factors of low back disorders. A high-end engineering solution could be to automate the task so that there is no human involved, for example, autopalletizing. However, full automation is not always desirable or feasible. The focus here is on general interventions at the interface of the human and the task to reduce the risk exposure of the employee. An effective intervention will improve the employee performance, usually through better quality or reduced time to do the task (Oxenburgh, 1991). Engineering solutions are favored over administrative or work practice approaches as engineering solutions are less susceptible to the variability of worker behavior. However, engineering solutions are rarely implemented alone without some administrative or work practice change as well.

The common occupational risk factors for low back disorders from an epidemiological perspective are heavy physical work, static work postures, frequent bending and twisting, lifting, pushing, and pulling, repetitive work, vibrations, and psychological and psychosocial factors (Andersson, 1998). More specifically, situations to avoid are: a fully flexed or bent spine; twisting of the spine, especially with bending; a large moment arm (distance of the load from the low back); and lifting soon after a prolonged static flexed posture, such as sitting or stooping (McGill, 1998).

16.2   Choosing the Right Solution

To derive a successful solution many employees should be involved, including but not limited to, engineers, safety and health personnel, production supervisors, and the operators. The following steps should be considered for identifying problems and determining effective improvements. The process of selecting an assistive device and implementing it are discussed in more detail in Chapter 17.

1. Identify the problems thoroughly, including productivity issues and systems concerns such as double handling, not just the risk factors of low back disorder
2. Identify the root causes of the problems so that there is a full list of issues to correct or improve
3. Aim for simple, elegant, and cost-effective solutions. Do not over engineer
4. Remember that solutions do not have to be all or nothing. The stresses of a job can be reduced with great benefit to the worker; the stress factors do not have to be eliminated
5. Keep the design objectives in mind to ensure they are all addressed by the solutions
6. Once a solution is decided and if equipment is to be purchased, research the market well to determine the best model with the criteria that are needed to make a difference
7. Identify additional features that may be important, such as mobility
8. Ensure that equipment is well engineered for the task and has good usability features
9. Ensure compatibility with other equipment and consider support needs such as preventive maintenance and specialized training
10. Conduct a trial to determine effectiveness prior to purchase, whenever possible
11. Confirm effectiveness of the interventions once fully installed

16.3   Objectives of Interventions

The main objectives to control the risk of low back disorders are to reduce:

- Excess weight, lifted or carried
- High forces, pulling or pushing
- Bending and twisting and awkward postures
- Large moment arms; in other words, extended reaches that increase the distance of the load from the back
- Sustained unsupported posture
- Exposure to whole body vibration

The remainder of this chapter gives some common examples of interventions to control low back disorders during manual materials handling in general industry. Specialized areas, such as patient handling and vibration attenuation, are not addressed.

16.4 General Solutions

16.4.1 Lifting

The following approaches are presented: keeping hands-off (the employee does not lift); using handling aids; reducing weight; sliding versus lifting; keeping the hands around waist height; and bringing the load close.

16.4.1.1 Keeping Hands-Off

- *Automation* (Figure 16.1)
- *Unit size or load principle.* Increase the size of the load so that it cannot be handled manually. This includes approaches such as automatic feed systems from or to containers, or bulk containment that reduces repeated handling (Figure 16.2a and Figure 16.2b).

16.4.1.2 Using Handling Aids

There are a variety of handling aids for lifting a load, some of which are industry specific, such as patient handling aids that are used in health care, or handling aids adapted for the construction industry. Specialty aids will not be presented; the following examples are common ones for many types of industries.

- *Hoists.* Hoists can be just as adept at lifting light loads as heavy loads and there is great versatility as they can carry loads vertically and horizontally. The key is to choose the appropriate system for the job, which is not always easy to do. The goals are ease of use, and moving fluidly and quickly to keep the current pace of the job or go even faster. The size of the load, lifting distance, and desired speed need to be matched to the parameters of the hoist specification. Hoist criteria are based on the frequency of use, power source, type of motor, controls, chain or wire rope, and mounting options (Forger, 2003). The basic options are shown in Table 16.1.
- Intelligent assist devices (IAD) is the newest technology that is computer driven. They are devices programmed to sense the weight and inertia of the load so that the operator needs only to apply a nominal push in any direction to guide the load and can control the load with speed and precision. IADs also detect changing weight of a load as it empties. These features provide great ergonomic advantages in reducing the forces exerted by operators and provide ease of

FIGURE 16.1 An autopalletizer for sacks eliminates manual lifting.

(a)

(b)

FIGURE 16.2   (a) Two people are manually emptying boxes. (b) The bulk container automatically dumps the contents on the line, instead of boxes of product being manually lifted and emptied.

use and fine control. More sophisticated IADs can sense where the load is in space and place it precisely and automatically (Nash, 2002).

- Types of end effectors of hoists include vacuum, hooks, belts, clamps, and magnetic attachments, all of which can be a single head or multi-head as the task dictates. The load can be lifted, rotated, and tilted according to the task needs and tailored for specific loads such as drums. Specialized ends to hoists can be custom-made (Figure 16.3).

TABLE 16.1   Hoist Features and Options for Each Feature

Feature	Options	
Duty class (degree of use)	H1–H5: infrequent or standby; light; medium; heavy; severe	
Power	Powered: electric; air	Manual: rachet lever; hand chain
Motor	Single, double, or variable speeds	—
Controls	Pendant, remote, or IAD	—
Rope	Chain or wire	
Mounting	Fixed point (up and down only)	
	Trolley along an I-beam	
	Overhead crane (jib, articulating or nonarticulating)	

FIGURE 16.3 Two examples of hoists: one on the left is a vacuum hoist and the other on the right is a simple chain and strap hoist.

- *Powered mechanical lifts.* Small lifts can be used to raise items. Often the lifts are transporters for short distances as well. Commonly, the lifts can be tailored for the items to be handled. The lifts can be either mechanical or powered (Figure 16.4).
- *Tilters and lifters.* There are many choices of lifters that can rotate and orient the product, as well as tip the product. Custom tailoring for the task is common (Figure 16.5a and Figure 16.5b).

16.4.1.3 Reducing Weight

Check that the weight is not outright too great, based on current guidelines, such as the Lifting Threshold Limit Value (ACGIH, 2001). Controlling the weight can be an option without investing in handling devices or other methods:

FIGURE 16.4 An employee is using a small mechanical lift to raise a roll of material to mount onto a machine.

- *Split the load.* Use smaller containers to reduce the load, but be aware that this approach may increase the frequency of handling. Consider the bulk approach mentioned above if the frequency of smaller loads is unacceptable.
- *Reduce dead weight.* Check if some of the weight being lifted is unnecessary, for example, using a stainless steel tray instead of aluminum or plastic. Consider lightening the material by punching holes.

16.4.1.4 Sliding versus Lifting

Avoid lifting when possible. Sliding on surfaces that reduce friction or using powered conveyors is generally preferable to lifting. There are many engineering choices for different circumstances. For example, simple roller or conveyor connectors can prevent lifting packed boxes to a shipping line that is 180° behind the packer. Air tables can help reduce friction for large items to move them across a surface:

- *Line connectors or multidirectional conveyors.* Ensure that there is no undue weight or force entailed if a connector is retractable or needs to be lifted by a person to pass by. The design must be very user

(a)   (b)

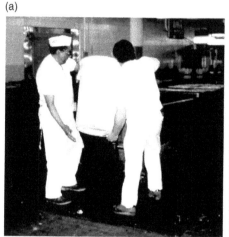

FIGURE 16.5   (a) Two men are needed to lift and dump the contents of the barrel. They also had to transport the barrel manually to the dumping area. (b) The drum was moved to the location by the mobile drum transporter. In addition, the transporter can tip the contents of the drum into the hopper. One person is involved in the task. While the drum drains, the person is free to perform other tasks.

friendly otherwise the employees are not likely to use the system. Multidirectional rollers are especially useful when cases or items need to be turned and oriented a particular way (Figure 16.6).

- *Air systems or friction reducing systems*. Air can be used to lift heavy objects and make them easier to handle. The technique is employed by mechanisms that move large machinery, as well as for moving heavy units on table-tops. The approach is effective for precise positioning of units. Slippery coatings can also be applied to work surfaces that reduce the force necessary to slide a product or item across the surface. One example of such coating is ultra-high molecular weight polyethylene (Figure 16.7).
- *Gravity principle*. Gravity is very effective in reducing forces exerted by the operator, yet it is underutilized in many processes (Figure 16.8a and Figure 16.8b).

16.4.1.5   Keeping the Hands Around Waist Height

Deep bending increases the load on the spine and the risk of low back disorders. Raising the load so that the hands are around waist height (above the hips and below the shoulders) not only reduces the risk of

FIGURE 16.6   Filled boxes are pushed across connecting roller conveyors onto the central shipping line.

FIGURE 16.7 An air system (note air hose to fixture) floats the heavy base of a unit so that it can be manually pushed with ease, into position under the press.

(a)

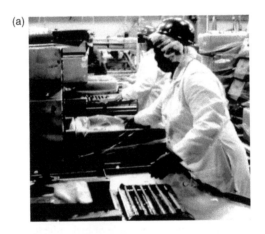

(b)

FIGURE 16.8 (a) Each packet of product is picked up by the left hand and passed to the right, which then stretches to place the product into the box. (b) Gravity brings each packet down a nonpowered roller conveyor to the box. The employee guides the product, letting it fall into position.

FIGURE 16.9   The pallet racks are permanently raised off the ground so that the hand height to lift the lowest case is between the hips and shoulders.

injury, but gains productive time that is otherwise lost when bending over. Because the concern during lifting is the height of the hands, it is easy to forget that the height changes with different-sized containers. Therefore, actual platform heights should be based on box size and where handles are located or how the hands are used to lift. Variable sizes of units would indicate the need for an approach to adjust surface height.

There are several common mechanisms to achieve height: permanently raise the height of the base of the load; use scissor lifts or lift tables; use forklifts as lift tables; retrofit hydraulic lifts on legs of the table or equipment; and install spring platform or self-leveling bases in containers. Lift tables are mechanical, hydraulic, or pneumatic, hydraulic being the most popular but the least clean method of control. They can come with customized tops to support specific products such as rolls.

High lifting places stress on the shoulders and neck, as well as on the back. The risk increases if the load is unstable. If the load cannot be lowered to around waist height, the person should be raised on a platform to the work height.

- *Raise height of the base* (Figure 16.9 through Figure 16.11).
- *Retrofit hydraulic leg lifts.* Retrofit kits are available of hydraulic leg systems that attach to existing table or equipment legs. There are also larger hydraulic poles for central support of a surface. A control raises or lowers the height of the surface. Such systems are commonly used to convert a fixed height work surface to one that is adjustable. Adjustability is especially useful when there are different-sized product runs or fine, sustained work being performed by people of different heights.
- *Self-leveling containers.* There are several types of self-leveling containers. A retrofit system is available to custom build a spring base for an existing container or on a cart frame. There are off-the-shelf carts with spring bottoms inside, such as those used in the laundry industry. In addition, other models exist, such as a platform truck with a spring actuated, self-leveling parts container. All these methods keep the load that is being lifted at around waist height.
- *Raise height of the person.* Platforms should be provided to raise the person up to the height of the work if the work cannot be lowered (Figure 16.12 and Figure 16.13).

16.4.1.6   Bringing the Load Close

Keeping the load close to the body reduces the moment arm and hence the forces on the low back. When possible, without undue stress on the shoulders, the item should be oriented so that the length of the part is across the body. This keeps the center of gravity of the item closest. There are several engineering approaches that can be taken to accomplish getting the load closer. Some main examples are tilting

(a)

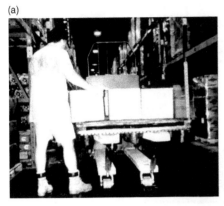

(b)

FIGURE 16.10 (a) Two illustrations show the pallet raised for loading. The picture on the left hand side in a distribution center shows the forks of a pallet jack raised up so that the first few layers of the pallet are stacked at around waist height. The picture on the right-hand side shows a simple platform truck being used to raise the pallet for palletizing from the conveyor belt. The platform truck reduced deep bending and more women were able to perform the job as the forces were reduced. (b) Two illustrations show the pallet raised up for unloading. The picture on the left-hand side shows traditional scissor lift tables that raise the pallets of paper, which are removed during the job. The picture on the right-hand side shows a forklift being used to raise the pallet for unloading.

containers, using gravity feed, reaching with devices such as a reach stick, and using turntables. Large containers or tanks can be tilted for access and there are models that have a drop gate to improve access as well:

- *Tilting containers.* Tilters can be used to angle containers to reduce the employee's reach and hence the moment arm. There are many variations of stationary container tilters as well as mobile carts that can lift the load and tilt it for access. For example, some tilters just lift and rotate the load, others may tilt to a limited degree such as 15°, and there are heavy-duty tilting systems. Tilters can be manual or motorized and hydraulic or pneumatic (Figure 16.14).

   Identify the needs of the task and be critical when assessing the equipment. Ensure there are minimal obstructions between the worker and the items to be reached so that the moment arm is as short as possible. For example, the handles of some tilt models protrude in the front where the person stands, which increases the horizontal distance to the bin even when it is tilted.
- *Gravity feed.* The angle of gravity rollers controls the speed of movement of the product. Gravity feed allows the product to move to the end of a line so that it is always convenient for handling (Figure 16.15).
- *Reach sticks.* Reach sticks aid in pulling items closer for handling.

FIGURE 16.11   One employee operates a stacking station with two feed conveyors, each going to a pallet on a high scissor lift. The stacking of the totes is at waist height. The pallet can be filled to the height of a truck without a change in stacking height for the employee. Note the person that is standing down at the ground level, illustrating how high these lifts are from the floor.

- *Turntables*. A rotation feature on a lift table reduces the moment arm of a lift as there is no need to reach to the back of the pallet. Turntables can be retrofitted to existing equipment. Multidirectional rollers or turning units on conveyors are effective to turn cases and reduce pushing forces or the need to lift (Figure 16.16).

16.4.2   Transporting

There are some common methods of transporting materials without carrying. Some methods are full automated, such as the automatic guided vehicles (AGVs) that do not have a human driver. Other methods do have a human interface. The following approaches for transporting are presented: lift trucks, pallet jacks, carts and hand trucks, dollies, pull/push assist devices, and conveyors. There are also carrying aids that improves the coupling between the object and the hand.

16.4.2.1   Lift Trucks

There is a great choice of lift trucks from walkies to reach trucks that should meet most criteria of the job, environment, serviceability, and performance. Some are designed to raise the operator up (turret trucks) while in others the operator stays down. The interfaces for the operators are becoming more ergonomic: effective to use, comfortable, and efficient. Look for a good line of sight of the load to perform the task, and in seated trucks choose high vibration absorption and cushioned seating.

The attachments for forklifts provide methods to handle materials that otherwise might be done manually or less efficiently (Forger, 2004; Trebilcock, 2004). Common attachments follow:

FIGURE 16.12   An employee stands on a platform while handling boxes.

FIGURE 16.13   A platform lift takes the operator up above the floor to the height of the truck for unloading.

- Side loaders or forklifts rotate up to 180° by having an articulating front.
- Side shifters move 45° left or right, which helps the operator align the load without backing up the forklift.
- Expanded fork lengths for reach.

FIGURE 16.14   The product is tilted so that it is closer to pick up. A tank tilter such as this also raises the product to waist height. The reduction in fatigue on the worker improves the performance of unloading the tank.

- Fork positioners automatically adjust the fork widths. Some lift truck models have straddle arms that help support the load instead of the truck as a counterbalance.
- Trucks designed to handle long loads such as piping or lumber.
- Carriages can handle up to two pallets deep and three loads wide and can be loaded from the side, all of which are very useful features for environments with a heavy volume of pallets.
- Clamping units for cartons eliminate pallets from the system. Sensors on the clamps provide the appropriate pressure to handle the load. Clamp units are also used for products that cannot be placed on pallets, such as rolls of paper and bales of wool (Figure 16.17).
- Load turner systems clamp and rotate a pallet that would otherwise be performed manually.
- Push/pull units attach to slip sheets, pulls the contents on to a platform, and removes the load off the platform once it is positioned.
- Layer picker can take off selected layers from a pallet.

FIGURE 16.15 The containers slowly roll to the end of the conveyor. One container had just been removed from the line.

- Bin dumping systems that attach to forklifts are common for transporting and dumping product into a hopper.
- Drum carrier and rotators stack and pour contents. Specialized fork ends can lift four drums at once.
- Lift trucks can be used as a lift table to raise the load for manual unloading.

16.4.2.2 Pallet Jacks and Platform Trucks

Pallet jacks are designed to lift pallets and typically have forks that slip under the pallet. Either a manual or mechanical mechanism slightly raises the forks to clear the ground. They are manually pushed and pulled or are powered. They are widely used by industry as they are versatile in facilities that are tight on space and they are cheaper than lift trucks. However, there is limited compatibility with lift tables. Zero lift tables are ones that have a platform down to the ground but may still need a ramp for a pallet jack. Thinner lift table models also may be served by a ramp but other models would need to be recessed into the floor or a small lift truck is needed to load and unload the table.

Pallet jacks that can lift the pallet and be used as a lift table are often referred to as walkie stackers. There is also a pallet jack that has a second set of forks that rises up to bring the load to waist height as shown earlier in this chapter. On many powered pallet jacks the operator can ride while standing. The base of a pallet jack can be a platform rather than forks and is used for nonpalletized loads and skid handling. It usually has a greater capacity than a pallet and there are walkie and rider versions.

Specialty pallet jacks includes one that folds for when space is limited. A sidewinder allows for the forks to be sideways from the direction of movement so that long loads can be carried down narrow aisles. Others have scales integrated into the jack. Low-profile jacks have forks that go especially low for passing under narrow clearances. There are specialized jacks for drum handling too. These jacks are mobile, have a hydraulic boom for the drum, and the "forks" are stabilizers.

All pallet jacks have general options such as weight capacity, fork length, width, spread and height, and type of power, usually manual or electric. The popularity of a pallet jack has provoked the design of other equipment to be compatible, such as a conveyor segment that can lower to the ground for a pallet jack to load on to it from either in line or sideways (Figure 16.18).

16.4.2.3 Carts and Hand Trucks

- *Characteristics.* Mack et al. (1995) conducted a survey of manual handling aids for transporting materials and determined the most important design features for two-wheeled and four-wheeled trucks and carts/trolleys.

FIGURE 16.16 A recessed lift table has a turntable so that the heavy units can be placed on the side close to the work area. Skirts are available, and should be present, to enclose the open area and reduce the risk of toes being caught under the lift.

FIGURE 16.18   A heavy duty-pallet truck used to pick up and transport a wagon.

FIGURE 16.17   A carton clamp is attached to a parked lift truck.

The characteristics are (Mack et al., 1995; Eastman Kodak, 2003; Jung, 2005):

- *Interface.* The type of handle, height, and orientation should be considered. Four-wheeled trucks should have handles at a height of about 91 cm (36 in.) but not higher than 112 cm (44 in.). If the truck in narrow and tall, vertical handles are preferable as it allows a comfort height for all users and easier steering. Vertical handles should be at least from 91 to 127 cm (36 to 50 in.) from the ground and no more than 46 cm (18 in.) apart. Two-wheeled hand trucks need to have handles that are at a comfortable height once tilted yet not too high to be unable to initiate the tilt to start off. A long hand truck may have a double handle, one set lower for initiating the tilt of the hand truck.
- *Size.* The size should be suitable for the task. Trucks and carts longer than 1.3 m (4 ft) or wider than 1 m (3 ft) are difficult to maneuver in standard aisles. The height of trucks should be kept to less than 127 cm (50 in.) so shorter people can see over the truck.
- *Weight.* Dead weight should be minimized and the cart or truck designed to tolerate the expected weight of the load.
- *Platform height and dimensions.* This influences the ease of loading and unloading. Shelf heights should be between 50 and 115 cm (20 and 45 in.) if possible.
- *Load securing system.* Ensure the load is stable when full and there is an easy and effective method to secure the load if necessary.
- *Wheelbase.* A wide base mounting of the wheels gives stability to a cart. Wheel bearings should be of good quality and easy to maintain. Brakes should be provided if the cart is taken on sloped floors or needs to be aligned with equipment.
- *Wheel type and size.* Small-diameter wheels help with maneuvering in small spaces, but large diameters make pushing and pulling easier. The best type of wheel and tire pressure depends on the floor surface. Starting forces increase with compressible tires when the loads are very heavy.
- *Swivel wheels.* Two directionally fixed, straight casters and two free moving (or swivel) casters are best for steering and controlling a cart or trolley. Having four wheels that can swivel helps with maneuverability, especially in tight areas.

Carts or floor hand trucks come in a variety of sizes and shapes depending on their purpose. Some are specialized in the form of racks for products such as frankfurters, store trays, carry panels, or shaped for drums. Most are conventional such as wire-caged carts in which to hold product, open two-level carts (or trolleys), or carts with several shelves. Stockpicker carts have a small ladder attached to reach higher shelves. A platform truck or cart can also be electric powered (Figure 16.19).

FIGURE 16.19   Pushing a platform cart with deliveries.

- *Hand trucks.* Hand trucks are two wheeled, tilted during travel, and used for small loads although the loads can be heavy. On occasion large loads are involved and a hand truck is used because of small areas or tight turning, such as encountered with furniture moving. Hand trucks vary in length, weight capacity, handle features (including the dual handle that provides a mechanical advantage when tipping the load), and type of wheels that are usually either pneumatic or hard rubber (Figure 16.20). Disc brakes for hand trucks are available for precision downhill braking and control of the load.

Types of hand trucks include folding and telescoping ones, as well as convertible ones that can become a four-wheeled platform cart. A three-position convertible cart can be used as a hand truck but partially unfolded so that a third support from two additional wheels can take the stress off the arms and hands. This has been recommended for beverage handling by the Occupational Safety and Health Administration (OSHA, 2001). Some models have handles at both ends of the hand truck so that if lifted down steps by two people, each has a handle. Appliance hand trucks are felted for protecting the items and are often wider than regular models. The "nose" or end of the hand truck can be specialized for cargo such as water bottles, cylinders or plated to slip under cases. A push out feature helps to remove product off the hand truck. Stair climbing is a feature that allows for leverage against a step to help ease a hand truck up. There are motorized stair climber hand trucks that are used for major transport of items such as pianos or other large furniture or equipment. These motorized hand trucks are also used for unloading from vehicles, or onto work platforms. Desk movers are a hybrid of a platform cart and pallet jack.

Drum handling is notorious for provoking injuries if manually handled. Apart from attachments to lift trucks, hoists, and jacks, there are specialized hand trucks in a variety of styles that may assist with dispensing or transporting. A drum caddy attaches at the bottom of the drum and is rather like a dolly with a long handle with which to push.

FIGURE 16.20   Loading a double-handled hand truck.

16.4.2.4   Dollies

Dollies are wheeled platforms that make it easier to push the item that is on the dolly. Some models provide a strap or tow handle on the dolly for pulling. There are many sizes and shapes including round for drums or pails or rectangle for other items. Some are minimal with only support for the wheels while others are solid with rims to catch any spills or to hold the container. Some round container dollies are versatile with ridges in concentric circles to hold containers of different diameters. There is a choice of dolly material, such as aluminum, plastic, wood, or steel, and choice of wheel type. Specialized dollies include one for pallets, machinery dollies that are low to the floor, ladder dollies which can help transport more than one ladder, and propel dollies that

FIGURE 16.21 Pushing a barrel on a dolly. Note the broken handle that occurred when pulling the barrel without a dolly.

each fit under a corner of very heavy crates and machinery. Dollies for carrying large sheets or panels have either A-frame or vertical rail supports and are usually referred to as panel or A-frame carts (Figure 16.21).

16.4.2.5 Pull/Push Assist Devices

An optimally designed cart may still be too heavy for manual pushing, or a piece of equipment or product may be too heavy for manual handling. There are power assist devices that can help maneuver a train of carts or a car, for example. Some assists are just for pulling while others will attach to push or pull. They are similar to the concept of a tug pulling a tanker. Stainless steel models are available for sterile environments. Small, nimble movers are available that can move very large items such as rolls of cable, paper, or pipes. These systems are usually air driven.

16.4.2.6 Conveyors

Conveyors are a means of transporting so that materials are not carried. The main flow of many manufacturing facilities will be based on some form of conveyor system. However, areas can still remain in which other modes of transport are used and therefore materials are double handled or items are carried. In such situations, conveyors can still be used, especially if they are temporary, versatile conveyors, such as flex ones, that can be positioned as needed and removed at the end of the job (Figure 16.22). A conveyor can also be added at the end of a line to extend it to reduce carrying. Boom conveyors are used when loading trucks (Figure 16.23). The boom extends to reach the back of a truck and is retracted as the truck fills. Slides and chutes are other methods of transporting, due to gravity or being motorized. These methods can reduce floor traffic from more traditional modes of transport.

FIGURE 16.22 Flex conveyor in position to bring cases closer for palletizing.

FIGURE 16.23   A boom conveyor is projecting into a truck. As the truck fills, the boom is moved back under the fixed roller conveyor.

16.4.2.7   Carrying Aids

Handles help reduce the forces on the hands, arms, and the low back when carrying is necessary. Ensure there are suitable handles when possible. If there are no handles there are carton grips available that are handles with fine spikes. These spikes can be put into the cardboard carton without affecting the integrity of the packaging or the contents. The grips are just as easily removed so that they can be used on the next case (Figure 16.24).

Hand suction cup handles can be attached to sheets of glass and other smooth materials, and handles can attach by magnetism on metal surfaces. Carpet grippers come with spiked pads. There is also a hook with a handle that is designed for quickly hooking under a sheet of material and has the handle to carry the load. A similar concept is available for hooking and pulling a hose where there is a handle included on the device so that it can be held with a strong power grip.

16.4.3   At the Workstation

Activity at a workstation can be dynamic with pulling and pushing product, often twisting to reach or lift materials. Bending over a work area can be driven by the need to see. Ensure that the work is brought closer to the worker, that there is good lighting, and, if necessary, magnification is provided. Some general guidelines follow with common engineering solutions to reduce exposures for low back disorders at the workstation.

16.4.3.1   Comfortable Height

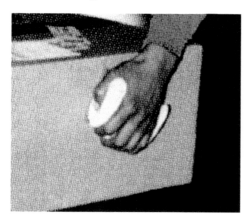

FIGURE 16.24   The lift aid spikes into the carton to provide a handle.

- *Adjustable work surface.* A good working height is ensuring the hands are between the waist and shoulder height for the task and to prevent a sustained bent back posture or elevated arms. Large materials would need to be on a lower surface to ensure work is not performed above shoulder height. If precision or good vision is required, the work height should be higher so that the worker can see. Especially large units may warrant adjustable height workstations. This can be achieved with adjustable workbenches or retrofit adjustable height legs, as mentioned earlier.

- *Raise the work.* Fine work may need to be especially close. This may be achieved by a platform on top of a traditional workbench and in some circumstances providing magnification. Supply bins should also be at the working height. Solutions such as using lift tables or lift trucks to raise work bins were mentioned earlier under lifting. Small, mobile lifts are available for parts bins. Hoists may be needed for heavy parts at a workstation.
- *Raise the worker.* An alternative solution to obtain a good work height may be to raise the worker. A platform can be at a fixed height or be adjustable depending on the duration and precision of the work. The more sustained the postures the more important to fine tune the working height of the person. A platform may be necessary for only one element of a task and put in place when needed (Figure 16.25).

16.4.3.2 Easy Reach

- *Access to work.* The work should be in comfortable reach with minimal twisting of the back. General guidelines recommend a 38 cm (15 in.) reach zone for primary frequent reaching and 50 cm (20 in.) for secondary, less frequent reach (Eastman Kodak, 2003) (Figure 16.26). An "U" or "L" workstation arrangement keeps supplies closer than a straight workbench. Keep supplies as close as possible to reduce twisting and tilt supply bins to reduce the reach. If the task is seated, ensure there is a good swivel seat if it is necessary to turn toward supplies. Supplies above the bench should be as low as possible to the bench for quick access.

  There should be caution when working on a paced assembly line that the worker can perform the task within the provided window without over reaching and twisting for the unit. A preferable approach that allows for individual differences is to work off line. Ensure that the product can be easily, without effort, reached and pulled off line, and also just as easily returned to the line. This can be achieved, for example, with low-friction multidirectional rollers. Diverters can bring the product close to a worker. The product could also be indexed automatically to an off line station (Figure 16.27).
- *Doing the work.* Working on a unit can lead to reaching out to the back of the unit, sometimes requiring the posture to be sustained. There are several approaches to bring the work closer. Positioners with universal joints can hold units and turn in many directions so that the work can be oriented for good access to the task. The work may benefit from being mounted and tilted on a frame and workbench turntables are easily available (Figure 16.28).

FIGURE 16.25 Adjustable work platforms raise the height of the worker.

FIGURE 16.26   Diverters direct the work close to worker so that reaching is prevented.

FIGURE 16.27   A connecting table with multidirectional rollers provides an easy method to pull and push product on and off line. The table is integrated with scales.

FIGURE 16.28   An adjustable frame for mounting a unit on which work is to be done is set toward the working edge. The reach for the unit is reduced and the unit does not need holding while work is performed.

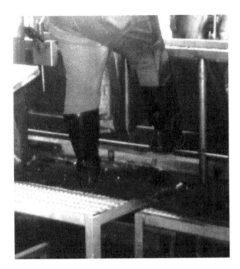

FIGURE 16.29 A railing supports a foot while standing at a workstation.

16.4.3.3 Supported Posture

- *Legs.* Footrests are important for both sitting and when standing in one position as they provide postural relief to the low back (Stuart-Buttle, 1995b). A footrest while seated is necessary if the feet are not well supported by the floor or if there is prolonged use of a chair footring that keeps the legs in a flexed position. Low back discomfort is commonly experienced during prolonged standing. A foot rail on which to raise a foot reduces the

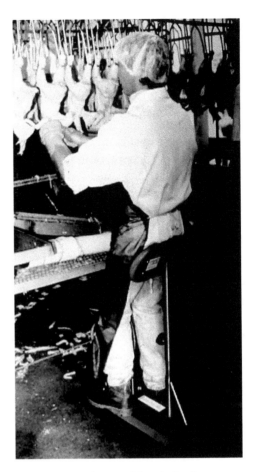

FIGURE 16.30 A lean stool that also has knee support. The stool is alternated with standing rather than used for a whole shift.

back lordosis and relaxes half the back muscles. For this reason, however, an operator should not favor one leg but alternate the legs on a footrail (Figure 16.29). Floor mats can also reduce back discomfort (Stuart-Buttle, 1994). Leaners or sit/stands change the pelvic position, which brings some relief to the back muscles. There should be caution that the reach distance is not increased when using these supports. Sit/stands or leaners should be alternated with standing, not used for a whole shift (Figure 16.30).

- *Torso.* There are occasions that a partial bend of the trunk cannot be avoided, such as leaning over a machine. When sustained this posture can rapidly lead to fatigue in the back muscles. Torso supports are designed to clamp on the edge of a machine or table and adjust to support the chest while leaning forward. Other supporting units provide mobile, horizontal support for the body and prop up the head to work under machinery or vehicles such as cars and smallaircraft.

16.5 Summary

General solutions for low back disorders include methods for lifting that are keeping hands-off (the employee does not lift), handling aids, reducing the weight, sliding versus lifting, keeping the hands around waist height, and bringing the load close. Transporting methods include lift trucks, pallet jacks and platform trucks, carts and hand trucks, dollies, pull/push assist devices, conveyors, and carrying aids. Solutions at the workstation include the general lifting and transport solutions plus additional methods to gain a comfortable height, easy reach, and supported posture.

References

ACGIH, 2001. *2001 Threshold Limit Values and Biological Exposure Indices.* American Conference of Governmental Industrial Hygienists, Ohio, 110–117.

Andersson, G. B. J. 1998. Epidemiology of back pain in industry. In *Handbook of Occupational Ergonomics*, W. Karwowski and W. S. Marras (Eds.), CRC Press, Boca Raton, FL, pp. 913–932.

Eastman Kodak Company. 2003. *Kodak's Ergonomic Design for People at Work*, 2nd edn., John Wiley, NewJersey.

Forger, G. 2003. Workhorse hoists. *Modern Materials Handling*, March: 31–33.

Forger, G. 2004. What's new in lift trucks. *Modern Materials Handling*, May: 35–40.

Jung, M.-C., Hiaght, J. M., and Freivalds, A. 2005. Pushing and pulling carts and two-wheeled hand trucks. *International Journal of Industrial Ergonomics*, 35: 79–89.

Mack, K., Haslegrave, C. M., and Gray, M. I. 1995. Usability of manual handling aids for transporting materials. *Applied Ergonomics*, 26(5): 353–364.

McGill, S. M. 1998. Dynamic low back models: theory and relevance in assisting the ergonomist to reduce the risk of low back injury. In *Handbook of Occupational Ergonomics*, W. Karwowski and W. S. Marras (Eds.), CRC Press, Boca Raton, FL, pp 945–965.

Nash, J. L. 2002. Can computers improve material handling safety? *Occupational Hazards*, July: 35–38.

OSHA. 2001. *E-Tool: Beverage Delivery.* Occupational Safety and Health Administration web page: http://www.osha.gov/SLTC/etools/beverage/index.html.

Oxenburgh, M. 1991. *Increasing Productivity and Profit Through Health and Safety*, CCH International, Australia.

Stuart-Buttle, C. 1994. The effects of mats on back and leg fatigue. *Applied Ergonomics*, 25: 29–34.

Stuart-Buttle, C. 1995a. A case study of factors influencing the effectiveness of scissor lifts for box palletizing. *American Industrial Hygiene Association Journal*, 56: 1127–1132.

Stuart-Buttle, C. 1995b. Why footrests? *Workplace Ergonomics*, July/August: 29–31.

Trebilcock, B. 2004. The right attachment for the job. *Modern Materials Handling*, May: 30–31.

17

Ergonomic Assist and Safety Equipment — An Overview

Gene Buer
*Crane Equipment and
Service Inc.*

17.1 Introduction

There is a wide variety of so-called "ergonomic lift assist" equipment available for industrial and commercial use. In the context of this chapter, we will address only those products that provide the worker a reduced risk of injury or CTDs in what is, or has been a totally manual task. This will eliminate any discussion of traditional lift trucks, high-speed conveyors, heavy-duty hoists, and other machinery that is employed due to speed, weight, or bulk volume considerations. We are also not addressing some of the other products that are marketed as ergonomic enhancements such as floor mats and wrist supports, among others. We will address the most commonly applied industrial ergonomic assist equipment in a very basic and practical way.

This chapter will provide a brief overview of "ergonomic assist and safety equipment" as an intervention device to reduce the risk of CTDs in the workplace. While this is not intended to be the definitive resource for all available mechanical solutions, it is a practical reference tool to understand both the potential benefits as well as the limitations of various types of equipment and systems. The generic description of these kinds of interventions frequently includes the word "ergonomic". In fact, many of the equipment applications are extremely successful in greatly reducing the risk of injury. However, it is extremely important to fully understand both the benefits as well as the potential limitations of equipment selection. Most often, dissatisfaction with the results of applying this type of equipment is most frequently because of one or more of the following reasons:

1. The existing process did not yield to a mechanical intervention without some modifications to make the equipment application successful.
2. The equipment selection was inappropriate for the application.
3. The limitations of the equipment were not well understood, so expectations were not in line with reality.
4. Lack of worker/operator training and education about the equipment.
5. Poor enforcement by management of its use has left many equipment applications in the workplace dormant and abandoned.

The following pages will provide a general summary of the basic types of lift assist equipment available and some important considerations for successful implementation into the workplace. We assume that the entire two-volume Ergonomic reference tool will be utilized for understanding the science of Ergonomics, and this chapter is recognized as a quick reference guide for specific equipment categories. We will cover the following topics:

- Introduction to manual lifting, material handling, and ergonomics
- Typical applications in manufacturing/assembly and warehousing/distribution environments
- Ergonomic movement classifications
- Listing and descriptions of basic types of "ergonomic material handling equipment"
- Tips for getting maximum benefit from your ergonomic investment.

17.2   Basic Types of Lift Assist Equipment and Systems

17.2.1   Industry Overview

Ergonomic lift assists all have a common denominator. They reduce the risk of injury or CTD by assisting to lift or position a payload or tool, or it positions the worker relative to the task in a way that reduces the risk of injury. This end benefit can and must be accomplished in many different ways. As a result, some types of lift assists are perfect for one situation, and totally inappropriate for another. So the wide variety of products available has been application driven by the diversity of need in the marketplace.

The "lift assist" industry is populated primarily with small- to medium-sized companies or business units of larger corporations. They offer a wide variety of solutions ranging from standard catalog equipment to custom-engineered integrated systems. Many lift assist products and systems are marketed through local, regional, or national sales channels such as manufacturer's representatives, distributors, or value-added integrators. User's needs will be application driven, and it is wise to work with an experienced and reputable supplier, since the process tends to very interactive and service oriented.

The next section contains information on basic lift assist equipment types as provided by the E.A.S.E. Council of the Material Handling Industry Trade Association. See www.mhi.org for more information and a listing of manufacturers.

17.2.2   List of Equipment Types and Typical Uses

- Industrial scissors lifts (lift tables)
- Adjustable worker elevation platforms
- Balancers
- Manipulators
- Vacuum assist devices
- Workstation cranes

- Adjustable workstations
- Conveyors
- Stackers
- Container tilters
- Pallet inverters or pallet rotators.

17.2.2.1 Industrial Scissors Lifts (Lift Tables)

Industrial scissors lifts are used in a wide variety of ergonomic applications where bottom-up movement is required. They are used in industries such as manufacturing, assembly, warehousing, distribution, sheet feeding, and printing, just to name a few.

Generally, scissors lifts, or lift tables (as they are sometimes called), may be used to position material so operators do not have to lift excessive loads, lift repetitively, or bend to do their jobs (Figure 17.1). These tables can include variations to facilitate horizontal movement on the deck such as conveyors or ball transfers. In addition, portability options for the base allow the complete unit to be moveable. Tilting devices can be added so loads can be positioned both vertically and angularly.

Scissors lifts can be adapted to any plant situation by choosing from different power options. For example, lifts can be powered with electric or air-powered hydraulic pump units, pneumatic lift systems or full mechanical lift systems.

17.2.2.2 Adjustable Worker Elevation Platforms

Adjustable worker elevation platforms (AWEP) are used to provide significant ergonomic benefits by positioning the worker to the job (Figure 17.2).

There are many defined workstations in manufacturing facilities with points of human interaction at an elevation above the plant floor, requiring that the operator be positioned on a raised platform. Often times, these platforms are makeshift and can present safety hazards It is also common for the plant maintenance department to fabricate non-adjusting platforms configured for the "average employee." These fixed-height platforms do not correctly accommodate the $5'-4''$ employee and the $6'-4''$ employee.

Multiple-shift operations, where three different operators interact with the same process machine in a 24-hours period, complicate the accommodation issue. Job rotation with operators changing jobs every couple of hours adds even more challenges to the workplace ergonomics issue. Job rotation may spread the risk or worker injury over a larger worker population, but it does not eliminate the risk. Modifying the characteristics of a workstation to reduce the number of bad choices an operator can make or providing operator control over the physical characteristics of workstation are proactive methods for addressing in-plant ergonomics issues.

AWEPs provide each operator with the ability to control their own position in relation to the task at hand, and position the task so that the operator can gain a biomechanical advantage over the task.

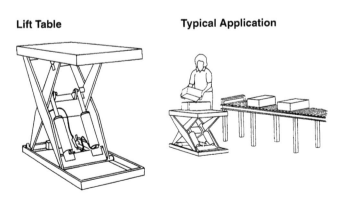

FIGURE 17.1   Industrial scissors lift. (Copyright 1996 E.A.S.E.)

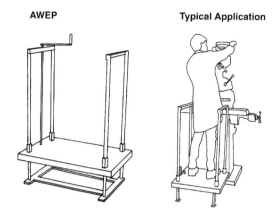

FIGURE 17.2   Adjustable worker elevation platform. (Copyright 1996 E.A.S.E.)

AWEPs are typically mechanically-driven devices that are manually adjusted. Platform sizes vary from $2' \times 3'$ one-person units to larger runways parallel to assembly lines (Figure 17.2). Heights start as low as $2\frac{1}{4}''$ above the shop floor. Most units have 12 inches of vertical adjustment. AWEPs can be pit-mounted, if necessary. Capacities are typically 400 lbs. Powered AWEPs can be activated electrically or using shop air.

17.2.2.3 Balancers

These overhead devices provide and perform functions similar to overhead hoists in that they can lift and lower a load. The balancer configurations and suspensions are similar to a hoist.

The balancer functions differently from a hoist. The functional difference is indicated in the name "balance." A balancer balances the load in a near weightless condition during the lifting operation. This feature allows the operator to maneuver the load easily (Figure 17.3). Balances are often used for awkward and/or rapid load movements. Balancers can also be used to suspend equipment which is used in repetitive operations. A variety of control options are available. A very broad range of load handling devices (end effectors) can expand this product's versatility. Balancers are typically supported on small jibs, light monorail systems or tubular track monorail systems, or as part of a workstation.

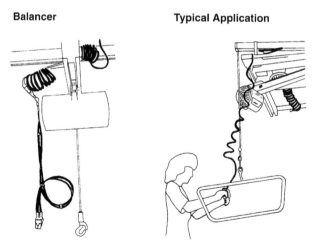

FIGURE 17.3   Balancer. (Copyright 1996 E.A.S.E.)

Manipulator   **Typical Application**

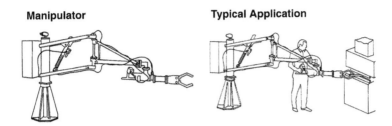

FIGURE 17.4   Manipulator. (Copyright 1996 E.A.S.E.)

Both air-powered and electronically driven balancers are commonly available. Balancers can weigh up to approximately 100 lbs. and have lifting capacities of 49 upto 1000 lbs.

17.2.2.4   Manipulators

A manipulator has a mechanical arm which can move a load horizontally as well as vertically. Manipulators, unlike the hoist or balancer, can provide more than vertical lifting and lowering. Manipulators combine mechanical arms, cylinders/motors and application-specific load handling devices. The devices are often called end effectors. The devices are generally dedicated to a single product.

Manipulators can be floor-mounted, column-mounted, mounted overhead on a rail system or attached to a ceiling.

The machines are generally operator-controlled at or near the load control devices. The operator can manually manipulate the load in the pick, move and place functions (Figure 17.4). The load is in a near weightless condition similar to the balance. Since the manipulator is often used to "reach" in the pick, move and place cycle, it is slower and may require more operator effort than a balancer.

Both products provide a wide range of ergonomic benefits. Balancers and manipulators provide a near weightless lifting situation, easy reach and orientation. They provide easy handling with difficult, rapid, repetitive or awkward applications.

Manipulators and balancers can be designed for a specific application. Before choosing a piece of equipment, you should consider its intended use. Either piece of equipment can adapt to different load sizes and shapes. Lift and lower distances are generally limited to the operator's reach.

Note: End effector types include vacuum cup, powered or manual clamp, probe, hook, forks, mechanical grab, magnet, or a nesting/cradling device.

17.2.2.5   Vacuum Assist Devices

Vacuum tube lifters are ergonomically designed to reduce back strain, injuries and accidents therefore reducing or eliminating workers compensation claims. When using a lifter on a daily basis, worker fatigue should be greatly reduced and productivity increased. The lifter impact on the work place, because of its versatility, should help reduce absenteeism and downtime. Using the vacuum lifter can accomplish many repetitive lifting tasks ergonomically, quickly and skillfully, without delay.

Vacuum tube lifters have a wide range of capacities from as little as 40 pounds up to 1,000 pounds. These units are powered by anywhere from 3.5 h.p. up to 7.5 h.p. regenerative blower.

Most vacuum lifters offer many standard vacuum heads. These heads usually twist or snap on or off to change one head to another and can be done in a matter of seconds. There are a variety of standard vacuum heads designed for specific applications from situations involving narrow spaces with little room to maneuver, to load destinations which are higher than the operator's head, loads from sacks, to cartons, to 55 gallon drums, or large wood or metal sheets. Other standard heads enable the handling of different package shapes and surface types. There are even heads which can enable the operator

Vacuum Assist Device   **Typical Application**

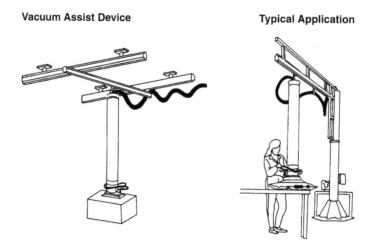

FIGURE 17.5 Vacuum assist device. (Copyright 1996 E.A.S.E.)

to reorient cargo vertically or horizontally, with integral vacuum pump or pneumatic assists devices (Figure 17.5).

17.2.2.6 Workstation Cranes (Enclosed Track Rail Systems)

Enclosed track workstation cranes are ergonomically designed for lighter loads, from 150 lbs. to 4,000 lbs. capacities. Horizontal movement is usually push, but can be powered. The high strength enclosed track design keeps rolling surfaces clean, which contributes to easier crane movement and longer wheel and track life. The tracks' low weight per foot reduces the dead weight, which makes for easier movement, increased worker safety and increased productivity (Figure 17.6).

Enclosed track workstation bridge cranes push very easily — typically 1 pound of force for every 100 pounds of load. This is much easier than an I-beam crane system that typically take 3 pounds of force to move every 100 pounds of load. Ergonomically designed, easy to move workstation cranes put less stress on operators. They offer precise load positioning, significant productivity improvements, fewer injuries and less fatigue.

Equipment configurations include single or double girder bridge cranes, jib cranes and monorails. Both steel and aluminum cranes are available. Suspension is varied. The options include free standing floor support, ceiling suspended cranes or a combination of the two.

Workstation Crane   **Typical Application**

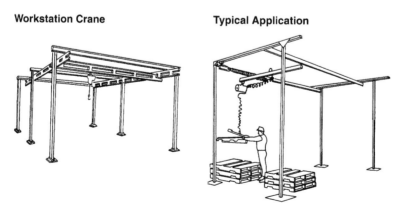

FIGURE 17.6 Workstation crane. (Copyright 1996 E.A.S.E.)

Load suspension trolleys, which ride on wheels inside the enclosed track, have devices that hang below the track opening to support vertical lift devices. These vertical lifting options include hoists, balancers, vacuum lifters and manipulators which contain holding and orienting devices such as hooks, slings, grabs, spreader bars, vacuum devices, custom end effectors and magnets. Enclosed track workstation cranes and monorails can offer ergonomic solutions to a single operation or an entire material handling system.

Note: Workstation cranes can also be powered and controlled with various drive packages that range from very basic to extremely sophisticated.

17.2.2.7 Adjustable Workstations

Workstations (Figure 17.7) meeting criteria relating to ergonomics should be designed to include the following:

- Vertical adjustment of the work surface. The surface should allow for persons of various heights and physiques to adjust the height to meet their physical characteristics. This would be especially true in multiple shift application.
- Vertical height adjustment is also necessary as manufacturing or assembly work changes. The height required to assemble or manufacture should change as the product or subassembly changes. Vertical adjustments should compensate for the size of the component.

Considerations for a well designed workstation:

1. Everything the worker needs for the task should be available without stretching or reaching and should be easy to handle
2. Items being handled should not require the hands to work at a level which averages more than 6″ above the work surface
3. Items in excess of 10 pounds should use a mechanical assist — nothing needs to be lifted from the floor
4. The ideal work area should be semicircular around the worker
5. The work surface should be at elbow height or slightly below it, so the forearm is horizontal or slightly slated down
6. Optimum height of work surface should be: For writing or light assemble — 27.5″ to 31″; For heavier manual work — range of 26″ to 28.5″
7. Work area should be well lighted with a glare eliminating system

Workstation

Typical Application

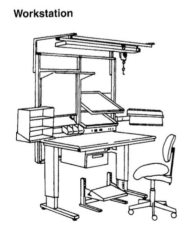
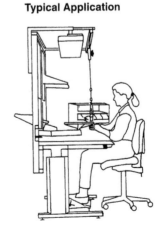

FIGURE 17.7 Workstation. (Copyright 1996 E.A.S.E.)

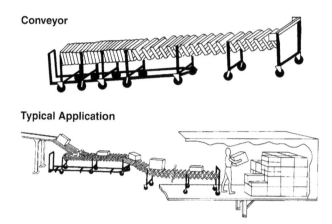

FIGURE 17.8   Conveyor. (Copyright 1996 E.A.S.E.)

17.2.2.8   Conveyors

Conveyors (Figure 17.8) are used in a variety of industries to transport unit/package loads from one location to the next. Conveyors in manufacturing operations are used primarily as part of the assembly process, while conveyors in warehousing and distribution are frequently used to process orders.

A conveyor provides an ergonomic advantage to the worker by providing a mechanized means of moving work to the worker. Expandable conveyors are ideal for assisting workers with tasks such as loading or unloading trucks and trailers. For order picking situations, conveyors provide workers with open cases at the proper height and in full view. This can mean fewer mistakes, less stooping, and less reaching.

Workers interfacing with conveyors are frequently engaged in bending, twisting, and reaching motions as they put on or take off materials from a conveyor. When using a conveyor system, it is important to evaluate the workstation design so that the equipment adjusts for the 6′ tall worker on the first shift and the 4′ 11″ worker on the second shift.

17.2.2.9   Stackers

Manually propelled lift trucks, commonly called stackers, can be an extremely versatile group of ergonomic assist products. They are designed to efficiently transport work from one work station to another, as well as, for elevating loads to comfortable ergonomic work heights.

Stackers are available with platforms, adjustable forks or fixed forks. Capacities range from 250 lbs. to 3,000 lbs., and load centers to 24″, to handle load sizes up to 48″ square.

Fork model stackers (Figure 17.9) offer vertical level from floor level (5″ for platform models) to 4′, 5′, 6′, 8′, 10′ and 12′ in lift height within a very small foot print. However, most ergonomic applications require only lower lift height models.

Hydraulics is the most frequently used method for lifting, although wire rope and winch are used with some lift weight manual types. Battery lifting power is most frequently used, but manual, AC electric, and air power models are available. A control lever is usually located at finger tip level on a control panel, but hand pendants and foot controls are also available.

Stackers effectively move, raise, and position wire baskets, tote boxes, crates, skids and pallets to proper ergonomic work levels for maximum efficiency, productivity, and worker safety. They are also available with numerous attachments and accessories to transport and position special work loads such as barrels, coils, rolls, etc.

Stacker   **Typical Application**

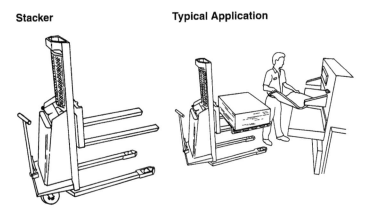

FIGURE 17.9   Stacker. (Copyright 1996 E.A.S.E.)

17.2.2.10   Container Tilters

Containers, boxes and baskets are widely used to store and transport parts and products. Although they are efficient, loading and unloading them requires motions that can be unsafe for workers. Regardless of how light the stored parts are, workers must repetitively bend, stoop, reach and lift as they work their way to the bottom of these containers.

To make the process safer more productive, the container should be moved up and toward the worker as it is unloaded. Proper positioning can eliminate the need for bending, stretching, reaching and unnecessary lifting.

Pictured are portable tilters (Figure 17.10) which allow containers to be picked up, moved into position, and tilted for easy access. There are similar stationary models which allow containers to be directly placed on the tilter by hand pallet trucks or fork trucks. In addition, there are floor height pivot point devices and tilters that mount on lift tables and other bases.

There are nearly as many tilter designs as there are container styles, and applications information must be carefully considered before equipment selection. Special attentions should be given to all container dimensions and the weight and configuration of the parts.

Container Tilter   **Typical Application**

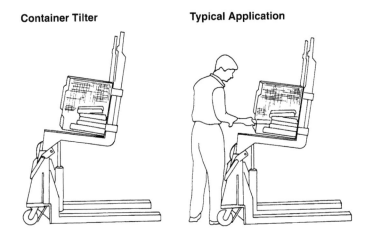

FIGURE 17.10   Container tilters. (Copyright 1996 E.A.S.E.)

Pallet Rotator/Invertor   **Typical Application**

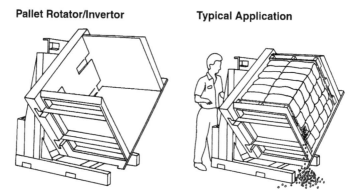

FIGURE 17.11   Pallet rotator or invertor. (Copyright 1996 E.A.S.E.)

17.2.2.11   Pallet Rotators/Invertors

Throughout industry, shipping and storing goods on pallets is a methodology for transferring inventory. This type of unit load handling can create numerous materials handling challenges Most of these difficulties are associated with restacking the pallet load and many can be handled by mechanically inverting the entire load instead.

Broken pallets can be easily removed and replaces. Crushed, damaged, or broken containers or bags can be pulled from the bottom of stacked loads by simply rotating the entire stack (Figure 17.11). Outgoing loads can be readily transferred from wood pallets to skids or slipsheets used for shipment. Incoming loads can be just as quickly transferred to permanent pallets used in racking systems or to special pallets for food handling and similar applications.

In operation, a pallet truck or fork truck is used to place palletized loads in the rotator. The pallet to which the load is to be transferred is placed on top of the stack. The rotator grips the load and rotates it 180 degrees, transferring the load from one pallet to the other.

17.3   Calibrating Expectations/Task and Process Assessment Considerations for Suitability and Impact of Lift Assists

The previous section on equipment types illustrate the wide variety of equipment designs and approaches to improving the ergonomics in the workplace. As one might suspect, the selection of the appropriate solution is application driven and frequently can be addressed successfully by more than one acceptable option.

Note: In many large organizations, specialists from various disciplines team up to drive these decisions. However, in small- and medium-sized companies, these decisions may be delegated to a few people, or even a single individual with little experience with this somewhat arcane equipment. The information in this section is designed to help guide the less experienced group or individual.

Sometimes, certain constraints make only one option available, and in some instances, the situation is so difficult that an equipment option is virtually impossible without a change in process or work cell configuration.

There are numerous potential drivers for determining what equipment options might be suitable under any given set of existing or future process parameters. Assuming that you have used the other chapters in this text properly and a lift assist is required, the following basic issues should be addressed when selecting a mechanical intervention to employed successfully:

- Define, refine and streamline the task and sequence
- Determine throughput and cycle times

- The importance of keeping it lean
- Measure and evaluate the ergonomic impact

17.3.1 Define, Refine and Streamline the Task and Sequence

While defining the sequence of operation appears elementary at first glance, it is critical to understand that the introduction of lift assist equipment into the workplace will be much more successful if the process is reviewed in light of maximizing the value of the new equipment. Depending upon the application and equipment employed, a change as minor as repositioning the worker relative to the payload might be the only change that is made. In other cases, changing the task sequence as well as upstream and downstream processes might be required to fully leverage the advantages of the lift assist equipment/system solution.

17.3.1.1 Defining the Present Task and Sequence with no Lift Assist

This is an extremely important step because many field studies have shown that the written standard operating procedure (SOP) for a given task or operation *is not, in fact, what is happening every day.* Workers frequently believe they are operating in one way, when in fact they perform their work even differently from what they believe they are doing. It would be wise to get a third party ergonomic and process assessment from a reputable company to fully define the task and the risks. Once the existing task sequence is well understood, it is time to investigate what mechanical interventions are suitable.

17.3.1.2 Refining the Task and Sequence for a Lift Assist

Among the first issues to be evaluated in this step is selecting a knowledgeable vendor or consultant to present the most suitable equipment options for any particular application. Several things can and will affect one's final selection, a few of which are listed as follow:

- Available floor space or headroom
- Flexibility to optimize presentation of payload
- Versatility required (variety of sizes, weights, shapes, etc.)
- Fixed or mobile lift assist required or desired
- Operator interface issues and cycle times
- Possible changes to this task to optimize ergonomics and productivity
- Possibility to modify upstream/downstream processes to optimize the task

The best equipment selection would most likely incorporate some or all of these considerations to provide the solution that will most likely deliver on the ergonomic promise of lower risk and higher productivity.

17.3.1.3 Streamlining and Worker Buy-In/Acceptance

Because of the human element implicit in lift assists, it is important to stay in tune with the workers who use them. The first step is training people to use these tools effectively. Like any other change, many people resist the introduction of lift assists into their space. Even after acceptance of these tools, many will believe that good ergonomics means not really working, making their job "easier," or that they "feel comfortable" while they are working. However, the real goal is to reduce the risk of injury and CTDs.

To that end, it is important to distinguish the difference between operator preference and good ergonomics. For example, many people stand and sit in poor posture because it is more comfortable to them at the time. It is hurting their spine in the long run, but "it feels more comfortable" at the moment. While it helps to have the operator "feel comfortable" with using the equipment, it should be noted that reducing the risk of injury is still the main objective. Operators will typically need training and time to adjust to the new method of doing his or her job.

Getting postinstallation feedback and ongoing task and process evaluation for continuous improvement in work cell ergonomics are an essential part of optimizing the benefit of lift assist equipment. One may discover that minor adjustments to controls or equipment placement are more convenient or attractive to the worker and do not adversely affect ergonomics or productivity. If the front-end work was done in a thorough manner, these changes should not be very radical or extensive.

However, it is imperative to let the science of ergonomics dictate modifications that affect the task risk, not what the worker perceives as good ergonomics or as more comfortable.

17.3.2 Cycle Times and Throughput

The introduction of a list assist product or system frequently suggest either faster or slower cycle times for the operator/worker. It is important to understand the probable impact of the intervention relative to this important economic parameter. Listed next are some examples of situations where the impact can be very different. The common denominator, however, is the reduced risk of injury and CTDs and the human and economic benefits obtained.

17.3.2.1 Lower Risk/Faster Cycle Time

In the manual handling of small part or boxes, a simple lift and tilt table can frequently have the dual advantage of reducing risk of injury while potentially increasing the number of units than can be moved per hour by each worker.

17.3.2.2 Lower Risk/Head Count Reduction/Same Cycle Time

There are many manual tasks in various industries that require two people due to the weight or awkward size of a payload. An overhead handling system or floor-mounted manipulator can provide the means for one person to perform the task in the same amount of time.

17.3.2.3 Lower Risk/Head Count Reduction/Faster Cycle Time

There many tasks where the introduction of semiautomated solutions such as conveyors, rotate–tilt–lift tables and stations, powered workstation cranes, adjustable worker elevation platforms, pallet rotators/invertors, stackers, and intelligent assist devices can frequently reduce headcount and increase cycle times.

17.3.2.4 Lower Risk/Perception of Slower Cycle Times

Occasionally, there will be instances when the fastest possible cycle time decreases with the introduction of a lift assist because the machine simply takes longer to complete the cycle than a human does with no assist. It is extremely important to understand that productivity is a marathon, not a sprint. Many high-risk tasks are not addressed with any interventions because the worker is not handling a heavy payload and in a short evaluation period, it seems to reduce productivity. However, in the long run, the total throughput is likely no better than if a risk-reduction system had been employed. And it will happen without the toll of frequent workplace injuries, lost days, and workers compensation claims.

17.3.3 Keeping it Lean

As all industries and enterprises embrace the lean operating model, it is important to recognize the benefit and importance of lift assist equipment as an integral part of a lean enterprise. A lean enterprise will plan its processes, or be willing to modify them, to embrace and incorporate lift assist systems where required into its operating model. This should drive access to truly lean solutions from the list of equipment suppliers and will ensure a productive workplace with a dramatically reduced risk of injuries and CTDs which threaten your most valuable asset . . . the employees.

17.3.4   Measure and Evaluate the Ergonomic Impact

One of the inherent dangers of addressing a specific risk in a task is that it is possible for the intervention to solve one problem while creating another. One of the classic examples of this is in overhead material handling applications where a workstation crane system is employed with a lifting or balancing device on the bridge and the combined weight and momentum of the bridge and the hoist creates an "over-travel" condition that makes it difficult for the worker to stop the entire mechanism. While the problem of spine loading originally being addressed had been mitigated effectively with the workstation crane and lifting or balancing device, the new problem areas were the neck and shoulders as a result of this over-travel condition.

This cautionary tale is presented not to dissuade one from implementing these products, but instead to encourage a complete assessment before and after implementation, and to take a more holistic approach to applying ergonomic mechanical interventions. Often, a supplier can help steer one in the right direction. However, independent third-party assessments by professional ergonomists are highly recommended. The ways in which one can inadvertently replace one problem with another are too numerous to be effectively addressed here. But the lift assist implementation strategy should include a method of proactive prevention and ongoing measurement and assessment to ensure that all of the risks are being identified, accounted for, and managed appropriately.

17.4   Ergonomic Motion Classifications

Figure 17.12 through Figure 17.14 illustrate Ergonomic motion classifications.

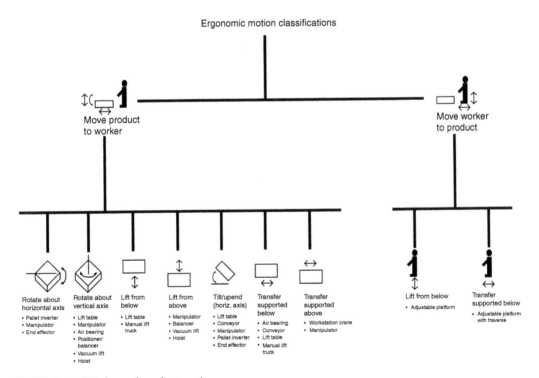

FIGURE 17.12   Worker and product motion.

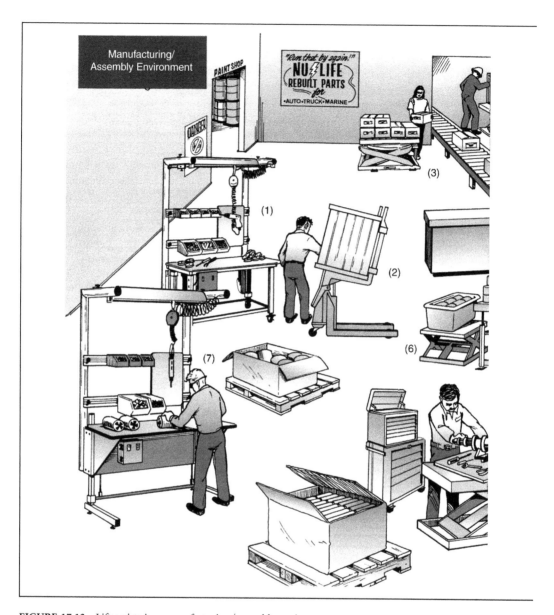

FIGURE 17.13   Lift assists in a manufacturing/assembly environment.

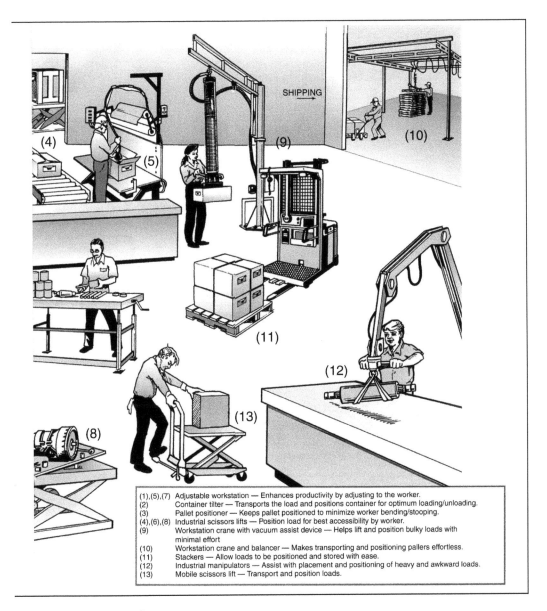

(1),(5),(7) Adjustable workstation — Enhances productivity by adjusting to the worker.
(2) Container tilter — Transports the load and positions container for optimum loading/unloading.
(3) Pallet positioner — Keeps pallet positioned to minimize worker bending/stooping.
(4),(6),(8) Industrial scissors lifts — Position load for best accessibility by worker.
(9) Workstation crane with vacuum assist device — Helps lift and position bulky loads with minimal effort
(10) Workstation crane and balancer — Makes transporting and positioning pallers effortless.
(11) Stackers — Allow loads to be positioned and stored with ease.
(12) Industrial manipulators — Assist with placement and positioning of heavy and awkward loads.
(13) Mobile scissors lift — Transport and position loads.

FIGURE 17.13 *(Continued)*

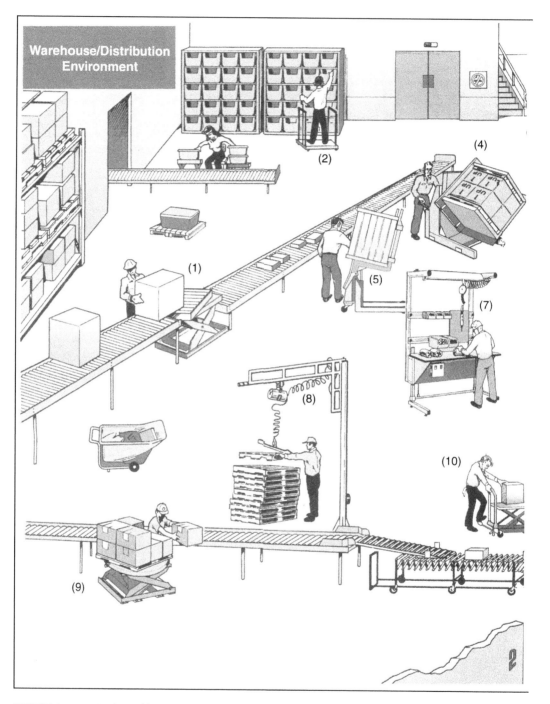

FIGURE 17.14   Warehouse/distribution environment.

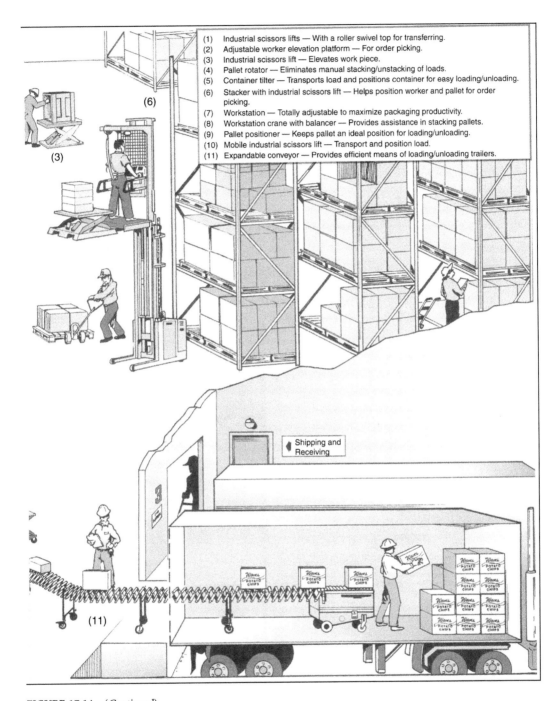

(1)   Industrial scissors lifts — With a roller swivel top for transferring.
(2)   Adjustable worker elevation platform — For order picking.
(3)   Industrial scissors lift — Elevates work piece.
(4)   Pallet rotator — Eliminates manual stacking/unstacking of loads.
(5)   Container tilter — Transports load and positions container for easy loading/unloading.
(6)   Stacker with industrial scissors lift — Helps position worker and pallet for order picking.
(7)   Workstation — Totally adjustable to maximize packaging productivity.
(8)   Workstation crane with balancer — Provides assistance in stacking pallets.
(9)   Pallet positioner — Keeps pallet an ideal position for loading/unloading.
(10) Mobile industrial scissors lift — Transport and position load.
(11) Expandable conveyor — Provides efficient means of loading/unloading trailers.

FIGURE 17.14   (*Continued*)

18

Gloves

R.R. Bishu

V. Gnaneswaran

University of Nebraska-Lincoln

A. Muralidhar

Auburn Engineers, Inc.

18.1 Importance of the Hand

The hand is probably the most complex of all anatomical structures in the human body. Along with the brain, it is the most important organ for accomplishing the tasks of exploration, prehension, perception, and manipulation, unique to humans. The importance of the hand to human culture is emphasized by its depiction in art and sculpture, its reference frequency in vocabulary and phraseology, and its importance in communication and expression (Chao et al., 1989). The human hand is distinguished from that of the primates by the presence of a strong opposable thumb, which enables humans to accomplish tasks requiring precision and fine control. The hand provides humans with both mechanical and sensory capabilities.

18.2 Prehensile Capabilities of the Hand

Napier (1956) divides hand movements into two main groups — prehensile movements, in which an object is seized and held partly or wholly within the compass of the hand, and nonprehensile movements, where no grasping and seizing is involved but by which objects can be manipulated by pushing or lifting motions of the hand as a whole or of the digits individually.

Landsmeer (1962) further classifies human grasping capabilities as power grip, where a dynamic initial phase can be distinguished from a static terminal phase, and precision handling, where there is no static terminal phase. The dynamic phase as defined by Landsmeer includes the opening of the hand, positioning of the fingers, and the grasping of the object. Westling and Johansson (1984) state that the factors that influence force control during precision grip are friction, weight, and a safety margin factor related to the individual subject. They also found that in multiple trials, the frictional conditions during a previous trial could affect the grip force. They also showed that the grip employed when holding small objects stationary in space was critically balanced such that neither accidental slipping between the skin and the object occurred, nor did the grip force reach exceedingly high values. This sense of critical balance as to the amount of force applied while gripping is important, as too firm a grip could result in the destruction of a fragile object, causing possible injury to the hand, or lead to muscle fatigue and interfere with further manipulative activity imposed upon the hand.

Sensory perception in the hand is due to the presence of mechanoreceptors distributed all over the palmar area, especially at the tips of the fingers. Thus, feedback from the hand is a critical component of the gripping task enabling the amount of force to be controlled. Anything that blocks the transmission of impulses from the hand interferes with the feedback cycle and affects grip force control.

18.3   Need for Protection of the Hand

Therefore, hand, which provides humans with both mechanical and sensory capabilities, needs to be protected from the environment. Protection is needed from mechanical trauma (abrasions, cuts, pinches, punctures, crush injuries), thermal extremes (heat and cold), radiation (nuclear, ultraviolet, x-ray, and thermal), chemical hazards, blood borne pathogens, electrical energy, and vibration.

There exist several forms of hand protection, which can be used as stand-alone protection, or in combination with other personal protective equipment. The commonly available hand protection are gloves, mittens, finger cots, and gauntlets, made of several materials, such as leather, cotton, rubber, nylon, latex, metal, and in combinations of the same, to provide maximum protection against the specific condition being guarded against.

It will be relevant to give a brief description of a variety of gloves that are available today. It will also be relevant to discuss performance effects of gloves, before detailing on the challenges of glove design.

18.4   Types of Gloves

There are a wide variety of gloves available today. Starting from a garden glove at 50 cents a pair from a local grocery stores to latex and vinyl gloves used by health care professionals to custom fit shuttle gloves donned by astronauts for extra vehicular activities (EVAs), which cost a few hundred thousand dollars a pair, the extent of variety among gloves can be overwhelming as to defy easy categorization. Gloves can be categorized along a number of dimensions such as materials, design, and location of use. According to the National Safety Council (1983) hand protection can be job-rated or general purpose. Job-rated hand protection is designed to protect against hazards of specific operations while general purpose gloves protect against many hazards. Materials used in gloves are: cotton, nylon, duck, jersey, canvas, terry, flannel, lisle, leather, rubber, synthetic rubber, wire mesh, aluminized fabric, asbestos, plastic and synthetic coatings impregnated fabrics, polyvinyl chloride, nitrile, neoprene, and many man-made fibers with identifiable brand names (Dionne, 1979; Riley and Cochran, 1988). Glove styles include liners, reversibles, open back, gloves or mittens with reinforced nubby palms and fingers and double thumb gloves. Certain tasks may need double or more gloves. For example, shuttle gloves are an assemblage of three layers of gloves, while latex sensitive people in the medical community wear an inner liner with an outer shell. The length of glove may be up to the wrist, elbow, or shoulder with exact dimensions depending on the manufacturer. In summary, the gloves range from easily available general purpose ones to highly task specific and job-rated ones.

The use of gloves, although a necessity in many work places, has some associated disadvantages. Gloves have been found to affect hand performance adversely, and the performance parameters affected are dexterity, task time and grip strength, and range of motion. Facilitation of these activities, with simultaneous

TABLE 18.1 Comparison of Bare Hand and Gloved Hand Capabilities

Indices	Bare Hand	Gloved Hand
Thermal tolerance	Poor	Good
Tactile perception	Excellent	Poor
Grip strength	Good	Reduced
Range of motion	Excellent	Poor
Manipulative ability	Excellent	Reduced
Prehension	Excellent	Poor
Torque capability	Poor	Improved
Vibration tolerance	Poor	Good
Dexterity	Excellent	Reduced
Chemical resistance	Poor	Excellent
Electrical energy	Poor	Excellent
Radiation (all kinds)	Poor	Excellent
Biohazard risk	Poor	Excellent
Abrasive trauma	Poor	Improved

protection from the hazards of the work environment, is often conflicting objectives of glove design. The conflicts associated with providing primary hand protection through the use of a glove while permitting adequate hand functioning has been widely recognized. Table 18.1 provides a summary of comparison of bare hand and gloved hand capabilities.

The glove effects on performance are enumerated in detail in the following section.

18.5 Glove Effects on Strength

18.5.1 Grip Strength

Published evidence exists for glove effect on grip strength, grasp strength, pinch strength, grasp at sub-maximal levels of exertion, torque capabilities, and on endurance time. Reduction in grip and grasp force when gloves are donned are perhaps the most common finding (Hertzberg, 1955, 1973; Lyman, 1957; Cochran et al., 1986; Wang et al., 1987; Sudhakar et al., 1988). Cochran, Albin, Bishu, and Riley (1986) examined the differences in grasp force degradation among five different types of commercially available gloves as compared to a bare-handed condition. The results indicated that the bare-handed condition was significantly higher in grasp force than any of the glove conditions. Wang, et al. (1987) performed an experiment on strength decrements with three different types of gloves. The authors showed that there was a reduction in grip strength when comparing gloved performance to bare-handed performance. Bishu et al. (1993a, b) studied the effects of NASA EVA gloves at different pressures on human hand capabilities. A factorial experiment was performed in which three types of EVA gloves were tested at five pressure differentials. A number of performance measures, namely grip strength, pinch strength, time to tie a rope, and the time to assemble a nut and bolt, were recorded. Tactile sensitivity was also measured through a two-point discrimination test. The salient results were that with EVA gloves strength is reduced by nearly 50%, and that performance decrements increase with increasing pressure differential. McMullin and Hallbeck (1991) studied the effect of wrist position, age, and glove type on the maximal power grasp force, and their findings indicate that a single layer glove is better than several layers, as the bunching of glove material at the joints could cause strength decrement. Muralidhar et al. (1996) evaluated two prototype gloves (contour and laminated) with a single layer, and a double-layered glove. Bare hand performance was measured to assess the exact glove effect. Considerable reduction in grip strength with gloves was found. Figure 18.1 shows the effect of gloves on grip strength. Similar results were also reported by Bronkema and Bishu (1996).

Kinoshita (1999) investigated the effect of gloves spatiotemporal characteristics of prehensile forces during lifting and holding tasks using two-finger precision grip. The author evaluated the effect of

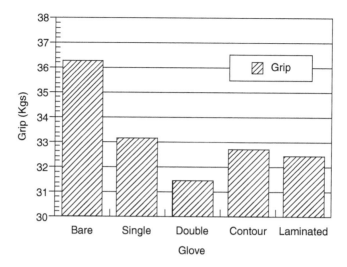

FIGURE 18.1   The effect of gloves on grip strength.

glove material (rubber and cotton) and glove thickness on slippery (rayon) and nonslippery (sandpaper) surfaces. It was identified that the grip force was influenced by glove thickness. The author also found a relatively lower grip force in rubber gloves for both slippery and nonslippery surfaces. The author suggests that rubber gloves provide better efficiency of force and temporal control than the cotton gloves in precision handling of small objects.

Fleming et al. (1997) determined the effect of wearing a work glove on hand grip fatigue and compared the effect of sustained grip contraction of concentric versus eccentric nature. They also determined the physiological muscle performance and subjective perceptual fatigue during concentric and eccentric gripping. The authors recorded the (1) time to limit of endurance (T_{lim}); (2) rate of perceived effort (RPE); (3) mean power frequency (MPF) derived from the electromyogram (EMG); and (4) the fatigue objective–subjective relationship (FOSR, which is the correlation coefficient between RPE and MPF). They found that the T_{lim} was greater for no glove and eccentric muscle action. They also found that the FOSR was the greatest for the glove condition and isometric muscle action. The authors conclude that the glove condition and the type of handgrip contraction have an effect on the physiological fatigue and subjective perception of fatigue.

Buhman et al. (2000) examined the grasp force at maximal and submaximal exertion. They conducted three experiments to examine the grasp force at the hand/handle interface for different performance conditions. In experiment 1, they examined the effect of glove type, pressure differential, and load lifted for submaximal condition. The effect of glove type, pressure differential, load lifted, handle size, and handle orientation at submaximal exertion was studied as experiment 2. The authors examined the effect of glove type, pressure differential, load lifted, handle size, and handle orientation for maximal exertion in experiment 3. It was identified that grasp force was affected by frictional and load tactile feedback. They found that the glove effect was strong at maximal exertions but marginal at submaximal exertions. From the findings they conclude that the neuromuscular mechanisms utilized during maximal exertions are differentially applied and are different from those used during submaximal or "just holding" types of exertion.

Shih et al. (2001) supported this study by investigating the effect of latex gloves on motor performances. They assessed the impact of multiple-layered gloves on tactile sensitivity using discriminating tests (two-point discrimination test and Von Frey hair test). It was observed that multiple layers of gloves impaired haptic sensitivity. The authors also determined if impaired sensation affected motor control. Grip and load forces were recorded for picking various masses (100, 150, and 200 g) using forces transducers. Greater grip and load forces were identified for multiple-layered gloves. It was also demonstrated that the gloves were more slippery than bare hand. The authors conclude that the increased grip force is due to lower friction between the object and glove surfaces.

Imrhan and Farahmand (1999) examined the effect of two glove conditions (dry and grease smeared) and selected handle task characteristics on tightening torques on cylindrical handles in simulated oil rig tasks. They found a 50% reduction of torque when using grease-smeared gloves compared to dry gloves. They also found a 15% increase with the long handle compared to the short one; a 25% increase with the medium-diameter handle compared to the small one; and a 12% increase with the horizontally oriented handle compared to the vertical one.

Bishu and Madhunuri (2004) in their evaluation of latex and vinyl gloves identified latex gloves enhanced grip strength than vinyl gloves. They determined that grip strength was high for latex powdered followed by vinyl powdered, latex nonpowdered, vinyl nonpowdered, and barehanded condition. They also found that males had better grip strength than females.

18.5.2 Torque Strength

Unlike grip strength, the overall findings on torque strength is somewhat muddled, with some studies suggesting that gloves increase torque strength, while others suggesting the reverse. Riley et al. (1985) examined forward handle pull, backward handle pull, maximum wrist flexion torque, and maximum wrist extension torque while using no glove, one-glove, and double-glove conditions. The results of this study showed that the one-glove condition was superior to both the no glove and two-glove conditions. Similar results have been reported by Adams and Peterson (1988) who investigated the effects of two types of gloves on torquing capabilities. In this study, a two-layer work glove and a three-layer chemical defense glove were found to enhance tightening performance, while only the work glove aided the loosening performance. Mital et al. (1994) have reported an increase in peak torque exertion capabilities when gloves are donned, with the extent of increase being dependent on the type of gloves donned. In contrast, Cochran et al. (1988) found that gloves reduce torquing force. They had subjects perform a flexion torquing task using four sizes of cylindrical handles (7, 9, 11, and 13 cm) while wearing three types of gloves (cotton glove, smooth leather glove, and suede leather glove). Using cotton gloves yielded the lowest torquing force, while barehanded had the highest, and the two leather gloves were in between and were not significantly different from each other. These results are supported by Chen et al. (1989), who found the forces generated using cotton gloves of all sizes were significantly lower than the leather or deerskin gloves of different sizes in a similar torquing task. In summary, the effect of gloves on torque capabilities is far less clear. However, it is reasonable to assume that gloves would aid torquing tasks.

18.5.3 Pinch Strength

As compared to grip or grasp capabilities, studies on glove effect on pinch strength are few and far between. Kamal et al. (1992) report that gloves do not affect lateral pinch capabilities. Hallbeck and McMullin (1991, 1993) found similar results for a three-jaw chuck pinch. Overall, gloves do not affect pinch strength.

Tsaousidis and Freivalds (1998) investigated the effect of gloves on maximum force and rate of force development in pinch, wrist flexion, and grip by creating force development profiles. They found that in the particular experimental task gloves did not affect pinching performance despite overcoming the pliability of gloves and loss of tactile information. It was also identified that grip force was negatively affected both in maximum value and rate of development. They determined that gloves had no effect on maximum torque production (contradicting finding from other studies).

Lowe and Freivalds (1997) investigated the pinch force required for handling windshield glass in a warehouse. Pinch force was measured using force sensitive resistors in the pulpar surfaces of the finger tips. They found that Spectra^{TM} gloves required 4–20% less grip force than Nitrile^{TM} gloves. It was found that for handling glasses on the lower shelf a palm press required the least force because of

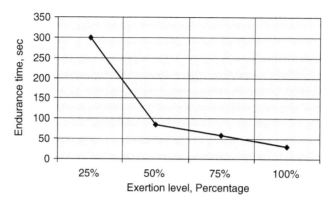

FIGURE 18.2 Endurance time vs. level of exertion.

the support from the thighs. The "over-under" pinch required the least force for handling glass on the middle shelf. For the top shelf retrieval (two-person operation) the authors identified no significant difference in grip forces between passer and catcher.

18.5.4   Endurance Time

Almost all activities with gloved hand involve certain levels of hand exertions for periods of time. Therefore, two issues are relevant here: the extent of exertion and the time of exertion. Most of the published studies on gloves have addressed the issue of extent of exertion. Bishu et al. (1994) addressed the question of how long can a person sustain a level of exertion in gloved hand condition. This deals with muscular fatigue and related issues. They reported that the endurance time at any exertion level depended just on the level of exertion expressed as a percentage of maximum exertion possible at that condition. Figure 18.2 shows the plot of the exertion level effect on the endurance time, across all glove and pressure configurations. The endurance time is least at 100% exertion level, while it is most at 25% exertion level.

18.6   Glove Effect on Dexterity

Bradley (1969) showed that control operation time was affected while wearing gloves. Banks and Goehring (1979), while studying the effects of degraded visual and tactile information in diver performance, found that the use of gloves increased task time by 50 to 60%. McGinnis et al. (1973) investigated the effect of six different hand conditions on dexterity and torque capability. They used bare hand, leather glove, leather glove with inserts, impermeable glove, impermeable glove with inserts, and an impermeable glove with built-in insulation. They found that under dry conditions, the impermeable glove had the best torque capability, and that the bare-handed dexterity performance was superior to that of gloved-hand performance. Plummer et al. (1985) studied the effects of nine glove combinations (six double and three single) on performance of the Bennett Hand Tool Dexterity Test apparatus. Results of the study indicated that subjects, with gloves donned, took longer times to complete the task, with the double glove causing longer completion times. Cochran and Riley (1986) found that gloves generally reduce dexterity and force capability. Bensel (1993) conducted an experiment in which the effects of three thicknesses (0.18, 0.36, and 0.64 mm) of chemical protective gloves on five dexterity tests (the Minnesota rate of manipulation-turning; the O'Connor finger dexterity test; a cord and cylinder manipulation; the Bennet hand-tool dexterity test; and a rifle disassembly/assembly task) were investigated. Mean performance times were shortest for the bare-handed condition and longest for the thickest (0.64 mm) glove. Nelson and Mital (1994) found no appreciable differences in dexterity and tactility among latex gloves of five different thicknesses: 0.2083, 0.5131, 0.6452, 0.7569, and 0.8280 mm. The

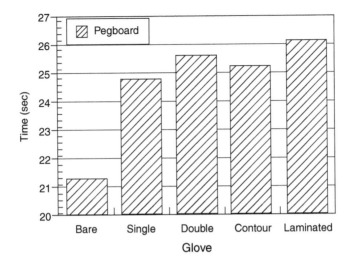

FIGURE 18.3 Glove effect on pegboard time (MMRT).

authors found the thickest latex glove (0.8280 mm) to be puncture resistant, with no loss in dexterity and tactility as compared to the thinner gloves. Bollinger and Slocum (1993) investigated the effect of protective gloves on hand movement and found that gloves decreased the range of motion in adduction/abduction and supination/pronation while extension/flexion were not affected. Their findings suggest that there is an overall reduction in the kinematics abilities of the hand while wearing gloves. Muralidhar et al. (1996) evaluated two prototype gloves (contour and laminated) with a single layer and a double-layered glove. Bare-hand performance was measured to assess the exact glove effect. A battery of tests comprising Pennsylvania Bi-Manual Worksample Assembly Test (PBWAT), Minnesota Rate of Manipulation Test-Turning (MRMTT), a rope tying task to evaluate dexterity for flexible object manipulation, and a manipulability test. Figure 18.3 shows the glove effect on MRMTT. Figure 18.4 shows the plot of the glove effect on PBWAT. Figure 18.5 shows the glove effect on the rope tying time, while Figure 18.6 shows the glove effect on the manipulation time.

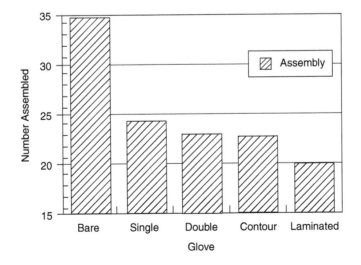

FIGURE 18.4 Glove effect on assembly time.

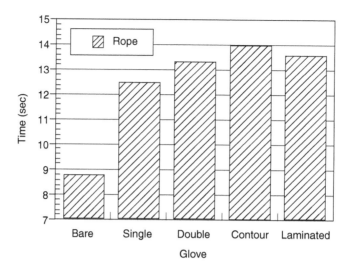

FIGURE 18.5   Glove effect on rope-tying task.

Geng et al. (1997a) studied the effect of gloves on manual dexterity in cold ($+19°C$ and $-10°C$) environments. They compared four different gloves and two different gloving (outer and inner) for bolt–nut and pick-up tasks. They found a significant difference in performance between the gloves in bolt–nut task. They also found that outer–inner combination gloving may be an approach to use for precision tasks.

Overall, gloves reduce finger dexterity and manipulability.

18.7   Glove Effect on Tactility

Although intuitively most obvious, the effect of gloves on tactile sensitivity has not been well documented. The evidence on this matter is somewhat confusing mainly due to inadequacies of measures and inadequacies of instruments. Monofilament test (Weinstein, 1993) is by far the most popular test to assess tactile sensitivity. Used in clinical testing, filaments with predetermined force are pressed against the fingers of the subjects by the experimenter till the sensation of touch is felt. The force is

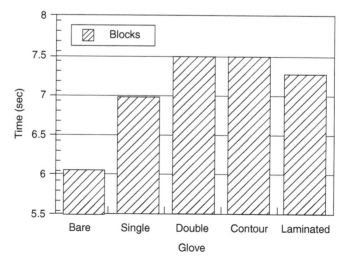

FIGURE 18.6   Effect of gloves on manipulation time.

recorded as the tactile sensitivity. The two-point discrimination test used by O'Hara et al. (1988) and by Bishu and Klute (1993) failed to give a clear indication of loss of tactile sensitivity. Bronkema et al. (1993) have used grasp force degradation at submaximal levels of exertion with gloves as a measure of loss of tactility. Their results indicate that gloves do reduce tactile sensitivity. Desai and Konz (1983) studied the effect of gloves on tactile inspection performance, and found that gloves had no significant effect on the inspection performance. In fact, they recommend that gloves be worn during tactile inspection tasks to protect the inspectors' hands from abrasion, and to help in the detection of small surface irregularities. Nelson and Mital (1994) found no appreciable differences in dexterity and tactility among latex gloves of five different thicknesses.

Geng et al. (1997b) investigated the tactile sensitivity of gloved hand in a cold (-12 and $-25°$C) operation. They measured the tactual performance using an identification task with various sizes of the objects over the percentage of misjudgment. They found that the tactual performance was affected both by the gloves and hand/finger cooling. They also identified that the effect of object size on tactile discrimination was significant and the misjudgement increased when similar sized objects were identified at $-25°$C.

Madhunuri et al. (2005) determined the effect of latex and vinyl gloves on hand performance. They developed a new test (Sponge test) to measure fine finger tactility. The evaluation of the effect of latex and vinyl gloves on human performance indicated that tactility, dexterity, and strength were better when subjects donned latex gloves than vinyl gloves. Results from functional tests showed that ability to perform was better when subjects donned latex gloves than vinyl gloves. However, the results showed that vinyl gloves generated less sweat than latex gloves. Figure 18.7 shows the effect of gloves on Tactility while Figure 18.8 shows the effect of gloves on sweat absorption.

In summary, although the evidence show that gloves reduce tactility, due to difficulties in measurement of tactility, its exact effect has not been clearly understood and should be the focus of glove research in the future.

18.8   Glove Effect on Vibration

Goel and Rim (1987) studied the role of gloves in reducing vibrations while using power tools like a pneumatic chipping hammer and found that vibration isolation gloves are useful in protecting the operator from power tool induced vibrations, and recommend the use of gloves.

Dong et al. (2002) developed a new test method based upon the total effectiveness acceleration transmissibility (TEAT) to study the vibration isolation performance of anti-vibration gloves. They compared the new methodology with the procedure outlined by ISO 101819 (Mechanical vibration and shock — Hand–arm vibration — Method for the measurement and evaluation of the vibration transmissibility of

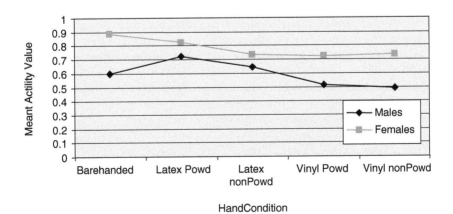

FIGURE 18.7   Effect of gloves on tactility.

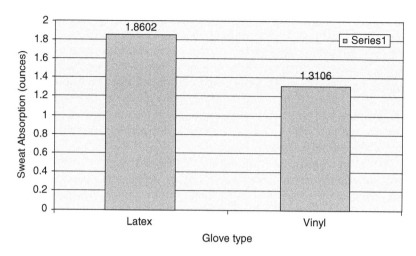

FIGURE 18.8 Effect of glove on sweat absorption.

gloves at the palm of the hand, International Organization for Standardization, Geneva, Switzerland, 1996). The vibration transmission characteristics of gloves were evaluated using the new method and compared with those derived from the standardized method to demonstrate the effectiveness of the TEAT approach. From the results, the authors conclude that the TEAT method, based upon vector sums of both the source and response accelerations, can effectively account for the majority of the measurement errors, and yield more repeatable and reliable assessments of gloves.

Rakheja et al. (2002) developed a new methodology based on the frequency response characteristics of the gloves to estimate the vibration isolation effectiveness of antivibration gloves as a function of handle vibration of specific tools. They synthesized handle vibration spectra of six different tools. The attenuation performances of two different gloves were characterized under tools vibration, and M- and H-spectra defined in ISO-10819. They found that the tool specific vibration isolation performance of a glove cannot be derived from standardized M- and H- spectra. They also identified that the frequency response characteristics of gloves are relatively insensitive to magnitude of vibration but strongly dependent upon viscoelastic properties of the glove materials. The mean measured frequency response characteristics were then applied to derive an estimate of tool-specific isolation effectiveness of the gloves. They conclude that the isolation effectiveness of gloves for selected tools can be effectively predicted using the proposed methodology.

Dong et al. (2003) evaluated the effectiveness on-the-hand measurement methods for assessing the vibration transmissibility of gloves (air gloves and gel-filled gloves) in conjunction with chipping hammer. They also used a transfer function method to predict the vibration transmission of gloves. It was found that the on-the-hand methods offered some unique advantages over palm adapter method outlined by ISO-10819. It was also found that the on-the-hand had poor repeatability for high degree of tool vibration variability. They found that the transmissibility values of the air glove obtained from on-the-hand measurement are consistent with those predicted from the transfer function method. They also identified that the on-the-hand measured transmissibility of the gel-filled glove may perform better than predicted by the transfer function method.

18.9   Glove Effect on Miscellaneous Performance

Chang et al. (1999) evaluated the effects of wearing a glove (bare-handed, cotton, nylon, and open-finger) and wrist support on hand–arm response while operating an in-line pneumatic screwdriver. They investigated the hand–arm responses in terms of triggering finger force, flexor digitorum EMG, and hand-transmitted vibration. They found wearing a nylon glove reduced 18.2% of the triggering force as compared with the bare-handed condition. In addition, wearing a nylon glove had comparatively low forearm muscular exertion, and reduced 16 and 15% of hand-transmitted vibration in the z-axis and the sum of three axes as compared with the bare-handed condition. They also found that the use of a wrist support required a greater triggering force and a 9.9% greater hand-transmitted vibration in the y-axis than when not using a wrist support.

Rybczynski and Fathallah (2002) investigated the effect of glove use in coating removal task. They monitored force exertion along with EMG readings from the finger flexors, finger extensors, biceps, and triceps. They found a significant increase in force output and muscle activities when using gloves as compared to bare hands. They identified that glove material and glove thickness are important in glove selection.

18.10   Liners

Today there is a growing trend towards the use of inner gloves or glove liners. For example, health care professionals often tend to use glove liners to prevent outer glove/skin interaction. Similarly meat processors usually wear glove liners, while astronauts use multiple layers of liners. Almost all of the research efforts on gloves have focused on the outer glove, while liners have drawn little research attention.

Branson et al. (1997) designed and evaluated prototype artificially cooled gloves that enhance thermal comfort as assessed by skin temperature, sweat rate, manual dexterity, and perceived comfort. They designed three multiple-layered gloves and tested them under simulated environmental conditions in a controlled laboratory chamber. They identified that artificial cooling and liner reduced skin temperature and perceived moisture related discomfort. They also found that one of their liner design hampered manual dexterity.

Using a standardized glove testing protocol, Bishu et al. (1997) investigated the effect of a number of glove liners. The study compared three types of liners; liners made from poly tetrafluoroethylene (PTFE), cotton, and latex. A battery of evaluation tests, comprising some standardized tests and certain functional tests, was designed. The tests assessed the following capabilities: tactile sensitivity, dexterity, manipulability, strength, and effect of continuous use. The actual tests performed were the pinch test, finger strength test, monofilament test, pain threshold test, rope tying test, pegboard test, and fatigue test. In the fatigue test the effect of continuous use of gloves was measured with Borg's RPE scale. Continuous use was simulated by making the subjects perform keyboard task and pegboard task alternately for an hour. Discomfort measures were recorded every 10 min. Important findings were as follows:

a. Liners had a distinct contribution toward the performance decrements with gloves
b. Liners had a significant effect on the overall comfort (discomfort) during extended periods of glove usage

Figure 18.9 shows the graph of effect of liners on tactility as measured with monofilament test. Figure 18.10 shows the liner effect on overall fatigue.

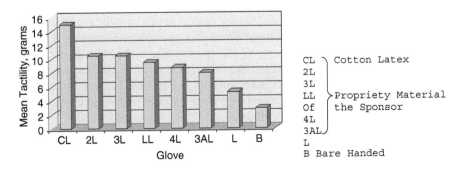

FIGURE 18.9   Liner effect on tactility.

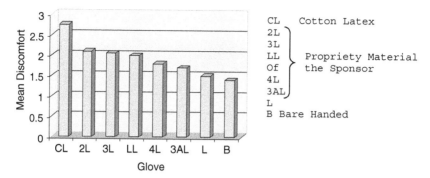

FIGURE 18.10   Liner effect on fatigue.

On a continuation study, Bishu and Chin (1998) evaluated a number of inner glove liners used to protect the skin from latex proteins and chemical skin sensitizers found in gloves. They used a battery of performance tests, objective measurements (skin temperature, skin conductance, and skin moisture content) and subjective comfort/discomfort tests to evaluate the inner gloves. They found that sweat, pegboard, and discomfort measures were the best discriminators of liners, while the other subjective and objective measures were not. They also identified that the liner size significantly affected performance in most of the experiments. They found that the extra large glove liner was better in fatigue test, pegboard test, range of motion test, scissor-cutting test, tactility test, and vibration test. The interesting finding of this study was that the pinch strength test did not significantly affect performance for inner liners.

Bishu and Chin (1998) evaluated glove liners using standardized protocol of short- (performance) and long-term (fatigue) tests. They suggest that the inner gloves can be evaluated using performance tests. They also identified that inner gloves pegboard and tactility tests along with long-term fatigue tests are two possible ways to accurately quantify differences between inner gloves. Consideration of the hand anthropometrics in design process is essential as the hand length, hand width, thumb, and index fingers contribute to hand prehension.

Moore et al. (1995) investigated the effects of hand condition on three-jaw chuck pinch strength, power grip strength, and manual dexterity. They compared performance of: (1) bare hand, (2) hand with a normal sized latex examination glove, and (3) hand with a tight-fitting latex examination glove. They found that the latex examination gloves do not have an effect on three-jaw chuck pinch strength or power grip strength. They also identified that ill-fitting latex examination gloves significantly reduce manual dexterity.

18.11   Glove Attributes

Bradley (1969) also investigated dexterity as a function of glove attributes such as snugness of fit, tenacity, and suppleness in a wide variety of 18 industrial gloves. The conclusions reached are that various glove attributes influence dexterity performance to varying extent. Bishu et al. (1987) found that glove attributes and the task performed had a significant effect on the force exertion. Wang et al. (1987) concluded that altered feedback from a gloved hand caused strength degradation. Batra et al. (1994) found grip strength reduction to be significantly correlated with glove thickness and subjective rating discomfort, and suggest that glove thickness should be minimized, while increasing the tenacity.

Purvis and Cable (2000) assessed the efficacy of phase control materials incorporated with soccer goalkeeper gloves to reduce heat load inside the glove. The phase control material glove was compared with normal foam material glove. They found that the particular specification of phase control material promoted heat gain and was inappropriate for soccer goalkeeper gloves.

Inspite of all these studies, no comprehensive model linking performance degradation with glove characteristics exists.

18.12 Challenges of Glove Design

In summary, gloves do reduce performance, but provide a vital protective function. Facilitation of performance, with simultaneous protection from the hazards of the work environment, is often conflicting objectives of glove design. The conflicts associated with providing primary hand protection through the use of a glove while permitting adequate hand functioning has been widely recognized. Looking at the glove attributes that cause performance differences, attributes or level of attributes that facilitate performance deteriorates safety function. This conflict poses certain challenges for the glove designer. Before attempting to design any kind of protection for the hand, it is necessary to first identify what it needs to be protected against. Human capability is limited to a narrow bandwidth of acceptable environmental conditions in which performance is not affected. There are a number of environmental hazards, often in combination, that are likely to pose a threat to the hands of workers interfacing with their work place.

Glove material is often fixed by the environment. Environments that expose the worker to radiation, electrical, biological, fire, chemical, and extreme thermal hazards warrant those specific materials be incorporated into the hand protection, irrespective of the design. The hazard-specificity of such materials also poses a problem for the glove designer, because the minimum thickness of the material required to provide adequate protection is usually a fixed value, limiting the designer's choice of variable parameters.

For example, a glove designed for use in an operating room or by a dental hygienist has to be capable of being sterilized either by steam or chemical disinfectants, impermeable to any potentially dangerous fluids, and of sufficient thickness and strength to maintain its integrity for a reasonable period of time.

However, in spite of the inflexibility in the materials' parameters, the glove is expected to enable the user to function without significant loss of desirable hand functions like grip strength, dexterity, range of motion, and tactile feedback. When these requirements are combined with multiple hazard condition protection requirements, glove design becomes a complex task. Table 18.2 shows the glove attributes as a function of design parameters. Muralidhar et al. (1999) suggest an ergonomic approach for glove design. Basing on published literature on force distribution in the hand during any task, they recommend that gloves should have variable thickness, with more thickness in regions were more force is exerted and

TABLE 18.2 Glove Attributes as a Function of Design Parameters

Prehension	Function (design, material)
Torque capability	Function (design, material)
Dexterity	Function (design, material)
Tactile feedback	Function (material, user)
Grip strength	Function (user, material, design)
Fit	Function (design, material)
Pressure-pain threshold	Function (user, material, design)
Range of motion	Function (user, design, material)
Abrasion resistance	Function (material)
Puncture resistance	Function (material)
Cut and tear resistance	Function (material)
Thermal protection	Function (material)
Chemical resistance	Function (material)
Biohazard protection	Function (material)
Radiation protection	Function (material)
Electrical protection	Function (material)
Vibration protection	Function (material, design)

less thickness on regions where force exertion is minimal. They argue that such an approach would yield gloves with minimal performance degradation and maximal protection.

18.13   Glove Evaluation Protocol

The question of how to evaluate a glove has always interested the designer, manufacturer, and the user of gloves alike. Standard evaluation protocols do not exist. Even in cases where they exist, as in cases of rubber gloves used by utility people, or fire fighters gloves, the protocols are inadequate. It is recommended that a typical glove evaluation protocol include the following:

1. Strength tests including grip and pinch tests.
2. A battery of standardized tests to assess dexterity, tactility, and manipulability. Typical standard tests for these include PBWAT, MRMTT, Purdue pegboard test, O'Connor dexterity test, and monofilament test.
3. A battery of functional tests. This is where most of the existing glove evaluation protocols lack. Functional tests are task specific and should be appropriately designed as to simulate actual tasks to be performed with the concerned gloves.

Evaluation protocols similar to the ones listed in the preceding text have been used by O'Hara et al. (1988) and Bishu and Klute (1993) in the evaluation of EVA gloves. Bishu and Goodwin (1997) determined the most appropriate test time for evaluating gloves. They examined the effect of time on glove comfort using three sets of experiment. Experiment one consisted of a battery of short time evaluation tests, while experiment two evaluated glove comfort when they were donned for an hour, and experiment three was a repeat of experiment two for 8 h. They found that gloves yield different levels of discomfort with use. A 2-h period was identified as the appropriate time to evaluate gloves.

Norton and Hignett (2001) developed a protocol for comparative testing of sterile and nonsterile gloves used by nurses. They developed tests represented typical nursing tasks considering the design features of gloves that may have impact on usability. A questionnaire was also used to receive subjective comments about each glove. The authors argue that the protocol is an ergonomic approach to ensure the provision of gloves that will not hinder the practice of nurses.

18.14   Glove Standards

Existing glove standards are of three types. The standards generally describe protective requirements as in some of the U.S. Occupational Safety and Health Administration (OSHA) standards or describe the protection of the gloves must provide for safety as in gloves in the chemical industry or for the utility personnel or specifically describe glove-testing requirements.

18.15   Conclusion

In summary, the following statements can be made with regards to gloves:

1. Gloves protect the hand from the environment, but affect the performance.
2. Gloves range widely in size, type, and cost. They range from general purpose gloves to highly specialized task-specific gloves.
3. Gloves reduce grip or grasp strength capabilities while they do not affect torque or pinch capabilities.
4. Gloves reduce hand dexterity, tactility, and manipulability.
5. Providing protection without compromising performance in a continuous challenge for glove designers.

References

Banks, W. W. and Goehring, G. S. (1979). The effects of degraded visual and tactile information on diver work performance. *Human Factors*, 21(4):409–415.

Batra, S., Bronkema, L. A., Wang, M., and Bishu, R. R. (1994). Glove attributes: can they predict performance? *International Journal of Industrial Ergonomics* (In Press).

Bishu, R. R., Batra, S., Cochran, D. J., and Riley, M. W. (1987). Glove effect on strength: an investigation of glove attributes, *Proceedings of the 31st Annual Meeting of the Human Factors Society*, pp. 901–905.

Bishu, R. R. and Chin, A. (1998). Inner gloves: how good are they? *Advances in Occupational Ergonomics and Safety 2*, S. Kumar. Ed. pp. 397–400.

Bishu, R. R. and Goodwin, B. (1997). Evaluation of gloves: short time tests vs. long time tests. *Proceedings of the Human Factors and Ergonomics Society 41st Annual Meeting*, pp. 692–696.

Branson, D. H., Simpson, L. S., Claypool, P. L., Chari, V., and Ruiz, B. M. (1997). Comparison of prototype artificially-cooled chemical protective glove systems. *Performance of Protective Clothing: Sixth Volume*, J. O. Stull and A. D. Schwope, Ed. pp. 314–325.

Bradley, J. V. (1969). Effect of gloves on control operation time. *Human Factors*, 11(1):13–20.

Buhman, D. C., Cherry, J. A., Bronkema-Orr, L., and Bishu, R. (2000). Effects of glove, orientation, pressure, load, and handle on submaximal grasp force. *International Journal of Industrial Ergonomics*, 25(3):247–256.

Caldwell, L. S., Chaffin, D. B., Dukes-Dobos, F. N., Kroemer, K. H. E., Laubach, L. L., Snook, S. H., and Wasserman, D. E. (1974). A proposed standard procedure for static muscle strength testing. *American Industrial Hygiene Association Journal*, 35(4):201–206.

Chang, C. H., Wang, M. J. J., and Lin, S. C. (1999). Evaluating the effects of wearing gloves and wrist support on hand–arm response while operating an in-line pneumatic screwdriver. *International Journal of Industrial Ergonomics*, 24(5):473–481.

Chen, S. C., Tarawneh, I., Goodwin, B., and Bishu, R. R. (1998). Evaluation of glove liners with objective, subjective and performance measures. *Proceedings of the Human Factors and Ergonomics Society 42nd Annual Meeting*, pp. 846–850.

Cherry, J., Christensen, A., and Bishu, R. (2000). Glove comfort vs. discomfort: are they part of a continuum or not? A multi-dimensional scaling analysis. *Proceedings of the XIV Triennial Congress of the International Ergonomics Association and the 44th Annual Meeting of the Human Factors and Ergonomics Society*.

Cho et al. (2002). A pair of Braille-based chord gloves. *Proceedings of the Sixth International Symposium on Wearable Computers*, pp. 154–155.

Cochran, D. J. and Riley, M. (1986). The effects of handle shape and size on exerted forces. *Human Factors*, 28(3):253–265.

Cochran, D. J., Albin, T. J., Bishu, R. R., and Riley, M. W. (1986). An analysis of grasp force degradation with commercially available gloves. *Proceedings of the 30th Annual Meeting of the Human Factors Society*, pp. 852–855.

Cochran, D. J., Batra, S., Bishu, R. R., and Riley, M. W. (1988). The effects of gloves and handle size on maximum torque. *Proceedings of the 10th Congress of the International Ergonomics Association*, pp. 254–256.

Desai, S. and Konz, S. (1983). Tactile inspection performance with and without gloves. *Proceedings of the Human Factors Society*, pp. 782–785.

Dong, R. G., Rakheja, S., Smutz, W. P., Schopper, A., Welcome, D., and Wu, J. Z. (2002). Effectiveness of a new method (TEAT) to assess vibration transmissibility of gloves. *International Journal of Industrial Ergonomics*, 30(1):33–48.

Dong, R. G. et al. (2003). On-the-hand measurement methods for assessing effectiveness of anti-vibration gloves. *International Journal of Industrial Ergonomics*, 32(4):283–298.

Ervin, C. A. (1988). A standardized dexterity test battery. *Performance of Protective Clothing: Second Symposium, ASTM STP989*. S. Z. Mansdorf, R. Sager, and A. P. Nielsen, Ed. American Society for Testing and Materials, Philadelphia, pp. 50–56.

Fleming, S. L., Jansen, C. W., and Hasson, S. M. (1997). Effect of work glove and type of muscle action on grip fatigue. *Ergonomics*, 40(6):601–612.

Geng, Q., Chen, F., and Holmer, I. (1997a). The effect of protective gloves on manual dexterity in the cold environments. *International Journal of Occupational Safety and Ergonomics*, 3(1–2):15–29.

Geng, Q., Kuklane, K., and Holmer, I. (1997b). Tactile sensitivity of gloved hands in a cold operation. *Applied Human Science*, 16(6):229–236.

Griffin, D. R. (1944). Manual dexterity of men wearing gloves and mittens. *Fatigue Lab., Harvard University, Report No. 22.*

Hallbeck, M. S. and McMullin, D. L. (1993). Maximal power grasp and three jaw chuck pinch as a function of wrist position, age and glove type. *International Journal of Industrial Ergonomics*, 11:195–206.

Hertzberg, T. (1955). Some contributions of applied physical anthropometry to human engineering. *Annals of New York Academy of Sciences*, 63:621–623.

Imrhan, S. N. and Farahmand, K. (1999). Male torque strength in simulated oil rig tasks: the effects of grease-smeared gloves and handle length, diameter and orientation. *Applied Ergonomics*, 30(5):455–462.

Kinoshita, H. (1999). Effect of gloves on prehensile forces during lifting and holding tasks. *Ergonomics*, 42(10):1372–1385.

Landsmeer, J. M. F. (1962) Power Grip and Precision Handling. *Annals of Rheumatic Disease*, 21:164–169.

Lowe, B. D. and Freivalds, A. (1997). Analyses of glove types and pinch forces in windshield glass handling. *Advances in Occupational Ergonomics and Safety*, B. Das and W. Karwowski, Ed. pp. 477–480.

Lyman, J. and Groth, H. (1958). Prehension force as measure of psychomotor skill for bare and gloved hands. *Journal of Applied Psychology*, 42(1):18–21.

McGinnis, J. S., Bensel, C. K., and Lockhar, J. M. (1973). Dexterity afforded by CB protective gloves. US Army Natick Laboratories, Natick, MA, Report No. 73-35-PR.

Muralidhar, A. and Bishu, R. R. (1994). Glove evaluation: a lesson from impaired hand testing. *Advances in Industrial Ergonomics and Safety VI*. F. Aghazadeh, Ed. Taylor & Francis.

Muralidhar, A. and Bishu, R. R. (2000). Safety performance of gloves using the pressure tolerance of the hand. *Ergonomics*, 43(5):561–572.

Muralidhar, A., Bishu, R. R., and Hallbeck, M. S. (1999). The development and evaluation of an ergonomic glove. *Applied Ergonomics*, 30(6):555–563.

Napier, J. R. (1956). The prehensile movements of the human hand. *The Journal of Bone and Joint Surgery*, 38B:902–913.

Norton, L. and Hignett, S. (2001). Examination gloves for nurses: a protocol for usability testing. *Contemporary Ergonomics 2001*. M. A. Hanson, Ed. pp. 535–539.

O'Hara, J. M., Briganti, M., Cleland, J., and Winfield, D. (1988). Extravehicular activities limitations study. Volume II: Establishment of physiological and performance criteria for EVA Gloves — Final report (Report no. AS-EVALS-FR-8701, NASA Contract no NAS-9-17702).

Purvis, A. J. and Cable, N. T. (2000). The effects of phase control materials on hand skin temperature within gloves of soccer goalkeepers. *Ergonomics*, 43(10):1480–1488.

Rakheja, S., Dong, R., Welcome, D., and Schopper, A. W. (2002). Estimation of tool-specific isolation performance of antivibration gloves. *International Journal of Industrial Ergonomics*, 30(2):71–87.

Riley, M. W. and Cochran, D. J. Ergonomic aspects of gloves: design and use. *International Reviews of Ergonomics*, 2:233–250. D. J. Oborne, Ed. Taylor & Francis.

Riley, M. W., Cochran, D. J., and Schanbacher, C. A. (1985). Force capability differences due to gloves. *Ergonomics*, 28(2):441–447.

Rybczynski, I. C. and Fathallah, F. A. (2002). Effects of glove use in a coating removal task. *Proceedings of the 46th Annual Meeting of the Human Factors and Ergonomics Society* pp. 1191–1195.

Shih, R. H., Vasarhelyi, E. M., Dubrowski, A., and Carnahan, H. (2001). The effects of latex gloves on the kinetics of grasping. *International Journal of Industrial Ergonomics*, 28(5):265–273.

Sudhakar, L. R., Schoenmarklin, R. W., Lavender, S. A., and Marras, W. S. (1988). The effects of gloves on muscle activity. *Proceedings of the 32nd Annual Meeting of the Human Factors Society*, pp. 647–650.

Tsaousidis, N. and Freivalds, A. (1998). Effects of gloves on maximum force and the rate of force development in pinch, wrist flexion and grip. *International Journal of Industrial Ergonomics*, 21(5):3535–3560.

Wang, M. J., Bishu, R. R., and Rodgers, S. H. (1987). Grip strength changes when wearing three types of gloves. *Proceedings of the Fifth Symposium on Human Factors and Industrial Design in Consumer Products*, Interface 87, Rochester, NY.

19

Wrist and Arm Supports

Carolyn M. Sommerich
The Ohio State University

19.1 Introduction

A concise overview of the literature on upper extremity supports is presented in this chapter. These are devices or elements that are incorporated into a workstation's design in order to reduce postural muscle activity and fatigue, joint and soft tissue loading, or the effects of physiologic tremor on precision tasks. The use of arm supports for some specific work activities, including keyboarding and microsurgery, are discussed in some detail. Where sufficient evidence exists, recommendations are provided for circumstance-specific use of upper extremity supports. A related chapter in this book, "Medical Management: Wrist Splints," describes another means by which wrists may be supported, a method that seems best reserved for people with medical conditions, such as rheumatoid arthritis or carpal tunnel syndrome.

19.2 Fundamentals

In this section, basic theories of engineered upper extremity support are explored, including the ways in which various designs provide support, and under what circumstances they may be employed.

Work in constrained postures is an acknowledged risk factor for neck and shoulder disorders,[1] particularly when joint angles deviate significantly from neutral (resting) orientations. In reviewing numerous studies of the role of posture in neck and upper limb disorders, Wallace and Buckle[2] concluded that posture may not always be a sufficient cause, but may interact with or add to other physical or psychosocial risk factors. Constrained postures impose continuous loads (stress) on muscles, tendons, ligaments, and joints. When they are of sufficient duration, even low levels of stress can lead to adverse health outcomes. Aarås[3] reported a significant reduction in lost time associated with musculoskeletal illness at an electronics assembly facility when static loads on workers' trapezius muscles were reduced from levels of 4–6% of maximum voluntary contraction (% MVC) to below levels of 2% MVC, through workstation modification.* For reference, simply positioning the arm for typing (upper arm vertical, elbow bent 90°) has been estimated to require trapezius muscle activation of 3% MVC.[5]

Potential pathologic reactions from sustained work-related stress on various musculoskeletal elements in the neck and shoulder include degenerative joint disease; tendon disorders, including degeneration and tendinitis; and problems with muscles, including myofascial syndrome.[1] Development of localized muscular fatigue may be an immediate consequence of sustained muscle loading. Symptoms of fatigue can include discomfort or pain, reduced strength capacity, increased time for hand–eye coordination, increased hand tremor, and, when severe, difficulties in hand positioning.[6] Low level muscle contractions (5% MVC) sustained for 1 h were shown to result in a 12% reduction in MVC, and shifts in the spectral frequency of the electromyographic (EMG) signal that are associated with localized muscle fatigue.[7] Considering typical work break schedules (morning and afternoon work blocks, each 4 h in length and each interrupted by a single 15 min break), maintaining a work posture for an hour would not be extraordinary, especially for individuals performing precise, hand-intensive work, or work requiring intense concentration.

Working with the upper arm unsupported and out of the vertical plane has been associated with symptoms of musculoskeletal disorders in the neck and shoulder region, with duration of exposure a key element.[8,9] Ergonomic hazard surveillance tools assign penalties for arms positioned out of the vertical or not supported.[10] Working with arms unsupported, significant levels of static activity have been recorded from muscles in the shoulder and forearm of typists performing typical keyboard operations.[11,12] Positioning an unsupported arm or arm segment in any posture away from vertical requires activation of shoulder and arm muscles in order to oppose gravitational forces. Concomitant tensile loads are imposed on tendons and ligaments in the upper extremity, including the shoulder. These active and passive internal loads contribute, along with any external loads at the hand, to compressive and shear loading of the joints of the upper extremity. It is the interaction of joint posture and arm segment weight and the weight of any handheld obejcts that affects muscle activity requirements. The moments that result from these interactions directly determine muscle activity requirements. These levels of muscle contraction, and associated soft tissue and joint loads, are what are necessary in order for an individual to accommodate to the physical conditions of his or her work, including workspace layout and tools. However, levels of muscle activity that exceed physical work requirements are not uncommon, and may occur due to operator training or technique, environmental conditions, task complexity, or personality.[13–15] The effects of these latter items are not likely to be altered by the presence of arm supports.

19.2.1   Joint and Soft Tissue Stress Reduction

The use of arm supports may be an effective method for reducing muscle activity requirements imposed by physical work conditions. Effectively, arm supports reduce arm segment weight. This reduces joint moments due to external forces, which, in turn, reduces moment contributions required from muscles and may reduce stress on other soft tissues. Frequently supporting hands and arms on the work surface was found to be associated with lower incidence of pain in the neck, shoulder, and arms in groups of professional keyboard operators performing different types of keyboarding tasks.[16] In a laboratory-based study where subjects performed a simulated soldering task, Schüldt et al.[17] demonstrated that supporting the upper extremity through suspending the elbow or resting the elbow on a support resulted in marked reductions in shoulder muscle activity. Muscle activity was reduced whether subjects sat upright or sat with trunks flexed over the work surface. Sitting in a slightly reclined position reduced muscle activity so significantly that arm support did not provide further reductions in activity in that condition.

Based on force plate measurements, Occhipinti et al.[18] found that 4 to 7% of body weight was supported through resting wrists on the keyboard support surface, while 7 to 14% was supported when

---

*Explanation of *static load*: If the static load for a muscle during a particular job was determined to be *X*% MVC, this means that 10% of the time, the muscle is active at or below *X*% of its maximum. A static load of 2 to 5% is desirable.[4]

resting forearms on the same surface. Ranges reflect differences due to changes in unsupported trunk postures. Based on biomechanical modeling, they estimated reductions in spinal loading at the level of the third lumbar vertebra (L3) ranging from approximately 25–100 kg, depending upon type of arm support and trunk posture. In a related study, Colombini et al.[19] reported 6.6% of body weight supported through a forearm support when the back was also supported.

19.2.2 Methods of Support

There are many ways to provide direct support for the upper extremity, and several points at which a support could contact the upper extremity including the elbow, along the forearm, at the wrist, or at the base of the palm. Objective effectiveness and user acceptance are both important issues in determining which methods may be appropriate for a particular situation. For example, as mentioned earlier, researchers have demonstrated reductions in shoulder muscle activity when the elbow was supported via suspension, or through fixed support located directly under the elbow during a soldering task.[20] Chaffin[6] demonstrated that time to fatigue was extended by two to four times with the use of an elbow support. In an early study of alternative keyboards, subjects tended to prefer a split keyboard with a large built-in forearm–wrist support over both a split keyboard with a small built-in support and a traditional keyboard with a large support.[21] Subjects were found to exert about twice as much force on the large support (about 35 N), compared to the small one (about 17 N), meaning that more upper body weight was supported by the larger support. A number of alternative keyboards currently on the market have built-in wrist/palm rests, as do most keyboards built into portable computers. Efficacy and preference for various types of supports will be examined in more detail in Section 19.3.

19.2.3 Influence of Support on Posture

Supports may be useful in helping muscles to maintain or improve the position of the supported limb. However, other body parts may be impacted by support, as well. For example, during a series of six 10 min long typing periods, Weber et al.[22] found arm abduction was greater when trained typists keyed with a forearm–wrist support, when compared to unsupported keying. Subjects also sat nearer to the keyboard support surface when arms were supported.

In theory, one effect of wrist pad usage should be a reduction of wrist extension (if the pad is positioned appropriately). However, in a study of 12 office employees, Paul and Menon[23] reported fairly pronounced wrist extension postures, ranging from 29° to 41° when subjects used five different wrist pads that varied in shape, width, and compressiveness. However, no unsupported condition was provided to show whether extension with any of the pads was different from extension when no pad was used. The pronounced extension angles reported in that study, may be due to the interactive effect, reported by Damann and Kroemer,[24] between the presence or absence of wrist support and the height of that support. Wrist extension increased markedly, both with and without support, when the height of the support was reduced from elbow level to 80% of elbow level.[24]

19.3 Applications

The theoretical benefits of various upper extremity supports were discussed in the previous section. Results from numerous studies are summarized in this section in order to demonstrate how and under what circumstances various types of upper extremity supports have been shown to be effective.

Arm support is typically considered for work that requires steady positioning of the hands for some period of time, such as electronics assembly, keyboarding, or surgery. One brief, but interesting mention was made by Jex and Magdaleno[25] regarding their efforts to model the potential vibration damping effects of elbow rests for pilots flying high-performance aircraft. Few intervention studies have been performed to examine the effects of arm support on the job. Much of the research is in the form of controlled studies conducted in laboratories or in work settings, on healthy subjects, with limited exposure to each testing condition, making it difficult to predict either short- or long-term effects from use of the various types of arm support. Results of many of those studies are summarized in the following sections.

19.3.1 Support during Specific Work Tasks

19.3.1.1 Assembly Work

Eighteen months after installing arm suspension devices at an electronics plant, Harms-Ringdahl and Arborelius[26] found that workers continued to choose to use the devices, which may have been due to the reductions in shoulder and neck discomfort experienced by the workers. Compared to before the suspension was introduced, average discomfort intensity was reduced by one-half to two-thirds. Before the suspension was introduced about 40% of the workers rated their afternoon shoulder and neck discomfort greater than 50 on a scale of 0–100, whereas after 18 months of suspension use, only about 14% rated their discomfort that high.

In making ergonomic modifications to several work stations in an electronics assembly plant, Aarås[3] recognized the importance of reducing shoulder load moments in workers. They achieved this through a number of engineering controls, including the provision of arm rests, either positioned on the chair or work surface, for any tasks carried out above elbow height.

19.3.1.2 Computer Use

When given an opportunity to work at an adjustable workstation, either with or without a forearm–wrist support, two-thirds of a sample of 67 keyboard operators preferred the support.[27] Seventy-eight percent of that sample did not find the support hindered their work on the keyboard. Based on observations of the operators, when the forearm–wrist support was present 80% of the subjects used the support. When no formal support was present, 50% of the subjects still rested their forearms or wrists on the keyboard support surface. More recently, a cross-sectional epidemiological study of video display terminal (VDT) work seemed to objectively confirm the importance of arm supports for keyboard operators. Bergqvist et al.[28] found neck/shoulder discomfort in a group of VDT operators was associated with a combination of three factors: more than 20 h per week of VDT work, limited rest break opportunity, and working with the lower arm unsupported. Further evidence of the benefits of forearm support was provided by Aarås et al.[29,30] In a lab-based study, Aarås et al.[30] placed the keyboard and monitor on the same support surface (adjustable height desk), and in the support condition, moved the keyboard back from the edge of the surface in order that the forearms could be fully supported on the desk while typing and using the mouse. They found that this method of support significantly reduced trapezius muscle activity for both typing and mousing. Aarås et al.[29] also tested the effectiveness of tabletop forearm support in a prospective intervention study, which lasted more than a year and included a control group. They found significant reductions in trapezius muscle activity and reductions in intensity and frequency of shoulder pain in the intervention groups.

In a study of the effects of palm rest height and profile, 40 typists were each allowed to work with nine different palm rests during a 1 week period, and were asked to use the one they preferred for at least half a day.[31] Though it is unclear how the supports compared physically, unlike the outcome of the study by Grandjean et al.,[27] in this study only four of the 40 typists found the rests to be useful, while seven commented about increased discomfort. Several operators suggested that arm rests on their chairs might have been more comfortable, because they would have provided support while enabling freedom of wrist and hand movement. Chair arm rests may not provide a long-term benefit, though. In a 3-year prospective study of computer users, Marcus et al.[32] did not find any protective benefit to the "presence of chair arm rests," in terms of development of neck/shoulder symptoms or disorders.

Relying on objective EMG data, two studies have demonstrated reduced activity in forearm extensor muscles with use of one particular alternative keyboard with a built-in wrist support,[33,34] although a portion of that reduction appeared to be due to reductions in typing speed which occurred with the alternative keyboard. However, palm rests were not shown to reduce activity in any of the forearm, shoulder, or back muscles studied by Fernström et al.[35] The authors discussed a couple of important limitations of their study which may have impacted their results: none of their subjects had prior experience with palm rests; and both practice (10–20 min) and testing (5 min) times in the study were limited. The authors suggested that people unfamiliar with palm rests tend to avoid using them (which may be important to remember when evaluating other support studies, or when introducing supports into a workplace). The palm rests that were built into a notebook computer were utilized by just over half of a small sample of novice notebook computer users in a laboratory study of factors affecting work posture during notebook computer use.[36] Though they were unable to assess effects on muscle activity, the researchers found very little difference in upper extremity, head, neck, or trunk posture when subjects typed on the computer with the built-in wrist rest and on one without.

Bendix and Jessen[37] also found that forearm muscle activity (extensor carpi radialis) was not affected by the use of a wrist support, in a situation wherein subjects typed on electric typewriters, although trapezius muscle activity actually increased somewhat with use of the support. Both performance and subjective assessments of conditions with and without the support were similar. All subjects in the study were professional secretaries who had experienced discomfort in the neck and shoulder or elbow region for a substantial portion of the 12-month period preceding the study. Prior to data collection, subjects had been allotted 2 weeks to adapt to the wrist support. In spite of the lack of objective or subjective beneficial effects of the support, 9 of the 12 subjects wanted to keep the support once the study was completed.

Unlike wrist pads which are fixed relative to the keyboard, full motion forearm supports are fixed relative to the user's forearms. They are designed specifically to facilitate unrestricted motion of the arm and hand in a fixed horizontal plane, while maintaining support of the forearm and hand. Nonetheless, Powers et al.[38] found that subjects, described as office workers with substantial computer experience, using full motion forearm supports thought that the supports slowed their typing and increased their errors, although objective measurements showed this not to be the case. The authors did not find any postural benefits to using this type of support, based on a limited assessment of wrist and elbow postures.

Erdelyi et al.[39] studied the effects of fixed arm supports and arm suspension on experienced keyboard operators, some of whom were experiencing shoulder and neck pain. Both support methods were effective in reducing trapezius activity in the group with pain. In contrast, the group of healthy subjects experienced either no reduction in muscle activity, or, in some conditions, an increase in activity with the supports. In each condition, pain sufferers tended to display higher levels of muscle activity than did healthy subjects. Neither group preferred working with the supports. Compared to these findings with experienced keyboard operators, Sihvonen et al.[40] found a significant reduction in trapezius activity with the use of moveable arm supports in a group of healthy, novice word processors.

A few studies have directly compared the effects of various support locations, though they have all been laboratory-based, and some exposed subjects to the various test conditions for very brief periods of time. In a study designed to find effective support for intensive mousing activities, Wells et al.[41] reported that trapezius muscle activity was reduced by any location of support, with a trend indicating that elbow support (from a chair arm rest) was most effective, followed by wrist on the work surface, and then forearm supported on work surface. Note that the finding, in this study, of beneficial effects of elbow support from a chair arm rest contrasts with the lack of long-term benefit of the presence of chair arm rests determined by Marcus et al.[32] This may be a function of the difference between "presence" and "use of," as well as the type of activity (intensive mousing[41] versus a mix of keying and mousing[32]). Wells, et al.[41] also found that activity in the infraspinatus muscle was reduced with any support; that the extrinsic finger extensor muscles were not affected by support condition; and that support at the wrist increased activity in the flexor digitorum superficialis and pronator teres muscles. After 1 h of continuous mousing activity, the trend in discomfort averaged across all body parts, was, from least to most, elbow support, forearm, wrist, and no support. At two hours and beyond, forearm and elbow were similar in total discomfort and less than wrist and no support, which were similar to each other. Other laboratory-based studies of simple data entry[42] or typing and mousing[43] also indicate that wrist support is either ineffective at reducing muscle activity or induces higher muscle activity in some upper extremity or shoulder muscles, compared with no support or support provided at the forearm, elbow, or both.

Table 19.1 provides a summary of the studies described in this section.

TABLE 19.1   Summary of Results from Studies that Examined Effects of Upper Extremity Support when Using Computers

Type of Work	Type of Support or Contact Point(s)	Effects	Effects of Support, on Balance[a]	Type of Study	Reference
Computer	Forearm–wrist	No hindrance of work; Used by 80% of subjects	+	Field, short-term trial (1 week)	[27]
Computer	Lower arm supported	Unsupported arms, limited rest breaks, and >20 h/week of computer work in combination elevated risk for neck/shoulder discomfort	+	Field, cross-sectional study	[28]
Computer	Table top support of forearm and wrist	Reduced trapezius muscle activity	+	Laboratory	[30]
Computer	Table top support of forearm and wrist	Reduced trapezius muscle activity; Reduced shoulder pain	+	Field, intervention, >1 yr	[29]
Computer	Full motion forearm supports	Performance, subjects' perceptions: slowed typing and increased errors. Performance, objective assessment: no impact; No postural benefit.	0/–	Laboratory	[38]
Computer	Fixed arm supports and arm suspension	Both supports reduced trapezius muscle activity in subjects with pain; no reduction or an increase in healthy subjects. Neither group preferred working with the supports.	+/0/–	Laboratory	[39]
Computer	Moveable arm supports	Reduced trapezius muscle activity	+	Laboratory	[40]
Computer	Wrist support and forearm support	Forearm support reduced activity in anterior deltoid; no effect on trapezius muscle. Wrist support had no effect on either muscle	+/0	Laboratory	[42]
Computer	Wrist support and forearm support	Wrist support: trapezius muscle activity increased or was similar to unsupported condition; Forearm support: trapezius muscle activity decreased or was similar to unsupported condition	+/0/–	Laboratory	[43]
Mousing	Chair arm support — elbow, work surface — wrist, work surface — forearm	Reduced trapezius and infraspinatus muscle activity with any support; No effect on extrinsic finger extensor activity; Wrist support increased flexor digitorum superficialis and pronator teres m. activity; Wrist support was similar to no support in total body discomfort during extended mousing (2–4 h)	+/0/–	Laboratory	[41]
Computer	Palm rest	Subjects did not find rests useful, and a few experienced increased discomfort using them	–	Field, short-term trial (1 week)	[31]
Computer	Palm rest	No reduction in any forearm, shoulder, or back muscles studied	–	Laboratory	[35]
Computer	Built-in keyboard wrist support	Reduced forearm extensor muscle activity reported, but confounded by reduced typing speed	?	Laboratory	[33]
Computer	Built-in keyboard wrist support	Reduced forearm extensor m. activity reported, but confounded by reduced typing speed	?	Laboratory	[34]
Electric typewriter	Wrist support	No effect on extensor carpi radialis muscle; Increase in trapezius muscle activity; $\frac{3}{4}$ of subjects wanted to keep the wrist support after the study	+/–	Laboratory	[37]
Computer	Chair arm rests	"Presence of chair arm rests" was not protective of development of neck/shoulder symptoms or disorders	0	Field, observational, prospective	[32]

[a]Symbols defined: +, positive effect of arm support; 0, no effect; –, negative effects; ?, effects are unclear; multiple symbols indicated mixed results within a study.

19.3.1.3 Medical Procedures

Several different wrist and forearm supports have been designed for use during microsurgical, plastic, and ophthalmic surgical procedures.[44,45] The common objectives of each are to provide fatigue relief for shoulder and back muscles during procedures that require surgeons to remain almost immobile for long periods of time, as well as to provide a steady base of support for control of postural tremor. Unfortunately, there are other restrictions in an operating room which may diminish the impact of arm supports, including the adjustability and height requirements of the surgical microscopes and the height and thickness of the patient support surface.

An EMG study of dentists performing a variety of procedures on patients revealed fairly high levels of activity in the trapezius and extensor carpi radialis muscles.[46] A second study was performed in order to determine the effect of several factors on muscle activity during dental procedures, including different types of upper extremity support.[47] Compared to a no support condition, arm support provided at the elbow significantly decreased muscle activity in the deltoid and trapezius muscles. Support at the elbow appeared to be more effective than support applied at the hand in reducing shoulder muscle activity. Whether such a support would be acceptable in practice is not clear from this study, since dentists were not used as subjects.

19.3.1.4 Possible Adverse Results

Most concerns are for the effects of localized pressure at contact sites between the support and the user's body. Direct pressure applied to the base of the palm was shown to raise the pressure in the carpal tunnel in cadaver specimens,[48] with point of application appearing to be the crucial factor. A 1 kg force applied just proximal to the distal wrist crease resulted in a mean increase in carpal tunnel pressure of 9 mmHg (median value of 1 mmHg), whereas that same force applied just distal to the distal wrist crease resulted in a mean pressure of 77 mmHg (median value of 29 mmHg).[48] Pressure in the range of 20–30 mmHg begins to impact nervous and circulatory function and performance. In live subjects, carpal tunnel pressure (CTP) has also been shown to increase when typing while resting the wrist, either on a wrist rest or on the keyboard support surface, in comparison to unsupported typing.[49] However, those authors did not report wrist posture in the three test conditions. If changes in wrist extension occurred between those three conditions, it would be difficult to determine whether the pressure change was due to external pressure on the wrist or change in wrist extension. The main points to take from this are: (1) contact stress should not be imposed at the locations(s) at which an operator supports his or her or her upper extremity while working, meaning the contact zone should not be concentrated on a very small area and the support surface should not "dig into" the skin, as would happen if the operator supported his or her arm on the edge, or, worse, the corner, of the work surface; and (2) support applied at the base of the palm or over the carpal tunnel, particularly if it is the sole point of contact (support), will likely raise the pressure on the median nerve within the carpal tunnel, and so, contact at the base of the palm should not be maintained continuously.

Pressure is also a concern at the elbow. Working with continuously flexed elbows, especially if supported at the elbow such that the ulnar nerve is compressed, may result in the development of cubital tunnel syndrome (compression of the ulnar nerve at the cubital tunnel).[50]

19.4 Summary

In reviewing the literature there seems to be little consensus on the effectiveness or acceptance of upper extremity supports. However, some summary statements and recommendations can still be made.

Wrist pads, forearm supports, and arm rests can all be categorized as engineering controls. Yet, based on the literature, their presence does not guarantee benefits or even usage. First, a worker's need for support must be balanced with the need for freedom of movement of the upper extremity. Continuous support methods would be appropriate if arms and hands are maintained in the same spatial location for extended periods of time. However, if movement is required, then the need is for a support that provides a resting point during mini-breaks (a few seconds in length) and longer pauses in activity. This resting support should not hinder the individual during work periods. Additionally, the worker's preference must be considered. For example, in modifying VDT workstations, once chair, monitor, and keyboard heights have all been established, operators should be provided with a variety of support options from which to choose. They should receive proper instructions in the ways the various supports should be used. Operators should also be encouraged to try out the supports at their own workstations.

There are a myriad of commercially available ergonomic support devices for keyboard workstations. The key to evaluating the efficacy of these products is to determine how and where the support is delivered, what postural changes might occur when using the device, and whether operators will find the support interferes with task performance. Based on the information in this chapter and some trial-and-error user-testing effective support devices or means may be found for those operators interested in utilizing them. For industrial settings, appropriate commercial supports may not exist. However, with some basic ergonomics training engineers, maintenance personnel, and operators may be able to devise unique support systems tailored to particular workstations or operators.

The goal of any upper extremity support is to aid an individual in achieving or maintaining a desired posture with less effort or less discomfort. Given that joints are designed to move, and most tasks require motion, there is often a conflict between the desire for support and the need for motion. Identifying a support which will provide support without interfering with motion requirements is important for the successful employment of upper extremity supports.

Acknowledgment

The author is grateful to Ms. Sahika Vatan for identifying and reviewing the most recent literature relevant to the topic of this chapter.

References

1. Hagberg, M., (1984) Occupational musculoskeletal stress and disorders of the neck and shoulder: a review of possible pathophysiology, *Int Arch Occup Environ Health* 53(3): 269–78.
2. Wallace, M. and Buckle, P., Ergonomic aspects of neck and upper limb disorders, in *International Reviews of Ergonomics*, Oborne, D. J. (ed.), Taylor & Francis, London, 1987, pp. 173–200.
3. Aarås, A., (1994) The impact of ergonomic intervention on individual health and corporate prosperity in a telecommunications environment, *Ergonomics* 37(10): 1679–96.
4. Jonsson, B., (1982) Measurement and evaluation of local muscular strain in the shoulder during constrained work, *J. Human Ergol* 11, 73–88.
5. Hagberg, M. and Sundelin, G., (1986) Discomfort and load on the upper trapezius muscle when operating a wordprocessor, *Ergonomics* 29(12): 1637–1645.
6. Chaffin, D. B., (1973) Localized muscle fatigue — definiton and measurement, *J Occup Med* 15(4): 346–354.
7. Jørgensen, K., Fallentin, N., Krogh-Lund, C., and Jensen, B., (1988) Electromyography and fatigue during prolonged, low-level static contractions, *Eur J Appl Physiol* 57, 316–321.
8. Melin, E., (1987) Neck–shoulder loading characteristics and work technique, *Ergonomics* 30(2): 281–285.
9. Jonsson, B. G., Persson, J., and Kilbom, Å. (1988) Disorders of the cervicobrachial region among female workers in the electronics industry, *Int J Inds Ergon* 3, 1–12.
10. McAtamney, L. and Corlett, E. N., (1993) RULA: a survey method for the investigation of work-related upper limb disorders, *Appl Ergon* 24, 91–99.
11. Onishi, N., Sakai, K., and Kogi, K., (1982) Arm and shoulder muscle load in various keyboard operating jobs of women, *J Human Ergol* 11, 89–97.

12. Sommerich, C. M., Marras, W. S., and Parnianpour, M., Activity of index finger muscles during typing, in *Proceedings of The Human Factors and Ergonomics Society 39th Annual Meeting*, San Diego, 1995, pp. 620–624.

13. Lundervold, A., (1958) Electromyographic investigations during typewriting, *Ergonomics* 1(3): 226–233.

14. Westgaard, R. H. and Bjørklund, R., (1987) Generation of muscle tension additional to postural muscle load, *Ergonomics* 30(6): 911–923.

15. Goldstein, I. B., (1964) Role of muscle tension in personality theory, *Psychol Bull* 61(6): 413–425.

16. Hünting, W., Läubli, T., and Grandjean, E., (1981) Postural and visual loads at VDT workplaces. I. Constrained postures, *Ergonomics* 24(12): 917–931.

17. Schüldt, K., Ekholm, J., Harms-Ringdahl, K., Németh, G., and Arborelius, U. P., (1987) Effects of arm support or suspension on neck and shoulder muscle activity during sedentary work, *Scand J Rehabil Med* 19(2): 77–84.

18. Occhipinti, E., Colombini, D., Frigo, C., Pedotti, A., and Grieco, A., (1985) Sitting posture: analysis of lumbar stresses with upper limbs supported, *Ergonomics* 28(9): 1333–1346.

19. Colombini, D., Occhipinti, E., Frigo, C., Pedotti, A., and Grieco, A., Biomechanical, electromyographical, and radiological study of seated postures, in *The Ergonomics of Working Postures*, Corlett, N., Wilson, J., and Manenica, I. (eds.), Taylor & Francis, London, 1986, pp. 331–344.

20. Schuldt, K., Ekholm, J., Harms-Ringdahl, K., Nemeth, G., and Arborelius, U. P., (1987) Effects of arm support or suspension on neck and shoulder muscle activity during sedentary work, *Scand J Rehabil Med* 19(2): 77–84.

21. Nakaseko, M., Grandjean, E., Hünting, W., and Gierer, R., (1985) Studies on ergonomically designed alphanumeric keyboards, *Human Factors* 27(2): 175–187.

22. Weber, A., Sancin, E., and Grandjean, E., The effects of various keyboard heights on EMG and physical discomfort, in *Ergonomics and Health in Modern Offices*, Grandjean, E. (ed.), Taylor & Francis, London, 1983, pp. 477–483.

23. Paul, R. and Menon, K. K., Ergonomic evaluation of keyboard wrist pads, in *Proceedings of 12th Triennial Congress of the International Ergonomics Association*, Toronto, 1994, pp. 204–207.

24. Damann, E. A. and Kroemer, K. H. E., Wrist posture during computer mouse usage, in *Proceedings of 39th Annual Meeting of the Human Factors and Ergonomics Society* The Human Factors and Ergonomics Society, San Diego, 1995, pp. 625–629.

25. Jex, H. R. and Magdaleno, R. E., Biomechanical models for vibration feedthrough to hands and head for a semisupine pilot, *Aviat Space Environ Med Jan*, 304–316, 1978.

26. Harms-Ringdahl, K. and Arborelius, U. P., One-year follow-up after introduction of arm suspension at an electronics plant, in *Proceedings of Tenth International Congress, World Confederation for Physical Therapy*, Syndey, 1987, pp. 69–73.

27. Grandjean, E., Hünting, W., and Pidermann, M., (1983) VDT workstation design: preferred settings and their effects, *Human Factors* 25(2): 161–175.

28. Bergqvist, U., Wolgast, E., Nilsson, B., and Voss, M., (1995) The influence of VDT work on musculoskeletal disorders, *Ergonomics* 38(4): 754–762.

29. Aarås, A., Horgen, G., Bjorset, H. H., Ro, O., and Thoresen, M., (1998) Musculoskeletal, visual and psychosocial stress in VDU operators before and after multidisciplinary ergonomic interventions, *Appl Ergon* 29(5): 335–354.

30. Aarås, A., Fostervold, K. I., Ro, O., Thoresen, M., and Larsen, S., (1997) Postural load during VDU work: a comparison between various work postures, *Ergonomics* 40(11): 1255–1268.

31. Parsons, C. A., Use of wrist rests by data input VDU operators, in *Contemporary Ergonomics*, Lovesay, E. J. (ed.), Taylor & Francis, London, 1991, pp. 319–321.

32. Marcus, M., Gerr, F., Monteilh, C., Ortiz, D. J., Gentry, E., Cohen, S., Edwards, A., Ensor, C., and Kleinbaum, D., (2002) A prospective study of computer users: II. Postural risk factors for musculoskeletal symptoms and disorders, *Am J Ind Med* 41(4): 236–49.

33. Smith, W. J. and Cronin, D. T., Ergonomic test of the Kinesis keyboard, in *Proceedings of The Human Factors and Ergonomics Society 37th Annual Meeting*, Seattle, 1993, pp. 318–322.

34. Gerard, M. J., Jones, S. K., Smith, L. A., Thomas, R. E., and Wang, T., (1994) An ergonomic evaluation of the Kinesis Ergonomic Computer Keyboard, *Ergonomics* 37(10): 1661–1668.

35. Fernström, E., Ericson, M. O., and Malker, H., (1994) Electromyographic activity during typewriter and keyboard use, *Ergonomics* 37(3): 477–484.

36. Moffet, H., Hagberg, M., Hansson-Risberg, E., and Karlqvist, L., (2002) Influence of laptop computer design and working position on physical exposure variables, *Clin Biomech (Bristol, Avon)* 17(5): 368–375.

37. Bendix, T. and Jessen, F., (1986) Wrist support during typing — a controlled, electromyographic study, *Appl Ergon* 17(3): 162–168.

38. Powers, J. R., Hedge, A., and Martin, M. G., Effects of full motion forearm supports and a negative slope keyboard support system on hand-wrist posture while keyboarding, in *Proceedings of the Human Factors Society 36th Annual Meeting*, 1992, pp. 796–800.

39. Erdelyi, A., Sihvonen, T., Helin, P., and Hanninen, O., (1988) Shoulder strain in keyboard workers and its alleviation by arm supports, *Int Arch Occup Environ Health* 60(2): 119–124.

40. Sihvonen, T., Baskin, K., and Hanninen, O., (1989) Neck-shoulder loading in wordprocessor use. Effect of learning, gymnastics and armsupports, *Int Arch Occup Environ Health* 61(4): 229–233.

41. Wells, R., Lee, I. H., and Bao, S., Investigations of upper limb support conditions for mouse use, in *Proceedings of the 29th Annual Conference of the Human Factors Association of Canada*, 1997.

42. Feng, Y., Grooten, W., Wretenberg, P., and Arborelius, U. P., (1997) Effects of arm support on shoulder and arm muscle activity during sedentary work, *Ergonomics* 40(8): 834–848.

43. Visser, B., de Korte, E., van der Kraan, I., and Kuijer, P., (2000) The effect of arm and wrist supports on the load of the upper extremity during VDU work, *Clin Biomech* 15 (Suppl 1), S34–S38.

44. Halliday, B. L., (1988) A new surgical head rest, *Br J Ophthal mol* 72, 284–285.

45. Bustillo, J. L., (1968) Hand and arm support in ophthalmic surgery, *Am J Ophthal mol* 66(2): 345–346.

46. Milerad, E., Ericson, M. O., Nisell, R., and Kilbom, A., (1991) An electromyographic study of dental work, *Ergonomics* 34(7): 953–962.

47. Milerad, E. and Ericson, M. O., (1994) Effects of precision and force demands, grip diameter, and arm support during manual work: an electromyographic study, *Ergonomics* 37(2): 255–264.

48. Cobb, T. K., An, K. N., and Cooney, W. P., (1995) Externally applied forces to the palm increase carpal tunnel pressure, *J Hand Surg [Am]* 20(2): 181–185.

49. Horie, S., Hargens, A., and Rempel, D., Effect of keyboard wrist rest in preventing carpal tunnel syndrome, in *Proceedings of Marconi Keyboard Research Conference* UC Berkeley Ergonomics Program, Marshall, CA, 1994.

50. McPherson, S. A. and Meals, R. A., (1992) Cubital tunnel syndrome, *Orthop Clin of North Am* 23(1): 111–123.

20

Lower Extremity Supports

Karen E.K. Lewis
*Honda of America
Manufacturing, Inc.*

20.1 Introduction

Musculoskeletal disorders of the back and upper extremity have drawn the vast majority of attention and scientific research because of well-documented high prevalence and severity. This is understandable because 56.4% of all injuries caused by overexertion affected the back and 76.2% of injuries caused by repetitive motion affected the upper extremity in 2001 according to the Bureau of Labor Statistics.[6]

The number of lower extremity injuries every year is not negligible. Injuries and illnesses to the lower extremity caused 21.0% of all incidents resulting in lost work days in 2001. In descending order, the primary causes of lower extremity injury are: (a) the "all other events" category; (b) falls to the same level; and (c) struck by object. There were a lesser number of total injuries, but "slips and trip without fall" results 62.5% of the time in injury to the lower extremity, especially to the knees and ankles.[6] A slip or a trip is more likely result in a lower extremity injury than an injury to any other body part.

Research into musculoskeletal disorders has focused on injuries caused by overexertion and repetitive motion. The lower extremity was affected by only 4.1% of overexertion injuries and 1.6% of repetitive motion injuries in 2001.[6] Although this is a relatively small percentage of total musculoskeletal disorders, it poses a very significant problem to the approximately 23,000 workers who experience these injuries every year.

In addition, the prevalence of lower extremity injuries can be expected to increase. The knees are particularly notorious for being susceptible to the effects of aging.[1] Current trends are predicting increasing proportions of older workers remaining in the workforce, and workers postponing retirement. These workers will be experiencing age-related degradation of the joints of the lower extremity through such mechanisms as osteoporosis and arthritis. The combination of age-related degradation and stresses from the occupational environment are likely to aggravate each other and result in early and more severe injury.

It has also been established that the most likely predictor of injury to any body part is past history of injury. Older workers are more likely to have a work-related or nonwork-related injury than younger workers, simply because they have had more years of opportunity.

20.2   Biomechanical Risks for Lower Extremity Injuries

Although musculoskeletal injuries of the lower extremity do not have a high frequency in the general working population, in some occupations they can be seen to be extremely common. High incidence of lower extremity symptoms have been noted in letter carriers and meter readers,[25] 30.9% of ballet dancers,[8] and carpet and floor laying workers.[12,26] Workers in these occupations experience long-distance walking, repetitive ankle movements, and knee mechanical stress, respectively, which have all been hypothesized as contributors to development of lower extremity injury.

As noted previously, 21.0% of all lost work day incidents in 2001 involved the lower extremity. Review of 2001 Bureau of Labor Statistics data indicates that some occupations have higher percentages of lower extremity injuries. For example, 88.7% of the injuries experienced by elevator operators involved the lower extremity. These data compared well with earlier data, showing that 48.1% of all injuries to meter readers; 62.3% of all injuries to dancers; and 30.4 and 38.9% of injuries to carpet installers and tile setters, respectively, affected the lower extremity (Table 20.1).

Outside of the athletic research community, injuries to the lower extremity have often been overlooked. There are few studies linking occupational risk factors to the occurrence of musculoskeletal disorders of the hips, knees, ankles, and feet. The studies that are available typically used occupation to measure exposure to a risk factor, rather than a direct measurement, so dose–responses relationship cannot be determined.

One study reviewed medical records of workers' compensation cases within a single practice for knee injuries. They identified 20.3% patellofemoral disorders, 20% meniscal pathology, 12.1% sprains, strains, contusions, 11.5% arthritis, 6.7% anterior cruciate ligament injuries, and 5.8% unexplained knee pain.[26] Of these diagnoses, arthritis has been linked to prolonged or repeated knee bending, as when squatting or kneeling more than 30 min per day or climbing more than ten flights of stairs.[7] This relationship held even when accounting for the effects of obesity. However, considering the frequency of arthritis of the knees in the general population, it is not yet clear whether occupational exposures will cause arthritis, or only accelerate the degenerative process in a susceptible individual.[15]

Bursitis was classified under the "meniscal pathology" cases. Peripatellar bursitis has been associated with carpet and floor laying occupations. Bursitis may result from extended time in a kneeling posture combined with mechanical stress.[12,26] The mechanical compression arising from kneeling on hard surfaces has been linked to knee disorders or symptoms in several studies.[11,14,22,23]

Tendinitis may have been included in the survey as one of the patellofemoral disorders. Tendinitis has been solidly linked in other body parts to repetitive overloading. Therefore, it can be expected more often in workers who frequently kneel, squat, or climb stairs. For example, the repetitive loading in ballet dancers has resulted in an extremely high incidence of achilles tendinitis.[8] It has also been speculated that types of work surface and footwear may contribute to the onset or exacerbation of tendinitis.

Walking for extended distances has also been linked to the development of joint discomfort in the lower extremity. Wells et al.[25] found that meter readers and letter carriers who walk more than 5 h per day have a more symptoms than postal clerks who are fairly stationary. Given the high percentage of lower extremity injuries that elevator operators, ushers, and various types of clerks experienced in 2001, it is also possible that extended time standing with little change of position may contribute to lower extremity injuries (see Table 20.1).

20.3   Prevention of Lower Extremity Injuries

Suggested methods for reducing the number of lower extremity injuries in the workplace have included changing work surfaces, retraining work methods, and implementing various types of personal protective

TABLE 20.1 Occupations with High Percentages of Injuries to the Lower Extremity Resulting in Days Away From Work

Occupation	Total Injuries	LE injuries	Percent LE injuries
Elevator operators	344	305	88.7
Forestry and conservation scientists	74	57	77.0
Ushers	265	199	75.1
Marine life cultivation worker	54	38	70.4
Postsecondary teachers, subject not specified	177	116	65.5
Dancers	448	279	62.3
Lawyers	682	412	60.4
Religious workers, n.e.c.	117	66	56.4
Operators, power plants	365	190	52.1
Engineers, mechanical	398	207	52.0
Social scientists, n.e.c.	76	39	51.3
Business and promotion agents	86	44	51.2
Meter readers	989	476	48.1
Athletes	1682	809	48.1
Real estate	226	108	47.8
Library clerks	155	72	46.5
Helpers, extractive	159	71	44.7
Supervisors, electricians, power transmission installers	863	382	44.3
Tailors	331	145	43.8
Timber cutting, logging	1238	537	43.4
Surveyors, mapping scientists	289	123	42.6
Hotel clerks	404	163	40.3
Agricultural and food scientists	199	80	40.2
Operators, paving, surfacing, tamping equipment	112	45	40.2
Engineers, civil	135	53	39.3
Tile setters, hard and soft	635	247	38.9
File clerks	936	361	38.6
Explosives workers	207	79	38.2
Statistical clerks	226	86	38.1
Artists, performers, n.e.c.	264	99	37.5
Physician's assistants	559	209	37.4
Railroad brake, signal, switch operators	689	256	37.2
Forestry workers, except logging	459	170	37.0
Engineers, industrial	419	151	36.0
Correctional institution officers	268	96	35.8
Protective service, n.e.c.	951	331	34.8
Guards and police, except public	9298	3232	34.8
Firefighting	127	44	34.6
Hoist and winch operators	830	286	34.5
Apprentices, plumber, pipefitter	1606	542	33.7
Supervisors, painters, paperhangers, plasterers	117	39	33.3
Stevedores	1044	344	33.0
Managers, properties, real estate	1654	539	32.6
Purchasing agents, buyers, farm products	117	38	32.5
Railroad conductors and yardmasters	1423	461	32.4
Supervisors, guards	323	104	32.2

(continued)

TABLE 20.1   *Continued*

Occupation	Total Injuries	LE injuries	Percent LE injuries
Teachers, except postsecondary, n.e.c.	4089	1292	31.6
Airplane pilots and navigators	1021	322	31.5
Pharmacists	140	44	31.4
Attendants, amusement, recreation	2475	775	31.3
Psychologists	482	149	30.9
Supervisors, personal service	699	216	30.9
Purchasing agents and buyers, n.e.c.	410	125	30.5
Managers, horticultural	276	84	30.4
Carpet installers	1697	516	30.4
Actors and directors	267	81	30.3
Roofers	5174	1567	30.3
Apparel	2979	901	30.2
Messengers	3054	922	30.2
Repairers, small engine	604	182	30.1
Supervisors, mechanics, repairers	2268	682	30.1

equipment. Undoubtedly, the most effective changes come with task redesign to eliminate the need to kneel, walk, squat, or climb steps in the workplace. Aside from the analysis of the using power stretchers instead of "knee-kickers" for carpet-laying occupations,[17] there have been few studies reporting such interventions.

There is a significant amount of research concerning various options for work surfaces. This topic is discussed in detail elsewhere. To summarize, concrete flooring likely contributes to symptoms of lower extremity discomfort that can be alleviated in part by the application of antifatigue floor matting or changing from a concrete to a wooden floor surface.

For the lower extremity, the subject of retraining work methods has only received limited attention, especially when compared with the work methods retraining for low back injuries. There is a theory that a significant contributor to osteoathrosis of the knees is "microklutziness," a result of individual incoordination causing cumulative microtraumas of the joints. Radin[19] cites unpublished work that biofeedback techniques may help retrain workers to eliminate this incoordination.

Given the difficulties and costs of redesigning work tasks, changing work surfaces, and retraining workers, many practitioners have turned to various kinds of personal protective equipment to attempt to prevent lower extremity injuries. This equipment may be applied to assist with immobilization of an injured joint, protect an uninjured joint, or correct movement problems.[18] Supports have been developed to assist with healing and to try to prevent injuries. Personal protective equipment for the lower extremity includes supports for the knee, ankle, and foot. As yet, there do not appear to be any feasible support devices for the hip on the commercial market.

20.3.1   Knee Supports

Personal protective equipment for the knee can consist of kneepads, knee straps, and knee braces. Kneepads are primarily recommended for individuals who experience mechanical compression through kneeling or using the knee to apply forces, such as using a "knee-kicker" during carpet laying. Kneepads act to cushion the knee from impact and should be replaced regularly. Sharrad[22] found that when worn-down or damaged, some styles of kneepads may contribute to the development of bursitis by degrading into hard ridges that press against the knee. He found that discomfort was reduced in miners with a previous knee injury by wearing kneepads, but this effect lasted less than a year as the kneepads became damaged and were not replaced.

Knee straps are clinically known as infrapatellar straps. They are believed to reduce the tensile force on the tendon and the compressive forces between the articulating surfaces during muscle contraction.[5] It is possible that knee straps may be helpful to individuals in occupations with repeated bending of the knee such as climbing stairs; however, there is little research to support the efficacy as a preventative tool for avoiding the development of tendinitis.

Knee braces primarily are used to aid in the recovery of an injured knee. Braces prevent excessive lateral shifting of the patella and absorb external forces away from the ligament, bones, and cartilage.[5] Most studies of the preventative capabilities of knee braces have taken place in an athletic context. This is motivated by the fact that 48.1% of the injuries to professional athletes occur to the lower extremity (see Table 20.1).

There has been speculation that the supportive properties of knee braces might be protective to individuals who work around unstable work surfaces, such as the construction industry. However, there have been conflicting studies on the ability of knee braces to prevent ligament injuries in athletes.[3] Furthermore, after an extensive review of the literature, in 1984 the American Association of Orthopaedic Surgeons determined that knee braces have not been proven to reduce knee injuries in athletes and may even contribute to injury.[27] Therefore, conclusive recommendation cannot be made for the use of knee braces to prevent occupational injuries.

Relatively recently a product has appeared on the commercial market called a "Knee Saver." The Knee Saver is a wedge-shaped pad that straps onto the lower leg and provides a cushion when an individual is working in a squatting or kneeling posture. This relieves muscular strain and prevents the knee from bending in extreme flexion. A similar device has been used recently by some catchers in professional baseball. There are no known studies on the efficacy of the Knee Saver or similar equipment in the peer-reviewed literature.

20.3.2 Ankle Supports

Ankle supports primarily consist of ankle braces and various techniques of ankle taping. Ankle braces are most often used as a postinjury method of returning an individual to work. The braces immobilize the ankle and are intended to provide pain relief.[24] Both ankle braces and ankle taping have been recommended as potential preventative measures for avoiding ankle sprains. In theory, the supports will restrict the amount of inversion of the ankle under conditions of sudden instability, such as uneven working surfaces. This theory is backed up by the finding that nonrigid ankle supports will decrease the inversion of the ankle during an unexpected fall.[2,13]

However there has been considerable debate in the athletic medicine community as to the best method of avoiding ankle inversion. One study found that both tape and a semirigid support would reduce the inversion and eversion of the ankle, but concluded that the support remained more effective after a period of exercise, because the tape often came loose or shifted, requiring frequent retaping.[10]

In a cadaveric study, both ankle taping and bracing were found to reduce the inversion moment of the ankle, particularly when wearing "high-top" shoes as opposed to "low-top" shoes.[21] Wearing an ankle brace with a low-top shoe was found to be equivalent to wearing a high-top shoe without any bracing. However, a review by Barrett and Bilisko[4] found contradictory studies on the effectiveness of shoe height and prevention of ankle strains.

20.3.3 Foot Supports

Shoe insoles are the most common support for the foot. Insole manufacturers have made sweeping claims that these products will "reduce foot, knee, and back pain," "reduce muscle fatigue," and "increase blood circulation." And, of course, "all this, while lowering your health care costs." However, there is only limited nonanecdotal evidence to support these claims.

Insoles have been prescribed to improve lower extremity alignment.[5] Theoretically, misalignment of the lower extremity can be caused by prior injury or a slightly abnormal foot shape. This misalignment then throws off ankle, knee, hip, and back coordination and causes inappropriate distribution of forces to

TABLE 20.2 Summary of Commonly Available Lower Extremity Supports

Body Part	Support Device	Evidence for Prevention of Injury or Discomfort
Knee	Kneepads	May reduce or prevent discomfort
	Kneestraps	Theoretical only
	Knee braces	Contradictory — may cause injury
Ankle	Ankle braces	May reduce injury
	Ankle taping	May reduce injury
	High-top shoes	Inconclusive
Foot	Shoe insoles	Theoretical

the joints, leading to discomfort and injury. However, foot shape has not been definitively proven to affect the frequency of injury and the effectiveness of insoles has not been shown.[20]

The type of shoe can also act as a foot support. Stress fractures have been found to be common in soldiers and may also be found in other occupation with extensive periods of walking.[9,16] Manufacturers claim that the viscoelastic properties of the shoe or insole can reduce the shock transmitted to the foot. Most studies of the efficacy of shoe type have been performed in military settings. Milgrom et al.[16] compared a high-top basketball shoe with a standard infantry boot during a 14-week infantry training period. No significant differences in the total number of stress fractures, knee fractures, or tendinitis were seen. Additionally, another study by Gardner et al.[9] found that an elastic polymer insole with good shock absorbency did not prevent stress fractures during 12 weeks of Marine training.

20.4 Conclusion

Injury and discomfort to the lower extremity present a significant challenge to the individual and the employer. There are plausible relationships between occupational factors and injury to the knee, ankle, and foot. However, the understanding of a dose–response relationship for specific occupational risk factors, if any, is still lacking and must be evaluated to distinguish the effects of aging from the effects of job tasks. Currently, many employers are implementing knee, ankle, and foot supports to affordably and quickly address some of the problems causing employees to experience discomfort or pain to the lower extremity (Table 20.2).

Existing evidence for the ability of lower extremity supports to prevent injury is sketchy at best. Individuals who work in a kneeling position may experience some relief from symptoms with the use of knee-pads or knee straps. Since the use of knee braces may increase the risk of knee injury in some situations, they should be discouraged as a preventative measure until more is known about this relationship.

Evidence for the use of ankle braces, ankle taping, or high-top shoes to prevent ankle injuries is too inconclusive to make any recommendations of the use or nonuse of this type of equipment. Insoles for the shoes have not been conclusively proven to reduce the frequency of lower extremity injuries during highly repetitive use. Clearly, there is a significant need for further investigation into this subject to clarify many of the contradictory findings and investigate the unanswered questions.

References

1. Allander, E. (1974). Prevalence, incidence and remission rates of some common rheumatic diseases or syndromes. *Scandinavian Journal of Rheumatology*, 3: 145–153.
2. Anderson, D.L., Sanderson, D.J., and Hennig, E.M. (1995). The role of external nonrigid ankle bracing in limiting ankle inversion. *Clinical Journal of Sports Medicine*, 5: 18–24.

3. Antich, T.J. (1997). Orthoses for the knee: the tibiofemoral joint. In D.A. Nawoczenski and M.E. Epler Eds., *Orthotics in Functional Rehabilitation of the Lower Limb*, pp. 57–76. Philadelphia, PA: W.B. Saunders Company.
4. Barrett, J. and Bilisko, T. (1995). The role of shoes in the prevention of ankle sprains. *Sports Medicine*, 20: 277–280.
5. Belyea, B.C. (1997). Orthoses for the knee: the patellofemoral joint. In D.A. Nawoczenski and M.E. Epler Eds., *Orthotics in Functional Rehabilitation of the Lower Limb*, pp. 31–56. Philadelphia, PA: W.B. Saunders Company.
6. Bureau of Labor Statistics (2002). Workplace injuries and illnesses in 2001. U.S. Department of Labor, Washington, D.C.
7. Cooper, C., McAlindon, T., Coggon, D., Egger, P., and Dieppe, P. (1994). Occupational activity and osteoarthritis of the knee. *Annals of the Rheumatic Diseases*, 53: 90–93.
8. Fernandez-Palazzi, F., Rivas, S., and Mujica, P. (1990). Achilles tendinitis in ballet dancers. *Clinical Orthopaedics and Related Research*, 257: 257–261.
9. Gardner, L.I., Dziados, J.E., Jones, B.H., Brundage, J.F., Harris, J.M., Sullivan, R., and Gill, P. (1988). Prevention of lower extremity stress fractures: a controlled trial of a shock absorbent insole. *American Journal of Public Health*, 78: 1563–1567.
10. Gross, M.T., Bradshaw, M.K., Ventry, L.C., and Weller, K.H. (1987). Comparison of support provided by ankle taping and semirigid orthosis. *Journal of Orthopaedic and Sports Physical Therapy*, 9: 33–39.
11. Jensen, L.K. and Eenberg, W. (1996). Occupation as a risk factor for knee disorders. *Scandinavian Journal of Work, Environment and Health*, 22: 165–175.
12. Jensen, L.K., Mikkelsen, S., Loft, I.P., and Eenberg, W. (2000). Work-related knee disorders in floor layers and carpenters. *Journal of Occupational and Environmental Medicine*, 42: 835–842.
13. Kimura, I.F., Nawoczenski, D.A., Epler, M., and Owen, M.G. (1987). Effect of the AirStirrup in controlling ankle inversion stress. *Journal of Orthopaedic and Sports Physical Therapy*, 9: 190–193.
14. Kivimaki, J., Riihimaki, H., and Hanninen, K. (1992). Knee disorders in carpet and floor layers and painters. *Scandinavian Journal of Work, Environment and Health*, 18: 310–316.
15. Kuorinka, I. and Forcier, L. Eds. *Work Related Musculoskeletal Disorders (WMSDs): A Reference Book for Prevention*, pp. 94–107. London: Taylor & Francis.
16. Milgrom, C., Finestone, A., Shlamkovicth, N., Wosk, J., Laor, A., Voloshin, A., and Eldad, A. (1992). Prevention of overuse injuries of the foot by improved shoe shock attenuation: a randomized prospective study. *Clinical Orthopaedics and Related Research*, 281: 189–192.
17. National Institute for Occupational Safety and Health (1990). *Preventing Knee Injuries and Disorders in Carpet Layers*. DHHS Publication No. 90-104.
18. Nawoczenski, D.A. (1997). Introduction to orthotics: rational for treatment. In D.A. Nawoczenski and M.E. Epler Eds., *Orthotics in Functional Rehabilitation of the Lower Limb*, pp. 2–13. Philadelphia, PA: W.B. Saunders Company.
19. Radin, E.L. (1995). Osteoarthrosis — the orthopedic surgeon's perspective. *Acta Orthopaedica Scandinavia*, 66(Suppl 266): 6–9.
20. Razeghi, M. and Batt, M.E. (2000). Biomechanical analysis of the effect of orthotic shoe inserts: A review of the literature. *Sports Medicine*, 29: 425–438.
21. Shapiro, M.S., Kabo, M., Mitchell, P.W., Loren, G., and Tsenter, M. (1994). Ankle sprain prophylaxis: an analysis of the stabilizing effects of braces and tape. *American Journal of Sports Medicine*, 22: 78–82.
22. Sharrad, W.J.W. (1963). Aetiology and pathology of beat knee. *British Journal of Industrial Medicine*, 20: 24–31.
23. Thun, M., Tanaka, S., Smith, A.B., Halperin, W.E., Lee, S.T., Luggen, M.E., and Hess, E.V. (1987). Morbidity from repetitive knee trauma in carpet and floor layers. *British Journal of Industrial Medicine*, 44: 611–620.

24. Trepman, E. and Yodlowski, M.L. (1996). Occupational disorders of the foot and ankle. *Orthopaedic Clinics of North America*, 27: 815–829.

25. Wells, J.A., Zipp, J.F., Schuette P.T., and McEleney, J. (1983). Musculoskeletal disorders among letter carriers: a comparison of weight carrying, walking and sedentary occupations. *Journal of Occupational Medicine*, 25: 814–820.

26. Westrich, G.H., Haas, S.B., and Bono, J.V. (1996). Occupational knee injuries. *Orthopaedic Clinics of North America*, 27: 805–814.

27. Wirth, M.A. and DeLee, J.C. (1990). The history and classification of knee braces. *Clinics in Sports Medicine*, 9: 731–741.

VI

Administrative Controls

21

Administrative Controls as an Ergonomic Intervention

Joseph M. Deeb
*ExxonMobil Biomedical
Sciences, Inc.*

21.1 Introduction

Workplace hazards that contribute to musculoskeletal disorders (MSDs) (injuries and illnesses) must be controlled in effective ways. Methods to control these hazards are available and have been practically tested NIOSH.[1] The most common approaches or methods used are:

1. *Engineering controls* — This is always considered to be the preferred solution to workplace hazards that may contribute to MSDs. In addition, it is the best solution to reduce and even prevent human errors. Engineering controls are discussed in part v of this book.
2. *Administrative controls* — This is the subject of this chapter.

It is very important to acknowledge that in the workplace, a combination of controls is used to reduce and prevent workers' exposure to MSDs. This means that these types of controls are not mutually exclusive. Given the fact that engineering controls are considered to be the optimum solution to reduce and prevent MSDs, they may not always be feasible given a workplace's current situation. This is so because of cost issues or technical solutions that are impractical or not yet feasible.

In this case, administrative controls offer approaches to contribute to the reduction in workplace MSDs. However, they are not as effective as engineering controls in eliminating hazards; administrative controls should not be relied on solely to solve ergonomic problems. In addition, since they do not fully eliminate workplace hazards, we must put continuous efforts to make sure that the prescribed administrative approaches are fully practiced, periodically evaluated, and well controlled. We will briefly discuss a number of these administrative approaches in this chapter.

21.2   Definitions

NIOSH[1] defines administrative controls as "management-dictated work practices and policies to reduce or prevent exposures to ergonomic risk factors." The commonly known risk factors that may contribute to MSDs are:

- *Repetitive tasks* — for example, rotating valves, flipping burgers.
- *Application of force* — for example, operating tools, performing manual handling tasks.
- *Awkward posture* — for example, working in confined spaces that can create poor body mechanics.
- *Contact stress* — for example, operating vibrating tools, supporting arms/forearms on sharp desk edges to perform keyboard activities.
- *Physical fatigue* — not enough rest during work hours. This is also true for leisure activities and hobbies such as gardening, playing the piano, violin, or guitar.

These ergonomic risk factors are not mutually exclusive. Usually, in the workplace, a combination of these factors is the norm. For example, performing keyboard activities can involve awkward posture of the wrist, repetition keying, contact stress between the forearm and the sharp edge of the desk, and no rest breaks.

21.3   Administrative Controls/Approaches

There are a number of administrative controls that have been used and have produced positive results. These are:

- Training
- Worker selection
- Worker rotation
- Overtime management
- Functional capacity assessment
- Physical ability testing
- Behavior-based safety

21.3.1   Training

Training should focus on preparing workers, supervisors, and managers to:

- Recognize and identify job ergonomic hazards
- Recognize signs and symptoms of related injuries and illnesses
- Participate in the development of strategies and design solutions to control and prevent workplace hazards and reduce stress and strain contributed by the work tasks

The goal is to have a classroom and on-the-job training using real case studies. The objectives should be to:

- Understand the basics of ergonomics
- Understand how muscles and bones work
- Identify different types of injuries
- Recognize warning signs of musculoskeletal injuries
- Identify factors of physical activities associated with CTDs (Cumulative Trauma Disorders) also referred to as RSIs (Repetitive Strain Injuries)

- Identify limitations and capabilities of people
- Increase awareness of the importance of the work environment
- Better recognize and perceive risk

Training must not be a one-time event. Refresher training should be part of any safety program and real case incidents must be shared in order to learn. NIOSH[1] provides details on the design of a training program.

21.3.2 Worker Selection

Worker selection to perform heavy physical tasks may be considered as an alternative to control MSDs on the job. This can also be referred to as "fit the human to the job." The goal here is to select the person with the ability and capability to perform heavy work and control the possibility of MSDs.

The variables that have been used for selection[2,3] are: age, gender, anthropometry, shoulder, arm, and back strength, to name a few. The capacity for a specific task is predicted using tables and regression equations.

The challenge in the worker-selection process is to ensure that it takes place within the guidelines set forth by the Equal Employment Opportunity Commission[4] and the American with disability act.[5] The guidelines make it very specific and clear that such selection tests:

1. Do not discriminate against any group of people
2. Are validated
3. Are reliable to ensure people who pass the selection tests can do that on a regular basis to continue doing the job

With increasing age, the third requirement mentioned above may become difficult to meet. This is so because muscles do atrophy and worker's capacity decreases with age leading to more risk of MSDs. For example, a worker is selected at age 30 to perform manual handling tasks at a warehouse. After 30 yr, he is still performing the same job, however, with less capacity. Therefore, at an older age, the probability of being injured is much higher.

Since selection is not the best solution, we need to concentrate on changing the design of the jobs to fit most workers instead of the selection process. However, some jobs may be difficult or impossible to improve ergonomically, and thus selection is warranted. For example, mining jobs where physical strength is a necessity.

Standardized strength testing for selection purposes has not shown much success in reducing or eliminating MSDs. Therefore, as mentioned earlier, the emphasis should be on the proper design of the workplace.

21.3.3 Worker Rotation

Rotating workers between tasks (also referred to as job rotation) has many benefits:[6–8]

1. Reducing physical fatigue and overexertion.
2. Reducing psychological fatigue such as boredom from performing monotonous jobs. Reducing boredom through job rotation is also referred to as job enlargement. Job enlargement is achieved through adding more variety of jobs/tasks. Conversely, adding more responsibility is referred to as job enrichment.
3. Widening the different skills a worker would have to:
   - Make job rotation more efficient, reliable, and productive.
   - Be able to cover for another worker who may be transferred to another job, sick, on vacation, etc.

Job rotation can produce very good results if applied properly. For example, in a supermarket environment, a worker may spend about 2 h in the bakery shop making bread, 2 h fixing, rearranging, and tiding up shelves, 2 h at the cashier desk, etc. In a petrochemical environment, a worker may spend 2 h in the field taking samples of products for laboratory tests, 2 h working with valves (open/close), 2 h in the maintenance shop, etc. Another example would be where workers who have been out on vacation, sick, or any type of leave of absence for a period of time. Rotation between jobs requiring different levels of force and repetition can help different muscle groups to progressively regain their tone and flexibility.

There are a number of guidelines or common practices that can be followed to achieve very good results in job rotation.[9] Some of these guidelines are:

1. There are tasks that should not be considered as part of the rotation process. They should be strong candidates for engineering rather than administrative controls. These tasks may have:
   • Had already recorded injuries associated with them.
   • Been identified to have the potential to cause MSDs either through task analysis or workers' complaints.
2. Task analyses should be performed for the different jobs/tasks that are candidates for rotation to ensure that:
   • Performing different activities require the involvement of different set of muscle groups. This will allow different muscle groups to alternate between work and recovery.
   • If some of the muscle groups used in task A are also used in task B, the muscle groups of task B are involved in light and dynamic work. This will allow the pumping action of the muscles under dynamic contraction–relaxation to provide more blood to the working muscle group to provide oxygen, remove heat and lactic acid and other by-products.[10]
   • Heavy physical work in a hot environment (i.e., middle of July in Texas, Louisiana) warrants the rotation of workers to lighter physical job. Of course, it would be extremely helpful if the worker can switch between hot and artificially cooler environment during the work day.

21.3.4   Overtime Management

Even though overtime is associated with workers paid by the hour, more and more people who are on the salary scale work longer and longer hours (above and beyond the 40-h week). Hourly workers are paid 1.5 to 2.0 times extra when their work exceeds the 40-h week. This is considered direct cost since it can be easily tracked. However, salaried employees working longer hours at the office, at home, and over the weekend make it a bit more difficult to track.

With the cost-cutting mode and the continuous job elimination companies have been going through, the amount of work that needs to be performed is now accomplished with fewer employees. Adding to that is the unplanned absenteeism, sickness, etc., and the number of employees becomes even less. This understaffing, at the end, leads to negative consequences in health and productivity, as well as litigation issues related to incidents, injuries, and death that are associated with overtime. In addition, age can become an issue where the average age of workers in the workforce has been on the rise over the years.[11] This brings questions about physical workload, fatigue, and MSDs when working longer and longer hours.

Best practices to manage overtime can be:

1. Employees (hourly or salaried) should respect their break times and stop doing work to allow recovery time to take place.
2. People should manage the number of hours worked per day. For example, in using the laptop computer, we spend 10 h or so at the office, take it home and use it for another 3 to 4 h, few hours over the weekend, at the airport while waiting for the plane, on the plane throughout the flight (except during take off and until reaching a certain altitude). No wonder that these people complain about eye fatigue and strain, and neck, arms, wrists, shoulders, and back pain and discomfort. The issue here is that there is no rest allowed in order for the involved muscle groups to rest and recover.
3. Supervisors and managers should be aware of people working long hours and should have the knowledge on how to deal with the situation.

21.3.5 Functional Capacity Assessment

Functional capacity assessment is a process to examine, evaluate, and match a person's capacity with the specific demands of a particular job. It is a range of assessment procedures to practically evaluate fitness for work. One of those assessment procedures is the physical ability testing discussed in the next section. Fraser[12] provides excellent guidelines in his book on fitness for work techniques.

To achieve accuracy and effectiveness of using the functional capacity assessment process, it must be done in conjunction with the physical demands analysis (PDA) of the job.[12] The PDA details the job activities or tasks and their related physical requirements such as carrying, lifting, lowering, pushing, pulling, arm strength and endurance, back flexibility, etc.[12] The PDA can also be used to identify and recommend different ways to execute the tasks or new engineering controls to eliminate any MSD- related issues and to increase productivity and efficiency. Indeed, without the use and context of the PDA, the functional capacity assessment process will amount to just a general medical assessment that does not have the full value of determining whether a potential match between a worker and a particular job does exist.[12] In addition, Fraser emphasizes that in order to achieve the full potential of the PDA and the functional capacity assessment process, they have to be done in detail to capture every activity/task and its related human factors/ergonomics issues and what to do about it. This kind of detail will provide a process to:

1. Assess workers capabilities and fitness to perform those job activities in a safe and efficient way
2. Establish a benchmark against which to assess workers
3. Envision solutions to reduce and even eliminate MSDs

The functional capacity assessment process has two main assessment components:[12]

1. Medical fitness
2. Work capacity

Medical fitness assessments concentrate on the following areas:

- General medical history
- Lifestyle history
- Occupational health history
- Physical examination

Work capacity assessments concentrate on the following physiological systems:

- Process and effectiveness of breathing and transfer of oxygen in the lungs
- Ability to match cardiac output with workload demands
- Efficient transportation of oxygen to working muscle groups
- Ability to exert and endure adequate muscular force to match workload demands

One thing to point out here is that when rotating a worker between jobs with different physical demands, this worker may be fit to work at a specific job but not the other. Therefore, it is of utmost importance to clearly know the different tasks and settings to properly assess fitness and to ensure that rotation does not lead to a misfit between workers and job.

A fitness for work program may involve the following components:

- Ergonomics/human factors awareness training for safety contacts, key supervisors, managers, and contractors
- Selection/evaluation of critical jobs based on medical information (absenteeism rates, clinic visits, etc.)
- Physical fitness evaluation of employees in critical jobs
- Job demands/fitness matching
- Job-specific physical conditioning training
- Ergonomic improvement implemented when possible

21.3.6 Physical Ability Testing

The goal of physical ability testing (PAT) is to screen employees (new hires and incumbents) on the basis of their physical capabilities and match that with the demands a job imposes on them.[13] However, in general, PAT is mainly applied to new hires and rarely for incumbents.[14] The reasons are possibly:

- Political
- Sensitive policies
- Lack of studies validating the baseline requirement of the job

To achieve success in the screening process, Chaffin and Andersson[15] suggest that when choosing a particular method of evaluation it needs to be:

- Safe to administer and practical
- Reliable, accurate, and valid
- Quantitative and able to predict risk of future injury or illness
- Related to the specific job requirements

Indeed, the age Discrimination in Employment Act,[16] American with Disabilities Act,[5] and Civil Rights Act[17] all emphasize that any test assessing employee's ability and capability to perform physically demanding tasks from entry to retirement has to be reliable, valid, and fair. The PAT procedure must show (by law) that the skills and abilities tested are those required by the job under consideration.

PAT requires a lesser degree of sophistication than the functional capacity assessment (discussed earlier). There are many cases where tests of the general fitness level of individuals are sufficient to establish their specific fitness level to do the job. However, there are other cases that need rigorous tests to establish whether individuals have the capabilities to perform job duties.

Most of the work in the area of PAT has been done in laboratory and under controlled conditions. The Canadian standardized test of fitness[18] provides simple sets of procedures for testing. These sets of procedures are explained in Fraser.[12] The test procedures include:

- Anthropometric measurement such as stature, body weight
- Aerobic fitness such as postexercise heart rate and blood pressure
- Muscular strength, endurance, and flexibility such as grip strength, trunk forward flexion

21.3.7 Behavior-Based Safety

Over the last 10 to 15 yr, many companies have selected to implement behavior-based safety (BBS) programs or processes. The common belief is that if they (companies) can change workers behavior to act safely, then accidents and incidents will disappear. The objective of all BBS programs is to reduce the number and severity of injuries. In general, safe behavior acts as a barrier to prevent an incident while unsafe behavior acts as the hazard that initiates the incident. The main goal of BBS programs is to identify safe behaviors for reinforcement, and correct unsafe behaviors. Even though our traits (attitudes, personalities, and values) can affect our behavior, what we observe, however, is only what the individual does or say and not the traits.[19] Therefore, BBS programs are based on field observations, providing immediate feedback to reinforce safe behavior and correct unsafe behavior. The recommended core elements for a BBS program[19] are:

- Implementation
  - The program has features that support continuous improvement
  - The program includes a plan for regular stewardship
- Buy-in
  - The program expects employee participation
  - The program has visible participation by site and line management

- Training
  - The program provides training for employees on:
    - Basic principles of behavioral reinforcement and correction
    - Observer's observation skills and improvement, if needed
    - How to deliver feedback and reinforcement
    - Considerations for human factors/ergonomics both in terms of physical activities and potential for human error
- Observation
  - Persons who are observed receive immediate feedback/reinforcement on safety behavior issues that were observed
  - Observers use a checklist or reminder list while observing work activities
- Tools
  - The program provides tools for consistent observations and analyses
  - The program provides plans/tools to effectively define and communicate solutions/learnings throughout the organizational levels and shifts/work groups
- Measurement
  - Initial implementation includes evaluation of "base case" to target needs
  - Measures and performance targets emphasize results and key activities and are used to prioritize interventions
- Contractors
  - The program can extend to appropriate contractors

BBS programs and processes are common administrative approaches that companies are using these days. Despite the advertisement and claims made by the owners of the different programs and processes, BBS has not led to, or created, a totally accident-free workplace. This proves that BBS programs should not be used alone as the main solution to incidents. They should be supported by feasible engineering controls. BBS programs have their positive and negative points. An excellent article by Eckenfelder[20] describes in detail these two points. The following are summaries from the article:

1. Positive points of BBS programs and processes:
   - Focuses on people
   - Define safe and unsafe behavior
   - Encourages continuing safe behavior and discourages unsafe behavior
   - Involves every worker in the plant
   - Workers see positive involvement from Management by spending the money to adopt BBS
   - Creates strong commitment at least in the early phases
2. Negative points of BBS programs and processes:
   - BBS is really not a new concept. Conducting observations and providing feedback are old processes and date back to the early years of Industrial Engineering.
   - In general, but not always, BBS focuses on the front-line employees (workers) and neglect others such as supervisors and managers who are critical to the process.
   - BBS focuses on changing behavior to change attitude (when we have positive attitude we do positive and safe behavior). However, BBS does not focus on changing beliefs and values, which are the key ingredients to change attitude permanently and thus behavior.
   - BBS focuses on behavior, which is considered as a symptom of the causes of incidents, and less on design issues. So, the question is usually "what more do we need to do to further change behavior so incidents do not occur again?"

References

1. NIOSH. *Elements of Ergonomics Programs — A Primer Based on Workplace Evaluations of Musculoskeletal Disorders.* DHHS (NIOSH) Publication No. 97–117, 1997.

2. Ayoub, M.M., Selan, J., and Jiang, B. Manual materials handling. In *Handbook of Human Factors*, Salvendy, G., Ed., Wiley, New York, 1987, Chapter 7.2.
3. Chaffin, D.B. and Andersson, G.B.J. *Occupational Biomechanics*. Wiley, 1991, Chapter 5.
4. Equal Employment Opportunity Commission, Civil Services Commission, Department of Justice, and Department of Labor. *Uniform Guidelines on Employment Selection Procedures*. Number 6570-08, Part 1607. Federal registry, 43(166), 1978, 38290–38345.
5. American With Disabilities Act, 1990. United States Equal Employment Opportunity Commission. http://www.eeoc.gov/laws/ada.html.
6. OSHA. *Ergonomics Program Management Guidelines for Meatpacking Plants*. OSHA 3123: Washington, DC: DOL/OSHA, 1990.
7. OSHA. *OSHA Proposed Ergonomic Protection Standard*. Federal Register 64(225): 65768, 1999.
8. OSHA. *Part II: 29 CFR part 1910, Ergonomics Program: Final Rule*. Federal Register 65(22): 68262–68870, 2000.
9. Eastman Kodak Company. *Ergonomic Design for People at Work*, 2nd edn. Wiley, New York, 2004, p. 448.
10. Konz, S. *Work Design*. Grid Publishing, Columbus, OH, 2001.
11. Bourdouxhe, M.A. et al. Aging and shiftwork: the effects of 20 years of rotating 12-hour shifts among petroleum refinery operators. *Experimental Aging Research*, 25(4), 323, 1999.
12. Fraser, T.M. *Fitness for Work*. Taylor & Francis, London, 1992, p. 81.
13. Karwowski, W. Occupational biomechanics. In *Handbook of Industrial Engineering*, 2nd edn, Salvendy, G., Ed. Wiley, New York, 1992, Chapter 39.
14. Sothmann, M.S. et al. Performance requirements of physically strenuous occupations: validating minimum standards for muscular strength and endurance. *Ergonomics*, 47(8), 864, 2004.
15. Chaffin, D.B. and Andersson, G.B.J. *Occupational Biomechanics*. Wiley, New York, 1991, p. 464.
16. Age Discrimination In Employment Act, 1967. United States Equal Employment Opportunity Commission. http://www.eeoc.gov/laws/adea.html.
17. Title VII of the Civil Rights Act of 1964, 1991. United States Equal Employment Opportunity Commission. http://www.eeoc.gov/laws/vii.html.
18. Ministry of Fitness and Amateur Sport. *Canadian Standardized Test of Fitness, Operations Manual*, 3rd edn., Document FAS 7378, Government of Canada, 1986.
19. Deeb, J.M., Danz-Reece, M.E., and Smolar, T.J. Industry experience with behavior-based safety programs: a survey. *Proceedings of the International Ergonomics Association 2000 and Human Factors and Ergonomics Society 2000 Congress*, San Diego, California, 2, 2000, 213.
20. Eckenfelder, D.J., 2003. http://www.occupationalhazards.com.

22

Worker Selection for Physically Demanding Jobs

Veikko Louhevaara
University of Kuopio and Kuopio
Regional Institute of
Occupational Health

Juhani Smolander
ORTON Research Institute and
ORTON Orthopaedic Hospital

Annina Ropponen
University of Kuopio,
Department of Physiology

22.1 Introduction

22.1.1 Ergonomics and Fitting the Job to the Worker

The fundamental principle of occupational ergonomics is to fit the job to the worker so that no worker selection is necessary. Fitting the job to the worker should assure that feasible ergonomics measures of job modification are done before the initiation of the selection process. According to the International Ergonomics Association (IEA), modern ergonomics is a scientific discipline concerned with understanding the interaction between humans and other elements of a system for optimizing the human well-being and overall system performance (IEA, 2000). In the definition of the IEA, "elements of a system" may be very diverse and related to technology and psychosocial issues that an individual utilizes or encounters during working and leisure time, and even when sleeping.

In Finland, occupational ergonomics is defined as follows: "(occupational) ergonomics is a multidisciplinary science, which is based on physiology, psychology, sociology, and technical science applications." It considers human capacities, needs, and limitations in the interaction of technical and organizational work systems. The integrated knowledge of ergonomics is used to develop work content and the environment through the use of job design and redesign measures. Ergonomics promotes health, work ability, and well-being by preventing work-related disorders, diseases, and injuries, and improves productivity and the quality of work by technical and organizational measures (Louhevaara, 1999). The impact of ergonomics may be enhanced when complemented by individual lifestyle measures and education and training for qualified professional competence (Louhevaara and Nevala, 2003). Ergonomics is both a scientific and practical discipline that always aims to improve the job and the work environment for meeting a worker's individual characteristics and capacities in the multidisciplinary context of working life.

22.1.2 Successful Job Placement

In successful job placement, the primary measures consist of the application of ergonomic technologies. Various health and fitness interventions as well as the implementation of assistive technologies may be considered secondary measures (http//umrerc.engin.umich.edu) (Figure 22.1). Successful job placement is considered to guarantee suitable individual characteristics and capacities for the job protecting both worker and employer against harmful consequences.

Successful job placement is supposed to alleviate individual health and safety risk associated with high-risk jobs in terms of physical, psychological, and social demands as well as to maintain efficient job performance in terms of productivity, quality of work, and effectiveness. Physically demanding jobs often require heavy dynamic muscle work, manual handling of materials, and static muscle work in poor or dangerous work environments. Typically, the intensity of the jobs may also unpredictably exceed near maximal or maximal levels (Lusa, 1994; Soininen, 1995; Vehmasvaara, 2004).

There are also various jobs that require specific psychosocial resources and professional competence, for instance, due to high demands for leadership position, difficult human relationships, high

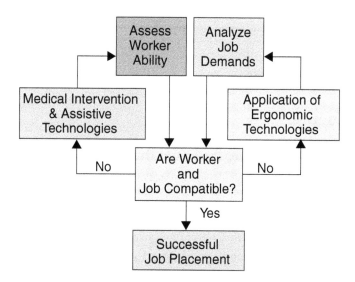

FIGURE 22.1   Model for successful job placement.

responsibility, or unfamiliar or hostile work conditions. The psychological suitability is often necessary to test for avoiding health problems, mental exhaustion or burnout, or economical risks. The testing of the abuse of drugs is also a current topic in pre-employment screening (Morland, 1993; Kraus, 2001; Matano et al., 2002).

Successful job placement requires a careful analysis of job demands and the assessment of individual abilities, capacities, and characteristics of the workers. For physically or psychosocially demanding jobs, successful job placement is practically impossible without a relevant process of worker selection. The main element of worker selection is the application of various tests for evaluating the match between the job demands and capacities of a worker.

Singleton (1989), Chaffin and Andersson (1991), Burke (1992), Jackson (1994), Konz (1995), Wickström (1996), Pheasant (1996), Gallagher (1999), and Louhevaara (1999) have published reports about the topic of worker selection, successful job placement, or human diversity at the textbook or review level.

22.1.3 Scope of the Chapter

The scope of this chapter is to consider worker selection, successful job placement, and the assessment of individual physical characteristics and capacities in pre-employment screening and in the context of physically demanding high-risk safety jobs. The focus is on the selection of healthy individuals with various capacities or job-related tests, which assess physical performance or the prerequisites of physical work ability. Examples of safety jobs include those of firefighters, police officers, paramedic workers, and security guards.

22.2 Validation Approaches in Worker Selection for Physically Demanding Jobs

22.2.1 Job Demands as a Basis for Worker Selection Procedures

Occupational ergonomics measures and the development of the job and the environment for matching the work to the worker needs to be based on reliable and accurate analysis of the critical and important elements of work. Many types of methods are available for job analysis including, for example, interviews, observations, video-taping, questionnaires, and direct physical or physiological measurements. When the relevant tasks and elements of the job have been identified and conceptualized, then the possible testing procedures can be validated against the constructs found in the job analysis.

The Dictionary of Occupational Titles (DOT, U.S. Department of Labor, 1991) is a widely used administrative system for matching occupational demands with workers' abilities. Although DOT has been developed for social security purposes (i.e., for evaluating persons usually seeking social security benefits and not jobs), it is interesting to look at the system with the aim of worker selection. DOT contains descriptions of 12,741 occupational titles. The classification of physical demands are descibed by 20 functional items, and for most of them the demand level is expressed in exposure times (not present, occasionally, frequently, constantly). The demand level for strength considers also the demand for force exertion, and is expressed in five levels (Table 22.1). As an example, the DOT physical demands for firefighter's occupation are shown in Table 22.2. As for the physical abilities required for firefighting, DOT lists different types of strength (explosive, dynamic, static, trunk), stamina, and gross body equilibrium.

Although widely used, especially in the rehabilitation area, the reliability and validity of the DOT system has been questioned (Frings-Dresen and Sluiter, 2003). And as can be seen in Table 22.2, the physical demands are expressed rather crudely, and thus DOT may not be a suitable basis for developing

TABLE 22.1 Dictionary of Occupational Titles System for Classifying the Strength Demands of Work

Physical Demand	Occasionally	Frequently	Constantly	Energy Required
Sedentary	<4.5 kg	Negligible	Negligible	1.5–2.1 METS
Light	<9.0 kg	<4.5 kg	Negligible	2.2–3.5 METS
Medium	9.0–22.5 kg	4.5–11.4 kg	<4.5 kg	3.6–6.3 METS
Heavy	22.5–45 kg	11.4–22.5 kg	4.5–9.0 kg	6.4–7.5 METS
Very heavy	>45.0 kg	>22.5 kg	>9.0 kg	>7.5 METS

test procedures for physically demanding high-risk jobs, or even other jobs (Frings-Dresen and Sluiter, 2003). The newer O*NET occupational information system will gradually replace the DOT system (Hanson et al., 2001).

The development of DOT has, however, created an "industry" of so-called functional capacity evaluation (FCE) systems. It has been estimated (Rudy et al., 1996) that there are several hundreds of FCE methods available on the market for the evaluation of physical capacity in relation to DOT physical demands (e.g., Blankenship, ErgoScience, WorkHab, ARCON, DOT-RFC). The duration of FCEs vary from 1 h to 2 days, and the prices are from US$400 up to 75,000 (King et al., 1998). Recently, Innes and Straker (1999a,b) evaluated the scientific basis of the available FCE systems and found that the psychometric properties of most of the methods had not been studied at all. Also, the reliability and validity of the studied FCE methods were mostly low. FCEs are also measures of performance and not capacity, which makes them less suitable for worker-selection purposes demanding near-maximal or maximal levels of effort.

According to Jackson (1994), the general methods commonly used to validate tests for physically demanding tasks usually come from combinations of psychophysical, biomechanical, and physiological data. The typical psychophysical method is the rating of perceived exertion (Borg, 1970). Biomechanical data include, for example, weights of the objects lifted or carried and forces needed in pushing and pulling objects. The physiological approach often concentrates on the endurance aspect of tasks (e.g., climbing stairs in fire fighting), and the measured variables are usually oxygen consumption

TABLE 22.2 Physical Demands of a Firefighter's Occupation According to Dictionary of Occupational Titles

Physical Demand Construct	Physical Demand Level
Strength	Very heavy
Climbing	Occasionally
Balancing	Occasionally
Stooping	Occasionally
Kneeling	Occasionally
Crouching	Occasionally
Crawling	Occasionally
Reaching	Frequently
Handling	Frequently
Fingering	Occasionally
Feeling	Occasionally
Talking	Frequently
Hearing	Frequently
Tasting/smelling	Not present
Near acuity	Frequently
Far acuity	Frequently
Depth perception	Frequently
Accommodation	Occasionally
Color vision	Frequently
Field of vision	Frequently

and heart rate. From these data it is possible to identify the underlying physical performance constructs and their demand levels, which are the prerequisites for successful job performance.

22.2.2 New Taxonomic Approaches to Evaluate Physical Functioning

The pioneer of taxonomic research linking human abilities to work behaviors is prof. Fleishman, whose work started in the 1960s with the structure of physical fitness (Fleishman, 1964), and continued with taxonomies of occupational performance (Fleishman and Reilly, 1992).

As a part of a recent project by U.S. Social Security Administration a functional assessment taxonomy (FAT) was developed based on existing taxonomies (Gaudino et al., 2001). The idea was to create a system that could link a broad variety of impairment taxonomies with taxonomies of occupational demands. The taxonomy includes five domains that contain 33 conceptual factors, which contain 133 constructs with definitions (Table 22.3). For example, in the physical domain one conceptual factor is "manual material handling," which contains constructs "reaching," "lifting and lowering," "pushing and pulling," and "carrying objects." In the DOT system, the last three constructs belong to the demand category "strength."

Another part of that project identified 800 valid instruments for evaluating work disability (Matheson et al., 2001). These measures were organized into a database that was linked to the taxonomy of FAT. Matheson (2003) gives examples of measures for evaluating constructs under conceptual factors "hand use" (nine constructs) and "manual material handling." Even thougth the construct is well defined, different protocols, scales, and equipment may be used to evaluate the same construct. For example, hand strength (one construct in "hand use") can be measured with different equipment and protocols.

WHO has recently published the International Classification of Functioning, Disability, and Health (known as ICF) (WHO, 2001). The ICF covers the components "body functions," "body structures," and "activity and participation." Each component consists of various domains, and within each domain, categories, which are the units of classification. Health and health-related states of an individual may be recorded by selecting the appropriate category. The ICF and FAT taxonomies have much in common, and both emphasize to focus on what a person can do (activity), especially in daily life in the usual environment (participation). It must be noted that ICF classification does not indicate any method at the category level.

The taxonomic approach may be a useful tool for occupational ergonomics, providing a standardized construct system for linking job demands, individual abilities, and the assessment methods of both of these.

22.2.3 Validation Strategies

Worker-selection procedures should be standardized, and the methods should have acceptable psychometric properties. According to Jackson (1994), validity depends on reliability and relevance. Relevance involves the qualities being tested, and it is defined by the job analysis. Validity depends on reliability and should be verified for those for whom the use of the instrument is intended (Jette, 1998). Basicly, three types of validation studies can be used for worker-selection tests: content validity, criterion validity, and construct validity. Sometimes the term face validity is used, but it can often be determined only after all other types of validity studies have been carried out. It indicates whether a test procedure is reasonable, acceptable, and useful for the intended purpose.

Content validity analysis should be enhanced at the development stage of tests and protocols. It is a process of collecting evidence that the worker selection tests represent important aspects of performance on the job. Professional judgment and job analysis form the basis for content validity. Content validation approaches often and naturally tend to result in work simulation tests, which usually have a high content validity because they represent the actual work tasks. A typical example is a ladder climbing test for firefighters.

TABLE 22.3   Relationship Between Domains and Conceptual Factors in the Functional Assessment Construct Taxonomy

Domain	Conceptual Factor	Ability within the Worker Role and Work Environment
Sensory-perceptual	Hearing	Ability to clearly perceive spoken and other sounds
	Vision	Ability to clearly perceive visual stimuli
	Cutaneus discrimination	Ability to discriminate light touch, pressure, vibration, pain, and temperature
	Proprioceptive discrimination	Ability to discriminate the position of the body in relation to the environment
	Body reactivity	Ability to tolerate the effect on the body of contact with the environment
	Psychomotor speed/reaction time	Ability to react quickly to a stimulus
	Visuospatial abilities	Ability to identify, generate, retain, and manipulate visual images
Physical	Digestive control	Ability to control nourishment and elimination of body wastes
	Lower extremity use	Ability to use legs and feet in coordinated and purposeful movements
	Upper extremity use	Ability to use shoulders and arms to perform coordinated and purposeful movement
	Head and trunk use	Ability to use hand and trunk to perform coordinated and purposeful movements
	Hand use	Ability to use wrists and fingers in coordinated and purposeful movements
	Whole body postural maintenance	Ability to maintain the body in a standing or sitting positions
	Whole body change of position	Ability to transition to and from lying, standing or sitting positions
	Whole body mobility	Ability to move oneself across space
	Manual material handling	Ability to lift, handle, and transport objects of various weights and sizes
Cognitive-intellectual	Consciousness	Ability to be awake and alert
	Orientation	Ability to identify oneself in relation to location and time
	Attention/concentration	Ability to focus on task to completion, managing symptom distractions
	Numerics	Ability to understand and apply numerical concepts
	Language and communication	Ability to communicate using language and alternate modes
	Memory	Ability to store and retrieve information
	Learning	Ability to develop conceptual, abstract, and practical knowledge, skills, and abilities

(continued)

TABLE 22.3 *Continued*

Domain	Conceptual Factor	Ability within the Worker Role and Work Environment
Interpersonal and emotional	Interpersonal interaction	Ability to interact constructively with others in the workplace
	Behavior modulation	Ability to modulate behavior associated with affective experience
	Tolerance for distractions	Ability to maintain activity despite distractions
	Adaptability	Ability to adapt to changes within the workplace
Vocational	Task performance	Ability to begin, follow through, and complete goal-directed activities
	Planning and organizing work	Ability to plan sequence, prioritize, and organize tasks
	Problem-solving and decision-making	Ability to adjust one's behavior according to changing contingencies
	Dependability	Ability to adhere to a structured work schedule
	Work safety	Ability to perform work in a manner that is safe for self and others
	Travel	Ability to travel to and from work and around work place

Source: From Matheson, L. 2003. The Functional Capacity Evaluation. In: Andersson, G., Demeter, S., Smith, G. (eds.) *Disability Evaluation*, 2nd ed. American Medical Association and Mosby Yearbook, Chicago, IL, pp. 748–768.

Criterion validity is the most important type of validity and should be assessed whenever feasible. Criterion validation data should indicate that a test significantly predicts (predictive validity) or correlates (concurrent validity) with important elements of job performance. If a "gold standard" of job performance exists, criterion validity should be assessed for those workers or applicants among whom the instrument will be used (Jette, 1998).

If the criterion validity cannot be assessed for a testing procedure, the construct validation approach should be used. The construct validation approach links theoretical and empirical relationships. Job analysis or professional judgment can indicate the key constructs (e.g., climbing a ladder) required for successful job performance, and the construct validation study indicates whether the selection instrument measures the same construct. Typically, correlation analysis is used to indicate to what extent intruments describe the same (convergent validity) or different (divergent validity) construct. For example, the ability to climb a ladder with a load could be hypothetized to correlate positively with trunk and leg strength, but not with hand dexterity. Another form of construction validation approach is the known-group method, where two or more groups should differ in a hypothetical manner assumed beforehand. For example, experienced firefighters should have better results than applicants when climbing a ladder with a load. Sensitivity and specificity determinations are also a form of construct validity, by which the study indicates, for example, how well a test classifies persons in poor and good performers.

Factor analysis is often used by psychologists to develop tests. It reduces several correlated variables to a smaller number of factors. The taxonomic approach of Fleishman and Reilly (1992) is partly based on factor analysis, and has been used create DOT and O*NET. The New York Fire Department developed tests in 1982 using the constructs from Fleishman's work (also in DOT). The tests were ruled discriminatory against women in the court, because strength, gross body equilibrium, and stamina were not shown to be related to the work of firefighters (Jackson, 1994). Thus, test developers should always try to use the criterion validity approach in their studies.

22.3   Assessment of the Prerequisites of Physical Work Ability

22.3.1   When Are the Assessments Needed?

There are a number of jobs whose demands include unavoidable health and safety risks. In these physically demanding high-risk jobs a worker with low physical work capacity is considered to have a high individual risk, and the assessment of physical work ability is particularly important (Taylor and Groeller, 2003). The jobs need to have workers who have capacities to control or cope with the risks, and the significance of individual abilities is strongly underlined when fitting the worker to the job. The assessment of physical work ability aims at preventing a worker's overstrain and improving safety at work (Shephard and Bonneau, 2002). The necessity of tests for the assessment of prerequisites of physical work ability is commonly agreed, and the use of tests has been required by employers, workers, and civil right organization (Jackson, 1994; Gonsoulin and Palmer, 1998; Shephard and Bonneau, 2002; Taylor and Groeller, 2003). The need for tests has been increased by the enrollment of women for jobs where the majority of workers have traditionally been men such as firefighters, paramedic workers, and police officers. In the recruitment phase for a job the tests may also serve as an orientation for the job and as learning aids. The tests help a worker to realize physical capacities that are required by the job (Shephard and Bonneau, 2002).

22.3.2   Types of Assessment and Interpretation

The worker selection for physically demanding jobs may be performed purely according to the subjective criteria without considering the validity and reliability of the selection process, but this approach may be very unfair for the applicants. More commonly the selection is carried out using different tests, which assess individual characteristics and capacities that are supposed to be necessary for successful work performance. The applied tests are typically tests of physical performance or simulations of work-relevant activities. The physical performance tests assess cardiorespiratory capacity, muscle strength and endurance, and motor coordination (Hogan, 1991). Previously mentioned validation approaches are important to indicate how well the measured constructs relate to the actual demands of work tasks.

The determination of cut scores aims at the prevention of overstrain and the improvement of safety at work (Shephard and Bonneau, 2002). The emphasis of the reference values has to be on the physical capacities, which are critical for the job performance (Glendhill and Jamnik, 1992; Jamnik and Glendhill, 1992; Pohjonen, 2001b; Shephard and Bonneau, 2002). It is supposed that in the determination of reference values work tasks and job demands are equal and independent of a worker's gender and age. Therefore, the reference values of the tests need to be as equal as possible for men and women with different ages (Chalal et al., 1992; Shephard and Bonneau, 2002). However, the reference values should as far as possible reflect the differences of physical capacity due to gender (Shephard and Bonneau, 2002).

22.3.3   Tests of Physical Work Ability

Physical performance or job-related tests can be used for the assessment of workers' prerequisites to cope with the demands of physically demanding safety jobs, such as firefighters, paramedic workers, and police officers. The reason for testing is that these jobs often include plenty of heavy dynamic muscle work, manual materials handling, and peak load situations (Gamble et al., 1991; Chahal et al., 1992; Jamnik and Gledhill, 1992; Jackson, 1994; Lusa, 1994; Soininen, 1995; Colledge et al., 1999; Shephard and Bonneau, 2002; Taylor and Groeller, 2003).

At their best job-related tests simulate as accurately as possible the key activity or activities of the job (Chahal et al., 1992; Gledhill and Jamnik, 1992; Jamnik and Gledhill, 1992; Taylor and Groeller, 2003). Occupationally relevant test batteries have been developed only for a few occupations (Jackson, 1994; Shephard and Bonneau, 2002; Taylor and Groeller, 2003). In Finland, a job-related test drill has been developed for firefighters that evaluates their cardiorespiratory capacity, muscle fitness, and motor coordination. The test drill includes five different tasks, which are carried out with fire protective clothing and a self-contained breathing apparatus (SCBA). The tasks are often required to be done in smoke-diving operations (entry into a smoke-filled space) (Louhevaara et al., 1994). The reliability of the test drill for firefighters was high ($r = 0.88$) (Lusa, 1994). The cardiorespiratory capacity of firefighters can be assessed with a treadmill test carried out with the fire protective clothing and SCBA used in smoke-diving operations (Lusa et al., 1993).

Commonly, physical performance tests have been used to measure a single construct of physical capacity. Such tests are, for instance, the submaximal incremental bicycle-ergometer test (Andersen et al., 1971; ACSM, 2000) and a 2-km walking test (Oja et al., 1991) for assessing cardiorespiratory fitness. Muscle strength and endurance have been evaluated with various laboratory tests or job-related tests in the field (Pohjonen, 2001b,c). In Finnish health care, the test battery of Alaranta et al. (1994) has been used most frequently. Its reference values are based on the results obtained from the workers of the City of Helsinki aged 35–54 yr ($n = 508$) in 1990. Recently, a health-related fitness test battery was launched (Suni et al., 1996). It also consists of tests for assessing flexibility and the control of movements. The test battery is based on the results obtained from the population sample of the City of Tampere ($n = 500$). Moreover, occupationally targeted test batteries have been developed, for instance, for firefighters (Lusa, 1994) and police officers (Smolander et al., 1985; Soininen, 1995; Shephard and Bonneau, 2002).

The cardiorespiratory fitness of firefighters, paramedic workers, police officers, and security guards has been measured with both a submaximal bicycle-ergometer test (Louhevaara, 1985; Gamble et al., 1991; Soininen, 1995; Laine et al., 2004) and a treadmill test (Gamble et al., 1991; Lusa et al., 1993). In addition, a running test has been developed in which a distance of 3.2 km was run as fast as possible using the sport wear and sneakers (Knapik et al., 1999) or the work clothing, boots, and the belt of work equipment (Laine et al., 2004). Also, the Cooper 12-min walking-running test has been used for the evaluation of the cardiorespiratory capacity of paramedic workers (Cooper, 1968; von Restoff, 2000).

The muscle strength and endurance of firefighters, paramedic workers, police officers, and security guards have been evaluated with a number of various physical performance tests (Gamble et al., 1991; Lusa, 1994; Soininen, 1995; Knapik et al., 1999; von Restroff, 2000; Laine et al., 2004). Handgrip strength has been measured most often using different dynamometers (Gamble et al., 1991; Knapik et al., 1999; von Restroff, 2000). Leg strength has been assessed with isokinetic equipment (Gamble et al., 1991), and static lifting strength with a dynamometer (von Restroff, 2000). The dynamic strength of trunk, leg, or arm muscles was assessed with various repetitive physical performance tests (Gamble et al., 1991; Lusa, 1994; Soininen, 1995; Knapik et al., 1999; von Restroff, 2000; Laine et al., 2004).

The control of body movements may be determined by single capacity tests, which evaluate, for instance, reaction time, balance, agility, or motor coordination. The balance has been evaluated with both functional/dynamic (Pohjonen, 2001b; Rinne et al., 2001; Laine et al., 2004) and static tests (Takala and Viikari-Juntura, 2000). The poor reliability and validity are often the problems of these tests (Punakallio, 2004).

22.4 Selection of Safety Workers in Pre-Employment Screening

22.4.1 Firefighters

22.4.1.1 Physical Job Demands

In the selection of firefighters, their occupational health and safety issues should be well recognized due to potential health risks in firefighting and rescue tasks. Firefighters face high physical and psychological demands in operative tasks. They are also exposed to risks of cardiovascular diseases related to hot work conditions and pulmonary diseases and cancer due to the release of toxic substances in fires, and, in addition, to noise-induced hear loss (Melius, 2001).

Professional firefighters ($n = 200$) rated the firefighting and rescue tasks according to their job demands on physical capacity in terms of cardiorespiratory capacity, muscle strength and endurance, and motor coordination (flexibility, agility, balance). Over three fourths of the firefighters (79%) rated smoke-diving with the use of fire protective clothing and SCBA as the task that demanded most cardiorespiratory capacity. Almost half of the firefighters (43%) considered that clearing of passages with heavy manual tools required the highest demands on muscle strength and endurance. About three fourths of the firefighters (72%) felt that the need for motor coordination was the greatest in firefighting and rescue operations on the roof. During the past 5 yr, 54 to 71% of the firefighters had carried out smoke-diving, clearing tasks, and roof work at least four times a year (Lusa et al., 1994).

22.4.1.2 Demands on Physical Work Ability

Several studies have recommended that a fire fighter should have a maximal oxygen consumption of 2.7 to 3.0 l/min/kg or 34 to 45 ml/min/kg of the body mass (Lemon and Hermiston, 1977; Sothmann et al., 1990; Louhevaara et al., 1994). In Finland, the Guide for Smoke-Diving (1991, 2002) gives a recommendation for a four-stage scale to classify firefighters' maximal oxygen consumption (Table 22.4). Male and female firefighters in the age range of 20 to 63 yr are considered to have a sufficient cardiorespiratory capacity for smoke-diving tasks when they attain a result of 3.0 l/min/kg or 36 ml/min/kg) in the tests for maximal oxygen consumption (Lusa, 1994).

In different firefighting and rescue operations with fire protective clothing and SCBA, such as smoke-diving, fire-suppression, ladder climbing, rescuing a victim, dragging a hose, and raising a ladder, the mean oxygen consumption levels were 2.1 to 2.8 l/min (Lemon and Hermiston, 1977; Louhevaara, 1985; Sothmann et al., 1990; Lusa, 1994). During peak loads the mean oxygen consumption was 3.8 l/min and the heart rate 180 beats/min in young and healthy firefighters (Lusa, 1994). In the heat, the effective regulation of body temperature is necessary for carrying out firefighting and rescue tasks without potential health hazards (Ilmarinen et al., 1996; Smith et al., 1996; Smith and Petruzzello, 1998; Richardson and Capra, 2001).

The biomechanical features of a simulated rescue-clearing task, in which a 9-kg power saw was lifted from the floor up to the ceiling level, were studied in seven young and six older firefighters by Lusa et al. (1991). The maximal muscular capacity of the firefighters did not differ with respect to their maximal isokinetic muscle strength or with respect to the results of repetitive muscle performance tests. In the task the mean dynamic compression force at the disc of L5/S1 was 6228 N. The peak torque for the back and knee extension were, on the average, 242 and 120 N m, respectively. The peak values corresponded to over 90% of their maximal isokinetic muscle strengths. The results showed that lifting and handling of a heavy power saw produced a high load on musculoskeletal system.

TABLE 22.4 Tests and Their Classification for Assessing Male and Female Firefighters' Cardiorespiratory Fitness and Muscular Performance

Test	Classification			
	Poor	Moderate	Good	Excellent
VO$_2$ max[a]				
l/min	≤2.4	2.5–2.9	3.0–3.9	≥4.0
ml/min/kg	≤29	30–35	36	≥50
Bench press (45 kg) (reps/60 sec)	≤9	10–17	18–29	≥30
Sit-up (reps/60 sec)	≤20	21–28	29–40	≥41
Squatting (45 kg) (reps/60 sec)	≤9	10–17	18–26	≥37
Pull-up (max. reps)	≤2	3–4	5–9	≥10

Note: Maximal oxygen consumption = *VO$_2$* max.
[a] Bicycle-ergometer or treadmill test.

Recent results by Punakallio et al. (2003) showed that the use of fire-protective clothing and SCBA, in particular, significantly impaired both postural and functional balance, and more negatively the balance of older firefighters than that of younger ones.

22.4.1.3 Assessment of Physical Work Ability

Gledhill and Jamnik (1992) developed a three-phase test procedure that could be applied for evaluating physical work capacity of firefighter applicants in pretraining screening and in the follow-up of the physical work capacity of firefighters. In the first phase, the applicants went through a health examination. In the second phase, the applicants performed a job-related test battery comprising seven key tasks. The test battery was carried out as fast as possible and a fixed maximal time was given for each task, except one task without time limit. In the third phase, the applicants performed tests for maximal oxygen consumption, dynamic strength and endurance of trunk flexors, and spine mobility. The reference value for the maximal oxygen consumption was 45 ml/min/kg, which was considered as the minimum for successful performance of fire and rescue tasks (Davis and Dotson, 1987). The reference values for muscle performance and mobility were adjusted according to the Canadian population values with respect to gender and age. The maximal score of the three-phase tests was 25. At maximum, it was able to have five points from the health examination, ten points from the test battery, and ten points from the performance tests (Gledhill and Jamnik, 1992).

A job-related test drill has been developed for Finnish firefighters (Louhevaara et al., 1994; Lusa, 1994). It simulates the physiologically heaviest work activities often required by smoke-diving operations, and the purpose is to assess the circulatory strain during the tasks with heart rate measurements. The test drill with fixed maximal working time of 14.5 min consists of five activities (walking and carrying, climbing/ascending stairs, hammering, moving over and under bars, and hose rolling) done with full personal protective clothing and SCBA of 25.5 kg (Figure 22.2). The test drill has to be carried out using a normal work pace within a fixed time. This is exceptional as compared to other job-related tests or test drills that are required to be performed as fast as possible (Misner et al., 1989; Gledhill and Jamnik, 1992;

FIGURE 22.2 Hammering is one task of the job-related test drill for firefighters.

Shephard and Bonneau, 2002). The test drill is submaximal performance for firefighters having a good cardiorespiratory capacity (Louhevaara et al., 1994). The correlation coefficient between repeated assessments in the test drill was, on the average, 0.88 (Lusa, 1994). A new version of the test drill is under development. It is physically more demanding, and the work pace is free. The passing of the drill within the allowed time of 12.5 min requires at least a maximal oxygen consumption of 3.0 l/min (Punakallio et al., unpublished data).

According to Williford et al. (1999), a test battery consisting of three items (1.5-mile run, fat-free body weight, and the number of pull-ups) had the best predictive value for physical performance of 91 firefighters in simulated firefighting tasks, which were stair climbing, hose forcible entry, hose advance, and victim rescue. In addition, several other physical performance tests showed significant associations with the work-related physical performance (Schonfeld et al., 1990; Williford et al., 1999).

The physical capacity tests in Table 22.4 are mainly designed to follow-up the firefighters' physical work capacity in Finland. These tests are also used for the selection of applicants for the training courses of the Emergency Service Institute of Finland. In the selection of applicants for basic training courses, the required minimum level has been between "good" and "excellent" in each test (Table 22.4). However, validation studies are not available for these tests and for the classification in terms of firefighter performance.

Body mass index (BMI) has been proved to be a useful characteristic for screening pre-employment status of firefighters in terms of health and physical capacity for duties. BMI showed a significant inverse association with blood pressure, maximal oxygen consumption, energy expenditure at work, and total cholesterol (Kales et al., 1999; Clark et al., 2002).

22.4.2 Paramedic Workers

22.4.2.1 Physical Job Demands

Paramedic (ambulance) work includes both physical and psychosocial demands related to emergency rescue and health care. Seriously injured or ill patients must often be lifted and carried without adequate equipment resulting in an increased risk for musculoskeletal overstrain and injuries. Also, early retirement due to musculoskeletal, cardiorespiratory, and mental reason is common among ambulance workers (Boreham et al., 1994; Rodgers, 1998a,b).

Vehmasvaara (2004) studied the job demands of Finnish paramedic workers with a questionnaire, and measured the physical load by simulating the most demanding physical tasks. In the questionnaire study, the respondents were 145 men and 24 women with a mean age of 36 yr. They worked as paramedics in fire and rescue departments and private ambulance services.

According to Vehmasvaara (2004), the paramedics working in the fire and rescue departments fetched patients most commonly from flats (83% of the cases) whereas those working in the private service fetched their patients mostly from houses or row houses (76% of the cases). About four fifths of the respondents (81%) estimated that the average distance of carrying a patient on a stretcher was 10 to 50 m. Almost every fifth (17%) considered the distance to be below 10 m. A few (2%) estimated that the distance was 50 to 100 m. The respondents perceived that the most demanding physical task was carrying the patient on a stretcher (57% of the respondents), the second demanding task was lifting and transferring the patient (37%), and the third was the carrying of equipment (2%).

In Sweden, eight work shifts of ambulance workers were followed-up. Station work occupied 30% of the working time. The corresponding values were 40% for ordered calls with low and medium priority, and 20% for the high-priority calls (Barnekow-Berqvist et al., 2002).

During the simulated tasks of paramedics the average heart rate (HR) ranged from 119 to 161 beats/min and the average circulatory strain was moderately high (63% of the maximal HR) (Vehmasvaara, 2004). The strain was highest during patient carrying being 80% of the maximal HR, and the rating of perceived exertion (Borg, 1970) was on the average 12.8 on the scale from 6 to 20. During patient carrying the average circulatory strain of female subjects (84% of the maximal HR) was higher that for male subjects (77% of the maximal HR). The peak HR averaged between 109 and 190 beats/min in the simulations.

22.4.2.2 Demands on Physical Work Ability

In the questionnaire study by Vehmasvaara (2004), the respondents were asked to rate the dimensions of physical capacity, which were needed mostly in the most demanding physical tasks of the paramedic work. Muscle strength (33% of the respondents), motor coordination and reaction ability (27%), and cardiorespiratory capacity (19%) were considered as the most important physical constructs.

Barnekow-Berqvist et al. (2002) simulated a work task where a patient was carried on a stretcher with a total weight of 92 kg. Fatigue development during the task could be explained (34 to 50% for men and 73 to 89% for women) by the level of maximal oxygen consumption, maximal static lifting strength, and balance.

22.4.2.3 Assessment of Physical Work Ability

Vehmasvaara (2004) developed a jobrelated test drill for assessing the prerequisites of physical work capacity for paramedic workers. The test drill is based on the results of a questionnaire study and work simulations (Vehmasvaara, 2004), and is used in the screening of applicants for paramedic education and training. The test drill is advised to be carried out using "a normal work pace" and correct work techniques. The performance can be followed with continuous monitoring and recording of HR and the rating of perceived exertion. The test drill includes three tasks carried out without breaks as follows:

1. *Carrying medical equipment.* The task simulates the situation in which a paramedic worker carries medical equipment needed in emergency care and climbs up stairs to a patient. In the beginning of the task the applicant takes two barbells, each weighing 12 kg, in his or her right and left hands. Then, the applicant walks forwards a distance of 7 m, and steps over three thresholds of 20 to 25 cm. The height of the fourth threshold is 14 cm, where the applicant performs 30 steps up and down.
2. *Cardiopulmonary resuscitation.* The second task simulates activities needed in cardiac emergency. After the first task is completed the applicant walks 8 m to the Anne Modular System Training Doll, PPE (Laerdal Oy, Helsinki, Finland), and starts pushing the chest 100 times/min for 4 min. The electronic unit of the Anne Doll controls the frequency and depth (force) of the pushes. After this task the applicant performs the Nine Hole Peg Test of Finger Dexterity (Mathiowetz et al., 1985).
3. *Transferring a patient on a stretcher.* The third task simulates the lifting and carrying of a patient. The applicant walks 6.5 m from the Anne Doll to a push stretcher weighing 39 kg (F035 Nordic Ambulance Stretcher). The other end of the stretcher is set down and the handles for carrying are 15 cm above the floor. A 40-kg weight is placed on that end of the stretcher to simulate the weight of a patient and equipment. The applicant lifts the stretcher up, and moves with side steps around a circle (diameter 3.6 m). Then the task is repeated in the reverse direction (Figure 22.3). Between circles the stretcher is layed down, and the body straightened. After the circles a bench (height 14 cm) is placed in front of the applicant and he or she steps up and down 20 times while holding the stretcher. After a short rest, the applicant walks backwards to the starting point (6.5 m) with the stretcher.

Vehmasvaara (2004) investigated the associations between physical capacity tests and the performance in the simulated physically most demanding tasks and in the job-related test drill. Handgrip strength, maximal strength of the leg extensor, local endurance strength of the leg extensors, and maximal oxygen consumption predicted significantly the level of circulatory strain both in the simulated tasks and job-related test drill. Also the physical fitness index correlated ($r = 0.61$ and 0.63) with the circulatory strain in the simulated paramedic tasks and job-related test drill. Moreover, Vehmasvaara (2004)

FIGURE 22.3   Transferring a patient on a stretcher is one task of the job-related test drill for paramedics.

correlated the average circulatory strain measured in the simulated paramedic tasks and job-related test drill. The correlation coefficient was 0.51.

Lagasse and Turcotte (1997) developed a job-related test battery for paramedic workers who operate in ambulances. The test battery included seven different activities, which imitated activities necessary in the paramedic work. The first activity includes running 150 m and stair climbing while carrying a load of 15 kg. It is followed by pushing a stretcher for 150 m. The third activity comprises stair climbing and carrying a load of 40 kg to a higher floor, which is followed by an activity that demands finger dexterity. The fifth activity includes stair climbing and pulling a two-wheel stretcher of 70 kg, which is followed by cardiopulmonary resuscitation. Finally, the two-wheel stretcher of 70 kg is pulled down the stairs. The test battery is carried out as fast as possible. The developed test battery was considered to describe well actual tasks of a paramedic worker, and also to predict his or her strain in the tasks (Lagasse and Turcotte, 1997).

22.4.3   Police Officers

22.4.3.1   Physical Job Demands

In operative police work, it is very difficult to affect job demands or to change the work environment. The health risks can be, however, decreased with continuous qualified training and adequate work equipment. Soininen (1995) confirmed the previous observations of Smolander et al. (1985) on the physical job demands associated with Finnish police work. The mean level of physical work demands was low, and police officers spent about 60% of their working hours in a sitting position. The most common work places were an office, a patrol car, and an alarm center. However, the questionnaire revealed that almost all police officers had a few physically extreme peak loads at their work every year. The peak loads were common when the police officer had to catch or transport criminals, and intoxicated or mentally ill persons. The situations leading to a physical peak load were usually very sudden and unexpected in nature. They most often demanded a high level of muscular strength and endurance (carrying, lifting, pressing, wrestling, running, wrenching), often in unfavorable (dark, cold, or cramped space) environmental conditions. Today clients are often abusers of drugs whose behavior may be extremely violent and unpredictable (Matano et al., 2002). Physical endurance is often needed in search tasks, for example, in forests and swamps.

22.4.3.2 Demands on Physical Work Ability

In order to carry out physical peak load situations successfully, the police officer needs to have a sufficient level of physical capacity. Smolander et al. (1984) suggested that the necessary level for those police officers operating in the street should be at least the same as in typical clients who were 20- to 40-yr-old men in Finland. Soininen (1995) concluded that the sufficient level of each physical capacity dimension (cardiorespiratory capacity, muscle strength and endurance, and motor coordination) should be higher than the mean gender- and age-related reference values obtained from the population studies.

22.4.3.3 Assessment of Physical Work Ability

Faranholz and Rhodes (1986) developed a job-related test battery for Canadian police officers. The tasks of the test battery simulated typical situations of catching and arresting of suspects. The successful passing of the test battery demanded good cardiorespiratory (aerobic and anaerobic) capacity and motor skills such good muscular coordination, agility, and speed.

There are no consistent practices in the assessments of physical work ability with performance tests. Cardiorespiratory capacity has been assessed with a submaximal bicycle-ergometer test, a step test, a 12-min walking–running test, and various long-distance running tests. There has been a large variation in the assessments of muscle strength and endurance. The common tests have been bench press, pull-up, sit-up, handgrip, squatting, and vertical jump. The capacity tests are often supplemented with the measurements of height and body mass, and sometimes the evaluations of the mobility of joints and low back (Moulson-Litchfied and Freedson, 1986; Shephard, 1991; Rhodes and Franholtz, 1992; Shephard and Bonneau, 2003). Smolander et al. (1985) recommended that the test battery for the evaluation of physical fitness of police officers should consist of the assessments of cardiorespiratory capacity (maximal oxygen consumption), muscle strength and endurance, and flexibility. The maximal oxygen consumption could be assessed with a maximal or submaximal bicycle-ergometer or treadmill tests. Field tests such as repetitive sit-up, pull-up, and push-up were recommended for the evaluation of muscle strength and endurance, and flexibility.

Based on the studies of Smolander et al. (1984, 1985) and Soininen (1995), the Finnish Ministry of Internal Affairs prepared a guide for testing physical work capacity of police officers. The cardiorespiratory capacity is determined with an indirect measurement of the maximal oxygen consumption using a submaximal bicycle-ergometer test or 2-km walking test; the strength and endurance of the trunk and arm muscles are evaluated with a sit-up test and a dynamic endurance and strength test for upper arms with 5- or 10-kg barbells. The determination of the flexibility of the hip and low back is done with a sit and reach test (Laukkanen, 1993; Pollock and Wilmore, 1994). Soininen (1995) stated that the assessment of physical capacity should be done with reliable and feasible tests possible to complete in the occupational health services.

22.4.4 Security Guards

22.4.4.1 Physical Job Demands

There are no scientific reports available in ergonomics or work physiology literature about the job demands and physical work capacity of security guards. In Finland, the study of Laine et al. (2004) was recently completed and they reported a 4-yr process for promoting the health and work ability of security guards in a small enterprise. A part of their study focused on the selection of relevant and feasible tests for assessing the prerequisites of the physical work capacity of the guards.

In Finland, the number of security guards is about 7000, and every fourth of them is a woman. Usually security guards are young, and they have a low level of education and training for the occupation. The length of the career often remains quite short. There are no commonly approved health or physical capacity qualifications for the security guards. However, they face the same risks and dangers in the field as, for instance, police officers: violent clients usually due to abuse of drugs, unpredictable and poor work conditions and situations, shift work and night work, and shortcomings in their protective equipment and weapons. During the work shifts the most common operative tasks of security guards are the catching of a suspect by running and controlling him or her by using arms and trunk. When the guard struggles with the suspect it is important to avoid falling down as far as possible. When the guard is down the suspect may use his or her legs for kicking and that may led to serious negative health consequences (Laine et al., 2004).

22.4.4.2   Demands on Physical Work Ability

In their job security guards face demands for muscle strength and endurance, cardiorespiratory capacity, and dynamic balance of the body. The importance of muscle capacity is evident but also cardiorespiratory capacity and balance are necessary in operative tasks (Laine et al., 2004).

22.4.4.3   Assessment of Physical Work Ability

Laine et al. (2004) suggested that the assessment of security guards' physical work ability would include two job-related tests. First, a 1.5-mile walking–running test with work clothing and equipment of 7 kg that consists of light protective clothing, a bullet vest, a weapon belt, and protective boots. The test is done on an indoor track. The performance time and HR are registered in the test. According to the time and HR the average and maximal oxygen consumptions are estimated and related to the mass of the body and equipment. The performance is accepted when the estimated maximal oxygen consumption is at least 36 ml/min/kg, which is also used as the qualified criteria for firefighters' smoke-diving operations (Lusa, 1994). The second test is a functional (dynamic) balance test. In the functional balance test, the applicant has to walk barefooted forwards and backwards across a wooden plank that is 2.5-m long, 9-cm wide, and 5-cm thick. He or she starts the test from footprints marked on the floor at one end of the plank. He or she walks forward to the marked 0.5-m-long mid-area of the plank, where he or she turns around 180° and walks backwards to the footprints at the opposite end of the plank. The task is then immediately repeated from the opposite end, with the applicant returning in reverse direction to the starting point. In the test the performance time and errors (touches to the floor) are registered. The accepted performance time may not exceed 12 sec, which can be considered the average result for men aged 20 to 55 yr (Punakallio et al., 2003).

According to Laine et al. (2004), suitable physical capacity tests for evaluating security guards' muscle strength and endurance are the sit-up test (Pollock and Willmore, 1990) and the pull-up test (Fleishman, 1964). The time limit is 60 sec, and the accepted performance requires at least 38 sit-ups/min. The pull-up test is done with a reverse grip without the time limit. The acceptable performance requires at least six pull-ups. The minimum limits require "good" performance according to the classification for firefighters (Lusa, 1994).

22.5   Concluding Remarks

22.5.1   Physical Fitness and Work Ability

The firefighters's job demands on physical work ability are occasionally very high during the entire occupational career. Their cardiorespiratory and muscular fitness should be tested regularly for guaranteeing adequate work capacity for physically extreme operations. The use of fitness tests is necessary in the planning and carrying out preventive measures for maintaining health and work ability of firefighters. The most effective and necessary measure, however, is regular physical training, which should be done either during the working hours or leisure time (Lusa, 1994).

Vehmasvaara (2004) stated that good physical fitness is one of the most important factors that influences work ability of the paramedics, and it can be considered a necessary tool in the job. The assessments of physical fitness must be based on analysis of the actual physical demands of the job. In the physical fitness of paramedics special attention must be paid to muscle strength and endurance of the arms and legs. Cardiorespiratory fitness is also needed when carrying the patient on a stretcher. The test drill developed by Vehmasvaara (2004) is a suitable screening procedure for applicants to degree programs in emergency care, and the assessments should be continued regularly throughout the occupational career of the paramedics.

Due to the strict selection criteria young police officers are healthy and physically fit when starting their careers. However, the life style of many middle-aged police officers is sedentary, and often associated with various health problems. These factors accelerate the unavoidable age-related reduction of physical fitness and, particularly, after the age of 40 yr physical capacity as well as work ability of police officers often decline rapidly. Obesity, elevated blood pressure, high serum lipid levels, and musculoskeletal disorders and diseases are the most common health problems that also reduce the physical fitness needed to carry out operative tasks without excessive strain and increased hazards on health or safety (Soininen, 1995).

According to Laine et al. (2004), the test battery developed for assessing the security guards' physical work ability is relevant and feasible both in pre-employment screening and in the follow-up of physical work capacity. The set "acceptable" limits of the tests are very strict and are needed to modify when older (over 30 yr) and female guards are evaluated. The tests have been used for 2 yr in a small enterprise of security guards and the results and experiences obtained have been positive.

22.5.2 Reliability and Validity

When various capacities or job-related tests are applied for worker selection, they should have acceptable reliability and validity (relevance) (Innes and Straker, 1999a; Shephard and Bonneau, 2003). Good reliability has been shown for the determination of the maximal oxygen consumption based on submaximal or maximal test protocols (Nevala-Puranen, 1997). The repetitive tests of physical fitness, handgrip, finger dexterity, and functional balance have good or excellent reliability (Alaranta et al., 1994; Suni et al., 1996; Pohjonen, 2001c; Rinne et al., 2001; Haward and Griffin, 2002; Punakallio, 2004). In addition, the reliability of the test drill for firefighters is good (Lusa, 1994).

The validity of single muscle fitness tests based on actual job demands is difficult to prove. The content and construct validities of various job-related tests are good if based on careful job and expert analysis. Thus, the criterion validity of the measurement of maximal oxygen consumption is good for all tasks that need intensive dynamic muscle work with large muscle masses such as smoke-diving and running (Lusa, 1994; Soininen, 1995; Laine et al., 2004).

22.5.3 Gender and Age

When the physical performance or job-related tests are relevant they should guarantee an acceptable work performance in physically demanding jobs that are equal for men and women of all ages (Pohjonen, 2001a). This leads to the situation where passing tests is easier for men than women, and for younger than older individuals (Davis and Dotson, 1987). For instance, about a half of the Finnish firefighters aged over 50 yr have serious problems in passing all physical fitness tests, and only one female firefighter is carrying out operative fire and rescue tasks. On the other hand, there are many female police officers, paramedics, and security guards who carry out operative tasks in the street.

The pension age is 65 yr in many safety jobs. The selection of safety workers based on physical performance and job-related tests seems to guarantee an acceptable level of physical fitness up to the age of 40 yr. After that there should be alternative respectful career paths for older workers who could not pass the tests. The reasons for the negative test outcome usually relate to various disorders and diseases, which prevent regular physical exercise needed for maintaining sufficient physical fitness in the middle-aged safety workers. Intensive and regular fitness training alleviates the evident decline of cardiorespiratory and muscular fitness due to age, and may maintain the physical work ability at the acceptable level for the age range of 55 to 60 yr. After that most of the safety workers have problems in meeting job demands in operative tasks.

Safety workers should be assessed by nondiscriminatory tests and standards at pre-employment screening, and when following-up their work ability (Shephard, 1991). However, certain differences exist in physical fitness between men and women, like gender differences in anthropometrics, cardio-respiratory capacity, haemoglobin levels, and muscular strength. Thus, the potential training effect for female safety recruits is likely to match that of male recruits. Therefore, equivalent criteria and capacity augmentation across gender should be provided for all recruits (Shephard and Bonneau, 2002).

The discrimination of the tests due to gender and age cannot be avoided in physically demanding high-risk jobs if the tests are relevant with respect to actual job demands. When the work safety is weighted against the discrimination aspect, the safety should be the first issue to be considered in worker selection.

References

ACSM. 2000. *ACSM Guidelines for Exercise Testing and Prescription*, 6th edition. American College of Sports Medicine. Lippincott Williams and Wilkins, Philadelphia.

Alaranta, H., Hurri, H., Heliövaara, M., Soukka, A., Harju, R. 1994. Non-dynamometric tests: reliability and normative data. *Scand. J. Rehabil. Med.* 26: 211–215.

Andersen, K.L., Shephard, R.J., Denolin, H., Varnauskas, E., Masironi, R. 1971. *Fundamentals of Exercise Testing*. World Health Organisation, Geneva.

Barnekow-Berqvist, M., Aasa, U., Ängquist, K.-A., Lyskov, E., Nakata, M., Johansson, H. 2002. Self-reported physical and psychosocial risk factors, physiological responses to job strain, and evaluation of individual factors that may facilitate heavy work tasks in ambulance work — implications for prevention. *Proceedings of the 34th Annual Congress of the Nordic Ergonomics Society.* 1–3 October, Kolmården, Sweden, pp. 144–149.

Borg, G. 1970. Perceived exertion as an indicator of somatic stress. *Scand. J. Rehabil. Med.* 2: 92–98.

Boreham, C.A., Gamble, R.P., Wallace, W.F., Cran, G.W., Stevens, A.B. 1994. The health status of an ambulance service. *Occup. Med. (Lond.)* 44: 137–140.

Burke, M. 1992. *Applied Ergonomics Handbook*. Lewis Publishers, Ann Arbor; CRC Press, Boca Raton, FL.

Chaffin, D.B., Andersson, G.B.J. 1991. *Occupational Biomechanics*, 2nd edition. John Wiley, New York, pp. 464–477.

Chahal, P., Lee, S., Oseen, M., Singh, M., Wheeler, G. 1992. Physical fitness and work performance standards: a proposed approach. *Int. J. Ind. Ergono.* 9: 127–135.

Clark, S., Rene, A., Theurer, W.M., Marshall, M. 2002. Association of body mass index and health status in firefighters. *J. Occup. Med.* 44: 940–946.

Colledge, A., Johns, R., Thomas, M. 1999. Functional ability assessment: guidelines for the workplace. *J. Occup. Environ. Med.* 41: 172–180.

Cooper, K. 1968. *Aerobics*. M. Evans and Company Inc., New York.

Davis, P.O., Dotson, C.O. 1987. Job performance testing: an alternative to age discrimination. *Med. Sci. Sports Exerc.* 19: 179–185.

Faranholz, D.W., Rhodes, E.C. 1986. Development of a physical abilities test municipal police officers in British Columbia. *Can. J. Appl. Sport Sci.* 11(abstract) .

Fleishman, E.A. 1964 *The Structure and Measurement of Physical Fitness*. Prentice-Hall, Englewood Cliffs, NJ.

Fleishman, E.A, Reilly, M. 1992. *Handbook of Human Abilities: Definitions, Measurements, and Job Task Requirements*. Consulting Psychologists Press, Palo Alto, CA.

Frings-Dresen, M.H.W., Sluiter, J.K. 2003. Development of a job-specific FCE protocol: the work demands of hospital nurses as an example. *J. Occup. Rehabil.* 13: 233–248.

Gallagher, S. 1999. Worker strength evaluation: job design and worker selection. In: Karwowski, W. Marras, W.S. Eds. *The Occupational Ergonomics Handbook*. CRC Press, Boca Raton, FL, pp. 279–380.

Gamble, R., Stevensen, A., McBrien, H., Black, A., Boreham, C. 1991. Physical fitness and occupational demands of the Belfast ambulance service. *Br. J. Ind. Med.* 48: 592–596.

Gaudino, E.A., Matheson, L.N., Mael, F.A. 2001. Development of the functional assessment taxonomy. *J. Occup. Rehabil.* 11: 155–175.

Gledhill, N., Jamnik, V. 1992. Development and validation of a fitness screening protocol for firefighter applicants. *Can. J. Sports Sci.* 17: 199–206.

Gonsoulin, S., Palmer, E. 1998. Gender issues and partner preferences among a sample of emergency medical technicians. *Prehospital. Disaster. Med.* 13: 34–40.

Guide for Smoke-Diving. 1991. Finnish Ministry of Internal Affairs, Helsinki.

Guide for Smoke-Diving. 2002. Finnish Ministry of Internal Affairs, Helsinki.

Hanson, M., Matheson, L., Borman, W. 2001. The O*NET occupational information system. In: Bolton, B. Ed. *Handbook of Measurement and Evaluation in Rehabilitation*, 3rd edition. Aspen Publishers, Gaithersburg, MD, pp. 281–309.

Haward, B., Griffin, M. 2001. Repeatability of grip strength and dexterity tests and the effects of age and gender. *Int. Arch. Occup. Environ. Health* 77: 111–119.

Hogan, J. 1991. Structure of physical performance in occupational tasks. *J. Appl. Psychol.* 76: 495–507.

Ilmarinen, R., Louhevaara, V., Griefahn, B., Kunemund, C. 1996. Thermal responses to consecutive strenuous fire-fighting and rescue tasks in the heat. In: Shapiro, Y., Moran, D.S., Epstein Y. Eds. *Environmental Ergonomics. Recent Progress and Environmental Frontiers*. ICEE 96. Freund Publishing House, London, pp. 295–298.

International Ergonomics Association (IEA). 2000. http://www.iea.cc/ergonomics/.

Innes, E., Straker, L. 1999a. Reliability of work-related assessments. *Work* 13: 107–124.

Innes, E., Straker, L. 1999b. Validity of work-related assessments. *Work* 13: 125–152.

Jackson, A. 1994. Pre-employment physical evaluation. *Exerc. Sports Sci. Rev.* 22: 53–90.

Jamnik, V., Gledhill, N. 1992. Development of fitness screening protocols for physically demanding occupations. *Can. J. Sports Sci.* 17: 222–227.

Jette, A. 1998. Desired characteristics of instruments to measure functional capacity to work. In: Wunderlich, G. Ed. *Measuring Functional Capacity and Work Requirements*. National Academy Press, Washington, DC, pp. 45–52.

Kales, S.N., Polychronopoulos, G.N., Aldrich, J.M., Leitao, E.O., Christiani, D.C. 1999. Correlates of body mass index in hazardous materials firefighters. *J. Occup. Environ. Med.* 41: 589–595.

King, P.M., Tuckwell, N., Barrett, T.E. 1998. A critical review of functional capacity evaluations. *Phys. Ther.* 78: 852–866.

Knapik, J., Harper, H., Crowell, H. 1999. Physiological factors in stretcher carriage performance. *Eur. J. Appl. Physiol.* 79: 409–413.

Konz, S. 1995. *Work Design. Industrial Ergonomics*, 4th edition. Publishing Horizons, Inc., Arizona, pp. 110–113.

Kraus, J.F. 2001. The effects of certain drug-testing programs on injury reduction in the workplace: an evidence-based review. *Int. J. Occup. Environ. Health* 7: 103–108.

Lagasse, P., Turcotte, F. 1997. Validity of requirements for ambulance technicians. *Med. Sci. Sports Exerc.* 29: S359(abstract).

Laine, K., Kolehmainen, M., Louhevaara, V. 2004. Development of the workplace health promotion enterprise of security guards. Work and human research reports 23. Safety book. *Action Program for Safety Occupations. Finnish Institute of Occupational Health*. Helsinki, pp. 33–44.

Laukkanen, R. 1993. Development and evaluation of a 2-km walking test for assessing maximal aerobic power of adults in field conditions. Kuopio University Publications, D. Medical Sciences 23, University of Kuopio, Doctoral dissertation.

Lemon, P.W. and Hermiston, R.T. 1977. Physiological profile of professional firefighters. *J. Occup. Med.* 19: 337–340.

Louhevaara, V., Tuomi, T., Smolander, J., Korhonen, O., Tossavainen, A., Jaakkola, J. 1984. Cardiorespiratory strain in jobs that require respiratory protection. *Int. Arch. Occup. Health* 55: 195–206.

Louhevaara, V. 1985. Effects of respiratory protective devices on breathing pattern, gas exchange, and heart rate at different work levels. University of Kuopio, Doctoral dissertation.

Louhevaara, V., Soukainen, J., Lusa, S., Tulppo, M., Tuomi, P., Kajaste, T. 1994. Development and evaluation of a test drill for assessing physical work capacity of firefighters. *Int. J. Ind. Ergon.* 13: 139–146.

Louhevaara, V. 1999. Job demands and physical fitness. In: Karwowski, W., Marras, W. Eds. *The Occupational Ergonomics Handbook.* CRC Press, Boca Raton, FL, pp. 261–273.

Louhevaara, V., Nevala, N. 2005. Ergonomics. In: Guidotti, T.L., Rantanen, J., Hua, F., Chan, G. Eds. *Basic Occupational Health.* World Health Organisation, Geneva (in press).

Lusa, S., Louhevaara, V., Smolander, J., Kinnunen, K., Korhonen, O., Soukainen, J. 1991. Biomechanical evaluation of heavy tool-handling in two age groups of firemen. *Ergonomics* 34: 1429–1432.

Lusa, S., Louhevaara, V., Smolander, J., Pohjonen, T., Uusimäki, H., Korhonen, O. 1993. Thermal effects of fire-protective equipment during a job-related exercise protocol. *SAFE J.* 23: 36–39.

Lusa, S., Louhevaara, V., Kinnunen, K. 1994. Are the job demands on physical work capacity equal for young and aging firefighters? *J. Occup. Med.* 36: 70–74.

Lusa, S. 1994. Job demands and assessment of physical work capacity of firefighters. Studies in Sport, Physical Education and Health, University of Jyväskylä, Jyväskylä, Doctoral dissertation. University Printing House.

Matano, R.A., Wanat, S.F., Westrup, D., Koopman, C., Whitsell, S.D. 2002. Prevalence of alcohol and drug use in a highly educated workforce. *J. Behav. Health Serv. Res.* 29: 30–44.

Matheson, L.N., Kaskutas, V., McCowan, S., Shaw, H., Webb. C. 2001. Development of a database of functional assessment measures related to work disability. *J. Occup. Rehabil.* 11: 177–199.

Matheson, L. 2003. The functional capacity evaluation. In: Andersson, G., Demeter, S., Smith, G. Eds. *Disability Evaluation*, 2nd edition. American Medical Association and Mosby Yearbook, Chicago, IL, pp. 748–768.

Mathiowetz, V., Weber, K., Kashman, N., Volland, G. 1985. Adult norms for the nine hole peg test of finger dexterity. *Occup. Ther. J. Res.* 5: 24–38.

Melius, J. 2001. Occupational health for firefighters. *Occup. Med.* 16: 101–108.

Misner, J., Boileau, R., Plowman, S. 1989. Development of placement tests for fire-fighting. A long term analysis by race and sex. *Appl. Ergon.* 20: 218–219.

Morland, J. 1993. Types of drug-testing programmes in the workplace. *Bull. Narc.* 45: 83–113.

Moulson-Litchfield, M., Freedson, P.S. 1986. Physical training programs for public safety personnel. *Clin. Sports Med.* 5: 571–587.

Nevala-Puranen N. 1997. Physical work and ergonomics in dairy farming. Effects of occupationally oriented medical rehabilitation and environmental measures. Studies in Sport, Physical Education and Health, University of Jyväskylä, Jyväskylä, Doctoral dissertation, University Printing House.

Oja, P., Laukkanen, R., Pasanen, M., Tyry, T., Vuori, I. 1991. A 2-km walking test for assessing the cardiorespiratory fitness of healthy adults. *Int. J. Sports Med.* 12: 356–362.

Pheasant, S. 1996. *Bodyspace. Anthropometry, Ergonomics and the Design of Work*, 2nd edn. Taylor and Francis, London, pp. 153–173.

Pohjonen, T. 2001a. Perceived work ability of home care workers in relation to individual factors in different age groups. *Occup. Med.* 51: 209–217.

Pohjonen, T. 2001b. Age-related physical fitness and the predictive values of fitness tests for work ability in home care work. *J. Occup. Environ. Med.* 43: 723–730.

Pohjonen, T. 2001c. Perceived work ability and physical capacity of home care workers. Effects of the physical exercise and ergonomic intervention on factors related to work ability. Kuopio University Publications D, Medical Sciences 260, Helsinki, Doctoral dissertation.

Pollock, M.L., Wilmore, J.H. 1990. Exercise in health and disease. *Evaluation and Prescription for Prevention and Rehabilitation.* W.B. Saunders Co., Philadelphia.

Punakallio, A. 2004. Trial-to-trial reproducibility and test-retest stability of two dynamic balance tests among male firefighters. *Int. J. Sports Med.* 25: 163–169.

Punakallio, A., Lusa, S., Luukkonen, R. 2003. Protective equipment affects balance abilities differently in younger and older firefighters. *Aviat. Space Environ. Med.* 74: 1151–1156.

von Restorff, W. 2000. Physical fitness of young women: carrying simulated patients. *Ergonomics* 43: 728–743.

Rhodes, E.C., Franholtz, D.W. 1992. Police officers' physical abilities tests compared to measures of physical fitness. *Can. J. Appl. Sport Sci.* 17: 228–233.

Richardson, J.E., Capra, M.F. 2001. Physiological responses of firefighters wearing level 3 chemical protective suits while working in controlled hot environments. *J. Occup. Environ. Med.* 43: 1064–1072.

Rinne, M., Pasanen, M., Miilunpalo, S. 2001. Test-retest reproducibility and inter-rater reliability of a motor skill test battery for adults. *Int. J. Sports. Med.* 22: 192–200.

Rodgers, L. 1998a. A five year study comparing early retirements on medical grounds in ambulance personnel with those in other groups of health service staff. Part I: Incidences of retirements. *Occup. Med.* 48: 7–16.

Rodgers, L. 1998b. A five year study comparing early retirements on medical grounds in ambulance personnel with those in other groups of health service staff. Part II: Causes of retirements. *Occup. Med.* 48: 119–132.

Rudy, T.E., Lieber, S.J., Boston, J.R. 1996. Functional capacity assessment: influence of behavioral and environmental factors. *J. Back Musculoskeletal Rehab.* 6: 277–288.

Schonfeld, B., Doerr, D., Convertino, V. 1990. An occupational performance test validation program for firefighters at Kennedy Space Center. *J. Occup. Med.* 32: 638–643.

Shephard, R. 1991. Occupational demand and human rights. Public safety officers and cardiorespiratory fitness. *Sports Med.* 12: 94–109.

Shephard, R., Bonneau, J. 2002. Assuring gender equity in recruitment standards for police officers. *Can. J. Appl. Physiol.* 27: 263–295.

Shephard, R.J., Bonneau, J. 2003. Supervision of occupational fitness assessments. *Can. J. Appl. Physiol.* 28: 225–239.

Singleton, W.T. 1989. *The Mind at Work: Psychological Ergonomics.* Cambridge University Press, Cambridge, pp. 88–93.

Smith, D.L., Petruzzello, S.J., Kramer, J.M., Misner, J.E. 1996. Physiological, psychophysical, and psychological responses of firefighters to firefighting training drills. *Aviat. Space Environ. Med.* 67: 1063–1068.

Smith, D.L., Petruzello, S.J. 1998. Selected physical and psychological responses to live-fire drills in different configurations of firefighting gears. *Ergonomics* 41: 1141–1154.

Smolander, J., Louhevaara, V., Oja, P. 1984. Policemen's physical fitness in relation to the frequency of leisure time physical exercise. *Int. Arch. Occup. Environ. Health* 54: 261–270.

Smolander, J., Louhevaara, V., Nygård, C.-H., Ylikoski, M. 1985. Job demands and assessment of physical working capacity in policemen's occupational health service. In: Brown, I.D., Goldsmith, R., Coombes, K., Sinclair, M.A. Eds. *Proceedings of the Ninth Congress of the International Ergonomics Association.* Taylor and Francis, Bournemouth, England, pp. 613–615.

Soininen, H. 1995. The feasibility of worksite fitness programs and their effects on the health, physical capacity and work ability of aging police officers. University of Kuopio, Kuopio, Finland, Doctoral dissertation.

Sothmann, M.S., Saupe, K.W., Jasenof, D., Blaney, J., Fuhrman, S.D., Woulfe, T., Raven, P.R., Pawelczyk, J.P., Dotson, C.O., Landy, F.J., Smith, J.J., Davis, P.O. 1990. Advancing age and the cardiorespiratory stress of fire suppression: determining a minimum standard for aerobic fitness. *Hum. Perfor.* 3: 217–236.

Suni, J., Oja, P., Laukkanen, R., Miilunpalo, S., Pasanen, M., Vuori, I., Vartiainen, T-M., Bös, K. 1996. Health-related fitness test battery for adults: aspects of reliability. *Arch. Phys. Med. Rehab.* 77: 399–405.

Takala, E.P., Viikari-Juntura, E. 2000. Do functional tests predict low back pain? *Spine* 15: 2126–2132.

Taylor, N., Groeller, H. 2003. Work-based physiological assessment of physically-demanding trades: a methodological overview. *J. Physiol. Anthropol. Appl. Human Sci.* 22: 73–81.

U.S. Department of Labor. 1991. *Dictionary of Occupational Titles*, 4th edition. U.S. Government Printing Office, Washington, DC.

Vehmasvaara, P. 2004. Physical load and strain of paramedics and development of a test drill for assessing the prerequisites of physical work capacity among paramedics. Kuopio University Publications D, Medical Sciences 324, Kuopio, Doctoral dissertation (in Finnish with English summary).

WHO. 2001. *International Classification of Functioning, Disability and Health.* World Health Organization, Geneva, http://www.who.int/classification/icf/intros/ICF-Eng-Intro.pdf.

Wickström, R.J. 1996. Evaluation physical qualifications of workers and jobs. In: Bhattacharya, A., McGlothlin, J.D. Eds. *Occupational Ergonomics. Theory and Applications.* Marcel Dekker, New York.

Williford, H.N., Duey, W.J., Olson, M.S., Howard, R., Wang, N. 1999. Relationship between fire fighting suppression tasks and physical fitness. *Ergonomics* 42: 1179–1186.

23

Training Lifting Techniques

Steven A. Lavender
The Ohio State University

23.1 Introduction

Occupational low back pain represents an enormous cost to society both financially and in terms of morbidity. Several occupational factors have been explored as to their contribution to low back disorders (LBDs). Frequently, combinations of lifting, bending, twisting, and general material handling tasks are described as the precursors of back injuries (Frymoyer et al., 1983; Damkot et al., 1984; Kelsey et al., 1984; Klein et al., 1984; Bigos et al., 1986; Punnett et al., 1991; Marras et al., 1993; Andersson, 1997; Kraus et al., 1997). Kumar (1990) showed that the cumulative loading experience of an individual is a predictor of low back pain. Andersson (1997) pointed out that low back pain represents the leading cause of activity limitation in individuals under 45 yr of age. This should not be surprising since this is the age group that more often have jobs with substantial manual material handling demands.

In sum, the epidemiological link between lifting and low back pain suggest a monotonically increasing relationship between LBDs and the biomechanical loading on the spine in material handling jobs (Chaffin and Park, 1973). High biomechanical loads most often result from the large muscle forces generated during lifting tasks. Many investigators have estimated the forces in the trunk muscles during lifting and material handling activities (Chaffin, 1969; Schultz and Andersson, 1981; Schultz et al., 1983; McGill and Norman 1985; Granata and Marras, 1993; Marras and Granata, 1997). In general, spine loading increases with asymmetry (lifts not directly in front of the body), with lifts from lower levels, lifts that require greater horizontal reaches, lifts that involve twisting, and with faster lifting speeds (Buseck et al., 1988; Gagnon and Gagnon, 1992; Lavender et al., 1992; Tsuang et al., 1992; De Looze et al., 1993; Gagnon et al., 1993; Dolan et al., 1994; Marras and Granata, 1995; Schipplein et al., 1995).

Injury prevention efforts typically include some combination of engineering and administrative control measures to reduce an individual's exposure to the stresses encountered during lifting. Engineering controls focus on designing the hazardous lifting out of the work either through changes in the worker interface, the work environment, or the nature of the work itself. In so doing, the exposure to the large spinal loads associated with lifting is eliminated or substantially reduced. Administrative controls comprise the second line of defense when the exposures of concern cannot be engineered out of the work in a cost effective manner. And while administrative controls can be effective at reducing exposure to the large and or repetitive loads associated with many occupational lifting tasks, they do not eliminate the exposure. Moreover, in many cases administrative controls require continued managerial support.

Snook et al. (1978) evaluated the effectiveness of three potentially viable preventative approaches taken toward controlling occupational LBDs: job design, training, and employee selection. These authors concluded that job design was the most promising of the three alternative approaches, and attempts to prevent the incidence of LBDs through ergonomic redesign have enjoyed good success in many manufacturing environments. However, in many work environments jobs are not easily redesigned due to constantly changing work demands. For example, employees in construction, warehousing, or delivery type jobs perform frequent lifting, but rarely under repeated circumstances. Warehouses may contain thousands of different items, thereby making the engineering changes for all but the heaviest or most frequently lifted items, cost prohibitive. In delivery jobs there is limited opportunity to engineer out the hazardous lifting, especially when half of the lifting may be occurring on the premises of the delivery's recipient. Even in manufacturing jobs, Bigos et al. (1986) reported that "improper lifting" was the most frequently cited cause of back injury in a retrospective study of Boeing employees. In these types of dynamic work environments, much of the effort aimed at back injury prevention has been directed towards training employees how to protect their backs while lifting with the belief that the trained techniques will be applied appropriately.

Historically, training employees in safe lifting techniques has appealed to organizations with many manual material handling jobs because it appears to be inexpensive relative to engineering controls, easy to implement, appears to be adaptable to a large variety of situations an employee may encounter, and can *potentially* protect a large number of employees. NIOSH (1981) recommends that employee training cover the following:

1. The risks to health of unskilled manual material handling
2. The basic physics of manual material handling
3. The effects of manual material handling on the body
4. Individual awareness of the body's strengths and weaknesses
5. How to avoid the unexpected
6. Handling skill
7. Handling aids

A typical training protocol involves teaching small groups of employees ($n < 25$) basic anatomy and biomechanics so that they will appreciate the stress put on the spine during lifting. Alternative techniques are then demonstrated and perhaps even practised in front of the instructor. NIOSH (1981) encourages the training to include lessons and demonstration outside of the classroom. The less expensive approach to training involves the purchase and showing of a videotape which illustrates the same concepts, however, generally lacks the practice component.

The effectiveness of training programs aimed at improving lifting technique has been questioned by many. Authors that have attempted to review the training efficacy literature have arrived at generally the same conclusion, namely, the effects are mixed, not well documented, and many of the studies lack randomization and control groups (Nordin et al., 1991; King, 1993; Karas and Conrad, 1996). Snook et al. (1978), in reviewing the experience of those insured by Liberty Mutual, concluded that such training programs were ineffective at reducing low back injury claims. Reddell et al. (1994), while investigating back support belts, included a comparison group that was provided with training in safe lifting techniques. In addition to finding no significant changes due to the back support belt, Reddell et al. (1994) reported that the training group did not differ with respect to the back injury incidence rate or the disability rate when compared with a nontrained control group. Walsh and Schwartz (1990) found no difference in productivity or time lost due to injury between a group of grocery ware-house workers who received a 1 h training session in back pain prevention and body mechanics and a control group who received no training. Others, however, have shown that the concepts taught in a presentation on spine anatomy, postural curves, exercises to "protect" their backs, and a demonstration of correct lifting technique were retained by navy recruits. But these same recruits were also more likely to report back problems than their noneducated counterparts (Woodruff et al., 1994). Daltroy et al. (1997) reported that a controlled trial of a "back-school" type program, designed to teach safe lifting techniques, was ineffective at controlling LBDs in their study of 4000 postal employees. Van Poppel et al. (1998) reported that education, which included: classroom lectures and demonstrations on spine anatomy and lifting mechanics at the inception of the program, repeated training on lifting mechanics 6 weeks later, and individual advice at the worksite 12 weeks into the intervention program, showed no benefit with regards to injury rates or lost time due to back injuries amongst airline cargo handlers.

23.2 Why Traditional Training Programs Are Not Successful

The results reported in the preceding section raise serious questions regarding the efficacy of teaching lifting techniques as a means to control occupational LBDs. The following four points summarize the issues:

1. The techniques taught may not be effective at reducing biomechanical stress (Stubbs et al., 1983) or applicable (Parnianpour et al., 1987; St.Vincent et al., 1989). Much of the literature on lifting techniques has focused on lifting style. The question of whether to stoop or squat when lifting low lying objects has been the topic of many research papers, the results of which, if the object cannot fit between the legs, are generally inconclusive (van Dieen et al., 1999). In some cases, for example, in specific patient handling tasks, variations in technique may reduce the biomechanical loads; however, not enough to substantially reduce LBD risk (Marras et al., 1999).

2. Most safe lifting programs rely on lecture, demonstration, and limited practice in front of the instructor. However, De Looze et al. (1994) demonstrated that trained observers, when compared to video analysis tools, failed to properly identify and categorize body mechanics during the simplest of forward lifts. These findings suggest that even if the student is given an opportunity to practice in front of the instructor, the instructor may not detect significant flaws in the demonstrated lifting technique. Moreover, Hall and Mason (1986), as reported by St. Vincent et al. (1989), showed that only 10% of a group of prospective trainers, after a week of training in how to teach lifting techniques, could perform all the movements taught correctly. Thus, the validity of the demonstrations is even questionable.

3. Most training programs fail to provide objective data by which performance improvements can be quantified and evaluated. For example, Arguss et al. (2004) found that providing feedback on objective performance measures the peak spine compressions could be reduced with training. Others have found that feeback combined with coaching could reduce the three-dimensional spine moments, although the magnitude of the change was dependent upon the willingness of the lifter to move their feet (Lavender et al., 2002).

4. Most training programs are initiated without a means to reinforce the desired behavior once it had been trained. Daltroy et al. (1997) noted that the employees understood the concepts covered in the instruction but failed to change their lifting behavior. Given that there was no behavioral change, the lack of injury reduction due to the training program should not be surprising. Komaki et al. (1980) clearly showed that just training employees in desired safety practices did

not result in the desired behavior being maintained. In the case of Komaki et al. the behaviors were only maintained when there was supervisory reinforcement.

23.3   What Is Needed to Make Training Programs Effective

It is clear that there are significant issues that must be considered in the development of a training program in order for it to be successful. First and foremost, the techniques taught need to be both biomechanically valid *and* useable. Second, lifting needs to be thought of and taught as a complex motor skill. Third, objective data must be supplied to validate the methods taught for each individual, and to determine whether a positive behavioral change has occurred. Fourth, the training process is incomplete without a means to maintain the desired behavior.

23.3.1   The Techniques Taught Need to Be Both Biomechanically Valid and Useable

This requires that a training program be job specific. In other words, the training must be based on an understanding of the job demands, including the perceived work pace. Thus, training the entire staff within a facility using a generic videotape is not likely to have much effect. Ultimately, the goal is to have workers adopt behaviors that are effective in reducing biomechanical loads. The trainees need to see how the techniques covered in the training program can be applied to their work environment given the wide range of constraints they typically report. One of the most frequently reported constraints is time. Many of the trainees in a large study conducted by myself and colleagues at Rush University in Chicago reported that they did not believe they had time to use "proper" lifting techniques. Most of the study participants worked in distribution centers where individual productivity rates are closely monitored. This prompted us to examine the time associated with three variations in lifting technique and the change in lift duration as a result of the 30-min individualized coaching session. For each of the 265 participants included in this analysis of the training process, we classified lifting technique based upon whether they voluntarily (without being told) took two steps, one step, or no steps during a simulated palletizing task in which boxes had to be moved from one position to another approximately 1 m away. Most would consider the two-step technique as approximating what is typically considered "proper" lifting technique in that the lifter faced the load when initially lifting and when placing the load at the destination point, thus, minimizing the lateral bending and twisting motions and moments. The one-step technique typically started as an asymmetric lift followed by a pivoting motion with one foot, thus allowing the lift to end more or less symmetric with regards to the mid-sagittal plane. The no-step technique, or what we called the "swing" technique, started and ended with asymmetric exertions and included no movement of the feet. In the training exercise the boxes were shifted from one scale to another, thus the computer monitoring the scales could detect both the initiation and termination of each lift. The lift duration, defined as the time from when an item was lifted from one scale and placed on a second scale, varied as function of technique used. On average, taking two steps took approximately 2 sec, one step took approximately 1.6 sec, and the swing approach took approximately 1.3 sec (Lavender et al., 2002). Thus, pushing workers to adopt a two-step technique will increase the amount of time required. In many jobs, the extra 0.75 sec may be of little consequence. In highly repetitive and tightly monitored jobs, the 0.75 sec, could accumulate if a worker lifts several times per minute. For example, in a distribution center where workers frequently lift 200 cases per hour, the extra 0.75 sec could result in an extra 20 min across an 8-h shift. Hence, the extra time works against encouraging changes in lifting technique both in the individual worker, who may be motivated by organizational policies and incentives, and by managers who are often looking at short-term indicators of system performance. Given this reluctance to change their stepping technique, we found it was best to direct the training towards optimizing whichever lifting technique workers were currently saying they used. When we did this, we found those trained on improving the two-step technique did slow down following the training by an average of 0.35 sec. There was no change in the time following the training of those using the one-step (pivot) technique. However, those using the swing technique learned behaviors that reduced the item transfer times by nearly 0.4 sec. The goal here was to show the workers that improvements could be attained without forcing them into a technique they did not believe was feasible in their environment. This increased the willingness to try behaviors that would still incrementally reduce the spine moments. Clearly, the magnitude of the reductions would be dependent upon the stepping style. Nevertheless, we did see reductions in the spine moments in behaviors that, if adopted outside the training, would reduce the moment exposure during the work process.

One of the challenges in teaching lifting techniques is determining what is to be taught. As Parnianpour et al. (1987) point out, it is a fallacy to think that there is a single correct lifting technique. In our training study (Lorenz et al., 2002) we identified several behaviors, summarized in Table 23.1, that could be emphasized during the training process. It should be noted that some of the behaviors listed are mutually exclusive and were dependent upon the stepping style used. In our training process, the selection of which behaviors were coached depended on each individual's stepping technique. What is conspicuously absent from this list is the classic "bend the knees" or the "keep the back straight" behaviors. As discussed previously, the value of techniques that use these behaviors is limited to when the lifted objects can be brought between the knees (van Dieen et al., 1999), which was one of the behaviors encouraged during our training process. Typically, if an object could be brought between the knees, and the trainee was encouraged to keep the box close, some degree of knee bending resulted. We found that pointedly emphasizing knee bending behaviors was not well received, as most of the people trained probably would be unable to use this technique throughout the day. The reader may notice that in some cases we discouraged trainees from "over-bending knees" (Table 23.1), given that in several cases the spine moments increased. When trainees used a two-step technique, the "head lift" instruction tended to help straighten the spine prior to lifting. Other behaviors, including the "body English" and "knee slide" also resulted in more upright spine postures during load placement, when the boxes were placed between mid-shank and hip level, thus reducing the forward bending moment. Another behavioral technique that addressed the forward bending moment when initiating lifts included rocking the box. This was particularly useful when handling taller and larger cartons, By tipping the box laterally prior to lifting handholds could be obtained at the corners with reduced forward bending, which in turn reduces the forward bending moment. At both the initiation and termination of the lift we strongly encouraged box sliding to minimize the reach distance. In so doing we found that it was important to emphasize a strategic approach both when selecting which carton to pick and where the cartons should be placed. In short, the trainees were encouraged to remove cartons by layer and build pallets by layer, thus eliminating the extended reaches to the back of the pallet.

Several behaviors were emphasized to minimize the twisting moments and lateral bending moments. Where the two-step technique was used, trainees were encouraged to keep the "box square," keep the "box close," to move the "box and the body together," and to focus on "finishing" the lift. Likewise, many of these same behaviors could be used with the pivoting technique. When lifting using the pivot and swing techniques, the emphasis was on swiveling the box prior to lifting such that the long axis of the carton was across the body. This allowed the carton to be slid laterally off the stack to a position in front of the body, which is biomechanically superior to rotating the carton after lifting. This swivel of the box reduced the lateral bending at the initiation of the lift, which depending on the height, reduced the twisting and lateral bending moments. All trainees were encouraged to swivel the box after it was placed if the final orientation required the box be rotated. When handing taller boxes, many of the trainees lifted with a hand positioned on the lower edges and on the other hand positioned up high around the vertical distal edge to stabilize the carton. This technique resulted in most of the load being carried by one arm, thus increasing the lateral bending moment. Instead, the "rock-the-box" or the "box level" behaviors were coached, depending on the size of the carton used.

As the training progressed, and new behaviors were practised, we realized it was important to push the trainees to increase the speed. Given that previous research has shown that spine load increase with increases in lifting speed (Buseck et al., 1988; Bush-Joseph et al., 1988; Tsuang et al., 1992; Lavender et al., 1999), the coaching emphasized the need to accelerate smoothly, thus, minimizing the peak

TABLE 23.1 A Description of the Behavioral Techniques Coached and Their Desired Effects on the Forward Bending Moment (FBM), Lateral Bending Moment (LBM), and Twisting Moments (TWM)

Behavioral Technique	Description	Desired Effect/Moment Reduced
Box and body together	Similar to "box square," this attribute refers to the *movement* phase between the pick and the placing actions. When turning, the trainees were instructed to keep the box in front and not allow the box to lead or lag the body's motion	LBM and TWM
Box between legs	Trainees were instructed to raise and lower boxes between the legs to minimize the horizontal reach distance	FBM
Box close	Trainees were instructed to keep the box resting against the body after the box was lifted box and until it was placed	LBM, FBM, and TWM
Box square	Trainees were instructed to orient this body such that box was picked up and placed in the mid-sagittal plane	LBM and TWM
Box swivel (begin/end)	Trainees using the swing or pivot technique were instructed to turn the boxes such that the long axis was parallel to the shoulders prior to lifting. Those using the swing technique were instructed to use the reverse of this technique if appropriate for box placement	LBM and TWM
Finishing	Trainees were instructed to complete the lift with the box directly in front and held close	TWM and LBM
Head lift	Trainees were instructed to lead the lift with their eyes, or in other words, look up as the lift was initiated	FBM
Hanging box	Trainees were instructed to keep the box hanging at or below pelvis height when carrying to keep the box close, minimize the physical work done, the moment arm, upper extremity fatigue, and the trunk motion	FBM
Knee slide	Trainees were instructed to bend their knees and slide the boxes down their thighs when placing the box at the end of the lift	FBM
Moving feet	Trainees were instructed to move their feet when turning the body	Reduced lateral bending and twisting motion
Not over-bending knees	Trainees were discouraged from performing deep squats as these would likely not be used on the job and were often associated with longer reaches as the load was initially lifted	FBM
Sliding boxes	Trainees were instructed to slide boxes picked from the far side of the pallet close to the body prior to lifting. Likewise, at the completion of the lift, trainees were instructed to slide into the final position without supporting its full weight	LBM and FBM
Rock-the-box	Trainees were instructed to tip large cartons laterally prior to lifting from below knuckle level to raise the bottom edge of one side of the carton, thus effectively raising the starting height of the lift. The raised edge provided one hand-hold while the upper edge rotated downward to provide the second hand-hold	FBM
Smooth	Trainees were instructed to lifted smoothly so to minimize the accelerations of the body and the box	FBM, LBM, and TWM
Speed	Trainees were encouraged to work at their normal work pace once they had practiced the new or the newly modified behaviors	Maintain worker productivity
Strategy	Trainees were instructed to plan out a set of lifts to facilitate box sliding and minimize reaches for all lifts in the set	FBM, LBM, and TWM

moments experienced during the quicker lifts. One of the five principal components identified by Wrigley et al. (2005) that differentiated techniques used between those who subsequently experienced low back pain and those who did not was related to the shape of the spine moment exposure function. More specifically, a moment exposure pattern that smoothed out the moment experienced over the course of the lift, for example by minimizing the peak exposure during the acceleration stages of the lift, was associated with the absence of reported back pain, even though characteristically this technique resulted in a prolonged moment exposure later in the lift (Wrigley et al., 2005).

23.3.2 Lifting Needs to Be Thought of and Taught as a Complex Motor Skill

Demonstrations and a few practice lifts in front of an instructor is no way to insure a beneficial behavioral change. It has been shown, for example, that observing a model does little to improve performance on a gymnastics skill (Magill and Schoenfelder-Zohd, 1996). Research suggests that complex motor skills are best learned through a coaching process rather than demonstrations. Kernodle and Carlton (1992) in their study of throwing behavior found that performance was enhanced when subjects were provided feedback as to specific attributes of their movement associated with task performance. Performance was even more enhanced when coaching was provided to direct the subjects towards improved performance. In motor control terms, the coach provided the "transitional cue" as to which changes in behavior should be tried. This situation is analogous to getting some "tips" from the golf pro on how we might improve our swing. Clearly, observing a video on golf technique, does not tell each of us what we are doing wrong and specifically what we need to adjust. Likewise, videos and generalized training on lifting techniques will likely do little to address the deficiencies in an individual's technique. The employees in the geriatric facility analyzed by Wood (1987) responded favorably to having a physiotherapist coach them on lifting techniques on the job. Barker and Atha (1994) reported that an interactive lifting instruction session with a personal tutor resulted in significantly improved lifting practices than a control group, and lower ratings of perceived exertion and predicted compressive force. The transitional cues, tips, target behaviors, from our study that have been described above were effective at reducing the dynamic spine moments when they were provided based on observed performance deficiencies.

23.3.3 Objective Data Must Be Supplied to Validate the Methods Taught for Each Individual and to Determine Whether a Positive Behavioral Change Has Occurred

One of the primary challenges in improving lifting technique is learning when techniques are being used correctly such that the biomechanical loads on the spine are minimized. This is because we lack a feedback system within the body that tells, at least in those individuals without back pain, how much load is being experienced by the spine during a lifting activity. When lifting heavy objects, we can feel the object is heavy and we may sense the stress placed on the bicep tissues. However, we typically have no feedback as to the compression of tissues within the elbow joint. With the spine, the muscle tension feedback is diffused across the many different muscles that work synergistically during lifting activities. Moreover, we have no feedback as to the compression or shear loads experienced by the intervertebral disc. This lack of feedback regarding lifting performance, hinders the learning and adoption of new behavioral techniques during a training process. It would be similar to trying to improve one's golf swing when you have no knowledge of the direction that the golf ball actually traveled or how far it went (Figure 23.1). Researchers have attempted to construct the feedback loop through various means for lifting tasks. For example, Stubbs et al. (1983) used intra-abdominal pressure to quantify the biomechanical stress associated with patient handling. Obviously, this methodology would not be suitable for a large work force.

The lack of feedback is also problematic for the trainer or coach. Given that there are typically no objectively defined quantitative measures that can evaluate a trainee's lifting behavior, the information

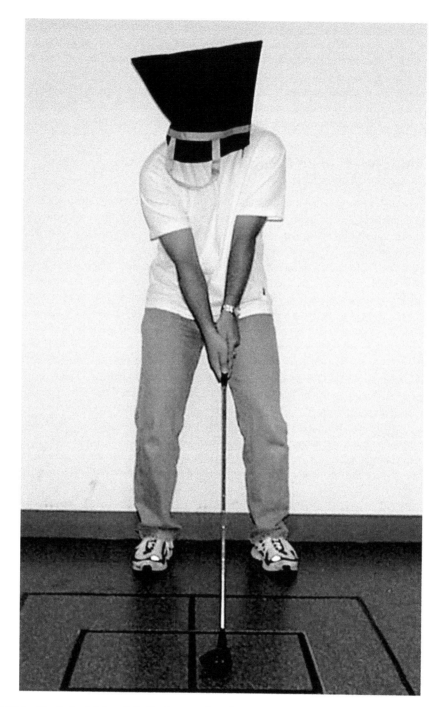

FIGURE 23.1 The lack of internal feedback conveying the spine biomechanical loading when learning lifting techniques is similar to trying to improve ones golf swing in the absence of information about the results of each swing.

provided by the trainer is typically based upon course behavioral observations which may or may not be valid. Hence, the lifting trainer's situation would be analogous to the golf pro giving advice where he or she has only limited knowledge of the trainee's performance and no knowledge of the results of the trainee's swing (Figure 23.2). In this analogy the golf pro, who observed the swing, can only ask how it felt given that neither the pro nor the trainee knows were the ball went. Thus, the tips offered by the golf pro

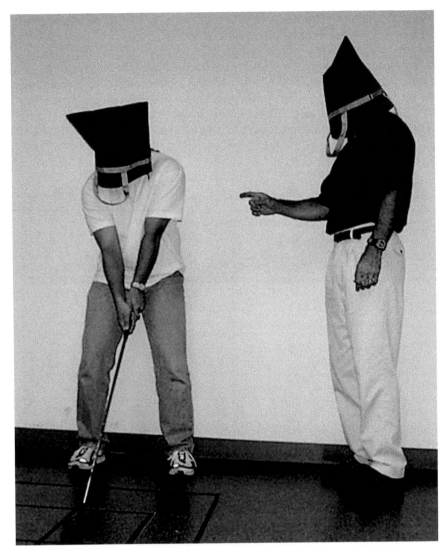

FIGURE 23.2　The lack of objective data describing lifting performance hampers the trainers efforts in a way similar to that of a golf pro that has only a little information available regarding the performance and no information regarding results of what changes have been made to the golf swing.

are of questionable value if the validity cannot be determined through objective performance measures. Barker and Atha (1994) measured improvements in lifting performance, spine compressive force, and ratings of perceived exertion, but failed to use these data in guiding the teaching process. Thus, while the validity of the training could be assessed after the fact, the trainee would not have the assurance that the trained techniques were effective. Each employee participating in a training process will likely be asked to change some aspect of their lifting behavior. Objective data showing, not just behavioral change, but improvement in the biomechanical consequences of the change, are necessary to document learning and validate that the behavioral adaptations are actually beneficial to the trainee. Essentially, we have to construct an external feedback loop that can be used to guide the process of learning or modifying the appropriate behaviors.

In our study (Lavender et al., 2002; Lorenz et al., 2002), we developed and tested an approach to training lifting techniques that essentially created an external feedback loop. The apparatus, which consisted of a magnetic motion measurement system for tracking motions in three dimensions and two scales for tracking the material handled, were combined with software that included a dynamic linked-segment model that could instantaneously compute the net moment at the base of the spine (L5/S1) in real time. The magnitude of the moment was used to adjust the pitch of a tone generated by the sound card in the computer, where higher pitches indicated higher moments. The tone provided instantaneous feedback on the biomechanical loading of the spine. Once trainees completed a set of lifts, typically between three to five lifts, data were displayed showing the peak forward bending, lateral bending, and twisting moments experienced during each lift (Figure 23.3). A coach interpreted the data for the

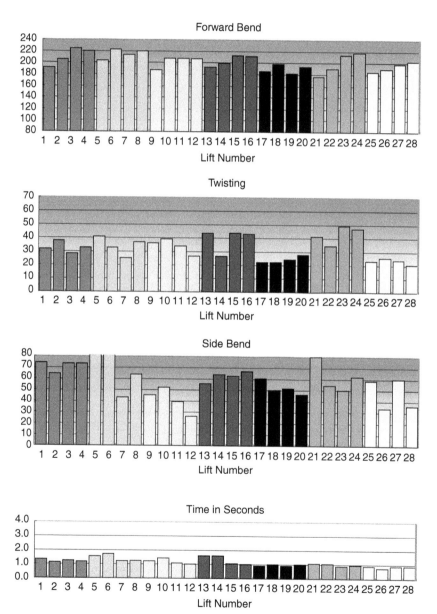

FIGURE 23.3   Charts showing the objective performance measures used in our study which comprised the peak directional moments acting on the spine during each lift. The top chart shows the peak forward bending moment for each lift in Newton meters (N m). Likewise, the second and third charts show the peak twisting and side bending moments, respectively. The bottom chart shows the lift duration in seconds.

participants and provided suggestions (transitional cues) as to how the lifting behaviors could be modified based on each individual's data. The training method produced significant changes in the directional spine loads from the beginning to the end of the 30-min training sessions (Lavender et al., 2002). By providing the objective data, trainees could assess their performance and know if modifications to their technique were successful in actually reducing spine loads.

23.3.4 The Training Process is Incomplete without a Means to Maintain the Desired Behavior

Probably most training programs in industry are conducted with no follow-up process built in. Thus, the potential for the training to be effective will likely diminish quickly with time. Investigators examining the retention of lifting behaviors taught during a 4-h training program found only selected behaviors were retained when the same workers were evaluated between 5 and 8 weeks after training (Chaffin et al., 1986). As noted above, the participants in Daltroy et al.'s study (1997) failed to use the behaviors even though they understood the concepts. Hence the challenge is how to reinforce the desired lifting behaviors such that the behavior is maintained? Komaki et al. (1978) report that simple feedback can be a powerful reinforcer. Following a training program to improve worker safety, observations were used to determine the percentage of operations performed safely and charted each week for all to see. Additionally, supervisors were made responsible for reinforcing safe behavior through positive feedback comments. In a later study, Komaki et al. (1980) isolated the impact of the training versus the feedback. It was shown that while training on safety practices had a positive effect, the gains were much larger once the feedback was instituted. These findings suggest that for training programs on lifting techniques to be effective, they may need to be combined with other behavioral safety initiatives within the facility such that workers remain cognizant of the desired behaviors and reinforced for their use. Alvero and Austin (2004) reported that the process of conducting behavioral observations improved the lifting behaviors used by the observer. These finding suggest a peer-review process where the employees at a facility evaluate each others lifting behaviors relative to what has been emphasized through a training process. The peer-review not only provides evaluation and reinforcement for the observed individuals, but also serves to reinforce the target behaviors in those currently serving as evaluators. Depending upon how this is approached, the process may be much better received than periodic evaluation by management.

23.4 Conclusions

While the evidence supporting the use of training programs as a means for preventing injury is very limited, the general belief is that they are of value. The underlying premise is that they provide individuals with the skills to protect themselves from the risks associated with manual materials handling across a wide range of situations that may be encountered on the job. The literature makes it clear that the training must go beyond the conceptual level and focus on behaviors. As such, these behaviorally oriented approaches should include techniques that are biomechanically valid *and* useable, provide objective data to validate that a positive behavioral change has occurred, and include a means to maintain or reinforce the desired behavior.

References

Agruss, C.D., Williams, K.R., and Fathallah, F.A. (2004). The effects of feedback training on lumbosacral compression during simulated occupational lifting. *Ergonomics* 47:1103–1115.

Alvero, A.M. and Austin, J. (2004). The effects of conducting behavioral observations on the behavior of the observer. *Journal of Applied Behavior Analysis* 37(4):457–468.

Andersson. G.B.J. (1997). The epidemiology of spinal disorders. In J.W. Frymoyer (ed.) *The Adult Spine: Principles and Practice, 2nd edition.* Lippincott-Raven Publishers, Philadelphia, 93–141.

Barker, K.L. and Atha, J. (1994). Reducing the biomechanical stress of lifting by training. *Applied Ergonomics* 25:373–378.

Bigos, S.J., Spengler, D.M., Martin, N.A., Zeh, J., Fisher, L., Nachemson, A., and Wang, M.H. (1986). Back injuries in industry: a retrospective study II. Injury factors. *Spine* 11:246–251.

Buseck, M. and Schipplein, O.D., et al. (1988). Influence of dynamic factors and external loads on the moment at the lumbar spine in lifting. *Spine* 13(8):918–921.

Bush-Joseph, C. and Schipplein, O., et al. (1988). Influence of dynamic factors on the lumbar spine moment in lifting. *Ergonomics* 31(2):211–216.

Chaffin, D.B. (1969). A computerized biomechanical model — development of and use in studying gross body actions. *Journal of Biomechanics* 2:429–441.

Chaffin, D.B., Gallay, L.S., Woolley, C.B., and Kuciemba, S.R. (1986). An evaluation of the effect of training program on worker lifting postures. *International Journal of Industrial Ergonomics* 1:127–136.

Chaffin, D.B. and Park, K.S. (1973). A longitudinal study of low-back pain as associated with occupational weight lifting factors. *American Industrial Hygiene Association Journal* 34:513–525.

Daltroy, L.H., Iversen, M.D., Larson, M.G., Lew, R., Wright, E., Ryan, J., Zwerling, C., Fossel, A.H., and Liang, M.H. (1997). A controlled trial of an educational program to prevent low back injuries. *The New England Journal of Medicine* 337:322–328.

Damkot, D.K., Pope, M.H., Lord, J., and Frymoyer, J.W. (1984). The relationship between work history, work environment and low-back pain in men. *Spine* 9:395–399.

De Looze, M.P., Toussaint, H.M., Van Dieen, J.H., and Kemper, H.C.G. (1993). Joint moments and muscle activity in the lower extremities and lower back in lifting and lowering tasks. *Journal of Biomechanics* 26:1067–1076.

De Looze, M.P., Toussaint, H.M., Ensink, J., Mangnus, C., and Van Der Beek, A.J. (1994). The validity of visual observation to assess posture in a laboratory-simulated, manual material handling task. *Ergonomics* 1335–1343.

Dolan, P., Earley, M., and Adams, M.A. (1994). Bending and compressive stresses acting on the lumbar spine during lifting activities. *Journal of Biomechanics* 27:1237–1248.

Frymoyer, J.W., Pope, M.H., Clements, J., Wilder, D.G., MacPherson, I.B., and Ashikaga, T. (1983). Risk factors in low-back pain: an epidemiological study. *The Journal of Bone and Joint Surgery* 65-A: 213–218.

Gagnon, D. and Gagnon, M. (1992). The influence of dynamic factors on triaxial net muscular moments at the L5/S1 joint during asymmetrical lifting and lowering. *Journal of Biomechanics* 25:891–901.

Gagnon, M., Plamondon, A., and Gravel, D. (1993). Pivoting with the load: an alternative for protecting the back in asymmetrical lifting. *Spine* 18:1515–1524.

Granata, K.P. and Marras, W.S. (1993). An EMG assisted model of loads on the lumbar spine during asymmetric trunk extensions. *Journal of Biomechanics* 26:1429–1438.

Hall, A.R. and Mason, I.D. (1986). L'evaluation du role d'une formation kinetique dans la prevention des accidents de manutention. *Le Travail Humain* 49:195–207.

Kelsey, J.L., Githens, P.B., White, A.A., III Holford, T.R., Walter, S.D. O'Connor, T., Ostfeld, A.M., Weil, U., Southwick, W.O., and Calogero, J.A. (1984). An epidemiologic study of lifting and twisting on the job and risk for acute prolapsed lumbar intervertebral disc. *Journal of Orthopaedic Research* 2:61–66.

Karas, B.E. and Conrad, K.M. (1996). Back injury prevention interventions in the workplace. An integrative review. *AAOHN Journal* 44:189–196.

Kernodle, M.W. and Carlton, L.G. (1992). Information feedback and the learning of multiple-degree-of-freedom activities. *Journal of Motor Behavior* 24:187–196.

King, P.M. (1993). Back injury prevention programs: a critical review of the literature. *Journal of Occupational Rehabilitation* 3:145–158.

Klein, B.P., Jensen, R.C., and Sanderson, L.M. (1984). Assessment of workers' compensation claims for back strains/sprains. *Journal of Occupational Medicine* 26:443–448.

Komaki, J., Barwick, K.D., and Scott, L.R. (1978). A behavioral approach to occupational safety: pinpointing and reinforcing safe performance in a food manufacturing plant. *Journal of Applied Psychology* 63:434–445.

Komaki, J., Heinzmann, A.T., and Lawson, L. (1980). Effect of training and feedback: component analysis of a behavioral safety program. *Journal of Applied Psychology* 65:261–270.

Kraus, J.F., Schaffer, K.B., McArthur, D.L., and Peek-Asa, C. (1997). Epidemiology of acute low back injury in employees of a large home improvement company. *American Journal of Epidemiology* 146:637–645.

Kumar, S. (1990). Cumulative load as a risk factor for back pain. *Spine* 15:1311–1316.

Lavender, S.A., Tsuang, Y.H., Andersson, G.B.J., Hafezi, A., Shin, C.C. (1992). Trunk muscle cocontraction: The effects of moment direction and moment magnitude. *Journal of Orthopedic Research* 10:691–700.

Lavender, S.A., Li, Y.C., et al. (1999). The effects of lifting speed on the peak external forward bending, lateral bending, and twisting spine moments. *Ergonomics* 42(1):111–125.

Lavender, S.A., Lorenz, E., Andersson, G.B.J. (2002). Training in lifting: do good lifting techniques adversely affect case handling times. *Professional Safety* 47(12):30–35.

Lorenz, E.P., Lavender, S.A., Andersson, G.B.J. (2002). Determining what should be taught during lift-training instruction. *Physiotherapy Theory and Practice* 18:175–191.

Magill, R.A., Schoenfelder-Zohd, B. (1996). A visual model and knowledge of performance as sources of information for learning a rhythmic gymnastics skill. *International Journal of Sport Psychology* 27:7–22.

Marras, W.S. and Granata, K.P. (1995). A biomechanical assessment and model of axial twisting in the thoracolumbar spine. *Spine* 20:1440–1451.

Marras, W.S. and Granata, K.P. (1997). Spine loading during trunk lateral bending motions. *Journal of Biomechanics* 30:697–703.

Marras, W.S., Lavender, S.A., Leurgans, S.E., Rajulu, S.L., Allread, W.G., Fathallah, F.A., and Ferguson, S.A. (1993). The role of dynamic three-dimensional trunk motion in occupationally related low back disorders: the effects of workplace factors, trunk position, and trunk motion characteristics on risk of injury. *Spine* 18:617–628.

Marras, W.S., Davis, K.G., et al. (1999). A comprehensive analysis of low-back disorder risk and spinal loading during the transferring and repositioning of patients using different techniques. *Ergonomics* 42(7):904–926.

McGill, S.M. and Norman, R.W. (1985). Dynamically and statically determined low back moments during lifting. *Journal of Biomechanics* 18:877–885.

National Institute for Occupational Safety and Health (NIOSH) (1981). Work practices guide for manual lifting, NIOSH Technical Report No. 81–122, US Department of Health and Human Services, National Institute for Occupational Safety and Health, Cincinnati, OH.

Nordin, M., Crites-Battie, M., Pope, M.H., and Snook, S. (1991). Education and training. In Pope, M.H., Andersson, G.B.J., Frymoyer, J.W., Chaffin, D.B. (eds.) *Occupational Low Back Pain: Assessment, Treatment, and Prevention*. Mosby, St. Louis, 266–276.

Parnianpour, M., Bejjani, F.J., et al. (1987). Worker training: the fallacy of a single, correct lifting technique. *Ergonomics* 30(2):331–334.

Punnett, L., Fine, L.J., Keyserling, W.M., Herrin, G.O., and Chaffin, D.B. (1991). Back disorders and nonneutral trunk postures of automobile assembly workers. *Scandinavian Journal of Work Environment and Health* 17:337–346.

Reddell, C.R., Congleton, J.J., Huchingson, R.D., and Montgomery, J.F. (1992). An evaluation of a weightlifting belt and back injury prevention training class for airline baggage handlers. *Applied Ergonomics* 23:319–329.

Schipplein, O.D., Reinsel, T.E., Andersson, G.B.J., and Lavender, S.A. (1995). The influence of initial horizontal weight placement on the loads at the lumbar spine while lifting. *Spine* 20:1895–1898.

Schultz, A.B. and Andersson, G.B.J. (1981). Analysis of loads on the lumbar spine. *Spine* 6:76–82.

Schultz, A.B., Haderspeck, K., Warwick, D., and Portillo, D. (1983). Use of lumbar trunk muscles in isometric performance of mechanically complex standing task. *Journal of Orthopaedic Research* 1:77–91.

Snook, S.H., Campanelli, R.A., and Hart, J.W. (1978). A study of three preventative approaches to low back injury. *Journal of Occupational Medicine*, 20:478–481.

Stubbs, D.A., Buckle, P.W., Hudson, M.P., and Rivers, P.M. (1983). Back pain in the nursing profession II. The effectiveness of training. *Ergonomics* 26:767–779.

St. Vincent, M., Tellier, C., and Lortie, M. (1989). Training in handling: an evaluative study. *Ergonomics* 32:121–210.

Tsuang, Y. H., Schipplein, O.D., et al. (1992). Influence of body segment dynamics on loads at the lumbar spine during lifting. *Ergonomics* 35(4):437–444.

van Dieen, J. H., Hoozemans, M.J., et al. (1999). Stoop or squat: a review of biomechanical studies on lifting technique. *Clin Biomechanics (Bristol, Avon)* 14(10):685–696.

Van Poppel, M.N.M., Koes, B.W., Van Der Ploeg, T., Smid, T., and Bouter, L.M. (1998). Lumbar supports and education for the prevention of low back pain in industry. A randomized controlled trial. *Journal of the American Medical Association* 279:1789–1794.

Walsh, N.E. and Schwartz, R.K. (1990). The influence of prophylactic orthoses on abdominal strength and low back injury in the workplace. *American Journal of Physical Medicine and Rehabilitation* 69:245–250.

Wood, D.J. (1987). Design and evaluation of back injury prevention program within a geriatric hospital. *Spine* 12:77–82.

Woodruff, S.I., Conway, T.L., and Bradway, L. (1994). The U.S. Navy healthy back program: Effects on back knowledge among recruits. *Military Medicine* 159:475–484.

Wrigley, A.T., Albert, W.J., Deluzio, K.J., and Stevenson, J.M. (2005). Differentiating lifting technique between those who develop low back pain and those who do not. *Clinical Biomechanics* 20:254–263.

24

Secondary Intervention for Low Back Pain

Stover H. Snook
Harvard School of Public Health

24.1 Introduction

Low back pain is defined as the subjective perception of pain in the lower back, buttocks, or leg (numbness and pain in the leg is known as sciatica). There is no objective measure of pain; it can only be measured subjectively through self-report (pain drawings, pain analogs, pain scales, pain words, pain diaries, etc.)

Low back pain is a very common problem; it affects about two thirds of the adult population.[1] Low back pain is also a very expensive problem. The estimated costs (direct medical and indirect costs) of low back pain in the United States are in the range of 50 billion dollars per year, and could be as high as 100 billion dollars per year.[2] Low back pain costs are about three times higher than the total cost of all forms of cancer diseases.[3]

From a research point of view, low back pain is a neglected problem. Only 0.2% of all randomized trials in medicine concern low back pain.[3] Since only about 15% of medical interventions are based on solid scientific evidence, one can begin to understand how little is known about low back pain.[4] According to Akeson and Murphy, two leading orthopedic surgeons, "Probably nowhere in the field of medicine has there been as much written with as few facts as in the field of 'discogenic low back pain'."[5]

Up to 85% of low back pain has no known cause.[6,7] Most back symptoms are of spontaneous and gradual onset, without a specific accident or unusual activity.[8-12] Therefore, most low back pain cannot be referred to as an "injury" — which, by definition, occurs at a specific time and place.

There are many possible sources of low back pain. However, the main suspected source, according to many researchers, is in the intervertebral disc.[13-19] The primary reason for this belief is that noxious stimulation of the disc produces symptoms of low back pain in many patients — whether by injections during discography, whether by probing during surgery under local anesthesia, or whether by electro-thermal heating of the disc (IDET). Anular tears and reduction in disc height are associated with a history of low back pain, but their sensitivity is poor (explaining only 6 and 7%, respectively, of the total variance).[20]

A clear distinction must be made between low back pain and low back disability.[21] Low back disability is the time lost on the job, or the restricted duty that results from low back pain. Low back disability is obviously related to low back pain, but includes many other psychological, social, and economic variables that influence disability. Examples of these variables are the type of job, management policies and practices, job dissatisfaction, personality differences, supervisor conflicts, domestic problems, compensation laws, unemployment rates, and litigation.

The relationship between low back pain and low back disability is not very good (the correlation coefficient is only about 0.3).[22–25] Some people experience high amounts of low back pain, but are not disabled. Others become disabled with only small amounts of low back pain.

There is no question that low back pain can be very painful. However, according to Waddell, less than 1% of low back pain is caused by a serious disease such as cancer, less than 1% is due to inflammatory disease such as arthritis, and less than 5% is true sciatica.[25] The rest of it is referred to as idiopathic or nonspecific low back pain.

24.2   Nature of Nonspecific Low Back Pain

Scientific evidence indicates that nonspecific low back pain is basically an age-related disorder that is affected by differences in occupation, genetics, and personal behavior.

24.2.1   Age

The prevalence of back pain increases from early adult life to the late 40s or early 50s and remains relatively constant thereafter, at least to the mid-60s. In those who continue to have back pain, it is more likely to be more frequent or more constant with increasing age.[25–27]

The body changes with increasing age, and the intervertebral disc is one of the earliest parts of the body to change. Boos and associates recently received the Volvo Award for demonstrating the detrimental effect of a diminished blood supply on the disc beginning at 11 to 16 yr of age.[28] Discs degenerate with age, and tears develop in the anulus of the disc. In their review of cadaver studies, Miller et al. found 7% of people in their 20s exhibit anular tears; 20% in their 30s; 41% in their 40s; 53% in their 50s; 85% in their 60s; and 92% for people over 70.[29] Videman and Battié reviewed the epidemiology of idiopathic back pain and concluded that "we currently cannot differentiate pain producing pathology from simple aging changes."[17]

The symptoms of low back pain also change with age.[11] People in their 20s and early 30s usually suffer from sudden acute attacks of short duration. During the mid- to late 30s, the pain often becomes more localized to one side or the other, listing to one side becomes evident, and there may be some residual, mild backache between attacks. During the 40s, radiation of pain into the buttocks, the thigh, and even down to the foot occurs more frequently. The pain often becomes more constant during the 50s, but less severe.

24.2.2   Occupation

The literature regarding the relationship between the physical demands of work and low back pain is contradictory.[8,26] One reason is that sedentary workers also suffer from low back pain. Vingård and Nachemson have reviewed the literature, and they conclude that "there seems to be a constant but weak relationship between workload factors and reports of back pain."[30] On the other hand, the National Research Council and the Institute of Medicine conclude that the evidence shows "a clear relationship between back disorders and physical load."[31] The question appears to be whether the physical demands cause low back pain, or whether they aggravate or exacerbate an underlying condition. From a workers' compensation point of view, it does not make any difference; both are compensable. The activities most commonly associated with low back pain are heavy lifting, frequent bending and twisting, and whole-body vibration.[30–33] With respect to lifting, the risk is greatest when lifting from the floor, and lifting bulky objects.[25,31,32,34]

Marras has suggested a load–tolerance relationship, where the tolerance of the spine must be considered in addition to the degree of occupational exposure.[35] Spine tolerance is believed to vary according to age, disc degeneration, repetitive motion, and time of day. This may help explain the contradictory literature.

24.2.3   Genetics

Low back pain tends to run in families. There are three studies that show a familial predisposition.[36–38] A fourth study using identical twins showed that genetics was more important than occupational factors in determining disc degeneration.[39] A fifth study using magnetic resonance imaging (MRI) on monozygotic and dizygotic twins concluded that the genetic contribution to lumbar disc degeneration was 74% (95% CI 64–81 for adjusted values).[40] The data suggest that disc height and disc bulging were the predominant factors contributing to the genetic determination of disc degeneration.

24.2.4   Personal Behavior

In a systematic review of the literature, Leboeuf-Yde concluded that body weight should be considered a possible weak risk factor for low back pain.[41] Deyo and Bass found a steady increase in back pain prevalence with increasing obesity, but most strikingly in the highest 20% of body mass index.[42]

Two reviews of the literature concluded that there is some evidence for a weak relationship between heavy smoking and low back pain, but no clear evidence for a relationship between sciatica and smoking.[43,44]

24.3   Diagnosis, Treatment, and Prognosis

About 85% of people with low back pain cannot be given a precise pathoanatomical diagnosis.[1,45] The most accurate diagnosis for these people is idiopathic or nonspecific low back pain. The common diagnosis of strain or sprain has never been anatomically or histologically characterized.[1] Bulging discs have little or no association with low back pain.[46] A herniated disc can be diagnosed in only about 4% of patients, and spinal stenosis in about 3%.[1] Herniated discs occur in 20 to 70% of people without low back pain, depending on age, selection, and definition of disc herniation.[46]

It is estimated that 40% of patients with low back pain fear that they have some serious disease.[25] One of the major roles of the physician is to provide reassurance that nonspecific low back pain is not a serious disease.[25] Overdiagnosis often leads to further anxiety, worry, dependence on medical care, unnecessary testing, and a conviction about the presence of disease.[1] Overdiagnosis leads directly to overtreatment.[25,47]

The treatment of low back pain varies greatly, depending upon the type of practitioner. Nachemson lists 52 different treatments for low back pain, which illustrates the lack of knowledge and the degree of uncertainty among clinicians who treat low back pain.[48] However, in spite of all the different treatments, the outcomes are about the same.[49–55] The scientific evidence shows that most of the treatments in routine use for low back pain are ineffective, and some are worse than no treatment at all.[25] Perhaps this is the reason why most adults with acute severe low back pain do not seek any health care during their most recent episode of pain.[56]

Clinical guidelines for the management of low back pain have been developed by many countries, based generally upon the same body of literature. Koes et al. have reviewed and compared the national guidelines of 11 countries that were published in English, German, or Dutch.[57] The guidelines are consistent in their recommendation that plain radiographs (x-rays) are not useful in acute nonspecific low back pain, and that the vast majority of cases should be managed in a primary care setting. There were also consistent recommendations for the early and gradual activation of patients, the discouragement of bed rest, and the recognition of psychosocial factors in the development of chronicity. However, there were differences in the recommendations for exercise, spinal manipulation, and medication.

There are indications that the body heals itself. Most patients improve considerably during the first 4 weeks — with or without medical treatment.[52,58,59] The pain from herniated discs also improves with time, but at a slower rate than low back pain alone.[1,60,61] The herniated portion of the disc tends to regress with time, with partial or complete resolution in two thirds of the cases after 6 months.[1]

However, the recurrence rate of low back pain is high. Recurrences range from 40 to 80%, depending upon the definition of recurrence, and the length of follow-up.[62] Other studies have consistently found that a history of low back symptoms is the best predictor for future low back pain.[17,32,33]

24.4 Secondary Intervention

There is no evidence that low back pain has decreased in recent years.[63] However, the disability from low back pain increased dramatically between the 1950s and 1970s without an increase in low back pain.[51] One study showed that low back disability from 1957 to 1976 increased at a rate that was 14 times greater than the rate of population growth.[64] Consequently, there are a growing number of investigators who believe that efforts at preventing low back pain are futile; that low back pain is an unavoidable consequence of life that will afflict most people at some point in their lives.[11,65–68] These investigators believe that programs aimed at reducing low back disability are likely to be much more effective and less costly. As Frymoyer and Cats-Baril pointed out in 1991, "The future challenge, if costs are to be controlled, appears to lie squarely with prevention and optimum management of disability, rather than perpetrating a myth that low back pain is a serious health disorder."[2]

Contrary to popular opinion, the majority of people with low back pain continue to work. Fifty percent of the U.S. population reports low back pain each year, but only 2% of the population is disabled from low back pain each year.[69,70] However, it is this small percentage that consumes most of the cost. Frymoyer and Cats-Baril estimated that 75% or more of the total costs of low back pain can be attributed to the 5% of people who become disabled temporarily or permanently from low back pain.[2] Data from the Liberty Mutual Insurance Company indicated that only 12.4% of the low back claims had a length of disability greater than 3 months, but these claims accounted for 88% of the costs.[71] It appears quite obvious that the major problem for industry today is low back disability, not low back pain.

24.4.1 Risk Factors for Low Back Disability

Disability risk factors refer to the increased risk of disability after the onset of pain, and the increased duration of disability. The major risk factors for low back disability are discussed next.

24.4.1.1 Increasing Age

Length of recovery and disability increase with age. The U.S. Bureau of Labor Statistics reported that workers in their 50s have twice the disability as workers in their 20s and 30s.[72] In a large prospective study of over 1000 disabled workers, Mayer et al. concluded that younger workers are far more likely to return and hold work after functional restoration.[73]

24.4.1.2 Occupational Factors (Type of Job)

Low back disability is closely related to the physical demands of the job. Waddell believes that there is little evidence to support the belief that heavy manual work causes low back pain, but there is almost unanimous agreement that people with heavy manual jobs lose more time from work because of back pain than workers with lighter jobs.[25] Lifting is more frequently associated with low back pain workers' compensation claims (49%) than any other task or movement;[12] particularly lifting from the floor, lifting bulky objects, lifting heavy objects, and lifting frequently.[25,34]

Bending, without lifting, is also a problem. Back disorders in an automobile assembly plant were associated with mild (20° to 45°) trunk flexion (odds ratio: 4.9), severe (>45°) trunk flexion (odds ratio: 5.7), and trunk twist or lateral bend (odds ratio: 5.9).[74]

24.4.1.3 Psychosocial Factors

Frymoyer and Cats-Baril have observed that "if the risk factors for low back disability are analyzed, it becomes clear that this is not a medical problem ... factors that are important are dominantly psychosocial ..."[2] Waddell et al. conclude that "there is now extensive evidence that psychosocial factors are more important than any physical changes in the back for the development and maintenance of chronic pain and disability."[75]

Psychosocial factors include the following:

- Job dissatisfaction
- Low supervisor support
- Monotonous work
- High work pace
- Personality conflicts
- Poor working conditions
- Poor health habits

A recent review of the literature by Linton concluded that "there is strong evidence that psychosocial variables are strongly linked to the transition from acute to chronic pain disability."[76] There is also strong evidence that psychological factors can be associated with the reporting of low back pain.

24.4.1.4 Health Care Providers

In many cases, low back disability is iatrogenic (i.e., caused by the physician or surgeon).[21,47,77–79] Deyo claims that physicians must share the blame for unnecessary disability claims because of alarming and inconsistent diagnostic labels, excessive testing, and unnecessary therapy.[80] Spitzer feels that low back disability is "aided and abetted by the health care provision system in general and by doctors and physiotherapists in particular."[68] Frymoyer and Cats-Baril ask the question: "... have medical professionals of all types become part of the problem, rather than the solution?"[2]

A study of workers' compensation claimants with acute, uncomplicated, disabling work-related low back pain found that increased disability was associated with increased utilization of specialty referrals and provider visits, use of MRIs, and use of opioids for more than 7 days.[52] Almost two thirds of the patients were prescribed various physical therapy interventions. Patients receiving physical therapy treatments such as back stretching exercises, back strengthening exercises, and ultrasound experienced longer lengths of disability than those who did not receive those treatments. Although x-rays are not recommended within the first month except for certain "red-flag" conditions, approximately 42% of all patients had x-rays taken in the first week, and 57% in the first month.[52]

24.4.1.5 Management

Management is also part of the problem by not responding well to low back pain when it does occur among workers. The problem is often ignored or delegated to lower level staff. For example, one survey reported that only about half of workers on disability were contacted by their immediate supervisor, despite evidence that supervisor concern and contact are associated with reduced disability.[81]

The Michigan Disability Prevention Study demonstrated that lower levels of disability are associated with management policies and practices, particularly safety diligence, safety training, and proactive return-to-work programs.[82] The probability of getting off disability and returning to work after various durations of disability is illustrated in Table 24.1 and Table 24.2.[71,83,84] Management plays an important role in keeping workers on the job, or in returning them to work quickly.

TABLE 24.1 Probability of Getting off Disability after 1 yr[71]

Disabled for 1 month	62%
Disabled for 3 months	44%
Disabled for 6 months	28%
Disabled for 9 months	14%

24.4.1.6 Organized Labor

Five years after the release of the ground-breaking report of the Quebec Task Force on Spinal Disorders in the Workplace,[6] Walter Spitzer, the Chairman of the Task Force, observed that the Quebec Workmens' Compensation Board had failed to implement any of the clinical or organizational recommendations of the task force. According to Spitzer, "It is generally acknowledged that this is largely because of strong opposition by organized labour. The unions never liked the key conclusion of the report: 'The best treatment for low back pain without radiation or objective clinical signs is work.'"[68]

Rigid work rules often prevent early return to work. Also detrimental are union referrals to "friendly" physicians who prolong disability, and referrals to "friendly" attorneys who press for lump sum settlements instead of rehabilitation.[85]

24.4.1.7 Compensation Insurance

Low back disability is closely related to the development of compensation insurance. However, according to Allan and Waddell, "it is wrong to infer that disability is caused by compensation. Indeed the converse is true: legislation for compensation was only passed after a need was recognised."[21]

The best available data, as reviewed by Loeser et al., suggests that a 10% increase in workers' compensation benefits produces a 1% to 11% increase in the number of claims, and a 2 to 11% increase in the average duration of claims.[86] However, Waddell et al. claim that the availability and ease or difficulty of getting benefits have a much greater impact on the number of claims and the number and duration of benefits paid than the financial level of the benefits.[75]

Compensation patients also have poorer results from back surgery, and do not respond as well to pain management and rehabilitation.[25] However, Waddell reminds us that health care professionals have benefited more from the compensation system than back pain patients have.[25]

24.4.2 Reducing Low Back Disability

Although low back pain may not be preventable at the present time, the good news is that we know how to reduce the disability from low back pain. There is sufficient evidence in the literature to suggest an approach consisting of the following components.

24.4.2.1 Workplace Redesign (Ergonomics)

Workplaces must be designed for people with low back pain as well as for people without low back pain. Good workplace design will permit employees with low back pain to remain on the job, or to return to the job sooner. Workers with low back pain have difficulty bending forward, and difficulty handling objects when excessive forward reaching is involved. The basic ergonomic principle is to get things up off the floor. Objects can often be handled with little stress if excessive bending and forward reaching are avoided. Benches, four-wheel carts, lift tables, hoists, balancers, two-wheel hand trucks, and conveyers

TABLE 24.2 Probability of Returning to Work

	McGill[83]	Rosen[84]
Off work for 6 months	50%	35–50%
Off work for 1 yr	25%	10–25%
Off work for 2 yr	Nil	2–3%

are often recommended to reduce bending and forward reaching. A large prospective study by Marras demonstrated that the introduction of lift tables and lift aids (e.g., overhead pulley systems, vacuum hoists) significantly reduced the incidence rate of reported low back disorders in repetitive manual handling jobs.[87]

Henry Ford offered good advice in the early 1900s when he said: "The work must be brought to the man waist-high. No worker must ever have to stoop to attach a wheel, a bolt, screw or anything else to the moving chassis."[88] Henry Ford was probably more interested in the extra time it takes to bend over, but the principle remains the same. Excessive bending is not good; it takes more time, it increases the probability of low back disability, and it increases the chance of aggravating an existing disorder.

Manual handling tasks should also be designed according to accepted guidelines for maximum weights and forces.[89,90] A study by Marras and his associates compared the strengths and weaknesses of the 1981 NIOSH Work Practices Guide, the 1993 NIOSH Revised Lifting Equation, and the Liberty Mutual psychophysical tables.[91] The results indicated that all three methods were predictive of low back disorders, but in different ways. The 1981 NIOSH Guide underestimated the risk, the 1993 NIOSH Equation overestimated the risk, and the psychophysical tables fell between the two. The 1993 NIOSH Equation was developed specifically for lifting tasks, and assumes that lifting and lowering tasks have the same level of risk for low back disorders. The psychophysical tables can be used to evaluate all types of manual handling tasks (i.e., lifting, lowering, pushing, pulling, and carrying).

Sit–stand workstations are also helpful. For jobs that require prolonged sitting, the sit–stand workstation allows the employee to change from a sitting posture to a standing posture (and vice versa) without interfering with the job. This is important for people with low back pain, because they cannot maintain the same posture for long periods of time.

A randomized clinical trial of 130 workers who had been absent from work for 6 weeks because of back pain investigated the effects of an occupational intervention that included ergonomic redesign and time limited light duties (the Sherbrooke Model). The median time off regular work was 67 days for workers with the occupational intervention compared to 131 days for workers without the occupational intervention.[92]

Another prospective study investigated one company that changed from standard care to an occupational management approach that included worker rotation schedules, reduced lifting loads, and ergonomic redesign of tasks. The total days lost per 100,000 h worked dropped from 60.9 days for standard care to 1.1 days for the occupational management approach.[93]

Management is often reluctant to redesign jobs because of the costs involved. However, committing capital to redesign jobs can often be a wise business investment. Decreases in compensation costs and increases in worker performance will return the cost of the initial investment over time. Determining the "payback period" will help convince management of the cost-effectiveness of redesigning jobs. Although ergonomics alone will not solve the low back disability problem, it can be an important part of the solution.

24.4.2.2 Proactive Return-to-Work Program

There is strong evidence that the longer a worker is off work with low back pain, the lower their chances of ever returning to work.[32,33,71,83,84,94] "Once a worker is off work for 4 to 12 weeks, they have a 10% to 40% risk (depending on the setting) of still being off work at one year; after 1 to 2 yr absence it is unlikely they will return to any form of work in the foreseeable future, irrespective of further treatment."[32] Furthermore, there is strong evidence that most clinical interventions are quite ineffective at returning people to work once they have been off work for a protracted period with low back pain.[32,33]

These statistics emphasize the importance of providing modified, alternative, or part-time work as a means of returning the disabled employee to the job as quickly as possible. The U.K. Occupational Health Guidelines recommend that the worker be encouraged to remain in his or her job, or to return at an early stage, even if there is still some low back pain — do not wait until they are completely pain-free.[32] There is moderate evidence that the temporary provision of lighter or modified duties facilitates return to work and reduces time off work.[32,33] A recent review of the literature concluded that modified work programs reduce the number of lost work days by half.[95]

A proactive return-to-work program is a supportive, company-based intervention for personally assisting the disabled employee from the beginning of the episode to its positive resolution. In a proactive program, the actions and responsibilities of individuals within the company and external providers are spelled out and related to the goal of resumption of employment.[82]

In the Michigan Disability Prevention Study, companies reporting a 10% greater level of achievement on the proactive return to work variable demonstrated a significant 13.6% lower rate of lost workday cases ($p < 0.01$).[82,96]

24.4.2.3 Communication

There is moderate evidence that communication, cooperation, and common agreed goals between the worker with low back pain, the occupational health team, supervisors, management, and primary health care professionals are fundamental for improvement in clinical and occupational health management and outcomes.[32,33] According to Burton, workers have come to believe that any back pain is likely caused by their work. Society has tended to reinforce this belief. "Attribution in particular is an important factor for compliance with intervention strategies, and it has been found that a simple educational program that detunes perception of attribution is capable of creating a positive shift in beliefs with a concomitant reduction in extended sickness absence."[63]

An uncontrolled study with hospital staff increased communications between the worker, employer, practitioner, and insurer.[97] When a compensation claim was received, the employer made immediate and repeated contacts with the worker and the insurer. The tone of the communication was always pleasant and the focus was always on the best interests of the worker. The message strongly communicated to the worker was: "You are a vital part of our team, your work is important, and your job is waiting for you." The program significantly reduced the proportion of long-term low back disability claims from 7.1 to 1.7%.

The power of the media in altering beliefs about back pain was recently demonstrated in an Australian state-wide public health intervention, entitled "Back Pain: Don't Take It Lying Down".[98-100] Television commercials aired during prime time offered advice on staying active and exercising, not resting for prolonged periods, and remaining at work. The television campaign was augmented by radio and printed advertisements, outdoor billboards, posters, seminars, workplace visits, publicity articles, and endorsements by sports celebrities and leading back pain researchers. An educational booklet (*The Back Book*)[120] was made available in 16 languages, and all doctors received evidence-based guidelines for treating low back pain. Results showed that population beliefs about back pain became more positive. Beliefs about back pain also improved among doctors. There was a clear decline in number of claims for back pain, rates of days compensated, and medical payments for back pain claims. The three million Australian-dollar investment led to an estimated 36 million dollar savings in disability claims, and a 5.7 million dollar reduction in medical expenses.[100]

24.4.2.4 Selection of Health Care Providers

Management should select health care providers who diagnose and treat low back pain according to accepted guidelines.[32,33,101-104] Nonspecific low back pain is benign. It should be treated and managed by primary care providers, and not referred to specialists without clear indications. Linton refers to "secondary prevention" as quality pain management at the primary care level. "The idea is to provide better multidimensional care, a little earlier, and with better coordination with other agents (e.g., the workplace, insurance companies, and authorities)."[105]

The making of a "specific" diagnosis at the beginning of a compensated episode carries the message that the condition is serious and that a "specific" clinical procedure must be carried out. One consequence of this "labeling effect" is to investigate and treat the lesion suspected of being the cause of the pain, rather than focusing on the functional recovery. Abenhaim et al. believe that "this situation encourages patients to believe there is a cure for their problem when it is known that only a small number will respond to a specific therapy."[77]

The U.K. Occupational Health Guidelines suggest that a simple clinical interview and examination can distinguish between simple back pain manageable at the primary care level and those pathological conditions requiring specialist referral. "Conventional clinical tests of spinal and neurological function are of limited value ... X-rays and MRI are primarily directed to the investigation of nerve root problems and serious spinal pathology. Much more relevant to occupational health management is the identification of individual and work-related psychosocial issues which form risk factors for chronicity."[33]

A recent Australian study demonstrated that evidence-based treatment guidelines really do work.[106] Thirteen special urban and rural clinics were established with trained medical practitioners who treated low back pain patients according to evidence-based guidelines. "The guidelines emphasize dealing with patients' fears and misconceptions, providing confident explanation, and empowering the patient to resume or restore normal activities of daily living through simple exercises and graded activity ... supplemented as necessary by simple measures such as analgesics and manual therapy for symptomatic relief."[106] The outcomes of these patients were compared with the outcomes of patients treated by their own general practitioners in their normal manner. A significantly greater reduction in pain, significantly fewer patients requiring continuing care, and a significantly greater proportion of patients fully recovered at 12 months were reported for patients treated according to the guidelines. These patients rated their treatment as extremely helpful, and offered positive unsolicited comments about their treatment.

24.4.2.5 Management Commitment

Occupational safety and health literature stresses the importance of management commitment in successful safety and health programs.[107–109] Safety must be managed in the same way as finance, sales, production, and advertising. The chief executive officer must also become the chief safety officer.[107]

Specific ways in which management commitment can be demonstrated include establishing goals, assigning staff, providing staff time, making resources available, evaluating results, communicating with employees, and encouraging employee involvement in the program. The intent is to create a corporate culture with a positive and supportive attitude regarding employees with low back pain. Trust building and employee advocacy are important ingredients.

The U.K. Occupational Health Guidelines report general consensus but limited scientific evidence that workplace organizational and/or management strategies may reduce absenteeism and duration of work loss.[32,33] High job satisfaction and good industrial relations are the most important organizational characteristics associated with low disability and sickness absence rates attributed to low back pain. One uncontrolled study reported a decrease in lost time low back pain cases from 20 per yr to 2 per yr, after a change in top management commitment.[107]

24.4.2.6 Supervisor Training

In general, supervisors do not respond well when workers experience low back pain. Employees are often accused of malingering, whether by innuendo or direct confrontation. Adversary situations are set up, and employees look for ways to retaliate. Prolonged disability is often the result, with increased costs for both management and the employee.

Supervisors must be trained in the true nature of nonspecific low back pain, that is, that low pain is a disorder of unknown cause that happens to practically everyone, that low back pain usually develops gradually and insidiously without an unusual or strenuous activity, that low back pain may recur frequently, and that low back pain does not respond well to treatment but usually resolves itself within a few days or weeks.[67]

The best way for a supervisor to respond to a disorder of this type is to show some concern for the needs of the employee, to avoid making judgments and setting up adversary relationships, to encourage the employee to seek appropriate medical treatment, and to consider adapting the workplace (or modifying the job) so the employee may continue working on the job.

An uncontrolled study conducted by Fitzler and Berger consisted of conservative, supportive, in-house treatment of low back pain while keeping the employee on the job.[110,111] The occupational health nurse treated the worker with heat, mild analgesics, and advice on proper postures and exercises, while the safety engineer looked at the job to see what modifications could be accomplished in the workplace. If necessary, light duty was available. A significant part of the program was the training given to management, supervisors, and workers on the true nature of low back pain, and the positive acceptance of low back pain when it does occur. Over a 3-yr period, the number of lost time low back pain cases was reduced by 32%, and the cost of low back pain compensation claims was reduced by over 90%. There was no reduction in low back pain, but a rather dramatic reduction in low back disability.

A disability management program for supervisors was developed by McLellan et al.[112] A total of 108 supervisors representing 7 employers were provided with a 1.5 h training session to reinforce a proactive and supportive response to work-related musculoskeletal symptoms and injuries among employees. Results showed improvements in supervisor confidence to investigate and modify job factors contributing to injury, to get medical advice, and to answer employees' questions. 38.5% of supervisors reported decreases in lost time within their department.

24.4.2.7 Coordination with Organized Labor

Traditionally, labor unions are opposed to early return to work after a disabling episode of low back pain. It is thought that the employee is entitled to time off for even a minor episode, despite sound medical evidence that activity and work will hasten healing.[68,113–115] Organized labor must be involved in the planning and execution of a disability reduction program. There must be agreement between labor and management on what constitutes the best interests of the employee.

24.5 Conclusions

At the present time, the prevention and treatment of low back pain have not been very successful. The bottom line is that we cannot really prevent low back pain — primarily because we cannot control aging, we cannot control genetics, and we cannot control personal behavior. However, we can control the job; and because we can control the job, we can reduce the disability from low back pain — and it is the disability that is the primary problem for industry.

Much of the responsibility for reducing low back disability rests in the lap of management.[68] Management must assume a more active role; they are the principal defense in the battle to reduce low back disability. Management must recognize that the primary responsibility for reducing low back disability belongs to them — not to the medical establishment, not to the insurance company, not to the Workers' Compensation Board, but to themselves.

Health care providers must also recognize that they have an important, but limited, role in reducing low back disability. Health care providers must first attend to the patient with relief or control of symptoms, a correct and objective diagnosis, referral to specialists only if clearly indicated, patient education and counseling, and early activation and return to work. The health care provider must also find ways to reassure and empower the patient. An equally important role for the health care provider is to work with management in developing training programs, ergonomics programs, and policies and practices that have been shown to be effective in reducing low back disability.[75]

According to the U.K. Occupational Health Guidelines report, "Helping and supporting the worker to remain at work, or in early return to work, is in principle the most promising means of reducing future symptoms, sickness absence, and claims."[33]

The Michigan Disability Prevention Study concluded that "disability can be managed; and those who do it well can expect to be rewarded with lower disability costs, more satisfied workers, greater productivity and, ultimately, higher profits."[82]

References

1. Deyo, R.A. and Weinstein, J.N., Low back pain, *N. Engl. J. Med.*, 344, 363, 2001.
2. Frymoyer, J.W. and Cats-Baril, W.L., An overview of the incidences and costs of low back pain, *Orthop. Clin. North Am.*, 22, 263, 1991.
3. Jonsson, E., Preface, in *Neck and Back Pain: the Scientific Evidence of Causes, Diagnosis, and Treatment*, Nachemson, A.F. and Jonsson, E., Eds., Lippincott Williams and Wilkins, Philadelphia, PA, 2000.
4. Smith, R., Where is the wisdom. . .? the poverty of medical evidence, *Br. Med. J.*, 303, 798, 1991.
5. Akeson, W.H. and Murphy, R.W., Editorial comment: low back pain, *Clin. Orthop.*, 129, 2, 1977.
6. White, A.A. and Gordon, S.L., Synopsis: workshop on idiopathic low-back pain, *Spine*, 7, 141, 1982.
7. Spitzer, W.O., LeBlanc, F.E., and Dupuis, M., et al., Scientific approach to the assessment and management of activity-related spinal disorders: a monograph for clinicians. Report of the Quebec Task Force on Spinal Disorders, *Spine*, 12(7S), S5, 1987.
8. Hall, H., McIntosh, G., Wilson, L., and Melles, T., Spontaneous onset of back pain, *Clin. J. Pain*, 14, 129, 1998.
9. Battié, M.C., Minimizing the impact of back pain: workplace strategies, *Sem. Spine Surg.*, 4, 20, 1992.
10. Hirsch, C., Etiology and pathogenesis of low back pain, *Israel J. Med. Sci.*, 2–3, 362, 1966.
11. Rowe, M.L., *Backache at Work*, Perinton Press, Fairport, NY, 1983.
12. Snook, S.H., Campanelli, R.A., and Hart, J.W., A study of three preventive approaches to low back injury. *J. Occup. Med.*, 20, 478, 1978.
13. Nachemson, A.L., The lumbar spine: an orthopaedic challenge, *Spine*, 1, 59, 1976.
14. Smyth, M.J. and Wright, V., Sciatica and the intervertebral disc: an experimental study, *J. Bone Joint Surg. (Am.)*, 40, 1401, 1958.
15. Kuslich, S. and Ulstrom, C., The tissue origin of low back pain and sciatica: a report of pain response to tissue stimulation during operations on the lumbar spine using local anesthesia, *Orthop. Clin. North Am.*, 22, 181, 1991.
16. O'Neill, C.W., Kurgansky, M.E., Derby, R., and Ryan, D.P., Disc stimulation and pattern of referred pain, *Spine*, 27, 2776, 2002.
17. Videman, T. and Battié, M.C., A critical review of the epidemiology of idiopathic back pain, in *Low Back Pain: A Scientific and Clinical Overview*, Weinstein, J.N. and Gordon, S.L., Eds., American Academy of Orthopedic Surgeons, Rosemont, IL, 1996, 317.
18. Kayama, S., Konno, S., Olmarker, K., Yabuki, S., and Kikuchi, S., Incision of the anulus fibrosus induces nerve root morphologic, vascular, and functional changes: an experimental study, *Spine*, 21, 2539, 1996.
19. Donelson, R., Aprill, C., Medcalf, R., and Grant, W., A prospective study of centralization of lumbar and referred pain: a predictor of symptomatic discs and anular competence, *Spine*, 22, 1115, 1997.
20. Videman, T., Battié, M.C., Gibbons, L.E., Maravilla, K., Manninen, H., and Kaprio, J., Associations between back pain history and lumbar MRI findings. *Spine*, 28, 582, 2003.
21. Allan, D.B. and Waddell, G., An historical perspective on low back pain and disability, *Acta. Orthop. Scand.*, 60 (Suppl. 234), 1, 1989.
22. Linton, S.J., The relationship between activity and chronic back pain, *Pain*, 21, 289, 1985.
23. Snook, S.H., Webster, B.S., McGorry, R.W., Fogleman, M.T., and McCann, K.B., The reduction of chronic, non-specific low back pain through the control of early morning lumbar flexion: a randomized controlled trial, *Spine*, 23, 2601, 1998.
24. Von Korff, M., Deyo, R., Cherkin, D., and Barlow, W., Back pain in primary care: outcomes at 1 year, *Spine*, 18, 855, 1993.
25. Waddell, G., *The Back Pain Revolution*, Churchill Livingstone, Edinburgh, 1998.
26. Waddell, G., *Epidemiology Review: The Epidemiology and Cost of Back Pain. The Annex to the Clinical Standards Advisory Group's Report on Back Pain*, HMSO, London, 1994.

27. Waxman, R., Tennant, A., and Helliwell, P., A prospective follow-up study of low back pain in the community, *Spine*, 25, 2085, 2000.

28. Boos, N., Weissbach, S., Rohrbach, H., Weiler, C., Spratt, K.F., and Nerlich, A.G., Classification of age-related changes in lumbar intervertebral discs, *Spine*, 27, 2631, 2002.

29. Miller, J.A.A., Schmatz, C., and Schultz, A.B., Lumbar disc degeneration: correlation with age, sex, and spine level in 600 autopsy specimens. *Spine*, 13, 173, 1988.

30. Vingård, E. and Nachemson, A., Work-related influences on neck and low back pain, in *Neck and Back Pain: The Scientific Evidence of Causes, Diagnosis, and Treatment*, Nachemson, A. and Jonsson, E., Eds., Lippincott Williams and Wilkins, Philadelphia, PA, 2000, 97.

31. National Research Council and Institute of Medicine, *Musculoskeletal Disorders and the Workplace: Low Back and Upper Extremities*, National Academy Press, Washington, DC, 2001.

32. Carter, J.T. and Birrell, L.N., Eds., *Occupational Health Guidelines for the Management of Low Back Pain at Work — Principal Recommendations*, Faculty of Occupational Medicine, London, 2000.

33. Waddell, G. and Burton, A.K., *Occupational Health Guidelines for the Management of Low Back Pain at Work — Evidence Review*, Faculty of Occupational Medicine, London, 2000.

34. Marras, W.S., Granata, K.P., Davis, K.G., Allread, W.G., and Jorgensen, M.J., Effects of box features on spine loading during warehouse order selecting, *Ergonomics*, 42, 980, 1999.

35. Marras, W.S., Occupational low back disorder causation and control, *Ergonomics*, 43, 880, 2000.

36. Matsui, H., Kanamori, M., Ishihara, H., Yudoh, K., Naruse, Y., and Tsuji, H., Familial predisposition for lumbar degenerative disc disease, *Spine*, 23, 1029, 1998.

37. Postacchini, F., Lami, R., and Pugliese, O., Familial predisposition to discogenic low-back pain: an epidemiologic and immunogenetic study, *Spine*, 13, 1403, 1988.

38. Richardson, J.K., Chung, T., Schultz, J.S., and Hurvitz, E., A familial predisposition toward lumbar disc injury, *Spine*, 22, 1487, 1977.

39. Battié, M.C., Videman, T., Gibbons, L.E., Fisher, L.D., Manninen, H., and Gill, K., Determinants of lumbar disc degeneration, *Spine*, 20, 2601, 1995.

40. Sambrook, P.N., MacGregor, A.J., and Spector, T.D., Genetic influences on cervical and lumbar disc degeneration: a magnetic resonance imaging study in twins, *Arthritis and Rheum.*, 42, 366, 1999.

41. Leboeuf-Yde, C., Body weight and low back pain: a systematic literature review of 56 journal articles reporting on 65 epidemiologic studies, *Spine*, 25, 226, 2000.

42. Deyo, R.A. and Bass, J.E., Lifestyle and low-back pain: the influence of smoking and obesity, *Spine*, 14, 501, 1989.

43. Goldberg, M.S., Scott, S.C., and Mayo, N.E., A review of the association between cigarette smoking and the development of nonspecific back pain and related outcomes, *Spine*, 25, 995, 2000.

44. Nachemson, A.L. and Vingård, E., Influences of individual factors and smoking on neck and low back pain, in *Neck and Back Pain: the Scientific Evidence of Causes, Diagnosis, and Treatment*, Nachemson, A.F. and Jonsson, E., Eds., Lippincott Williams and Wilkins, Philadelphia, PA, 2000, 79.

45. Nachemson, A.L. and Vingård, E., Assessment of patients with neck and back pain: a best-evidence synthesis, in *Neck and Back Pain: the Scientific Evidence of Causes, Diagnosis, and Treatment*, Nachemson, A.F. and Jonsson, E., Eds., Lippincott Williams and Wilkins, Philadelphia, PA, 2000, 189.

46. Jarvik, J.G. and Deyo, R.A., Imaging of lumbar intervertebral disk degeneration and aging, excluding disk herniations, *Radiol Clin. of North Am.*, 38, 1255, 2000.

47. Roland, M. and van Tulder, M., Should radiologists change the way they report plain radiography of the spine? *Lancet*, 352, 229, 1998.

48. Nachemson, A.L., Introduction, in *Neck and Back Pain: the Scientific Evidence of Causes, Diagnosis, and Treatment*, Nachemson, A.F. and Jonsson, E., Eds., Lippincott Williams and Wilkins, Philadelphia, PA, 2000, 1.

49. Van Tulder, M.W. and Waddell, G., Conservative treatment of acute and subacute low back pain, in *Neck and Back Pain: the Scientific Evidence of Causes, Diagnosis, and Treatment*, Nachemson, A. and Jonsson, E., Eds., Lippincott Williams and Wilkins, Philadelphia, PA, 2000.

50. Carey, T.S., Garrett, J., and Jackman, A., et al. The outcomes and costs of care for acute low back pain among patients seen by primary care practitioners, chiropractors, and orthopedic surgeons, *N. Engl. J. Med.*, 333, 913, 1995.
51. Waddell, G., A new clinical model for the treatment of low-back pain, *Spine*, 12, 632, 1987.
52. Mahmud, M.A., Webster, B.S., Courtney, T.K., Matz, S., Tacci, J.A., and Christiani, D.C., Clinical management and the duration of disability for work-related low back pain, *J. Occup. Environ. Med.*, 42, 1178, 2000.
53. Andersson, G.B.J., Lucente, T., Davis, A.M., Kappler, R.E., Lipton, J.A., and Leurgans, S., A comparison of osteopathic spinal manipulation with standard care for patients with low back pain, *N. Engl. J. Med.*, 341, 1426, 1999.
54. Cherkin, D.C., Deyo, R.A., Battié, M., Street, J., and Barlow, W., A comparison of physical therapy, chiropractic manipulation, and provision of an educational booklet for the treatment of patients with low back pain, *N. Engl. J. Med.*, 339, 1021, 1998.
55. Shekelle, P.G., What role for chiropractic in health care? *N. Engl. J. Med.*, 339, 1074, 1998.
56. Carey, T.S., Evans, A.T., and Hadler, N.M., et. al., Acute severe low back pain: a population-based study of prevalence and care-seeking, *Spine* 21, 339, 1996.
57. Koes, B.W., van Tulder, M.W., Ostelo, R., Burton, A.K., and Waddell, G., Clinical guidelines for the management of low back pain in primary care: an international comparison, *Spine* 26, 2504, 2001.
58. Von Korff, M. and Saunders, K., The course of back pain in primary care, *Spine*, 21, 2833, 1996.
59. Roland, M., Waddell, G., Moffat, J., Burton, K., Main, C., and Cantrell, T., *The Back Book*, The Stationery Office, London, 1996.
60. Ito, T., Takano, Y. and Yuasa, N., Types of lumbar herniated disc and clinical course, *Spine*, 26, 648, 2001.
61. Postacchini, F., Editorial. Lumbar disc herniation: a new equilibrium is needed between nonoperative and operative treatment, *Spine*, 26, 601, 2001.
62. Wasiak, R., Pransky, G.S., and Webster, B.S., Methodological challenges in studying recurrence of low back pain, *J. Occup. Rehabil.*, 13, 21, 2003.
63. Burton, A.K., Spine update: back injury and work loss, *Spine*, 22, 2575, 1997.
64. Frymoyer, J.W. and Durett, C.L., The economics of spinal disorders, in *The Adult Spine: Principles and Practice*, 2nd ed., Frymoyer, J.W., Ed., Lippincott-Raven Publishers, Philadelphia, PA, 1997.
65. Frank, J.W., Kerr, M.S., Brooker, A.S., DeMaio, S.E., Maetzel, A., Shannon, H.S., Sullivan, T.J., Norman, R.W., and Wells, R.P., Disability resulting from occupational low back pain. Part I: what do we know about primary prevention? A review of the scientific evidence on prevention before disability begins, *Spine*, 21, 2908, 1996.
66. Frank, J.W., Brooker, A.S., DeMaio, S.E., Kerr, M.S., Maetzel, A., Shannon, H.S., Sullivan, T.J., Norman, R.W., and Wells, R.P., Disability resulting from occupational low back pain. Part II: what do we know about secondary prevention? A review of the scientific evidence on prevention after disability begins, *Spine*, 21, 2918, 1996.
67. Snook, S.H., The control of low back disability: the role of management, in *Manual Material Handling: Understanding and Preventing Back Trauma*, Kroemer, K.H.E., McGlothlin, J.D., and Bobick, T.G., Eds., American Industrial Hygiene Association, 1989, p. 97.
68. Spitzer, W.O., Low back pain in the workplace: attainable benefits not attained, *Br. J. Ind. Med.*, 50, 385, 1993.
69. Fordyce, W.E., Ed., *Back Pain in the Workplace: Management of Disability in Nonspecific Conditions*, International Association for the Study of Pain. IASP Press, Seattle, WA, 1995.
70. Lawrence, R.C., Helmick, C.G., and Arnett, F.C., et. al., Estimates of the prevalence of arthritis and selected musculoskeletal disorders in the United States, *Arthritis Rheum.* 41, 778, 1998.
71. Hashemi, L., Webster, B.S., Clancy, E.A., and Volinn, E., Length of disability and cost of workers' compensation low back pain claims, *J. Occup. Environ. Med.*, 39, 937, 1997.
72. Bureau of Labor Statistics, *Issues in Labor Statistics: Older Workers' Injuries Entail Lengthy Absences from Work*, Summary 96-6, U.S. Department of Labor, 1996.

73. Mayer, T., Gatchel, R.J., and Evans, T., Effect of age on outcomes of tertiary rehabilitation for chronic disabling spinal disorders. *Spine*, 26, 1378, 2001.

74. Punnett, L., Fine, L.J., Keyserling, W.M., Herrin, G.D., and Chaffin, D.B., Back disorders and non-neutral trunk postures of automobile assembly workers, *Scand. J. Work Environ. Health*, 17, 337, 1991.

75. Waddell, G., Aylward, M., and Sawney, P., *Back Pain, Incapacity for Work and Social Security Benefits: An International Literature Review and Analysis*, The Royal Society of Medicine Press, London, 2002.

76. Linton, S.J., Psychological risk factors for neck and back pain, in *Neck and Back Pain: The Scientific Evidence of Causes, Diagnosis, and Treatment*, Nachemson, A.F. and Jonsson, E., Eds., Lippincott Williams and Wilkins, Philadelphia, PA, 2000, 57.

77. Abenhaim, L., Rossignol, M., Gobeille, D., Bonvalot, Y., Fines, P., and Scott, S., The prognostic consequences in the making of the initial medical diagnosis of work-related back injuries, *Spine*, 20, 791, 1995.

78. Frank, J.W., Sinclair, S., Hogg-Johnson, S., Shannon, H., Bombardier, C., Beaton, D., and Cole, D., Preventing disability from work-related low-back pain, *Can. Med. Assoc. J.*, 158, 1625, 1998.

79. Hrudey, W.P., Overdiagnosis and overtreatment of low back pain: long term effects, *J. Occup. Rehabil.*, 1, 303, 1991.

80. Deyo, R.A., Editorial: Pain and public policy, *N. Engl. J. Med.*, 342, 1211, 2000.

81. Workers' Compensation Monitor, *Disability leave: only half of injured workers were contacted by supervisor*, 11, 9, Sept. 1998.

82. Hunt, H.A. and Habeck, R.V., *The Michigan Disability Prevention Study: Research Highlights*, W. E. Upjohn Institute for Employment Research, Kalamazoo, MI, 1993.

83. McGill, C.M., Industrial back problems: a control program, *J. Occup. Med.*, 10, 174, 1968.

84. Rosen, N.B., Treating the many facets of pain, *Business and Health*, 3, 7, May 1986.

85. Snook, S.H., Approaches to the control of back pain in industry: job design, job placement and education/training, *Spine S.O.A. Rev.*, 2, 45, 1987.

86. Loeser, J.D., Henderlite, S.E., and Conrad, D.A., Incentive effects of workers' compensation benefits: a literature synthesis, *Med. Care Res. Rev.*, 52, 34, 1995.

87. Marras, W.S., Allread, W.G., Burr, D.L., and Fathallah, F.A., Prospective validation of a low-back disorder risk model and assessment of ergonomic interventions associated with manual materials handling tasks, *Ergonomics*, 43, 1866, 2000.

88. Burlingame, R., *Henry Ford: A Great Life in Brief.* Alfred A. Knopf, New York, 1966.

89. Snook, S.H. and Ciriello, V.M., The design of manual handling tasks: revised tables of maximum acceptable weights and forces, *Ergonomics*, 34, 1197, 1991.

90. Waters, T.R., Putz-Anderson, V., Garg, A., and Fine, L.J., Revised N.I.O.S.H. equation for the design and evaluation of manual lifting tasks. *Ergonomics*, 36, 749, 1993.

91. Marras, W.S., Fine, L.J., Ferguson, S.A., and Waters, T.R., The effectiveness of commonly used lifting assessment methods to identify industrial jobs associated with elevated risk of low-back disorders, *Ergonomics*, 42, 229, 1999.

92. Loisell, P., Abenhaim, L., Durand, P., Esdaile, J.M., Suissa, S., Gosselin, L., Simard, R., Turcotte, J., and Lemaire, J., A population-based, randomized clinical trial on back pain management, *Spine*, 22, 2911, 1997.

93. Lemstra, M. and Olszynski, W.P., The effectiveness of standard care, early intervention, and occupational management in worker's compensation claims, *Spine*, 28, 299, 2003.

94. Andersson, G.B.J., Svensson, H.O., and Odén, A., The intensity of work recovery in low back pain, *Spine*, 8, 880, 1983.

95. Krause, N., Dasinger, L.K., and Neuhauser, F., Modified work and return to work: a review of the literature, *J. Occup. Rehabil.*, 8, 113, 1998.

96. Hunt, H.A., Habeck, R.V., VanTol, B., and Scully, S.M., *Disability Prevention Among Michigan Employers*, Upjohn Institute Technical Report No. 93-004, W.E. Upjohn Institute for Employment Research, Kalamazoo, MI, 1993.

97. Wood, D.J., Design and evaluation of a back injury prevention program within a geriatric hospital, *Spine*, 12, 77, 1987.

98. Buchbinder, R., Jolley, D., and Wyatt, M., Population based intervention to change back pain belief and disability: three part evaluation, *Br. Med. J.*, 322, 1516, 2001.

99. Buchbinder, R., Jolley, D., and Wyatt, M., Effects of a media campaign on back pain beliefs and its potential influence on management of low back pain in general practice, *Spine*, 26, 2535, 2001.

100. The BackLetter, Lippincott Williams and Wilkins, Philadelphia, PA 16, 73, 2001.

101. *Clinical Guidelines for the Management of Acute Low Back Pain.* Royal College of General Practitioners, London, 1996 (Revised 1999).

102. Bigos, S., Bowyer, O., Braen, G., et. al., *Acute Low Back Problems in Adults. Clinical Practice Guideline No. 14*, AHCPR Publication No. 95-0642, Agency for Health Care Policy and Research, Public Health Service, U.S. Department of Health and Human Services, Rockville, MD, 1994.

103. Low Back Complaints, in *Occupational Medicine Practice Guidelines*, Harris, J.S., Ed., American College of Occupational and Environmental Medicine, OEM Press, Beverly, MA, 1997, Chapter 14.

104. *Report of a CSAG Committee on Back Pain*, Clinical Standards Advisory Group, National Health Service, HMSO, London, 1994.

105. Linton, S.J., Editorial. The socioeconomic impact of chronic back pain: is anyone benefiting? *Pain*, 75, 163, 1998.

106. McQuirk, B., King, W., Govind, J., Lowry, J., and Bogduk, N., Safety, efficacy, and cost effectiveness of evidence-based guidelines for the management of acute low back pain in primary care, *Spine*, 26, 2615, 2001.

107. Griffiths, D.K., Safety attitudes of management, *Ergonomics*, 28, 61, 1985.

108. *Ergonomics program management guidelines for meatpacking plants.* OSHA 3123, Occupational Safety and Health Administration, U.S. Department of Labor, 1990.

109. *Private Sector Ergonomics Programs Yield Positive Results*, GAO/HEHS-97-163, U.S. General Accounting Office, Washington, DC, 1997.

110. Fitzler, S.L. and Berger, R.A., Attitudinal change: the Chelsea back program, *Occup. Health Safety*, 51, 24, 1982.

111. Fitzler, S.L. and Berger, R.A., Chelsea back program: one year later, *Occup. Health Safety*, 52, 52, 1983.

112. McLellan, R.K., Pransky, G., and Shaw, W.S., Disability management training for supervisors: a pilot intervention program, *J. Occup. Rehabil.* 11, 33, 2001.

113. Derebery, V.J. and Tullis, W.H., Delayed recovery in the patient with a work compensable injury, *J. Occup. Med.*, 25, 829, 1983.

114. Nachemson, A.L., Work for all, *Clin. Orthop. Relat. Res.*, 179, 77, 1983.

115. Malmivaara, A., Hakkinen, U., and Aro, T., et. al., The treatment of acute low back pain — bed rest, exercises, or ordinary activity? *N. Engl. J. Med.* 332, 351, 1995.

25

Psychosocial Characteristics of Work as Targets for Change

Catherine A. Heaney
Stanford University

Kaori Fujishiro
University of Illinois at Chicago

25.1 Introduction

During the last few decades, there has been a tremendous growth in our understanding of the causes of occupational injuries and illnesses, resulting in a significant decline in fatal and severe injuries experienced by workers. The U.S. Centers for Disease Control and Prevention has recognized the importance of these improvements in occupational safety and health by designating "safer workplaces" as one of the ten great public health achievements of the 20th century.[1]

One component of this increased understanding is a broadening of the scope of occupational risk factors being addressed. Although continuing to address the physical and chemical hazards in the workplace that have been the traditional focus of occupational safety and health (OSH) efforts, researchers and OSH professionals have also recognized the role of work organization factors and their potential effects on worker health. This trend is particularly evident in the literature addressing the prevention of work-related musculoskeletal disorders (WMSDs). All of the recent major reviews and critical assessments of this literature have devoted considerable energy to critiquing the potential role of work organization and psychosocial work characteristics in the etiology of WMSDs.[2-4]

This chapter will address the implications of this growing emphasis on psychosocial work characteristics for the prevention of WMSDs. More specifically, Section 25.2 will define work organization factors and psychosocial work characteristics and examine conceptual models linking them to

WMSDs. Section 25.3 will describe implications for ergonomic interventions and assess the relevant intervention literature. Section 25.4 will present an agenda for future intervention research in this area. It should be noted that the scope of this chapter is limited to the role of psychosocial work characteristics in the initial experience or the incidence of WMSDs. Psychosocial factors also play a major role in the clinical management, recovery and recurrence of WSMDs; however, this role is complex[5-7] and will not be described here.

25.2 Psychosocial Work Characteristics and WMSDs

The ergonomic literature has tended to use the term "psychosocial" quite indiscriminately. All of the following have been labeled a "psychosocial factor" in recent studies of WMSDs: job satisfaction; satisfaction with specific aspects of work including pay and opportunities for advancement; job security; job task characteristics such as concentration demands and task variety; organizational climate; control over work; relationships with supervisors and coworkers; fatigue and exhaustion; socioeconomic characteristics of the workers; job demands; work pace; job stress; worker personality traits; and employee depression and anxiety (see, for reviews, Ref. 8–10). In short, every aspect of work that does not fit under the umbrella of physical or biomechanical job demands and every aspect of the worker that is not a reflection of physical capability appears to have the potential of being classified as a "psychosocial" factor.

While some of the early definitions of psychosocial work factors reflect this diffuse and broad conceptualization,[11] we would like to offer a more focused framework. This framework begins with the concept of work organization. This is a multilevel concept that includes "the work process [the way jobs are designed and performed] and the organizational practices [management and production methods and accompanying human resource policies] that influence job design" (Ref. 12, p. 2). The work process includes how work is scheduled and paced, the content and structure of the work (e.g., complexity and repetition), how job-related decisions are made, and how employees are supervised. Organizational practices and policies that influence the work process include quality and process management initiatives (e.g., lean production and total quality management), alternative employment arrangements (e.g., contract employees and job sharing), pay systems (e.g., incentive-based pay), communication systems and norms, opportunities for career advancement, and human resource policies (e.g., employee benefits).[12]

Work organization can be measured without relying on the reports of the employees themselves: work cycles can be counted, rest breaks can be quantified, interactions with supervisors can be observed, and written policies can be reviewed. However, the organization of work gives rise to employees' experience of work. Some of these experiences are physical in nature, some are psychological, and some are social. The latter two categories can be combined and constitute "psychosocial characteristics of work." These characteristics of work are defined or created through the workers' cognitive appraisals and affective processing of the job. For example, organizing work so that employees specialize in one task and have no job rotation can result in repetitive work (a physical characteristic). However, for the job to be deemed monotonous or boring (a psychosocial characteristic), employees must appraise the repetitive nature of the job and assess the extent to which it violates their own expectations or preferences. Another example is work pace. Setting high production standards (a work organization factor) can lead to a rapid pace of work (a physical work characteristic). For employees to experience work overload or perceive that they have too little time to complete the required work, they must appraise the demands of the rapid work pace and compare them to their own capabilities and expectations.

Psychosocial work characteristics are based on employees' perceptions, and people's perceptions of their work are likely to be a reflection of both the work itself and the person's expectations, previous work history, resources for coping with work, and general mood and well-being. Thus, measures of psychosocial work characteristics provide somewhat uncertain cues as to what is contributing to

WMSDs — is it the demands of work or the attitudes and resources of the worker (or both)? Psychosocial work characteristics can only be measured through self-report. However the amount of subjectivity involved in a given self-report depends on how the questions are worded. For example, questions that ask the worker to make judgments or interpretations (e.g., "How heavy is your usual workload?") introduce more subjectivity than do questions that ask for more descriptive, quantifiable assessments (e.g., "On how many days a week are you unable to complete the work assigned to you?"). But it is not unusual for psychosocial work characteristics to be only weakly or moderately correlated with the work organization factors that give rise to them.[13] For example, in a study of police dispatchers, self-reported workload was only moderately correlated ($r = 0.35$) with workload as measured by the observed number of work activities completed per hour.[14] However, extreme or intense work organization factors are likely to be perceived similarly by employees exposed to them, thus reducing variation and increasing the correlation between the work organization measure and the psychosocial work characteristic.[15]

Figure 25.1 presents four models of how work organization factors can be linked with WMSDs. Model 1 indicates that work organization factors can give rise to both physical and psychosocial characteristics of work, which, in turn, contribute to the development of WMSDs. For example, an organization that subscribes to Taylor's[16] principles of scientific management may create jobs that exact high physical exertion and provide little control over the work process to employees. High exertion and low job control have been associated with increased risk of WMSDs.[10,17–19] Models 2 and 3 illustrate that the influence of physical characteristics of work on MSDs can be partially mediated by psychosocial characteristics and vice versa. For example, a fast work pace can directly contribute to WMSDs (i.e., without any cognitive appraisal by employees) and can also give rise to a sense of time urgency among employees. Time urgency has been consistently associated with WMSDs.[2,8] Likewise, when employees have little say over how their work is accomplished, the work process may entail higher physical demands or more awkward work postures — both of which are well-established risk factors for WMSDs.[2] Model 4 indicates that models 1, 2, and 3 can operate in the workplace simultaneously. All four models are consistent with, and potentially explain, the empirical evidence for strong covariation between physical work characteristics and certain psychosocial characteristics.[20]

Obviously, these models are simplified schematics to show the potential relationships between work organization, physical characteristics of the job, psychosocial work characteristics, and WMSDs. Another level of detail could be added to explain the pathways linking both physical and psychosocial work characteristics with WMSDs. Indeed, much research has investigated potential biological pathways including muscle recruitment patterns, neuroendocrine responses to stressors, increased muscle tension, and hyperventilation theories.[2,21,22] These biological pathways are supported by varying amounts of empirical evidence, ranging from the relatively speculative to the well-established.

While progress in identifying the biological injury mechanisms is scientifically engaging and intellectually satisfying, it is unlikely to prove particularly helpful in guiding the development of effective prevention programs. As Westgaard and Winkel state in their review of ergonomic interventions for musculoskeletal health (Ref. 23, p. 492), "injury mechanisms are likely to be multicausal" and the importance of any one pathway may vary across individuals and occupations. However, in anticipation of our focus on interventions in the next section, it is important to note that global work attitudes such as job satisfaction and perceived job stress levels partially mediate the relationship between work characteristics (both physical and psychosocial) and WMSDs (see Figure 25.2). Physical characteristics such as heavy lifting and psychosocial characteristics such as lack of autonomy have been shown to consistently predict employee job dissatisfaction.[24,25] In turn, job satisfaction and global job stress are two of the "psychosocial factors" included in previous reviews of the literature that have been most consistently and strongly related to WMSDs.[8,10]

The models depicted in Figures 25.1 and Figure 25.2 illustrate the process of mediation where the causal transmission of an effect is described. Physical and psychosocial characteristics of work may also moderate each other's influence on risk for WMSDs. In other words, the presence of one may alter the strength of the relationship between the other and WMSDs.[26] Only a few studies have

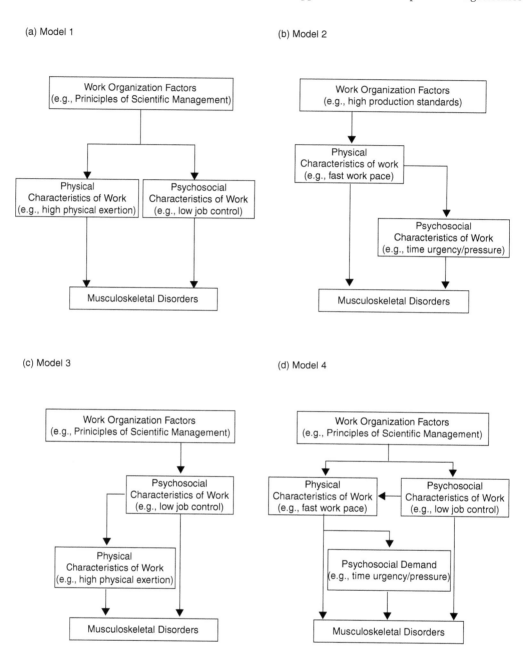

FIGURE 25.1 (a) Parallel effect model. (b) Psychosocial characteristics partially mediate the effect of physical demands on musculoskeletal disorders. SDs. (c) Physical demands partially mediate the effect of psychosocial characteristics of work on musculoskeletal disorders. (d) Mixed model.

investigated potential moderating effects and the results have been inconsistent. However, some of the studies have demonstrated significant moderating effects. Barnekow-Bergkvist and colleagues[27] found that the influence of heavy lifting on the risk for low back pain was stronger when employees were performing work that employees experienced as monotonous. A study of employees in nursing homes in Germany found that psychological demands (measured as time pressure and concentration demands) had a stronger association with musculoskeletal complaints for employees with high physical work

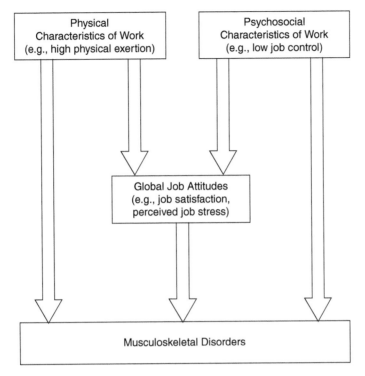

FIGURE 25.2 Effects of physical and psychosocial characteristics of work on WMSDs Partially mediated by global job attitudes.

loads (as measured by trunk postures and number of times heavy weights are lifted) than for employees with low physical demands.[28] This same moderating effect was found in a study of manual material handlers and office workers in the United Kingdom.[29] Moderator effects are important because they shed light on *when* (or under what conditions) WMSDs are likely to occur, and thus *when* certain interventions are likely to be effective.

25.3 Implications for Preventive Intervention

The cascading causal processes presented in the models in Figure 25.1 and Figure 25.2 suggest potential points of intervention for reducing the incidence of WMSDs. Interventions can target global work attitudes, physical characteristics of work, psychosocial characteristics of work, and work organization factors. Global work attitudes, psychosocial work characteristics, and physical work characteristics are the more proximal, "downstream" targets for intervention. Moving "upstream," work organization provides a context for and gives rise to these downstream factors. The "downstream/upstream" continuum is not to be confused with the "top down/bottom up" continuum; they are independent dimensions. Upstream factors such as work organization can be modified using an approach that emphasizes action of top management or through a grassroots movement of entry-level employees.

25.3.1 General Intervention Principles

In general, effectiveness of an intervention is likely to be optimized by targeting the more distal or upstream risk factors in the causal chain. As can be seen in models 1–4, modifications in work organization have the potential to influence multiple risk factors for WMSDs. For example, the introduction of job rotation can reduce an employee's exposure to excessive weights, awkward postures, and repetitive motions. In addition, job rotation may reduce monotony and boredom on the job. As another example, increasing employee participation and influence in decision-making may lead to job redesign that reduces the physical demands of the job and may also create enhanced feelings of employee job control.

Work organization interventions should be evaluated for their effects on these multiple risk factors. At present, this is rarely accomplished. Participatory ergonomics programs are a case in point; they are likely to be effective not only because of the introduction of well-designed biomechanical interventions that are appropriate to the local context, but also because the participatory process of-dentifying priority problems and generating feasible, useful solutions may generate among employees a sense of increased involvement and control over work processes. Although assessing the impact of these interventions on employee perceptions of control would be instructive, few participatory ergonomic studies have actually included measures of job control or other psychosocial work characteristics (see, e.g., Ref. 30–32 and see Ref. 33 as an exception). Similarly, few studies have examined the effectiveness of job rotation in preventing WMSDs, and the ones that are in the published literature tend to address only biomechanical risk factors.[34] Without comprehensive impact evaluations, the potential of work organization interventions is likely to be underestimated, and the added benefit of using participatory processes as opposed to outside expert-guided processes will be poorly quantified.

In addition to creating a potential cascade of change in employees' experience of work, modifying work organization creates an organizational context that enhances the likelihood of sustaining changes in physical and psychosocial characteristics of work. For example, it is likely to be more difficult to convince employees to take rest breaks if they are paid on a piece rate basis than if they are paid an hourly wage. If employees actually end up with smaller paychecks, the rest break initiative is unlikely to be sustained. Similarly, the introduction of new equipment is not likely to be successful if using it slows workers down and they are paid under a piece-rate system. Work organization modifications (in this case, changes in the reward system) can support work process changes.

When attempting to modify a psychosocial characteristic of work, the program designer has the choice of focusing on changing the worker or changing the workplace. Since psychosocial work characteristics are created in part through the cognitive and affective processing of employees, they theoretically can be modified without making actual changes to work processes. Interventions that focus on changing the worker fall into two categories: (1) those that attempt to engage the worker in a more positive reappraisal of the relevant work organization factors and (2) those that attempt to reduce worker vulnerability to demanding or otherwise risk-increasing work organization factors. The first category includes educational strategies and persuasive messages to help the workers reduce their expectations or modify their standards of comparison. For example, management can suggest that their employees compare their own situation with that of workers who have to work harder and under less felicitous conditions. Management can introduce new reasons why higher production standards are both necessary and more easily accomplished than employees think they will be. Such interventions may be construed by employees as attempts at "attitude adjustment" and can foster reactance and other negative reactions. Thus, caution should be exercised in mounting such interventions. Interventions in the second category might include stress management or time management programs to address perceived work overload. Recent research has shown that programs that rely solely on changing workers have not proven highly effective in reducing WMSDs.[35,36]

It is important to note that the various available targets of intervention are not mutually exclusive; interventions may incorporate several targets for change. For example, the introduction of flexible scheduling for employees may be complemented by an employee training program in time management skills. A social ecological perspective suggests that multiple targets of change at multiple levels (individual, job, work team, organization) will be most likely to reduce the incidence of WMSDs.[37] The dynamic nature of social ecological systems allows for cross-level linkages, where changes at one level bring about changes in another.

25.3.2   Identifying Specific Work Factors to Target

In contrast to the emphasis on upstream factors in the preceding discussion of general principles, when gathering assessment data to guide intervention planning it is helpful to begin with a focus on the most proximal or downstream factors and work up to the more distal or upstream factors. One of the most consistent findings from the epidemiological studies investigating the relationships of psychosocial work characteristics to WMSDs is that job satisfaction and job stress levels have the strongest associations with WMSDs.[8–10] This finding can serve as the basis of an organizational assessment strategy for determining program priorities for preventing WMSDs. A cross-sectional assessment of the correlates of job satisfaction and job stress among the employees at the worksite can be conducted. These correlates are likely to include both physical and psychosocial characteristics of work, but are also likely to vary across worksites, industries, and occupations. Once the work characteristics most strongly associated with satisfaction and stress levels have been identified, best practices for ergonomic interventions can be followed.[38] Employees can be engaged in a dialogue to discover how best to make changes in these work characteristics.

Organizations rarely have the time and resources to conduct prospective studies to assess which work characteristics are most strongly associated with incident WMSDs, but they may be more willing to conduct a one-time employee survey to guide intervention development. With the growing availability of easily accessed, valid and reliable questionnaire measures of psychosocial work characteristics,[39] job satisfaction,[25,40] and job stress,[41] it is feasible for even smaller organizations to implement such an assessment.

An alternative assessment approach was recently used by Huang and Feuerstein[42] to identify targets for intervention to prevent WMSDs. They also conducted a cross-sectional employee survey, but identified those work characteristics (both physical and psychosocial) that were most strongly associated with musculoskeletal symptoms among the employees. The problem with such an approach is that the possibility of "reverse causation" is strong once symptoms are manifest. In other words, the experience of pain or a decrement in mobility is likely to change the employee's experience of work in certain ways. And thus, the work characteristics identified through such an approach may not be those that contribute to WMSDs, but those that are most affected by WMSDs.

25.4   Intervention Research Agenda for the Prevention of WMSDs

Although knowledge about both physical and psychosocial risk factors for WMSDs is accumulating, our ability to develop effective worksite interventions to address these risk factors has lagged.[43] This is not unique to the field of ergonomics. In the public health arena, knowledge about how to change identified risk factors does not keep pace with epidemiological breakthroughs and other basic science research findings. For example, smoking has long been identified as the major preventable cause of premature mortality in the United States, yet health professionals are still struggling to devise effective strategies for helping people choose to be nonsmokers.

In an effort to enhance the ability of the occupational safety and health community to identify effective injury and illness prevention strategies, the NIOSH-supported National Occupational Research Agenda (NORA) chose *intervention effectiveness research* as one of its 21 priority research areas.[44] The NORA Intervention Effectiveness Research Implementation Team created a framework for understanding the intervention research process.[45] This framework posits three phases of intervention research: developmental, implementation, and effectiveness (see Figure 25.3). For each of these phases, there is a set of five central research tasks that need to be carried out. This framework can be applied to the specific area of preventing WMSDs.

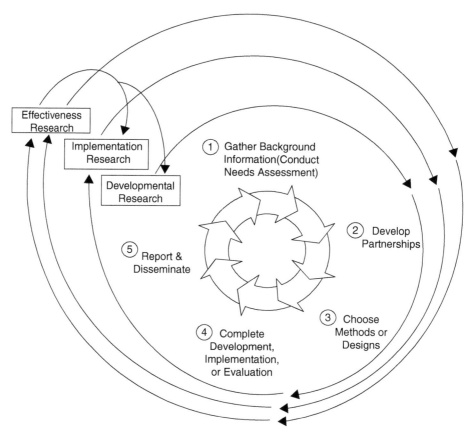

FIGURE 25.3   A conceptual framework for intervention research in occupational safety and health. (From Goldenhar, L.M., LaMontagne, A.D., Katz, T., Heaney, C.A., and Landsbergis, P. 2001 *Journal of Occupational and Environmental Medicine*, 43, pp. 616–622. With permission.)

25.4.1   Developmental Phase

Developmental studies are intended to gather the information necessary to develop an intervention that has the greatest chance of reducing occupational injuries and illnesses. These studies are the key components of evidence-based practice. Goldenhar and colleagues[46] state that developmental studies are intended to answer the following questions:

- What changes are needed to enhance the health of the workers at this site?
- What are the best ways to bring about these changes?
- What are the challenges in bringing about these changes?
- To what extent do the workers understand the need for and desire the changes?
- To what extent might models and theories in occupational safety and health and related fields help guide these change efforts?

These studies can include epidemiological studies to identify the physical and psychosocial work characteristics at a particular site that are the most promising targets for change in that if changed, the incidence of WMSDs would be likely to decline. The literature is replete with this type of study, although they are of varying scientific rigor.[2,3] However, these studies have rarely incorporated measures of workers' perceptions of the need for change. If a proposed intervention conflicts with many of the workers' perspectives on what needs to be changed and how that change should be implemented, it is less likely to succeed. Thus, it is beneficial to include measures of the workers' perspectives.

Developmental studies also include efficacy studies of the proposed intervention components. When the proposed intervention is a change to the physical work station or the physical work process, they are typically tested in the laboratory under highly controlled conditions before being introduced in the worksite (see, e.g., Ref. 36). Similar studies can be carried out for psychosocial characteristics of work, but rarely are. For example, if high concentration demands have been shown to be a strong predictor of job dissatisfaction among a group of workers, then strategies for reducing concentration demands should undergo efficacy tests (either in the lab or in one carefully observed work unit) before widespread introduction into the workplace.

Occupational safety and health interventions, and ergonomic interventions more specifically, rarely incorporate a clear theoretical basis for selecting one intervention strategy or approach over another.[46] For interventions that attempt to modify work organization factors, organizational change theories might be particularly helpful. The process of translating results from the organizational change literature to the field of ergonomics is not straightforward (see Ref. 47, for a discussion). However, there are strategies for increasing the likelihood of success of organizational change efforts that stem from Kurt Lewin's classic model for conceptualizing change.[48] These strategies include: analyzing the current situation (as described earlier), creating a perceived need for change among the employees, creating a shared vision for change, providing leadership and assigning responsibility for change, communicating the change to all stakeholders, providing necessary education and training, modifying organizational structures and policies to support the change effort, involving employees in all steps of the process, and evaluating the change effort to provide feedback for additional change and sustainability.[47] It is interesting to note that these strategies overlap considerably with the typical steps in a participatory ergonomics intervention (see, e.g., Ref. 30) and best practices in ergonomic interventions.[38] Nevertheless, a more explicit examination and consideration of organizational change theory has potential for enhancing the effectiveness of all WMSD prevention programs.

Another type of developmental study focuses on the development of reliable and valid measures of all relevant work characteristics, as well as the desired musculoskeletal health outcomes. Unfortunately, the research literature linking psychosocial work characteristics and WMSDs has been plagued by the use of unreliable and nonvalidated measures.[8] This is true even though there are several well-validated multiple item self-report measures of various psychosocial work characteristics available.[41] Measurement development is an important area of endeavor, and recent efforts have spawned new and creative measurement techniques for psychosocial work characteristics.[49–51]

25.4.2 Implementation Research

Implementation research studies systematically document how an intervention is carried out. This phase of the intervention research process serves multiple purposes: provides feedback for improving the intervention implementation; assesses the appropriateness of mounting a large effectiveness study (only well-implemented interventions are appropriate candidates for effectiveness studies); aids in the interpretation of the results of effectiveness studies; and helps guide effective replication and dissemination of an intervention that has proven effective.[45]

Intervention processes are typically not well described in the ergonomics literature. One major exception is the participatory ergonomics literature. Perhaps because participatory ergonomics is conceptualized as a process rather than as the introduction of a specific innovation, several studies have provided detailed descriptions and critiques of the implementation of the participatory intervention.[30,31,52–54] These studies established the feasibility of implementing participatory approaches in a wide variety of industries and also identified some of the major obstacles encountered.

By documenting the experience of workers participating in the intervention, implementation research can help identify reasons for why desired outcomes were or were not achieved. For example, in a study of nurses' aides,[55] an exercise program was implemented in order to reduce vulnerability to physical demands of the job. The program was carried out twice a week during work hours, but was poorly attended. Fewer than half of the nurses' aides who had been invited to participate did so with any regularity. In addition, those who did participate reported a deterioration in their ability to plan their work effectively. Because the workers were not relieved of any of their regular work duties, participation in the exercise program during work hours actually intensified the demands of the job. When making changes in work organization, implementation studies are indispensable because of the potential for unintended effects on other aspects of work.

25.4.3   Effectiveness Studies

Once an intervention has been carefully developed and demonstrated feasible implementation, an effectiveness study can be conducted to establish the extent to which the intervention brings about the desired reduction in WMSDs under real-world conditions. Effectiveness studies are difficult to conduct in a dynamic context where market forces and other constantly changing external influences drive organizational innovation and employee turnover. Researchers have little control over the working conditions to which study participants are being exposed over the course of an intervention study. In addition, random assignment of employees or work units to intervention and control is rarely feasible. Once again, examples from the participatory ergonomics literature are instructive.

Most of the effectiveness studies of participatory ergonomics interventions are case studies. While these studies are instructive, they do not provide strong evidence that changes in the incidence of WMSDs are due to the intervention being conducted. Other studies used quasi-experimental designs where comparison groups similar to the intervention group were included in the data collection (see, e.g., Ref. 56,57). Only one randomized controlled trial of a participatory ergonomics program (where work units were randomly assigned to intervention or control) was identified in the literature.[58] Although this study found that the intervention work units had made changes that appeared to reduce their risk for WMSDs, the intervention is not well described, appears to focus solely on biomechanical risks, involves only minimal participation on the part of workers, and includes no planned changes in work organization. To the best of our knowledge, there are no randomized controlled effectiveness studies of interventions that aim to reduce WMSDs through changes in work organization.

Randomized controlled trials are the accepted standard of scientific rigor for determining the effectiveness of interventions, and such studies would be a welcome addition to the literature. However, logistical constraints and the cost of such studies make the use of other study designs necessary. Well-designed quasi-experimental studies that collect multiple types of data (self-report, observational, physiological, and archival) can provide the basis for reasonably strong causal inferences. Various discussions of and guides to methodological strategies for effectiveness studies in occupational safety and health can be consulted.[59–61]

25.5   Conclusions

Even as our understanding of how physical and psychosocial work characteristics play a role in the etiology of WMSDs has increased, the prevalence and cost of WMSDs has remained unacceptably high. Understanding, in and of itself, does not prevent the occurrence of injuries. It is imperative that our ability to intervene effectively to reduce injuries keep pace with new knowledge about risk factors. A strong emphasis on intervention research is needed. Such research will help develop optimally successful interventions through the identification of effective change agents and change strategies.

The potential benefit of incorporating work organization interventions into ergonomic practice appears to be substantial. This chapter has reviewed how targeting an upstream factor such as work organization can bring about multiple changes in work characteristics that reduce the risk for WMSDs. Unfortunately, there are numerous obstacles to mounting such interventions. Many ergonomists, and OSH professionals more generally, may not perceive themselves to be appropriately trained and adequately skilled in modifying work organization. They may also feel that developing interventions that go beyond the traditional scope of health and safety is outside the prescribed purview of their jobs.

Formal degree programs and continuing education programs should include curricula on the relationships among work organization, physical work characteristics, psychosocial work characteristics, and WMSDs. In addition, they should include opportunities for building skills in organizational change strategies. At the same time, drawing on the expertise of organizational scientists and organizational change consultants will be helpful. As the scope of ergonomic interventions broadens, interdisciplinary intervention teams will be best suited to developing ergonomic initiatives.

Ergonomists must not only be the recipients of educational efforts, they must be the teachers as well. Upper management and labor leaders need to be educated about the potential importance of work organization in preventing WMSDs. Professional organizations can also play a role in terms of: (1) advocating for broadening the scope of ergonomic training, (2) increasing awareness among industry leaders, and (3) raising the visibility of work organization intervention research. A concerted effort is needed to overcome the obstacles and resistance to developing multifaceted interventions that reduce both physical and psychosocial risk factors for WMSDs.

References

1. Center for Disease Control and Prevention, Ten Great Public Health Achievements — United States, 1900–1999, *Morbidity and Mortality weekly Report*, 48, pp. 241–243, 1999.
2. Panel on Musculoskeletal Disorders and the Workplace, *Musculoskeletal Disorders and the Workplace*. Washington, DC: National Academy Press, 2001.
3. Bernard, B.P. Musculoskeletal Disorders and Workplace Factors, DHHS (NIOSH) Publication No. 97–141, 1997.
4. NORA Musculoskeletal Disorders Team, National Occupational Research Agenda for Musculoskeletal Disorders: Research Topics for the Next Decade, Cincinnati, OH: Department of Health and Human Services, Centers for Disease Control and Prevention, National Institute for Occupational Safety and Health, 2001.
5. Akerlind, I., Hornquist, J.O., and Bjurulf, P. 1992 Psychological factors in the long-term prognosis of chronic low back pain patients, *Journal of Clinical Psychology*, 48, pp. 596–605.
6. Gatchel, R.J., Polatin, P.B., and Mayer, T.G. 1995 The dominant role of psychosocial risk factors in the development of chronic low back pain disability, *Spine (Philadelphia Pa: 1986)*, 20, pp. 2702–2709.
7. Wright, A.R. 1998 Psychosocial factors involved in the progression from acute to chronic low back pain disability, in *Dissertation Abstracts International: Section B: The Sciences & Engineering*, 58, p. 4479.
8. Davis, K.G. and Heaney, C.A. 2000 Relationship between psychosocial work characteristics and low back pain: underlying methodological issues, *Clinical Biomechanics*, 15, pp. 389–406.
9. Bongers, P.M., de Winter, C.R., Kompier, M.A., and Hildebrandt, V.H. 1993 Psychosocial factors at work and musculoskeletal disease, *Scandinavian Journal of Work, Environment & Health*, 19, pp. 297–312.
10. Bongers, P.M., Kremer, A.M., and ter Laak, J. 2002 Are psychosocial factors risk factors for symptoms and signs of the shoulder, elbow, or hand/wrist? A review of the epidemiological literature, *American Journal of Industrial Medicine*, 41, pp. 315–42.
11. International Labor Office, *Psychosocial Factors at Work: Recognition and Control*. Geneva, Switzerland: International Labor Office, 1986.
12. NORA Organization of Work Team, *The Changing Organization of Work and the Safety and Health of Working People: Knowledge Gaps and Research Directions*. Cincinnati, OH: Department of Health and Human Services, Centers for Disease Control and Prevention, National Institute for Occupational Safety and Health, 2002.

13. Rau, R. 2004 Job strain or healthy work: a question of task design, *Journal of Occupational Health Psychology*, 9, pp. 322–338.

14. Kirmeyer, S.L. and Dougherty, T.W. 1988 Workload, tension, and coping: Moderating effects of supervisor support, *Personnel Psychology*, 41, pp. 125–139.

15. Frese, M. and Zapf, D. 1999 On the importance of the objective environment in stress and attribution theory. Counterpoint to Perrewe and Zellars, *Journal of Organizational Behavior*, 20, pp. 761–765.

16. Taylor, E. *The Principles of Scientific Management.* New York: Norton and Company, 1911.

17. Hoogendoorn, W.E., van Poppel, M. N., Bongers, P.M. Koes, B.W., and Bouter, L.M. 2000 Systematic review of psychosocial factors at work and private life as risk factors for back pain, *Spine (Philadelphia Pa: 1986)*, 25, pp. 2114–2125.

18. Ferguson, S.A. and Marras, W.S. 1997 Literature review of low back disorder surveillance measures and risk factors, *Clinical Biomechanics*, 12, pp. 211–226.

19. Weiser, S. and Cedraschi, C. 1992 Psychosocial issues in the prevention of chronic low back pain—a literature review, *Baillieres Clinical Rheumatology*, 6, pp. 657–684.

20. MacDonald, L.A., Karasek, R.A., Punnett, L., and Scharf, T. 2001 Covariation between workplace physical and psychosocial stressors: evidence and implications for occupational health research and prevention, *Ergonomics*, 44, pp. 696–718.

21. Huang, G.D., Feuerstein, M., and Sauter, S.L. 2002 Occupational stress and work-related upper extremity disorders: concepts and models, *American Journal of Industrial Medicine*, 41, pp. 298–314.

22. Schleifer, L.M., Ley, R., and Spalding, T.W. 2002 A hyperventilation theory of job stress and musculoskeletal disorders, *American Journal of Industrial Medicine*, 41, pp. 420–432.

23. Westgaard, R.H. and Winkel, J. 2000 Ergonomic intervention studies for improved musculoskeletal health: a review of the literature and some implications for practitioners, presented at *Proceedings of the XIV Triennial Congress of the International Ergonomics Association and 44th Annual Meeting of the Human Factors and Ergonomics Association*, Ergonomics for the New Millennium.

24. Wilson, M.G., DeJoy, D.M., and Vandenverg, R.J. 2004 Work characteristics and employee health and well-being: test of a model of healthy work organization, *Journal of Occupational and Organizational Psychology*, 17, pp. 565–588.

25. Spector, P.E. 1997 *Job Satisfaction: Applications, Assessment, Causes, and Consequences.* Thousand Oaks, CA: Sage.

26. Baron, R. and Kenny, D. 1986 The moderator–mediator variable distinction in social psychological research: conceptual, strategic, and statistical considerations, *Journal of Personality and Social Psychology*, 51, pp. 1173–1182.

27. Barnekow-Bergkvist, M., Hedberg, G.E., Janlert, U., and Jansson, E. 1998 Determinants of self-reported neck-shoulder and low back symptoms in a general population, *Spine (Philadelphia Pa: 1986)*, 23, pp. 235–243.

28. Hollmann, S., Heuer, H., and Schmidt, K.-H. 2001 Control at work: a generalized resource factor for the prevention of musculoskeletal symptoms?, *Work & Stress*, 15, pp. 29–39.

29. Devereux, J.J., Vlachonikolis, I.G., and Buckle, P.W. 2002 Epidemiological study to investigate potential interaction between physical and psychosocial factors at work that may increase the risk of symptoms of musculoskeletal disorder of the neck and upper limb, *Occupational and Environmental Medicine*, 59, pp. 269–277.

30. de Jong A.M. and Vink, P. 2002 Participatory ergonomics applied in installation work, *Applied Ergonomics*, 33, pp. 439–448.

31. Moore J.S. and Garg, A. 1996 Use of participatory ergonomics teams to address musculoskeletal hazards in the red meat packing industry, *American Journal of Industrial Medicine*, 29, pp. 402–408.

32. Vink, P., Urlings, I. J.M., and van der Molen, H. F. 1997 Participatory ergonomics approach to redesign work of scaffolders, *Safety Science*, 26, pp. 75–85.

33. Evanoff, B.A., Bohr, P.C., and Wolf, L.D. 1999 Effects of a participatory ergonomics team among hospital orderlies, *American Journal of Industrial Medicine*, 35, pp. 358–365.

34. Paul, P., Kuijer, F.M., de Vries, W. H. K., van der Beek, A. J., van Dieen, J. H., Visser, B., and Frings-Dresen, M.H. 2004 Effect of job rotation on work demands, workload, and recovery of refuse truck drivers and collectors, *Human Factors*, 46, pp. 437–448.

35. Horneij, E., Hemborg, B., Jensen, I., and Ekdahl, C. 2001 No significant differences between intervention programmes on neck, shoulder and low back pain: a prospective randomized study among home-care personnel, *Journal of Rehabilitation Medicine: Official Journal of the UEMS European Board of Physical and Rehabilitation Medicine*, 33, pp. 170–176.

36. Lincoln, A.E., Vernick, J.S., Ogaitis, S., Smith, G.S., Mitchell, C.S., and Agnew, M. 2000 Interventions for the primary prevention of work-related carpal tunnel syndrome, *American Journal of Preventive Medicine*, 18, pp. 37–50.

37. Stokols, D. 1996 Translating social ecological theory into guidelines for community health promotion, *American Journal of Health Promotion*, 10, pp. 282–298.

38. Cohen, A., Gjessing, C.C., Fine, L., Bernard, B.P., and McGlothlin, J.D. 1997 Elements of Ergonomics Programs: A Primer Based on Workplace Evaluations of Musculoskeletal Disorders, Cincinnati, OH: US Department of Health and Human Services, Centers for Disease Control and Prevention, National Institute for Occupational Safety and Health.

39. Wealleans, D. 2003 *The People Measurement Manual: Measuring Attitudes, Behaviors, and Beliefs in Your Organization*. Aldershot, UK: Gower.

40. van Saane, N., Sluiter, J.K., Verbeek, J.H.A.M., and Frings-Dresen, M.H.W. 2003 Reliability and validity of instruments measuring job satisfaction: a systematic review, *Occupational Medicine (Oxford England)*, 53, pp. 191–200.

41. Hurrell, J.J.J., Nelson, D.L., and Simmons, B.L. 1998 Measuring job stressors and strains: Where we have been, where we are, and where we need to go, *Journal of Occupational Health Psychology*, 3, pp. 368–389.

42. Huang G.D. and Feuerstein, M. 2004 Identifying work organization targets for a work-related musculoskeletal symptom prevention program, *Journal of Occupational Rehabilitation*, 14, pp. 13–30.

43. Pransky, G., Robertson, M.M., and Moon, S.D. 2002 Stress and work-related upper extremity disorders: implications for prevention and management, *American Journal of Industrial Medicine*, 41, pp. 443–455.

44. National Institute for Occupational Safety and Health, National Occupational Research Agenda. Cincinnati, OH: US Department of Health and Human Services, Centers for Disease Control and Prevention, 1996.

45. Goldenhar, L.M., LaMontagne, A.D., Katz, T., Heaney, C.A., and Landsbergis, P. 2001 The intervention research process in occupational safety and health: An overview from the National Occupational Research Agenda Intervention Effectiveness Research Team, *Journal of Occupational and Environmental Medicine*, 43, pp. 616–622.

46. Goldenhar, L.M. and Schulte, P.A. 1994 Intervention research in occupational health and safety, *Journal of Occupational and Environmental Medicine*, 36, pp. 763–775.

47. Heaney, C.A. 2003 Worksite health interventions: targets for change and strategies for attaining them, in *Handbook of Occupational Health Psychology*, J. C. Quick and Tetrick, L.E. Eds. Washington, DC: American Psychological Association, pp. 305–323.

48. Lewin, K. 1951 Field theory and learning, in *Field Theory in Social Science: Select Theoretical Papers*, D. Cartwright, Ed. New York: Harper-Collins, pp. 60–86.

49. Waldenstrom, M., Theorell, T., Ahlberg, G., Josephson, M., Nise, P., Waldenstrom, K., and Vingard, E. 2002 Assessment of psychological and social current working conditions in epidemiological studies: experiences from the MUSIC-Norrtalje study, *Scandinavian Journal of Public Health*, 30, pp. 94–102.

50. Dane, D., Feuerstein, M., Huang, G.D., Dimberg, L., Ali, D., and Lincoln, A. 2002 Measurement properties of a self-report index of ergonomic exposures for use in an office work environment, *Journal of Occupational and Environmental Medicine/American College of Occupational and Environmental Medicine*, 44, pp. 73–81.

51. Mairiaux, P. and Vandoorne, C. 2000 A simple risk assessment tool for use in ergonomics participatory processes, presented at *Proceedings of the XIV Triennial Congress of the International Ergonomics Association and 44th Annual Meeting of the Human Factors and Ergonomics Association*, Ergonomics for the New Millennium.

52. Rosecrance, J.C. and Cook, T.M. 2000 The use of participatory action research and ergonomics in the prevention of work-related musculoskeletal disorders in the newspaper industry, *Applied Occupational and Environmental Hygiene*, 15, pp. 255–262.

53. Haims, M.C. and Carayon, P. 1998 Theory and practice for the implementation of 'in-house', continuous improvement participatory ergonomic programs, *Applied Ergonomics*, 29, pp. 461–472.

54. Moore, J.S. and Garg, A. 1997 Participatory ergonomics in a red meat packing plant. Part II: Case studies, *American Industrial Hygiene Association Journal*, 58, pp. 498–508.

55. Skargren, E. and Oberg, B. 1999 Effects of an exercise programme on organizational/psychosocial and physical work conditions, and psychosomatic symptoms, *Scandinavian Journal of Rehabilitation Medicine*, 31, pp. 109–115.

56. Wickstrom, G., Hyytiaeinen, K., Laine, M., Pentti, J., and Selonen, R. 1993 A five-year intervention study to reduce low back disorders in the metal industry, *International Journal of Industrial Ergonomics*, 12, pp. 25–33.

57. Carrivick, P., Lee, A., and Yau, K. 2002 Effectiveness of a workplace risk assessment team in reducing the rate, cost, and duration of occupational injury, *Journal of Occupational and Environmental Medicine*, 44, pp. 155–159.

58. Straker, L., Burgess-Limerick, R., Pollock, C., and Egeskov, R. 2004 A randomized and controlled trial of a participative ergonomics intervention to reduce injuries associated with manual tasks: physical risk and legislative compliance, *Ergonomics*, 47, pp. 166–188.

59. Goldenhar, L.M. and Schulte, P.A. 1996 Methodological issues for intervention research in occupational health and safety, *American Journal of Industrial Medicine*, 29, pp. 289–294.

60. Robson, L.S., Shannon, H.S., Goldenhar, L.M., and Hale, A.R. 2001 Evaluating the Effectiveness of Strategies for Preventing Work Injuries: How to Show Whether a Safety Intervention Really Works, Cincinnati, OH: US Department of Health and Human Services, Centers for Disease Control and Prevention, National Institute for Occupational Safety and Health.

61. Shannon, H.S., Robson, L., and Guastello, S. 1999 Methodological criteria for evaluating occupational safety intervention research, *Safety Science*, 31, pp. 161–179.

26

Assessment of Worker Functional Capacities

Glenda L. Key
KEY Method Assessments, Inc.

26.1 Introduction

There are nearly 500,000 U.S. workers each year who, as a result of injury, are unable to resume their jobs for long periods of time. Low back pain will afflict roughly 80% of the population at some time in their lives (Wheeler and Hanley, 1995). Workers' compensation costs have reached $70 billion per year, tripling since 1980. The direct cost of the situation is staggering and in each case the degree of loss is unpredictable.

The knowledge of what a worker's functional capabilities are, is one of the most valuable pieces of information that a professional can have for reducing workers' compensation costs and prevention of injury in the workplace today. There are two occasions in the employment of a worker when it is critical to know just what the physical work capability of that worker is. One occurs during the decision to hire, and the other is upon return to work of an injured employee. The focus of this chapter will be on these two occasions as highlighted in Figure 26.1, the Worker Care Spectrum representing assessment of worker functional capabilities.

26.2 Worker Assessments

26.2.1 Definitions

26.2.1.1 Functional Capacity Assessment

The functional capacity assessment (FCA) is a return-to-work testing process that determines an individual's physical functional work-related capabilities through measuring, recording, and analyzing

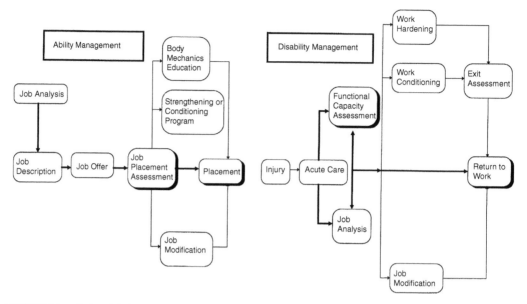

FIGURE 26.1   Industrial therapy worker care spectrum.

data gathered during a standardized physical testing procedure. There are many different formats of functional capacity assessments and almost as many names for them. The term used may be physical capacity assessment, functional capacity evaluation, work capacity evaluation, or work assessment. For the purposes of this chapter, the term functional capacity assessment is used.

The activities of an FCA include materials handling functions of lifting, carrying, pushing, and pulling. It also includes tolerances of posture such as standing, sitting, kneeling, crawling, and activities of walking and stair climbing (Key, 1995a).

Simulation of the movements and postures during the weighted activities and standardizing for consistency of instruction, weight loading protocols, and determining termination points are primary issues that need to be considered in a functional capacity assessment. An FCA allows for the testing of the individual to confirm meeting or not meeting the minimum physical requirements of the job.

As quickly as possible, the workers should be returned to their jobs and workplaces — even if they are unable to be fully productive. Athletes do not wait until they are fully recovered before they start back in practice or training. They work their bodies to the point where they still maintain safety and will not reinjure themselves. Their capabilities are built up gradually until they are back in their "game." It does not mean that they play without pain. It does not mean that they can do everything as well as or in the same manner as before. This approach applies to the injured worker as it does to the athlete.

26.2.1.2   Job Placement Assessment

The job placement assessment (JPA) provides data prior to hiring that assists managers in making decisions resulting in significantly decreased incident rates in the future. When armed with the information of the requirements of the job as provided through a job analysis and job description, and the physical abilities of the individual as provided through a job placement assessment, it is relatively easy to determine whether persons should be hired into a specific job or not. If they meet the requirements of the job, the answer is "Yes." If they do *not* meet the requirements of the job, the answer is "No."

"The job placement assessment is a series of specific, objective, and standardized protocols followed in a consistent progression to allow for objective, accurate, and repeatable results" (Frey, 1995). These results identify the prospective employee's work capabilities in the areas of lifting, carrying, pushing, pulling, and other job-specific activities.

The continuous and significant increase cost of health care and workers' compensation has added to the already inherent need to have healthy, productive employees. This process begins with hiring individuals who are able to physically perform the requirements of the job. How can employers tell if their next hire will be their best employee or become their worst workers' compensation and litigation nightmare?

The employee has the most to gain and lose if decisions are made based on anything other than accurate data. At the time of hire, the prospective employee has the right to be judged through a nondiscriminatory process. Generally, his or her personal need is to be employed and be a productive part of society. The expectation is that the hiring process will be fair and that the job activity will not be a precursor of an injury. Once injured, whether on the job or not, the worker relies on the medical system and those individuals carrying out company policies to establish the return-to-work process. The desire to be productive, the fear of reinjury, the unfamiliar role of patient, and the acceptability of not working can all be contributors to the complex course of action. It is of substantial importance that the match of the worker to the work be accurate to prevent any reinjury yet place the worker in the highest level of productivity possible — for economic and social reasons.

At these two points in the Worker Care Spectrum, the employee is especially vulnerable to the decisions of others. Employees often report that their experience in a worker assessment, JPA or FCA, is the first time that decisions are being made based on their actual participation, with themselves and their activities in control of the end points. To have their capabilities objectively tested in a participatory manner is important to them.

26.2.2 Components

Recommendations of an individual's physical work ability are made based on assessing components in the following three categories.

1. Weighted capabilities
2. Tolerance and endurance parameters
3. Validity of participation determinants

26.2.2.1 Weighted Capabilities

These tests include analyzing an individual's ability to perform specific work-related materials handling functions such as lifting at three standard heights, carrying, pushing, and pulling. A detailed list of components is found in Figure 26.2. Cardiac responses, kinesiological changes and repetition of posture and posture changes are documented and included in the formulas of decision of the individual's capabilities. This further confirms cardiovascular and body mechanics safety of the employee or prospective employee at that level of physical activity. Important questions are answered through these thorough testing techniques. For example, can the employee lift the 48 lbs. that are required for the job? How can one know unless the worker is tested?

26.2.2.2 Tolerances and Endurance

Testing includes and recommendations are given in the categories of standing tolerance, sitting tolerance, and workday tolerance. Kneeling, crawling, and activities such as walking and stair climbing are also covered. A detailed list of components is found in Figure 26.2. For example, can your employee return to an eight hour workday or is he or she limited to a lesser amount? And if so, what is the safe limitation of a workday?

26.2.2.3 Validity of Participation Determinants

It is important to objectively, scientifically, and statistically determine and demonstrate when individuals are not being forthright in the representation of their capabilities (Grossman, 1985; Osterweis et al.,

Weighted Capabilities	
Lifting desk to chair 30" to 18"	Lifting chair to floor 18" to 0"
Carrying	Lifting above shoulder 30" to 60"
Pulling	Pushing

Tolerances and Endurance	
Balancing	Bending
Cervical mobility	Circuit board tolerance
Climbing	Grip strength
Crouching	Crawling
Fine manipulation	Fastener board tolerance
Keyboard tolerance	Firm grasping
Simple grasping	Kneeling
Repetitive foot motion	Reaching
Sitting	Simple grasping
Standing	Squatting
Tool station work tolerance	Stooping
Walking	Work day tolerances

FIGURE 26.2   Components of physical ability.

1987). It is known that some individuals will demonstrate less than their full capability. They may be attempting to prolong their time off work or may be attempting to falsely maximize their case closure settlement.

The percentage of people participating in this kind of exaggeration and falsification is less than the public generally perceives it to be. Figure 26.3 presents the percentages in each category of participation determinant based on a sample of 43,000 injured clients (Worker Data Bank, 1994–1997).

It is important to the valid participators to have their honest attempts proven to be full effort demonstrations. The injustice is not only when the conscious and deceptive low participator is not revealed, but also when the honest, full effort, low capability level participator is misjudged.

Validity of participation determinants have been developed by one vendor of FCA equipment and protocols (Gilliand et al., 1986; Personnel Decisions, Inc., 1994). The individual's level of participation is determined through the use of algorithms and decision science objectively identifying the degree of an individual's consistency or inconsistency. The formula includes data from a database made up of over 100,000 cases of this standardized, protocolized, FCA (Worker Data Bank, 1994–1997). These

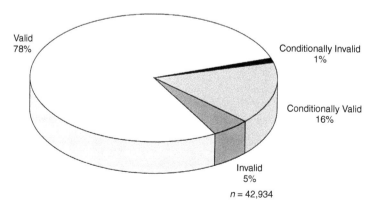

FIGURE 26.3   Functional capacity assessment validity determinants.

data have provided valuable correlations, one of which is between the amount someone can lift at specific heights and how much they can push and pull. Scientific literature and the same database have provided statistics on certain vital signs, such as heart rate, which respond predictively as the individual approaches and reaches exertion level (Astrand and Rhyming, 1954; Gilliand et al., 1986; Worker Data Bank, 1994–1997).

The delineations of performance levels as introduced by KEY Method in Minneapolis, Minnesota are (Gilliand et al., 1986; Key, 1984; Worker Data Bank, 1994–1997):

1. Valid participation. This determinant means that the individual participated with full effort and the results and recommendations reflect safe capabilities.
2. Invalid participation. This as a determinant means that the participant consciously and intentionally provided less than full effort. The numbers produced by this individual reflect less than full capability. The results can be used in addressing the issue directly with the participant or, if necessary, in litigation.
3. Conditionally valid participation. An individual with this determinant has demonstrated less than full capability. The results reflect their *perception* of capability even though they can physically do more. The numbers and capabilities are, therefore, safe, but may also reflect an unconscious psychological barrier.
4. Conditionally invalid participation. The individual with this determinant has demonstrated work levels that are beyond what would be considered full, safe levels for extended work periods. When left to their own end point determinations of work activity, they exceed the safe levels.

The importance of standardization, statistical analysis of data, and carrying the level of data in a database to produce determinants of participation is paramount.

26.2.3 Standardization

When looking for an assessment, standardization and validity of participation determinants are of consummate importance. They are the primary elements of defense against the occurrence of litigation and the primary elements of defense should the case eventually go to court. One of the standardized systems maintains a complete data bank of all assessments performed (Worker Data Bank, 1994–1997).

To be assured of such level of standardization, studies should be provided to those using the assessment that demonstrate reliability and consistency. The consistency of the KEY Method assessment results was analyzed and is represented in Table 26.1. Analysis using 2-way analysis by ranks found "no differences" in providers' results across the United States. Statistically, any differences were found to be *extremely not significant* (Aitken, 1996; Portney and Watkins, 1993, Worker Data Bank, 1994–1997).

TABLE 26.1 Consistency of KEY Method Assessment Results

	Validity by Percent in Each United States Territory			
	West	Central	North East	South East
Valid	78	78	78.5	76
Invalid	5	3	5	5
Conditionally Valid	15	17	15	17
Conditionally Invalid	1	2	1.5	1

Freidman two way analysis of variance by ranks. Region by participation (%) $k = 4$, $n = 4$; $X^2 = 7.5$ $X_r^2 + 0.9$

Another analysis demonstrating consistency in the assessment results, representing standardization, is the graph in Figure 26.4. This confirms the consistency of the results of the assessment when administered in three different countries: the United States, Canada, and Australia.

26.2.4　Worker Assessment Principles

Principles to look for in selecting a functional capacity assessment include:

1. The assessment must contain standards for identifying validity of participation of the individual being tested
2. The methodology administered must be consistent from tester to tester and test to test
3. Standardized equipment must be used and the same procedure followed with each assessment
4. The administrator of the assessment must be thoroughly trained and objective
5. The processing of the information gathered during testing must be standardized

The psychology of the individual needs to recognized in the assessment as well as the kinesiology of the activities (Schmidt et al., 1989). There is a growing body of literature supporting the theory that a statistical relationship does exist between low back pain and an individual's psychological factors as tested by the Minnesota Multiphasic Personality Inventory (MMPI) (Block et al., 1996). The MMPI has also been able to predict the occurrence of job-related low back pain or when poor response to surgery would be the outcome (Bigos et al., 1991; Blair et al., 1994; Block et al., 1996; Keller and Butcher, 1991; Schmidt et al., 1989).

26.2.5　Assessment Reports

A report should provide details of the results of the assessment and then compare those results with the physical demands of the job. Through this matching exercise, the decision maker can make an informed, unbiased, and defensible decision relating to the return to work of a previously injured employee.

It is also helpful to receive a visual display of the major highlights of the results, as in Figure 26.5. Graphic displays of these major areas can be most helpful in assisting the decision-making process for rehire or case management. The individual's assessment results are compared with the same profile of individual in the data bank. They also may be compared with the job requirements. Another option is to compare norms from the injured database with similar profiles of individuals in the uninjured database.

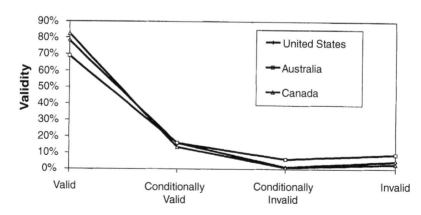

FIGURE 26.4　Descriptive summary validity.

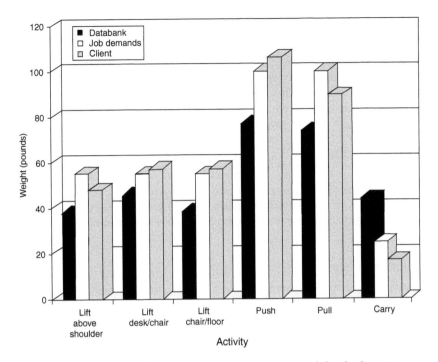

FIGURE 26.5 Injured worker FCA results compared to job requirements and data bank norms.

26.3   The Provider of Worker Assessments

The provider of these assessments can be instrumental in increasing return-to-work percentages, reducing reinjury rates, decreasing lost workdays, and yielding other short and long-term cost reductions.

Physical and occupational therapists are the primary providers of worker assessments, job placement assessments, and return-to-work functional capacity assessments. The motive for providing quality results in functional assessments lies primarily in the accuracy of the results and the defensibility in the courts. The therapist needs to provide the most objective, unbiased data. These data are then used as an assist in the hiring and returning-to-work decisions. As a result, therapists need to be able to support their method with objectivity, standardized equipment, and consistent protocols and procedures.

26.4   Outcomes

Outcome surveys or studies should be available for review for one to trust the results and the legitimacy of the recommendations of a worker assessment (Dobrzykowski, 1995; Key, 1995b). The predictive ability of an FCA needs to be demonstrated based on a track record of return-to-work without reinjury.

The primary outcomes that one should be looking for include:

1. Decreased reinjury rates
2. Decrease lapse of time from date of injury to date of return to work
3. Decrease incidents and cost of litigation

The State of Colorado studied the impact of FCAs on shortening the amount of time required for vocational evaluation of injured workers. H. D. Waite found that the inclusion of the KEY Method FCA resulted in a median of 18 fewer days of rehabilitation. This would save the State of Colorado over $200,000 annually (Waite, 1987).

26.4.1   JPA Outcomes

Having the data and using them to safely match workers' capability levels to the job demands, or being in a position to know the modifications necessary to facilitate a match results in fewer injuries and subsequent lower health care costs, lower workers' compensation complications, preserved productivity, worker morale, and job satisfaction. These issues along with that of the assessments' predictive capabilities regarding future injuries are the primary issues in outcome reports. Examples assist in demonstrating the effects (Karwowski and Salvendy, 1998).

26.4.1.1   Outcome — Case A

In 1988 a paper manufacturer in Minnesota instituted job placement assessments to help stem the costs related to workers' compensation claims and lost workdays. In analysis two years later, the experiences of 70 employees hired before the use of JPAs were compared to 70 hired after the initiation of administering JPAs. Use of the JPAs lowered both lost workdays and workers' compensation costs (see Figure 26.6) (Frey, 1995).

26.4.1.2   Outcome — Case B

A state transportation authority discovered that job placement assessments enable them to predict if an employee is at risk of injury. Of the 36 employees injured from 1985 to 1992, 75% had been categorized as "at risk" by JPAs performed when they were hired. Fourteen providers across the state administered the same system of JPA. Personnel Decisions, Inc. reports, "While the analysis is based on a relatively few cases, the results are statistically significant. The chi square for this cross-tabulation is 15.4, which is significant at $p = .00045$" (Personnel Decisions, Inc., 1994).

26.4.1.3   Outcome — Case C

A major trucking firm used the JPA for two separate hiring locations over a continuous eighteen-month period. Since implementing the administration of the JPA on each candidate and hiring based on the results, there were no new injuries (Worker Data Bank, 1994–1997).

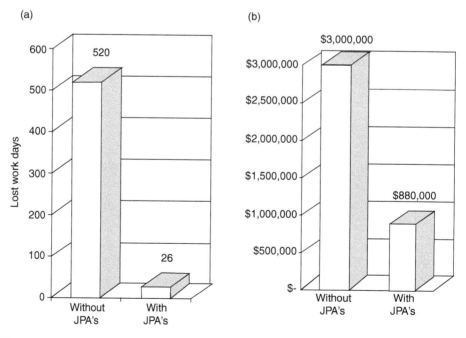

FIGURE 26.6   (a) Lost work days; (b) Workers' compensation costs.

26.5   After the Assessment Results Are In

The results of job placement assessments and functional capacity assessments offer options for pathways to follow, depending on the data assembled. The decision tree layout of Figure 26.7 and Figure 26.8 demonstrate optional pathways available.

26.6   Diversification Options

As change occurs in industry so will industry's approach to minimizing work-related injuries. To meet these changing needs medical providers are diversifying the services they now offer to industry and the mode of delivery of those services. Today's marketplace offers some examples of successful diversification.

26.6.1   Mobile Assessments

For those who offer functional capacity assessments or job placement assessments, a beneficial adaptation has been to perform these assessments at the job site rather than in a clinic. This is done by utilizing a mobile testing unit that uses the same standardized test procedures as those which are practiced in the clinical setting. By having the JPAs performed onsite, the individual responsible for the "hire/no-hire" decision has immediate access to the assessment results. Because of this, no administrative or production time is lost waiting for reports. They can be printed up immediately upon completion of the JPA. With

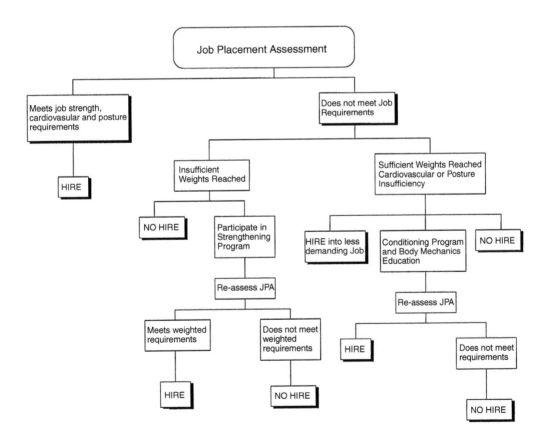

FIGURE 26.7   Flow chart for job placement assessment.

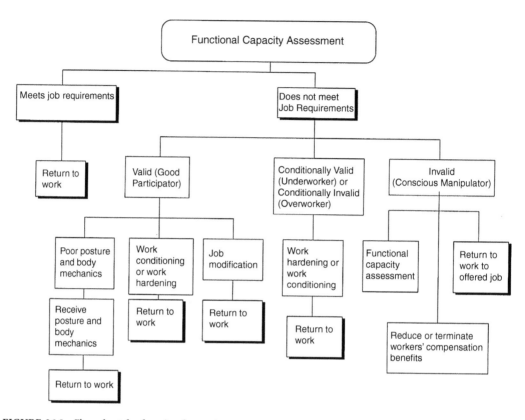

FIGURE 26.8 Flow chart for functional capacity assesment.

today's tight economic issues and tight worker market it is important that the hire be made quickly. Onsite JPAs provide a system by which many employees can test through and begin work the same day.

Having the return-to-work assessment, the FCAs, performed onsite promotes the bond between the injured worker and the workplace. It assists in the process of employee acceptance, co-worker to coworker. Management and nonmanagement observe that the individual is indeed participating in the return-to-work program. Bringing the worker back to the site for testing also allows the employee to reunite with other colleagues and co-workers.

Additionally, therapists are able to develop a closer working relationship with management and thereby better understand and meet the company's needs.

26.6.2 Mobile Occupational Health Clinic

Taking this idea a step further, some have adapted large vans or buses to house a mobile occupational health clinic. This includes facilities for examinations and physicals, functional capacity assessments, job placement assessments, job analysis, ongoing treatment, and educational materials, as well as facilities for drug, hearing, and vision screenings. This mobility presents industry with a broad offering for delivery of services. One such mobile unit travels up to 200 miles to deliver such services to hundreds of industries in a rural region.

26.6.2.1 Onsite Services

Onsite industrial therapist services are becoming a very cost-effective way for industries to more thoroughly meet the needs of their company. When the provider of services shows up for work every day at the plant site, there is a deepened knowledge level of the problems and a quicker pathway to the solutions. Services onsite may be delivered by contracting with a local clinic or therapist. It may be to the company's advantage to have the individual or individuals be employees of the company.

Onsite services can vary from full-time positions to a few hours per week depending on the size of the company and the workers' risk for injury. This approach requires that the company designate space for the medical provider to assess and treat workers. The most apparent benefits of onsite therapy services are the immediacy of services and the decreased worker down time. Musculoskeletal injuries (i.e., repetitive motion injuries, back injuries, sprain/strain injuries) receive treatment faster and therefore allow the patient to return to work faster.

Prompt treatment is the key to returning the worker to work as quickly as possible. In addition to helping the worker overcome injuries, prompt treatment can also impact the bottom line by cutting the costs of finding a replacement worker and reducing workers' compensation premiums. An intangible benefit of onsite services is the therapist's ability to effectively communicate the workers' situations to administration. This independent opinion can alleviate tensions which often develop between injured workers and their supervisors. It is these tensions that can lead to exaggerated symptoms and litigation.

The services that can be provided to industry and the worker through this approach are limited only by the willingness of the parties involved. Most services offered in a medical clinic can be offered at the plant site. The number of employees and the incident rate usually dictate the choice of services brought in-house.

The entire Worker Care Spectrum as represented in Figure 26.1 can be provided along with drug, hearing and vision screening, medical physicals, ergonomics consulting, fitness and wellness programs, inoculations, and traditional general medical treatment. The company also experiences economic benefits through the immediacy of evaluation and treatment of resultant earlier return-to-work, less time off work, and maintaining the bond between the worker and the worksite. Diversification can take many forms. These are but a few examples which are being offered to industry in today's market and demonstrate insight of how challenges will be met in the future.

26.7 Conclusion

Financial reward for losses as result of work activities is evidenced as far back as Greek warriors and pirates. More that 2000 years ago, the families of warriors who lost their lives in battle were compensated. Pirates were also provided a scheduled award for loss of a limb, but only after a successful attack and after the captain had taken his share of the recovered booty. Today's workers' compensation system is much more complicated. In 1986, Liberty Mutual reported $6807 as the mean cost per case for *low back pain*. In 1989, this cost had risen to $8321. This 1989 amount is more than twice the amount for the average workers' compensation claim which was $4075 (Webster and Snook, 1990). Liberty estimates the cost of time lost by injured workers to be $18.7 billion. Liberty also reports the cost of disability for corporations to be 6 to 12 percent of their corporate payroll. In 1993, 8.5 million people had disabling injuries (Liberty Mutual, 1998). It has been shown that the majority of costs incurred when an employee is injured comes from nonmedical costs (Quebec Task Force, 1987). A 1987 study revealed that only 14% of the total cost incurred was direct medical. The remaining 86% was attributed to indirect costs including the costs surrounding the replacement of an employee. Considering that industry directly pays for lost time, it is incumbent that industry know what an individual's capability is upon hiring and upon return-to-work from an injury.

Armed with accurate data, the human resource department can be sure that potential employees who do not have the capacity for the work will not be hired. Also, when an injured worker returns to work, it will be as soon as they are able to do so without risk of reinjury.

The employer needs increasingly more objective determination of job requirements and of worker capabilities to defensibly support their decisions of hire and of return-to-work once injured. By focusing on what the worker *can* do, job placement assessments and functional capacity assessments can trigger a more accurate and proactive response from the employer. This includes proper placement upon hiring and easier placement and accommodation, if necessary, once injured (Key, 1995a).

References

Aitken MJ, Creighton University School of Pharmacy and Allied Health, Department of Occupational Therapy, Omaha, Nebraska June, 1996.

Astrand PO, Ryhming I, A nomogram for calculation of aerobic capacity from pulse rate during sub-maximal work, *J of Applied Physiology* 7:218–221, 1954.

Bigos SJ, Battié MC, Spengler DM, et al. A prospective study of work perceptions and psychosocial factors affecting the report of back injury. *Spine* 16:1–6, 1991.

Blair JA, Blair RS, Rueckert P, Pre-injury emotional trauma and chronic back pain — an unexpected finding. *Spine*, 19: 10, May 15, 1994.

Block AR, Vanharanta H, et al. Discogenic pain report, influence of psychological factors, *Spine* 21(3):334–338, February 1996.

Dobrzykowski E, Data collection and use in industrial therapy, in *Industrial Therapy*, Key GL (Ed.) St. Louis, Mosby, 1995, p 42–60.

Frey DH, Job placement assessments and pre-employment screening, in *Industrial Therapy*, Key, GL (Ed.) St. Louis, Mosby, 1995, pp 110–122.

Gilliand RG, Sevy BA, Ahlgren A, *A Study of Statistical Relationships Among Physical Ability Measures of Injured Workers Undergoing KEY Functional Assessments*, Minneapolis, Minnesota 1986.

Grossman P, Respiration, stress, and cardiovascular function, *Psychophysiology* 20(33):284–300, 1983.

Karwowski E, Salvendy G, Ed. *Ergonomics and Manufacturing: Raising Productivity Through Work Place Improvement*, Society of Manufacturing Engineers, 1998.

Keller LS, Butcher JN, *Assessment of Chronic Pain Patients with the MMPI-2*. Minneapolis, MN: University of Minnesota Press, 1991.

Key GL, Functional capacity assessment, in *Industrial Therapy*, Key GL (Ed.) St. Louis, Mosby, 1995a.

Key GL, The impact and outcomes of industrial therapy, in *Industrial Therapy*, Key GL (Ed.) St. Louis, Mosby, 1995b, p 220–254.

Key GL, *Key Functional Assessment Policy and Procedures Manual and Training and Resource Manual*, Minneapolis, MN, 1984.

Liberty Mutual, For Your Employees: Workers' Compensation, available at: www.libertymutual.com/business/workcomp.html, accessed 10-6-98.

Osterweis M, Kleinman A, Mechanic D, *Pain and Disability, Clinical, Behavioral, and Public Policy Perspectives*, Washington, DC, National Academy Press, 1987.

Personnel Decisions Inc. *Key Functional Assessment Pre-employment Screening Battery as a Predictor of Job Related Injuries*, Minneapolis, MN, January, 1994.

Portney LG, Watkins MP, *Foundations of Clinical Research*, Norwalk, Conn., Appleton & Lange, 1993.

Quebec Task Force Study: Scientific approach to the assessment and management of activity-related spine disorders, *Spine*, 12:7S, 1987.

Schmidt AJM, Gierlings EH, Madelon LP, Environmental and interoceptive influences on chronic low back pain behavior. *Pain* 38:137–43, 1989.

Waite HD, *Use of a New Physical Capacities Assessment Method to Assist in Vocational Rehabilitation of Injured Workers*, University of Colorado, Thesis, 1987.

Webster BS, Snook, SH, The cost of compensable low back pain, *J Occup Med*, 32:13–15, 1990.

Wheeler AH, Hanley EN, Spine update: nonoperative treatment for low back pain — rest to restoration, *Spine*, 20(3): 375–378, February 1, 1995.

Worker Data Bank, KEY Method, Minneapolis, Minnesota. 1994, 1995, 1996, 1997.

27

Physical Ability Testing for Employment Decision Purposes

Charles K. Anderson
Advanced Ergonomics, Inc.

27.1 Introduction

Much of ergonomic activity is focused on designing or altering the demands of the job so that there is better match with the capabilities of the workforce. Sometimes, this approach reaches a point where further change in the job is either cost-prohibitive or technically infeasible at the moment. An alternative approach is to consider matching workers to the job demands on the basis of their physical abilities. For instance, if the job requires individuals lift cases weighing 60 lbs and there is no way to reduce the case weight, one approach would be to assess job candidates' ability to lift 60 lbs as part of the process of determining whether the individual is able to perform the job.

A number of studies have documented the effectiveness of physical ability testing as a means of identifying individuals who will be able to safely perform a given job (Chaffin et al., 1977; Cady et al., 1979; Keyserling, 1979; Reilly et al., 1979; Anderson and Herrin, 1980; Keyserling et al., 1980a, b; Arnold et al., 1982; Herrin et al., 1982; Ayoub, 1983; Jackson et al., 1984, 1991a, b, 1992; Laughery and Jackson, 1984; Liles et al., 1984; Anderson and Catterall, 1987; Laughery et al., 1988; Anderson, 1989a, b, 1992a, b; Craig et al., 1998). The experience of the author is that injury rates for new hires typically fall 20% to 40% with the implementation of testing (Anderson, 2003).

There are numerous issues that need to be considered when implementing a physical ability test battery to assure that it will be effective. One of the first concerns is to assure that the test battery will truly assess what is intended to be assessed — the individual's ability to perform the job. This means that a thorough job analysis must be performed, tests carefully chosen, and efforts taken to validate that the battery truly predicts job performance. Without this foundation, the employer may find that the battery provides no useful information while being a source of additional hiring cost and time delay.

In many countries, including the United States, there is a legal mandate that test batteries be valid predictors of job performance, particularly if it can be anticipated that protected groups, such as females and older individuals, will be less likely to pass the battery. In the United States, lack of compliance with the various pieces of legislation addressing employment testing can result in a company having to pay back wages to all individuals denied employment on the basis of the test, being required to offer those individuals employment, and perhaps in being liable for punitive damages of up to $300,000. Hence, there are moral, financial and legal considerations when implementing a physical ability test battery.

The courts prefer that the emphasis be on the ability to perform the job rather than the risk of an injury or illness that may occur sometime in the future. The basis for this preference is the observation that employers have sometimes decided not to hire individuals with particular health conditions, such as a "bad back," because of the employer's perception that the individual is at increased risk of injury. If the basis for establishing the battery is risk of injury, the technical assistance manual for Title I of the Americans With Disabilities Act, which deals with employment discrimination, emphasizes that injury must be expected to occur in the near future, be severe and have a significantly higher likelihood in the population to be denied employment than the risk in the general population (EEOC, 1992).

All of these considerations can be met with a carefully constructed implementation and ongoing management practice. The implementation consists of the job analysis, test battery design, and validation. Ongoing practices refer to the steps taken to assure that all individuals are given equal consideration and the manner in which other information, such as prior experience on a similar job, is integrated in the decision-making process.

27.2 Implementation

27.2.1 Job Analysis

One of the most critical aspects of the job analysis is to identify the essential functions of the job. These are the functions that define the purpose of the job and are typically the elements that have to be performed in order to be considered a satisfactory employee. In most cases, the essential functions are the ones that are performed frequently, though there may be situations where a function that is performed infrequently is still considered essential. For example, the essential function of a warehouse selector is to lift cases from storage racks to a pallet that will be shipped to a client. A second essential function is to drive a motorized vehicle that transports the pallet throughout the warehouse. These tasks comprise the bulk of the selector's activity over the course of the shift. As a counter-example, one of the essential functions of a firefighter is to respond to emergencies. The amount of time actually spent in emergency response may be small, but it is essential that the firefighter be able to perform these duties. An example of a nonessential function for the warehouse selector would be the task of cleaning up broken cases since selectors rarely need to perform that task and it is an essential function of the janitor.

When developing a physical ability test battery, the degree to which it can be anticipated that there will be a substantial portion of individuals who will be unable to perform the task is also an important consideration. If the task can be performed by virtually everyone, then there is little economic justification for testing applicants since virtually everyone would pass the test.

The cost–benefit analysis can be quantified by calculating the cost of not being able to perform the task and the probability of an individual not being able to perform the task, and then balancing these against the cost of testing all applicants for the ability to perform the task. For instance, consider a situation where 100 employees are hired per year. If there is a 1% probability that an individual could not perform the task, and the inability to perform the task results in a $2000 loss associated with hiring and training the individual, then there is an expected cost of $2000 per year for not using the test (100 new-hires × 1% chance of not being able to perform the job × $2000 cost). If the cost of the test is $30 per applicant and 101 applicants must be tested to find 100 who pass (99% pass rate), the expected cost of testing would be $3030 per year. Hence, in this example, it is less costly to not test and accept the risk that 1% of all the workers will not be able to perform the job (expected cost of $2000 per year) rather than test all candidates ($3030 per year). In contrast, if it can be anticipated that only 80% of new-hires truly have the ability to perform the job, then the cost of not testing rises to $40000 per year (100 new-hires × 20% chance of not being able to perform × $2000 cost). The cost of testing would now be $3750 ($30 per test × 125 applicants tested to find 100 who pass). Under this scenario there is better than a 10-to-1 benefit–cost ratio for implementation of testing ($40000 without testing vs. $3750 with testing).

A third consideration during the job analysis is whether the essential functions that are potentially difficult for a substantial portion of the population can be modified through some form of ergonomic intervention (i.e., reasonable accomodation). For instance, wholesale grocery delivery drivers have to lift cases weighing up to 60 lbs on a routine basis. In addition, to access the cases, they must raise the rear door on their trailer. When brand-new trailers arrive from the manufacturer, virtually no force is required to open the door. However, the door track can become bent from forklifts running into the track while loading pallets onto the trailer. This can lead to a situation where a force of up to 200 lbs has to be exerted to open the door. Obviously, an essential function of the driver's job is to open the trailer door, but is it reasonable to design a test that assesses the driver's ability to exert 200 lbs to open a jammed door? An alternative is to require preventative maintenance rather than require the driver to exert 200 lbs. This would reduce the demand for this essential function back into the region where it can be anticipated that virtually all candidates will be able to perform the task. Hence, in this example, the critical task for the purposes of test battery design for these drivers would be the ability to lift 60-lb pound cases.

27.2.2 Test Battery Design

It has already been mentioned that tests should be included only for those essential functions that will be difficult for a substantial portion (5 to 10% or more) of the applicant population. There are three additional considerations in designing the test battery.

27.2.2.1 Job Relatedness

First, it is critical that the tests bear a high degree of job-relatedness to the essential functions of the job that are used as the basis for the test battery design. For instance, when testing the ability of an individual to lift a 60-lb case, the most job-related test would be to have the individual literally demonstrate the ability to lift a 60-lb case by having them move the case through the region (e.g., floor to table level) that is found on the job. In contrast, having the applicant perform a series of exercises such as push-ups, chin-ups, and sit-ups would not be a job-related test of the ability to lift a 60-lb case.

27.2.2.2 Basis for Establishing the Cutoff Score

The second consideration in test selection is the ease with which the cutoff score can be determined. The cutoff score is the level of test performance required in order to be considered as having demonstrated the ability to perform the job. Ideally, there is a clear relationship between the cutoff score and the performance of the job. For instance, if the test for the ability to lift a 60-lb case consists of lifting a case from floor level to table level, the most job-related cutoff for weight lifted would be 60 lbs.

In contrast, an alternative method is to use normative data. For example, some companies strap the applicant into a fixture that isolates their movement to flexion and extension about the low back. The test consists of measuring the moment, or torque, that the applicant can then generate with these muscles. It is difficult to directly measure the moment about the low back required for a given individual to lift a 60-lb case, since the moment will be a function of the posture used, how rapidly the individual lifts the case, the anthropometry of the individual, and a host of other factors. The only alternative is to measure the moment created by a number of incumbents, and develop a distribution of their scores. Conceivably, one could then choose the lowest score of the incumbents and require that the applicant at least demonstrate the ability to demonstrate that level of strength. A problem with this approach is that there is no guarantee that the scores demonstrated by the incumbents have a bearing on what the job requires. For instance, if all of the incumbents are regular participants in heavy weight training, their strength may be a reflection of the weight training rather than the job requirement.

A second example illustrating the problem of using cutoff scores based on normative data comes from test batteries used for evaluating police officer candidates. Some batteries for police officers require that applicants perform a run over a short distance to simulate the essential function of chasing suspects fleeing on foot. The score for the test is the time taken to cover the distance. The question becomes "How fast must an applicant complete the distance in order to be able to perform the job?" One approach that has been used is to determine whether the applicant's time is above the median for individuals of their sex and age. The justification has been that one would want their police officers to be at least as fit as average person of their sex and age. Unfortunately, this does not speak to whether the officer can perform the essential function of the job.

Most jobs lend themselves to determining the cutoff scores directly from the job requirements, as described in the first example given in this section — handling 60-lb cases. Likewise, there typically are testing formats available that will allow the test administrator to assess the ability at issue in a manner that can be directly compared to the job requirement. Hence, the problems that arise with utilizing the "normative data" approach can generally be avoided once the issue is recognized.

27.2.2.3 Accuracy, Reliability, and Safety

The third consideration is that the tests have a high degree of accuracy, reliability, and safety. An accurate test is one that precisely measures the attribute it purports to measure. A reliable test is one that yields the same results when it is repeated over time and by different test administrators. Reliability is enhanced when objective test measures are used rather than subjective assessments or opinions of the test administrator.

Safety is an important consideration, particularly if maximal strength tests are being used. Typically, steps can be taken to reduce the risk of injury during the test. For instance, an alternative to having the participant demonstrate the maximum amount of weight they can lift is to ask them to only demonstrate the ability to lift the weight required by the job. A second safety modification would be to have the participant first demonstrate the lift with an empty case, and then add weight to the box in fixed increments so that there is the chance to detect inability to lift the required weight without actually having the individual attempt the lift with the full weight required on their first effort.

27.2.3 Validation

27.2.3.1 Statistical Validation Studies

The strongest form of validation is a prospective statistical analysis. The typical form of this type of analysis involves administering the test to a group of individuals who are about to be hired into the position in question, although the test scores are not used as part of the hiring decision. The performance of these individuals is then tracked over the course of their employment. Measures of performance might be productivity, retention, injury rate, or supervisor evaluation. Care must be taken when selecting a measure that it reflect the important aspects of job performance and be reliable. Performance for individuals who pass the test would then be compared to performance for individuals who fail. As an example, Figure 27.1 shows the worker compensation injury rate for new-hires who failed a physical test battery compared to the injury rate for their peers who passed the battery designed for grocery warehouse selectors (Anderson, 1989a). The injury rate for those who failed the test was about double the rate for those who passed.

The effectiveness of the battery can be projected by comparing the performance of the entire group of new-hires, which reflects the performance anticipated without implementation of the battery, to the performance of those individuals who passed the battery. This latter group reflects the performance that would be expected with implementation of the battery. Table 27.1 summarizes the results calculated

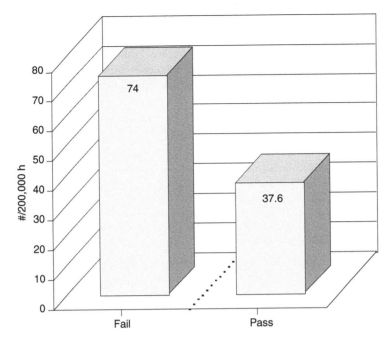

FIGURE 27.1 Worker compensation incidence rate by pass/fail status.

in this manner for three prospective statistical studies performed by the author (Anderson, 1989b, 1992a, b).

A less strong form of statistical validation is to compare performance for new-hires who began employment in time periods prior to and after implementation of the battery. The primary shortfall of this approach is that there is no control for factors that did not stay constant between the two time periods, such as work load, job structure, or management policies. The benefit of this type of analysis is that there is at least some indication of the effectiveness of implementing the battery when it is technically or economically infeasible to perform a prospective validation study. As noted earlier, the author has found that worker compensation incidence rates for new-hires typically fall 20% to 40% with implementation of a well-designed test battery (Anderson, 2003).

27.2.3.2 Content Validation Studies

The alternative to a statistical validation is a content validation. A content validation study consists of demonstrating that the content of the job is reflected in the content of the test battery, and that the cutoffs on the tests reflect the job demands. It is valuable to be able to also make reference to any

TABLE 27.1 Summary of Expected Impact of Test Battery Implementation

	Grocery Distribution	Soft Drink Distribution	Retail Distribution
Decrease in incidence rate	18%	26%	17%
Decrease in lost time due to injuries	n/a	11%	27%
Decrease in workers' compensation cost	n/a	55%	n/a
Increase in retention	10%	15%	7%
Increase in productivity	n/a	n/a	7%

Note: n/a: data not available for all locations in the study.
Source: From Charles Anderson/Physical Ability Tests/Workplace Ergonomics/1995. With permission.

statistical validation studies in similar industries that further support the content validity of the battery at issue.

In the United States, the Equal Employment Opportunity Commission (EEOC) states that a content validation is acceptable when a statistical validation is technically infeasible (EEOC, 1978). For example, this could be the case when there is a small number of new-hires on which to base the study, or the job demands are not homogeneous.

27.3 Ongoing Practices

It is critical to not only carefully construct the test battery, but also the management policies that surround its use. Four aspects are particularly crucial to assure that the manner in which the test battery is used is fair to all applicants.

27.3.1 Equal Treatment

The fundamental consideration is that all applicants must be treated equally. For instance, all applicants for the job for which the test battery has been designed must be given the same opportunity to test. It is not considered appropriate to only give the battery to females, individuals over the age of 40, or those considered disabled under the premise that these are individuals who are most likely to be unable to perform the job. Furthermore, they should all be given the same information prior to the test. This information might include instructions to wear loose-fitting clothes, avoid a heavy meal prior to the test, and have adequate rest the night prior to their appointment.

Another dimension of equal treatment is that the same scoring process must be used for all individuals. It is generally recommended that the same cutoff be used for all individuals without exception. Problems arise when managers attempt to factor in their own judgement by allowing some individuals who fall below the cutoff to be employed, particularly when the decision as to who shall benefit is arbitrary.

Finally, all individuals who fail should be given the same opportunity to be reconsidered. This opportunity could be structured on the basis of their performance, but must be equally applied. For instance, all individuals who score within 95% of the cutoff might be offered the chance to retest immediately. Those who score between 80% and 95% might be allowed to retest after 4 weeks, whereas those who score less than 80% have to wait at least 8 weeks to retest. There should be a valid basis for the time periods, such as the percent of improvement in performance that could be expected at particular points in time with a well-designed physical conditioning program.

27.3.2 Prior Experience

There is only one exception to the policy of rigid interpretation of the cutoff score, which occurs when an individual who fails the test has demonstrated the ability to perform the job during prior experience in similar employment. It is important to remember that the best indicator of an individual's ability to perform a job is actual demonstration of that performance. Lacking direct evidence of ability to perform the job, one might consider performance on a test battery designed to provide information about that ability. The test battery is not necessarily a perfect indicator, so it is possible that an individual who can perform the job might fail the battery. Clearly, the battery should be designed so that this possibility is minimized. This is the reason for selecting tests that bear the closest resemblance to the way the job is actually performed, carefully selecting the cutoff score and taking all possible steps to assure that the test scores are valid predictors of job performance.

In the circumstance where there is additional information available about an individual's ability to perform the job, such as prior experience, this information can be used to override a failing test result. It is critical, though, that all individuals in similar circumstances be treated the same way. This can be promoted by establishing clear definitions of what would be considered adequate previous experience that can be consistently applied in each case that arises. For instance, if an individual applying for a warehouse selector position has previous experience as a selector yet fails the battery, the manager might ascertain if the previous experience involved handling cases of similar weight at a similar pace for a similar shift length using similar equipment (e.g., manual lifting vs. picking full pallets with a forklift). If all aspects of the previous job and the job at issue are similar, then the prior experience could be used as a legitimate basis for hiring the individual in spite of the failing test results.

27.4 Summary

Physical ability testing can be an effective way to improve the match between job demands and worker ability, thereby enhancing performance and reducing the risk of injury. Thorough job analysis is required to guarantee that the test battery selected is a valid predictor of the ability to perform the essential functions of the job. Furthermore, it is important to assure that the management policies related to the utilization the battery guarantee equal application of its use and interpretation. This includes to whom the battery is administered, the information that applicants are given prior to the test, the way their results are interpreted and the opportunity for retesting that is extended to those who fail the test. When the test battery is designed with these factors in mind, it can meet the legal requirements and be an effective management tool. Results from a wide range of applications indicate that there typically is a 20% to 40% reduction in worker compensation injuries associated with implementation of a well-designed test battery.

References

Anderson, C.K. 1989a. Strength and endurance testing for pre-employment placement. In K. Kroemer, ed. *Manual Material Handling: Understanding and Preventing Back Trauma*, American Industrial Hygiene Association, Akron.

Anderson, C.K. 1989b. Impact of physical ability screening for grocery warehousing. Technical Report, Advanced Ergonomics, Inc.

Anderson, C.K. 1992a. Impact of physical ability testing on injuries and retention for the Coca-Cola Bottlers association. Technical Report, Advanced Ergonomics, Inc.

Anderson, C.K. 1992b. Impact of physical ability testing on workers' compensation injury rate and severity: target distribution centers. Technical Report, Advanced Ergonomics, Inc.

Anderson, C.K. 2003. The advanced ergonomics physical ability testing program: a fourteen year review, Technical Report, Advanced Ergonomics, Inc.

Anderson, C.K. and Catterall, M.J. 1987. The impact of physical ability testing on incidence rate, severity rate and productivity. In S.S. Asfour, ed. *Trends in Ergonomics/Human Factors IV*, Elsevier Science Publishers, Amsterdam.

Anderson, C.K. and Herrin, G.D. 1980. Validation study of pre-employment strength testing at Dayton Tire and Rubber Company. Technical Report, University of Michigan Center for Ergonomics.

Arnold, J.D., Rauschenberger, J.M., Soubel, W.G., and Guion, R.M. 1982. Validation and utility of a strength test for selecting steel workers. *Journal of Applied Psychology*, 67(5):588–604.

Ayoub, M.A. 1983. Design of a pre-employment screening program. In Kvalseth, T.O., ed. *Ergonomics of Workstation Design*, Butterworths & Co., London.

Cady, L.D., Bischoff, D.P., O'Connell, E.R., Thomas, P.C., and Allan, J.H. 1979. Strength and fitness and subsequent back injuries in fire fighters. *Journal of Occupational Medicine*, 21(4):269–272.

Chaffin, D.B., Herrin, G.D. Keyserling, W.M., and Foulke, J.A. 1977. *Pre-Employment Strength Testing in Selecting Workers for Materials Handling Jobs*. NIOSH Physiology and Ergonomics Branch. Contract No. CDC 99-74-62, Cincinnati, OH.

Craig, B.N., Congleton, J.J., Kerk, C.J., Lawler, J.M., and McSweeney, K.P. 1998. Correlation of injury occurrence data with estimated maximal aerobic capacity and body composition in a high-frequency manual materials handling task. *American Industrial Hygiene Association Journal*, 59:25–33.

Equal Employment Opportunity Commission (EEOC). *Uniform Guidelines on Employee Selection Procedures*. Title 29, CFR, Part 1607, 1978.

Equal Employment Opportunity Commission (EEOC). *Title I of the Americans With Disabilities Act: EEOCs Technical Assistance Manual*. 1992.

Herrin, G.D., Kochkin, S., and Scott, V. 1982. Development of an employee strength assessment program for United Airlines. Technical Report, The University of Michigan, Center for Ergonomics.

Jackson, A.S., Osburn, H.G., and Laughery, K.R. 1984. Validity of isometric strength tests for predicting performance in physically demanding tasks. *Proceedings of the Human Factors Society — 28th Annual Meeting*, 452–454.

Jackson, A.S., Osburn, H.G., and Laughery, K.R. 1991a. Validity of isometric strength tests for predicting endurance work tasks of coal miners. *Proceedings of the Human Factors Society — 35th Annual Meeting*, 753–767.

Jackson, A.S., Osburn, H.G., Laughery, K.R., and Vaubel, K.P. 1991b. Strength demands of chemical plant work tasks. *Proceedings of the Human Factors Society — 35th Annual Meeting*, 758–762.

Jackson, A.S., Osburn, H.G., Laughery, K.R., and Vaubel, K.P. 1992. Validity of isometric strength tests for predicting the capacity to crack, open, and close industrial valves. *Proceedings of the Human Factors Society — 36th Annual Meeting*, 688–691.

Keyserling, W.M. 1979. *Isometric Strength Testing in Selecting Workers for Strenuous Jobs*. Ph.D. Dissertation, University of Michigan Center for Ergonomics.

Keyserling, W.M., Herrin, G.D., and Chaffin, D.B. 1980a. Isometric strength testing as a means of controlling medical incidents on strenuous jobs. *Journal of Occupational Medicine*, 22(5):332–336.

Keyserling, W.M., Herrin, G.D., Chaffin, D.B., Armstrong, T.A., and Foss, M.L. 1980b. Establishing an industrial strength testing program. *American Industrial Hygiene Association Journal*, 41:730–736.

Laughery, K.R. and Jackson, A.S. 1984. Pre-employment physical test development for roustabout jobs on offshore platforms. Technical Report, Kerr McGee Corporation.

Laughery, K.R., Jackson, A.S., and Fontenelle, G.A. 1988. Isometric strength tests: predicting performance in physically demanding transport tasks. *Proceedings of the Human Factors Society — 32nd Annual Meeting*, pp. 695–699.

Liles, D.H., Deivanayagam, S., Ayoub, M.M., and Mahajin, P. 1984. A job severity index for the evaluation and control of lifting injury. *Human Factors*, 26(6):683–693.

Reilly, R.R., Zedeck, S., and Tenopyr, M.L. Validity and fairness of physical ability tests for predicting performance in craft jobs. *Journal of Applied Psychology*, 64(3):262–274.

28

Human Resources Management

Thomas J. Slavin
*International Truck and Engine
Corporation*

28.1 Role of Human Resources

The role of human resources (HR) in industrial environments is to bring different people and functions together to achieve common business goals. HR provides the glue that holds various business functions together to make the company work smoothly and translate the business vision into action. The goal of HR is to hire and retain the right people, create business systems that support and motivate them, and help align departmental objectives with overall business objectives. An effective HR function uses ergonomics tools to accomplish all of these goals. In addition, HR must ensure compliance with several employment laws and ergonomics can be a key element in a compliance strategy.

Ergonomics practitioners can assist HR accomplish HR goals, and can work through HR to accomplish ergonomics goals. Many of the HR objectives can provide support for workplace ergonomics improvements. For example, diversity and Americans with Disability Act (ADA) compliance goals may help support changes that allow a greater percentage of the population to perform a given job. And ergonomics job improvements can increase morale and help retain and motivate people.

Some of the areas on which HR and ergonomics must work most closely together are preemployment functional capacity screening, return to work of injured and disabled employees, and implementation of ergonomic solutions such as job rotation that may involve labor relations issues. Perhaps most important to the business, HR and ergonomics professionals can work together to make the workforce more productive.

TABLE 28.1   Core Functions of Human Resources

Employee recruitment, selection, retention, and termination	Compliance with laws and regulations
Compensation and benefits (pension, disability, health care)	Federal Equal Employment Opportunity laws and
Labor/employee relations	regulations including:
Discipline	Title VII of the Civil Rights Act of 1964
Diversity Promotion	Age Discrimination Employment Act of 1967
Compliance	Americans with Disabilities Act (ADA)
Violence prevention, sexual harassment prevention	Executive Order 11246 as amended
Leadership development	Uniform Guidelines on Employee Selection
Performance evaluation	Procedures
Training/tution refund	National Labor Relations Act
Absenteeism	Fair Labor Standards Act
Employee assistance program	Wage and Hour laws
Drug-free program	Family and Medical Leave Act

28.2   Human Resources Functions

HR in most organizations typically performs the set of core functions listed in Table 28.1. Responsibility for other areas such as those listed in Table 28.2 is commonly assigned to HR, but is also often assigned to other functional areas such as treasury, finance, engineering, manufacturing, or legal.

HR plays a key role in any effective ergonomics program, and conversely ergonomics is a vital part of an effective HR function. A look at some of the functions that HR performs in many organizations shows how critical it is for the HR function to work well with an ergonomics program.

28.2.1   Employee Selection, Placement, and Retention Functions

Ergonomics is an important tool for optimizing productivity. Ensuring a good fit for worker and job is a key to avoiding many musculoskeletal disorders and other problems such as absenteeism and turnover. It is simply not possible to design every job to fit every potential worker. Having well-defined physical and mental demands for each job is critical to candidate evaluation, selection, placement, and exclusion. If ergonomic requirements of a job are available, companies can use the criteria to perform functional capacity tests to ensure candidates are capable of performing the job. A poor match not only increases risk of injury but can also result in poor performance, poor quality, poor morale, and costly turnover. Conversely, a good match provides a valuable and reliable resource for the company and a safe, healthy job for the worker.

In the selection of new employees a documented assessment of job demands helps find those employees who can perform the job. It may be an ideal of ergonomics to design all jobs for all people, but in practice some people cannot do certain jobs without injury. In those cases, unless and until the jobs can be redesigned it is best to identify such mismatches before making a hiring mistake.

TABLE 28.2   Functions Commonly But Not Always Performed by Human Resources

Safety	Insurance programs
Industrial hygiene	Workers' compensation
Environmental compliance	Fleet
Ergonomics	General liability
Medical	Products liability
Security	Long-term planning/corporate strategy
Fire prevention	Government affairs
Workers' compensation	
Communication	Culture transformation
Travel services/relocation programs	Wellness and health initiatives

The Americans with Disabilities Act prohibits discrimination against people with disabilities, but the ADA does allow employers to reject applicants who cannot perform essential functions of the job. The use of screening procedures such as functional capacity testing may be shown to be job related and consistent with business necessity provided the physical or mental demands of the job are properly evaluated and documented. Before an individual deemed "disabled" under the ADA is rejected on the basis of the functional capacity test, it may be necessary for the employer to demonstrate that it is not possible to provide "reasonable accommodation" to enable the disabled individual to perform the essential functions of the job by altering the tasks or duties. No one is better qualified to make such determinations than the ergonomics professional.

28.2.2 Diversity Function

Many companies have moved beyond the goal of compliance with federal equal employment opportunity laws and regulations to internalize the spirit and objectives of the laws and to recognize that diversity of the work force can provide strategic advantages. Diversity initiatives may increase variation in the employee population and create challenges to design jobs to accommodate a greater range of size, strength, and cognitive functions. Designing jobs to fit a wider range of people and to accommodate disabilities can also help place workers who are temporarily disabled due to a workplace injury or who have other limitations associated with short-term nonoccupational disabilities.

28.2.3 Labor Relations Function

For those operations represented by a union, certain ergonomic activities must be considered against the background of the collective bargaining agreement. Seniority, a key element of most collective bargaining agreements, determines who has rights to select which jobs through a carefully defined process. In general, the most senior person gets first choice of jobs in the bargaining unit. In fact, one simple way to find jobs that are the best candidates for ergonomic intervention is to find jobs with the lowest seniority. These are the most challenging jobs that nobody wants, often because of musculoskeletal stresses and risk factors.

Job rotation, an important administrative control for reducing musculoskeletal stresses may be difficult to implement in a seniority system. Employees who have worked many years to get enough seniority to claim the best available job may resist the idea of sharing elements of their job with people who have less seniority and less desirable jobs. Even when rotation may introduce variety and make a job more interesting, the near sacred regard for the seniority principle may make it difficult to persuade employees to accept rotation. With education and cooperation and in an environment of trust it is sometimes possible to obtain agreement to implement a rotational scheme. When new work processes are created, it is possible to introduce job rotation in the new job definitions. Labor contracts reserve certain functions for management including the right to manage the affairs of the business and to direct the working forces of the company. Management has the right to determine the work to be done, but defining jobs in a new manufacturing system is much easier than redefining old jobs in an existing system.

Many collective bargaining agreements contain provisions for joint ergonomic programs. Typically, these specify the formation and membership of a joint union management committee and may even specify a process for evaluating jobs and making improvements. Bargaining ergonomics provisions into labor contracts should not be seen as a win–lose proposition or as management giving something up or labor winning something from the company. Rather this should be an opportunity to express a company wide philosophy and commitment to ergonomics and to engage both management and labor in a common purpose. It is important for ergonomics specialists on both sides of the bargaining table to be involved in the design of contract language to make sure it is a suitable framework for the ergonomics program.

One key to a successful joint ergonomics program is the technical and educational support to make design changes. Programs can become overburdened by committee meetings and bureaucracy unless there are real results to show for the efforts. Improvements may take a few iterations to find a solution that is both effective and embraced by the operators. For example, a new hoist may seem like the best solution for lifting a 70-lb part from a pallet to an assembly line. A hoist could reduce the stress for a lifting task, but the initial installation may be awkward or difficult to use, may restrict the range of motion, or may slow the job down. The operator may like to work ahead and install parts before they reach the designated work station in order to create some buffer time for personal use. If the hoist does not permit that flexibility, the operator may ignore it. Persevering to overcome obstacles and modify solutions until they work for the operators, or working with the operators to accept a change that may seem awkward at first, requires committed support to keep working toward the solution. It is easy for a joint ergonomics committee to become frustrated by initial failure or resistance. Engineering and educational support to keep refining a solution and earning operator acceptance is critical to the success of a program.

28.2.4   Workers' Compensation Function

Before 1910, an employee injured while working had to prove employer neglect to obtain compensation. Employers could limit their liability through three legal theories. The fellow-servant doctrine applied if the injury was caused by the negligence of a coworker. The assumption of risk concept held that the employee knew there were risks when taking the job. The contributory negligence argument applied where an employee was partly responsible. Cases took a long time to settle and the judgment amounts were unpredictable. New Jersey was the first state to enact a workers' compensation law in 1911, and several other states soon followed. Workers' compensation began as a trade-off between the certainty of quick compensation payments regardless of fault for employees and limited liability for employers. Under the workers' compensation system payments are somewhat less than regular wages (usually 65 to 80%, tax free) to provide an incentive to return to work.

Workers' compensation is a state-by-state program with different rules and benefit levels in each state. It is mandatory for employers in every state except Texas, which allows employers to opt out of the workers' compensation system and choose the liability system. Most states allow employers to become insured or, if they meet certain requirements related to the ability to pay claims, to self-insure. A few states (e.g., Ohio, North Dakota, Washington, West Virginia, and Wyoming) have exclusive state funds that take the place of private insurance. Most of these still allow self-insurance. Most employers who choose to be self-insured use a third party administrator (TPA) to pay claims and administer the program.

There are a myriad of financial arrangements for workers' compensation and some of the terms can be confusing. Because ergonomics programs often involve discussions of workers' compensation costs, it is important to explain some of the key terms.

- *Premium rate* — This is the rate of payment usually expressed in percent of payroll, or dollars per hundred dollars of payroll. It varies by industrial classification and by state. The more dangerous occupations pay a higher premium rate. In 2002, according to a study by the Oregon Department of Consumer and Business services the state by state average rates ranged from \$1.24 per \$100 of payroll in North Dakota to \$5.23 per \$100 in California. (Reinke, 2003) Within each state the more hazardous industries may pay rates that are several times higher than the average.
- *Experience modification (EM) factor* — This is a premium adjustment based on prior years experience compared with the experience of other employers in the same classification. A company with an EM factor of 0.92, for example, would pay 92% of the standard premium for that classification. The EM calculation is usually based on the prior 3 yr experience.
- *Incurred loss* — This is the total of payments already made on a claim plus the amount of estimated future payments. When discussing or comparing incurred costs it is important to specify the "as of" date. Payments and reserves vary over time as cases develop and estimates of future payments become clearer. For example, to meaningfully compare the results of 2001 and 2002 it is necessary

to specify an "as of" date such as February 1 of the following year (February 1, 2002 for the 2001 year and February 1, 2003 for the 2002 year). Even if performance is similar for both years, using the same date (2/1/2003 in this example) to value claims from both years may make the 2001 year look 50% higher because of an extra year of case development.

- *Paid loss* — This is the amount of money actually paid out. When discussing workers' compensation costs it is necessary to specify the period of injuries for which payments are made. Without clarification "paid losses for 2003" could refer to any of the following:
  - All payments made up to the current date on only those injuries that occurred in 2003
  - All payments made during 2003 on injuries from 2003 and all prior years
  - All payments made during 2003 on injuries from only 2003
- *Reserves* — The case reserve is the estimate of future payments for a given claim. Reserves tend to "develop" over time for open cases. As with the other payment related terms, care is needed in specifying the period covered by the reserve as well as the "as of" date for valuing the reserve.
- *Indemnity/disability* — There are several categories of payment that make up a workers' compensation claim. Medical and legal expenses are fairly self-explanatory. The other major category of expense is payment for lost wages while the claimant is away from work. This type of payment is made directly to the employee and is referred to either as an indemnity payment or a disability payment. Disability in this sense must be distinguished from nonoccupational disability programs that cover short- or long-term disability.

Workers' compensation laws have evolved over time. It was not until the 1970's that cumulative trauma cases began to be recognized as work related and compensable. Prior to that the theory was that gradual onset conditions were personal conditions that merely surfaced at work. Heart attacks that occur at work are generally treated this way currently. In order to be compensable an injury such as a back injury had to be associated with a specific work-related event: "I bent over to pick up a bolt and heard a pop."

Keep in mind changes in workers' compensation laws when looking at historical trends. Work places are much safer now than in prior decades when judged by fatality rates and rates of traumatic injuries (cuts, fractures, amputations). However, workers' compensation costs are higher and overall injury rates, while lower, have not declined as much as traumatic injury rates have. Soft tissue injuries now account for about one third of all industrial cases and two thirds of workers' compensation costs. Such soft tissue injuries were probably common in prior decades, but were not recorded because they were not considered work related or compensable.

When the transcontinental railroad was being built in the 1860s detailed accident records were compiled. Many deaths, amputations, fractures, and other serious injuries incurred as laborers struggled to lay track across the continent. Interestingly, no back injuries were recorded. It is unlikely that the people of the 19th century had much better backs than people today. Rather it is likely that such injuries were not considered industrially related. It is much different today. One can only wonder what conditions are today considered part of the aging process (arthritis comes to mind) and therefore not compensable, but which could become so in the future.

Workers' compensation laws are only one of the many factors that come into play in the recognition of musculoskeletal disorders. Figure 28.1 is from the National Institute for Occupational Safety and Health (NIOSH) National Occupations Research Agenda (NORA) musculoskeletal disorder team. It shows the many work related, personal, and external factors that can impact the development of a musculoskeletal disorder and subsequent disability. Physical stressors are only one part of the process that determines whether a musculoskeletal disorder will result in a disability with loss of time from the job.

At the individual level, there are important interventions that can make a difference between disease and disability. One successful approach to back pain called "The Back Power Program" observes that 90% of back pain patients have no structural abnormalities or disease. (Imrie, 1990) The majority of back pain is caused by muscle injury and need not be permanently disabling. Diagnosing which muscle groups have sustained injury and developing stretching exercises for those muscles can prevent additional episodes

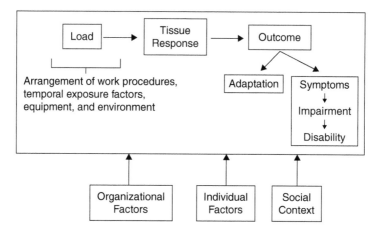

FIGURE 28.1   Conceptual model of factors that potentially contribute to musculoskeletal disorders. (CDC-NIOSH, 2001).

of pain. Even without stretching, just knowing the nature of the pain and realizing that it is not caused by a structural problem can give a person confidence to continue working. Working with pain is often discretionary and depends on whether the work environment is positive enough to entice the worker to come to work despite discomfort or stressful enough to encourage the worker to stay home.

An important part of effective workers' compensation administration and cost control is a return-to-work program that gets injured employees back to work as soon as they are able. In some cases, this may be a light duty program that occupies an injured employee in some productive endeavor until recovery. Ideally, this would be a transitional work program that assigns work that is increasingly challenging, consistent with the recovery process. Ergonomics professionals have an important role to play in identifying work opportunities that are consistent with an employee's transition to recovery. The longer a person is off work, the harder it is to reenter the workforce. Sometimes the workers' compensation process introduces an adversarial climate that alienates the injured employee and makes it hard to feel part of the workforce and reluctant to come back to work.

28.2.5 Disability Function

Nonoccupational disability (short term or long term) is often considered separately from workers' compensation and administered by a different organization under a different insurance program. However, the need to return people to productive work consistent with recovery is the same for both functions. Whether the disability is work-related or not, there are similar direct costs to the company for the absence whether paid through a short-term disability program or workers' compensation. Indirect costs, such as replacement workers, training, impact on morale of coworkers, etc. are also similar. Therefore, it makes sense to focus as much on returning to work those employees who have nonoccupational disabilities as on returning those with occupational injuries.

28.2.6 Compensation Function

Organizations have many functions with different, often opposing goals. Marketing, production, accounting, engineering, and legal departments all have different goals and need to be brought together in a common purpose. The reward structure needs to recognize activities and results that are good for the organization as a whole. HR plays a central role in designing the reward system through performance evaluation and incentive compensation criteria. Ergonomics professionals can take advantage of the evaluation and incentive system by helping define goals and objectives for managers and supervisors. For example, goals based on the number or percent of people in a department with ergonomics training, or percent of jobs evaluated, or number of category x risks reduced to category y, or percent of injured workers returned to work can be used to align ergonomics goals with individual or department goals.

28.2.7   Wellness and Absenteeism Function

Ergonomics professionals often focus only on job-related musculoskeletal stressors, but a big part of any company's health care cost and absenteeism is related to musculoskeletal disorders (MSDS). Even though they may not be occupational in origin they still have an impact on the individual, their family, and the company's bottom line. Job modifications that permit people to be productive despite musculoskeletal limitations are worth pursuing. In addition, increasing the health and conditioning of the working population will help reduce both occupational and nonoccupational injuries and will lower health care costs.

28.2.8   Training and Development Function

Many companies maintain a training and development function that can be a useful resource for ergonomics professionals. Professional trainers may not know much about ergonomics, but ergonomics professionals are not often experts in training. Training is essential for an ergonomics program and training professionals can help design, conduct, evaluate, and revise training programs. Scheduling and recordkeeping are other functions that can be made much easier with the help of training professionals.

28.3   Human Resources Vision

HR is sometimes perceived as a function devoted to rigid enforcement of policies, reactive response to problems, bureaucratic interpretation of benefit plans, and stingy administration of salary budgets. It does not have to be that way. The difference between a reactive and bureaucratic HR function and one that is a key strategic force in implementing business objectives lies in the vision of HR. Vision can also help determine the direction of ergonomics activities. The importance of vision can be illustrated by a principle from a 1947 book on traffic safety by Maxwell Halsey. The book was written as a collection of principles and the first one was the following:

> Efficient transportation, not accident reduction is the fundamental purpose. Accidents, like congestion, are only indices of inefficiency.

If you think about traffic accident reduction, the main solutions that come to mind may be speed limits, traffic lights, stop signs, and tougher laws against drunk driving. However, if you were to plan a trip across the country in the safest fashion possible you would not necessarily choose the route with the most stop signs and traffic lights, or the lowest speed limits, or through the greatest number of dry counties. Instead, you would choose the interstate highways that are designed for efficient transportation and also happen to be the safest way to travel by car.

The point of Halsey's principle #1 is that if you think in terms of accident reduction you may be led to a different set of solutions than if you think in terms of efficient transportation. That is a good lesson for human resources and ergonomics. For HR, a vision of making the workforce more effective may lead to different priorities than a vision of minimizing benefit costs or ensuring compliance with employment laws and policies.

For ergonomics, a vision of employee productivity may be a better driver for workplace improvement than reduction in MSD risk. Better outcomes may result from process redesign for improved productivity than from changes designed only to reduce stress factors. In fact, a total productivity vision may be a more powerful and compelling driver because it can usually engage more support for change and often can provide greater cost justification. It is often easier to justify productivity improvement than MSD relief.

For example, in one engine assembly plant, tightening bolts with a power hand tool and setting torque for a V-8 engine head led to symptoms and cases of carpal tunnel syndrome (CTS). Installing a multiple spindle bolt runner (Figure 28.2) for that job cost $200,000. To try to justify the cost on the basis of CTS reduction would have required some assumptions about causation, projected future incidence, and amount of projected savings. Sometimes the questions and debate over assumptions creates additional work, requires more data, and adds delay to the project approval process. In this case, quality and productivity improvements (installing 10 bolts at once and with precisely controlled torque) were relatively easy to support and could easily justify the change.

28.4   Total Productivity, An Integrated Approach

Organizations can operate as many functional silos or stove pipes — areas such as engineering or benefits or safety that function independently with little communication or cooperation. The silo approach is often counterproductive and produces suboptimal results.

Take the example of modified job placements. Returning injured workers to work in a modified capacity makes good sense for the organization because it speeds recovery, keeps workers feeling a part of the work group, and gets productive work from employees who would otherwise be collecting 65–80% of their normal pay while on workers' compensation. An employee on light duty or restricted activity who is doing some productive work is a greater asset than one who is on workers' compensation or short-term disability. With a silo mentality several issues may arise:

- *Supervisor silo* — Supervisors who are judged on productivity (defined as hours per unit) want employees who are capable of maximum effort. An injured employee who needs a light duty or modified job assignment for optimum recovery and return to work progress may reduce the supervisor's productivity metric. Supervisors prefer a replacement worker.
- *Medical silo* — Meanwhile, medical may be too busy to find a job that is consistent with the injured employee's restrictions and with rehabilitation and recovery.
- *Labor silo* — If the plant is represented by a union finding a job may involve considerable negotiations over the nature of the work to be done, whether it is part of someone else's existing work assignment, and how long the modified assignment will last. Labor relations may be reluctant to create an issue over job assignment and seniority. If the modified work assignment looks like a desirable job other employees with more seniority may contend that it is a permanent job and demand the right to bid on it.
- *Disability/workers' compensation silos* — If the placement issues cannot be resolved and the injured worker must take time off work the workers' compensation and short-term disability managers may try to shift the employee into the other program. Conditions that may originate either on

FIGURE 28.2   A multiple spindle bolt runner replaces hand tools used for tightening and torqueing bolts.

or off the job, including many musculoskeletal disorders, are more prone to this internal cost shifting debate.

In this return-to-work scenario, no one is looking out for what is best for the employee's recovery or the company's bottom line. An integrated approach, on the other hand, would tie all the functions together in a common purpose, ideally with a set of performance metrics that encourage behavior that benefits the employee as well as the company.

An integrated approach works best when functions share a common vision and goal. Total productivity is one such vision: helping people be as productive as possible at work, off the job in their private life, and ultimately in retirement. A positive vision can engage a broader range of functions. For example, supervisors who are accountable for absenteeism (including absenteeism by employees on workers' compensation or short-term disability) in addition to standard productivity metrics may be more interested in cooperating with return to work programs.

28.5   Valuing Human Assets

"People are our most important asset" is an old bromide that may have some truth, but too often rings hollow because of a lack of meaningful measurement. Human resources is responsible for the "asset" but seldom calculates or gives any meaning to the term relative to management of other corporate assets. Ergonomics has a big role to play not only in protecting the human assets but in optimizing their use. The following exercise to help put the human asset in perspective may be useful.

It may be more accurate to consider people as leased assets rather than as owned assets. Strictly speaking, of course, people are not owned by a company, although permanent liabilities can accrue for pension and other postretirement obligations. A typical employee may earn $50,000 per year in salary that we could consider the annual lease cost, and another $10,000 in benefits (insurance, vacation, etc.) that we could consider as an operating expense. Over a 20-yr career the lease cost would total $1 million and the operating expenses would be another $200,000. In this simplistic example we do not discount for present value nor account for other operating costs such as training. A company of 1000 employees would have $1 billion in human assets with a $50 million annual lease cost and $10 million per year operating cost.

Thinking of people as leased human assets may seem insensitive, but the exercise has a purpose. Think of how a company would approach the care of a $1 million machine tool. Anyone who consistently abused, misapplied, or carelessly overused the machine tool causing it to break down would not be considered a good manager. The same should be true for human machine tools that get misused, misapplied, or overused to the point of breakdown. Moreover, this financial model may show that it is wise to invest in ergonomic improvements that can extend the life or improve the function of the human assets.

The simple calculation of human asset value above produces a number that is similar to an accounting book value rather than the productive value of the asset. This is the same distinction that may be made for a building or machine that may have a production value that is greater or lesser than its book value. The ability of the employees at all levels of the company to produce value from their abilities and intellect is a human capital asset. It is not easy to quantify this aspect of human capital. One way to look at this on an individual basis may be to imagine the most productive and highly motivated worker, one with tremendous initiative and customer focus. Now imagine the least productive worker. An organization made up of employees who were all like the most productive worker would be much more successful than an organization made up of employees like the least productive worker.

It is difficult to quantify this aspect of human capital, that is, managing the potential difference between highly motivated, effective people and others who are not motivated or effective. The job of human resources is to provide the policies, work environment, and support systems to move people along the value continuum toward the highly productive state. Some of the tools to accomplish this include employee selection, placement, training and development, communication, morale, compensation, and benefits. Ergonomic improvements are another tool that can be used to make people more effective and motivated.

References

CDC — NIOSH. 2001. National Occupational Research Agenda for Musculoskeletal Disorders. DHHS (NIOSH) Publication No. 2001-117. Cincinnati, OH.

Imrie, D. and Barbuto, L. 1990. *The Back Power Program.* John Wiley & Sons Canada, Limited. Mississauga, ON.

Reinke, D. and Manley, M. 2003. 2002 Oregon workers' compensation Premium Rate Ranking. Oregon Department of Consumer and Business Services. Salem, OR.

29

Workday Length and Shiftwork Issues

Peter Knauth
University of Karlsruhe

29.1 Introduction

The recent trend toward extended and irregular work hours as well as the increase in night work and shiftwork in some fields may give grounds for concern.[1,2]

This chapter discusses both the potential problems of extended work hours and ergonomic recommendations (Section 29.2) as well as shiftwork issues and the corresponding recommendations (Section 29.3).

29.2 Workday Length

Extended workdays (of 9, 10, 12, or more hours, for example) can have many negative effects, such as increased fatigue, sleepiness, burnout symptoms, increased absenteeism, reduced performance, higher risk of accidents, and longer toxic exposure.

29.2.1 Increased Fatigue, Drowsiness, or Sleepiness

Although many authors found a connection between increased fatigue, drowsiness or sleepiness, and extended daily working hours,[3–9] some authors did not establish increased sleepiness or decreased alertness after 12-h shifts when compared to 8-h shifts.[10–12]

Not only the duration of the working day, but also many other factors can contribute to workers' level of fatigue; for example, time of day, work-rest scheduling, job task and workload, the environment, social/domestic demands and support, and long commuting time.

The effect of the combination of high workload and longer working hours is shown in Figure 29.1. Spencer et al.[5] analysed 98 air traffic controllers' diaries during a period of high workload. Fatigue increased from 3–5 h to 5–6 h ($p < 0.05$), to 6–7 h ($p < 0.01$), and to greater than 7 h ($p < 0.001$). As Spencer et al. [5, p. 254] stated, "Mean levels of fatigue increased from a rating of

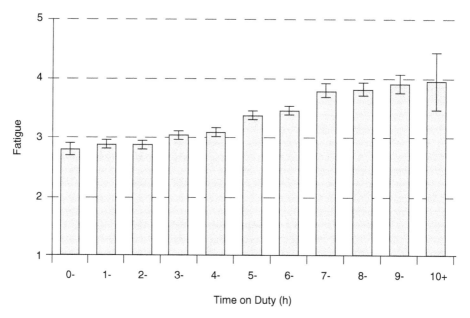

FIGURE 29.1 The effect of time on duty on fatigue. (From Spencer, M.B., Rogers, A.S., Birch, C.L., and Belyavin, A.J. In: *Shiftwork in the 21st Century*, Hornberger, S., Knauth, P., Costa, G., and Folkard, S., Eds., p. 254, Peter Lang Verlag, Frankfurt am Main, 2000. With permission.)

3.00 following periods classified as low-workload periods to 3.17 following periods of medium workload and to 3.36 following periods of high workloads."

29.2.2   Reduced Performance

Worksite studies measuring the association between performance and extended work shifts are rare. In a study of the automotive industry, poorer performance on neuropsychological tests, such as trailmaking and card sorting, was attributed to workers' recent overtime.[13] In a worksite study at a natural gas utility, there were decrements in reaction time and subjective alertness even 10 months after the change from an 8-h shift to a 12-h shift schedule.[4] Axelsson et al.[11] also compared 12-h shifts during weekends with 8-h shifts during weekdays. These performance tests (reaction time and alertness), however, did not show any significant main effects for length of shift.

With respect to productivity, equal numbers of oil refinery managers interviewed reported increases, decreases, or no change after the introduction of 12-h shifts.[14] However, a major U.S. electronics manufacturer and a majority of U.S. government sites abandoned long work shifts because of reduced productivity.[15]

29.2.3   Increased Accident Risk

As shown in Figure 29.2, Nachreiner[16] compared four studies that support the interpretation of increasing accident risk if workers worked beyond a "normal" 8-h working day, whereas Folkard[17] and Haenecke et al.[18] found that accident risk increased exponentially after 7 to 8 h at work. Åkerstedt[19] and Nachreiner[16] observed a later beginning, that is, after 9 or 10 h. Laundry and Lees[20] carried out a long-term study of accidents during 8-h and 12-h shifts at a yarn manufacturer over a 10-yr period. They found that lower rates of most minor injuries on the job occurred during the 12-h shifts, but higher rates of more major injuries occurred off the job.

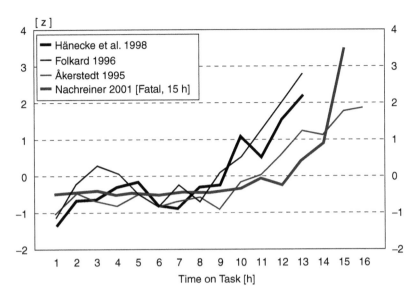

FIGURE 29.2 Accident risk as a function of time on task for four studies using aggregated data. (From Nachreiner, F., *Journal of Human Ergology*, 30, 101, 2001. With permission.)

29.2.4 Longer Toxic Exposure

Chemicals that are not eliminated from the body can quickly accumulate during compressed working hours. As Jung et al.[21] stated, "The prolongation of daily or weekly occupational exposure maintaining the yearly work time essentially influences the toxicity of substances with medium half-lives (10–1000 h)" (p. 6).

Figure 29.3[21] shows an example of the simulated effects of 8-, 10-, and 12-h shifts on blood toxin levels.

29.2.5 Ergonomic Recommendations

Having reviewed the literature on extended work shifts, Rosa [22, p. 51] concluded that "extended work shift schedules should be instituted cautiously and evaluated carefully, with appropriate attention [...] to staffing, levels, workload, job rotation, environmental exposure, emergency contingencies, rest breaks, commuting time, and social or domestic responsibilities."

Tucker and Rutherford[23] studied the factors moderating the impact of long work hours (per week) on well-being and health of train drivers. They found the following negative effects:

- Overtime was worked as a result of external factors/financial pressure
- Train drivers had little control over how much overtime had to be worked
- There was a lack of social support from colleagues or friends
- There was a high level of interference between hours worked and life outside work (e.g., performing household duties, nondomestic chores, leisure activities, maintaining close relationships, exercising, eating healthily, and sleeping adequately)

Based on train drivers' comments, Tucker and Rutherford[23] concluded that "the way in which their hours were arranged may have been at least as salient as the overall quantity of those hours."

To summarize the complex picture, extended work shifts (> 8 h) should only be instituted if:[24,25]

- The nature of work and workload are suitable for extended work hours (and individual resources)
- Adequate rest breaks are allowed
- The whole working time arrangement is designed to minimize the accumulation of fatigue (see, for instance, Section 29.3.3.6)

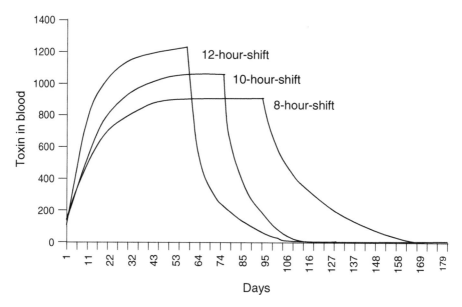

FIGURE 29.3 Simulated effects of a polongation of the daily occupational exposure (half-live = 168 h, toxin exposure = 10/h). (Modified from Jung, D., Wilhelm, E., Rose, D.M., and Konietzko, J., Schriftenreihe der Bundesanstalt fÿr Arbeitsschutz und Arbeitsmedizin, Ld. 10, Wirtschaftsverlag NW, Bremen, 1998. With permission.)

- Staffing levels are sufficient to cover absenteeism of colleagues[26]
- Exposure to toxic chemicals is limited (decreased exposure proportionate to prolonged working hours[27])
- Workers can recover completely after work

29.3 Shiftwork Issues

29.3.1 Definition of Shiftwork

Shiftwork may be defined as "work at alternating times of the day (e.g., morning shift, evening shift, night shift) or at permanent unusual times of day (e.g., permanent night shift, permanent evening shift)."

29.3.2 Disruptions of Physiological and Social Rhythms

Shiftwork — in particular nightwork — results in a phase shift of work and sleep in relation to the "normal" time structure of the body's biological rhythms.

In contrast to the subjective assessment of shiftworkers in general, the daily ("circadian") rhythms of psychological functions cannot adjust completely to nightwork within a week of night shifts.[28]

Dayworkers' circadian rhythms are organized in such a way that the body is "programmed" to perform during the day and to sleep at night. An incomplete adjustment of shiftworkers' circadian rhythms to nightwork means, for instance, that they have to sleep after night shifts at a time of the day that the body is not well prepared for sleep. Besides the incomplete adjustment of body rhythms to nightwork, there are also other stresses on shiftworkers. Shiftworkers live partly disconnected from their social environment. In particular, evening shifts and weekend work are incompatible with the leisure activities of day working family members and friends.[29]

Furthermore, shiftworkers are more frequently exposed to unfavorable environmental conditions (such as noise, unfavorable climate, or light conditions) than dayworkers.[30] As discussed in Section 29.2, other factors such as inadequate staffing levels, high workloads, and a lack of control over work or working time may increase the level of stress on shiftworkers.

All this stress can lead to disturbances to well-being, sleep, appetite, alertness, performance, and social life and may even lead to illnesses. However, individual workers all react differently to shiftwork (see Section 29.3.3.3). The following chapters discuss the typical problems that shiftworkers experience.

29.3.3 Shiftwork-Related Problems

29.3.3.1 Sleep Problems and Fatigue

Between 40 and 50% of night and shiftworkers complain of insomnia or insufficient sleep.[31] In particular, shiftworkers do not sleep as long after night shifts as they would usually (Figure 29.4)[32] and the quality of this day sleep also suffers.[33] An accumulation of deficits has to be expected if workers work several night shifts in succession.[34] In a study by Dumont et al.,[35] shiftworkers who had worked permanent night shifts for many years had persistent disturbances of sleep, even after changing to permanent day shifts.

However, not only the sleep after night shifts, but also the sleep before morning shifts may be reduced.[36] The earlier the morning shift starts (Figure 29.5)[37] and the longer the commuting, the shorter the sleep before the morning shift will be.

Most shiftworkers are sleepy by the end of their night shift. Dawson and Reid[38] have shown that this sleepiness corresponds to the effect of moderate alcohol consumption (1) in terms of psychomotor performance. There are several factors that contribute to this form of sleepiness: first, nightworkers have to work during their body rhythm's circadian trough (in the early morning hours). Second, workers on the first night shift, who have not slept before work, have been awake a long time since last sleeping. Third, sleep deficits can accumulate after several successive night shifts. Fourth, fatigue built up during the work hours of extended shifts may contribute to overall tiredness.[6]

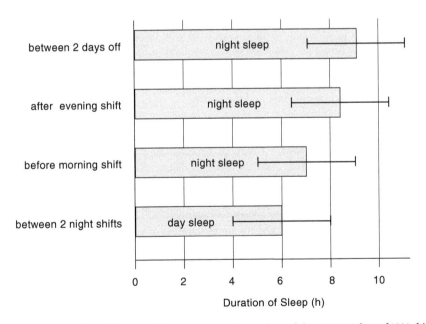

FIGURE 29.4 Mean durations of sleep by shift type based on the analysis of about 10 000 days of 1230 shift workers. (Modified from Knauth, P., Landau, K., Drš̌ge, C., Schwittek, M., Widynski, M., and Rutenfranz, J., *International Archives of Occupational and Environmental Health*, 46, 169, 1980. With permission.)

Morning Shift Starting

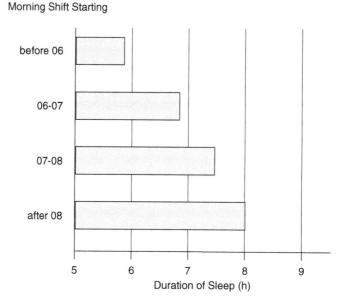

FIGURE 29.5 Duration of sleep dependent on the starting time of morning shift. (Based on data from Folkard, S. and Barton, J., *Ergonomics*, 36 (1–3), 85–91, 1993. With permission.)

If sleep before morning shifts is reduced, shiftworkers are also sleepy during this shift.[39–41] Furthermore, evening types will have more problems getting up to work a morning shift with an early start than morning types.

29.3.3.2 Disturbances to Individuals' Dietary Habits

Shiftwork may upset the eating habits of shiftworkers, promoting digestive trouble. This can be explained by a perturbed link between mealtimes and circadian phases of gastrointestinal functions (such as gastric secretion, enzyme activity, intestinal motility) associated with unbalanced food content (e.g., cold food and prepacked food containing fat), shortage of time available during work breaks, nibbling, and sometimes increased intake of caffeinated drinks, tobacco smoking, and alcohol consumption.[29,42–44]

Shiftworkers frequently complain about disturbances of appetite, dyspepsia, heartburn, abdominal pains, constipation, stomach grumbling, and flatulence.[45]

29.3.3.3 Health Disorders

Shiftwork — in particular nightwork — may be regarded as a risk factor for health. As shown in Figure 29.6,[46] three pathways leading from shiftwork to disease are discussed: a mismatch of circadian rhythms, a behavioral change, and social disturbances.[46–48]

Most epidemiological studies have shown a higher incidence of gastrointestinal disorders, for example, chronic gastritis, gastroduodenitis, peptic ulcers, and colitis.[45,48,49] Recently, the discussion of the causative role of *Helicobacter pylori* has changed the understanding of gastrointestinal pathology.[50] As Costa and Pokorski state,[51] it "does not mean that shiftwork is no more associated with it, but that it is one of the numerous factors which can favour the manifestation of ulcers" (p. 74).

More recent epidemiologic studies have focussed on the relationship between shiftwork and coronary heart disease (CHD). A review article by Boggild and Knutsson[52] concluded that shiftworkers' risk of developing CHD was 40% greater than for dayworkers (possible mechanisms that might explain the relation between shiftwork and cardiovascular disease are discussed in Knutsson).[47]

Furthermore, shiftworkers may be at potential risk to reproductive health, for example, in that there is an increased risk of spontaneous abortion.[53] Significantly higher occurrences of breast cancer have been

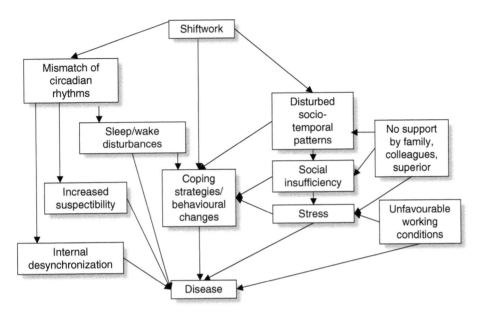

FIGURE 29.6 Disease mechanisms in shift workers. (Modified from Knutsson, A, *Scandinavian Journal of the Society of Medicine*, suppl. 44, 23, 1989. With permission.)

established in nurses and flight attendants.[54] Recent studies have attributed the higher risk of breast cancer in women working night shifts to the additional light they are exposed to at night.[55,56]

There have not been many studies on the mortality rates of shiftworkers. Taylor and Pocock[57] observed 8603 male manual workers between 1956 and 1968. A total of 1578 deaths in three groups (dayworkers, shiftworkers, and ex-shiftworkers) were notified. The dayworkers had slightly fewer deaths than expected in comparison to the general population (736 vs. 756.4). Shiftworkers and ex-shiftworkers had slightly more (722 vs. 711.4 and 120 vs. 100.9, respectively). As the differences were not statistically significant, the authors concluded that shiftwork would appear to have no adverse effect upon mortality. In a Danish cohort study[58] conducted over 22 yr, the relative death risk was 1.1 (95% confidence interval, 0.9–1.3).

More recent studies found an equally blurry relation between shiftwork and mortality.[59,60]

29.3.3.4 Disturbances of Social and Family Life

Our society values free time in the evening and on weekends more highly than free time at other times of the day or week. Evening shifts, night shifts, and work at weekends are incompatible with these value judgments. The severity of the problems created by this discrepancy depends on a variety of factors such as the type of shift system, shiftworkers' ages, sex, personal characteristics, coping behavior, marital status, family (e.g., the number and age of children), and the kind of community they live in.[29,61,62]

Each shift and shift system has its own advantages and disadvantages with respect to social and family life as well as personal preferences.

Mott et al.[61] found that male workers on fixed afternoon shifts reported the most dissatisfaction with the amount of time available for companionship with their wives and for relaxation and diversions from household duties. Seventy-nine percent of steel workers on an afternoon shift included in a study by Wedderburn[63] felt that their shift restricted their social and family life, compared to 17% of workers on a dayshift. Shiftworkers on a fixed afternoon shift have problems seeing their school-age children.[61,64]

Compared to the fixed afternoon shift, the fixed night shift schedule affords more time for family and social interaction. As per Colligan and Rosa,[62] "the cost of this extra time, however, may be health problems for the worker and lifestyle disruption for family members."

The effects of rotating shift systems vary according to the type of shift system. Shiftworkers who had changed from a slow, backwards rotating shift system to a more rapid, forward rotating shift system complained less about disturbances of social and family life under the new system.[65,66]

Some studies have looked at the effects of parents' shiftwork on the school careers of their children. Diekmann et al.[67] found a clear tendency of children of shiftworkers achieving lower grades of education. They were able to show that this effect was — at least in part — a shiftwork effect and not due to the higher education of dayworkers as compared to shiftworkers.

29.3.3.5   Impaired Performance and Accidents

Considering the circadian variations of body rhythms and shiftworkers' difficulty with sleep, it is evident that human performance and accident rates cannot be constant 24 h a day. In their summary of the results of three earlier studies with "real-job" measures of speed and accuracy over a period of 24 h, Folkard and Tucker[68] found a clear "dip" in these efficiency measures from about 10 p.m. to 6 a.m., with the through occurring at 3 a.m., during the night shift. A secondary dip in efficiency measurements was observed after 12 p.m. More recent studies have also found a dip in efficiency in the early morning hours. Assis et al.[69] reported that most of the 340 occurrences of error in maintenance operations of high-capacity airlines observed during their study occurred during the early morning. According to Bonnefond et al.,[70] who studied worker performance in an aircraft technical maintenance unit, performance lapses increased twice to threefold during the night shift among the workers aged between 35–49 and 50–58 yr when compared to the youngest group (25 to 34 yr).

Single-vehicle accidents in Texas also had a major peak in the early morning hours.[71]

The data in Figure 29.7[68] are based on eight studies of 8-h shift systems by different authors. As the figure shows, the relative risk of incidents was lowest in the morning shift and highest in the night shift. It is also possible that night shift accidents have greater severity than during morning and evening shifts.[72]

Folkard and Tucker[68] also summarized seven studies that reported incident frequencies separately for each night over a span of at least four successive night shifts. Figure 29.8[68] shows the increase of the relative risk from the first to the fourth night shift. The risk is 36% higher on the fourth night shift relative to the first night shift.

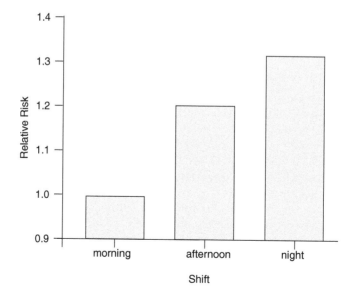

FIGURE 29.7   The relative risk across the three shifts. (From Folkard, S. and Tucker, P., *Occupational Medicine*, 53 (2), 97, 2003. With permission.)

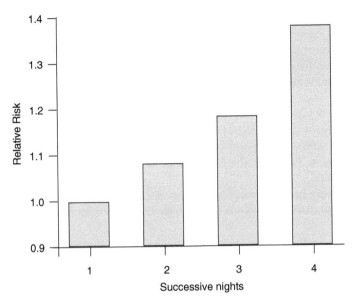

FIGURE 29.8 The relative risk over four successive night shifts. (From Folkard, S. and Tucker, P., *Occupational Medicine*, 53 (2), 97, 2003. With permission.)

Based on five studies, Folkard and Tucker[68] calculated the average hourly relative risk values over successive morning or day shifts (Figure 29.9).[68] On an average, the risk was 17% higher on the fourth morning/day shift than on the first shift.

The main conclusion to be drawn is that both safety and productivity are reduced at night. Therefore, it is important to design adequate shift systems that take into consideration, for instance, the number of successive night shifts, the length of the night shift, and the provision of breaks within them.

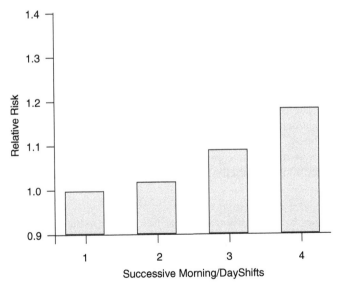

FIGURE 29.9 The relative risk over four successive morning/day shifts. (From Folkard, S. and Tucker, P., *Occupational Medicine*, 53 (2), 98, 2003. With permission.)

29.3.3.6 Ergonomic Recommendations for the Design of Shift Systems

Different shift systems may cause different problems; some workers may have problems under a given shift system, whereas others may not. Therefore, there is no optimum solution. An adequate combination of countermeasures will have the greatest effect.[73,74]

However, one of the most effective preventive countermeasures is the ergonomic design of shift systems, which is discussed in this chapter.

As shown in Table 29.1, there are two contrary basic strategies of ergonomic shift system design. The aim of the first strategy is to accelerate circadian adjustment to night work with the help of an exposure to bright light at night, the use of "goggles" (special dark sunglasses) after the night shift, and melatonin medication, for example. Following the first strategy, it is recommended that workers work several night shifts in succession.

The second, alternative strategy aims at limiting the disturbance that night work does to the circadian rhythms. Therefore, this strategy advocates few night shifts in succession in quick, forward-rotating shift systems.

Table 29.2 lists some pros and cons of the first strategy. The first strategy aims at realizing adequate phase shift of the circadian rhythms to improve alertness and performance during night shifts and sleep afterwards. While there have been many laboratory studies on this aspect, only a few studies have been conducted involving real shiftwork. Whereas some studies showed that bright light exposure during the night or goggles during the daylight caused a significant phase of endogenous circadian rhythms on most of the subjects,[75–79] other studies established no significant phase shift.[80,81]

In some studies, alertness and performance during bright light exposure in the night shift were enhanced.[75,77,82,83] However, some other studies found no significant positive effects of bright light exposure on alertness/fatigue and performance and some even noted deteriorations.[78,80,81,84,85]

Czeissler et al.,[75] Dawson and Campbell,[83] Eastman et al.,[76,77] and James and Boivin[79] observed positive effects associated with exposure to bright light at night or use of goggles after the night shift while sleeping during the day in all subjects — or at least of those with a larger phase shift.

Lewy et al.[86] state that, "the fact that the human melatonin and light PRCs (phase-response curves) are nearly opposite in phase suggests that endogenous melatonin may function to augment entrainment of the endogenous circadian pacemaker by the light-dark cycle" (p. 389). Crowly and Eastman[87] have used melatonin pills in combination with dark sleep episodes and black sunglasses or bright light. The combination of all four induced the greatest phase delay in the circadian clock.

Although the first strategy may have positive effects on alertness, performance, and sleep, there are also some negative aspects to it (Table 29.2). The most important argument against the first strategy is the fact that long-term effects on health in particular are unknown. Enhanced breast cancer risk in women working in night shifts has been attributed to exposure to additional lighting at night.[55,56] However, these authors state that "it will be necessary to further explore the relationship and cancer risk through the melatonin pathway" [55, p. 1657]. Up to now, there are no longitudinal studies on the long-term effects of bright light exposure on the health of shiftworkers.

Most studies on bright light have been conducted in laboratories. However, social contacts may have phase-resetting capabilities;[88] that is, shiftworkers' social contacts might influence the results outside the laboratoy. It is much more difficult to convince real shiftworkers to participate in well-controlled field studies than to conduct laboratory studies with healthy young (nonshift working) men. Czeissler and

TABLE 29.1 Two Contrary Basic Strategies of Ergonomic Shift System Design

Strategy	1. Accelerate circadian adjustment	2. Limit disturbances of circadian rhythms
Measures	Bright light Dark sunglasses Melatonin medication	Quickly rotating shifts Forward rotation of shifts
Recommendation	Many night shifts in succession	Few night shifts in succession

TABLE 29.2 Some Pros and Cons of Strategy 1 (i.e., Accelerate Circadian Adjustment to Night Work, e.g., With Help of Bright Light)

Arguments Supporting Strategy 1	Critical Aspects of Strategy 1
Enhanced alertness	Long-term effects of health are unknown
Better performance	Limited experience outside laboratories
Better sleep	Limited knowledge of adequate light management
	Social isolation
	Not applicable, e.g., for long distance train, truck, and bus drivers

Dijk[89] conclude that there is "an urgent need for longitudinal studies of bright light application in real-life settings."

As mentioned in Table 29.2, we do not know enough to be able to recommend the ideal light situation for each individual and shift system (e.g., intensity, wavelength, duration, starting and ending of bright light, day sleep without daylight).

If bright light is used to support and extend permanent nightwork, the concerned workers are relatively isolated compared to normal dayworkers.

Strategy 1 may help workers on an oil rig, who are socially isolated anyway, to adjust to nightwork and subsequently to readapt to day life.[90] Strategy 1 is not applicable to some professions, such as that of train drivers, long-distance truck drivers, or bus drivers.

In summary of the arguments contained in Table 29.2: until the results of additional research are available, we will assume for the moment that the second strategy is preferable.

The ergonomic recommendations for the design of shift systems in Table 29.3 are based on strategy 2. Several studies have shown that the problems experienced by shiftworkers may be reduced when these recommendations are applied.[65,66,91–96]

As Boggild and Jeppesen[97] have shown, when more ergonomic recommendations for the design of new shift systems are realized, the better the positive changes of risk factors for cardiovascular diseases will be.

The recommendations contained in Table 29.3 are explained in more detail in Knauth.[24,25]

Figure 29.10, Figure 29.11, and Figure 29.12 show examples of continuous and discontinuous shift systems, corresponding to the recommendations in Table 29.3. The continuous shift system "A" (Figure 29.10),[28] which needs five teams, is in operation in the chemical industry, steel industry, mechanical industry, power plants, and some service sector companies. In general, shiftworkers work more than 33.6 h per week. Therefore, additional shifts have to be worked in this shift system. These shifts may be used, for example, for continuation training, maintenance, and to cover for absentees. As recommended in Table 29.3, there are short periods of morning, evening, and night shifts, forward rotation of the shifts, blocks of free weekends, and at least 48 h off between the end of the last night shift and start of the following morning shift.

TABLE 29.3 Ergonomic Recommendations for the Design of Shift Systems

Nightwork should be reduced as much as possible. If this is not possible, quickly rotating shift systems are preferable (i.e., not more than three night, three morning, and three evening shifts in succession)

The forward rotation of shifts (first morning, then evening, thereafter night shifts) is recommended

An early start of the morning shift should be avoided (e.g., 06.30 better than 06.00, 06.00 better than 05.30)

Quick changeovers (e.g., from night to afternoon shift at the same day or from afternoon to morning shift) must be avoided

Every shift system should include some free weekends with at least two successive full days off

If the shift system has to be flexible from the employer's point of view it also should allow individuals some flexibility (the highest flexibility is possible in "time autonomous groups")

Length of shift, see Section 29.2.5

Source: From abridged version of Knauth, P. and Hornberger, S. Occupational Medicine, 53, 109–116, 2003.

Week	Mo	Tu	We	Th	Fr	Sa	Su	Mo	Tu	We	Th	Fr	Sa	Su
1+2	M	M	E	E	N	N					M	M	E	E
3+4	N	N				M	M	E	E	N	N			
5+6			M	M	E	E	N	N					M	M
7+8	E	E	N	N					M	M	E	E	N	N
9+10					M	M	E	E	N	N				

☐ = day off
M = morning shift
E = evening shift
N = night shift

Average weekly working hours:
33.6 h/week without additional shift
35.2 h/week with 2 additional shifts/10 weeks
36.8 h/week with 4 additional shifts/10 weeks
38.4 h/week with 6 additional shifts/10 weeks
40.0 h/week with 8 additional shifts/10 weeks

FIGURE 29.10 Continuous shift system "A," worked by five crews over a 10-week cycle. (From Knauth, P., Design of shift work systems. In: *Shiftwork. Problems and Solutions*, Colquhoun, W.P., Costa, G., Folkard, S., and Knauth, P., Eds., Peter Lang, Frankfurt am Main, 1996. With permission.)

The continuous shift system "B" (Figure 29.11) has been chosen by shiftworkers because of the larger leisure time block. Between the last morning shift and the first evening shift (and also between the last evening shift and the first night shift), workers always have 24 h off work. Therefore, the period of seven consecutive working days seems to be acceptable.

In Figure 29.12, the principle of shifts rotating forward quickly is applied to a discontinuous shift. In one company, shiftworkers swapped the first week's Saturday night shift to the Sunday of the second week in order to gain an additional free Saturday evening.

Week	Mo	Tu	We	Th	Fr	Sa	Su
1+2	M	M	E	E	N	N	N
3+4	E	E	N	N			
5+6	N	N					
7+8					M	M	M
9+10			M	M	E	E	E

☐ = day off
M = morning shift
E = evening shift
N = night shift

Average weekly working hours:
33.6 h/week without additional shift
35.2 h/week with 2 additional shifts/10 weeks
36.8 h/week with 4 additional shifts/10 weeks
38.4 h/week with 6 additional shifts/10 weeks
40.0 h/week with 8 additional shifts/10 weeks

FIGURE 29.11 Continuous shift system "B," worked by five crews over a 5-week cycle. (From Knauth, P., Design of shift work systems. In: *Shiftwork. Problems and Solutions*, Colquhoun, W.P., Costa, G., Folkard, S., and Krauth, P., Eds., Peter Lang, Frankfurt am Main, 1996. With permission.)

Week	Mo	Tu	We	Th	Fr	Sa	Su
1	M	M	E	E	N	N	
2			M	M	E	E	
3	N	N			M	M	
4	E	E	M	M			

  □ = day off
M = morning shift
E = evening shift
N = night shift

FIGURE 29.12 Discontinuous shift system, worked by four crews over a 4-week cycle (36 h/week).

If the system has to be operated with only three teams, a 40-h working week, and no work on Saturdays or Sundays, it is generally not possible to introduce quick-rotating shifts.

The implementation of a new shift system is always a very difficult process. However, if the following factors are taken into consideration, there is a greater chance that workers will accept a new shift system: workers' participation in planning the new shift system, information being made available to everybody concerned, training, commitment of promoter, professional project management, tailor-made solutions, regular discussions during a trial period, and a ballot after it.[98]

References

1. Härmä, M. New work times are here — are we ready? *Scandinavian Journal of Work & Environmental Health*, 24 (suppl. 3), 3–6, 1998.
2. Costa, G. Shift work and occupational medicine: an overview. *Occupational Medicine*, 53, 83–88, 2003.
3. Rosa, R.R. Performance, alertness, and sleep after 3–5 years of 12 h shifts: a follow-up study, *Work and Stress*, 5, 107–116, 1991.
4. Rosa, R.R. and Bonnet, M.H. Performance and alertness on 8-hour and 12-hour rotating shifts at a natural gas utility. *Ergonomics*, 36, 1177–1193, 1993.
5. Spencer, M.B. et al. A diary study of fatigue in air traffic controllers during a period of high workload. In: *Shiftwork in the 21st Century*, Hornberger, S., Knauth, P., Costa, G., and Folkard, S., Eds., Peter Lang, Frankfurt am Main, 2000, pp. 201–206.
6. Fischer, F.M. et al., Alterness and sleep after 12-hour shifts: differences between day and night work. In: *Shiftwork in the 21st Century*, Hornberger, S., Knauth, P., Costa, G., and Folkard, S., Eds., Peter Lang, Frankfurt am Main, 2000, pp. 43–48.
7. Sato, S., Taoda, K., Kaemura, M., Wakaba, K., Fukuchi, Y., and Nishiyama, K., Heart rate variability during long truck driving work, *Journal of Human Ergology*, 30, 235–240, 2001.
8. Ekstedt, M. et al. Non-scheduled overtime and sleep. *Shiftwork International Newsletter*, 20 (2), 64, 2003.
9. Son, M. et al. The relationship of working hours and work intensity with sleep disturbance among 12-hour shift workers in the automobile factory in Korea. *Shiftwork International Newsletter*, 20 (2), 175, 2003.

10. Peacock, B., Glube, R., Miller, M., and Clune, P., Police officers' response to 8 and 12 hour shift schedules. *Ergonomics*, 26, 479–493, 1983.

11. Axelsson, J. et al. Effects of alternating 8- and 12-hour shifts on sleep, sleepiness, physical effort and performance. *Scandinavian Journal of Work & Environmental Health*, 24 (Suppl. 3), 62–68, 1998.

12. Tucker, P. et al. Shift length as a determinant of retrospective on-shift alertness. *Scandinavian Journal of Work & Environmental Health*, 24 (suppl. 3), 49–54, 1998.

13. Proctor, S.P. The influence of overtime on cognitive function as measured by neurobehavioral tests in an occupational setting. Doctoral thesis, Boston University, School of Public Health, 1998.

14. Campell, L.H. Can new shifts motivate. *Hydrocarbon Processing*, April, 249–256, 1980.

15. Tepas, D.I. and Tepas, S.K. Alternative work schedules practice in the United States. International report prepared for the National Institute for Occupational Safety and Health, 1981.

16. Nachreiner, F. Time on task effects on safety. *Journal of Human Ergology*, 30, 97–102, 2001.

17. Folkard, S., Effects on performance efficiency. In: *Shiftwork — Problems and Solutions*, Colquhoun, W.P., Costa, G., Folkard, S., and Knauth, P., Eds., Peter Lang, Frankfurt am Main, 1996, pp. 65–87.

18. Hänecke, K., Tiedemann, S., Nachreiner, F., and Grzech Sukalo, H. Accident risks as a function of hour at work and time of day as determined from accident data and exposure models for the German working population. *Scandinavian Journal of Work & Environmental Health*, 24 (suppl. 3), 43–48, 1998.

19. Åkerstedt, T. Work injuries and time of day — national data. *Shiftwork International Newsletter*, 12, 2, 1995.

20. Laundry, B.R. and Lees, R.E.M. Industrial accident experience of one company on 8- and 12-hours shift system. *Journal of Occupational Medicine*, 33, 903–906, 1991.

21. Jung, D., Wilhelm, E., Rose, D.M., and Konietzko, J. *Circadiane Rhythmen und ihr Einfluß auf Toxizität von Arbeitsstoffen*. Schriftenreihe der Bundesanstalt für Arbeitsschutz und Arbeitsmedizin, Ld. 10, Wirtschaftsverlag NW, Bremerhaven, 1998.

22. Rosa, R.R. Extended workshifts and excessive fatigue. *Journal of Sleep Research*, 4 (suppl. 2), 51–56, 1995.

23. Tucker, P. and Rutherford, C. Factors moderating the impact of long work hours on well-being. *Shiftwork International Newsletter*, 20 (2), 186, 2003.

24. Knauth, P. The design of shift systems. *Ergonomics*, 36 (1–3), 15–28, 1993.

25. Knauth, P. Hours of work. In: *Encyclopaedia of Occupational Health and Safety*, 4th ed, Stellman, J., Ed., International Labour Office, Genf, 1998, pp. 43.1–43.15.

26. Bourdouxhe, M., Queinnec, Y., and Guertin, S. The interaction between work schedule and workload: case study of 12-hour shifts in a Canadian refinery, In: *Shiftwork in the 21st Century*, Hornberger, S., Knauth, P., Costa, G., and Folkard S., Eds., Peter Lang, Frankfurt am Main, 2000, pp. 61–66.

27. OSHA, *Modification of PELs for Prolonged Exposure Periods*. Occupational Safety and Health Administration, Washington, DC, 1979.

28. Knauth, P. Design of shiftwork systems. In: *Shiftwork. Problems and Solutions*, Colquhoun, W.P., Costa, G., Folkard, S., and Knauth, P., Eds., Peter Lang, Frankfurt am Main, 1996, pp. 155–173.

29. Knauth, P. and Costa, G. (1996). Psychosocial effects. In: *Shiftwork. Problems and Solutions*, Colquhoun, W.P., Costa, G., Folkard, S., and Knauth, P. Eds., Peter Lang, Frankfurt am Main, 1996, pp. 89–112.

30. Knauth, P. Ergonomische Beiträge zu Sicherheitsaspekten der Arbeitszeitorganisation. *Fortschr.-Ber. VDI-Z.*, Reihe 17, Nr. 18, VDI-Verlag, Düsseldorf, 1983.

31. Härma, M. et al. Combined effects of shift work and life-style on the prevalence of insomnia, sleep deprivation and daytime sleepiness. *Scandinavian Journal of Work & Environmental Health*, 24, 300–307, 1998.

32. Knauth, P. et al. Duration of sleep depending on the type of shift work. *International Archives of Occupational and Environmental Health*, 46, 167–177, 1980.

33. Åkerstedt, T. Shiftwork and disturbed sleep/wakefulness. *Occupational Medicine*, 53, 98-94, 2003.

34. Holmes, A.L. et al. Daytime cardiac autonomic activity during one week of continuous night shift. *Journal of Human Ergology*, 30, 223–228, 2001.

35. Dumont, M., Montplaisir, J., and Infante-Rivard, C. Sleep quality of former night-shift workers. *International Journal of Occupational and Environmental Health, Supplement*, 3 (3), 10–14, 1997.

36. Ingre, M. et al. Train drivers work-hours, health, and safety: a summary of results from the TRAIN-project. *Shiftwork International Newsletter*, 20 (2), 96, 2003.

37. Folkard, S. and Barton, J. Does the "forbidden zone" for sleep onset influence morning shift sleep duration? *Ergonomics*, 36 (1–3), 85–91, 1993.

38. Dawson, D. and Reid, K. Fatigue, alcohol and performance impairment. *Nature*, 388, 235, 1997.

39. Hak, A. and Kampmann, R. Working irregular hours: complaints and state of fitness of railway personnel. In: *Night and Shift Work, Biological and Social Aspects*, Reinberg, A., Vieux, N., and Andlauer, P., Eds., Pergamon Press, Oxford, 1981, pp. 229–236.

40. Moors, S.H. Learning from a system of seasonally-determined flexibility: beginning work earlier increases tiredness as much as working longer days. In: *Shiftwork: Health, Sleep and Performance*, Costa, G., Cesana, G., Kogi, K., and Wedderburn, A. Eds., Peter Lang, Frankfurt am Main, 1989, pp. 310–315.

41. Kecklund, G., Åkerstedt, T., and Lowden, A. Morning work: effects of early rising on sleep and alertness. *Sleep*, 20, 215–223, 1997.

42. Lennernäs, M.A. *Nutrition and Shift Work*. Acta Universitatis Upsaliensis, Uppsala, 1993.

43. Cristofoletti, M.F., Rocha, L.E., and Moreno, C.R.C. Nutritional status among call center operators working in three shift fixed schedule. *Shiftwork International Newsletter*, 20 (2), 57, 2003.

44. Latzer, Y., Tzischinsky, O., and Epstein, R. Eating habits and attitudes among shift-wok nurses in Israel. *Shiftwork International Newsletter*, 20 (2), 123, 2003.

45. Vener, K.J., Szabo, S., and Moore, J.G. The effect of shift work on gastrointestinal (GI) function: a review. *Chronobiologia*, 16, 421–439, 1989.

46. Knutsson, A. Shift work and coronary heart disease. *Scandinavian Journal of the Society of Medicine*, Suppl. 44, 1989.

47. Knutsson, A. Association between shiftwork and coronary heart disease: a review of recent findings and mechanisms of action. In: *Shiftwork 2000, Implications for Science, Practice and Business*, Marek, T., Oginska, H., Pokorski, J., Costa, G., and Folkard, S., Eds., Chair of Managerial Psychology and Ergonomics, Institute of Management Jagiellonian University, Krakow, 2000, pp. 99–117.

48. Knutsson, A. Health disorders of shift workers. *Occupational Medicine*, 53 (2), 103–108, 2003.

49. Costa, G. Effects on health and well-being. In: *Shiftwok. Problems and Solutions*, Colquhoun, W.P., Costa, G., Folkard, S., and Knauth, P., Eds., Peter Lang, Frankfurt am Main, 1996, pp. 113–139.

50. Ott, M.G. et al. Gastrointestinal illness relative to shiftwork and *Helicobacter pylori* infection. In: *Shiftwork in the 21st Century* Hornberger, S., Knauth, P., Costa, G., and Folkard, S., Eds., Peter Lang, Frankfurt am Main, 2000, pp. 201–206.

51. Costa, G. and Pokorski, J., Effects on health and medical surveillance of shiftworkers. In: *Shiftwork 2000, Implications for Science, Practice and Business*, Marek, T., Oginska, H., Pokorski, J., Costa, G., and Folkard, S., Eds., Chair of Managerial Psychology and Ergonomics, Institute of Management Jagiellonian University, Krakow, 2000, pp. 71–97.

52. Boggild, H. and Knutsson, A. Shiftwork risk factors and cardiovascular disease. *Scandinavian Journal of Work & Environmental Health*, 25, 85–99, 1999.

53. Nurminen, T. Shift work and reproductive health. *Scandinavian Journal of Work & Environmental Health*, 24 (suppl. 3), 28–34, 1998.

54. Welp, E.A. et al. Environmental risk factors of breast cancer. *Scandinavian Journal of Work & Environmental Health*, 24 (1), 3–7, 1998.

55. Schernhammer, E.S. et al. Rotating night shifts and risk of breast cancer in women participating in the nurses' health study. *Journal of the National Cancer Institute*, 93 (20), R1563–R1568, 2001.

56. Davis, S., Mirick, D.K., and Stevens, R.G. Night shift work, light at night, and risk of breast cancer. *Journal of the National Cancer Institute*, 93 (20), 1557–1562, 2001.

57. Taylor, P.J. and Pocock, S.J. Mortality of shift and day workers 1956–68. *British Journal of Industrial Medicine*, 29, 201–207, 1972.

58. Boggild, H. et al. Shift work, social class, and ischaemic heart disease in middle aged and elderly men; a 22 year follow up in the Copenhagen male study. *Occupational Environmental Medicine*, 56, 640–645, 1999.

59. Åkerstedt, T., Fredlund, P., and Gillberg, M. Mortality and shiftwork/hours — a prospective study. *Shiftwork International Newsletter*, 20 (2), 31, 2003.

60. Karlsson, B., Knutsson, A., and Alfredsson, L. Mortality of Swedish shift and day workers in pulp and paper industry between 1952–1998. *Shiftwork International Newsletter*, 20 (2), 109, 2003.

61. Mott, P.E. et al. *Shift Work: The Social, Psychological and Physical Consequences.* The University of Michigan Press, Ann Arbor, 1965.

62. Colligan, M.J. and Rosa, R.R. Shiftwork effects on social and family life. In: *Shiftwork*, Scott, A.J., Ed., Hanley and Belfus, Philadelphia, 1990, pp. 315–322.

63. Wedderburn, A.A.I. Some suggestions for increasing the usefulness of psychological and sociological studies of shiftwork. *Ergonomics*, 21, 827–833, 1978.

64. Tasto, D.L. et al. *Health Consequences of Shift Work.* National Institute for Occupational Safety and Health, Behavioral and Motivational Factors Branch, (NIOSH) Publication no. 78–154, 1978.

65. Hornberger, S. and Knauth, P. Follow-up intervention study on effects of a change in shift schedule in shiftworkers. *Shiftwork International Newsletter*, 12, 27, 1995.

66. Härmä, M. et al. Controlled intervention study of a quickly forward rotating shift system among young and elderly maintenance workers. *Shiftwork International Newsletter*, 20 (2), 86, 2003.

67. Diekmann, A., Ernst, G., and Nachreiner, F. Auswirkung der Schichtarbeit des Vaters auf die schulische Entwicklung der Kinder. *Zeitschrift für Arbeitswissenschaften*, 35 (7 NF), 174–178, 1981.

68. Folkard, S. and Tucker, P. Shift work, safety and productivity. *Occupational Medicine*, 53 (2), 95–101, 2003.

69. Assis, M.R. et al. Shiftwork and performance in aircraft maintenance: developing programs to reduce human error. *Shiftwork International Newsletter*, 20 (2), 36, 2003.

70. Bonnefond, A. et al. Interaction of age with shift-related sleep-wakefulness, sleepiness, performance and social life. *Shiftwork International Newsletter*, 20 (2), 43, 2003.

71. Bruno, G., Smolensky, M.H., and Griffin, L. Temporal patterns of single-vehicle accidents (SVA) specifically due to fatigue of drivers of trucks and passenger cars in Texas. *Shiftwork International Newsletter*, 20 (2), 49, 2003.

72. Cordova, V. Shiftwork accidentability pattern in 7,007 Chilean companies. *Shiftwork International Newsletter*, 20 (2), 54, 2003.

73. Kogi, K. Ergonomic guidelines for managers and workers in shiftworking plants: experiences from power plant maintenance work. In: *Shiftwork 2000, Implications for Science, Practice and Business*, Marek, T., Oginska, H., Pokorski, J., Costa, G., and Folkard, S., Eds., Chair of Managerial Psychology and Ergonomics, Institute of Management Jagiellonian University, Krakow, 2000, pp. 173–186.

74. Knauth, P. and Hornberger, S. Preventive and compensatory measures for shift workers, *Occupational Medicine*, 53 (2), 109–116, 2003.

75. Czeisler, Ch.A. et al. Exposure to bright light and darkness to treat physiologic maladaptation to night work. *The New England Journal of Medicine*, 322 (18), 1253–1259, 1990.

76. Eastman, Ch.I. et al. Dark goggles and bright light improve circadian rhythm adaptation to night-shift work. *Sleep*, 17 (6), 535–543, 1994.

77. Eastman, Ch.I., Liu, L., and Fogg, L.F. Circadian rhythm adaptation to simulated night shift work: effect of nocturnal bright light duration. *Sleep*, 18 (6), 399–407, 1995.

78. Daurat, A. et al. Detrimental influence of bright light exposure on alertness, performance, and mood in the early morning. *Neurophysiology Clinics*, 26, 8–14, 1996.

79. James, F.O. and Boivin, D.B. A light/darkness intervention to realign the cortisol rhythm to night shift work. *Shiftwork International Newsletter*, 20 (2), 99, 2003.

80. Costa, G. et al. Effect of bright light on tolerance to night work. *Scandinavian Journal of Work & Environmental Health*, 19 (6), 414–420, 1993.

81. Whitmore, J.N., French, J., and Fischer, J.R., Psychophysiological effects of a brief nocturnal light exposure. *Journal Human Ergology*, 30, 267–272, 2001.

82. French, J., Hannon, P., and Brainard, G.C. Effects of bright illuminance on body temperature and human performances. *Annual Reviews of Chronopharmacol* 7, 37–40, 1990.

83. Dawson, D. and Campbell, S.S. Timed exposure to bright light improves sleep and alertness during simulated night shifts. *Sleep*, 14 (6), 511–516, 1991.

84. Dollins, A.B. et al. Effect of illumination on human nocturnal serum melatonin levels and performance. *Physiological Behaviour*, 53, 153–160, 1993.

85. Iwata, N., Ichii, S., and Egashira, K. Effects of bright artificial light on subjective mood of shift work nurses. *Industrial Health*, 35, 41–47, 1997.

86. Lewy, A.J. et al. Melatonin shifts human circadian rhythms according to a phase–response curve. *Chronobiology International*, 9 (5), 380–392, 1992.

87. Crowley, S.J. and Eastman, C.I. Black plastic and sunglasses can help night workers. *Shiftwork International Newsletter*, 18 (1), 65, 2001.

88. Honma, K.J. et al. Differential effects of bright light and social cues on reentrainment of human circadian rhythms. *American Journal of Physiology/Regulatory Integrative and Comparative Physiology*, 268, R528–R535, 1995.

89. Czeisler, Ch.A. and Dijk, D.J. Use of bright light to treat maladaptation to night shift work and circadian rhythm sleep disorders. *Journal of Sleep Research*, 4 (Suppl. 2), 70–73, 1995.

90. Bjorvatn, B., Kecklund, G., and Åkerstedt, T. Bright light treatment used for adaptation to night work and re-adaptation to night work back to day life — a field study at an oil platform in the North Sea. *Journal of Sleep Research*, 8, 105–112, 1999.

91. Williamson, A.M. and Sanderson, J.W. Changing the speed of shift rotation: a field study. *Ergonomics*, 29, 1085–1096, 1986.

92. Knauth, P. and Kiesswetter, E. A change from weekly to quicker shift rotations: a field study of discontinuous three-shift workers. *Ergonomics*, 30, 1311–1321, 1987.

93. Knauth, P. and Schönfelder, E. Effects of a new shift system on the social life of shiftworkers. In: *Shiftwork: Health, Sleep and Performance*, Costa, G., Cesana, G., Kogi, K., and Wedderburn, A., Eds., Peter Lang, Frankfurt am Main, 1990, pp. 537–545.

94. Ng-A-Tham, J.E.E. and Thierry, H.K. An experimental change of the speed of rotation of the morning and evening shift. *Ergonomics*, 36, 51–57, 1993.

95. Knauth, P. Designing better shift systems. *Applied Ergonomics*, 27 (1), 39–44, 1996.

96. Knauth, P., Hornberger, S., and Scheuermann, G. Wahlarbeitszeit für Schichtarbeiter. In: *Good Practice, Ergonomie und Arbeitsgestaltung*, Landau, K., Ed., Ergonomie Verlag, Stuttgart, 2003, pp. 387–395.

97. Boggild, H. and Jeppesen, H.J. Intervention in shift schedule and changes in risk factors of cardiovascular disease. *Shiftwork International Newsletter*, 16 (2), 78, 1999.

98. Knauth, P. Strategies for the implementation of new shift systems. *Journal of Human Ergology*, 30, 9–14, 2001.

30

Back Belts

Stuart M. McGill
University of Waterloo

30.1 Introduction

The use of abdominal belts in industrial settings continues to be the topic of lively debate. The premier question still remains "Should abdominal belts be prescribed to workers in industry to perform manual materials handling tasks?" In the first edition of this handbook published a few years ago (McGill, 1999), I reviewed the available scientific literature pertaining to the use of back belts in industry with the objective of formulating a policy for belt prescription. The intent was not to take a position, either unconditionally for or against belt usage, but rather to interpret the scientific evidence and weigh the potential assets and liabilities for the development of belt prescription guidelines. The purpose of this chapter is to briefly review the literature previously reported, together with the most recent scientific data to see if our position on belt prescription has changed.

Abdominal belts and lumbar supports continue to be sold to industry in the absence of a regulatory requirement to conduct controlled clinical trials similar to that required of drugs and other medical devices. Many claims have been made as to how abdominal belts could reduce injury. For example, the suggested mechanisms have included the notion that belts remind people to lift properly. Some have suggested that belts may possibly support shear loading on the spine that results from the effect of gravity acting on the hand-held load and mass of the upper body when the trunk is flexed. Compressive loading of the lumbar spine has been suggested to be reduced through the hydraulic action of increased intra-abdominal pressure associated with belt wearing. Belts have been suspected of acting as a splint, reducing the range of motion, and thereby decreasing the risk of injury. Still other hypotheses as to how belts may affect workers include (1) providing warmth to the lumbar region, (2) enhancing proprioception via pressure to increase the perception of stability, and (3) reducing muscular fatigue. Some liabilities have been suggested such as belt wearing modulates blood pressure and long tidal volume and may stimulate adaptations that are detrimental once the belts are no longer worn. These issues, together with others, will be addressed in this chapter.

An influential publication from National Institute of Occupational Safety and Health (NIOSH, 1994), entitled *Workplace Use of Back Belts*, contained critical reviews of a substantial number of scientific reports evaluating back belts and concluded at that time that back belts do not prevent injuries among uninjured workers nor do they consider back belts to be personal protective equipment. While many more studies have been published since that time, this position remains generally consistent with our position stated in 1999, although my personal position for belt prescription is somewhat more moderate in specific cases.

The following sections have subdivided the scientific studies into clinical trials and those that examined biomechanical, psychophysical, and physiological changes from belt wearing. Finally, based on the evidence, guidelines are recommended for the prescription and usage of belts in industry.

30.2   Clinical Trials

Many clinical trials that have been reported in the literature were fraught with methodological problems and suffered from the absence of a matched control group, no posttrial follow-up, limited trial duration, and insufficient sample size. While the extreme difficulty in executing a clinical trial is acknowledged, only a few trials that have received notable attention will be reviewed in this chapter.

The first trial reviewed here was reported by Walsh and Schwartz (1990), in which 81 male warehouse workers were divided into three groups: a control group ($n = 27$); a group that received a 0.5-h training session on lifting mechanics ($n = 27$); and a group that received the 1-h training session and wore low-back orthoses while at work for the subsequent 6 months ($n = 27$). Instead of using more common types of abdominal belts, this research group used orthoses with hard plates that were heat molded to the low-back region of each individual. Given the concern that belt wearing was hypothesized to cause the abdominals to weaken, the abdominal flexion strength of the workers was measured both before and after the clinical trial. The control group and the training-only group showed no changes in abdominal flexor strength nor any change in lost time from work. The third group, which received both training and wore the belts, showed no changes in abdominal flexor strength or accident rate, but did show a decrease in lost time. However, it appears that the increased benefit was only to those workers who had a previous low-back injury.

In a larger clinical trial reported by Reddell et al. (1992), 642 baggage handlers who worked for a major airline were divided into four treatment groups: a control group ($n = 248$); a group that received only a belt ($n = 57$); a group that received only a 1-h back education session ($n = 122$); and a group that received both a belt and a 1-h education session ($n = 57$). The trial lasted 8 months and the belt used was a fabric weight lifting belt, 15 cm wide posteriorly and approximately 10 cm wide anteriorly. There were no significant differences between treatment groups for total lumbar injury incident rate, lost workdays, or workers' compensation rates. While the lack of compliance by a significant number of subjects in the experimental group was cause for consideration, those who began wearing belts but discontinued their use had a higher lost-day case injury incident rate. In fact, 58% of workers belonging to the belt-wearing groups discontinued wearing belts before the end of the 8-month trial. Further, there was an increase in the number and severity of lumbar injuries following the trial of belt wearing.

The clinical trial reported by Mitchell et al. (1994) was a retrospective study administered to 1316 workers who performed lifting activities in the military. While this study relied on self-reported physical exposure and injury data over 6 yr prior to the study, the authors did note that the costs of a back injury that occurred while wearing a belt were substantially higher than if injured otherwise.

A controversial study, reported by Kraus et al. (1996), surveilled the low-back injury rates of nearly 36,000 employees of the Home Depot Stores in California from 1989 to 1994. It was controversial because of the promotion and widespread reporting in the popular press and the subsequent political implications and debate that ensued. As reported, the company implemented a mandatory back belt use policy during this time period. Even though the authors claimed that belt wearing reduced the incidence of low-back injury, analysis of the data and methodology suggested a much more cautious interpretation was warranted. The data showed that while belt wearing reduced the risk in younger males and those older than 55 yr, belt wearing appeared to *increase* the risk of low-back injury for men working longer than 4 yr by 27% (although the large confidence interval required an even larger increase for statistical significance) and in men working less than 1 yr. However, of greatest concern was the lack of scientific control to ferret out the true belt wearing effect — there was no comparable nonbelt wearing group which was critical given that the belt wearing policy was not the sole intervention at Home Depot. For example, over the period of the study, the company increased the use of pallets and forklifts, changing the physical demands of the work, they installed mats for cashiers, implemented post-accident drug testing, and enhanced worker training. In fact, a conscious attempt was made to enhance safety in the corporate culture. This was a large study and the authors deserve credit for the massive data reduction and logistics. However, despite the title and claims that back belts reduced low-back injury, this uncontrolled study could not answer the question about the effectiveness of belts. It is sometimes interesting how social forces shape progress of science. Many politically based columns appeared following this study which questioned some of the positions taken by groups like NIOSH for example, who had stated previously that there was insufficient evidence to warrant the promotion of belts. The debate motivated a large study, funded by NIOSH, to control extraneous variables and test the effectiveness of belt wearing — it was published as the Wassell et al. (2000) study.

The Wassell et al. (2000) study was an important one since it, in many ways, replicated the controversial Kraus et al. (1996) study but with better scientific control. Surveilling over 13,873 employees at newly open stores of a major retailer, where 89 stores required employees to wear belts and 71 stores had only voluntary use, the belts failed to reduce the incidence of back injury claims (Wassell et al., 2000). Specifically, neither voluntary nor mandatory back belt use reduced the incidence of back injury claims or reported pain.

In summary, difficulties in executing a clinical trial are acknowledged: the Hawthorne effect is a concern, as it is difficult to present a true double-blind paradigm to workers since those who receive belts certainly know so; and there are logistical constraints on duration, diversity in occupations, and sample size. However, the data reported in the better-executed clinical trials cannot support the notion of universal prescription of belts to all workers involved in manual handling of materials to reduce the risk of low-back injury. There is weak evidence to suggest that those already injured may benefit from belts (or molded orthoses) with a reduced risk of injury recurrence. However, there does not appear to be support for uninjured workers wearing belts to reduce the risk of injury, and, in fact, there appears to be an increased risk of injury during the period following a trial of belt wearing. Finally, there appears to be some evidence to suggest that cost per back injury may be higher if the worker was wearing a belt than if injured otherwise.

30.3 Biomechanical Studies

Biomechanical studies have examined changes in low-back kinematics, posture, and issues of specific tissue loading. Two studies in particular have suggested that wearing an abdominal belt can increase the margin of safety during repetitive lifting: Lander et al. (1992) and Harman et al. (1989). Both of these papers reported ground reaction force and measured intra-abdominal pressure while subjects performed repeated lifting of barbells. Both reports observed an increase in intra-abdominal pressure when abdominal belts were worn. These researchers assumed that intra-abdominal pressure is a good indicator of spinal forces, which is highly contentious. Nonetheless, they assumed the higher recordings of intra-abdominal pressure indicated an increase in low-back support, which, in their view, justified the use of wearing belts. Spinal loads were not directly measured or calculated in these studies.

Several studies have questioned the hypothesized link between elevated intra-abdominal pressure and reduction in low-back load. For example, using an analytical model and data collected from three subjects lifting various magnitudes of loads, McGill and Norman (1987) noted that a build-up of intra-abdominal pressure required additional activation of the musculature in the abdominal wall, resulting in a net increase in low-back compressive load and not a net reduction of load as had been previously thought. In addition, Nachemson et al. (1986) published some experimental results that directly measured intradiscal pressure during the performance of valsalva maneuvrers documenting that an increase in intra-abdominal pressure increased, not decreased, the low-back compressive load. Therefore, it would seem erroneous to conclude that an increase in intra-abdominal pressure due to belt wearing reduces compressive load on the spine. In fact, it may have no effect or may even increase the load on the spine.

In another study, McGill et al. (1990) examined intra-abdominal pressure and myoelectric activity in the trunk musculature while six male subjects performed various types of lifts both wearing and without wearing an abdominal belt (a stretch belt with lumbar support stays, velcro tabs for cinching, and suspenders for when subjects were not lifting). Wearing the belt increased intra-abdominal pressure by approximately 20%. Further, it was hypothesized that if belts were able to help support some of the low-back extensor moment, one would expect to measure a reduction in extensor muscle activity. There was no change in activation levels of the low-back extensors nor in any of the abdominal muscles (rectus abdominis or obliques).

One study that examined the affect of belts on muscle function Reyna et al. (1995), examined 22 subjects for isometric low-back extensor strength and found belts provided no enhancement of function (although this study was only a 4-day trial and did not examine the affects over a longer duration of time). Ciriello and Snook (1995) examined 13 men over a 4-week period lifting 29 metric tons in 4 h twice a week both with and without a belt. Median frequencies of the low-back electromyographic signal (which is sensitive to local muscle fatigue) were not modified by the presence or absence of a back belt, strengthening the notion that belts do not significantly alleviate the loading of back extensor muscles. Once again this trial was not conducted over a very long period of time.

Lantz and Schultz (1986) observed the kinematic range of gross body motions while subjects wore low-back orthoses. While they studied corsets and braces rather than abdominal belts, they did report restrictions in the range of motion, although the restricted motion was minimal in the flexion plane. Subsequently, McGill et al. (1994) tested flexibility and stiffness of the lumbar torsos of 20 male and 15 female adult subjects, both while they wore and did not wear a 10-cm leather abdominal belt. The stiffness of the torso was significantly increased about the lateral bend and axial twist axes but not when subjects were rotated into full flexion. Woldstad and Sherman (1999) confirmed the restriction about the twist axis, but not the other two. Thus, it would appear from these studies that abdominal belts assist to restrict the range of motion about the lateral bend and axial twist axes but do not have the same effect when the torso is forced in flexion, as in an industrial lifting situation. Posture of the lumbar spine is an important issue in injury prevention for several reasons, but in particular Adams and Hutton (1988) have shown that the compressive strength of the lumbar spine decreases when the end range of motion in flexion is approached. Therefore, if belts restrict the end range of motion one would expect the risk of injury to be correspondingly decreased. While the splinting and stiffening actions of belts occur about the lateral bend and axial twist axes, stiffening about the flexion–extension axes appears to be less. A later data set presented by Granata et al. (1997) supports the notion that some belt styles are better in stiffening the torso in the manner described above, namely the taller elastic belts which span the pelvis to the rib cage. Furthermore, they also documented that a rigid orthopedic belt generally increased the lifting moment, while the elastic belt generally reduced spinal load but a wide variety in subject response was noted (some subjects experienced increased spinal loading with the elastic belt). A more recent study (Giorcelli et al., 2001) suggested that torso angles were reduced with belts although hips were flexed more and that workers moved at a slower pace while belted. Even in well-controlled studies, it appears that belts can modulate lifting mechanics in some positive ways in some people and in negative ways in others.

30.4 Physiological Studies

Blood pressure and heart rate were monitored by Hunter et al. (1989) while five males and one female subject performed dead lifts, bicycle riding, and one-armed bench presses, while wearing and not wearing a 10-cm weight belt. A load of 40% of each subject's maximum weight in the dead lift was held in a lifting posture for 2 min. The subjects were required to breathe throughout the duration so that no valsalva effect occurred. During the lifting exercise blood pressure was significantly higher (up to 15 mmHg), while the heart rate also was higher when the belt was worn. Given the relationship between elevated systolic blood pressure and an increased risk of stroke, Hunter et al. (1989) concluded that individuals who may have cardiovascular system compromise are probably at greater risk when undertaking exercise while wearing back supports.

Later work conducted in our own laboratory (Rafacz and McGill, 1996) investigated the blood pressure of 20 young men performing sedentary and very mild activities both with and without a belt (the belt was the elastic type with suspenders and velcro tabs for cinching at the front). Wearing this type of industrial back belt significantly increased diastolic blood pressure for quiet sitting and standing both with and without a hand-held weight, during a trunk rotation task, and during a squat lifting task. There is increasing evidence to suggest belts increase blood pressure!

Over the past 15 yr I have been asked to deliver lectures, and participate in academic debate on the back belt issue. On several occasions, occupational medicine personnel have approached me after hearing the effects of belts on blood pressure and intra-abdominal pressure, and have expressed suspicions that long-term belt wearing at their particular workplace may possibly be linked with higher incidents of varicose veins in the testicles, hemorrhoid, and hernia. At this point in time, there has been no scientific and systematic investigation of the validity of these claims and concerns. Rather than wait for strong scientific data to either lend support to these conditions, or dismiss them, it may be prudent to simply state concern. This will motivate studies in the future to track the incidents and prevalence of these pressure-related concerns to assess whether they are indeed linked to belt wearing.

Two recent studies have examined the effect of belt wearing on other physiological phenomenon — the Bobick et al. (2001) study reported lower mean oxygen consumption while the Parker et al. (2000) study reported reductions in lung ventilation tidal volumes in higher abdominal fat groups but increased tidal volumes in lower fat groups. Collectively these data could have several implications regarding lower work rates while belt wearing or modulated lung function that would impact performance. Once again, it appears to be a function of the individual worker making it difficult to justify a single, universal belt wearing policy or guideline.

30.5 Psychophysical Studies

Studies based on the psychophysical paradigm allow workers to select weights that they can lift repeatedly using their own subjective perceptions of physical exertion. McCoy et al. (1988) examined 12 male college students while they repetitively lifted loads from floor to knuckle height at the rate of three lifts per minute for a duration of 45 min. They repeated this lifting bout three times, once without a belt and once each with two different types of abdominal belts (a belt with a pump and air bladder posteriorly, and the elastic stretch belt previously described in the McGill et al., 1990. study). After examining the various magnitudes of loads that subjects had self-selected to lift in the three conditions, it was noted that wearing belts increased the load that subjects were willing to lift by approximately 19%. There has been some concern that wearing belts fosters an increased sense of security, which may or may not be warranted. This evidence may lend some support to this criticism of wearing abdominal belts.

30.5.1 What About Athletes Using Belts?

It is common to hear a worker request a belt because they saw a weightlifter wearing one, perceiving that it would help them avoid injury. Thus, highly trained athletes perform maximum effort lifts and employ several techniques and approaches in order to lift more weight and win. Belts assist in elastic recoil contributions to the low back and hip extensor moment together with adding stability. These athletes also hold their breath to enhance these mechanisms. But these are done to enhance ultimate performance to win — not for health reasons. Workers must realize that these special athletes have entirely different objectives and they approach their tasks to win a competition — not to achieve optimal health. This is not a desirable objective to be emulated by workers. For more information on athletic use see McGill (2004).

30.6 Back Belt Prescription

My earlier report (McGill, 1993) presented data and evidence that neither completely supported, nor condemned, the wearing of abdominal belts for industrial workers.

Definitive laboratory studies that describe how belts affect tissue loading and physiological and biomechanical function have yet to be performed. In addition, clinical trials of sufficient scientific rigor to comprehensively evaluate the epidemiological risks and benefits from exposure to belts must be done. The challenge remains to arrive at the best strategy for wearing belts; therefore, the available literature will be interpreted and given placement, and also combined with "common sense" to derive the most sensible position on prescription.

Given the available literature, it would appear the universal prescription of belts (i.e., providing belts to all workers in a given industrial operation) is not in the best interest of globally reducing both the risk of injury and compensation costs. Uninjured workers do not appear to enjoy any additional benefit from belt wearing and in fact may be exposing themselves to the risk of a more severe injury if they were to become injured and may have to confront the problem of weaning themselves from the belt. However, if some *individual* workers perceive a benefit from belt wearing then they may be allowed to conditionally wear a belt, but only on trial. The mandatory conditions for prescription (*for which there should be no exception*) are as follows:

1. Given the concerns regarding increased blood pressure and heart rate, and issues of liability, all those who are candidates for belt wearing should be screened for cardiovascular risk by medical personnel.
2. Given the concern that belt wearing may provide a false sense of security, belt wearers must receive education on lifting mechanics (back school). All too often belts are being promoted to industry as a quick fix to the injury problem. Promotion of belts, conducted in this way, is detrimental to the goal of reducing injury as it redirects the focus from the cause of the injury. Education programs should include information on how tissues become injured, techniques to minimize musculoskeletal loading, and what to do about feelings of discomfort to avoid disabling injury.
3. No belts will be prescribed until a full ergonomic assessment has been conducted of the individual's job. The ergonomic approach will examine, and attempt to correct, the cause of the musculoskeletal overload and will provide solutions to reduce the excessive loads. In this way, belts should only be used as a supplement for a few individuals while a greater plant-wide emphasis is placed on the development of a comprehensive ergonomics program.
4. Belts should not be considered for long-term use. The objective of any small-scale belt program should be to wean workers from the belts by insisting on mandatory participation in comprehensive fitness programs and education on lifting mechanics, combined with ergonomic assessment. Furthermore, it would appear wise to continue vigilance in monitoring former belt wearers for a period of time following belt wearing, given that this period appears to be characterized by elevated risk of injury.

References

Adams, M.A. and Hutton, W.C. (1988) Mechanics of the intervertebral disc. In *The Biology of the Intervertebral Disc*, Ed. P. Ghosh, Boca Raton, FL: CRC Press.

Bobick, T.G., Belard, J.-L., Hsiao, M., and Wassell, J.T. (2001) Physiological effects of back belt wearing during asymmetrical lifting. *Appl. Ergon.* 32:541–547.

Ciriello, V.M. and Snook, S.H. (1995) The effect of back belts on lumbar muscle fatigue. *Spine* 20(11):1271–1278.

Giorcelli, R.J., Hughes, R.E., Wassell, J.T., and Hsiao, H. (2001) The effect of wearing a backbelt on spine kinematics during asymmetric lifting of large and small boxes. *Spine* 26:1794–17798.

Granata, K.P., Marras, W.S., and Davis, K.G. (1997) Biomechanical assessment of lifting dynamics, muscle activity and spinal loads while using three different style lifting belts. *Clin. Biomech.* 12(2):107–115.

Harman, E.A., Rosenstein, R.M., Frykman, P.N., and Nigro, G.A. (1989) Effects of a belt on intra-abdominal pressure during weight lifting. *Med. Sci. Sports Exerc.* 2(12):186–190.

Hunter, G.R., McGuirk, J., Mitrano, N., Pearman, P., Thomas, B., and Arrington, R. (1989) The effects of a weight training belt on blood pressure during exercise. *J. Appl. Sport Sci. Res.* 3(1):13–18.

Kraus, J.F., Brown, K.A., McArthur, D.L., Peek-Asa, C., Samaniego, L., and Kraus, C. (1996) Reduction of acute low back injuries by use of back supports. *Int. J. Occup. Environ. Health* 2:264–273.

Lander, J.E., Hundley, J.R., and Simonton, R.L. (1992) The effectiveness of weight belts during multiple repetitions of the squat exercise. *Med. Sci. Sports Exerc.* 24(5):603–609.

Lantz, S.A. and Schultz, A.B. (1986) Lumbar spine orthosis wearing. I. Restriction of gross body motion. *Spine* 11(8):834–837.

McCoy, M.A., Congleton, J.J., Johnston, W.L., and Jiang, B.C. (1988) The role of lifting belts in manual lifting. *Int. J. Ind. Ergon.* 2:259–266.

McGill, S.M. (1993) Abdominal belts in industry: A position paper on their assets, liabilities and use. *Am. Ind. Hyg. Assoc. J.* 54(12):752–754.

McGill, S.M. (1999) Update on the use of back belts in industry: more data — same conclusion. In: *Interventions, Controls and Applications in Occupational Ergonomics*, Ed. W. Karwowski and W. Marres, CRC Press.

McGill, S.M. and Norman, R.W. (1987) Reassessment of the role of intra-abdominal pressure in spinal compression. *Ergonomics* 30(11):1565–1588.

McGill, S., Norman, R.W., and Sharratt, M.T. (1990) The effect of an abdominal belt on trunk muscle activity and intra-abdominal pressure during squat lifts. *Ergonomics* 33(2):147–160.

McGill, S.M., Seguin, J.P., and Bennett, G. (1994) Passive stiffness of the lumbar torso in flexion, extension, lateral bend and axial twist: The effect of belt wearing and breath holding. *Spine* 19(6):696–704.

McGill, S.M. Ultimate back fitness and performance, Wabuno Publishers, www.backfitpro.com.

Mitchell, L.V., Lawler, F.H., Bowen, D., Mote, W., Asundi, P., and Purswell, J. (1994) Effectiveness and cost-effectiveness of employer-issued back belts in areas of high risk for back injury. *J. Occup. Med.* 36(1):90–94.

Nachemson, A.L., Andersson, G.B.J., and Schultz, A.B. (1986) Valsalva maneuver biomechanics. Effects on lumbar trunk loads of elevated intra-abdominal pressures. *Spine* 11(5):476–479.

NIOSH. *Workplace Use of Back Belts*. U.S. Department of Health and Human Services, Centres for Disease Control and Prevention. National Institute for Occupational Safety and Health, July, 1994.

Parker, P.L., Crumpton-Young, L.L., and Brandon, K.M. (2000) Does abdominal body composition modulate the effects of back belts on the respiratory system. *Int. J. Ind. Ergon.* 26:561–567.

Rafacz, W. and McGill, S.M. (1996) Abdominal belts increase diastolic blood pressure. *J. Occup. Environ. Med.* 38(9):925–927.

Reddell, C.R., Congleton, J.J., Huchinson R.D., and Mongomery J.F. (1992) An evaluation of a weightlifting belt and back injury prevention training class for airline baggage handlers. *Appl. Ergon.* 23(5):319–329.

Reyna, J.R., Leggett, S.H., Kenney, K., Holmes, B., and Mooney, V. (1995) The effect of lumbar belts on isolated lumbar muscle. *Spine* 20(1):68–73.

Walsh, N.E. and Schwartz, R.K. (1990) The influence of prophylactic orthoses on abdominal strength and low back injury in the work place. *Am. J. Phys. Med. Rehab.* 69(5):245–250.

Wassall, J.T., Gardneux, L.T., Landsittal, D.P., Johnston, J.J., and Johnston, J.M. (2000) A prospective study of back belts for prevention of back pain and injury. *JAMA* 284:2727–2732.

Woldstad, J.C. and Sherman, B.R. (1999) The effects of a back belt on posture, strength, and spinal compressive force during static lift exertions. *Occup. Health Ind. Med.* 40(1):2.

.

31

Job Rotation

David Rodrick
Florida State University

Waldemar Karwowski
Peter M. Quesada
University of Louisville

31.1 Introduction

The concept of job rotation is not new. Job rotation has been extensively studied and applied in business organizations over the last five decades. In the 1960s and 1970s, job rotation was utilized to motivate employees. The main objectives of job rotation were to provide task variability, enhance socialization, assist in management and executive development, and enhance employee career development. Job rotation can be defined as a systematic lateral transfer of employees between jobs, tasks, assignments, or projects within an organization to achieve many different objectives of the organization. These objectives include, but are not limited to, orienting new employees; preventing job boredom or burnout; reducing stress, absenteeism, and turnover rates; preventing fatigue and exposure to risks/hazards of work-related musculoskeletal disorders (WMSDs); training employees; involving managers in training; rewarding employees; enhancing career development; exposing employees to various environments, etc. Specifically, job rotation has been generally recognized among practitioners and researchers as one of the effective administrative controls for preventing WMSDs. Several studies have reported that job rotation is a commonly used redesigning approach for improving an organization's health.

31.2   Job Rotation and Physical Tasks

Physical tasks are considered to be among the major causes of musculoskeletal disorders, illnesses, or disabilities, which generally result in high costs, reduced production, and most importantly, aggravated quality of life (Ayoub et al., 1997; Karwowski and Marras, 1997). There is a general consensus among researchers and practitioners that many variables and their interactions affect the magnitude of severity, risk, or hazards due to physical tasks. In general, physical tasks produce forces on connective tissues of the musculoskeletal system. High forces on discs, joints, ligaments, and muscles have long been considered to expose individuals to high injury risk (Frazer et al., 2003). Norman et al. (1998) argued that high peak compression and shear forces, as well as cumulative effects of such forces should be considered independent WMSD risk factors. The adverse effects of both high peak and high cumulative loading on spinal tissues have been well documented through laboratory research (e.g., Adams and Hutton 1985; Hansson et al., 1987) and epidemiological studies (e.g., Norman et al., 1998; Kerr et al., 2001). According to the U.S. Department of Labor (1993), the underlying principle of job rotation for physical tasks is to mitigate physical fatigue and stress for a particular set of muscles by rotating employees among other jobs that use primarily different muscle groups. Accordingly, some fundamental assumptions can be drawn that are deemed appropriate for job rotation and physical tasks:

1. Decreasing loading time for the same muscle groups and joints can reduce an individual's physical workload (Putz-Anderson, 1988; Wands and Yassi, 1993)
2. Exposure to risk can be averaged over the workforce and overall risk can be reduced by spreading high loads over many workers, rather than having the same worker continuously exposed to high risk (Frazer et al., 2003)
3. In order to determine the effects of rotation on a task's injury risk, it is necessary to quantify the injury risk for performing the task, as well as the injury risk of the tasks in the rotation scheme.

31.3   Empirical Evidence of the Effect of Job Rotation on Physical Tasks

Several research efforts have addressed the role or effectiveness of job rotation as an administrative control for reducing physical workload or fatigue. For instance, in his study, Henderson (1992) developed a rotation scheme for workers engaged in poultry processing tasks. Tasks were rated by workers on a scale that ranged from low physical stress to unacceptably high physical stress. The rotation scheme recommended redesign of high physical stress tasks. The scheme also recommended that workers should not perform consecutive high stress tasks, and that high stress tasks should be preceded and followed by low stress tasks. The study observed that the rotation scheme resulted in reduced number of musculoskeletal complaints. In another study, Hinnen et al. (1992) examined the effect of job rotation on check-out system design. The study evaluated 152 female supermarket cashiers using a self-administered questionnaire and a medical exam. The study found a very beneficial impact of job rotation on the prevalence of musculoskeletal disorders among the cashiers.

Kuijer et al. (1999) studied the effect of job rotation on refuse collecting tasks. The study involved three kinds of tasks: refuse collecting, street sweeping, and driving. The study compared two nonrotation groups with two rotation groups. The nonrotation groups were refuse collectors and street sweepers and the rotation groups were refuse collectors/street sweepers and street sweepers/drivers. Physical workload was measured via perceived load, energetic load, and postural load during a full working day. The study reported that job rotation significantly decreased perceived load and energetic load, and slightly decreased the postural load. Results demonstrated that job rotation reduced the overall physical workload of the refuse collectors with regard to the total amount of work performed by the refuse collectors.

Kuijer et al. (2002) recently conducted a 1 yr prospective investigation to examine effects of job rotation on need for recovery, prevalence of musculoskeletal complaints, and sick leave due to musculoskeletal complaints among 130 male refuse collectors. The study utilized a self-administered questionnaire that was taken by three groups, at two time intervals. The groups consisted of a reference group of refuse collectors that did not rotate at either interval (NR–NR); a group that rotated between truck driving and refuse collecting at both testing intervals (R–R); and a group of refuse collectors that were nonrotating at t_0 but rotated between truck driving and refuse collecting at t_1 (NR–R). The study found that the R–R group's need for recovery was slightly lower than the NR–NR group's need, and that there was no difference between the NR–R and NR–NR groups. Unexpectedly, it was found that the prevalence of low back complaints was significantly greater for the R–R and NR–R groups relative to the NR–NR group. There was no difference in sick leave between the groups. Kuijer et al. (2002) argued that several factors, such as, the presence of a healthy worker effect in the NR–NR group, the presence of an unhealthy worker effect in the two job rotation groups, exposure to different risk factors (e.g., whole body vibration) for those who performed both truck driving and refuse collecting, and different levels of peak and cumulative spinal loading risk factors might have played significant roles in study outcomes.

In another study, Frazer et al. (2003) utilized a custom software package to analyze two jobs in the automotive industry. The aim of the study was to estimate levels of exposure to factors known to be associated with risk of reporting low back pain. These exposures were then utilized in conjunction with the software to evaluate the effects of job rotation implementation on both predicted risk of reporting low back pain and redistribution of predicted risk for the two workers involved in the rotation schedule. The study found that job rotation produced a greater overall risk of reporting low back pain than working without rotation. It was found that a worker in a less demanding job who "rotated in" to a more demanding job experienced an increase in the risk of reporting low back pain that was substantially greater than the reduction experienced by a worker "rotating out" of a more demanding job to a less demanding one. The study also found that the redistribution of risk was not uniform across jobs of different risk levels. It showed that when "rotating in," the worker was immediately exposed to the peak loading parameters associated with the more demanding job resulting in a step increase in the risk of reporting low back pain.

The preceding discussion of the empirical evidence of the effect of job rotation on physical tasks indicates that job rotation can be an effective administrative control methodology for reducing exposure to physical stress and fatigue in different types of physical tasks.

31.4 Job Rotation and Cognitive Tasks

As stated earlier, apart from the highly strenuous physical tasks, there are tasks that are highly demanding with respect to employees' mental (cognitive) capacities. Tasks with cognitive demand level can affect human functional processes (e.g., information processing, mental execution, communication, learning), resource processes (e.g., attention, memory), as well as employee behavioral states (e.g., motivation, perception of work environment). In a recent study, Wei and Salvendy (2003) illustrated the utilization of the Purdue Cognitive Job and Task Analysis methodology, and the Human-Centered Cognitive Performance model based Purdue Cognitive Task Analysis Questionnaire (PCTAQ). The study showed that the PCTAQ allowed to analyze jobs and tasks, and it provided a mechanism for improving cognitive job and task performance. The authors concluded that the utilization of this methodology could assist in job evaluation, job design and job rotation, as well as personnel selection and training.

Allwood and Lee (2004) proposed that utilization of job rotation could address operator's problem solving skills in lean manufacturing settings. The authors postulated that increasing efficiency of operators faced with repeating tasks was widely accepted and was characterized by the well-known learning curve. This study modified the learning curve to describe problem solution times, incorporate forgetting effects, and treat both general skills and specific skills related to a particular problem. The model was tested in a simulation of a serial flow shop, subject to a range of interruptions. The efficiency of the flow shop was characterized by its run-ratio, and the effects of eight key variables on the run-ratio were tested through simulation. The results showed that the run-ratio generally increased as operators learned more rapidly and forgot more slowly, and decreased as the number of problem types increased. The study also observed that the run-ratio always declined with introduction of job rotation schemes, however implemented. The study indicated that this decline occurred because job rotation mitigated any possible specialization advantage related to problems, which did not occur uniformly at all workstations.

In another study by Harris et al. (2002), a cognitive mapping method was used to elicit mental models of psychosocial hazards at work. This study demonstrated the utilization of cognitive mapping method as an assessment tool for representing mental models of psychosocial hazards at work. Using cognitive maps of 35 individuals from eight organizations and a detailed example of one of the participating organizations, the study showed how understanding mental models of psychosocial hazards at work could aid the assessment of psychosocial risk and the development and implementation of intervention programs to reduce psychosocial hazards and harms.

31.5 Quantitative Methods to Determine Proper Job Rotation Scheme

Although the existing body of literature shows clear evidence that job rotation is an effective control strategy for reducing physical, cognitive, or environmental hazards, the effectiveness of job rotation relies on how well designed the rotation scheme is. In general, to develop a job rotation plan may involve: (1) determining the set of tasks/jobs to be included in the rotation scheme; (2) the rotation sequence; and (3) the length of the rotation interval (Tharmmaphornphilas and Norman, 2004). It can be seen from previous studies that utilizing formal quantitative methods to develop a proper job rotation plan has been ignored in most instances. Most of these previous studies assumed that tasks/jobs to be rotating, as well as task sequencing were already known. However, a few have studied different methods/policies to develop a job rotation plan. For example, in the study by Henderson (1992) the specific policy included choosing not to assign workers to successive high stress tasks.

Carnahan et al. (2000) argued that job rotation plan development would be facilitated if a planner could utilize scheduling rules. According to these investigators, such rules would have to provide guidance as to the maximum number of hours per day a worker could safely perform a physical task. Carnahan et al. (2000) proposed that the rules for a material handling task should take the following factors into account: (1) the physical demands of the task; (2) the gender of the worker; (3) the lifting capacities of the worker; and (4) the demands and time spent performing other tasks. As a set, rules would interact with each other. Subsequently, Carnahan et al. (2000) developed a job rotation method implementing integer programming and genetic algorithm methods. This allowed them to generate multiple job rotation schedules to reduce lower back pain/injury due to manual lifting tasks. Following the generation of various job rotation schedules, the study used a clustering method to determine a general set of rules governing task exposure for each group of workers. In another study, Tharmmaphornphilas et al. (2003) utilized an integer programming technique to develop a job rotation schedule to reduce noise exposure for sawmill workers. Recently, Tharmmaphornphilas and Norman (2004) utilized integer programming approach to determine job rotation length in a similar work setting.

In light of the preceding discussion, a generic set of steps could be proposed to develop a job rotation schedule as follows:

1. Determine and quantify level of risk for each job/task
2. Determine the set of jobs to be included in the job rotation schedule
3. Utilize a quantitative (or a qualitative/subjective when relationship between tasks and the level of risk for each task is too obvious) technique to determine the task sequence
4. Utilize a quantitative technique to determine length of rotation
5. Follow up the different schedules and measure the effectiveness

31.6 A Generic Strategy for Implementing Job Rotation in Physical Tasks

A good job rotation scheme begins with identification of problem jobs. Problem jobs are those that contain one or more risk factors for musculoskeletal disorders. The effects of upper extremity and low back pain/disorders on industry and service workers have been well documented in the existing body of literature. A considerable number of studies have so far been conducted to identify possible risk factors that induce WMSDs. Such risk factors can be work-related, psychosocial, or individual factors. Although there are debates as to which category of risk factors contribute more to development of WMSDs, it is clearly understood that there are direct relationships between physical stresses (e.g., force, joint angle, recovery, vibration, and temperature) and factors, such as, duration, repetition, and stress magnitude.

31.6.1 Identification of Problem Jobs

According to Konz (2001), problem jobs can be identified from a number of sources and methods. These include: (1) records/statistics of the medical/safety department (e.g., OSHA logs); (2) worker discomfort levels (e.g., body discomfort map, Borg category scale), (3) interview with the workers, (4) expert opinion, and (5) checklists (Lifshitz and Armstrong, 1986; Rogers, 1992). For illustration, Figure 31.1 and Figure 31.2 illustrate the body discomfort map and Borg category scale, respectively, while Table 31.1 and Table 31.2 show two checklists for generic problem job identification and jobs requiring utilization of upper extremities.

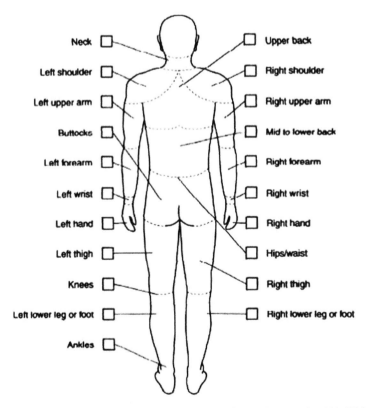

FIGURE 31.1 An illustration of body discomfort map. (Taken from Dimov et al., 2000. With permission.)

0	Nothing at all
0.5	Very, very weak (just noticeable)
1	Very weak
2	Weak (light)
3	Moderate
4	Somewhat strong
5	Strong (heavy)
6	
7	Very strong
8	
9	Very, very strong (almost maximal)
10	Extremely strong (maximal)

FIGURE 31.2 Borg category scale (CR-10) as a measure of discomfort. (Adapted from Konz, S. In G. Salvendy, ed., *Handbook of Industrial Engineering: Technology and Operations Management*. New York: Wiley, 2001, 1353–1390. With permission.)

31.6.2 Quantification of the Level of Risks

Following the identification of the problem jobs, it is important that the risks are quantified. Over the years, several quantitative and semiquantitative methods have been proposed. Recently, Moore and Garg (1995) have developed a semiquantitative job analysis methodology for identifying industrial jobs associated with distal upper extremity (wrist/hand) disorders. This proposed strain index methodology involves the measurement or estimation of six task variables, including: (1) exertion intensity, (2) exertion duration per cycle, (3) efforts per minute, (4) wrist posture, (5) exertion speed, and (6) task duration per day. An ordinal rating is assigned for each of the variables according to the exposure data. The proposed strain index is the product of these six multipliers assigned to each of the variables (see Moore and Garg, 1995 for details).

Another well-studied methodology is the revised NIOSH (1991) lifting equation (Waters et al., 1993). The revised lifting equation is

TABLE 31.1   A Generic Checklist to Prioritize Potential Problem Jobs

Job title————————————————————————Analyst————————————————————————
Specific task——————————————————————Phone————————————————————————
Job number———————————————Dept——————————————————Date of analysis————————————
Location————————————————————————————————

Body part		Effort	Continuous effort time	Effort/min	Priority	Effort 1 = light 2 = moderate 3 = heavy
Back		———	———	———	———	
Neck/shoulders	R	———	———	———	———	
	L	———	———	———	———	Cont effort time 1 = <6 sec 2 = 6 to 20 sec 3 = >20 sec
Arms/elbow	R	———	———	———	———	
	L	———	———	———	———	
Wrists/hands/fingers	R	———	———	———	———	Efforts/min
	L	———	———	———	———	1 = <1
Legs/knees	R	———	———	———	———	2 = 1 to 5
	L	———	———	———	———	3 = >5
Ankles/feet/toes	R	———	———	———	———	
	L	———	———	———	———	

Priority for change:
After completing the effort (force), continuous effort time (duration), and efforts per minute (repetition), determine the priority for change.
Verbal descriptors for any combination (e.g., 333, 212, or 123) of the ratings for effort, continuous effort time, and efforts per min can range from Extremely high to not at all.

Source: Adapted from Konz, S. In G. Salvendy, ed., *Handbook of Industrial Engineering: Technology and Operations Management*. New york: Wiley, 2001, 1353–1390. With permission.

TABLE 31.2 A Sample Checklist for the Upper-Extremity Problem Jobs

No	Yes	Risk factors
		Physical stress
—	—	1. Can the job be done without hand/wrist contact with sharp edges?
—	—	2. Is the tool operating without vibration?
—	—	3. Are the worker's hands exposed to temperatures >70°F (20°C)?
—	—	4. Can the job be done without using gloves?
		Force
—	—	1. Does the job require exerting less than 10 lbs (4.5 kg) of force?
—	—	2. Can the job be done without using a finger pinch grip?
		Posture
—	—	1. Can the job be done without wrist flexion or extension?
—	—	2. Can the tool be used without wrist flexion or extension?
—	—	3. Can the job be done without deviating the wrist from side to side?
—	—	4. Can the tool be used without deviating the wrist from side to side?
—	—	5. Can the worker be seated while performing the job?
—	—	6. Can the job be done without a clothes-wringing motion?
		Workstation hardware
—	—	1. Can the work surface orientation be adjusted?
—	—	2. Can the work surface height be adjusted?
—	—	3. Can the tool location be adjusted?
		Repetitiveness
—	—	1. Is the cycle time longer than 30 sec?
		Tool design
—	—	1. Are the thumb and finger slightly overlapped in a closed grip?
—	—	2. Is the tool handle span between 2 and 2.75 in. (5 and 7 cm)?
—	—	3. Is the tool handle made from material other than metal?
—	—	4. Is the tool weight below 9 lbs (4 kg)? Note exceptions to the rule.
—	—	5. Is the tool suspended?

Source: Adapted from Konz, S. In G. Salvendy, ed., *Handbook of Industrial Engineering: Technology and Operations Management.* New york: Wiley, 2001, 1353–1390. With permission.

based on three main components: (1) the standard lifting location, (2) load constant, and (3) risk factor multipliers. The standard lifting location (SLL) serves as the three-dimensional reference point for evaluating the parameters that define a worker's lifting posture. The load constant (LC) refers to a maximum weight value for the SLL (23 kg). The recommended weight limit (RWL) is the product of the load constant and six multipliers:

$$RWL(kg) = LC * HM * VM * DM * AM * FM * CM$$

The multipliers (M) are defined in terms of related risk factors, including horizontal location (HM), vertical location (VM), vertical travel distance (DM), coupling (CM), frequency of lift (FM), and asymmetry angle (AM). The multipliers for frequency and coupling are defined using relevant tables. In addition to lifting frequency, work duration and vertical distance factors are used to compute the frequency multiplier (see Waters and Putz-Anderson, 1999 for more details).

31.6.3 Determination of Jobs in the Job Rotation Scheme

Developing a good job rotation scheme requires determination of a job set to be included in the rotation, the rotation sequence, and the proper rotation interval length (Tharmmaphornphilas and Norman, 2004). Repetition, duration, and magnitude of risk factors present in a job determine whether exposure to those risk factors would lead to development of any WSMDs. Certainly job rotation as an administrative control will reduce repetition of a problem job. The fundamental question that remains, however, is what is the maximum duration that a worker can be exposed to a given risk factor magnitude without any detrimental effects from the exposure. One strategy would be to select a set of jobs with a wide range of risk levels (high to no risk), and with jobs requiring a mix of both static and dynamic physical abilities, as well as exertion of different muscle groups. It is also important to select jobs that require different cognitive abilities, even in small magnitude.

31.6.4   Quantification of Job Sequence and Length of Rotation

Following the determination of a set of jobs for rotation, the job sequence and length should be designed. The job sequence determines the repetition rate at which a worker will be exposed to a given risk, while length of rotation determines exposure to the magnitude of such a risk. For example, if there are three jobs in a job set: A, B, and C, and the risk for these jobs is assessed to be high, low, and low, respectively, then the quantification of the job sequence would determine how many times a worker would be performing Job A (repetition rate for A). If the job sequence is set to A–B–C–B–C–A, then the worker is exposed to the high-risk job (Job A) for two times. The length of rotation determines how long a worker would perform on a high-risk job (Job A). The duration would be a standard time frame (e.g., typically 2 h) derived computationally or by mathematical modeling. Computationally driven estimation of duration of rotation would allow for estimating a suitable time for each job in the job set with respect to worker's physiological and biomechanical capacities and limitations.

One very simple way of job sequencing would be to rotate workers randomly through the jobs. Another simple method is to have workers move from the first workstation to the second, and so on, until the worker at the last workstation moves to the first one. A third possibility is to rotate workers based upon a specific policy, such as choosing not to assign a worker to two stressful tasks in succession (Henderson, 1992). As mentioned earlier, Carnahan et al. (2000) recently proposed a method for reducing low back injury in a manual lifting environment by implementing a genetic algorithm to provide multiple good job rotation schedules. They then used a clustering method to determine a general set of rules governing task exposure for each group of workers. Tharmmaphornphilas et al. (2003) used an integer programming method to develop a proper rotation plan to reduce noise exposure for workers working in a sawmill plant. In a subsequent study, Tharmmaphornphilas and Norman (2004) utilized an integer programming approach to determine the job rotation length in a similar work setting.

31.6.5   Evaluation of the Job Rotation Scheme

A specific job rotation scheme should be evaluated to ascertain its effectiveness. Incident rates of the jobs before and after implementing the job rotation scheme can be compared to assess its effectiveness. However, the existing body of literature does not have information on the specific time interval at which this evaluation should be done.

31.7   Job Rotation and Manufacturing Systems

According to Brandt (2003), as the concept of lean manufacturing job rotation is implemented for increasing worker's motivation, improving performance and thus, achieving better efficiency in the company. As reported by Womack et al. (1991), this concept originates from Japanese culture in which a firm is regarded as a "dojo," which is literally a training place where one practices the martial arts of life. In a lean production system, employees are placed on a job rotation scheme, in which they work in different departments of the company for a specified period of time. Thus, job rotation allows employees to learn and to acquire considerable expertise in a number of areas of the company. Consequently, the employees become well-rounded company representatives who comprehend the company as a holistic system. This synergistic understanding is a central concept of human-centered design (Brandt, 2003). Furthermore, as argued by Brandt (2003), job rotation enables employees to see, understand, and benefit from the whole process of the company more clearly; and they feel themselves truly as a part of it. In turn, employee motivation and interest would always be kept at a high level, which results in increased productivity.

In one case study by Harryson (1997) on Canon and Sony, the author identified and illustrated the key mechanisms that these companies utilized to foster product innovation. The study showed that Canon and Sony used a combination of external and internal networking mechanisms to identify and acquire key technologies and related skills, gain market knowledge, improve the results of internal research and development efforts, and ensure the successful transfer of these results to efficient production processes. This study identified four key mechanisms underlying successful product innovation at Canon and Sony: (1) strategic training and job rotation for engineers, (2) application-driven research and development, (3) direct transfer of development teams from research and development to production, and (4) extensive networking with external centers of excellence and key suppliers. With regard to job rotation, the study concluded that rather than viewing this job rotation strategy as a drain on the technological expertise in their laboratories, both companies relied on strong external networks with key suppliers and university-based researchers as important sources for acquiring new technologies and the competencies needed to support them. Brandt (2003) argued that due to job rotation, knowledge and skills are shared among all the workers and the workers are always in a continuous learning process. This learning process of the employees diminishes the company's dependence or reliance on a few highly specialized workers.

31.8   Job Rotation and Management

From the management perspective, job rotation is valued by employees because of its causal relationship with promotion and salary growth. Nevertheless, job rotation is traditionally considered to be a training and development tool that improves employees' job/organizational knowledge and skills. According to Cheraskin and Campion (1996), job rotation contributes to improving three different kinds of skills, that is, technical (accounting, finance, and operating procedures), business (financial knowledge, employee support, international issues, knowledge of company operation), and administrative (planning, communication, interpersonal, leadership, cognitive, computer). Cheraskin and Campion (1996) proposed eight recommendations that an organization should consider when implementing a job rotation program. These recommendations are listed in Table 31.3.

31.9   Job Rotation and Knowledge Management

In recent years, several studies reported important implications of job rotation for knowledge management in organizations. More specifically, job rotation is found to play a very important role in organizational learning, problem solving skills, technological innovations, and employee career development. Osterman (1994), in a 1992 U.S. survey, reported that about 26% of organizations involved more than half of their core employees in job rotation. In 1997, Osterman (2000), reported an increase to 50% of these organizations. As argued by Ortega (2001), job rotation has important implications on organizational learning in at least two respects. First, by rotating employees through different tasks/departments/projects, the organization gains information about the quality of various job–employee matches (Jovanovic, 1979). Second, job rotation also allows the organization to be knowledgeable about profitability of different jobs within the organization. These implications are particularly important for companies where innovative production processes are being implemented or new products are launched (Ortega, 2001).

Several studies have shown significant relationships between job rotation and other job related attributes. It was found that there is a negative correlation between frequency of rotation and job tenure (Campion et al., 1994). Studies also showed that there is a positive correlation between job rotation

TABLE 31.3   Recommendations to be Considered for Job Rotation Program

1. Management
   Proactively manage job rotation as a component of the training and career development system. Job rotation may be especially valuable for organizations that require firm-specific skills because it provides an incentive to organizations to promote from within
2. Skills
   Have a clear understanding of exactly which skills will be enhanced by placing an employee into the job-rotation process. Address skills that are not enhanced by job rotation through specific training programs and management coaching.
3. Primary application
   Use job rotation for employees in nonexempt jobs, as well as for those in professional and managerial jobs. Job rotation may be of great value for developing employees in all types of jobs
4. Secondary applications
   Use job rotation with later-career and plateaued employees, as well as with early-career employees.
   Some organizations may have the tendency to rotate employees too fast in early-career stages and too slow in later-career stages. Job rotation can be a good way to reduce the effects of the plateauing process by adding stimulation to employees' work
5. Career development
   Job rotation can be used as a means of career development without necessarily granting promotions — so it may be especially useful for downsized organizations because it provides opportunities to develop and motivate employees
6. Equal employment opportunity
   Special attention should be given to the job rotation plans for female and minority employees
   Recent federal equal employment opportunity legislation has recognized the importance of job rotation to promotional opportunities when examining the limited representation of minorities and females in executive jobs (called the "glass ceiling" effect). Title II of the Civil Rights Act of 1991 has ordered a commission to study the barriers and opportunities to executive advancement, specifically including job-rotation programs (Section 204, paragraph a5).
7. Career development planning
   Link rotation with the career development planning process so that employees know the developmental needs addressed by each job assignment
   Both job-related and development-related objectives should be defined jointly by the employee and the manager when the employee assumes a new position. The rate of rotation should be managed according to the time required to accomplish the goals of the job and the time required to achieve the developmental benefits of the job. The advantage of this approach is that both the employee and the manager will have a clear understanding of expectations and the required tenure on the job will be related to predetermined outcomes. Job rotation should be perceived as voluntary from the employee's point of view if it's going to have the intended developmental effects
8. Benefits and costs of rotation
   Implement specific methods of maximizing benefits and minimizing costs of rotation.
   Examples include increasing the benefits of organizational integration and stimulating work by carefully selecting jobs, increasing career and awareness benefits by ensuring that they are reflected in the development plans, decreasing workload costs by managing the timing of rotations, decreasing learning-curve costs by having good operating procedures, and decreasing the dissatisfaction of coworkers by helping them understand the role of job rotation in their own development plans

Source: Modified after Cheraskin, L. and Campion, M.A. *Personnel Journal*, 75(11), 31–38, 1996. With permission.

and utilization of new technologies (Gittleman et al., 1998). However, Ortega (2001) argued that the employee learning theory is not that consistent with the relationship between job rotation and technological innovation. From his mathematical formulations, Ortega (2001) showed that, in fact, organizations learn more from rotation when employees and technologies are relatively new.

Brandt (2003) also argued that job rotation is very important in terms of information transmission, and management within the company. It is important that with the rapid development of information and communication technologies, the problem of using the available information appropriately has to be solved. Moreover, beyond formal knowledge, such as technical maintenance or process procedures, there is an inherent tacit knowledge of employees that cannot be written down and is situated under the visible and documented processes (Hunecke and Preuschoff, 2003). According to Brandt (2003), job rotation offers a unique opportunity in terms of both sharing knowledge and passing tacit knowledge of senior/experienced employees to new employees.

31.10   Concluding Remarks

Job rotation can be an effective administrative tool to alleviate the work-related musculoskeletal stresses due to physical tasks. Job rotation has also been found to be a predictor of some job-related behavior, such as, tenure, promotion, and salary growth. Recent research efforts showed that a synergetic approach based on the anthropometric, biomechanical, and mathematical modeling would guide to a working job rotation scheme to reduce the risk of musculoskeletal injury or pain. However, further research should be conducted to formulate the effective risk assessment techniques with respect to epidemiological and/or mathematical modeling approach. Second, an efficient method is needed to determine the number or proportion of the workers to be included in the set of jobs for job rotation. This issue should also address the proportion of workers with respect to different percentile groups to be included in the job rotation scheme. Third, along with genetic programming, other soft computing techniques such as, neural networks, fuzzy logic, adaptive neuro-fuzzy techniques should be applied in the future research to estimate the duration of job rotation.

References

Adams, M.A. and Hutton, W.C. (1985). Gradual disc prolapse. *Spine*, 15, 311–316.

Allwood, J.M. and Lee, W.L. (2004). The impact of job rotation on problem solving skills. *International Journal of Production Research*, 42(5), 865–881.

Ayoub, M.M., Dempsey, P.G., and Karwowski, W. (1997). Manual materials handling. In G. Salvendy (Ed.), *Handbook of Human Factors and Ergonomics*. New York: Wiley, 1085–1123.

Brandt, D. (2003). Reflections on Human-Centred Systems and Leadership, Summer Academy SAC, Eger, Hungary.

Carnahan, B.J., Redfern, M.S., and Norman, B. (2000). Designing safe job rotation schedules using optimization and heuristic search. *Ergonomics*, 43(4), 543–560.

Campion, M., Cheraskin, L., and Stevens, M. (1994), Career-related antecedents and the outcomes of job rotation. *Academy of Management Journal*, 37(6), 1518–1542.

Cheraskin, L. and Campion, M.A. (1996), Study clarifies job-rotation benefits. *Personnel Journal*, 75(11), 31–38.

Dimov, M., Bhattacharya, A., Lemasters, G., Atterbury, M., Greathouse, L., and Ollila-Glenn, N. (2000). Exertion and body discomfort preceived symptoms associated with carpentry tasks: an on-site evaluation. *AIHA Journal*, 61(5), 685–691.

Frazer, M.B., Norman, R.W., Wells, R.P., and Neumann, W.P. (2003). The effects of job rotation on the risk of reporting low back pain. *Ergonomics*, 46(9), 904–919.

Gittleman, M., Horrigan, M. and Joyce, M. (1998). Flexible workplace practices: evidence from a nationally representative survey. *Industrial Labor Relations Review*, 52(1), 99–115.

Hansson, T.H., Keller, T.S., and Spengler, D.M. (1987). Mechanical behaviour of the human lumbar spine. II. Fatigue, strength during dynamic compressive loading. *Journal of Orthopaedic Research*, 5, 479–487.

Harris C., Daniels, K., and Briner, R.B. (2002). Using cognitive mapping for psychosocial risk assessment, *Risk Management*, 4(3), 7–21.

Harryson, S.J. (1997). How Canon and Sony drive product innovation through networking and application-focused R&D. *Journal of Product Innovation Management*, 14(4), 288–295.

Henderson, C.J. (1992). Ergonomics job rotation in poultry processing. In S. Kumar (Ed.), *Advances in Industrial Ergonomics and Safety*. London: Taylor & Francis, 443–450.

Hinnen, U., Laubli, T., Guggenbuhl, U., and Krueger, H. (1992). Design of check-out systems including laser scanners for sitting work posture. *Scandinavian Journal of Work Environment and Health*, 18, 186–194.

Hunecke, H. and Preuschoff, E. (2003). The competitive advantage of locally embedded knowledge the internal enterprise networking. In D. Brandt, (Ed.), *Human-Centred System Design - First: People, Second: Organisation, Third: Technology. 20 Case Reports*. Aachen: Aachener Reihe Mensch und Technik, 42, 44–53.

Jovanovic, B. (1979). Job matching and the theory of turnover. *Journal of Political Economics*, 87(6), 972–990.

Karwowski, W. and Marras, W. (1997). Work-related musculoskeletal disorders of the upper extremities, in G. Salvendy (Ed.), *Handbook of Human Factors and Ergonomics*. New york: Wiley, 1124–1173.

Kerr, M.S., Frank, J.W., Shannon, H.S., Norman, R.W.K., Wells, R.P., Neumann, W.P., Bombardier, C., and The Ontario Universities Back Pain Study Group (2001). Biomechanical and psychosocial risk factors for low back pain at work. *American Journal of Public Health*, 91, 1069–1075.

Konz, S. (2001). Methods engineering. In G. Salvendy (Ed.), *Handbook of Industrial Engineering: Technology and Operations Management*, New york: Wiley, 1353–1390.

Kuijer, P.P.F.M., Visser, B., and Kemper, H.C.G. (1999). Job rotation as a factor in reducing physical workload at a refuse collecting department. *Ergonomics*, 42(9), 1167–1178.

Kuijer, P.P.F.M., Van Der Beek, A.J., Van Dieën, J.H., Visser, B., and Frings-Dresen, M.H.W. (2002). Effect of job rotation on need for recovery and (sick leave due to) musculoskeletal complaints: a prospective study among refuse collectors. In P.P.F.M. Kuijer (Ed.), *Effectiveness of Interventions to Reduce Workload in Refuse Collectors*, Wageningen, The Netherlands: Ponsen & Looijen BV, 133–147.

Lifshifz, Y. and Armstrong, T.J. (1986). A design checklist for control and prediction of cumulative trauma disorders in hand intensive manual jobs. *Proceedings of the Human Factors Society 30th Annual Meeting*.

Moore, J.S. and Garg, A. (1995). The strain index: a proposed method to analyze jobs for risk of distal upper extremity disorders. *AIHA Journal*, 56, 443–458.

Norman, R., Wells, R., Neumann, P., Frank, J., Shannon, H., and Kerr, M., and the Ontario Universities Back Pain Study (OUBPS) Group, (1998), A comparison of peak versus cumulative physical work exposure risk factors for the reporting of low back pain in the automotive industry. *Clinical Biomechanics*, 13, 561–573.

Ortega, J. (2001). Job rotation as a learning mechanism. *Management Science*, 47(10), 1361–1370.

Osterman, P. (1994). How common is workplace transformation and who adopts it? *Industrial Labor Relations Review*, 47(2), 173–188.

Osterman, P. (2000). Work reorganization in an era of restructuring: trends in diffusion and effects on employee welfare. *Industrial Labor Relations Review*, 53(2), 179–196.

Putz-Anderson, V. (1988). *Cumulative Trauma Disorders: A Manual For Musculoskeletal Diseases of the Upper Limbs*. London: Taylor & Francis.

Rogers, S.A. (1992). A functional job analysis technique. *Occupational Medicine: State of the Art Review*, 7(4), 679–711.

Tharmmaphornphilas, W. and Norman, B.A. (2004). A quantitative method for determining proper job rotation intervals. *Annals of Operations Research*, 128, 251–266.

Tharmmaphornphilas, W., Green, B., Carnahan, B.J., and Norman, B.A. (2003). Applying mathematical modeling to create job rotation schedules for minimizing occupational noise exposure. *AIHA Journal*, 64, 401–405.

U.S. Department of Labor (1993). Ergonomics program management guidelines for meatpacking plants. Occupational Safety and Health Administration Publication No. 3123, Washington, DC: Occupational Safety and Health Administration.

Wands, S.E. and Yassi, A. (1993). Modernization of a laundry processing plant: is it really an improvement. *Applied Ergonomics*, 24, 387–396.

Waters, T.R. and Putz-Anderson, V. (1999). Revised NIOSH lifting equation, in W. Karwowski and W. Marras (Eds.), *Occupational Ergonomics*. CRC Press: Boca Raton, 1037–1061.

Waters, T.R., Putz-Anderson, V., Garg, A., and Fine, L.J. (1993). Revised NIOSH equation for the design and evaluation of manual lifting tasks. *Ergonomics*, 36(7), 749–776.

Wei, J. and Salvendy, G. (2003). The utilization of the Purdue cognitive job analysis methodology. *Human Factors and Ergonomics in Manufacturing*, 13(1), 59–84.

Womack, J.P., Jones, D.T., and Roos, D. (1991). *The Machine That Changed the World: The Story of Lean Production*, First Harper Perennial Publishers: New york.

32

Epidemiology of Upper Extremity Disorders

Bradley Evanoff
Washington University School of Medicine

David Rempel
University of California

This chapter summarizes findings from epidemiologic studies that address workplace and individual factors associated with upper extremity musculoskeletal disorders. These disorders are not new: epidemics and clinical case series of work-related upper extremity problems were reported throughout the 1800s and early 1900s (Conn, 1931; Thompson, 1951). Although there are almost no prospective studies in this area, within the last 20 years a number of well-designed, cross-sectional studies have focused on disorders of the hand, wrist, and elbow as related to work. These studies point to the multi-factorial nature of work-related upper extremity disorders. The severity of these disorders is influenced not only by biomechanical factors, but also by other work organizational factors, the worker's perception of the work environment, and medical management.

From an epidemiologic point of view, this topic is problematic because there are many specific disorders that can occur in the hand, arm, and shoulder, ranging from arthritis to nerve entrapments. To complicate the matter further, there are few accepted criteria for case definitions for these many disorders. In their early stages, these disorders usually present with nonspecific symptoms without physical examination or laboratory findings. In fact, the only laboratory tests consistently of value in diagnosing these disorders are nerve conduction studies for nerve entrapment disorders and radiographs for osteoarthritis. Finally, symptoms at the hand or wrist may be due to nerve compression or vascular pathology in the neck or shoulder.

32.1   Frequency, Rates, and Costs

Rates of hand and wrist symptoms and associated disability among working adults were assessed by a 1988 national interview survey of 44,000 randomly selected U.S. adults (National Health Interview Survey) (Park, 1993). Of those who had worked anytime in the past 12 months, 22% reported some finger, hand, or wrist discomfort that fit the category "pain, burning, stiffness, numbness, or tingling" for one or more days in the past 12 months. Only one-quarter were due to an acute injury such as a cut, sprain, or broken bone. Nine percent reported having prolonged hand discomfort that was not due to an acute injury; that is, discomfort of 20 or more days or 7 or more consecutive days during the last 12 months. Of those with prolonged hand discomfort, 6% changed work activities and 5% changed jobs due to the hand discomfort.

Elbow pain and epicondylitis are common in working populations. Symptoms of elbow pain are reported by 7 to 21% of workers in industrial populations (Chiang, 1993; Ohlsson, 1989; Buckle, 1987). Epicondylitis is seen in 0.7 to 2.0% of workers in jobs with low levels of physical demands to the arms and hands, and in 2 to 33% of worker groups with high levels of demands.

In the U.S., hand and wrist disorders account for 55% of all work-related repeated motion disorders reported by U.S. private employers (Bureau of Labor Statistics, 1993). This category excludes low back pain. A similar percentage is also reported in industrial (McCormick, 1990) and other national studies (Kivi, 1984). A similar rise in work-related hand/forearm problems has been observed in other countries such as Finland (Kivi, 1984), Australia (Bammer, 1987), and Japan (Ohara, 1976).

Costs for work-related musculoskeletal disorders are difficult to estimate reliably. Webster and Snook (1994) analyzed 1989 insurance claims data from 45 states, restricting their analysis to upper extremity claims classified as cumulative trauma disorders. They estimated that the total compensable cost for upper extremity cumulative trauma disorders in the U.S. was $563 million in 1989. The National Institute for Occupational Safety and Health has estimated that the annual workers' compensation costs for neck and upper extremity disorders is $2.1 billion, plus $90 million in indirect costs (NIOSH, 1996).

32.2 Disorder Types and their Natural History

Table 32.1 lists the most common workplace hand, wrist, and elbow problems. Nonspecific hand/wrist pain is the most common problem, followed by tendinitis, ganglion cysts, and carpal tunnel syndrome (Silverstein, 1987; McCormack, 1990; Hales, 1994). In many workplace studies, rates of nonspecific symptoms, tendinitis, and CTS appear to track each other, that is, a number of specific disorders typically occur together. For example, in a pork processing plant, the rank order of hand and wrist problems, as a percentage of all morbidity, was: nonspecific hand/wrist pain (39%), CTS (26%), trigger finger (23%), trigger thumb (17%), and DeQuervain's tenosynovitis (17%) (Moore, 1994). Similar ratios of disorders have been observed in manufacturing (Armstrong, 1982; Silverstein, 1986; McCormack, 1990), food processors (Kurppa, 1991; Luopajarvi, 1979), and among computer operators (Hales, 1994; Bernard, 1993).

Tendinitis is the most common specific, work-related hand disorder (McCormack, 1990; Luopajarvi, 1979). For the purposes of this chapter tendinitis will include hand, wrist, and distal forearm tendinitis or tenosynovitis, and trigger finger. Tendinitis occurs at discrete locations; the most common site is the first extensor compartment (De Quervain's Disease), followed by the five other pulley sites on the extensor side of the hand and three on the flexor side. The diagnosis is based on history, symptom location, and palpation and provocative maneuvers on physical exam. There has been no association of tendinitis with age or gender, but work-related tendinitis is higher among workers with less than 3 years of employment (McCormack, 1990).

TABLE 32.1 Examples of Disorders of the Hand, Wrist, and Elbow Observed in Workplace Studies

Non-specific hand and wrist pain	Hand Arm Vibration Syndrome
Tendinitis	Osteoarthritis
Tenosynovitis	Hypothenar hammer syndrome
Finger tendinitis	Gamekeeper's thumb
Wrist tendinitis	Digital neuritis
Stenosing tenosynovitis	Nerve entrapments
Lateral epicondylitis	Carpal Tunnel Syndrome
Medial epicondylitis	Ulnar neuropathy at the wrist
Ganglion cysts	Ulnar neuropathy at the elbow

Lateral epicondylitis is the most common specific elbow disorder; medial epicondylitis is less common. The diagnosis is based on pain and tenderness over the lateral or medial elbow and pain on movement of the wrist or fingers against resistance. Other disorders of the elbow which may be related to occupational activities include olecranon bursitis, triceps tendinitis, and osteoarthritis.

Studies of carpal tunnel syndrome have generated considerable controversy. While there is agreement that this disorder results from compression of the median nerve at the wrist, there are no universally accepted diagnostic criteria for carpal tunnel syndrome. Some consider an abnormal nerve conduction study a gold standard (Katz, 1991; Nathan, 1992; Heller, 1986). However, relying exclusively on nerve conduction studies can lead to reporting very high prevalence rates — 28% (Nathan, 1992) and 19% (Barnhart, 1991) in low-risk working populations. A case definition incorporating typical symptoms and signs has been proposed by NIOSH for surveillance purposes (CDC, 1989); however, the usual signs have relatively poor sensitivities and specificities (Katz, 1991; Heller, 1986; Franzblau, 1993). Therefore, this definition may have limited value in distinguishing CTS from other hand disorders. Hand diagrams completed by patients are reproducible and sensitive, but may lack specificity (Katz, 1990; Franzblau, 1994). Only in the later stages are weakness and thenar atrophy a noticeable feature. In approximately 25% of cases, CTS is accompanied by other disorders of the hand or wrist (Phalen, 1966).

Few studies have evaluated the work-relatedness of osteoarthritis of the hand and wrist (Hadler, 1978; Williams, 1987). Hadler et al. (1978) assessed the hands of 67 workers at a textile plant in Virginia. Significant differences in finger and wrist joint range of motion, joint swelling, and X-ray patterns of degenerative joint disease were observed between three different hand intensive jobs; the observed differences matched the pattern of hand usage.

Hand arm vibration syndrome or Vibration White Finger disease occurs in occupations involving many years of exposure to vibrating hand tools (NIOSH, 1989). This is a disorder of the small vessels and nerves in the fingers and hands presenting as localized blanching at the fingertips with numbness on exposure to cold or vibration. The symptoms are largely self-limited if vibration exposure is eliminated at an early stage (Ekenval, 1987; Futatsuka, 1986).

Hypothenar hammer syndrome or occlusion of the superficial palmar branch of the ulnar artery has been associated in clinical series and case-control studies with habitually using the hand for hammering (Little, 1972; Nilsson, 1989) and with exposure to vibrating hand tools (Kaji, 1993). The mean years of exposure before presentation were 20 to 30 years.

Small case-control studies or clinical series have described factors associated with less common disorders such as Gamekeeper's thumb (Campbell, 1955; Newland, 1992), digital neuritis, and ulnar neuropathy at the wrist (Silverstein, 1986).

32.3   Individual Factors

Some data on individual risk factors, such as age and gender, are available for carpal tunnel syndrome but not for other disorders of the hand and wrist. The risk of CTS increases with age (Stevens, 1988), but in a cross-sectional study of an industrial cohort, age explained only 3% of the variability in median nerve latency (Nathan, 1992). Although CTS is more common among women in the general population, in workplace studies, when employees perform similar hand activities, the ratio of female to male rates is close to 1.2 : 1 (Franklin, 1991; Nathan, 1992; Silverstein, 1986). Certain female specific factors, such as pregnancy (Eckman-Ordeberg, 1987) are clearly associated with pregnancy; however, the role of other female factors such oophorectomy, hysterectomy (Cannon, 1981; Bjorkquist, 1977; de Krom, 1990), or use of oral contraceptives (Sabour, 1970), is less certain. Other individual factors have strong associations with carpal tunnel syndrome based on multiple studies: diabetes mellitus (Phalen, 1966; Yamaguchi, 1965; Stevens, 1987), rheumatoid arthritis (Phalen, 1966; Yamaguchi, 1965; Stevens, 1987), and obesity (Nathan, 1992; DeKrom, 1990; Falck, 1983; Vessey, 1990; Werner, 1994). For some putative risk factors, the associations are based on single studies on studies presenting conflicting results: thyroid disorders (Phalen, 1966; Hales, 1994), vitamin B6 deficiency (Amadio, 1985; Ellis,

TABLE 32.2　Work-Related Factors Associated with Disorders of the Hands and Wrists

Repetition	Mechanical contact
Force	Duration
Posture extremes	Work organization
Vibration	

1982; McCann, 1978), wrist size and shape (Johnson, 1983; Armstrong, 1979; Bleeker, 1985), and general de-conditioning (Nathan, 1988, 1992).

32.4　Work-Related Factors

Table 32.2 summarizes the characteristics of work that have been associated with elevated rates of upper extremity symptoms and specific disorders, including carpal tunnel syndrome and tendinitis. These associations have been observed in multiple studies and in different population groups, while dose–response trends have been seen in several studies. Most studies have been cross-sectional in design, limiting our ability to draw conclusions about causation. The preponderance of evidence, however, suggests strongly that there is a causal relationship between work exposures and upper extremity disorders. Carpal tunnel syndrome and hand–wrist tendinitis have been the best studied; several recent reviews have evaluated the work-relatedness of these disorders and concluded that there is a causal relationship (Stock, 1991; Hagberg, 1992; Kuorinka and Forcier, 1995). Table 32.3, Table 32.4, and Table 32.5 summarize selected studies of wrist and hand tendinitis, carpal tunnel syndrome, and epicondylitis.

Studies using crude measures of exposure have reported associations between repetition and hand/wrist pain and disorders. In a study relying exclusively on nerve conduction measurements, median nerve slowing occurred at a higher rate among assembly line workers than among administrative controls (Nathan, 1992; Hagberg, 1992). Although no systematic assessment of exposure was carried out, the

TABLE 32.3　Controlled Epidemiologic Workplace Studies Evaluating the Association between Work and Wrist, Hand or Distal Forearm Tendinitis[f]

Authors	Exposed Population	Control Population	Rate in Exposed	Rate in Control
Luopajarvi et al., 1979[e]	152 bread packaging	33 shop attendants	53%[a]	14%
Silverstein et al., 1986[b,c]	Industrial	Industrial		
	143 low force/high repetition	136 low force/low repetition	3%	1.5%
	153 high force/low repetition	136 low force/low repetition	4%[a]	1.5%
	142 high force/high repetition	136 low force/low repetition	20%[a]	1.5%
McCormack et al., 1990	Manufacturing	Manufacturing		
	369 packers/folders	352 knitting workers	3.3%[a]	0.9%
	562 sewers	352 knitting workers	4.4%[a]	0.9%
	296 boarding workers	352 knitting workers	6.4%[a]	0.9%
Kurppa et al., 1991[d,e]	102 meat cutters	141 office workers	12.5%	0.9%
	107 sausage makers	197 office workers	16.3%[a]	0.7%
	118 packers	197 office workers	25.3%[a]	0.7%

[a]Significant difference from control.
[b]Adjusted for age, sex, and plant.
[c]Analysis includes other disorders, although tendinitis was most common.
[d]Cohort study with 31-month follow-up.
[e]All exposed and control subjects are female.
[f]Case criteria are based on history and physical examination.
Source: From Rempel, D. and Punnet, L., in *Musculoskeletal Disorders in the Workplace: Principles and Practice*, eds. M. Nordin et al., Mosby-Year Book, Inc., St. Louis, Missouri, 1997. With permission.

TABLE 32.4 Selected Controlled Epidemiologic Workplace Studies Evaluating the Association between Work and Carpal Tunnel Syndrome[e]

Authors	Exposed Population	Control Population	Criteria	Rate in Exposed	Rate in Control
Silverstein et al., 1987[b]	Industrial high force, high repetition	Industrial low force, low repetition	History & physical exam	5.1%[a]	0.6%
Nathan, 1988[c,d]	22 keyboard operators	147 administrative/ clerical	Electrodiagnostic	27%	28%
	164 industrial assembly line	147 administrative/ clerical	Electrodiagnostic	47%[a]	28%
	115 general plant	147 administrative/ clerical	Electrodiagnostic	38%	28%
	23 grinders	147 administrative/ clerical	Electrodiagnostic	61%[a]	28%
Chiang, 1990[b]	Frozen food factory	Frozen food factory	History and signs		
	37 high repetition	49 low repetition & cold		46%	4%
	121 high repetition & cold	49 low repetition & cold		47%[a]	4%
Barnhart, 1991[c]	106 ski manufacturing repetitive jobs	67 ski manufacturing nonrepetitive jobs	Electrodiagnostic and signs	15.4%[a]	3.1%
Osorio, 1994[b,c]	Supermarket workers High exposure	Supermarket workers Low exposure	History & signs Electrodiagnostic	63%[a] 33%[a]	0.0% 0.0%

[a]Significantly different from control group.
[b]Control for age, gender, years on job.
[c]Control for age and gender.
[d]Low participation rate and limited exposure assessment.
[e]Diagnosis based on history and physical exam or nerve conduction study.
Source: From Rempel, D. and Punnet, L., in *Musculoskeletal Disorders in the Workplace: Principles and Practice*, eds. M. Nordin et al., Mosby-Year Book, Inc., St. Louis, Missouri, 1997. With permission.

assembly line work was considered more repetitive than the control group. Rate of persistent wrist and hand pain was higher in garment workers performing repetitive hand tasks than in the control group, hospital employees (Punnett, 1985). Persistent wrist pain, or that lasting most of the day for at least one month in the last year, occurred in 17% of garment workers and 4% of hospital controls, while persistent hand pain occurred in 27% of garment workers and 10% of controls. Others have observed a similar link between high hand/wrist repetition and carpal tunnel syndrome (Chiang, 1990; Barnhart, 1991) and tendinitis (Kurppa, 1991). The link to repetition may be that these are jobs that require high velocity or accelerations of the wrist (Marras and Schoenmarklin, 1993).

Rates of wrist tendinitis among scissors makers was compared to shop attendants in department stores in Finland. Examinations and histories were systematized and performed by one person. The rates between the groups were not significantly different; however, among the scissors makers the rate of tendinitis increased with increasing number of scissors handled (Kuorinka, 1979). Luopajärvi et al. (1979) compared packers in a bread factory to the same control group. The packers' work involved repetitive gripping, up to 25,000 cycles per day, with maximum extension of thumb and fingers to handle wide bread packages. Approximately half of the packers had wrist/hand tenosynovitis compared to 14% among the controls. The most common disorder of the hand or wrist was thumb tenosynovitis followed by finger/wrist extensor tenosynovitis. CTS was diagnosed in four packers and no controls.

The force applied to a tool or materials during repeated or sustained gripping are also predictors of risk for tendinitis and carpal tunnel syndrome. For example, in a study of the textile industry the risk of hand and wrist tendinitis was 3.9 times higher among packaging and folding workers than among knitters (McCormack, 1990). The packing and folding workers were considered to be performing physically demanding work compared to the knitting workers. Armstrong et al. (1979) observed that women with carpal tunnel syndrome applied more pinch force during production sewing than did their

TABLE 32.5 Selected Epidemiologic Workplace Studies Evaluating the Association between Work and Epicondylitis[a]

Authors	Exposed Population	Control Population Criteria	Rate in Exposed	Rate in Controls
Kurppa, 1991[b]	107 female sausage makers	197 female office workers and supervisors	11.1	1.1
	118 female meatpackers	197 female office workers and supervisors	7.0	1.1
	102 male meat cutters	141 male office workers, maintenance men and supervisors	6.4	0.9
Chiang, 1993[c]	28 fish processors with high repetition and high force movements of the arms	61 fish processors without high repetition or high force	21.4%	9.8%
	118 fish processors with high repetition or high force movements of the arms	61 fish processors without high repetition or high force	15.3%	9.8%
Roto and Kivi, 1984[c]	90 male meat cutters	77 male construction foremen	8.9%	1.4%
McCormack, 1990[c]	369 manufacturing workers	352 knitting workers	2.2%	1.4%
	562 manufacturing workers	352 knitting workers	2.1%	1.4%
	468 manufacturing workers	352 knitting workers	1.9%	1.4%
	296 manufacturing workers	352 knitting workers	1.0%	1.4%
Viikari-Juntura, 1991[c]	332 meat plant workers	288 office workers, maintenance workers and supervisors	0.6%	0.5%
Luopajärvi, 1979[c]	152 female packers	133 female shop assistants	2.6%	2.3%

[a]Diagnosis based on history and physical exam.
[b]Prospective cohort study: rates are incidence of epicondylitis per 100 workers/yr.
[c]Cross-sectional study: rates are prevalence of epicondylitis observed in active workers.

job- and sexmatched controls. It is possible that those with carpal tunnel syndrome altered their working style as the carpal tunnel syndrome progressed; however, it is unlikely that they would increase the pinch force because this would also trigger symptoms. In a study by Moore et al. (1994) at a pork processing plant, the jobs that involved high grip force or long grip durations, such as Wizard knife operator, snipper, feeder, scaler, bagger, packer, hanger, and stuffer, affected almost every employee. Others have observed a similar relationship with work involving sustained or high-force grip in grinders (Nathan, 1992), meatpackers and butchers (Kurppa, 1991; Falck, 1983), and other industrial workers (Thompson, 1951; Welch, 1972).

The most comprehensive study of the combined factors of repetition and force was a cross-sectional study of 574 industrial workers by Silverstein et al. (1986, 1987; Armstrong, 1987). Disorders were assessed by physical exam and history and were primarily tendinitis followed by carpal tunnel syndrome, Guyon tunnel syndrome, and digital neuritis. Subjects were classified into four exposure groups based on force and repetition. The "high-force" work was that requiring a grip force on average of more than 4 kg-force, while "low-force" work required less than 1 kg of grip force. The "high-repetition" work involved a repetitive task in which either the cycle time was less than 30 seconds (greater than 900 times in a work day) or more than 50% of cycle time was spent performing the same kind of fundamental movements. The high-risk groups were compared to the low-risk group after adjusting for plant, age, gender, and years on the job. The odds ratio of all hand/wrist disorders for just high force was 4.9, and it increased to 30 for jobs which required both high-force and high-repetition. The identical analysis of just carpal tunnel syndrome revealed an odds ratio of 1.8 for force and 14 for the combined high-force and highrepetition group. A meta-analysis of Silverstein's data and Luopajarvi study concluded that for high-force and high-repetition work the common odds ratio for carpal tunnel syndrome was 15.5 (95% C.I. 1.7–141) and for hand/wrist tendinitis it was

9.1 (95% C.I. 5–16) (Stock, 1991). Estimates of the percentage of CTS cases among workers who perform repetitive or forceful hand activity that can be attributed to work range from 50 to 90% (Hagberg, 1992; Cummins, 1992; Tenaka, 1994).

With regard to epicondylitis, the individual roles played by force and repetition are less clear. One cohort study and six cross-sectional studies have evaluated the incidence or prevalence of epicondylitis in relation to specific jobs, which were characterized by high force, high repetition, or both. Kurppa (1991) found a relative risk of 6.4 for epicondylitis in jobs with high repetition, some of which also involved high force. One cross-sectional study found a significantly elevated risk of epicondylitis only among recently employed workers in high-repetition or high-repetition/high-force jobs (Chiang, 1993). Another cross-sectional study found an odds ratio of 6.9 epicondylitis in a high-repetition, high-force job (Roto and Kivi, 1984). This odds ratio was not statistically significant. Four other cross-sectional studies found little or no increase in risk for epicondylitis in workers involved in jobs characterized by high force and high repetition. (McCormack, 1990; Viikari-Juntura, 1991; Luopajärvi, 1979; Dimberg, 1987.)

Work involving increased wrist deviation from a neutral posture in either the extension, flexion or ulnar, radial direction has been associated with carpal tunnel syndrome and other hand and wrist problems (Thompson, 1951; Hoffman, 1981; Tichauer, 1966). De Krom et al. (1990) conducted a casecontrol study of 156 subjects with carpal tunnel syndrome compared to 473 controls randomly sampled from the hospital and population registers in a region of the Netherlands. After adjusting for age and sex, a dose–response relationship was observed for increasing hours of work with the wrist in extension or flexion. No risk was observed for increasing hours performing a pinch grasp or typing. Some studies of computer operators have linked awkward wrist postures to severity of hand symptoms (Faucett, 1994), risk of tendinitis or carpal tunnel syndrome (Seligman, 1986), arm and hand discomfort (Sauter, 1991; Duncan, 1974; Hunting, 1981).

Prolonged exposure to vibrating hand tools, such as chain saws, has been linked in prospective studies to Hand Arm Vibration Syndrome (Ekenval, 1987; Futatsuka, 1986). The risks are primarily vibration acceleration amplitude, frequency, hand coupling to tool, hours per day of exposure, and years of exposure. However, based on existing studies, there is no clear vibration acceleration/frequency/duration threshold that would protect most workers. Therefore, medical surveillance is recommended to identify cases early while the disease can still be reversed (NIOSH, 1989). Use of vibrating hand tools may also increase the risk of CTS (Seppalainen, 1970; Cannon, 1981; Rothfleisch, 1978) indirectly by increasing applied grip force through a reflex pathway (Radwin, 1987).

Prolonged or high-load localized mechanical stress over tendons or nerves from tools or resting the hand on hard objects have been associated with tendinitis (Tichauer, 1966) and nerve entrapments (Phalen, 1966; Hoffman, 1985) in case studies.

The average total hours per day that a task is repeated or sustained has been a factor in predicting hand problems (Margolis, 1987; Macdonald, 1988). Among computer operators increasing self-reported hours of computer use has been a predictor of symptom intensity or disorder rate in all (Faucett, 1994; Burt, Bernard, 1993; Oxenburgh, 1985; Maeda, 1982; Hunting, 1981). DeKrom et al. (1990) did not observe a relationship between CTS and hours of computer use.

Work organizational (work structure, decision control, work load, deadline work, supervision) and psychosocial factors (job satisfaction, social support, relationship with supervisor) appear to have some influence on hand and wrist symptoms among computer users. Among newspaper reporters and editors, work organizational factors modified the expected relationship between workstation design and hand and wrist symptoms. Symptom intensity increased as keyboard height increased among those with low decision latitude but not among those with high decision latitude (Faucett, 1994). In another study of newspaper employees, the risk of hand and wrist symptoms was increased among those with increasing hours on deadline work and less support from the immediate supervisor (Bernard, 1993). Among directory assistance operators at a telephone company, high information processing demands were associated with an elevated rate of hand and wrist disorders (Hales, 1994). On the other hand, in the industrial setting, Silverstein et al. (1986) observed no effect on job satisfaction.

32.5 Summary

The lack of prospective studies and an uncertainty about the precise pathophysiologic mechanisms involved limits our ability to definitively identify causative factors. Nonetheless, current studies point to a multifactorial relationship between work exposures and disorders of the hand, wrist, and elbow. Symptom severity and disorder rate appear to be influenced by work organizational factors, such as decision latitude and cognitive demands. Some disorders, such as tendinitis and carpal tunnel syndrome, are clearly associated with work involving repetitive and forceful use of the hands. It seems likely that there is a causal relationship between some work exposures and these disorders. For other disorders, such as epicondylitis and osteoarthritis, the relationship to work exposures is less clear, although current data are suggestive. Carpal tunnel syndrome has been linked to individual factors in population-based studies and in clinical case series. However, in workplace studies where workplace exposures are adequately quantified, individual factors play a limited role relative to workplace factors (Cannon, 1981; Silverstein, 1987; Armstrong, 1979; Franklin, 1991; Faucett, 1994; Hales, 1994).

References

Adams ML, Franklin GM, Barnhart S. Outcome of carpal tunnel surgery in Washington State workers' compensation. *Am J Ind Med* 1994; 25:527–536.

Al-Qattan MM, Bowen V, Manktelow RT. Factors associated with poor outcome following primary carpal tunnel release in non-diabetic patients. *J Hand Surg* (Br Volume) 1994; 19B:622–625.

Amadio PC. Pyridoxine as an adjunct in the treatment of carpal tunnel syndrome. *J Hand Surg* 1985; 10A:237–241.

Armstrong TJ, Langolf GD. Ergonomics and occupational safety and health, in *Environmental and Occupational Medicine*, ed WN Rom, Little, Brown Co, Boston, 1982, pp. 765–784.

Armstrong TJ, Chaffin DB. Carpal tunnel syndrome and selected personal attributes. *J Occup Med* 1979; 21:481–486.

Armstrong TJ, Foulke JA, Joseph BS, Goldstein SA. Investigation of cumulative trauma disorders in a poultry processing plant. *Am Ind Hygiene Assoc J* 1982; (43)2:103–116.

Armstrong TJ, Buckle P, Fine LJ et al. A conceptual model for work-related neck and upper-limb musculoskeletal disorders. *Scand J Work Environ Health* 1993; 19:73–84.

Bammer G. VDUs and musculoskeletal problems at the Australian National University, in *Work with Display Units 86*, Eds Knave B and Wideback PG, Elsevier Science Publishers B.V. North-Holland, 1987.

Barnhart S, Demers PA, Miller M, Longstreth WE, Rosenstock L. Carpal tunnel syndrome among ski manufacturing workers. *Scand J Work Environ Health* 1991; 17:46–52.

Bernard B, Sauter S, Peterson M, Fine L, Hales T. Health Hazard Evaluation Report: Los Angeles Times. U.S. Department of Health and Human Services, Public Health Service, Centers for Disease Control, National Institute for Occupational Safety and Health, NIOSH Report No. 90–013–2277. 1993.

Birkbeck MQ, Beer TC. Occupation in relation to the carpal tunnel syndrome. *Rheumatol Rehabil* 1975; 14:218–221.

Bjorkqvist SE, Lang AH, Punnonen R, Rauramo L. Carpal tunnel syndrome in ovariectomized women. *Acta Obstet Gynecol Scand* 1977; 56:127–130.

Bleeker MQ, Bohlman M, Moreland R, Tipton A. Carpal tunnel syndrome: role of carpal canal size. *Neurology* 1985; 35:1599–1604.

Burt S, Hornung R, Fine L, Silverstein B, Armstrong T. Health hazard evaluation report: Newsday. U.S. Department of Health and Human Services, Public Health Service, Centers for Disease Control, National Institute for Occupational Safety and Health, NIOSH Report No. 89–250–2046. 1990.

Campbell, CS. Gamekeeper's Thumb. *Journal of Bone and Joint Surgery* 1955; 37(B) 1:148–149.

Cannon LJ, Bernacki EJ, Walter SD. Personal and occupational factors associated with carpal tunnel syndrome. *J Occup Med* 1981; 23:255–258.

Centers for Disease Control: Occupational disease surveillance — carpal tunnel syndrome. *MMWR* 1989; 38:485–489.

Cheadle A, Franklin G, Wolfhagen C, Savarino J, Liu PY, Salley C, Weaver M. Factors influencing the duration of work-related disability: a population-based study of Washington State Workers' Compensation. *Am J Pub Health* 1994; 84:190–196.

Chiang HC, Chen SS, Yu HS, Ko YC. The occurrence of carpal tunnel syndrome in frozen food factory employees. *Kaohsiung J Med Sci* 1990; 6:73–80.

Chiang HC, Ko YC, Chen SS, Yu HS, Wu TN, Chang PY. Prevalence of shoulder and upper-limb disorders among workers in the fish-processing industry. *Scand J Work Environ Health* 1993; 19:126–131.

Conn HR. Tenosynovitis. *Ohio State Med J* 1931; 27:713–716.

de Krom M, Kester A, Knipschild P, Spaans F. Risk factors for carpal tunnel syndrome. *Am J Epi* 1990, 132:1102–1110.

Duncan J, Ferguson D. Keyboard operating posture and symptoms in operating. *Ergonomics* 1974; 17:651–662.

Ekenvall L, Carlsson A. Vibration white finger: a follow up study. *Br J Ind Med* 1987; 44:476–478.

Ekman-Ordeberg G, Salgeback S, Ordeberg G. Carpal tunnel syndrome in pregnancy: A prospective study. *Acta Obstet Gynec Scand* 1987; 66:233–235.

Ellis J, Folkers K, Watanabe T et al. Clinical results of a cross-over treatment with pyridoxine and placebo of the carpal tunnel syndrome. *J Clin Nutr* 1979; 2046–2070.

Falck B and Aarnio P. Left-sided carpal tunnel syndrome in butchers. *Scand J Work Environ Health* 1983; 9:291–297.

Faucett J and Rempel D. VDT-related musculoskeletal symptoms: Interactions between work posture and psychosocial work factors. *Am J Ind Med* 1994; 26:597–612.

Fine LJ, Silverstein BA, Armstrong TJ et al. Detection of cumulative trauma disorders of upper extremities in the workplace. *J Occup Med* 1986; 28:674–678.

Franklin GM, Haug J, Heyer N, Checkoway H, Peck N. Occupational carpal tunnel syndrome in Washington State, 1984–1988. *Am J Pub Health* 1991, 81:741–746.

Franzblau A, Werner R, Valle J, Johnston E. Workplace surveillance for carpal tunnel syndrome: a comparison of methods. *J Occup Rehab* 1993; 3:1–14.

Franzblau A, Werner RA, Albers JW, Grant CL, Olinski D, Johnston E. Workplace surveillance for carpal tunnel syndrome using hand diagrams. *J Occup Rehab* 1994; 4:185–198.

Futatsuka M, Ueno T. A follow-up study of vibration-induced white finger due to chain-saw operation. *Scand J Work Environ Health* 1986; 12:304–306.

Hadler N, Gillings D, Imbus H et al. Hand structure and function in an industrial setting. *Arthritis and Rheum* 1978, 21:210–220.

Hagberg M, Morgenstern H, Kelsh M. Impact of occupations and job tasks on the prevalence of carpal tunnel syndrome. *Scand J Work Environ Health* 1992; 18:337–345.

Hales TR, Sauter SL, Peterson MR, Fine LJ, Putz-Anderson V, Schleifer LR, Ochs TT, Bernard BP. Musculoskeletal disorders among visual display terminal users in a telecommunications company. *Ergonomics* 1994; 10:1603–1621.

Heller L, Ring H, Costeff H, Solzi. Evaluation of Tinel's and Phalen's signs in diagnosis of carpal tunnel syndrome. *Eur Neurol* 1986; 25:40–42.

Hoffman J, Hoffman PL. Staple gun carpal tunnel syndrome. *J Occup Med* 1985; 27:848–849.

Hünting W, Läubli T, Grandjean E. Postural and visual loads at VDT workplaces. *Ergonomics* 1981; 24:917–931.

Johnson EW, Gatens T, Poindexter D, Bowers D. Wrist dimensions: correlation with median sensory latencies. *Arch Phys Med Rehab* 1983; 64:556–557.

Katz JN, Stirrat CR, Larson MG, Fossel AN, Eaton HM, Liang MH. A self-administered hand symptom diagram for the diagnosis and epidemiologic study of carpal tunnel syndrome. *J Rheumatol* 1990; 3:1–14.

Katz JN, Larson MG, Fossel AH, Liang MH. Validation of a surveillance case definition of carpal tunnel syndrome. *Am J Public Health* 1991; 81:189–193.

Kaji H, Honma H, Usui M, Yasuno Y, Saito K. Hypothenar Hammer Syndrome in workers occupationally exposed to vibrating tools. *J Hand Surg* (Br Volume) 1993; 18B:761–766.

Kivi P. Rheumatic disorders of the upper limbs associated with repetitive occupational tasks in Finald in 1975–1979. *Scand J Rheum* 1984; 13:101–107.

Kuorinka I, Koskinen P. Occupational rheumatic diseases and upper limb strain in manual jobs in a light mechanical industry. *Scand J Work Environ Health* 1979, 5:39–47.

Kourinka I, Forcier L. (eds.) *Work Related Musculoskeletal Disorders: A Reference Book for Prevention.* London: Taylor & Francis, 1995.

Kurppa K, Viikari-Juntura E, Kuosma E, Huuskonen M, Kivi P. Incidence of tenosynovitis or peri-tendinitis and epicondylitis in a meat processing factory. *Scand J Work Environ Health* 1991; 17:32–37.

Little JM, Ferguson DA. The incidence of Hypothenar Hammer Syndrome. *Arch Surg* 1972; 105:684–685.

Luopajarvi T, Kuorinka I, Virolainen M, Holmberg M. Prevalence of tenosynovitis and other injuries of the upper extremities in repetitive work. *Scand J Work Environ Health* 1979. 5:48–55.

Marras WS, Schoenmarklin RW. Wrist motions in industry. *Ergonomics* 1993; 36:341–351.

Maeda K, Hunting W, Grandjean E. Factor analysis of localized fatigue complaints of accounting-machine operators. *J Human Ergol* 1982; 11:37–43.

Magnusson M, Ortengren R. Investigation of optimal table height and surface angle in meatcutting. *Applied Ergonomics* 1987; 18.2:146–152.

Masear VR, Hayes JM, Hyde AG. An industrial cause of carpal tunnel syndrome. *J Hand Surg* 1986; 11A:222–227.

McCormack RR Jr., Inman RD, Wells A, Berntsen C, Imbus HR. Prevalence of tendinitis and related disorders of the upper extremity in a manufacturing workforce. *J Rheumatol* 1990; 17:958–964.

Moore JS, Garg A. Upper extremity disorders in a pork processing plant: relationships between job risk factors and morbidity. *Am Ind Hyg Assoc J* 1994, 55:703–715.

Muffly-Elsey D, Flinn-Wagner S. Proposed screening tool for the detection of cumulative trauma disorders of the upper extremity. *J Hand Surg* 1987, 12A:2(2), 931–935.

Nathan PA, Keniston RC, Myers LD, Meadows KD. Obesity as a risk factor for slowing of sensory conduction of the median nerve in industry. *J Occup Med* 1992; 34:379–383.

Nathan PA, Meadows KD, Doyle LS. Occupation as a risk factor for impaired sensory conduction of the median nerve at the carpal tunnel. *J Hand Surg* 1988; 13B:167–170.

NIOSH Criteria for a recommended standard. Occupational exposure to hand-arm vibration. DDHS Publication No. 89–106. 1989. National Institute for Occupational Safety and Health. Cincinnati, Ohio.

NIOSH National Occupational Research Agenda. DDHS Publication No. 96–1115. 1996. National Institute for Occupational Safety and Health. Cincinnati, Ohio.

Newland, CC. Gamekeeper's Thumb. Orthopedic Clinics of North America 1992; 23(1):41–48.

Nilsson T, Burström L, Hagberg M. Risk assessment of vibration exposure and white fingers among platers. *Int Arch Occup Environ Health* 1989; 61:473–481.

Ohara H, Aoyama H, Itani T. Health hazards among cash register operators and the effects of improved working conditions. *J Human Ergology* 1976; 5:31–40.

Osorio AM, Ames RG, Jones JR, Rempel D, Castorina J, Estrin W, Thompson D. Carpal tunnel syndrome among grocery store workers. *Am J Ind Med* 1994, 25:229–245.

Oxenburgh M, Rowe S, Douglas D. Repetitive strain injury in keyboard operators. *J Occup Health and Safety — Australia and New Zealand* 1985; 1:106–112.

Park CH, Wagener DK, Winn DM, Pierce JP. Health conditions among the currently employed: United States, 1988. National Center for Health Statistics. *Vital Health Stat* 1993. 10(186).

Phalen GS. The carpal-tunnel syndrome. *J Bone Joint Surg* 1966; 48A:211–228.

Punnett L, Robins JM, Wegmen DH, Keyserling WM. Soft tissue disorders in the upper limbs of female garment workers. *Scand J Work Environ Health* 1985; 11:417–425.

Radwin RG, Armstrong TJ, Chaffin DB. Power hand tool vibration effects on grip exertions. *Ergonomics* 1987; 30:833–855.

Rempel D, Punnet L. Epidemiology of wrist and hand disorders, in *Musculoskeletal Disorders in the Workplace: Principles and Practice*, eds M Nordin et al., Mosby-Year Book, Inc., St. Louis, Missouri, 1997.

Roto P, Kivi P. Prevalence of epicondylitis and tenosynovitis among meatcutters. *Scand J Work Environ Health* 1984; 10:203–205.

Sauter SL, Schleifer LM, Knutson SJ. Work posture, workstation design, and musculoskeletal discomfort in a VDT data entry task. *Human Factors* 1991; 33:151–167.

Seligman P, Boiano J, Anderson C. Health Hazard Evaluation of the Minneapolis Police Department. NIOSH HETA 84-417–1745, 1986. U.S. Department of Commerce, NTIS, Springfield, Virginia.

Seppalainen AM. Nerve conduction in the vibration syndrome. *Scand J Work Environ Health* 1970; 7:82–84.

Silverstein BA, Armstrong T, Longmate A, Woody D. Can in-plant exercise control musculoskeletal symptoms? *J Occup Med* 1988; 30:922–927.

Silverstein BA, Fine LJ, Armstrong TJ. Hand wrist cumulative trauma disorders in industry. *Br J Ind Med* 1986. 43:779–784.

Silverstein BA, Fine LJ, Armstrong TJ. Occupational factors and carpal tunnel syndrome. *Am J Ind Med* 1987. 11:343–358.

Silverstein BA, Fine LJ, Stetson D. Hand-wrist disorders among investment casting plant workers. *J Hand Surg* 1987; 12A(5 part 2):838–844.

Stevens JC, Sun S, Beard CM, O'Fallon WM, Kurland LT. Carpal tunnel syndrome in Rochester, Minnesota. 1961 to 1980. *Neurology* 1988; 38:134–138.

Stock SR. Workplace ergonomic factors and the development of musculoskeletal disorders of the neck and upper limbs: A meta-analysis. *Am J Ind Med* 1991; 19:87–107.

Tanaka S, Wild DK, Seligman PJ, Behrens V, Cameron L, Putz-Anderson V. The U.S. prevalence of self-reported carpal tunnel syndrome. *Am J Public Health* 1994; 84:1846–1848.

Thompson A, Plewes L, Shaw E. Peritendinitis crepitans and simple tenosynovitis: a clinical study of 544 cases in industry. *Br J Ind Med* 1951; 8:150–160.

Tichauer E. Some aspects of stress on forearm and hand in industry. *J Occup Med* 1966; 8:63–71.

Viikari-Juntura E. Neck and upperlimb disorders among slaughterhouse workers. *Scand J Work Environ Health*, 1983; 9:283–290.

Vessy MP, Villard-MacIntosh L, Yeates D. Epidemiology of carpal tunnel syndrome in women of childbearing age. Finding in a large cohort study. *Am J Epi* 1990; 19:655–659.

Webster BS, Snook SH. The cost of compensable upper extremity cumulative trauma disorders. *J Occup Med* 1994; 36:713–717.

Welch R. The causes of tenosynovitis in industry. *Indust Med* 1972; 41:16–19.

Werner RA, Albers JW, Franzblau A, Armstrong TJ. The relationship between body mass index and the diagnosis of carpal tunnel syndrome. *Muscle & Nerve* 1994; 17:632–636.

Williams, Cope et al. Metacarpo-phalangeal arthropathy associated with manual labor (Missouri metacarpal syndrome). *Arthritis and Rheum* 1987; 30:1362–1371.

Yamaguchi D, Liscomb P, Soule E. Carpal tunnel syndrome. *Minn Med J* 1965; January 22–23.

VII

Medical Management

33

Medical Management of Work-Related Musculoskeletal Disorders

Thomas Hales
National Institute for Occupational Safety and Health

Patricia Bertsche
Ross Laboratories

33.1   Introduction

The Bureau of Labor Statistics (BLS) reports that in 1994 nearly two thirds of the workplace illnesses were disorders associated with repeated trauma (one category of musculoskeletal disorders) (BLS 1995). These figures do not include low back disorders associated with overexertion, which accounted for 380,000 lost time cases in 1993. The number of repeated trauma cases reported in 1994 was 332,000, a 10% increase from the 1993 figure. In fact, since 1982, the number of reported disorders associated with repeated trauma has been increasing each year (BLS 1995). Not surprisingly, many health care providers (HCPs) find evaluating and treating these employees consumes an increasing proportion of their time and energy.

To prevent or reduce symptoms, signs, impairment, or disability associated with work-related musculoskeletal disorders (WRMSDs), employers, in collaboration with HCPs, should develop a medical management program which is outlined in Figure 33.1. This chapter provides assistance to employers setting up a medical management program and to HCPs managing these cases in two ways — first, by outlining the general principles and listing the components of a program needed to adequately evaluate and treat affected employees; second, by providing HCPs with practical guidance and forms to collect the appropriate information. These forms can then be incorporated into the employee's medical record.

33.2 Terminology

Before addressing the various components of a medical management program, the term musculoskeletal disorder must be defined. MSDs are disorders of the muscles, tendons, peripheral nerves, or vascular system not directly resulting from an acute or instantaneous event (e.g., slips or falls). These disorders are considered to be work-related when the work environment and the performance of work contribute significantly, but as one of a number of factors, to the causation of a multifactorial disease (WHO 1985). Physical risk factors that cause or aggravate MSDs and that may be present at the workplace include, but are not limited to: repetitive, forceful, or prolonged exertions; frequent or heavy lifting; pushing, pulling, or carrying of heavy objects; fixed or awkward work postures; contact stress; localized

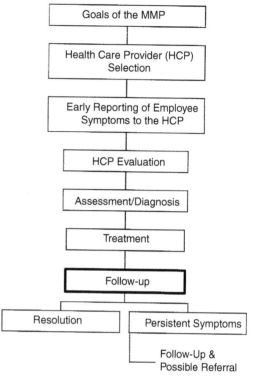

FIGURE 33.1 Overview of a medical management program (MMP).

or whole-body vibration; cold temperatures; and poor lighting leading to awkward postures. These workplace risk factors can be intensified by work organization characteristics, such as inadequate work–rest cycles, excessive work pace and/or duration, unaccustomed work, lack of task variability, machine-paced work, and piece rate.

33.3 Selection of a Health Care Provider (HCP)

An HCP is a practitioner operating within the scope of his or her license, registration, certification, or legally authorized practice. The evaluation and treatment of employees with WRMSDs should be performed by an HCP with experience and training in managing these disorders. Many HCPs are capable of providing these services, including physicians, occupational health nurses, physical therapists, occupational therapists, and hand therapists. Employers and employees may be more familiar with the services of physicians, therefore Table 33.1 provides information regarding some of the other HCPs who might be directly providing the care or coordinating the care of employees with WRMSDs. Considerations for the employer to use in selecting an HCP include:

- Specialized training and experience in ergonomics and the treatment of work-related musculoskeletal disorders
- Current working knowledge of the worksite and the specific industry
- Willingness to periodically tour the worksite
- Willingness to communicate with the employer and employees (Louis, 1987; Haig et al., 1990)
- Experience in the case management of work-related musculoskeletal disorders
- Willingness to consider conservative therapy prior to surgery
- History of successful treatment of work-related musculoskeletal disorders

TABLE 33.1 Non-Physician Health Care Providers Who Might Be Involved in the Medical Management of Work-Related Musculoskeletal Disorders (Not Intended to Be All-Inclusive)

Profession	Scope of Practice	Training/Experience	Services They Provide
Occupational Health Nurse (OHN)	An OHN is a Registered Nurse (RN), independent licensure with scope defined by individual state boards of nursing; certification is voluntary (COHN); Advanced practice nurses (nurse practitioners) treat independently or provide medical treatment with protocol depending on requirements of state licensing board. RNs refer to physicians and other health care providers when treatment beyond their scope of practice is required.	Basic education includes complete assessment (history and physical examination) of all body systems; OHNs have academic and continuing education in assessment of the musculoskeletal and nervous systems and diagnosis, treatment, and rehabilitation of work-related musculoskeletal disorders.	Assessment, treatment of common work-related musculoskeletal disorders, particularly in early stages (under protocol when required by state statute), referral to other appropriate health care providers as needed, and rehabilitation including case management; Preventive services include trend analysis, education and training, and involvement in the job improvement process including job analysis.
Occupational Therapist (OT)	49 states, the District of Columbia, Guam, and Puerto Rico have laws regulating the profession; The American Occupational Therapy Certification Board's national certification exam is a basic requirement in the states/jurisdictions that license or certify OTs. Generally, an OT may independently provide services, however, in certain states, occupational therapy laws/regulations require physician referral for services for specific medical conditions.	OTs have either a bachelor's or master's degree and pursue continuing education and extensive on-the-job training to specialize in work-related musculoskeletal disabilities; OTs have a comprehensive background in the biological and behavioral sciences; knowledge and application of the components of human performance including psychosocial, neurological, cognitive, perceptual, and motor function.	OTs use standardized tests, observational skills, activities and tasks designed to evaluate specific work-related skills, functional abilities, physical abilities, and behaviors. Examples of assessments include: functional capacity evaluation, physical capacity testing, examination of essential functions of a job. Other services include work hardening and involvement in the job improvement process such as job analysis and workstation and tool modification.

(*continued*)

TABLE 33.1 *Continued*

Profession	Scope of Practice	Training/Experience	Services They Provide
Physical Therapist (PT)	PTs licensed in all states, the District of Columbia, Puerto Rico, and the U.S. Virgin Islands; Direct physician oversight is not required. Of the 53 jurisdictions, 44 permit physical therapy evaluation without physician referral.	PTs have either a bachelor's or graduate degree and pursue continuing education to specialize in prevention and rehabilitation of work-related musculoskeletal disorders. PTs' basic education includes courses in anthropometrics, biomechanics, ergonomic interventions, kinesiology, movement and posture analysis, the components of human psychophysical performance, orthotic prescription, fabrication, and application of supportive devices.	PTs evaluate a variety of conditions such as abnormalities of body alignment and movement patterns; impaired motor function and learning; impaired sensation; limitations of joint motion; muscle weakness; and pain. PTs perform tests and measures such as batteries of work performance; assessment of work hardening or conditioning; determination of dynamic capabilities and limitations during specific work activities. Involvement in the job improvement process including analysis of jobs or activities, and workstation or tool modifications.
Hand Therapist (HT)	A Hand Therapist is either an OT or PT who voluntarily becomes certified by the Hand Therapy Certification Commission. Certified HTs specialize in upper extremity rehabilitation.	HTs have specialized training and experience in assessment and rehabilitation of work-related musculoskeletal disorders.	Services include diagnostic work up of quantitative sensory testing to determine peripheral neuropathy, grip strength, and motor testing to determine the localization of muscular tenderness areas of inflammation; physical or functional capacity evaluations. HTs apply treatments such as thermotherapy, ultrasound, and electric stimulation; reeducation home exercise programs, splintage, pain management, soft tissue mobilization and myofascial release. HTs are skilled in work task analysis and therefore are well suited for involvement in the job improvement process.

33.4 Early Reporting of Symptoms and Access to Health Care Providers

The case management process begins with an employee informing his or her employer of the presence of musculoskeletal symptoms or signs. Generally, the earlier that symptoms are identified, an evaluation completed, and treatment initiated, the likelihood of a significant disorder developing is reduced. Early treatment of many MSDs has been shown to reduce their severity, duration of treatment, and ultimate disability (Haig et al., 1990; Wood, 1987; Wiesel et al., 1984; Mayer et al., 1987). There can be various workplace situations influencing an employee's decision to report symptoms. These situations can result in employees over-reporting, or under-reporting, symptoms. In either case, to prevent severe disorders from occurring, employees must not be subject to reprisals or discrimination based on reporting symptoms to their supervisors.

Supervisors and foremen are not trained to evaluate and assess MSDs. To prevent supervisors or other plant personnel from performing triage, employees reporting persistent musculoskeletal symptoms (e.g., symptoms lasting seven days from onset, or symptoms that interfere with the employee's ability to perform the job) should have the opportunity for a prompt HCP evaluation. If an HCP is available at the workplace, this initial assessment should be offered when the employee reports symptoms or at least within two days. If the HCP is offsite, the employer should make available an assessment to the employee promptly, but no later than a week after the signs or symptoms are reported. This is not meant to imply that employers should wait seven days from onset of all employee's symptoms before referring the employee to an HCP. There are foreseeable circumstances where immediate evaluation by an HCP would be warranted. For example, an employee who reports to the supervisor that he/she is experiencing severe low back pain with numbness and tingling radiating down his/her leg, an inability to sleep due to the pain, and obvious difficulty walking should immediately be referred to the HCP.

33.5 Health Care Providers Being Familiar with the Employee's Job

HCPs who evaluate employees, determine an employee's functional capabilities, and prepare opinions regarding work-relatedness and work-readiness, must be familiar with employee jobs and job tasks. Being familiar with employee jobs not only assists HCPs in making informed case management decisions, but also demonstrates to employers and employees the importance HCPs place on making informed decisions, assists with the identification of workplace hazards that cause or aggravate MSDs, assists with the identification of alternate duty jobs, and can help establish the proper diagnosis for the employee's condition.

Critical to this process is open lines of communication with the employer, employee, and the HCP. The employer should appoint a contact person who is familiar with plant jobs and workplace risk factors to communicate and coordinate with the HCP. In addition, HCPs should perform a plant walk-through. Once familiar with plant operations and job tasks, the HCP can periodically revisit the facility to remain knowledgeable about working conditions. Other approaches to become familiar with jobs and job tasks include review of job analysis reports, job surveys or risk factor checklists, detailed job descriptions, job safety analyses, photographs and videotapes accompanied by narrative or written descriptions, and interviewing the employee.

33.6 Evaluation of the Employee

The HCP evaluation of the symptomatic employee should contain a relevant occupational and health history, a physical examination, laboratory tests appropriate to the reported signs or symptoms, and conclude with an initial assessment/diagnosis. If the HCP providing the initial evaluation does not have the training or experience to make a preliminary assessment or diagnosis, the employee should be referred to an HCP with such training and experience. The content of the evaluation is outlined below with a recording form available (see Form 1).

1. Characterize the symptoms and history
   - Onset (date; circumstance; abrupt vs. gradual, etc.)
   - Duration and frequency
   - Quality (pain; tingling; numbness; swelling; tenderness, etc.)
   - Intensity (mild; moderate; severe; other rating scales)
   - Location
   - Radiation
   - Exacerbating and relieving factors or activities (both on-the-job and off-the-job)
   - Prior treatments
2. Relevant considerations:
   - Demographics (e.g., age; gender; hand dominance)
   - Past medical history (e.g., prior injuries or disorders related to the affected body part)
   - Recreational activities, hobbies, household activities
   - Occupational history with emphasis on the (a) job the employee was performing when the symptoms were first noticed, (b) prior job if the employee recently changed jobs, (c) amount of time spent on that job, and (d) whether the employee was working any other "moonlighting" or part-time jobs.
3. Characterize the job:

   Becoming familiar with an employee's job is a critical component of the HCP evaluation and treatment process. In addition to collecting the information from the plant contact person and plant walk-through (described above), employees should be interviewed regarding their work activities. The employee should be asked to describe their required job tasks with respect to known workplace risk factors for MSDs and the duration of exposure such as hours per day, days per week and shift work. Workplace risk factors for MSDs include repetitive, forceful, or prolonged exertions; frequent or heavy lifting or lifting in awkward postures (e.g., twisting, trunk flexion, or lateral bending); pushing, pulling, or carrying of heavy objects; fixed or awkward work postures; contact stress; localized or whole-body vibration; cold temperatures; and others. The employee should also be asked if there has been any recent changes in their job, such as longer hours, increased pace, new tasks or equipment, or new work methods which may have caused or contributed to the current illness.
4. Physical examination:

   The physical examination should be targeted to the presenting symptoms and history. Components of the exam include inspection (redness, swelling, deformities, atrophy, etc.), range of motion, palpation, sensory and motor function (including functional assessment), and appropriate maneuvers (e.g., Finkelstein's). It is important to note that clinical examinations may not identify the specific structure affected, nor find classic signs of inflammation (e.g., redness, warmth, swelling). This should not be surprising since the role of inflammation in the pathophysiology of these disorders is unclear (Nirschl 1990). For further information on the content of an appropriate exam, or the technique to perform the exam, please consult the following references: (AHCPR, 1994; ASSH, 1990; Hoppenfield, 1976; Tubiana et al., 1984).
5. Assessment and diagnosis:

   For each employee referred for an assessment, the HCP should make a specific diagnosis consistent with the current International Classification of Diseases, or the HCP should summarize the findings of his or her assessment. Terms such as repetitive motion disorders (RMDs), repetitive strain injury (RSI), overuse syndrome, cumulative trauma disorders (CTDs), and work-related musculoskeletal disorder (WRMSD) are not ICD diagnoses and, although useful as general terms, should not be used as medical diagnoses. Given the difficulty in establishing the specific structure affected, many diagnoses should describe the anatomic location of the symptoms without a specific structure diagnosis (e.g., unspecified neck symptoms or disorders should be listed as ICD-9 723.9; unspecified disorders of the soft tissues should be listed as ICD-9 729.9). When a specific anatomical structure can be ascertained, most of these conditions involve the muscles or tendons (unspecified disorders of muscle, ligament, and fascia should be listed at ICD-9 728.9; unspecified disorders of synovium, tendon, and bursa should be listed as ICD-8 727.9). Table 33.2 provides a listing of ICD-9 codes.

The HCP should assist in determining whether occupational risk factors are suspected to have caused, contributed to, or exacerbated the condition. Factors helpful in making this determination are:

- Is the medical condition known to be associated with work?
- Does the job involve risk factors (based on job surveys or job analysis information) associated with the presenting symptoms?
- Is the employee's degree of exposure consistent with those reported in the literature?
- Are there other relevant considerations (e.g., unaccustomed work, overtime, etc.)?

33.7 Treatment of the Employee

Before initiating treatment, the HCP should document the specific treatment goals (e.g., symptom resolution or restoring of functional capacity), expected duration of treatment, dates for follow-up evaluations, and time frames for achieving the treatment goals. Resting the symptomatic area, and treatment of soft tissue and tendon disorders are the mainstays of conservative treatment. Despite the wide application of some therapeutic modalities, many are untested in controlled clinical trials.

33.7.1 Resting the Symptomatic Area

Reducing or eliminating employee exposure to musculoskeletal risk factors through engineering and administrative controls in the workplace is the most effective way to rest the symptomatic area while allowing employees to remain productive members of the workforce (Upfal 1994). Until effective controls are installed, employee exposure to workplace risk factors can be reduced through restricted duty and temporary job transfer. The specific amount of work reduction for employees on restricted duty must be individualized; however, the following principles apply: the degree of restriction should be proportional to the condition severity and to the frequency and duration of exposure to relevant risk factors involved in the original job. HCPs are responsible for determining the physical capabilities and work restrictions of the affected worker. The employer is responsible for finding a job consistent with these temporary restrictions. The employer's contact person (who is knowledgeable about the employee's job requirements and their associated risk factors) is critically important to this process. The contact person should communicate and collaborate with the HCP so that appropriate job placement of the employee occurs during the recovery period. Written return-to-work plans ensure that the HCP, the employee, and the employer all understand the steps recommended to promote recovery, and ensure that the employer understands what his or her responsibility is for returning the employee to work. A form is included to collect and distribute this written plan (Form 2). The HCP is also responsible for employee follow-up to document a reduction in symptoms during the recovery period.

Complete removal from the work environment should be avoided unless the employer is unable to accommodate the prescribed work restrictions. Research has documented that the longer the employee is off work, the less likely he/she will return to work (Vallfors 1985). In these cases, the employer's contact person and the employee should be in day-to-day contact, and the employee can be encouraged to participate in a fitness program that does not involve the injured anatomical area.

TABLE 33.2   Specific ICD-9 Diagnoses Referred to as Musculoskeletal Disorders by ICD-9 Numbers

Tendon, synovium, and bursa disorders	**727**
Trigger finger (acquired)	727.03
Radial styloid tenosynovitis (deQuervain's)	727.04
Other tenosynovitis of hand and wrist	727.05
Specific bursitides often of occupational origin	727.2
Unspecified disorder of synovium, tendon, and bursa	727.9
Peripheral enthesopathies	**726**
Rotator cuff syndrome, supraspinatus syndrome	726.10
Bicipital tenosynovitis	726.12
Medial epicondylitis	726.31
Lateral epicondylitis (tennis elbow)	726.32
Unspecified enthesopathy	726.9
Disorders of muscle, ligament, and fascia	**728**
Game-Keepers thumb	728.8
Muscle spasm	728.85
Unspecified disorder of muscle, ligament, and fascia	728.9
Other disorders of soft tissues	**729**
Myalgia, myositis, fibromyositis	729.1
Swelling of limb	729.81
Cramp	729.82
Unspecified disorders of soft tissue	729.9
Osteoarthritis	**715**
Mononeuritis of upper limb	**354**
Carpal tunnel syndrome (median nerve entrapment)	354.0
Cubital tunnel syndrome	354.2
Tardy ulnar nerve palsy	354.2
Lesions of the radial nerve	354.3
Unspecified mononeuritis of upper limb	354.9
Peripheral vascular disease	**443**
Raynaud's syndrome	443.0
Hand-Arm Vibration Syndrome	443.0
Vibration White Finger	443.0
Arterial embolism and thrombosis	**444**
Hypothenar hammer syndrome	444.2
Ulnar artery thrombosis	444.21
Nerve root and plexus disorders	**353**
Brachial plexus lesions:	353.0
Cervical rib syndrome	353.0
Costoclavicular syndrome	353.0
Scalenus anticus syndrome	353.0
Thoracic outlet syndrome	353.0
Unspecified nerve root and plexus disorder	353.9
Spondylosis (inflammation of the vertebrae)	**721**
Cervical without myelopathy	721.0
Cervical with myelopathy	721.1
Thoracic without myelopathy	721.2
Lumbarsacral without myelopathy	721.3
Thoracic or lumbar with myelopathy	721.4
Intervertebral disc disorders	**722**
Displacement of cervical disc	722.0
Displacement of thoracic or lumbar disc	722.1
Degeneration of the cervical disc	722.4
Degeneration of the thoracic or lumbar disc	722.5
Intervertebral disc disorder with myelopathy	722.17
Disorders of the cervical region	**723**
Cervicalgia (pain in neck)	723.1
Cervicobrachial syndrome (diffuse)	723.3
Unspecified neck symptoms or disorders	723.9
Unspecified Disorders of the Back	**724**
Low back pain	724.2

FORM 1 — Occupational and Health History Recording Form for Musculoskeletal Disorders

Name: _____ Dept: _____ Job Title: _____

Age: ____ yrs   Gender: ____ F ____ M   Length of time at the plant: _____ mo/yrs

Dominate Hand: ____ R ____ L ____ Both   Length of time on-the-job: _____ mo/yrs

Symptom Characterization:

Onset: Date: _____ Abrupt vs. Gradual: _____

Quality: (let employee describe, check all that apply)

___ pain   ___ tenderness ___ weakness   ___ soreness   ___ numbness

___ tingling ___ burning   ___ swelling   ___ cramping   ___ throbbing

Duration: _____   Frequency: _____

Intensity: (mild, moderate, or severe) _____

Location: (R = right, L = left) ___ neck   ___ upper arm ___ lower arm ___ back ___ upper leg ___ foot

(Check all that apply)   ___ shoulder ___ elbow   ___ hand/wrist ___ hip ___ lower leg

Radiation: (R = right, L = left) ___ neck   ___ upper arm ___ lower arm ___ back ___ upper leg ___ foot

(Check all that apply)   ___ shoulder ___ elbow   ___ hand/wrist ___ hip ___ lower leg

Exacerbating or relieving activities (both on-the-job and off-the-job):

Exacerbating: 1) _____ 2) _____ 3) _____

Relieving: 1) _____ 2) _____ 3) _____

Past Medical History (prior injuries or disorders):

1) _____ 3) _____

2) _____ 4) _____

Recreational Activities, Hobbies, Household Activities:

1) _____ 3) _____

2) _____ 4) _____

Occupational History:

1) _____ 3) _____

2) _____ 4) _____

Characterize the Job:

Forceful, repetitive or sustained **exertions** can be estimated from production standards, employee ratings of efforts required to complete job tasks, descriptions of work objects and tools, weights of work objects and tools, and length of the workday. Extreme, repetitive or sustained **postures** can be estimated from a description of work methods and equipment. Employees can demonstrate the posture required for each step of the job task, or simulate the workstation in the examining room. Insufficient rest, pauses, or **recovery time** and be estimated from a description of rest breaks, production standards, work flow, and work organization factors. Extreme levels, repeated or long exposure to **vibration** can be estimated from a description of hand tools, or equipment. **Cold temperatures**, repeated or long exposure to cold can be based on

Physical Stress	Property		
	Magnitude	Repetition Rate	Duration
Force			
Joint Angle			
Recovery			
Vibration			
Temperature			

temperature measurements, estimated from a description of the work environment, and the duration of time spent in cold areas.

FORM 2 — Musculoskeletal Disorder Management Plan
Forward Only Work Related Medical Information to the Employer

Date of Assessment: _____

Name: _____ Date of Birth: _____
Employer: _____ Contact Person: _____ Phone: _____ FAX: _____

Diagnosis/Assessment:
_____
_____

–

Treatment Plan: (e.g., medications/dosage, splints, physical or occupational therapy including frequency and duration   of   treatment,   etc.)
_____
_____

–

Next Appointment: _____

Other   Scheduled   Appointments:
_____

WORK STATUS

Is the Employee able to perform his/her regular work?

____ Yes, Full duty
____ No, Remove from Work Environment until _____
____ No, Modified or Alternate Work until _____

(Complete Activity Checklist below for Job Modifications)

Name: _____

Description of Restricted Work Activity

Activity	Duration	Frequency
a. Sitting	____ Hrs. Per Day	____ Hrs. at a Time
b. Standing	____ Hrs. Per Day	____ Hrs. at a Time
c. Walking	____ Hrs. Per Day	____ Hrs. at a Time
d. Lift/Carry:____ lbs.	____ Hrs. Per Day	____ Times Per Hr.
e. Climbing Stairs	____ Hrs. Per Day	____ Times Per Hr.
f. Climbing Ladders	____ Hrs. Per Day	____ Times Per Hr.
g. Kneeling	____ Hrs. Per Day	____ Times Per Hr.
h. Bending at Waist	____ Hrs. Per Day	____ Times Per Hr.
I. Squatting	____ Hrs. Per Day	____ Times Per Hr.
j. Twisting	____ Hrs. Per Day	____ Times Per Hr.
k. Pull/Push: ____ lbs.	____ Hrs. Per Day	____ Times Per Hr.
l. Reach Above Shoulder	____ Hrs. Per Day	____ Times Per Hr. ____ L ____ R
m. Extended Reaching	____ Hrs. Per Day	____ Times Per Hr. ____ L ____ R
n. Neck bend/twisting	____ Hrs. Per Day	____ Times Per Hr.
o. Elbow/Forearm Twist	____ Hrs. Per Day	____ Times Per Hr. ____ L ____ R
p. Hand/Wrist Bending	____ Hrs. Per Day	____ Times Per Hr. ____ L ____ R
q. Pinch Gripping	____ Hrs. Per Day	____ Times Per Hr. ____ L ____ R
r. Forceful Grasping	____ Hrs. Per Day	____ Times Per Hr. ____ L ____ R
s. Continuous Keyboard Use	____ Hrs. Per Day	____ Times Per Hr.
t. Vibrating Tool/Equip Use	____ Hrs. Per Day	____ Times Per Hr. ____ L ____ R
u. Ankle/Foot Bend/Twist	____ Hrs. Per Day	____ Times Per Hr. ____ L ____ R
v. Cold Temperature	____ Hrs. Per Day	

Other   Restricted   Job   Tasks   (including   frequency   and   duration):
_____

–

| Other | Specific | Job | Recommendations: |

–

Health Care Provider Name: _____
Address: _____
City/State/Zip: _____
Phone: (___)_____ FAX: (___)_____
Copy of Form Given to Employee: _____ Yes   _____ No
Health Care Provider Signature: _____   Date: _____

Wrist immobilization devices, such as wrist splints or supports, can help rest the symptomatic area in some cases. These devices are especially effective off the job, particularly during sleep. They should be dispensed to individuals with MSDs only by HCPs with the training and experience in the positive and potentially negative aspects of these devices. Wrist splints, typically worn by patients with possible carpal tunnel syndrome, should not be worn at work unless the HCP determines that the employee's job tasks do not require wrist deviation or bending. Struggling against a splint can exacerbate the medical condition due to the increased force needed to overcome the splint. Splinting may also cause other joint areas (elbows or shoulders) to become symptomatic as work technique is altered. Recommended periods of immobilization vary from several weeks to months depending on the nature and severity of the disorder. Immobilization should be prescribed judiciously and monitored carefully to prevent iatrogenic complications (e.g., disuse muscle atrophy).

The *prophylactic* use of immobilization devices worn on or attached to the wrist or back is not recommended. Research indicates wrist splints have not been found to prevent distal upper extremity musculoskeletal disorders (Rempel 1994). Likewise, there is no rigorous scientific evidence that back belts or back supports *prevent* injury, and their use is not recommended for prevention of low back problems (NIOSH 1994; Mitchell et al., 1994). Where the employee is allowed to use a device that is worn on or attached to the wrist or back, the employer, in conjunction with a HCP, should inform each employee of the risks and potential health effects associated with their use in the workplace, and train each employee in the appropriate use of these devices (McGill 1993).

The HCP should advise affected employees about the potential risk of continuing non-modified work, or spending significant amounts of time on hobbies, recreational activities, and other personal habits that may adversely affect their condition (e.g., requires the use of the injured body part). However, as mentioned above, the employee should engage in a fitness program designed for exercise and aerobic conditioning that does not involve the injured anatomical area.

33.7.2 Thermal (More Frequently Cold) Therapy

Such treatment is generally considered useful in the acute phase of some MSDs. Cold therapy may be contraindicated for other conditions (e.g., neurovascular).

33.7.3 Oral Medications

Aspirin or other nonsteroidal anti-inflammatory agents (NSAIA) are useful in reducing the severity of symptoms either through their analgesic or anti-inflammatory properties. Their gastrointestinal and renal side effects, however, make their prophylactic use among asymptomatic employees inappropriate, and may limit their usefulness among employees with chronic symptoms. In short, NSAIAs should not be used prophylactically.

It must be noted that the effectiveness of Vitamin B-6 for treatment of musculoskeletal disorders has not been established (Amadio 1985; Stransky et al., 1989; Spooner et al., 1993). Additionally, at this time there is no scientifically valid research that establishes the effectiveness of Vitamin B-6 for *preventing* the occurrence of musculoskeletal disorders.

33.7.4   Stretching and Strengthening

A valuable adjunct in individual cases, this approach should be under the guidance of an appropriately trained HCP (e.g., physiatrists, physical and occupational therapists). Exercises that involve stressful motions or an extreme range of motions, or that reduce rest periods may be harmful.

33.7.5   Hot Wax

At this time there is no scientific evidence regarding the effectiveness of hot wax treatments as a preventative measure or as a therapeutic modality.

33.7.6   Steroid Injections

For some disorders resistant to conservative treatment, local injection of a corticosteroid by an experienced physician may be indicated. The addition of a local anesthetic agent to the injection can provide valuable diagnostic information.

33.7.7   Surgery

With an effective ergonomics and medical management program, surgery for work-related MSDs should be needed rarely. Surgical intervention should be used for objective medical conditions and should have proven effectiveness. While the indications for prompt or emergency surgical intervention may still be present (e.g., ulnar artery thrombosis), surgery should be reserved for severe cases (e.g., very high levels of pain resulting in significant functional limitations) not responding to an adequate trial of conservative therapy.

33.8   Follow-Up and Return to Work

33.8.1   Follow-Up

Many, if not most, WRMSDs improve with conservative measures. HCPs should follow up the symptomatic employee to document improvement, or to reevaluate employees who have not improved. The time frame for this follow-up depends on the symptom type, duration, and severity. A clinical exam or telephone contact with the employee should be made once a week, followed by a complete reevaluation within ten days from the last examination if the employee's symptoms are not improving. Where HCPs are available at the workplace, monitoring the symptomatic employee should occur every 3 to 5 working days depending on the clinical severity of the disorder (Wiesel et al., 1984; Wiesel et al., 1994).

In reassessing employees who have not improved, the following should be considered:

- Is the diagnosis correct?
- Are the treatment goals appropriate?
- Have the MSD risk factors on and off the job been addressed?
- Is referral appropriate?

If the job's relevant risk factors have been eliminated but the employee's symptoms persist, it is important for the HCP to realize that employee reactions to pain and functional limitations may prolong the recovery period. Strategies to help the employee cope with the pain and stress associated with these disorders should be incorporated into the employee's treatment plan. The time frames for considering referral depends on the primary HCP's training and expertise, in addition to the type, duration, and severity of the condition. In general, severe symptoms with objective physical examination findings interfering with an employee's ability to perform his/her job should be referred to an appropriate HCP specialist sooner than milder symptoms without objective findings.

33.8.2 Return to Work

If an employee's treatment plan required time away from work, the next step is to return the employee to work in a manner that will minimize the chance for re-injury. Employees returning to the same job without a modification of the work environment are at risk for a recurrence. Key to the return to work process is open communication among the employee, the HCPs, and management. This will allow: (1) prompt treatment, (2) an expedient return to work consistent with the employee's health status and job requirements, and (3) regular follow-up to manage symptoms and modify work restrictions as appropriate. The principles guiding the return to work determination include the type of MSD condition, the severity of the MSD condition, and the MSD risk factors present on the job.

Employees with MSDs who have difficulty remaining at work or returning to work in the expected timeframes are candidates for rehabilitation therapy. Rehabilitation refers to the process in which an injured worker follows a specific program that promotes healing and helps him or her return to work. During the rehabilitation process, psychosocial factors (factors present both on the job, and off the job, that can compromise an individual's ability to cope with symptoms, physical disorders, and functional limitations) should be addressed.

33.9 Screening

Currently there is no scientific evidence that validates the use of preassignment medical examinations, job simulation tests, or other screening tests as a valid predictor of which employees are likely to develop MSDs (Frymoyer 1992; Werner et al., 1994; Cohen et al., 1994). Literature findings are mixed on the use of preplacement strength testing as a valid predictor of back injury.

33.10 Conclusion

The financial and human costs of work-related musculoskeletal disorders to our society are staggering. This chapter on the medical management of these disorders should help employers and HCPs wishing to prevent or reduce the severity of these disorders, resulting in a healthier, more productive workplace.

Acknowledgments/Disclaimer

We would like to thank the individuals and professional associations who contributed to the draft ANSI Z-365 standard, and the draft OSHA Ergonomic Protection Standard. Ms. Bertsche's contribution to this chapter occurred while a visiting scientist at the National Institute for Occupational Safety and Health (NIOSH), from the U.S. Department of Labor Occupational Safety and Health Administration (OSHA). This chapter represents the views of the authors and does not constitute official policy of NIOSH.

References

AHCPR (1994). Acute Low Back Problems in Adults: Assessment and Treatment. U.S. Dept of Health and Human Services, Public Health Service, Agency for Health Care Policy and Research. Rockville, MD. Publication No. 95-0643.

Amadio PC (1985). Pyridoxine as an adjunct in the treatment of carpal tunnel syndrome. *J Hand Surg;* 10A:237–241.

ANSI (1995). ANSI Z-365 Control of Work-Related Cumulative Trauma Disorders Part 1: Upper Extremities. American National Standards Institute. Chicago, IL: Working draft 4/17/95.

ASSH (1990). *The Hand: Examination and Diagnosis*, 3rd ed. American Society for Surgery of the Hand. New York, NY; Churchill Livingstone.

BLS (1995). Workplace injuries and illnesses in 1994. U.S. Department of Labor, Bureau of Labor Statistics. Washington, D.C.

Bongers PM, De Winter CR, Kompier MA, Hildebranndt VH (1993). Psychosocial factors at work and musculoskeletal disease. *Scand J Work Environ Health;* 19:297–312.

Cohen JE, Goel V, Frank JW, Gibson ES (1994). Predicting risk of back injuries, absenteeism, and chronic disability. *J Occup Med;* 36(10):1093–1099.

Day DE (1987). Prevention and return to work aspects of cumulative trauma disorders in the workplace. *Seminars in Occup Med;* 2(1):57–63.

Frymoyer JW (1992). Can low back pain disability be prevented? *Bailliere's Clinical Rheumatology;* 6(3):595–607.

Haig A, Linton P, McIntosh M, Moneta L, Mead P (1990). Aggressive early medical management by a specialist in physical medicine and rehabilitation: effect on lost time due to injuries in hospital employees. *J Occup Med;* 32(3):241–244.

Hoppenfield S (1976). *Physical Examination of the Spine and Extremities.* Norwalk, CT: Appleton Century Crofts.

Kiefhaber TR, Stern PH (1992). Upper extremity tendinitis and overuse syndromes in the athlete. *Clinics in Sports Med;* 11(1):39–55.

Louis DS (1987). Cumulative trauma disorders. *J Hand Surg;* 12A(5):823–825.

Mayer TG, Gatchel RJ, Hayer H et al. (1987). A prospective two-year study of functional restoration in industrial low back injury: an objective assessment procedure. *JAMA;* 258:1763–1767.

McGill SM (1993). Abdominal belts in industry: A position paper on their assets, liabilities, and use. *Am Indust Hygiene Assoc;* 54(12):752–554.

Mitchell LV, Lawler FH, Bowen D, Mote W, Ajundi P, Purswell J (1994). Effectiveness and cost-effectiveness of employer issued back belts in areas of high risk for back injury. *J Occup Med;* 36(1):90–94.

NIOSH (1994). Workplace use of back belts. Review and recommendations. U.S. Department of Health and Human Services, Public Health Service, Centers for Disease Control and Prevention, National Institute for Occupational Safety and Health. Cincinnati, OH; Publication No. 94–122.

Nirschl RP (1990). Patterns of failed healing in tendon injury, in Leadbetter WB, Buckwalter JA, Gordon SL (eds). Sports-Induced Inflammation. Park Ridge, IL: American Orthopaedic Society, pp 577–585.

OSHA (1995). Draft Ergonomics Protection Standard. U.S. Department of Labor, Occupational Safety and Health Administration. Washington, D.C. March, 1995.

Ranney D (1993). Work-related chronic injuries of the forearm and hand: Their specific diagnosis and management. *Ergonomics;* 36(8):871–880.

Rempel D, Manejlovic R, Levinsohn DG, Bloom T, Gordon L (1994). The effect of wearing a flexible wrist splint on carpal tunnel pressure during repetitive hand activity. *J Hand Surg;* 19(1): 106–110.

Spooner GR, Desai HB, Angel JF, Reeder BA, Donat JR (1993). Using pyridoxine to treat carpal tunnel syndrome. Randomized control trial. *Canadian Family Practice;* 39:2122–2127.

Stransky M, Rabin A, Leva NS, Lazaro RP (1989). Treatment of carpal tunnel syndrome with vitamin B-6: A double blind study. *Southern Med J;* 82(7):841–842.

Tubiana R, Thomine JM, Mackin E (1984). *Examination of the Hand and Upper Limb.* Philadelphia, PA: W.B. Saunders Co.

Upfal M (1994). Understanding medical management for musculoskeletal injuries. *Occup Hazards;* Sept:43–47.

Vallfors B (1985). Subacute and chronic low back pain; Clinical symptoms, absenteeism, and work environment. *Scand J Rehab Med;* Suppl 11:1–99.

Werner RA, Franzblau A, Johnston E (1994). Quantitative vibrometry and electrophysiological assessment in screening for carpal tunnel syndrome among industrial workers. *Arch Phys Med Rehabil*; 75:1228–1232.

WHO (1985) Identification and control of work-related diseases. World Health Organization Geneva: WHO Technical report; 174:7–11.

Wiesel SW, Feffer HL, Rothman RH (1984). Industrial low back pain, a prospective evaluation of a standardized diagnostic and treatment protocol. *Spine*; 9:199–203.

Wiesel SW, Boden SD, Feffer HL (1994). A quality based protocol for management of musculoskeletal injuries. *Clinical Orthopaedics and Related Research*; 301:164–176.

Wood DJ (1987). Design and evaluation of a back injury prevention program within a geriatric hospital. *Spine*; 12:77–82.

34

Systems Approach to Rehabilitation

Ann E. Barr
Temple University

34.1 Introduction

Work-related musculoskeletal disorders (WMSDs) accounted for one in three lost work time illnesses in U.S. private industry in 2001 (BLS, 2003a). While the total number of these disorders represented a 10% reduction compared to the number reported in the previous year (BLS, 2002), illnesses in the U.S. workplace due to repeated trauma accounted for 62 to 67% of all reported illnesses in U.S. private industry since 1992 (BLS, 1989–2001). Although occupational injury and illness data are available for 2002, changes in the reporting requirements by the United States Occupational Safety and Health Administration (OSHA) have made comparisons with more recent calendar years impossible. (BLS, 2003b). Based upon such trends over the last decade, WMSDs continue to represent the majority of all occupational illnesses, cause substantial worker discomfort and disability, and impose a heavy economic burden.

Although efforts to manage these disorders through government regulation have been fraught with controversy, research concerning the etiology and treatment of WMSDs in the past decade provides employers and health-care providers (HCPs) with guidelines for efficient and effective rehabilitation of affected workers. The most important tenet of WMSD management is the need for a multidisciplinary team, involving stakeholders in the workplace and the clinic that uses cooperative problem solving to address both the physical and nonphysical aspects of this problem. Clearly, the most effective rehabilitation programs for WMSD are based on the biopsychosocial model of musculoskeletal pain and disability and forge active and clear communication between the health-care team, the affected worker, and the employer (Feuerstein et al., 2003a; Gatchel, 2004).

This chapter will describe the relationship of HCPs to other members of the ergonomics program team in the rehabilitation of workers with WMSD. Selected medical management approaches from the literature will be summarized from which a generalized model for medical management will be derived. Specific assessment tools will be highlighted to assist HCPs in identifying workplace and nonworkplace risk factors that contribute to WMSDs. Clinical findings that help to direct appropriate treatment will be summarized. Specific treatments that have been determined to be effective will be recommended, and treatment pitfalls will be identified. Because new assessment tools and treatment approaches are constantly being reported and refined in the clinical literature, employers should rely on the expertise of its HCP team members in selecting the most appropriate assessment tools and treatments when establishing a rehabilitation program for WMSD.

34.2　Definition and Scope of WMSDs

WMSDs are defined by the United States Department of Labor (USDL) as injuries or disorders of the muscles, nerves, tendons, joints, cartilage, and spinal discs associated with exposure to risk factors in the workplace (BLS, 2003a). WMSDs may be cases where the nature of the injury or illness is sprains, strains, tears; back pain, hurt back; soreness, pain, hurt (except the back); carpal tunnel syndrome; hernia; or musculoskeletal system and connective tissue diseases and disorders and when the event or exposure leading to the injury or illness is bodily reaction, bending, climbing, crawling, reaching, twisting, overexertion; or repetition (BLS, 2003a). While the USDL has excluded Raynaud's phenomenon, tarsal tunnel syndrome, and herniated spinal disc from its definition of WMSD, this was done for reporting purposes and does not preclude these disorders from being related to work. The workplace risk factors that have been associated with WMSDs include highly repetitive or forceful exertions, particularly in the presence of awkward or extreme postures, vibration or cold temperatures, and work-related psychosocial factors (e.g., rapid work pace, monotonous work, low job satisfaction, low decision latitude, and job stress) (NRC, 2001). While the definition of WMSD is quite broad, it emphasizes two important aspects of these disorders: (1) they develop over time with prolonged exposure to risk factors and (2) they are "work related," which is to say that they may be caused or exacerbated by the performance of work in the context of a complex and multifactorial web of causal factors that include nonworkplace as well as workplace elements.

Possible ICD-9 diagnoses for WMSDs are listed in Table 34.1. In 2001, the most common (80%) of all lost workday injuries and illnesses due to lifting were sprains, strains, or tears regardless of the anatomical regions affected (BLS, 2001). The most common (41%) of all lost workday injuries and illnesses due to repetitive motion was carpal tunnel syndrome, followed by sprains, strains, or tears (18%), tendonitis (13%), and soreness, pain, hurt, except back (7%) (BLS, 2001). Ninety-six percent of lost workday disorders of the peripheral nervous system due to repetitive motion were diagnosed as carpal tunnel syndrome in 2001 (BLS, 2001). Of nonspecified injuries and disorders caused by lifting, 60% were classified as back pain or hurt back (BLS, 2001). In nonspecified cases involving soreness, pain, hurt, except the back, 82% were caused by repetitive motion (BLS, 2001).

34.3　Components of a Multidisciplinary WMSD Management Program

In 1997, the National Institute for Occupational Safety and Health (NIOSH) published three documents on the subject of WMSD: a comprehensive review of the literature concerning the epidemiology of

TABLE 34.1 Specific ICD-9 Codes Corresponding to Common Diagnoses Referred to as Work-Related Musculoskeletal Disorders (Not Intended to Be All-Inclusive)

Disorder	ICD-9 Code
Soft tissue disorders	
Tendon, synovium, and bursa	727
Trigger finger	727.03
DeQuervain's (radial styloid) tenosynovitis	727.04
Other tenosynovitis of hand and wrist	727.05
Specific bursitides often of occupational origin	727.2
Unspecified disorder of tendon, synovium, and bursa	727.9
Peripheral enthesopathies and allied syndromes	726
Rotator cuff syndrome	726.1
Bicipital tenosynovitis	726.12
Medial epicondylitis (golfer's elbow)	726.31
Lateral epidcondylitis (tennis elbow)	726.32
Unspecified enthesopathy	726.9
Disorders of muscle, ligament, and fascia	728
Muscle spasm	728.85
Unspecified disorder of muscle, ligament, and fascia	728.9
Other disorders of soft tissues	729
Myalgia and myositis, unspecified	729.1
Neuralgia, neuritis, and radiculitis, unspecified	729.2
Fasciitis, unspecified	729.4
Pain in limb	729.5
Swelling of limb	729.81
Cramp	729.82
Unspecified disorders of soft tissue	729.9
Bone and joint disorders	
Osteoarthritis	715
Unspecified arthropathies	716
Unspecified disorders of joint	719
Effusion of joint	719
Pain in joint	719.4
Stiffness of joint	719.5
Other specified disorders of joint	719.8
Peripheral nerve disorders	
Mononeuritis of upper limb	354
Carpal tunnel syndrome (median nerve entrapment)	354
Cubital tunnel syndrome	354.2
Ulnar nerve palsy	354.2
Lesions of the radial nerve	354.3
Unspecified mononeuritis of upper limb	354.9
Mononeuritis of lower limb	355
Tarsal tunnel syndrome	355.5
Nerve root and plexus disorders	353
Brachial plexus lesions	353.0
Cervical rib syndrome	353.0
Costoclavicular syndrome	353.0
Scalenus anticus syndrome	353.0
Thoracic outlet syndrome	353.0
Cervical root lesions, not elsewhere classified	353.2
Lumbosacral root lesions	353.4
Unspecified nerve root and plexus disorder	353.9
Spinal disorders	
Spondylosis (inflammation of the vertebrae)	721
Cervical without myelopathy	721
Cervical with myelopathy	721.1

(continued)

TABLE 34.1 *Continued*

Disorder	ICD-9 Code
Thoracic without myelopathy	721.2
Lumbosacral without myelopathy	721.3
Thoracic or lumbar with myelopathy	721.4
Intervertebral disc disorders	722
Displacement of cervical disc	722.0
Displacement of thoracic or lumbar disc without myelopathy	722.1
Degeneration of cervical disc	722.4
Degeneration of thoracic or lumbar disc	722.5
Intervertebral disc disorder with myelopathy	722.17
Disorders of the cervical region	723
Cervicalgia (neck pain)	723.1
Cervicobrachial syndrome (diffuse)	723.3
Unspecified neck symptoms or disorders	723.9
Unspecified disorders of the back	724
Pain in thoracic spine	724.1
Low back pain (lumbago)	724.2
Sciatica	724.3
Disorders of the sacrum	724.6
Vascular disorders	
Peripheral vascular disease	443
Raynaud's syndrome	443.0
Hand-arm vibration syndrome	443.0
Vibration white finger	443.0
Arterial embolism and thrombosis	444
Hypothenar hammer syndrome	444.2
Ulnar artery thrombosis	444.21
Unspecified arteries and arterioles	444.9

WMSD (Bernard, 1997), a compendium of presentations from industry representatives who have established successful workplace ergonomics programs (NIOSH, 1997a), and guidelines for workplace ergonomics programs (NIOSH, 1997b). Although the OSHA Ergonomics Program Rule (Federal Register, 2000) was repealed in 2001, OSHA is presently developing industry- and task-specific ergonomic guidelines for voluntary use by U.S. employers (OSHA, 2002, 2003d). Currently, final guidelines exist for meatpacking plants (OSHA, 1993) and nursing homes (OSHA, 2003c), and draft guidelines are under review for poultry processing (OSHA, 2003b) and retail grocery stores (OSHA, 2003a). All of these guidelines embrace the same basic program components, which are delineated in the OSHA Guidelines for Nursing Homes as follows: (1) management support, which consists of clearly developed program goals, clearly identified program responsibilities for participants, adequate provision of resources, and oversight to ensure program success; (2) employee involvement, which relies on employees to report problems early, provide insight (suggestions and evaluation) into work design issues and solutions, and share responsibility for program development and implementation; (3) problem identification, which requires systematic procedures utilizing various sources of information such as OSHA injury and illness logs, workers' compensation claims data, surveillance and workplace analysis; (4) solution implementation, which typically involves changes to workplace hazards, work practices, or both; (5) injury management, which should emphasize early detection and intervention; (6) training of employees and managers in the risks and detection of WMSD and the procedures required to respond to an WMSD incident; and (7) regular program evaluation in which quantifiable measures, such as number and severity of WMSDs, are determined (OSHA, 2003c).

NIOSH schematized these major program components in its Primer on the Elements of Ergonomics Programs (NIOSH, 1997b). This schematic emphasizes not only the importance of an effective response to cases of WMSD, but also that of prevention of these disorders in presently unaffected employees. It also emphasizes the need for the program to become an integral part of the workplace culture in order to be successful. It has been this author's experience that the commitment shown by employers who spearhead the development of a workplace ergonomics program has a positive effect on employee morale and, therefore, job satisfaction. In other words, the mere presence of a company-based ergonomics program may, in and of itself, reduce one of the risk factors for WMSD. The remainder of this chapter will elaborate upon the relationship of the injury management component of a comprehensive ergonomics program within the context of a workplace system. The role of HCPs will be clearly defined, and their responsibilities to injured workers as well as to other component members of the ergonomics program will be delineated. Specific treatment models will be suggested for secondary prevention of WMSDs.

34.4 Injury Management in a Comprehensive Ergonomics Program

Whether WMSD management is workplace or clinic based, HCPs, in addition to delivering treatment to injured workers, also must interact directly with the workplace. Such interaction may range from communication with management about the worker's limitations while undergoing treatment to recommending surveillance or job changes based on patterns of injury and illness among employees. Certainly, HCPs have a major role in the evaluation of ergonomics program effectiveness. HCPs who serve this patient population will fall short of their obligation if they diagnose and treat these patients in clinical isolation. Furthermore, the most successful rehabilitation programs for WMSD include multidisciplinary HCP teams with various expert perspectives to address the multifactorial causes of these disorders. Representative multidisciplinary programs will be highlighted in the paragraphs that follow.

34.4.1 The Biopsychosocial Model of WMSD Management

The biopsychosocial model of WMSD management has been clearly demonstrated to represent the most successful approach to WMSD management. In this model, the interactions between physical impairments, personal psychological factors, and psychosocial factors are recognized and targeted for treatment in a more holistic approach to care than in the traditional biomedical reductionist model (Gatchel, 2004). Given recent evidence for the contribution of workplace psychosocial as well as personal psychological factors in the development and severity of WMSD (Leino and Hanninen, 1995; Symonds et al., 1995; Strasser et al., 1999; Tittaranonda et al., 1999; Haufler et al., 2000; Warren et al., 2000; Kerr et al., 2001; MacDonald et al., 2001; Clauw and Williams, 2002; Dersh et al., 2002; Lundberg, 2002; Shaw et al., 2002), the biopsychosocial model of management is intuitively as well as empirically sound.

34.4.2 Recent Examples of Multidisciplinary WMSD Management Programs

One example of such a biopsychosocial approach is the PREVention of work handICAP, or PREVICAP, program developed in Quebec, Canada, for the secondary prevention of long-term WMSD disability (Loisel et al., 2003). This program evolved from an earlier such program, the Sherbrooke Model, for low back pain (Loisel et al., 1994). The PREVICAP program has the following components: (1) a disability diagnosis step, in which the precise causes (e.g., physical, psychosocial, occupational, and administrative) for disability are determined; and (2) a progressive return to work rehabilitation phase involving an interdisciplinary HCP team that is capable of addressing the various causes of disability and that interacts closely with the workplace. The PREVICAP program is a person–environment interaction model in which interventions are directed toward the worker (physical and psychosocial), the workplace, and the interaction between the two. The interdisciplinary HCP team includes a medical doctor with skills in musculoskeletal disorders and rehabilitation, an occupational therapist, a kinesiologist, an ergonomist, and a team coordinator (i.e., case manager). The team members meet on a weekly basis so that consensus regarding progressive rehabilitation is achieved and to ensure that consistent information is being communicated between the HCP team, the worker, and the workplace. To date, the PREVICAP program has been implemented from a clinical site at a teaching hospital. In a prospective evaluation of 127 workers with work absences from low back pain or upper extremity disorders, approximately 75% were working 1 yr after completing the PREVICAP rehabilitation phase and there was a high work retention rate after 3 yr. The evolution of this program has proven valuable in identifying two key concepts in medical management of WMSD: (1) the need for coordination between worker- and workplace-focused interventions through an interdisciplinary HCP team and team coordinator; (2) an intervention delivered at the secondary level of care (i.e., subacute or early chronic stage of injury) can still result in successful return to work as long as the long-term goals of the program are disability management as opposed to medical cure of the disorder. Although the Sherbrooke Model was developed for workers with low back pain disability, preliminary evaluation of the PREVICAP program was conducted for workers with all WMSDs (low back pain, 67%; neck/upper extremity, 30%; and other WMSDs, 3%), which demonstrates that the organizational components of the program may be adapted to different diagnoses provided that appropriate assessment tools are used to examine workers and the workplace.

In the United States, Feuerstein et al. (2000) investigated the effects of a multicomponent intervention program on upper extremity (UE) WMSD in sign language interpreters. The program had five components: (1) reduce physical work load and improve flexibility and endurance through exercise; (2) train employees in job stress management; (3) change work organization and work style to reduce biomechanical risk exposure; (4) improve managerial skill in addressing UE WMSD and increase supervisor support through training; and (5) educate workers and supervisors regarding optimal health-care utilization. The intervention was conducted by various team members consisting of a clinical psychologist, a physical therapist, an exercise physiologist, and a physician with board certification in occupational medicine and rehabilitation medicine. In the 3 yr following the intervention, there was a 69% reduction in the number of UE WMSD cases reported as compared to the 3 yr prior to the intervention. There was a 30% reduction in workers' compensation health-care costs in the 2 yr following the intervention as compared to the 2 yr prior to the intervention. There was a 64% decline in workers' compensation indemnity costs in the 3 yr following the intervention. Finally, there was an increase in the number of work hours logged following the intervention, which coincides with the decreased number of reported cases of UE WMSD. This study not only demonstrated the effectiveness of a multidisciplinary approach to WMSD management, but also illustrated the feasibility of a workplace management program tailored to the needs of the specific employer and employees, and is one of the few studies to analyze health-care cost and indemnity outcomes. The latter findings illustrate clearly the cost benefits of workplace ergonomics management in terms of both cost savings and increased worker productivity.

In a later study, Feuerstein et al. (2003a) demonstrated the effectiveness of an integrated case management (ICM) approach involving ergonomic and problem-solving components in a group of federal workers with UE WMSD. This study examined patient satisfaction, future symptom severity, function, and return to work in 205 subjects randomly assigned to either usual care or ICM care. Both patient groups received the services of a case manager, but in the case of the ICM group, the case managers, who were occupational nurses, received 2 days of formal training in systematic methods in the integration of ergonomic and psychosocial assessment and intervention into the UE WMSD care and recovery process. This included such topics as initial interview and examination, problem solving, ergonomic assessment and accommodation, re-injury prevention, and clinical follow-up. Workers receiving the ICM intervention had higher patient satisfaction, which predicted decreased symptom severity, increased function, and earlier return to work at 6-month follow-up. This ICM approach is part of an ongoing experimental program with joint sponsorship from the U.S. Department of Labor/Office of Workers' Compensation Programs, Georgetown University, and the Robert Wood Johnson Foundation (Feuerstein et al., 1999a). As in the previous programs already cited, the ICM project also emphasizes the need for coordination of all interdisciplinary team members both within the HCP team as well as within the context of the comprehensive ergonomics program. Furthermore, usual case management practices are not as effective as those in which specific ergonomic and problem-solving skills, which can be obtained through case manager training, are brought to bear on each WMSD case.

Although the three studies illustrated represent a small cross section of the available reports in the literature, they are representative of many such programs for several reasons: (1) they emphasize the importance of a multidisciplinary HCP team to address the multiple causal factors for any given WMSD case; (2) they illustrate the need for clear communication and cooperation between HCPs, injured workers, and various stakeholders in the workplace; (3) they show that such communication is greatly enhanced by trained case managers, who oversee all aspects of health-care and workplace management; and (4) they demonstrate that effective health-care interventions for WMSD do not necessarily result in a cure, but rather lead to long-term worker- and workplace-focused management through effective problem solving on the part of the entire WMSD management team.

34.4.3 Relationship of HCPs to Other Members of an Ergonomics Program

The lines of communication between the HCP team and other member components of an ergonomics program are depicted in Figure 34.1. The different types of professionals who may participate as members of the HCP team are indicated in the figure and the scope of their practice and the services they provide is elaborated in Table 34.2. The nature of the information flowing between the individual ergonomics program components is also indicated in Figure 34.1. One of the most important types of information, both delivered to and received from the HCP team, is feedback about the success of implemented solutions. Such feedback directs adjustments needed for each individual worker–workplace system.

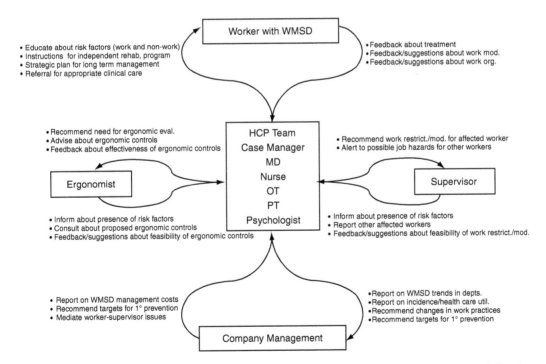

FIGURE 34.1 Lines of communication between the HCP team in a WMSD management program and the other members of a comprehensive, multidisciplinary ergonomics program. The nature of the information flow is specified in order to illustrate the ways in which the HCP responsibilities extend beyond clinical treatment in the workplace system. The types of health professionals that serve as HCP team members are indicated and their specific scope of practice and training relevant to WMSD are elaborated in Table 34.2. MD, medical doctor; OT, occupational therapist; PT, physical therapist.

TABLE 34.2   Health Care Professionals Who Might Serve as Members of the Health-Care Provider Team in a Comprehensive Ergonomics Program for WMSD (Not Intended to Be All-Inclusive)

Profession	Scope of Practice and Specialist Training	Relevant Services They Provide
Medical Doctor (MD)	The following specialist areas have expertise relevant to WMSD management: occupational and environmental medicine, rheumatology, physical medicine and rehabilitation, orthopedics, neurology, pain medicine	Evaluation, diagnosis, and medical treatment or referral for nonmedical treatment. Role in diagnosis, conservative and specialist (level III) treatment in acute, subacute, and chronic cases
Registered Nurse (RN)	Certification in Occupational Health Nursing (COHN) available on a voluntary basis with academic or continuing education. Nurse practitioners have advanced training and treat independently or provide medical treatment by protocol based on practice scope defined by jursidiction	Initial screening assessment and conservative treatment as well as job/workstation analysis. Makes referrals to other HCPs as needed. Most common HCP to act as Case Manager. Role in treatment of acute and subacute cases (unless as Case Manager in chronic cases)
Occupational Therpist (OT)	Entry-level graduate education in psychosocial, neurological, cognitive, perceptual, and motor systems in context of disability rehabilitation. Ergonomics (including task and job analysis) may be included in entry-level education. Postprofessional academic and continuing education in ergonomics and WMSDs available on a voluntary basis. Hand specialist certification (CHT) available on a voluntary basis	Asessment of motor and psychosocial/cognitive aspects of work-related skills and treatment to maximize function. Performs job and workstation analyses and modification, functional capacity analysis, and work hardening/simulation. Provides adaptive technology, assistive devices, orthotics (splints), and pain management. Role in conservative treatment and rehabilitation of acute, subacute, and chronic cases. Makes referrals to other HCPs as needed. CHT will have extensive continuing education and clinical expertise relevant to treatment of upper extremity WMSDs
Physical Therapist (PT)	Entry-level graduate education in biomechanics, kinesiology, neurology, motor control, exercise physiology, and psychophysical performance measures in context of disability rehabilitation. Ergonomics (including job analysis and work hardening) included in entry-level education. Postprofessional continuing education in ergonomics and WMSDs available on a voluntary basis. Hand specialist certification (CHT) and orthopedics certification (OCS) available on a voluntary basis	Evaluation, diagnosis, and treatment of impairments of the musculoskeletal, neuromuscular, and cardiovascular systems. Performs job and workstation analyses and modification, functional capacity analysis, and work hardening/simulation. Provides orthotics (splints, braces), pain management, and progressive strength, and cardiovascular fitness training. Role in conservative treatment and rehabilitation of acute, subacute, and chronic cases. Makes referrals to other HCPs as needed. May act as Case Manager. Clinical specialists (i.e., hand and orthopedics) will have extensive continuing education and clinical expertise relevant to treatment of WMSDs, including management of low back pain and upper extremity disorders
Psychologist	Entry-level graduate education in cognitive and behavioral therapeutic techniques for individual and group counseling settings	Assessment and treatment of psychosocial and individual psychological aspects of WMSD cases. Provides individual or group counseling regarding stress management, relaxation techniques, cognitive coping strategies, assertiveness training, problem solving, and behavioral modification. Usually receives referral from another HCP team member in subacute or chronic cases

Without such feedback, program success cannot be ensured or evaluated. Thus, the iterative nature of the problem-solving process must be appreciated by all members of the ergonomics program team, and follow-up by all team members to monitor implemented solutions is obligatory.

The discussion up to this point has emphasized secondary prevention interventions for WMSD. While it is more typical for HCPs to participate in secondary prevention strategies, they also play an important role in alerting employers to potentially hazardous work practices. In a truly integrated ergonomics program, where a particular workplace is covered by a single HCP team, trends in injuries within the workplace can be appreciated soonest by HCPs and referred to other members of the program for investigation (see Figure 34.1). In this way, the HCP team may contribute to primary prevention of WMSD. Once again, such a contribution is only possible with clear and frequent communication between the HCP team and other member components of a comprehensive ergonomics program.

Finally, when treating a worker with WMSD, the HCP team must investigate the extent to which non-workplace factors contribute to or exacerbate WMSD symptoms. While this investigation is inherent to the biopsychosocial perspective, HCPs need to be reminded to investigate all possible sources of causality. In a third-party payer system, such as the U.S. health-care system, in which causality must sometimes be partitioned between workers' compensation and private insurance payers, HCPs must sometimes proceed delicately in trying to determine the existence of nonworkplace risk factors. One approach to this problem is to maintain worker confidentiality regarding nonworkplace factors (i.e., communicate only work-related information to the ergonomics program team) and to educate the worker to be aware of potential contributing risks in a generalized enough way that he or she is able to recognize such factors. The latter approach is especially helpful in focusing a portion of WMSD management on the worker, thereby rendering him or her a primary responsible party in long-term management of these disorders.

34.5 Algorithm for Secondary Prevention of WMSD in an Occupational Setting

Now that the role of the HCP team has been clarified with respect to the entire ergonomics program, it is necessary to delineate the specific roles of HCPs in caring for injured workers (see Table 34.2). The steps in the clinical care of such cases are as follows: (1) initial assessment; (2) referral for specialist care if needed; (3) immediate intervention; (4) long-term management; and (5) follow-up.

For many companies, supporting a comprehensive, multidisciplinary HCP team is prohibitively expensive. Furthermore, not all affected workers will require the services of every member of the HCP team. In other words, companies have more modest needs than recommended for an ideal medical management program, and more pragmatic alternatives need to be determined.

The algorithm for medical management of WMSD in Figure 34.2 is based on a workplace ergonomics program at a music and entertainment company with approximately 1500 employees where computer work was highly prevalent (Barr et al., 1999; Barr, 2002). As part of this program, a clinical algorithm was established in which an employee-initiated report of WMSD is handled in a triage procedure. Occupational Health Nurses perform an initial screening examination that helps to determine the appropriate level of care. Three levels of care are available: level I (or conservative nursing management), level II (on-site physician diagnosis and management), and level III (specialist diagnosis and management). Because this is a company-based program, there were limitations as to the expertise of the company-based HCPs in treatment delivery. However, initial assessment for triage purposes is well within the scope of this small Health and Safety Department, as is acute and long-term conservative management of affected employees. Workers with more severe symptoms may be referred to a clinic-based practice that specializes in rehabilitation of WMSD and utilizes the multidisciplinary HCP team model and with which the company has a strong relationship, or they may choose another outside practitioner for further evaluation. In either case, employees receiving level III care are assigned a nurse case manager who is a

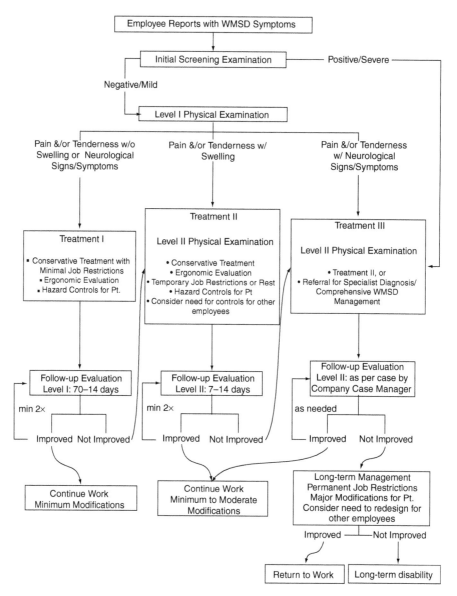

FIGURE 34.2 Algorithm for company-based medical management of WMSDs. The algorithm is triggered by an employee report of WMSD symptoms. An on-site occupational health nurse conducts an initial screening examination, which takes approximately 10 to 15 min. Based on the result of the screening, the employee proceeds to either a Level I physical examination, which is a more thorough nursing examination, or to a Level II physical examination, which is conducted by an occupational health physician. In the case of a Level I physical examination, employees with mild signs and symptoms may be treated conservatively by nursing and fitness center staff (Treatment I). For moderate to severe signs and symptoms, employees will go on for a Level II physical examination. Based on the Level II examination results, employees will be recommended for either Treatment II, in which temporary job restrictions may be imposed, or Treatment III, which may require specialist referral for further diagnosis and treatment outside the company Health and Safety Department or permanent job restrictions or redesign. For all treatment options, the algorithm includes at least two time points for follow-up. In cases where the employee's condition either does not improve or worsens, the employee can be referred for the next highest level of care in the algorithm. The use of such an algorithm empowers health-care providers within the company to clearly identify and track progress of affected employees and provides guidelines for clinical decision-making. (Reprinted from: Barr AE (2002). In *Rehabilitation of the Hand: Surgery and Therapy*, 5th edition (Vol. 1), pp. 996–1004, Mosby, Philadelphia. With permission).

member of the company Health and Safety Department. In this way, those patients receiving outside care are benefited by the establishment of effective lines of communication between outside HCPs and other members of the workplace ergonomics program. While this algorithm was developed for a specific setting, it possesses the five aforementioned components of WMSD clinical care and may be utilized in many occupational health and safety settings. Others have recently reported on the success of this approach in a larger workplace (Bernacki errociz).

34.6   Clinical Evaluation of Workers with WMSD

34.6.1   Questionnaires

Questionnaires are frequently used to obtain subjective information from workers with WMSD regarding both their present complaints as well as relevant medical history, and exposure to workplace and non-workplace risk factors for WMSD. Such tools may be administered as a pen-and-paper questionnaire that the affected worker completes prior to an examination by an HCP (e.g., Kourinka et al., 1987; Hales and Bertsche, 1992; NIOSH, 1997b; McConnell et al., 1999) or through structured interview by an HCP prior to physical examination (e.g., Ohlsson et al., 1994; Barr et al., 1999). The primary elements of a questionnaire tool for a WMSD management program are as follows:

- Demographic information (date of birth, height, weight, gender, hand dominance).
- Job characteristics (job title; time on the job; if less than 12 months in present job, title and time for last job; shift/work hours per week; primary job activities with regard to the duration of repetitive motions, forceful exertions, awkward or extreme postures, contact stress on body part(s), presence of vibration, poor lighting, and cold temperatures; symptoms within the last year believed by the worker to be work related).
- Current symptoms (anatomical regions affected; nature of symptoms, such as pain, numbness, aching, tingling, burning, swelling, stiffness, or other as described by the worker; temporal behavior of symptoms, particularly in relation to the work schedule).
- Leisure activities that may be comorbid risk factors for WMSD (e.g., athletic activities/exercise, gardening, sewing/knitting, computer use, playing a musical instrument).
- Past medical history (i.e., surgery or trauma; medical conditions such as diabetes, arthritis, inflammatory muscle disease, thyroid disease, hypertension, heart disease; chronic pain conditions; menopause or pregnancy; vision correction; smoking history).
- Psychosocial/work stress factors (e.g., work schedule control, group cohesion, time pressure, supervisor support, decision latitude).

Various literature references for questionnaire tools available to assess these primary elements are provided in Table 34.3. HCPs should be aware that employees are sometimes reluctant to share certain types of information with workplace occupational safety and health departments due to fear of reprisal for expressing negative opinions or feelings about work practices or supervisory personnel, or for providing personal information, such as past medical history, that may adversely affect access to workers' compensation resources. These are not necessarily unfounded fears, and they should be recognized and addressed in an open and forthright manner by HCPs. It is vital to disclose privacy policies in writing to affected employees at the time of their evaluation (i.e., in the introductory section of any forms or information sheets) in which explicit detail is given regarding what information will be shared and with whom. It may also be necessary to reassure employees verbally of confidentiality practices. In many cases, follow-up of a self-report survey with structured interview to fill in gaps in information may be necessary. Finally, assessments of psychosocial status can be particularly invasive and HCPs ultimately must respect an affected worker's right to privacy. It may be advisable in certain circumstances to refer an employee to an outside practitioner with no connection to the workplace for such assessments.

TABLE 34.3 Examples of Questionnaires Used in Epidemiological, Clinical, and Field Studies of Upper Extremtiy and Low Back Work Related Musculoskeletal Disorders

Reference	Name of Questionnaire	Purpose of Questionnaire
Fairbank (1980)	Oswestry Back Pain Disability Scale	Pain intensity and disability in activities of daily living caused by low back pain
Kourinka (1987)	Standardized Nordic Questionnaire	Musculoskeletal symptoms in different anatomical regions. Number of items depends on number of anatomical regions included
Sauter (1991)	Musculoskeletal Discomfort Questionnaire	Musculoskeletal discomfort (pain, tenderness, stiffness, numbness, or tingling) in 18 body regions using body diagram
Hales and Bertsche (1992) and NIOSH (1997b)	Symptoms Survey: Ergonomics Program	Job characteristics, symptoms in all anatomical regions, employee estimation of cause, treatments in the past year, effects on work, employee suggestion for improvement
Ware and Sherbourne (1992) and Ware et al. (1995)	MOS 36-item short-form health survey (SF-36); SF-12 Physical and Mental Health Summary Scales of the SF-36	SF-36: Eight subscales/dimensions (physical functioning, role-physical, bodily pain, general health perceptions, social functioning, vitality, mental health, and role-emotional) SF-12: Two subscales from SF-36 (physical functioning and mental health)
Leino and Hanninen (1995)	Psychosocial Workload and Musculoskeletal Symptom Survey	Four indices of psychosocial workload (work content, work control, social relationships, overstrain); musculoskeletal symptoms in all anatomical regions
Stock et al. (1990) and Durand et al. (2002)	Neck and Upper Limb Functional Status Index (NULI)	Impact of upper limb musculoskeletal disorders on five subscales (physical activities, work, psychosocial factors, sleep, and finances)
McConnell et al. (1999)	Disabilities of the Arm, Shoulder and Hand (DASH) Questionnaire	Four subsections (functional ability, symptom severity, sports/performing arts activities impact, and work impact)
Haufler et al. (2000) and Moos and Moos (1992)	Work Stressors and Work Resources Subscales of the Life Stressors and Social Resources Inventory	Items examining work stressors (relationship with supervisor, work pressure, and unpleasant physical conditions) and work resources (relationships with fellow employees, recognition for work, clearly defined responsibilities, work challenge, and possibility for initiative)
Halpern (2000) and Halpern et al. (2001)	Job Description Questionnaire (JDQ)	Perceptions of exposure to physical (exertion, equipment, and environment) and work organization (work control, work pace, resources) risk factors associated with low back pain
MacDonald et al. (2001)	Psychosocial Stessor Measure	Ten stressor measures (job strain, psychological demands, decision latitude, mental work load, poor work schedule control, lack of group cohesion, group pressure, lack of supervisor support, lack of coworker support, opinions not accepted by work group)
Salerno et al. (2001) Strasser et al. (1999)	Upper Extremity Questionnaire	Part 1: demographic, medical and exercise history, and symptom reports in three body regions (neck/shoulder/upper arm, elbow/forearm; wrist/hand/finger). Part 2: subtests of psychosocial conditions at work (perceived stress and job content)
Dane et al. (2002) Shaw et al. (2002)	Job Tasks Section of the United States Air Force Job Requirements and Physical Demands Survey (JR-PDS)	Perceived level of exposure to workplace risk factors (administrative, workstation, environmental, equipment-related, and exertion)

(continued)

TABLE 34.3 *Continued*

Reference	Name of Questionnaire	Purpose of Questionnaire
Durand et al. (2002)	Work Disability Diagnosis Interview (WoDDI)	Information about clues of a work handicap situation obtained through structured interview by HCP in eight catagories (demographic characteristics and work history, pain syndrome, general health and previous health history, family and social hisotry, medical history, work environment, patient perception of disabilty status)
Haldorsen (2002)	Prognosis for Return to Work, Psychosocial Factors Questionnaire	Psychological and motivational factors shown to predict return to work (ability to work, pain, belief in personal control over pain, general well-being, expectations regarding return to work)

34.6.2 Physical Examination

After an affected worker completes the self-report questionnaire, the HCP conducts a physical examination that should be guided by the subjective report of symptoms provided by the worker. Depending on the expertise of the HCP, this examination may be geared toward making an appropriate diagnosis for a specific set of symptoms, or it may be geared toward ruling out, or screening for, certain severe conditions requiring further expert diagnosis. This latter triage process, which fits the algorithm depicted in Figure 34.2, relies on the detection of "flags" that are summarized in Table 34.4.

A thorough physical examination for a WMSD should include the following key elements:

- Provocative tests for peripheral nerve function (e.g., Phalen's sign for median nerve irritation, straight leg raise for sciatic nerve irritation).
- Inspection for signs of swelling, inflammation, or infection (e.g., red pallor, increased skin temperature, suppuration, wounds, pitting edema, rash, deformity).
- Passive, active, and functional range of motion of affected joints and immediately adjacent joints with recording of symptoms on motion.
- Resisted efforts with recording of weakness and symptoms on resistance.
- Palpation of musculotendinous, skeletal, peripheral nerve, and ligamentous structures with recording of symptoms on palpation.

There is no gold standard for the physical examination of workers with WMSD, and workplace ergonomics programs should rely on the expertise of the various HCP team members to determine what tests are most appropriate for a particular WMSD case.

34.6.3 Assessment and Diagnosis

Although the OSHA Ergonomics Program Rule was repealed, the recommendations contained in that document regarding assessment and diagnosis represent sound practice for WMSD management. These recommendations state that an HCP written opinion about a WMSD case must contain the following information (Federal register, 2000, p. 68851):

1. The HCPs assessment of the employee's medical condition as related to the physical work activities, risk factors and [W]MSD hazards in the employee's job;
2. Any recommended work restrictions, including, if necessary, time off work to recover, and any follow-up needed;

TABLE 34.4   Summary of Indicators or "Flags." Found on Worker Evaluation, Which Help to Direct Outside Referral, Diagnosis, and Management of WMSD

Category	Indicators
"Red" flags: indicators of serious physical conditions requiring special tests or referral for specialist diagnosis[a–c]	Neurological symptoms or signs (sensory loss, motor weakness, muscle atrophy parasthesiae, positive provocative tests, incoordination, impaired speech, or hyperreflexia)
	Severe swelling, edema, redness, increased skin temperature, or skin discoloration
	Joint or structural deformity
	History of violent trauma
	Constant, progressive, nonmechanical pain
	Age less than 20 yr or greater than 55 yr
	Chest pain
	History of cancer, substance abuse, HIV, inflammatory arthritis or myositis, diabetes, or vascular disease
	Fever
	Rapid, unexplained weight loss
	Being in poor health or physically unwell
	Pain that increases when lying down or that is severe and unremitting, especially at night
	Persistent, severe restriction of lumbar flexion
	Bowel or bladder control problems
	Severe tenderness to palapation
	Palpable mass (whether painful or not)
	Tenderness with motion
Personal psychosocial risk factors for developing chronic WMSD[c,d]	Counterproductive attitudes (fear-avoidance, passive attitude toward rehabilitation, negative beliefs, catastrophizing, unrealistic goals, or low expectations for rehabilitation)
	Significant reduction in activity level or withdrawal from activities of daily living
	Excessive reliance on aids or appliances
	Reporting excessively high levels of pain
	Increased use of alcohol or other substances
	Compensation issues (lowest or highest compensation benefits, history of filing injury claims, history of extended leave for injuries or pain, or past adversarial experience with work injury case manager)
	High frequency of health-care visits
	Negative emotions (hopelessness, hyperattentiveness to pain or health problems, depression, anxiety, irritability, disinterest, high levels of stress, or feeling out of control)
	Family issues (overly solicitous or punitive family members, poor communication, single or divorced, or having dependants)
	Long work absence
	Social isolation
	Poor coping strategies
Occupational psychosocial risk factors for developing chronic WMSD[b,c]	Physically demanding job
	Unsupportive or unhelpful work environment (supervisors and fellow workers)
	Low decision latitude/control
	High pressure to produce
	Light duty unavailable
	Low eductional level or socioeconomic status
	Low job satisfaction
	Patterns of frequent job change or short job tenure

(continued)

TABLE 34.4 *Continued*

Category	Indicators
Organizational risk factors that may influence worker disability[c,d]	National policies/laws/programs (minimum wage rate, workers' compensation benefits, disability benefits)
	Company policies (sick leave policy, occupational health procedures, work accommodations)
	Health-care practices (Iatrogenesis due to incompatible information or diagnoses from different HCPs, unsound medical advice for prolonged rest or leave of absence from work, provision of passive treatment; stigmatization)

[a]Barr et al. (1999).
[b]Durand et al. (2002).
[c]Gatchel (2004).
[d]Cedraschi et al. (1998).

3. A statement that the HCP has informed the employee of the results of the evaluation, the process to be followed to effect recovery, and any risk factors and [W]MSD hazards in the employee's job; and
4. A statement that the HCP has informed the employee about work-related or other activities that could impede recovery from the injury.

Important aspects of these recommendations are the obligation of the HCP to identify a specific medical condition and asses its relationship to workplace risk factors, to clearly communicate suggested interventions, whether they are clinical or ergonomic in nature, and to educate the affected worker about both workplace and nonworkplace risk factors that may be contributing to his or her symptoms.

Specific diagnoses should be made by HCPs whenever possible according to the ICD-9 criteria (Table 34.1). Umbrella terms, such as WMSD or RSI, are ambiguous and they should not be used as medical diagnoses. For findings on patient evaluation that defy specific diagnostic criteria, categories for unspecified disorders are available based on the likely tissues affected (see Table 34.1). To determine the relationship of the employee's work to his or her diagnosis, HCPs must consider the employee's report concerning the temporal behavior of symptoms and the presence of probable risk factors (see Section 34.6.1). The HCP should also rely on any job analyses available from other ergonomics program team members. If such an analysis has not been performed, the HCP should recommend that it be done (see Figure 34.1). Although the HCP is obligated to explore the presence of nonworkplace risk factors and educate the employee accordingly, such factors are generally not communicated to the ergonomics program members as a whole, since doing so would constitute a breach of confidence and possible invasion of the employees' privacy. With a thorough assessment, the specific treatment options and the goals for treatment outcomes are clarified. Some of these treatment options and goals will be highlighted in the next section.

34.7 Treatment of Workers with WMSD

Recent research in an animal model of repetitive motion has revealed some of the tissue pathophysiology that occurs in the early stages of WMSD development. These changes are consistent with the diagnoses summarized in Table 34.1. For example, in rats performing highly repetitive or forceful reaching and grasping, microtrauma of muscle, tendon, ligament, peripheral nerve, and bone tissues is evident after 4 to 6 weeks of exposure (Barbe et al., 2003; Barr et al., 2003). After 9 to 12 weeks of exposure to either a high repetition–low force or a high repetition–high force task, nerve conduction velocity slowing of the median nerve at the carpal tunnel can be measured (Clark et al., 2003, 2004). Furthermore, animals exhibit both avoidance behaviors as well as develop uncoordinated movement patterns that indicate the presence of pain and somatosensory degradation (Barbe et al., 2003; Clark et al., 2004). Reduction of exposure level (i.e., decreasing task repetitiveness and forcefulness) attenuates the severity of these changes, thereby proving that a primary underlying cause of the tissue and behavioral changes is exposure to physical risk factors (Barr and Barbe, 2004). These findings help to direct the development of effective interventions for WMSD, which will be the focus of this section.

Clinical studies support the presumption that early intervention reduces long-term disability and improves rates of return to work (e.g., Jordan et al., 1998; Linz et al., 2002; Gatchel et al., 2003; Gatchel, 2004). Furthermore, multidisciplinary medical management programs, such as described in previous sections, may also aid in the identification of workers at high risk for chronicity, so that appropriate treatment intensity and emphasis can be implemented to optimize recovery and return to work (Kasdan and Lewis, 2002; Haldorsen et al., 2003; Gatchel et al., 2003). This section will summarize how the emphasis on the various aspects of treatment in a multidisciplinary WMSD management program may change for cases in the acute, subacute, and chronic phases. Finally, common treatment pitfalls that may contribute to iatrogenesis will be identified.

34.7.1 Acute-Phase Treatment

Acute-phase WMSDs are characterized by symptoms or work disability of less than 4 weeks duration. As the animal studies cited earlier suggest, such symptoms result from an inflammatory reaction to tissue microtrauma. The goal of intervention, therefore, is to promote tissue recovery and restoration of full function through the resolution of the acute inflammatory episode before permanent tissue changes or chronic inflammation ensues. The following treatments may be included in such a conservative, acute-phase management program:

- Reduce inflammation (e.g., temporary rest/immobilization, nonprescription nonsteroidal anti-inflammatory drugs [NSAIDs], application of ice for the first week followed by application of heat as needed).
- Reduce workplace and nonworkplace exposure to physical ergonomic risk factors.
- Restoration of function (e.g., gentle stretching and range of motion exercises for the first week followed by graded strengthening to tolerance).
- Graded increase in work activity with return to normal duty.
- Regular follow-up (per Figure 34.2) to assess effectiveness of interventions.

34.7.2 Subacute-Phase Treatment

Subacute-phase WMSDs are characterized by symptoms or work disability of 4 to 12 weeks and may include absence from work. This is a critical period in the development of a WMSD case, in that failure to resolve work disability before prolonged work absence occurs may lead to long-term disability (Gatchel, 2004). Therefore, the emphasis of treatment in this phase shifts toward determining predictors of chronicity, managing rather than completely curing symptoms, and addressing psychosocial as well as physical issues. It is with this group of affected workers that the full benefits of a comprehensive WMSD management program must be brought to bear. Treatment options in the subacute phase of WMSD include the following:

- Conservative management as in acute phase and supplemented as appropriate for specific diagnoses (e.g., prescription anti-inflammatory medication, iontophoresis, steroid injection; Pilligian et al., 2000).
- Screen for predictors of chronicity (Haldorsen et al., 2002; Gatchel et al., 2003) and adjust treatment intensity accordingly.
- Work restrictions if possible. If employee is not working, engage in simulated work activities (e.g., work hardening) with an emphasis on functional recovery.

- Improve general physical fitness through aerobic exercise two to three times per week and graded strengthening with an emphasis on a home exercise regimen (Jacobson and Aldana, 2001; Linz et al., 2002; Silverstein and Clark, 2004).
- Psychosocial interventions, such as stress management.
- Regular follow-up (per Figure 34.2) to assess effectiveness of interventions.

34.7.3   Chronic-Phase Treatment

Chronic-phase WMSDs are characterized by symptoms and severe work disability or absence of greater than 12 weeks. If affected workers have experienced significant work absence due to their WMSD, they are far less likely to return to work without intensive, multidisciplinary intervention (Pransky et al., 1999; Gatchel et al., 2003; Gatchel, 2004). While the treatment options are similar to those used for subacute WMSD cases, chronic WMSD treatment should be more frequent and intense with clear emphasis on psychosocial interventions (Feuerstein et al., 1993, 2003b; Pransky et al., 1999). Treatment options in the chronic phase of WMSD management include the following:

- Conservative management as in acute and subacute phases and supplemented as appropriate for specific diagnoses (e.g., prescription anti-inflammatory medication, iontophoresis, steroid injection; Pilligian et al., 2000).
- Daily physical conditioning, including aerobic exercise, individualized exercises for specific WMSD diagnosis, and graded work simulation with an emphasis on optimizing functional recovery (Feuerstein et al., 1993).
- Daily sessions with clinical psychologist in pain and stress management (e.g., relaxation techniques, cognitive coping strategies, assertiveness training to facilitate workplace communication, problem-solving skills for use in workplace and nonworkplace situations; Feuerstein et al., 1993; Gatchel, 2004).
- Additional specialized psychological assessment and treatment for significant mental health conditions (Pransky et al., 1999; Schade et al., 1999).
- Home exercise program with an emphasis on maintaining aerobic fitness (Linz et al., 2002).
- Regular follow-up, through a case manager, to assess the effectiveness of interventions.

34.7.4   Treatment Pitfalls

In the subacute and chronic phases of WMSD rehabilitation, interventions in which the worker is a passive recipient of the treatment should be minimized. Examples of such passive treatments include massage and manual therapy. Although such treatments may be effective in symptom reduction, they tend to encourage patient dependence and undermine patient responsibility in long-term WMSD management.

Stigmatization of workers with WMSD by HCPs should be discouraged. Terms such as "malingerer" or "symptom amplifier" may apply as much to the frustration of clinicians in their inability to cure a WMSD as to the patient's attempts to "milk the system." In either case, these terms cause breakdown in effective communication between the affected worker and the HCP team, which has been shown to impede treatment progress and ultimate outcome (Cedraschi et al., 1998). Furthermore, HCPs should remember that they as well as their patients gain financially from maximizing and prolonging workers' compensation benefits for health-care services. In cases where worker's compensation or litigation issues contribute to a WMSD case, these issues should be handled as mere components in the biopsychosocial gestalt of that case and interventions should be directed toward reducing their impact on worker disability.

While the use of NSAIDs and immobilization devices, such as splints, have a place in conservative WMSD management, workers should be advised not to continue work at full capacity while using such treatments without frequent follow-up. The point of these conservative measures is to promote tissue recovery. Continued performance of work at a level of exposure associated with tissue injury may be counterproductive to such recovery or may contribute to further injury. Again, frequent follow-up is needed to ensure that additional tissue damage is not taking place.

Finally, in spite of current knowledge, some patients with WMSD may present with puzzling symptoms or examination findings that are either inconsistent with diagnostic criteria or are unresponsive to conventional treatment. In such cases, HCPs tend to blame the worker (see comments on stigmatization given previously). Recent evidence in both animal models and patients suggest that changes to the central nervous system in response to chronic, painful conditions may lead to abnormal movement patterns, heightened pain sensations, or both (Byl and Melnick, 1997; Byl et al., 1997; Byl and McKenzie, 2000). HCPs who encounter such clinical findings should consult the recent literature regarding emerging treatment alternatives for such WMSD cases.

34.8 Summary

As integral members of a multidisciplinary ergonomics program, HCPs play an essential role in the diagnosis and clinical management of WMSDs. However, the duties of HCPs extend beyond the clinic to maintaining clear communication with other ergonomics program team members, particularly through a clinical case manager, so that appropriate accommodations in the workplace are implemented. Follow-up of affected workers with regard to treatments as well as other accommodations is necessary to monitor progress and program success. While most HCPs deliver secondary prevention for WMSDs, their participation as a member of a comprehensive ergonomics program may guide primary prevention strategies as well. Early intervention is tantamount in reducing chronicity and maintaining worker employment status. Treatments should emphasize a biopsychosocial approach in which workers are encouraged to take an active role in the long-term management of their own disorders. As WMSDs progress from acute to subacute and chronic phases, treatments should become more frequent and more focused on the psychosocial aspects of the disorders. Care must be taken to avoid pitfalls that may undermine successful treatment and contribute to long-term disability.

References

Barbe MF, Barr AE, Amin A, Gorzelany I, Gaughan JP, Safadi F (2003). Repetitive reaching and grasping causes motor decrements and systemic inflammation. *J Orthop Res* 21:7–16.

Barr AE (2002). Approach to workplace management of work-related musculoskeletal disorders. In *Rehabilitation of the Hand: Surgery and Therapy*, 5th edition (Vol. 1), pp. 996–1004, Mosby, Philadelphia.

Barr AE, Barbe MF (2004). Inflammation reduces physiological tissue tolerance in the development of work related musculoskeletal disorders. *J Electromyogr Kinesiol* 14:77–85.

Barr AE, Badenchini IT, Forsyth-Bee M, Duff JM, Herring KM, Covit AB, Nordin M (1999). Development of a physical examination for a company-based management program for work-related upper extremity cumulative trauma disorders. *J Occup Rehabil* 9:63–77.

Barr AE, Safadi FF, Gorzelany I, Amin M, Popoff SN, Barbe MF (2003). Repetitive, negligible force reaching in rats induces pathological overloading of upper extremity bones. *J Bone Mineral Res* 18:2023–2032.

Bernacki EJ, Tsai SP (2003). Ten years' experience using and integrated workers' compensation management system to control workers' compensation costs. *J Occup Environ Med* 45:508–516.

Bernard BP, ed. (1997). *Musculoskeletal Disorders (MSDs) and Workplace Factors: A Critical Review of Epidemiologic Evidence for Work-Related Musculoskeletal Disorders of the Neck, Upper Extremity, and Low Back*. DHHS (NIOSH) Publication No. 97–141.

Bureau of Labor Statistics (1989–2001). Occupational injuries and illnesses: industry data, http://www.bls.gov/iif/home/htm, accessed 1/30/04.

Bureau of Labor Statistics (2001). Table R15: number of nonfatal occupational injuries and illnesses involving days away from work by nature of injury or illness and selected events or exposures leading to injury or illness, http://www.bls.gov/iif/oshwc/osh/case/ostb1170.pdf, accessed 1/30/04.

Bureau of Labor Statistics, News (2002). United States Department of Labor, USDL 02-687, December 19, http://www.bls.gov/iif/home.htm, accessed 5/6/03.

Bureau of Labor Statistics, News (2003a). United States Department of Labor, USDL 03-138, March 27, http://www.bls.gov/iif/home.htm, accessed 5/6/03.

Bureau of Labor Statistics, News (2003b). United States Department of Labor, USDL 03-913, December 18, http://www.bls.gov/iif/home.htm, accessed 12/30/03.

Byl NN, McKenzie A (2000). Treatment effectiveness for patients with a history of repetitive hand use and focal dystonia: a planned, prospective follow-up study. *J Hand Ther* 13:289–301.

Byl NN, Melnick M (1997). The neural consequences of repetition: clinical implications of a learning hypothesis. *J Hand Ther* 10:160–174.

Byl NN, Merzenich MM, Cheung S, Bedenbaugh P, Nagarajan S, Jenkins WM (1997). A primate model for studying focal dystonia and repetitive strain injury: effects on primary somatosensory cortex. *Phys Ther* 77:269–284.

Cedraschi C, Nordin M, Nachemson Al, Vischer TL (1998). Health care providers should use a common language in relation to low back pain patients. *Baillieres Clin Rheumatol* 12:1–15.

Clark BD, Barr AE, Safadi FF, Beitman L, Al-Shatti T, Barbe MF (2003). Median nerve trauma in a rat model of work-related musculoskeletal disorder. *J Neurotrauma* 20:681–695.

Clark BD, Al-Shatti TA, Barr AE, Amin M, Barbe MF (2004). Performance of a high-repetition, high-force task induces carpal tunnel syndrome in rats. *J Orthop Sports Phys Ther* 34:244–253.

Clauw DJ, Williams DA (2002). Relationship between stress and pain in work-related upper extremity disorders: the hidden role of chronic multisymptom illnesses. *Am J Ind Med* 41:370–382.

Dane D, Feuerstein M, Huang G-D, Dimberg L, Ali D, Lincoln A (2002). Measurement properties of a self-report index of ergonomic exposures for use in an office work environment. *J Occup Environ Med* 44:73–81.

Dersh J, Gatchel RJ, Polatin P, Mayer T (2002). Prevalence of psychiatric disorders in patients with chronic work-related musculoskeletal pain disability. *J Occup Environ Med* 44:459–468.

Durand M, Loisel P, Hong Q, Charpentier N (2002). Helping clinicians in work disability prevention: the work disability diagnosis interview. *J Occup Rehabil* 12:191–204.

Fairbank JCT, Couper J, Davies JB, O'Brien JP (1980). The Oswestry low back pain disability questionnaire. *Physiotherapy* 66:271–273.

Federal Register (2000). 29 CFR Part 1910: ergonomics program; final rule. Federal Register; 65(220, Part II):68262–68870.

Feuerstein M, Callan-Harris S, Dyer D, Armbruster W, Carosella AM (1993). Multidisciplinary rehabilitation of chronic work-related upper extremity disorders: long-term effects. *J Occup Med* 35:396–403.

Feuerstein M, Shaw W, Miller VI, Wood P, Berger R, Lincoln AE, Hickey PF (1999). Integrated case management for work-related upper extremity disorders provider manual, http://www.georgetown.edu/research/rwj-icm/provman.html/, accessed 1/20/04.

Feuerstein M, Marshall L, Shaw WS, Burrell LM (2000). Mulitcomponent intervention for work-related upper extremity disorders. *J Occup Rehabil* 10:71–83.

Feuerstein M, Huang GD, Ortiz JM, Shaw WS, Miller VI, Wood PM (2003a). Integrated case management for work-related upper extremity disorders: impact of patient satisfaction on health and work status. *J Occup Environ Med* 45:803–812.

Feuerstein M, Shaw WS, Lincoln AE, Miller VI, Wood PM (2003b). Clinical and workplace factors associated with a return to modified duty in work-related upper extremity disorders. *Pain* 102:51–61.

Gatchel RJ (2004). Musculoskeletal disorders: primary and secondary interventions. *J Electromyogr Kinesol* 14:161–170.

Gatchel RJ, Polatin PB, Noe C, Gardea M, Pulliam C, Thompson J (2003). Treatment- and cost-effectiveness of early intervention for acute low-back pain patients: a one-year prospective study. *J Occup Rehabil* 13:1–9.

Hales TR, Bertsche PK (1992). Management of upper extremity cumulative trauma disorders. *AAOHN J* 40:118–128.

Haldorsen EMH, Grasdal AL, Skouen JS, Risa AE, Kronholm K, Ursin H (2002). Is there a right treatment for a particular patient group? Comparison of ordinary treatment, light multidisciplinary treatment, and extensive multidisciplinary treatment for long-term sick-listed employees with musculoskeletal pain. *Pain* 95:49–63.

Halpern M (2000). The effects of low back pain on the perception of job demands. Proceedings of the International Ergonomics Association 2000 and Human Factors and Ergonomics Society Congress 2000; 4:765–768.

Halpern M, Hiebert R, Nordin M, Goldsheyder D, Crane M (2001). The test-retest reliability of a new occupational risk factor questionnaire for outcome studies of low back pain. *Appl Ergon* 32:39–46.

Haufler AJ, Feuerstein M, Huang GD (2000). Job stress, upper extremity pain and functional limitations in symptomatic computer users. *Am J Ind Med* 38:507–515.

Jacobson BH, Aldana SG (2001). Relationship between frequency of aerobic activity and illness-related absenteeism in a large employee sample. *J Occup Environ Med* 43:1019–1025.

Jordan KD, Mayer TG, Gatchel RJ (1998). Should extended disability be an exclusion criterion for tertiary rehabilitation? Socioeconomic outcomes of early versus late functional restoration in compensation spinal disorders. *Spine* 23:2110–2116.

Kasdan ML, Lewis K (2002). Management of carpal tunnel syndrome in the working population. *Hand Clin* 18:325–330.

Kerr MS, Frank JW, Shannon HS, Norman RWK, Wells RP, Neumann WP, Bombardier C (2001). Biomechanical and psychosocial risk factors for low back pain at work. *Am J Public Health* 91:1069–1075.

Kourinka I, Jonsson B, Kilbom A, Vinterberg H, Biering-Sorensen F (1987). Standardized nordic questionnaire for the analysis of musculoskeletal symptoms. *Appl Ergon* 18:233–237.

Leino PI, Hanninen V (1995). Psychosocial factors at work in relation to back and limb disorders. *Scand J Work Environ Health* 21:134–142.

Linz DH, Shepherd CD, Ford, LF, Ringley LL, Klekamp J, Duncan JM (2002). Effectiveness of occupational medicine center-based physical therapy. *J Occup Environ Med* 44:48–53.

Loisel P, Durand P, Abenheim L, Gosselin L, Simard R, Turcotte J, Esdaile JM (1994). Management of occupational back pain: the sherbrooke model. Results of a pilot and feasibility study. *Occup Environ Med* 51:597–602.

Loisel P, Durand M, Diallo B, Vachon B, Charpentier N, Labelle J (2003). From evidence to community practice in work rehabilitation: the Quebec experience. *Clin J Pain* 19:105–113.

Lundberg U (2002). Psychophysiology of work: stress, gender, endocrine response, and work-related upper extremity disorders. *Am J Ind Med* 41:383–392.

MacDonald LA, Karasek RA, Punnett L, Scharf T (2001). Covariation between workplace physical and psychosocial stressors: evidence and implications for occupational health research and prevention. *Ergonomics* 44:696–718.

McConnell S, Beaton DE, Bombardier C (1999). *The DASH Outcome Measure: A User's Manual*, Institute for Work & Health, Toronto, Ontario.

Moos RH, Moos BS (1992). *Life Stressors and Social Resources Inventory: Adult Form Manual*. Center for Health Care Evaluation, Stanford University and Department of Veterans Affairs Medical Centers, Palo Alto, CA.

National Institute of Occupational Safety and Health (1997a). Ergonomics: effective workplace practices and programs. Transcripts of Presentations for Conference Sponsored by NIOSH and OSHA, Chicago, IL, http://www.cdc.gov/niosh/homepage.html.

National Institute of Occupational Safety and Health (1997b). *Elements of Ergonomics Programs: A Primer Based on Workplace Evaluations of Musculoskeletal Disorders.* DHHS (NIOSH) Publication No. 97–117.

National Research Council and Institute of Medicine (2001). *Musculoskeletal Disorders and the Workplace: Low Back and Upper Extremities,* National Academy Press, Washington, DC.

Ohlsson K, Attewell RG, Johnsson B, Ahlm A, Skerfving S (1994). An assessment of neck and upper extremity disorders by questionnaire and clinical examination. *Ergonomics* 37:891–897.

OSHA (1993). *Ergonomics Program Management Guidelines for Meatpacking Plants.* U.S. Department of Labor, Occupational Safety and Health Administration, OSHA 3123 (reprinted).

OSHA (2002). National News Release: OSHA Announces Comprehensive Plan To Reduce Ergonomic Injuries: Targeted Guidelines and Tough Enforcement Two Key Elements, USDL 02-201, April 5.

OSHA (2003a). *Ergonomics for the Prevention of Musculoskeletal Disorders: Draft Guidelines for Retail Grocery Stores,* http://www.osha.gov/ergonomics/guidelines/grocerysolutions/, accessed 12/31/03.

OSHA (2003b). *Ergonomics for the Prevention of Musculoskeletal Disorders: Draft Guidelines for Poultry Processing,* http://www.osha.gov/ergonomics/guidelines/poultryprocessing/, accessed 12/31/03.

OSHA (2003c). *Guidelines for Nursing Homes: Ergonomics for the Prevention of Musculoskeletal Disorders,* www.osha.gov/ergonomics/guidelines/nursinghome/index.html.

OSHA (2003d). Trade News Release: Ergonomics Guidelines Announced for the Nursing Home Industry: OSHA Recommends Eliminating Manual Lifting Of Residents When Feasible, USDL 03-128, March 13.

Pilligian G, Herbert R, Hearns M, Dropkin J, Landsbergis P, Cherniack M (2000). Evaluation and management of chronic work-related musculoskeletal disorders of the distal upper extremity. *Am J Ind Med* 37:75–93.

Pransky G, Benjamin K, Himmelstein J, Mundt K, Morgan W, Feuerstein M, Koyamatsu K, Hill-Fotouhi C (1999). Work-related upper extremity disorders: a prospective evaluation of clinical and functional outcomes. *J Occup Environ Med* 41:884–892.

Salerno DF, Franzblau A, Armstrong TJ, Werner RA, Becker MP (2001). Test-retest reliability of upper extremity questionnaire among keyboard operators. *Am J Indust Med* 40:655–666.

Sauter SL, Schleifer LM, Knutson SJ (1991). Work posture, workstation design, and musculoskeletal discomfort in a VDT data entry task. *Human Factors* 33:151–167.

Schade V, Semmer N, Main CJ, Hora J, Boos N (1999). The impact of clinical, morphological, psychosocial and work-related factors on the outcome of lumbar discectomy. *Pain* 80:239–249.

Shaw WS, Feuerstein M, Lincoln AE, Miller VI, Wood PM (2002). Ergonomic and psychosocial factors affect daily function in workers' compensation claimants with persistent upper extremity disorders. *J Occup Environ Med* 44:606–615.

Silverstein B, Clark R (2004). Interventions to reduce work-related musculoskeletal disorders. *J Electromyogr Kinesiol,* 14:135–152.

Stock S, Streiner D, Tugwell P, Loisel P, Reardon R, Durand MJ (1996). Validation of the neck and upper limb index (NULI), a functional status instrument for work related musculoskeletal disorders. In: 25eme Congres annuel de la commission internationale de sante au travail (CIST), Stockholm.

Strasser PB, Lusk SL, Franzblau A, Armstrong TJ (1999). Perceived psychosocial stress and upper extremity cumulative trauma disorders. *AAOHN J* 47:22–30.

Symonds TL, Burton AK, Tillotson KM, Main CJ (1995). Absence resulting from low back trouble can be reduced by psychosocial intervention at the work place. *Spine* 20:2738–2745.

Tittiranonda P, Burastero S, Rempel D (1999). Risk factors for musculoskeletal disorders among computer users. *Occup Med State of the Art Rev* 14:17–38.

Ware JE, Sherbourne CD (1992). The MOS 36-item short-form health survey (SF-36). *Med Care* 30:473–483.

Ware JE, Kosinski M, Keller SD (1995). *SF-12: How to Score the SF-12 Physical and Mental Health Summary Scales.* The Health Institutes, New England Medical Center, Boston.

Warren N, Dillon C, Morse T, Hall C, Warren A (2000). Biomechanical, psychosocial, and organizational risk factors for WRMSD: population-based estimates from the Connecticut Upper-Extremity Surveillance Project (CUSP). *J Occup Health Psychol* 5:164–181.

35

Postinjury Rehabilitation/ Management

Tom G. Mayer
*University of Texas
  Southwestern Medical Center
  at Dallas and PRIDE Research
  Foundation*

Robert J. Gatchel
*University of Texas at Arlington
and University of Texas
Southwestern Medical Center
at Dallas*

Skye Porter

Brian R. Theodore
PRIDE Research Foundation

35.1 Introduction

Postinjury rehabilitation for musculoskeletal disorders can be classified into three distinct levels of treatment.[1] Diagnosis and the anticipated healing time from an inciting event dictate the timing of administering the treatment, as the following delineates:

(a) *Primary care.* The term "rehabilitation" is used guardedly in this stage of postinjury non-operative care, because patients may be severely limited by acute inflammatory changes from active participation in treatment. It is provided during the acute stage of an injury, focusing primarily on pain control but also taking into consideration the avoidance of deconditioning. Duration of primary care is determined by the type of injury. Thermal (heat/cold) modalities, medication, immobilization, bed rest, traction, and injection methods constitute some of the treatments in primary care.

(b) *Secondary rehabilitation.* This is provided in the postacute phase of an injury, with a focus on preventing chronic disability by avoiding physical deconditioning, medication habituation, and disruptive psychological reactions before they occur. This phase begins when there is sufficient tissue healing and stabilization of the injury to permit progressive motion and strengthening exercises. Major treatment modalities consist of exercise regimens that emphasize active joint mobilization and strengthening of the para-articular muscles, which may be reinforced by pain control methods.

(c) *Tertiary rehabilitation.* This is provided for a small percentage of cases that go on to chronic disabling musculoskeletal pain not responding to early surgical and nonoperative intervention. The main focus is the prevention of permanent disability for the patients who have experienced deconditioning and chronic distress. There are two approaches to tertiary care. Functional restoration emphasizes intensive physical training within an interdisciplinary cognitive behavioral disability management program. On the other hand, interventional pain management consists of injections, often with long-term opiates accompanied by a psychosocial focus on de-emphasizing the pain experience, with patients playing a more passive role. Neuroablative procedures and devices are sometimes utilized for palliation in end-stage cases. Tertiary care begins when chronicity is established, usually 4–6 months following injury. However, a functionally based program may be appropriate if significant psychosocial or treatment resistance problems emerge, with total disability, in an early intervention approach at 6–8 weeks.

35.2 Primary Nonoperative Care

Primary care is focused on controlling pain symptoms, and preparing the musculoskeletal system for proper healing from the injuries sustained. Duration of primary care is determined by the type of injury; typically, 10–14 days in the case of mild sprains, strains and lacerations, and ranging from 8 to 12 weeks when dealing with complex fractures and dislocations. As such, primary rehabilitation is concerned with the treatment of acute stages of pain and disability. Treatment modalities include thermal (heat/cold) application, medication, immobilization, bed rest, traction, and injection methods. The literature does not conclusively demonstrate the superiority of any one modality over others.[1] Upon receiving primary care, patients have a 50–60% positive response to early symptomatic treatment within 8 weeks, provided there is no significant surgically treatable pathology. They should be able to return to work as symptoms decrease, or stay at work during healing.

35.3 Secondary Rehabilitation

The goals of secondary care are restoring function to the musculoskeletal system, preventing chronic disability, and avoiding physical deconditioning, medication habituation, and disruptive psychological reactions before they occur. This program is administered to patients who do not recover function or overcome symptoms as the tissue heals. As such, secondary rehabilitation is concerned with patients who may be at risk for progressing into the chronic stage of pain and disability. Secondary rehabilitation can take a single discipline or an interdisciplinary approach (Guidelines for Low Back Pain), as reviewed in the following text.

35.3.1 Physical Therapy

The main modality in secondary care is an exercise regimen to stimulate active joint mobilization and strengthening of the para-articular muscles. However, most disorders require multiple components to provide more comprehensive care. The simplest type of secondary rehabilitation simply restores mobility during the early post-acute process, even before muscle wasting and recruitment problems arise. Later stages of secondary rehabilitation prevent early deconditioning from progressing. They may include a single discipline (usually physical therapy) exercise treatment, including aerobic training, aquatic therapy, muscle strengthening, yoga-type exercise, balance and positional tolerance training, or stabilization exercises. Patients are unlikely to benefit simply from "home exercises," because the pain experience is generally associated with fear of movement or *kinesophobia*. Pain control modalities, including bracing, manual medicine, thermal modalities, medications, and injections, may be commonly used during secondary rehabilitation to provide pain control associated with the later stages of tissue healing. Postoperative rehabilitation usually utilizes secondary care approaches. *Interdisciplinary programs*, such as work hardening, may be useful for a smaller cohort of more difficult patients reaching higher levels of chronicity and the postoperative period. Such programs should be reserved for patients unlikely to respond to single discipline care, who demonstrate good effort and consistency on functional testing, and who lack complicating psychosocial issues, such as medication dependence or psychiatric mood disorders.

Preventing immobility through activation of the patient and exercise to restore muscle flexibility, muscle balance, and coordination are paramount in restoring function. Ultimately, these exercise programs should emphasize the strengthening of weaker structures while not aggravating pain symptoms. Prior to starting focused strengthening programs, emphasis should be on exercise in nonpainful ranges and planes of motion. At this stage, injections for pain control may be utilized to facilitate active participation in the therapy by controlling the pain. As rehabilitation progresses, the exercise regimens should be focused on endurance and return to preinjury function. Studies focusing on acute low back pain have shown that these types of individually tailored, graded exercise regimens are associated with improvements in several critical socioeconomic outcomes such as return to work and reduced sick leave.[1-3]

35.3.2 Psychosocial Interventions

In addition to physical therapy, psychosocial interventions may be introduced as adjuncts on an as-needed basis.[4] These interventions are used when patients exhibit an unwillingness to engage in therapy due to severe pain or secondary gain factors. The first step involves an evaluation of the socioeconomic and psychosocial barriers to recovery. When socioeconomic barriers are present, a disability case manager may be assigned to discuss and clarify the issues. In mild cases, the patient may still continue the secondary rehabilitation program with supportive counseling. When the evaluation uncovers severe levels of emotional distress, psychotropic medication, and cognitive–behavioral counseling are usually added as part of a comprehensive tertiary approach.

35.3.3 Evidence for Interdisciplinary Approaches to Secondary Care

Interdisciplinary approaches to secondary care are usually termed work hardening programs. In addition to restoring function, these programs also aim to improve work status. Various studies have documented the increased effectiveness of an interdisciplinary approach to secondary care relative to a single discipline approach, especially in terms of return-to-work outcomes.[5-7] A systematic review of randomized and nonrandomized controlled clinical trials, that evaluated the effectiveness of interdisciplinary rehabilitation for subacute low back pain among working age adults, indicated moderate evidence for its effectiveness for subacute low back pain, with increased effectiveness if a workplace visit is incorporated into the treatment program.[8] Additionally, a number of large-scale reviews and prospective studies showed that there is no effect at all on return-to-work in low back pain patients when only physical exercise was utilized.[9,10]

35.4 Tertiary Rehabilitation

Prevention of permanent disability for the person who has experienced deconditioning and chronic distress is the goal of tertiary rehabilitation. Tertiary programs are multidisciplinary and interdisciplinary in nature and require medical direction. The multidisciplinary aspect of this level of care takes into account the view that pain and disability are perpetuated by a complex interaction of physiological, psychosocial, and socioeconomic factors.[11] Functional restoration and palliative pain management are two types of tertiary programs. Unlike secondary rehabilitation, psychosocial care is incorporated into the core curriculum of the tertiary rehabilitation program. Prior to starting the program, multidisciplinary medical (neurologist, orthopedist, rehabilitation and pain specialists, etc.) assessments are conducted to ensure that tertiary rehabilitation is the last resort after all reasonable surgical options have been exhausted or ruled out.

35.4.1 Interdisciplinary Assessment

Rehabilitation often begins with some form of physical, functional, and psychosocial assessment. The assessments then lead to treatment, and the treatment, in turn, often leads to recovery with measured outcomes, even in complex cases. The rationale for the assessments is that, in order to meet the treatment needs of the individual, the various aspects that contribute to his or her pain/disability deconditioning must be examined. As stated earlier, assessments involve physical, functional, and psychosocial components to determine the type and dose of treatment needed. It is important to note that, in tertiary rehabilitation, the assessment process is more intense and comprehensive than in primary or secondary care programs.

At the physical level, an individual's strength, motion, and endurance of the injured "weak link" are examined. This raises the issue of the difference between the *deconditioning syndrome* and an athlete simply "getting out of shape in the off season." In the latter situation, the whole body loses tone, endurance, and coordination to a relatively small degree, and generalized reconditioning prior to the athletic season is usually sufficient to prepare the athlete for sports again. In the deconditioning syndrome, an injured area becomes a "weak link," which becomes far more limited in mobility, strength and endurance than other body regions, are more painful, and become progressively more debilitated with the passage of time. Patients need to "work around the injured area" to perform functional tasks, such as lifting, squatting, climbing, and bending. Physical capacity is measured by assessing motion, strength, and endurance in an injured area. Functional capacity is assessed by measuring the patient's residual ability to perform functional tasks. These areas are often measured in quantitative terms by various types of reliable measurement approaches that assess spine performance. The physical examination will also provide useful information, but inclinometers (for measuring mobility) as well as isometric, isokinetic, and isoinertial devices are essential for functional and strength measurements. Hand function tests, activities of daily living tests (pushing, pulling, carrying, squatting, etc.), upper body ergometry assessment, and submaximal cardiovascular endurance tests are also used to assess the body as a whole. One evaluation used often after secondary or tertiary care programs is the functional capacity evaluation (FCE). These protocols assess function after treatment, and may provide useful data on the subject's ability to perform work or recreational activities.

Psychosocial assessment is also an important prerequisite for undergoing tertiary rehabilitation. It is inaccurate to assume that the pain is either physical or psychological. Instead, it is more effective and accurate to identify a variety of psychosocioeconomic barriers to recovery or risk factors for persistent disability. At the psychosocial level, screening tests for symptom magnification and somatization are used, and include such tests as the quantified pain drawing, the Pain Disability Questionnaire (PDQ), the Million Visual Analog Scale (MVAS), and the Oswestry Disability Inventory (ODI). Depression is screened by using one of several instruments, such as the Beck Depression Inventory (BDI), as well as the Hamilton Rating Scale for Depression. Other measures used to assess psychosocial functioning are the Minnesota Multiphasic Personality Inventory (MMPI) and the Structured Clinical Interview for the Diagnostic and Statistical Manual of Mental Disorders, fourth edition (the SCID). The psychiatric diagnosis can be made using the SCID. A psychosocial interview by a clinical psychologist is helpful and can identify risk factors for disability and resistance to treatment.[1]

During the comprehensive assessment, the disability system should not be ignored, and it is essential to know if the individual is dealing with workers' compensation (state or federal) or other compensation injuries. Workers' compensation can create a variety of incentives and deterrents that affect response to treatment. Some factors to take into consideration are whether the patient is receiving current total disability benefits, impairment or disability benefits, vocational rehabilitation benefits, or is approaching an array of financial/medical end-points. It is important to know about any third party claims that a patient may be involved in, as well as any disability (short- or long-term) or Supplemental Security or Social Security Disability (SSI/SSDI) the patient may receive.[1]

In physically intensive functional restoration programs (tertiary programs), more quantitative methods of assessment are often needed. Fear avoidance, chronicity, and complex high-cost cases are clearly more common in these programs. Therefore, it is essential for the assessments to be more accurate and discriminating in order to guide the treatment team in restoring function. Tertiary programs use interdisciplinary team (PT, OT, psychologist, etc.) assessments, in which several health care professionals are needed to evaluate the patients. The interdisciplinary assessment is mostly about determining the duration and intensity of the program for the individual patient, based on their degree of physical deficit and psychosocial involvement. The multidisciplinary evaluation by specialist physicians is individualized to be certain that all reasonable single discipline nonoperative, or surgical, options have been offered or provided. If indicated by history and physical examination, appropriate imaging and electrodiagnostic tests are performed before tertiary rehabilitation, and surgical care offered if medically necessary, before "last resort" tertiary rehabilitation is provided. In cases of an active surgical option, which the patient wishes to avoid, tertiary care can be provided. However, patients are commonly assessed at the midpoint of treatment to see if they still have confidence in pursuing this approach, or wish to reconsider the open surgical option before completing all treatment at an end-point generally termed *maximum medical improvement (MMI)*.

Often, patients have undergone weeks or months of physical therapy guided by their acceptance of assistance from a variety of health professionals. The quantification of physical functioning by using objective measures helps physical therapists set parameters for appropriate exercise dose. Physical therapists often assess patients with acute low back pain using the McKenzie assessment.[12] These assessments are also being used in secondary and tertiary rehabilitation and, during the evaluation, patients are required to perform repetitive lumbar end-range test movements and positions while monitoring their immediate subjective pain response.

After a successful rehabilitation intervention, a single functional capacity evaluation for work tolerance is often performed. These evaluations assist in determining work restrictions in patients returning to duty after treatment. In select primary and secondary treatment cases, human performance testing can be used to identify specific joints or regional deficits that need continued reconditioning in a fitness maintenance program, and to record improvement. Such human performance testing is always standard in tertiary rehabilitation. A major form of tertiary care — functional restoration — is discussed next.

35.4.2 Functional Restoration

Developed in 1983, and based on the biopsychosocial model of musculoskeletal pain and disability, functional restoration was a variant of chronic pain management specifically intended for compensation injuries.[13] This program is based on a quantitatively directed exercise progression approach, combined with a multimodal disability management program incorporating psychosocial and case management interventions. An underlying assumption in functional restoration is that the source of much of the disability consists of both physical and psychosocial components.

35.4.3 Physical and Occupational Therapy

In terms of the physical components, functional restoration programs utilize the measurement of physical and functional capacity to quantitatively guide the rehabilitation. Quantification allows for the bracketing of the "dose" and intensity of exercise, thus preventing patients from excessively exerting themselves, while at the same time encouraging them to utilize the benefits of the exercise regimen by avoiding suboptimal levels of performance that would be a waste of their time. Serving as valuable adjuncts to the intensive exercise regimens are various medical modalities, such as intra-articular injections, TENS units, and anti-inflammatory medication.

The initial stages of the program emphasize mobility exercises that lead ultimately to muscle training and endurance training. This initial stage of physical therapy ensures that the "weak link" area of the joint or spine is isolated and focused on by the therapist in a supervised environment.[14] As the patients add occupational therapy to the rehabilitation, the focus shifts to coordinating the injured "weak link" with other body parts with the goal of achieving close to preinjury functional status. These activities are designed to simulate common tasks found in a work environment.[15]

35.4.4 Psychosocial Interventions and Disability Management

Chronic pain and disability define the patients in a functional restoration setting. As a result, various psychosocial barriers to positive treatment outcomes may have developed prior to admission to such a program.[16] To overcome these barriers and induce active participation in the rehabilitation process, psychosocial interventions such as education and counseling are standard in a functional restoration program.[16] These interventions consist of individual and group counseling, behavioral stress management training, and family counseling. Additionally, support is provided for overcoming disability, reintegrating into the workforce, acquiring relevant vocational placement/training and helping with case settlement issues.

35.4.5 Evidence for Functional Restoration

The success of the functional restoration method has been documented by patient outcomes measured in terms of function and socioeconomic parameters. These parameters include work status (return-to-work and work retention), future health utilization (additional surgery to the injured area, persistent health care seeking behavior, and number of visits to new providers), recurrent injury after work return (new claims and lost time) and case closure. Both 1-yr follow-up[13,17,18] and 2-yr follow-up studies[19] have shown that functional restoration is associated with improvement in critical socioeconomic outcomes such as increased return to work, decrease in health utilization, decrease in additional surgery and higher percentages of case settlement. Replications of the functional restoration program outside the United States, when evaluated using randomized controlled trials, yielded similar positive findings.[20–23] In addition to treatment effectiveness, comprehensive program reviews of rehabilitation programs have also demonstrated the cost effectiveness of functional restoration and other multidisciplinary pain management programs.[15,24,25]

Behavioral and cognitive–behavioral treatment approaches, a key component of the psychosocial interventions within a functional restoration program, have also been subjected to studies that have demonstrated their therapeutic effectiveness. A systematic review of randomized controlled trials of behavioral and cognitive behavioral therapy for chronic pain in adults found significantly less pain experience, increased cognitive coping and appraisal and reduced behavioral expression of pain.[26] A similar review of published literature indicated that the behavioral and cognitive–behavioral approaches also improved daily functioning.[27]

35.5 Palliative Pain Management

Another form of tertiary care is palliative pain management. The ultimate goal of palliative pain management programs is reduction of pain. Pain palliation techniques in these programs include neuroablative procedures, radiofrequency neurotomies in the spine, spinal cord stimulation, or intrathecal drug pumps. In contrast to functionally oriented programs, palliative or interventional pain management techniques depend on narcotics for pain control. Due to the absence of modalities aimed at overcoming disability, efforts are focused on helping patients accept relatively nonfunctional lifestyles, such as dealing with the stress and tension associated with inactivity and the lack of productivity. Outcomes are difficult to measure in these programs due to the possibility of patients' self-assessments of pain and health status being confounded by financial secondary gain issues, the subjectivity of pain self-report, and the reliance on narcotics.

35.6 Conclusions

The three distinct categories of postinjury nonoperative care are primary, secondary, and tertiary rehabilitation. Primary care is concerned with controlling pain while early healing progresses. Secondary rehabilitation is concerned with avoiding deconditioning and preventing chronic disability, and consists mainly of exercise regimens that may be accompanied by psychosocial interventions as an adjunct. A small percentage of patients that develop chronic disabling musculoskeletal disorders are treated using tertiary rehabilitation. Prevention of permanent disability and return to productivity with lower pain levels are the main goals of these types of treatment programs. Tertiary rehabilitation can consist of intensive physical training with a cognitive behavioral disability management program (e.g., functional restoration) or an interventional pain management approach.

References

1. Mayer, T.G. and Press, J. Musculoskeletal rehabilitation. In Vacarro, A., Freedman, M., and Mayer, T., eds. *Orthopedic Knowledge Update (8)*. Chicago, IL: AAOS Press, 2003.
2. Gatchel, R.J., Polatin, P., Noe, C., Gardea, M., Pulliam, C., and Thompson, J. Treatment- and cost-effectiveness of early intervention for acute low back pain patients: a one-year prospective study. Journal of Occupational Rehabilitation 2003; 13:1–9.
3. Moffett, J.K., Torgenson, D., Bell-Syer, S., et al. Randomized controlled trial of exercise for low back pain: clinical outcomes, costs and preferences. British Medical Journal 1999; 319:279–283.
4. Polatin, P.B. and Gatchel, R.J. Psychosocial care for spinal disorders. In Fardon, D.F., Garfin, S.R., et al. eds. *Orthopedic Knowledge Update (8)*. Chicago, IL: AAOS Press, 2002.
5. Indahl, A., Haldorsen, E.H., Holm, S., Reiheras, O., and Ursim, H. Five-year follow-up study of a controlled clinical trial using light mobilization and an informative approach to low back pain. *Spine* 1998; 23:2625–2630.
6. Lindstrom, I., Ohlund, C., Eek, C., et al. The effect of graded activity on patients with subacute low back pain: a randomized controlled clinical study with an operant-conditioning behavioral approach. *Physical Therapy* 1992; 72:279–290.
7. Hagen, E.M., Eriksen, H.R., and Ursin, H. Does early intervention with a light mobilization program reduce long term sick leave for low back pain? *Spine* 2000; 25:1973–1976.
8. Karjalainen, K., Malmivaara, A., van Tulder, M., et al. Multidisciplinary biopsychosocial rehabilitation for subacute low back pain in working age adults: a systematic review within the framework of the Cochrane Collaborative Back Review Group. *Spine* 2001; 26:262–269.
9. Abenhaim, L., Rossignol, M., and Valat, J.P., et al. The role of activity in the therapeutic management of back pain: Report of the International Paris Task Force on Back Pain. *Spine* 2000; 25(4 Suppl): 1S–335.
10. van Tulder, M., Malmivaara, A., Esmail, R., and Koes, B. Exercise therapy for low back pain: a systematic review within the framework of the Cochrane Collaborative Back Review Group. *Spine* 2000; 25:2784–2796.
11. Gatchel, R.J. and Turk, D.C. Interdisciplinary treatment of chronic pain patients. In Gatchel, R.J. and Turk, D.C. eds. *Psychosocial Factors in Pain: Critical Perspectives*. New York: Guilford, 1999, pp. 435–444.
12. McKenzie, R. *Lumbar Spine Mechanical Diagnosis and Therapy*. Waikanae, Newzealand: Spinal Publications Ltd., 1981.
13. Mayer, T., Gatchel, R.J., Kishino, N., et al. Objective assessment of spine function following industrial injury: a prospective study with comparison group and one-year follow-up. 1985 Volvo Award in Clinical Sciences. *Spine* 1985; 10:482–493.
14. Mayer, T. Functional restoration program characteristics in chronic pain tertiary rehabilitation. In Slipman, C., Derby, R., Simeon, F., Mayer, T., eds. *Interventional Spine* London, UK: Elsevier Global Publications. (in press).

15. Deschner, M. and Polatin, P.B. Interdisciplinary programs: chronic pain management. In Mayer, T.G., Gatchel, R.J. and Polatin, P.B., eds. *Occupational Musculoskeletal Disorders.* Philadelphia, PA: Lippincott, Williams and Wilkins, 2000, pp. 629–637.
16. Mayer, T.G. and Polatin, P.B. Tertiary nonoperative interdisciplinary programs: the functional restoration variant of the outpatient chronic pain management program. In Mayer, T.G., Gatchel, R.J., and Polatin, P.B., eds. *Occupational Musculoskeletal Disorders.* Philadelphia, PA: Lippincott, Williams and Wilkins, 2000, pp. 639–649.
17. Hazard, R.G., Fenwick, J.W., Kalisch, S.M., et al. Functional restoration with behavioral support. A one-year prospective study of patients with chronic low back pain. *Spine* 1989; 14(2):157–161.
18. Mayer, T., Gatchel, R.J., Kishino, N., et al. Study of chronic low back pain patients utilizing novel objective functional measurements. *Pain* 1986; 25:53–68.
19. Mayer, T., Gatchel, R.J., Mayer, H., Kishino, N., Keeley, J., and Mooney, V. Prospective two-year study of functional restoration in industrial low back injury. *Journal of the American Medical Association* 1987; 258:1181–1182.
20. Bendix, A.E., Bendix, T., Vaegter, K., et al. Multidisciplinary intensive treatment for chronic low back pain: a randomized, prospective study. *Cleveland Clinic Journal of Medicine* 1996; 63:62–69.
21. Corey, D.T., Koepfler, L.E., Etlin, D., and Day, H.J. A limited functional restoration program for injured workers: a randomized trial. *Journal of Occupational Rehabilitation* 1996; 6:239–249.
22. Hildebrandt, J., Pfingsten, M., Saur, P., and Jansen, J. Prediction of success from a multidisciplinary treatment program for chronic low back pain. *Spine* 1997; 22:990–1001.
23. Jouset, N., Fanello, S., Bontoux, L., et al. Effects of functional restoration versus 3 hours per week physical therapy: a randomized controlled study. *Spine* 2004; 29(5):487–493.
24. Turk, D.C. and Gatchel, R.J. Multidisciplinary programs for rehabilitation of chronic low back pain patients. In Kirkaldy-Williams, W.H. and Bernard, T.N. eds., *Managing Low Back Pain*, 4ᵗʰ ed. New York: Churchill & Livingston, 1999, pp. 299–316.
25. Okifuji, A., Turk, D.C., and Kalavokalani, D. Clinical outcomes and economic evaluations of multidisciplinary pain centers. In Block, A.R., Kremer, E.F., and Fernandez, E., eds. *Handbook of Pain Syndromes*. Mahwah, NJ: Lawrence Erlbaum Associates, 1999, pp. 169–191.
26. Morley, S., Eccleston, C., and Williams, A. Systematic review and meta-analysis of randomized controlled trials of cognitive behavior therapy and behavior therapy for chronic pain in adults, excluding headache. *Pain* 1999; 80:1–13.
27. McCracken, L.M. and Turk, D.C. Behavioral and cognitive–behavioral treatment for chronic pain. *Spine* 2002; 27:256–257.

36

Wrist Splints

Carolyn M. Sommerich
The Ohio State University

36.1 Introduction

A concise overview of the literature on wrist splints is presented in this chapter. Wrist splints are primarily considered to be a form of medical treatment to relieve symptoms or improve functionality. For example, wrist splints are commonly prescribed in the conservative treatment of carpal tunnel syndrome (CTS). Splints are sometimes also utilized by workers as personal protective equipment, similar to the use of back belts, though their efficacy in this role is yet to be demonstrated. Both aspects of wrist splint use are discussed in this chapter. A related chapter in this book, "Engineering controls: wrist and arm supports," describes other methods by which wrists, and more generally the upper extremity, may be effectively supported while working.

36.2 Fundamentals

In this section, basic theories of wrist splinting are explored, including the ways in which various designs provide support, and under what circumstances they may be employed.

In the treatment of fractures, splinting had long been known to provide rest and relief from pain and disability. By analogy, splinting has been applied to arthritis,[1] where the treatment is typically prescribed to relieve pain and inflammation, but may also be prescribed to prevent deformities (contracture), or increase function.[2] Wrist splinting is also used in the conservative treatment of CTS. When CTS is attributed to pregnancy or to work that is highly repetitive or requires sustained nonneutral wrist postures, splinting may reduce pain and paraesthesia.[3] Symptoms would not be expected to respond to splinting when CTS is due to vibration exposure, lumbrical muscles entering the distal carpal tunnel during pinch grips, pathologic encroachment (carpal bone fracture or dislocation), or external pressure applied to the palm (from hand tools with sharp edges or short handles, or from repetitive activation of palm buttons).

Splinting the wrist in extension (using a cock-up splint) is a common practice. This practice is based on studies of wrist position during common activities, which have shown the centroid of motion for most common tasks to be an extended position.[4] However, minimum pressure in the carpal canal, and therefore on the median nerve, has been shown to occur when the wrist is in a neutral orientation (along both flexion–extension and radial–ulnar axes).[5] Splinting the wrist in extension may be effective for treating early stage (fully reversible inflammatory) lateral epicondylitis.[6] Passive extension of the wrist relieves tensile stress at the common origin of the wrist and finger extensor muscles, the loci of pain in epicondylitis. Pain from epicondylitis is also addressed effectively by wearing a brace (a support band about 5-cm wide) just distal to the elbow, which covers the upper part of the forearm.[6-8]

Wrist splints are categorized as either rigid or flexible. Illustrations and descriptions of several different rigid and flexible splints, each designed to accommodate deficits due to specific nerve lesions (radial, ulnar, or median), are provided in Paternostro-Sluga et al.[9] Rigid splints do not permit any wrist motion, while flexible splints permit a limited range of motion, thereby preserving more hand function. Rigid splints may also be referred to as immobilizing, positioning, resting, or static splints. Flexible splints are also referred to as activity, functional, or working splints, wrist orthoses, or wrist supports. Splints cover some aspect of the forearm, hand, and finger(s), and usually have a plate or bar in or across the palmar region which can interfere with grasping activities. Both prefabricated and custom-made splints are used in treatment protocols, depending on the preference of the medical provider. Wearing recommendations may be for nighttime use, usage during painful activity, or continuous usage, depending on the protocol recommended by the medical provider.

36.3   Applications

The theoretical benefits of various upper extremity supports were discussed in the previous section. Results from several studies are summarized in this section in order to demonstrate how and under what circumstances wrist splints have been shown to be effective.

36.3.1   Wrist Splints as Medical Treatment for Disorders Affecting the Wrist

There is a consensus that use of a splint should be based on medical opinion, and that the usage should be supervised by a medical provider.[10] However, usage protocols and treatment effectiveness seem to vary widely across providers. Results of several studies of various protocols are provided in the following paragraphs.

Use of a working splint for 1 week, at night and during stressful activities, was found to reduce, at the end of the week, subjective assessments of pain, numbness, and tingling in a group of CTS patients with abnormal nerve conduction study results, when compared with symptoms in a similar group of patients who did not receive splints.[11] Wrists were splinted at 0 to 5° of flexion.

The efficacy of conservative treatment of CTS based on steroid injection followed by continuous use of a neutral position, rigid splint for 4 weeks was tested prospectively by Weiss et al.[12] No patients with advanced CTS or associated medical conditions (such as diabetes, pregnancy, or arthritis) were admitted to the study. The authors found that only 13% of the treated hands were cured (defined as symptom free) at final follow-up, which occurred between 6 and 18 months after the injection.

A similar success rate (17%) was achieved by Banta,[13] with a regimen of ibuprofen and 3 weeks of continuous, neutral positioned, wrist immobilization for 23 hands with early-mild CTS. A second-stage treatment consisting of 1 week of iontophoresis and splinting followed by two more weeks of continuous splinting resulted in an improvement in 58% of the remaining hands. The author's overall success rate was 65%, based upon absence of symptoms at 6 months follow-up.

From a retrospective review of 105 CTS patients treated with a neutral angle rigid splint designed to permit hand function, Kruger et al.[14] determined that 67% of the patients received symptom relief attributable to the splint. There was a significant decrease in the median sensory latency group average between pre- and posttreatment measurements. Splinting seemed to be most effective for patients who had experienced symptoms for a short period of time (1 to 3 months). Patients were told to wear the splints at night and during the day as much as possible; however, the authors were not able to confirm the wearing patterns of the patients. Though no lower limit was provided for post-treatment measurements, some occurred as long as 17 months after receiving the splint. In a prospective, randomized clinical trial with a 6-week follow-up, Walker et al.[15] also found reductions in median sensory latency, as well as median motor latency, in a group of CTS patients who were instructed to wear, full time, a custom-molded thermoplastic, volar, neutral wrist splint with radial bar. The other patient group in that study received nighttime only wearing instructions, and only experienced a decrease in sensory distal latency. That decrease was significantly less than the reduction in the full-time wear instruction group.

Based on a review of 363 hands with CTS, Kaplan et al.[16] identified five factors that predicted outcome success or failure from medical management of CTS. The five factors were age over 50 yr, symptom duration exceeding 10 months, constant paresthesia, stenosing flexor tenosynovitis, and a positive Phalen's test in less than 30 sec. Treatment success was defined as a patient remaining symptom free for 6 months following conservative treatment with a rigid wrist splint (neutral wrist position; worn at night and during the day when symptomatic) and anti-inflammatory medication. Follow-up averaged 15.4 months, but ranged from 6 to 48 months. Although their overall success rate was only 18.4%, there was a 66.7% success rate in the subgroup of patients who did not have any of the five factors. In the subgroup of patients who had one of the five factors, the success rate dropped to 40.4%; in the group with two factors the rate was 16.7%, and was only 6.8% in the group with three factors. No patients with more than three factors were successfully treated.

Splinting may be specifically prescribed to relieve nocturnal CTS symptoms. This is based on the theory that patients sleep with wrists curled (flexed), which results in elevated pressure on the median nerve. Luchetti et al.[17] found no statistically significant differences in carpal tunnel pressure due to splinting, either during the day or throughout the night, between two groups of CTS patients (one splinted in 20° of extension, and the other not splinted). It is not known whether splinting would have had an effect if splints had maintained the wrists in neutral rather than extended postures.

The question of appropriate splint angle was addressed in a clinical trial in which patients' subjective responses indicated that a 2-week regime of splinting in a neutral position was more effective in relieving CTS symptoms than was splinting in 20° of extension.[18] Both splint designs appeared to be more effective for relief of nocturnal symptoms than daytime symptoms. Also of interest was the finding that only 8% of patients who continued to use their splints for 2 months experienced further symptom relief.

Compliance is an important part of any medical treatment. In their study of splint usage in the treatment of rheumatoid arthritis, Spoorenberg et al.[2] found that patients were less likely to adhere to wearing advice for an immobilization splint than for an activity splint, although none of the patients that were told to wear their splints at night did so. Only 17% of the patients with immobilization splints who rested often (and therefore had the opportunity to wear the splint often), did so. In contrast, 57% of the patients with activity splints wore them often. This is, undoubtedly, at least partly due to patients' perceptions of the splints. While 96% of the rheumatologist responding to the survey prescribed both immobilization and activity splints for reduction of pain, only 44% of patients with immobilization splints found they relieved pain, and only 26% found they improved hand function. Seventy-eight percent of those patients found the rigid splints both unwieldy and ugly. In contrast, 75% of the patients with activity splints found they relieved pain and improved hand function. Only 29% thought they were ugly, but 63% found even the activity splints to be unwieldy. Patients' perceptions of pain relief and other perceived benefits of wearing splints during various tasks were the primary factors associated with wear compliance in another study of patients with rheumatoid arthritis.[18]

36.3.2   Wrist Splints as Medical Treatment for Disorders Affecting the Elbow

As is the case with most musculoskeletal disorders and their associated risk factors, unless exposure to the risk factors associated with tennis elbow are reduced or eliminated after immobilization, treatment pain returns once a patient resumes the activities that lead to the development of tennis elbow.[6] Therefore, Nirschl[6] advocated for alteration or elimination of abusive activity rather than for immobilization of either the elbow or wrist joint. He cautioned that rigid immobilization could result in muscle atrophy. He also recommended changes in training techniques and equipment, along with counterforce bracing once activities were resumed. Little[19] did not find splints to be effective in the treatment of tennis elbow. In two groups of subjects that both received cortisone and ultrasound treatment for tennis elbow, he found essentially no difference in recovery time (time when subjects became symptom-free) though the patients in one group were given splints and told to rest, while members of the other group were told to continue their normal activities.

36.3.3   Wrist Splints Used during Task Performance

The science of ergonomics seeks to tailor workplace requirements to human capabilities and limitations. The preferred method of accomplishing this is through engineering controls, by which work or the work environment is altered to fit the worker. Use of wrist splints on the job is, instead, a method of altering the worker to fit the work. Falkenburg[10] described wrist splinting as "an industrial treatment to keep the employee on a job." She specifically mentioned, however, that splinting interferes with hand grasp, thereby reducing hand strength and interfering with holding or grasping tasks. Such functionality problems are documented in some of the splint research studies that are summarized in the following paragraphs. Most of the research on splints has been conducted in laboratory settings using healthy subjects, so those results are only suggestive of what may occur in actual employment situations.

36.3.3.1   Hand-Intensive Tasks

In an early field study of splinting as an intervention for CTS, results from a battery of objective and performance-based tests demonstrated that CTS was actually aggravated by the use of a rigid splint during work.[20] In contrast, time off and light duty were found to be effective intervention methods.

One reason splints might be thought to be useful on the job for patients with CTS would be to reduce pressure on the median nerve by restricting patients from deviating wrists too far from a neutral posture. However, in a laboratory-based study using only healthy subjects, Rempel et al.[21] were not able to demonstrate any difference in carpal tunnel pressure during a hand-intensive task between conditions when subjects wore a flexible wrist splint and when they did not, even though wrist motion, in both flexion–extension and radial–ulnar directions, was restricted by the splint. Rempel et al.[21] did find, however, that simply putting on the splint significantly raised carpal tunnel pressure in each subject.

Perez-Balke and Buchholz[22] also observed reduced wrist motion in healthy subjects performing a repetitive pick and place task while wearing a rigid splint (with a malleable metal bar across the forearm, wrist, and palm), compared to performing the same task without the splint. They found reductions in peak grip force when subjects were splinted. Female subjects were particularly handicapped by the splint. When wearing the splint, their grip strength was reduced by a greater percentage than the males in the study.[22,23] Their results might be explained, at least in part, by the work of Fransson-Hall and Kilbom,[24] who recorded significantly lower hand pain threshold tolerances and shorter times to experience pain in female subjects who experienced pressure applied to various locations on the palm, fingers, and thumb, compared with results from male subjects who experienced the same test protocol. Another important finding from Perez-Balke and Buchholz[22] was the increase in activity in the deltoid muscle and forearm flexor muscle group during the pick and place task when the splint was worn. In other words, subjects were less efficient when wearing the splint (more input for the same output). As a counterpoint, it should be noted that 75% of the 42 subjects in a study of splint use and preference in patients with rheumatoid arthritis reported that a splint made their wrist or hand feel stronger.[25] Patients used the metal bar "to rest things on" or used the stay as a lever — in other words, stresses that would ordinarily have to be resisted by the wrist were, instead, resisted by the metal stay, which facilitated performance of heavier tasks that required lifting, pushing, or pulling.

In another study using healthy subjects, Carlson and Trombly[26] found that wearing a rigid splint slowed the performance of several manual tasks. Increased performance times were recorded for each of seven different tasks in the Jebsen Hand Function Test (which includes writing, manipulating small- and medium-sized objects, picking up heavy cans, and feeding) when subjects wore a rigid splint, compared to when no splint was worn. The authors suggested that increased task performance times would be particularly problematic for an employee wearing a splint on a job with speed requirements (such as a highly repetitive assembly task) or for an employee who fatigued easily. They also made note of remarks of fatigue in the shoulder and upper trunk from several subjects, following task performance with the splint. Whether or not a person is wearing a wrist splint, the hand still needs to be in the same spatial location to perform a given task. If the wrist is not able to position the hand, the positioning task falls to the more proximal joints (elbow, shoulder, spine), and the muscles that control those joints. This is another example of the worker adapting to the work place.

Stern et al.[27] also employed the Jebsen Hand Function Test to evaluate the impact on performance due to splint usage in a group of healthy subjects. In contrast to the previous study, these authors tested several styles of commercial, nonrigid splints, which they referred to as "static wrist extensor orthoses," devices designed to support rather than immobilize the wrist. Only in manipulating small objects was there a significant difference in performance between the splinted and free hand conditions. In the other six tests, performance when wearing at least one of the five test splints did not differ from performance in the free hand condition, though it was not the same commercial splint that matched barehanded performance in each test. Subjects also wore each splint for a day, and reported differences in subjective experiences among the different splints. Splints differed in terms of the types of daily tasks with which they interfered (such as hygiene, housekeeping, or driving). Splints also differed in terms of comfort, temperature, and pain experienced by the subjects.

36.3.3.2 Vibration Exposure

In a laboratory study with healthy subjects, Chang et al.[28] investigated the effects of wearing a wrist support, with and without gloves of different materials, while using an inline pneumatic screwdriver. The wrist support was a band wrapped about the distal end of the forearm and which did not extend over the hand. Their dependent measures were force applied to the tool's trigger, muscle activity in the forearm (labeled "flexor digitorum"), and vibration transmission. The researchers found a significant interaction between glove and wrist support. Specifically, in the no-wrist-support conditions, subjects pressed the tool's trigger with less force when wearing a nylon or cotton glove than with a bare hand, but this benefit was not realized in the wrist-support condition. There was also a significant main effect of wrist support on vibration transmission: transmission to the hand was greater with the wrist band than without it. Khalil et al.[29] also studied the effect of a wrist band on vibration transmission. Subjects grasped a handle that was excited at 10 Hz. They were instructed to maintain a "uniform comfortable grip." Vibration was measured on the forearm, and transmission was found to be reduced when the wrist band was worn. The differences in these two studies of the effects of a wrist band on vibration transmission[28,29] include characteristics of the vibration (wide spectrum from a tool versus single 10 Hz), nature of the task (supporting and aiming a tool to perform a targeted task versus holding onto a fixed handle), and location of vibration measurement (hand versus forearm). The differing results from these two studies signal the need for further study of the effects of wrist support on vibration transmission, in light of the strong associations of vibration exposure with CTS[30–32] and hand-arm vibration syndrome.[31]

36.3.3.3 Wheelchair Propulsion

The prevalence of upper extremity MSDs is elevated in people who use wheelchairs.[33] Carpal tunnel pressure has been shown to be greater in neutral and extended wrist postures in wheelchair users when compared with non-wheelchair users.[34] This is thought to be the result of repeated exposure to the primary exertions performed in a wheelchair (chair propulsion and pressure relieving maneuvers), which are both performed with pronounced wrist extension. In a lab-based study utilizing "13 subjects with prior wheeling experience," Malone et al.[35] sought to determine if wrist extension could be reduced during wheelchair propulsion, through the use of wrist splints or gloves. They found wrist extension and range of motion were reduced when wearing the splint. As far as compensation for this reduction showing up in other joints, elbow angle range of movement was not affected by the splint, and shoulder motion was not measured. The researchers also reported "no detrimental effects in median nerve function under any" test condition, based on electrodiagnostic testing (motor conduction velocity and amplitude). However, further studies should be conducted to determine the full short- and long-term effects of splint use on wheelchair users prior to making a recommendation for splint use to this group of individuals.

36.3.3.4 Risk of Acute Injury Due to Splint Use

No reports were found in the literature which associated wrist splint usage with accidental injury during work. However, one report described a fracture of the distal radius in a gymnast wearing a standard gymnast's wrist support, which caught while he worked on the high bar. This experience should cause workers, employers, and medical providers to consider whether a wrist support could pose a danger in some work situations. Wearing a splint might increase the chances of an employee becoming caught in rotating or other machinery with accessible moving parts, either directly by catching the splint, or indirectly by restricting the employee's movements and thereby restricting his or her ability to avoid contact with the equipment.

36.4   Summary

In reviewing the literature there seems to be consensus among authors on the demonstrated effectiveness of wrist splints in treating symptoms of certain medical conditions (at certain stages) and compensating for a patient's weakness of the hand or wrist while performing tasks. However, the use of wrist splints as personal protective equipment at work for healthy workers is consistently not supported.

Wrist splints are a form of medical treatment, which varies in usage recommendation and splint design, based on the type of disorder (e.g., nerve lesion, rheumatoid arthritis, or epicondylitis) and location of disorder (e.g., radial, median, or ulnar nerve lesion). As such, the use of a wrist splint is an issue that should be decided between a medical provider and patient. Individuals should not self-prescribe these devices. If an individual's symptoms are the result of physical exposures, either in the workplace or outside of work, there is no reason to believe the symptoms will not continue or return if the individual continues to be exposed to the same risk factors. Wrist splints may, in fact, cause a disorder to migrate to a more proximal joint that is forced to compensate for the reduced functionality imposed on the injured wrist by the splint. Wrist splints used at work would, at best, be considered an administrative control measure. However, no where are splints categorized as personal protective equipment to be used by healthy workers, and they may actually make hand-intensive tasks more difficult to perform or cause other problems. The decision to wear a splint on the job should be made by an occupational health professional who has firsthand knowledge of the patient's workstation, tools, and tasks. Wrist splints may treat symptoms, but they do not address the root cause of an individual's discomfort. Thoughtful engineering controls are the most effective method for reducing an individual's exposure to physical risk factors associated with work-related upper extremity symptoms and disorders, such as carpal tunnel syndrome.

Acknowledgment

The author is grateful to Ms Sahika Vatan for identifying and reviewing the most recent literature relevant to the topic of this chapter.

References

1. Ehrlich, G.E. Splinting for arthritis, *Med Times* 96 (5), 485–489, 1968.
2. Spoorenberg, A., Boers, M., and van der Linden, S., Wrist splints in rheumatoid arthritis: a question of belief? *Clin Rheum* 13 (4), 559–563, 1994.
3. Ditmars, D.M., Jr., Patterns of carpal tunnel syndrome, *Hand Clin* 9 (2), 241–52, 1993.
4. Palmer, A.K., Werner, F.W., Murphy, D., and Glisson, R., Functional wrist motion: a biomechanical study, *J Hand Surg* 10A (1), 39–46, 1985.
5. Weiss, N.D., Gordon, L., Bloom, T., So, Y., and Rempel, D.M., Position of the wrist associated with the lowest carpal-tunnel pressure: implications for splint design, *J Bone Joint Surg* 77A (11), 1695–1699, 1995.
6. Nirschl, R.P., Muscle and tendon trauma: tennis elbow, in *The Elbow and its Disorders*, 2nd ed, Morrey, B.F., ed., W.B. Saunders, Philadelphia, 1985, pp. 537–552.
7. Froimson, A.I., Treatment of tennis elbow with forearm support band, *J Bone Joint Surg* 53A (1), 183–184, 1971.
8. Valle-Jones, J.C. and Hopkin-Richards, H., Controlled trial of an elbow support ("Epitrain") in patients with acute painful conditions of the elbow: a pilot study, *Curr Med Res Opin* 12 (4), 224–233, 1990.
9. Paternostro-Sluga, T., Keilani, M., Posch, M., and Fialka-Moser, V., Factors that influence the duration of splint wear in peripheral nerve lesions, *Am J Phys Med Rehabil* 82 (2), 86–95, 2003.
10. Falkenburg, S.A., Choosing hand splints to aid carpal tunnel syndrome recovery, in *Occup Health Saf.* 56(5), 60, 63–64, 1987.
11. Dolhanty, D., Effectiveness of splinting for carpal tunnel syndrome, *Con J Occup Ther* 53 (5), 275–280, 1986.
12. Weiss, A.-P.C., Sachar, K., and Gendreau, M., Conservative management of carpal tunnel syndrome: a reexamination of steroid injection and splinting, *J Hand Surg* 19A (3), 410–415, 1994.
13. Banta, C.A., A prospective, nonrandomized study of iontophoresis, wrist splinting, and anti-inflammatory medication in the treatment of early-mild carpal tunnel syndrome, *J Occup Med* 36 (2), 166–168, 1994.
14. Kruger, V.L., Kraft, G.H., Deitz, J.C., Ameis, A., and Polissar, L., Carpal tunnel syndrome: objective measures and splint use, *Arch Phys Med Rehabil* 72, 517–520, 1991.
15. Walker, W.C., Metzler, M., Cifu, D.X., and Swartz, Z., Neutral wrist splinting in carpal tunnel syndrome: A comparison of night-only versus full-time wear instructions, *Arch Phys Med Rehab* 81 (4), 424–429, 2000.
16. Kaplan, S.J., Glickel, S.Z., and Eaton, R.G., Predictive factors in the non-surgical treatment of carpal tunnel syndrome, *J Hand Surg* 15B (1), 106–108, 1990.
17. Luchetti, R., Schoenhuber, R., Alfarano, M., DeLuca, S., De Cicco, G., and Landi, A., Serial overnight recordings of intracarpal canal pressure in carpal tunnel syndrome patients with and without wrist splinting, *J Hand Surg* 19B (1), 35–37, 1994.
18. Burke, D.T., Burke, M.M., Stewart, G.W., and Cambré, A., Splinting for carpal tunnel syndrome: in search of the optimal angle, *Arch Phys Med Rehabil* 75 (Nov), 1241–1244, 1994.
19. Agnew, P.J. and Maas, F., Compliance in wearing wrist working splints in rheumatoid arthritis, *Occup Ther J R* 15 (3), 165–180, 1995.
20. Little, T., Tennis elbow — to rest or not to rest? (Letter to the editor), *The Practitioner* 228, 457, 1984.
21. Armstrong, T., Terminal Progress Report, NIOSH Report No. 2 R01 OH 00679, 1981.

22. Rempel, D., Manojlovic, R., Levinsohn, D.G., Bloom, T., and Gordon, L., The effect of wearing a flexible wrist splint on carpal tunnel pressure during repetitive hand activity, *J Hand Surg* 19A (1), 106–110, 1994.

23. Perez-Balke, G. and Buchholz, B., A study of the effect of a wrist splint on extrinsic flexor and anterior deltoid electromyography during a pick and place task, in *Proceedings of The Human Factors and Ergonomics Society 39th Annual Meeting*, San Diego, 1995, p. 958.

24. Perez-Balke, G. and Buchholz, B.O., A study of the effect of a "resting splint" on peak grip strength, in *Proceedings of The Human Factors and Ergonomics Society 38th Annual Meeting*, 1994, pp. 544–548.

25. Fransson-Hall, C. and Kilbom, A., Sensitivity of the hand to surface pressure, *Appl Ergon* 24 (3), 181–189, 1993.

26. Stern, E.B., Ytterberg, S.R., Krug, H.E., Larson, L.M., Portoghese, C.P., Kratz, W.N., and Mahowald, M.L., Commercial wrist extensor orthoses: a descriptive study of use and preference in patients with rheumatoid arthritis, *Arthritis Care Res* 10 (1), 27–35, 1997.

27. Carlson, J.D. and Trombly, C.A., The effect of wrist immobilization on performance of the Jebsen Hand Function Test, *Am J Occup Ther* 37 (3), 167–175, 1983.

28. Stern, E.B., Sines, B., and Teague, T.R., Commercial wrist extensor orthoses: hand function, comfort, and interference across five styles, *J Hand Ther* 7, 237–244, 1994.

29. Chang, C.-H., Wang, M.-J. J., and Lin, S.-C., Evaluating the effects of wearing gloves and wrist supports on hand arm response while operating an in-line pneumatic screwdriver, *Int J Ind Ergon* 24 (5), 473–481, 1999.

30. Khalil, T.M., Zaki, A., and Kassem, F., The effectiveness of a wrist band on activities involving the hand and wrist, in *Advances in Industrial Ergonomics and Safety IV*, Kumar, S., Ed., Taylor & Francis, London, 1992, pp. 1209–1216.

31. Wieslander, G., Norbäck, D., Göthe, C.-J., and Juhlin, L., Carpal tunnel syndrome (CTS) and exposure to vibration, repetitive wrist movements, and heavy manual work: a case-referent study, *Br J Ind Med* 46, 43–47, 1989.

32. Bernard, B.P., Report No. DHHS (NIOSH) Publication No. 97–141, 1997.

33. Nathan, P.A., Meadows, K.D., and Istvan, J.A., Predictors of carpal tunnel syndrome: an 11-year study of industrial workers, *J Hand Surg (Am)* 27 (4), 644–651, 2002.

34. Boninger, M.L., Robertson, R.N., Wolff, M., and Cooper, R.A., Upper limb nerve entrapments in elite wheelchair racers, *Am J Phys Med Rehabil* 75 (3), 170–176, 1996.

35. Gellman, H., Chandler, D.R., Petrasek, J., Sie, I., Adkins, R., and Waters, R.L., Carpal tunnel syndrome in paraplegic patients, *J Bone Joint Surg Am* 70 (4), 517–519, 1988.

36. Malone, L.A., Gervais, P.L., Burnham, R.S., Chan, M., Miller, L., and Steadward, R.D., An assessment of wrist splint and glove use on wheeling kinematics, *Clin Biomech (Bristol, Avon)* 13 (3), 234–236, 1998.

37

Clinical Lumbar Motion Monitor

Sue A. Ferguson
William S. Marras
The Ohio State University

37.1 Introduction

Low back disorders continue to be a widespread problem. In a recent review of the literature, the National Research Council (2001) found 22.4 million Americans had back pain for at least a week during the past 12-month period and estimated that back pain accounted for 149 million lost workdays. In the state of Ohio alone, the total cost of back injury claims was over 200 million dollars in 1996 (Hamrick, 2000). Given the rising costs of medical care and inflation, the economic impact due to low back injury would only be expected to increase. Appropriate medical management may minimize both the human and economic impacts of these injuries.

Medical management programs were developed to prevent or reduce symptoms, impairment, and disability associated with work-related musculoskeletal disorders (Hales and Bertsche, 1999). One of the components of a medical management program is a functional assessment. Traditionally, low back functional assessment has included such evaluations as range of motion, trunk strength testing (isometric and isokinetic), lifting capacity, and aerobic capacity (Waddell, 1998; Polatin and Mayer, 2001). This chapter focuses on the development of a dynamic functional assessment tool to quantitatively assess low back functional status.

An objective quantitative assessment of low back function is needed for several reasons. First, it would establish a quantitative benchmark of the severity of low back impairment (impairment is a deviation from normal in a body part and its function, AMA, 1993). Severity of low back injury is often measured by the patient's reported perception of pain. The shortcoming of using pain as a parameter is that it is subjective, influenced by many factors (Fordyce, 1976), and the same stimulus may cause different reports of pain (Gatchel et al., 1986). Second, an objective assessment would provide a mechanism to quantify the effectiveness of treatments. Third, a quantitative functional assessment may provide a tool to objectively determine readiness to return to work. Such a tool would minimize the subjective nature of disability determination. Disability is the degree to which the individual is affected by the impairment (Waddell, 1998). Finally, since low back disorders are recurrent in nature quantification of maximum medical improvement may provide a new benchmark for a specific patient when an exacerbation occurs.

Even though quantification is needed and there are several techniques available, no gold standard for low back assessment has been adopted. A standard low back assessment tool must have several qualities. First, the tool must be valid (high sensitivity and high specificity) (Andersson, 1991; Nachemson and Vingard, 2000; Polatin and Mayer, 2001). Second, the tool should be reliable and reproducible (Andersson, 1991; Polatin and Mayer, 2001). Andersson (1991) suggests several qualities for a functional assessment tool including that it causes minimal discomfort, be inexpensive, easy to administer, and safe. Nachemson and Vingard (2000) believe that the tool should have some predictive qualities. Thus, any tool developed for functional assessment must be scientifically tested.

37.2 Development of the Clinical Lumbar Motion Monitor (CLMM) Functional Assessment

37.2.1 Equipment

Three-dimensional trunk kinematics are measured with the lumbar motion monitor (LMM). The LMM is an exoskeleton of the spine and has been validated previously (Marras et al., 1992). The LMM measures position, velocity, and acceleration. Figure 37.1 illustrates the LMM on a person with the appropriate fit. In addition to the LMM, a shoulder and waist harness are required. As illustrated the shoulder harness plate should fit in the middle of the back between the shoulder blades. The waist harness should fit so that the top of the harness is on the iliac crests (hip bones). Finally, a computer is required for display and data storage. The computer should be placed directly in front of the person for the initial task.

37.2.2 Functional Performance Testing Protocol

The clinical lumbar motion monitor (CLMM) assessment was developed to document the

FIGURE 37.1 Lumbar motion monitor.

symmetric and asymmetric trunk motion kinematics of both low back pain patients and asymptomatic individuals. The test protocol required the participants to flex and extend their trunk repeatedly in five different trunk asymmetries including (1) zero or symmetric, (2) 15° clockwise, (3) 15° counterclockwise, (4) 30° clockwise, and (5) 30° counterclockwise. The trunk asymmetry was controlled within $\pm 2°$ of the desired asymmetry. In addition to the five control tasks the evaluation included two tasks of maximum twisting clockwise and counterclockwise. The motion profile observed during repeated flexion and extension of the torso at different trunk asymmetries is believed to be a reflection of the trunk's musculoskeletal motor program or "central set" (Horak and Diener, 1994). In asymptomatic participants, it is theorized that this motor control program has been developed during the participant's lifetime. However, in patient's it is hypothesized that the musculoskeletal control program must be adjusted to compensate for limitations related to muscle function, structural restrictions, and guarding behaviors. It is believed that the patient's motor control program would be repeatable even though it may be adapting due to the injury.

37.2.3 Instructions to the Subject

The instructions to the subject are critical. This fact is pointed out by the work of Marcel et al. (1999), who found that the LMM and Fastrak systems resulted in the same kinematic results when using the same instruction set. Thus, the specific type of equipment used may not be as critical as the instruction set to the subjects. The instructions to the subjects were to: (1) cross their arms in front of their chest; (2) stand with their feet shoulder width apart; and keep them in the same location for all the conditions; (3) flex and extend their trunks repeatedly in the sagittal plane as fast as they can comfortably while keeping the transverse plane position in the target zone; (4) watch the display at all times; (5) if the transverse plane position fell outside the control zone the gray region of the computer screen will turn red and the trial will be repeated; and (6) move continuously until instructed to "relax." Data were collected for 8 to 14 sec. The length of data collection time was depended upon having five flexion/extension cylces.

37.2.4 Trunk Kinematic Measures

Custom software has been developed to convert the electrical signal from the LMM to position using a calibration equation. The velocity and acceleration are derived from the position data. The software displays the position data for each plane of the body separately. In the sagittal plane, the first flexion and extension motions were considered a warm-up motion and were discarded. The following four flexion and extension cycles were analyzed. If extra cycles were collected then the software automatically discarded them. If less than two cycles of trunk motion were collected then the software would not complete the analysis due to insufficient data. The analysis program quantifies the average range of motion, average peak flexion velocity, average peak extension velocity, average peak flexion acceleration, and average peak extension acceleration for the sagittal plane. The same measures were calculated for the frontal and transverse planes; however, instead of flexion and extension it would be right and left. The frontal and sagittal plane data are evaluated for musculoskeletal status but not the transverse plane motion since it is being controlled within the asymmetry requirement for each task. Thus, a total of 50 trunk motion measures are evaluated from the control tasks.

In addition to the data from the control tasks, there is an ability measure. This measure was an integer from one to five indicating the number of control tasks the person successfully completed. Finally, there was a twisting range of motion measure that quantifies both maximum twisting clockwise and counterclockwise. Thus, a total of 52 motion parameters may be evaluated after completing all tasks in the CLMM functional assessment protocol.

37.2.5 Repeatability of Trunk Kinematic Measures

One of the qualities of a good functional assessment tool is that it should be repeatable (Andersson, 1991). Marras et al. (1994) evaluated the repeatability of the five CLMM functional performance control tasks. Twenty subjects with no history of back pain evenly split between males and females were evaluated five times, on the same day of the week and same time of the day. The results showed that sagittal performance measures of range of motion, flexion and extension velocity, and flexion and extension acceleration all had Cronbach alpha correlation above 0.90 for all five control tasks. The lateral motion measures had Cronbach alpha correlations above 0.80 for the symmetric control task but decreased to between 0.61 and 0.89 for the asymmetric conditions. The Cronbach alpha correlations showed that all the sagittal plane trunk motion measures were repeatable; however, the lateral motion at the 15 and 30° asymmetries were not as repeatable. Hence, the sagittal plane measures for all control tasks and the lateral plane measures at the symmetric control task all have good reliability and may be used to assess low back functional status of neuromuscular control.

37.2.6 Database of Asymptomatic Population

In order to understand the functional performance capabilities of the healthy back performing the five control tasks a database of those without history of low back injury was developed (Marras et al., 1994). At least 25 males and females were recruited in each 10-yr age category between 20 and 70. The asymptomatic low back population study evaluated the influence of age, gender, and task asymmetry on low back functional performance. As expected there were significant differences between males and females. The males had significantly greater trunk motion performance measures than females. As one would expect all the functional performance measures decreased with an increase in age. Interestingly, there was a significant difference in the influence of age for males and females. The velocity performance measures of the female population decreased the most between the 20s and 30s whereas the males decreased the most between the 50s and 60s.

In the asymptomatic population, performance in the sagittal plane decreased as task asymmetry increased. Figure 37.2(a)–(c) shows the range of motion, flexion velocity, and flexion acceleration performance measures as a function of asymmetry for the asymptomatic individuals, respectively. There was no significant difference between flexion and extension velocity and acceleration; therefore, only the flexion results are presented. In addition, there were no significant differences between the clockwise and counterclockwise measures. All three figures illustrate a decrease in functional performance as task asymmetry increased. The reduced performance at increased task asymmetry is theoretically due to the motor recruitment patterns necessary to perform the tasks. The symmetric task requires the recruitment of the large flexor and extensor muscles of the trunk whereas the 15° asymmetry tasks require the recruitment of the smaller oblique muscles in combination with the large flexor and extensor muscles, and thus resulting in lower performance at the increased asymmetry tasks. The 30° conditions require a greater contribution of oblique muscle recruitment compared to the 15° condition, again resulting in significantly lower performance. The figures illustrate the neuromuscular motion signature of the asymptomatic population in all age categories performing the control asymmetry tasks.

The asymptomatic functional performance database provides a benchmark of performance as a function of age and gender for each asymmetry. Figure 37.3 illustrates a healthy individual's data for all control tasks as a percentage of normal for their age and gender groups. The top left panel shows that the range of motion for all five tasks was above 90% of normal. The flexion and extension velocity measures in the top middle and right of the figure were at least 70% of normal. The results of the flexion and extension accelerations in the bottom panels show similar results to the velocities. By normalizing the functional performance measures, the age and gender effects are eliminated. The functional performance measures from patient evaluations would quantify the extent of impairment for each trunk kinematic measure as opposed to any age or gender influences. The one caveat is that the patient must provide the correct age information.

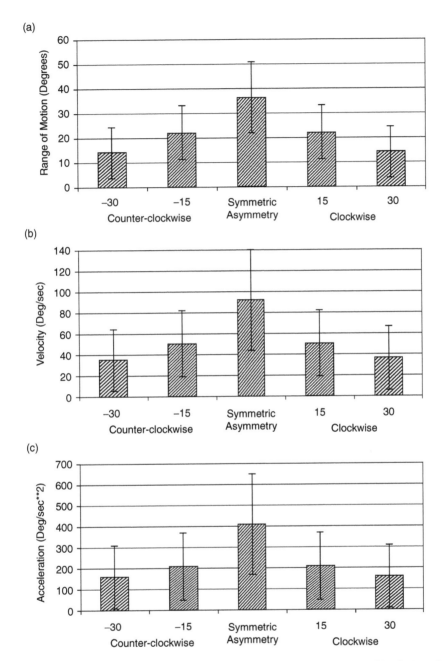

FIGURE 37.2 (a) Range of motion as function of asymmetry for asymptomatic database. (b) Velocity as function of asymmetry for asymptomatic database. (c) Acceleration as a function of asymmetry for asymptomatic database.

37.2.7 Database of Low Back Pain Patients

A database of low back pain patients was created using the same protocol as described above (Marras et al., 1993, 1995, 1999). The patients were recruited from a secondary referral practice and were considered chronic low back pain cases. The patients were in one of ten low back disorder categories listed in Table 37.1. The categories are a combination of structural pathology and pain location categories from the Quebec Task Force Study (Spitzer et al., 1987).

FIGURE 37.3   Percentage normal charts for each trunk motion parameter for an asymptomatic subjects.

37.2.8   Sensitivity and Specificity of Trunk Kinematics

In order for a quantitative assessment to be validated it must have good sensitivity and specificity (Andersson, 1991). Sensitivity is defined as the proportion of those with the disease who are correctly identified. Specificity is the proportion of those without disease who are correctly identified. Typically, there is a trade-off between sensitivity and specificity. The CLMM functional performance evaluation generated a total of 52 motion parameters. In order to evaluate sensitivity and specificity of the CLMM evaluation, Marras et al. (1993, 1995, 1999) have evaluated several different multivariate models using four different statistical methods. All of the multivariate models developed showed excellent sensitivity and specificity. The series of articles by (Marras et al., 1993, 1995, 1999) illustrate a growing database and refinement of the model.

The most recent Marras publication used a database of 374 asymptomatic individuals and 335 low back pain patients (Marras et al., 1999). The large database allowed the data to be examined using

TABLE 37.1   Low back disorder categories

Quebec 1 — Local low back pain
Quebec 2 — Local low back pain with proximal radiculopathy
Quebec 3 — Local low back pain with distal radiculopathy
Spondylolisthesis
Herniated disc pain >3
Herniated disc pain <3
Stenosis
Quebec 9.2 — postsurgical still having symptoms
Nonorganic
Quebec 11 — Scoliosis

---

cross-validation procedures as well as splitting the data randomly into training and test data sets. Univariate discriminant function analysis procedures showed that velocity and acceleration measures had better sensitivity and specificity than range of motion (Marras et al., 1999). The development of multivariate models had the best sensitivity and specificity. Marras et al. (1999) developed a six-variable model consisting of (1) ability to perform each task, (2) twisting range of motion, (3) sagittal range of motion at zero, (4) sagittal extension velocity at zero, (5) sagittal extension acceleration at zero, and (6) lateral range of motion at zero. The discriminant function analysis procedure as well as other statistical methods used generated a summary functional performance probability from 0.0 to 1.0. A performance of 0.5 to 1.0 would classify as an asymptomatic performance and a score of 0.0 to 0.5 would classify in a patient functional performance category. The sensitivity and specificity of this model using discriminant function analysis was 85 and 95%, respectively, using cross-validation techniques. The training and test set data were examined using four statistical techniques. The training set sensitivity range from 83 to 88% and the specificity range from 93 to 96%. The test set sensitivity ranged from 90 to 92% and specificity ranged from 92 to 97%. Regardless of the specific statistical technique the sensitivities and specificities were all excellent. The high sensitivity and specificity indicate that the functional assessment technique is valid for quantitative assessment of asymptomatic and impaired low back function, which is one of the qualities of a good assessment tool (Anderson, 1991; Nachemson and Vingard, 2000; Polatin and Mayer, 2001).

The multivariate model generates a functional performance probability ranging from 0.0 to 1.0, which is dichotomized at 0.5 to evaluate sensitivity and specificity. Dichotomizing the measure is necessary to evaluate sensitivity and specificity; however, it eliminates much of the information in the functional performance probability score. For medical management purposes the medical provider should use the functional performance probability as a continuous measure. For example, a patient with a score of 0.45 would classify as impaired; however, that person is doing much better than someone with a score of 0.12. Medical providers may make different decisions for these two patients regarding medical care whereas from a categorical standpoint both would have impaired functional performance. Thus, using the score as a continuous measure provides much more information on the severity of injury than dichotomizing the score.

37.2.9 Independent Validation of Functional Performance Probability

Cherniack et al. (2001) performed a study to independently assess the correlation between clinical examination and trunk motion characteristics from the CLMM. A multi-specialist physician panel administered a structured physical examination. The CLMM functional performance assessment was completed. The study included 19 patients recruited based on symptoms of chronic recurrent low back pain. The findings showed that the CLMM functional performance probability score and physician panel were in agreement on the presence or absence of abnormality. The CLMM functional performance probability tended to be more consistent with clinical history than the clinical examination. Hence, the functional performance probability from the CLMM has clinical relevance, another important quality of function assessment tools according to Polatin and Mayer (2001).

37.2.10 Functional Performance Probability and Spine Loading

The risk of low back injury in the workplace is assessed via spine loading. Spine loading is a function of the workplace design as well as worker lifting techniques. It was hypothesized that the impaired worker may develop alternative lifting techniques for performing tasks that may further compromise the spine. Therefore, it may be important to understand how spine loading in impaired workers changes as compared to workers with a healthy low back functional performance scores. The functional performance probability, which measures impairment, may be predictive of changes in spine loading.

Marras et al. (2001, 2004a) evaluated 62 low back pain patients and 61 asymptomatic individuals performing a variety of lifting exertions that varied in horizontal distance, vertical height, lift asymmetry, and weight. An EMG-assisted model was used to evaluate spine loading. The asymptomatic population had a functional performance probability of 0.82 (0.24) compared to the low back patient population, which had an average of 0.12 (0.17). The results showed that low back pain patients had greater spine compression and anterior/posterior (a/p) shear forces in all lifting regions compared to the asymptomatic individuals. The shoulder and waist region lifting tasks resulted in the greatest difference between low back pain patients and asymptomatic individuals. The greatest spine loading occurred in the knee far region for both groups. The increased level of spine loading was due to increased levels of antagonistic muscle activity in the low back pain patients. The patients activated muscles unnecessarily, thereby increasing the loading on the spine.

Marras et al. (2005) also investigated how well the functional performance probability predicted spine loading. Using a multivariate mixed modeling procedure with a model that included lift region, task asymmetry, weight, gender, and functional performance probability spine loads were predicted. The results of the model had pseudo-R^2 values of 0.87, 0.61, and 0.65 for compression, a/p shear, and lateral shear, respectively. This study demonstrated that functional performance probability is predictive in nature, one of the qualities of a good functional assessment tool recommended by Nachemson and Vingard (2000).

37.2.11 Interpretation of CLMM Results

The functional performance probability is a summary measure indicating whether or not the person evaluated has impaired low back function. However, the results of the CLMM may be viewed much more extensively. The range of motion, flexion velocity, extension velocity, flexion acceleration, and extension acceleration may be compared by using the asymptomatic data for the patient's age and gender. Figure 37.4 illustrates that the range of motion in the top left for this patient's age and gender is between 50 and 75% of the average asymptomatic group performance. The figure shows that flexion and extension velocity (top middle and right, respectively) are at best 35% for this patient's age and gender. Flexion acceleration at the symmetric task was only 10% of the patient's age and gender group. Overall the range of motion was the least impaired and the acceleration performance

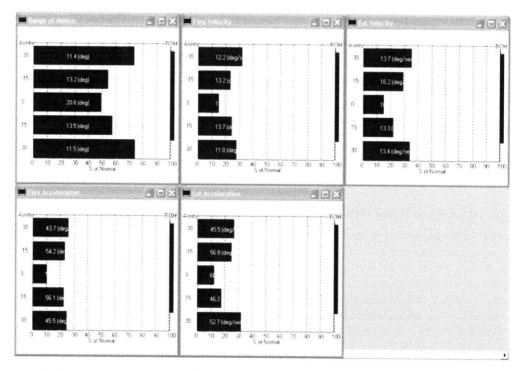

FIGURE 37.4 Percentage normal charts for a low back pain patient.

had the greatest impairment level. Medical providers have traditionally only examined range of motion as a functional performance measure. The additional information gleaned from velocity and acceleration motion parameters may provide insight for better decisions on treatment, extent of functional recovery or residual impairment, and return to work, thereby reducing the risk of recurrent back injury.

The functional performance results in Figure 37.4 may assist in providing information regarding treatment. Notice that all the performance measures show a U-shaped pattern indicating that the patient is most impaired at the symmetric task and less impaired at the asymmetric tasks. The medical management provider may use this information to direct treatment toward strengthening the primary flexor and extensor muscles of the trunk. Even though a specific diagnosis may not be known the functional performance outcome would provide information for targeted treatment. Such information may allow more appropriate treatment thereby reducing the number of treatments necessary as well as the length of time treatment is required, consequently reducing the extent of lost time.

37.3   Streamline Functional Performance Protocol

The CLMM functional performance protocol requires five control tasks to be performed, which may be considered rather time consuming (30 min). Therefore, a streamline protocol was developed and assessed for sensitivity and specificity. The goal of the streamline evaluation was to maintain high sensitivity and specificity and reduce the time required for testing. From a theoretical standpoint the streamline protocol evaluates the function of the primary flexor and extensor muscles of the trunk but not the smaller oblique muscles. In order to get a more complete understanding of musculoskeletal status the full protocol would be necessary.

The streamline protocol requires the same equipment as the full protocol. There is only one control task required, which is the sagittally symmetric task. In addition to the one control task, two tasks are required: a maximum twist clockwise and maximum twist counterclockwise. The two maximum twist scores are combined into twisting range of motion. The instructions to the subject were the same as in the original protocol for the control task. The data analysis was the same as for the full protocol. The total number functional performance measures generated from the streamline protocol were 11.

In order to assess the sensitivity and specificity of the streamline protocol a multivariate discriminant function model was developed (Ferguson and Marras, 2004). The model consisted of (1) twisting range of motion, (2) sagittal range of motion at zero, (3) sagittal extension velocity at zero, and (4) sagittal extension acceleration at zero. These variables are four of the original six measures in the Marras et al. (1999) multivariate model. The original database of 335 patients and 374 asymptomatic individuals was randomly split into training and test sets. The training and test sets had a sensitivity of 85 and 90% and a specificity of 90 and 92%, respectively. For the test set data, the sensitivity was the same as in the Marras et al. (1999) model and the specificity decreased 2%. Overall, the streamline protocol showed outstanding sensitivity and specificity.

The excellent sensitivity and specificity from the streamline protocol multivariate indicates that the protocol is valid. Since the streamline model uses factors that have been shown to be repeatable from the original full protocol the streamline protocol would be reliable. The streamline protocol is much more time efficient (10 min or less) and easier to administer than the full protocol, qualities suggested by Andersson (1991). However, from a musculoskeletal status perspective the streamline protocol only provides information on the status of the flexor and extensor muscles of the trunk and not the smaller oblique muscles. It should be noted that many times low back pain patients are only able to perform the symmetric task at an initial evaluation. As the patient recovers the more extensive full protocol with all five control tasks would provide a more comprehensive assessment of the neuromuscular recovery of the individual, including the oblique muscles. The medical management provider may decide that certain patients need a full evaluation whereas other patients may require only a streamline evaluation. This decision may be a function of the patient's injury, job demands, or other factors.

37.4　Sincerity of Effort

In order for a functional performance assessment to provide useful information to medical management the patient must be performing with a true effort. One troublesome aspect of low back functional performance assessment is that patients may magnify the impairment for several reasons including fear, mistaken beliefs, maladaptive coping strategies, and active attempts to seek treatment (Luoto et al., 1996). If a patient with low back pain does not perform the tasks to the best of his or her ability during the functional evaluation session, then the quantitative measure may erroneously document the musculoskeletal status of the trunk. Thus, in assessing trunk status, it is important to objectively quantify whether the patient is magnifying his or her low back impairment, which is a challenge (Chegalur et al., 1990). Simonsen (1995) and Luoto et al. (1996) have assessed impairment magnification of the trunk, using coefficient of variation measures with varying levels of success. This may be a promising avenue for evaluating sincerity of effort during trunk functional assessment. In any case it is crucial to evaluate sincerity of effort objectively in order to provide the medical management an indicator of the quality of information reported from the functional assessment evaluation.

37.4.1　Sincerity of Effort Protocol

Marras et al. (2000) developed a sincerity of effort evaluation using the LMM to be used in conjunction with the CLMM functional performance evaluation. In theory, when a low back injury occurs, the motor control pattern that has been established is either relearned or a new motor program is established. In either case a sincere effort would result in repeatable motion because the same motor program is engaged to elicit the motion each time. In an insincere effort the patient would rely on a "mental model" to override the recruitment pattern, therefore resulting in a less repeatable motion. Hence, a protocol was developed to evaluate both sincere effort as well as insincere effort. The sincerity of effort protocol required the symmetric control task as well as three additional tasks. The tasks were uncontrolled sagittal flexion extension, lateral bending side to side, and twisting side to side. A total of four tasks were required and each task was performed with a sincere effort and an insincere effort.

Marras et al. (2000) evaluated 100 asymptomatic and 100 low back pain patients performing all four tasks. There were at least ten asymptomatic individuals in four age categories (<30, 30–39, 40–49, and 50–59 yr) for both males and females. Each participant performed each task twice. The instructions were to "cross your arms in front of your chest and stand with feet shoulder width apart." For the sagittal tasks participants were instructed to flex and extend the torso. For the transverse tasks, participants were instructed to twist at the waist to the left and right. For lateral bending, participants were told to bend from side to side continuously. To elicit a sincere effort participants were instructed to "move as fast as you can comfortably." To obtain an insincere effort asymptomatic participants were instructed to "move as fast as you can pretending you have back pain" and patients were instructed to "move as fast as you can comfortably pretending your pain is worse."

37.4.2　Sensitivity and Specificity of Sincerity of Effort

Data analysis included quantification of position, velocity, acceleration, and jerk in all three planes. Phase plane diagrams were constructed for all combinations of position, velocity, acceleration, and jerk, which allow for the examination of the relationship between two variables to be expressed in a value of rho. Rho is the distance between the point in the phase plane and the centroid of the phase plane. The mean, standard deviation, and coefficient of variation for all rho values were calculated. Fifty-four trunk motion performance measures were extracted from each task for a total of 216 measures from all tasks.

Discriminant function analysis was used to evaluate how well each individual motion parameter distinguished between sincere and insincere efforts. The analysis showed that higher-order derivatives of motion classified the experimental conditions better than position. The coefficient of variation of rho for the velocity–acceleration phase plane distinguished well between sincere and insincere. Figure 37.5

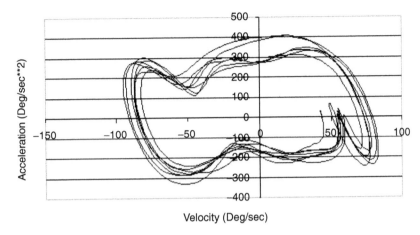

FIGURE 37.5 Velocity–acceleration phase plane plot for high sincerity score.

shows a high sincerity velocity–acceleration phase plane with consistent elliptical shapes. Figure 37.6, on the other hand, shows a low sincerity phase plane, which appears more like a scatter plot. Thus, it appears that these two figures support the theory of the protocol that sincere effort is repeatable especially with higher-order derivatives of motion and that insincere effort is less repeatable.

In order to achieve the maximum sensitivity and specificity possible multivariate models were developed. Marras et al. (2000) developed models with at least one measure from each of the three planes and higher-order derivatives of position were used because they represent the musculoskeletal recruitment pattern. Three models were developed, one for the low back pain group, one for the asymptomatic group, and one for all the data combined. Table 37.2 lists the variables in each of the models. Each model contains at least one motion parameter from each plane of the body. All the models have at least one coefficient of variation parameter using a higher-order derivative of motion. The models were selected with a goal of having equal sensitivity and specificity. The low back pain group model had a sincerity and specificity of 75%. The asymptomatic group model produced a sincerity and specificity of 92%. The model for both the patients and asymptomatic groups had a sensitivity and specificity of 81.5%. All three models have good sensitivity and specificity, which is one of the qualities of a good assessment tool (Andersson, 1991; Nachemson and Vingard, 2000).

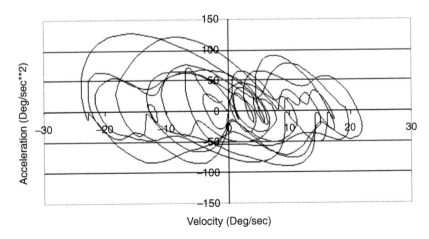

FIGURE 37.6 Velocity–acceleration phase plane plot for low sincerity score.

TABLE 37.2   Model Components for Sincerity of Effort

Sample Population	Model Variables
LBD patient group	Lateral standard deviation of position
	Uncontrolled sagittal standard deviation of velocity
	Lateral coefficient of variation of acceleration
	Twisting coefficient of variation of rho for velocity–acceleration phase plane
	Twisting coefficient of variation of rho for acceleration–jerk phase plane
Asymptomatic group	Twisting coefficient of variation of position
	Twisting peak velocity during extension
	Twisting peak velocity during twisting task
	Controlled sagittal coefficient of variation of acceleration
	Lateral coefficient of variation of rho for the position–acceleration phase plane
Combined LBD patients and asymptomatic individuals	Uncontrolled sagittal coefficient of variation of position
	Lateral peak velocity during trunk flexion
	Lateral peak velocity during lateral bending task
	Uncontrolled sagittal coefficient of variation of rho of velocity–acceleration phase plane
	Lateral coefficient of variation of acceleration
	Twisting range of rho for the position–jerk phase plane

Source: Adapted from Ferguson et al. (2000). *Spine* 25(15):1950–1956, Lippincott. With permission.

37.4.3 Repeatability and Validation

Ferguson et al. (2003) validated the sincerity of effort and evaluated inter-rater reliability as well as test–retest reliability. The validation was a blinded randomized study to evaluate whether or not the probability of sincere effort accurately identified the participants providing a sincere effort. The test–retest phase of the study evaluated intra- and inter-rater reliability.

Sincerity of effort is a continuous measure; however, in order to evaluate sensitivity and specificity of the measure it has must be dichotomized. The initial discriminant function analysis used a cutoff of 0.50. The results of the validation study indicated that a cutoff of 0.60 would create greater sensitivity and specificity. The sensitivity was 90% and the specificity was 100% in the blinded study. However, the score would be best used as a continuous measure to indicate the quality of data. The inter- and intra-rater evaluation results indicated that the sincerity of effort protocol was reliable both between raters and across trials.

37.4.4 Interpretation of Sincerity of Effort

The probability of sincere effort was developed for use in conjunction with the functional performance probability, which is used to evaluate the musculoskeletal status of a low back pain patient. The probability of sincere effort was not intended to identify malingering but to indicate the quality of functional performance data. If an individual has a low sincerity of effort score the functional performance evaluation should be repeated. As pointed out by Main and Waddell (1988), the misuse and misinterpretation of this type of measure may have unintended consequences. The objective of the sincerity of effort score is to provide medical management with an understanding of the quality or believability of the functional performance information.

37.5 Quantification of Recovery

Impairment is an objective measure of anatomical or functional loss (Mayer and Gatchel, 1988; AMA, 1998; Waddell, 1998). The percentage of normal chart provides an objective quantitative measure of the extent of functional loss for each kinematic measure. Multiple assessments quantify the initial extent of impairment as well as the amount of recovery at each evaluation as shown in Figure 37.7. The figure shows the results of three evaluations from a patient with a muscular strain/sprain. The figure illustrates that initially the patient was not able to perform the two 30° conditions. The range of motion was initially 100% for the three tasks performed, whereas velocity was 40 to 62% and acceleration was 15 to 55% of normal function. Initially, based on range of motion medical management may decide that this patient was not impaired; however, the velocity and acceleration measures indicate impairment. The second visit shows the patient performing all five tasks. The velocity improved to between 50 and 95% and the acceleration was between 28 and 80%. The third visit illustrates that the flexion velocity on the left-hand side was 100% whereas to the right-hand side it was 70 to 90% of normal. The flexion acceleration on the right-hand side was 85% of normal but only 60% on the left-hand side. The pattern of recovery over time was that range of motion recovered first followed by recovery of velocity and finally acceleration. This pattern has been shown in the control CLMM functional performance results as well as in all three planes of the body (Marras et al., 1995; Ferguson, 1998). Notice that at the third evaluation the flexion acceleration performance was better than the extension performance. It is hypothesized that this discrepancy between the flexion and the extension may be indicative of underlying cognitive fear of pain with forward flexion.

The probability of sincere effort was developed to provide a quality check of the data. The probability of sincere effort for the three visits shown in Figure 37.7 were 0.85, 0.92, and 0.91, respectively. These probabilities indicate that the patient was giving a true effort when performing the functional evaluation. Thus, the medical provider knows that the musculoskeletal results in Figure 37.7 document the true functional recovery of the low back function of the patient.

The CLMM functional performance results from multiple evaluations illustrated in Figure 37.7 may be used in several ways. First, it provides medical management quantitative documentation of the improvement or lack of improvement in low back function of the patient, which may determine treatment

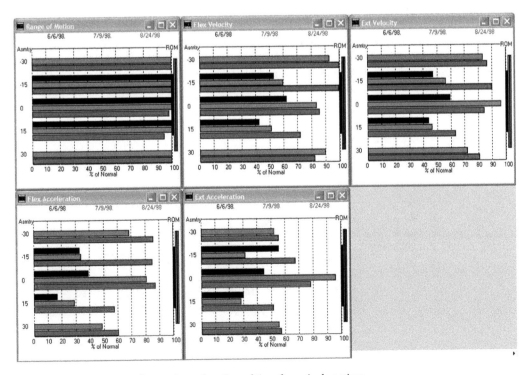

FIGURE 37.7 Percentage of normal as a function of time for a single patient.

termination. Second, it may be used to quantify the effectiveness of various types of treatments or combinations of treatment. Medical providers may be able to determine which type of treatment most effectively restores function as well as quantitatively assess the optimal number of treatments. Third, multiple evaluations may be used to quantify maximum medical improvement and establish a new benchmark of functional performance for the patient. If the patient has an exacerbation, then the functional restoration goals for the patient would be their own benchmarks. Finally, the CLMM functional assessment may be utilized in conjunction with job assessment information to aide in the decision for return to work.

37.6 CLMM Functional Performance Recovery Compared to Traditional Recovery Measures

The most common outcome measures for low back pain are return to work, symptoms, and activities of daily living. Ferguson et al. (2000) assessed the natural course of recovery using the CLMM functional performance probability, return to work, symptoms measured by the McGill Pain Questionnaire (MPQ), and impairment of activities of daily living measured with the Million Visual Analog Scale (MVAS). The study evaluated 32 acute low back pain patients with muscular strain sprains. There were 16 occupational and 16 nonoccupational patients. The patients were assessed every 2 weeks for 3 months. The four outcome measures all showed significant improvement over time. However, none of the outcome measures showed significant differences between the two groups. The functional performance probability was the only measure to show a significant difference in the rate of recovery between the two groups. The nonoccupational group recovered faster than the occupational group over time. On an average, the occupational group was recovered at 12 weeks whereas the nonoccupational group was recovered at 8 weeks.

In comparing the four outcome measures illustrated in Figure 37.8, the CLMM functional performance probability, MPQ symptoms score, and MVAS of impairment of activities of daily living demonstrate similar trends of recovery for the first 12 weeks. The CLMM functional performance probability provides an objective quantitative assessment of impairment that is independent of the subjective impression of the patient, which is the drawback of traditional symptoms and activity of daily living questionnaires. At 14 to 18 weeks, 32% of patients were impaired based on CLMM results whereas only 20% were impaired according to symptom and MVAS scores. This indicates that musculoskeletal function is recovering after symptoms have been eliminated. It is hypothesized that at the point when

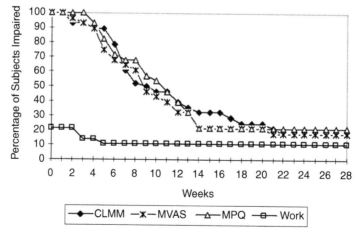

FIGURE 37.8 Percentage of patients impaired as a function of time. (Adapted from Marras et al. (2000) *Spine* 25(15):587–595, Lippincott. With permission.)

symptoms have recovered but musculoskeletal function remains impaired a patient may be at highest risk for recurrent injury. Using multiple assessment measures provides a more complete understanding of the severity of impairment and extent of recovery over time, which may allow medical providers to better determine readiness to return to full duty work or termination of treatment.

As shown in Figure 37.8, work status initially indicated that 21% of the population was impaired whereas the other three outcome measures indicated that 100% of the population was impaired. The figure clearly illustrates that study participants were continuing to work even though they may be impaired based on symptoms, impairment of activities of daily living, and CLMM functional performance probability. Baldwin et al. (1996) suggested that return to work was a misleading outcome measure of low back pain recovery. The discrepancy between return to work and the other three outcome measures appears to support the idea. Return to work or disability is a complex issue that is dependent not so much on functional impairment but rather on psychosocial work environment factors, psychological factors, and individual factors (Feuerstein et al., 1999). Ferguson et al. (2000) found that the correlation between the other three outcome measures and work status were all less than 0.5. The weak correlation indicates that work status is independent of the worker's functional performance probability, impairment of activity of daily living, or severity of symptoms.

The independence of CLMM functional performance probability and work status is illustrated in Figure 37.9, which shows the improvement in CLMM functional performance probability as a function of time for two patients. The figure illustrates that both patients improved during the 12 weeks of functional performance evaluations; however, one worker continued to work and the other was disabled. Ferguson et al. (2000) discussed these findings indicating the disabled patient did not return to work during the entire study because the employer was not willing to accommodate any type of work restriction. The employer required that workers be able to lift at least 50 lb in this patient's work department. The employer was not willing to allow the worker to temporarily change departments to accommodate work restrictions. The patient who continued working even though her back functional performance was less than 0.1 from the CLMM had an employer who was willing to accommodate her medical provider's work restrictions. The CLMM functional assessment results of functional performance probability provide a quantitative assessment of impairment but not necessarily disability. The CLMM results may assist the medical provider in assessment of the severity of impairment and the need for work

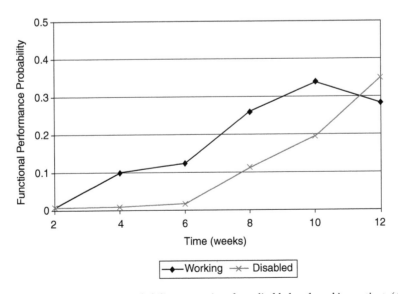

FIGURE 37.9 Functional performance probability across time for a disabled and working patient. (Adapted from Marras et al. (2000) *Spine* 25(15):587–595, Lippincott. With permission.)

restrictions. However, disability may be a function of whether or not the employer is willing to accommodate the restrictions prescribed by the medical providers.

37.7 Summary

The CLMM low back functional assessment has been scientifically developed and shown to be repeatable, reliable, and valid. There are two functional performance protocols that measure low back functional impairment. The full protocol is a complete assessment of the trunk's musculoskeletal status at five controlled asymmetries. This protocol provides quantification for function of primary flexor and extensor muscles as well as oblique muscles. The full protocol may require up to 30 min to complete. The second streamline protocol is more efficient. The streamline protocol requires only one controlled asymmetry task and the entire protocol requires at most 10 min. The streamline protocol provides quantitative results for only the primary flexors and extensors. Both protocols are valid; however, it is clear that the full protocol provides more clinical information on patient recovery. The CLMM results objectively quantify the extent of impairment given the patients' age and gender. In conjunction with the sincerity of effort protocol the medical provider can determine if the results of the CLMM evaluation provide insight into the patient's true musculoskeletal status. The CLMM may be used to quantify improvement or lack thereof in the patient's functional impairment status when multiple evaluations are conducted. The medical management may use the CLMM results in conjunction with job assessment information to make a determination of disability for a specific job or recommendations for light duty. However, it would be a corporate decision as to whether or not these recommendation for light duty would be accommodated.

References

AMA (1993) *Guides to the Evaluation of Permanent Impairment*, 4th edn. American Medical Association, Chicago.

Andersson, G.B.J. Sensitivity, specificity, and predictive value a general issue in the screening for disease and in the interpretation of diagnostic studies in spinal disorders. In: Frymoyer, J.W. (ed.) *The Adult Spine: Principles and Practice*, 1st edn. New York: Raven Press, 1991, pp. 277–287.

Baldwin, M., Johnson, W., Butler, R. (1996) The error of using return-to-work to measure the outcome of health care. *Am J Ind Med* 29:632–641.

Chengular, S., Smith, G., Nelson, R., Sadoff, A. (1990) Assessing sincerity of effort in maximal grip strength test. *Am J Phys Med Rehabil* 69:148–153.

Cherniack, M., Dillon, C., Erdil, M., Ferguson, S., Kaplan, J., Krompinger, J., Litt, M., Murphy, M. (2001) Clinical and psychological correlates of lumbar motion abnormalities in low back disorder. *Spine J* 1:290–298.

Ferguson, S. (1998) Quantification of low back pain recovery. Doctoral dissertation, The Ohio State University.

Ferguson, S., Marras, W. (2004) Revised protocol for the kinematic assessment of impairment. *Spine J*, 4:163–169.

Ferguson, S., Marras, W., Gupta, P. (2000) Longitudinal quantitative measure of the natural course of low back pain recovery. *Spine* 25(15):1950–1956.

Ferguson, S., Gallagher, S., Marras, W. (2003) Validity and reliability of sincerity test for dynamic trunk motion. *Disability and Rehabilitation* 25:236–241.

Feuerstein, M., Berkowitz, S., Huang, G. (1999) Predictors of occupational low back disability: implications for secondary prevention. *J Occup Environ Med* 41(12):1024–1031.

Fordyce, W. (1976) *Behavioral Methods for Chronic Pain and Illness.* C.V. Mosby Company, St. Louis.

Gatchel, R., Mayer, T., Capra, P., Diamond, P., and Barnett, J. (1986) Quantification of lumbar function Part 6. The use of psychological measures in guiding physical functional restoration. *Spine* 11:36–42.

Hales, T., Bertsche, P. Medical management of work-related musculoskeletal disorders. In: Karwowski, W., Marras, W. (eds.) *The Occupational Ergonomics Handbook.* New York: CRC Press, 1999, pp. 1255–1267.

Hamrick, C. (2000) CTDs and ergonomics in Ohio. Paper presented at International Ergonomics Association 2000/Human Factors and Ergonomics Society 2000 Congress, July 29–August 4, San Diego, CA.

Horak, F., Diener, H. (1994) Cerebellar control of postural scaling and central set in stance. *J Neurophysiol* 72:479–493.

Luoto, S., Hulpi, M., Alaranta, H., Hurri, H. (1996) Isokinetic performance capacity of trunk muscles: Part II. Coefficient of variation in isokinetic measurement in maximal effort and in submaximal effort. *Scand J Rehabil Med* 28:207–210.

Main, C., Waddell, G. (1998) Behavioural response to examination a reappraisal of the interpretation of "nonorganic signs." *Spine* 23:2367–2371.

Marcel, M.A., Costigan, P.A., Stevenson, J.M. (1999) Comparative study of the lumbar motion monitor and fastrak in assessing spine kinematics. In: *Advances in Occupational Ergonomics and Safety.* IOS Press, pp. 89–94.

Marras, W., Fathallah, F., Miller, R., Davis, S., Mirka, G. (1992) Accuracy of a three-dimensional lumbar motion monitor for recording dynamic trunk motion characteristics. *Int J Ind Ergon* 9:75–87.

Marras, W., Parnianpour, M., Ferguson, S., Kim, J., Crowell, R., Simon, S. (1993). Quantification and classification of low back disorders based on trunk motion. *Eur J Phys Med* 3:218–235.

Marras, W., Parnianpour, M., Kim, J., Ferguson, S., Crowell, R., Simon, S. (1994). A Normal database of dynamic trunk motion during repetitive trunk flexion and extension as function of task asymmetry, age and gender. *IEEE* 2(3):137–146.

Marras, W., Parnianpour, M., Ferguson, S., Kim, J., Crowell, R., Bose, S., Simon, S. (1995). The classification of anatomic and symptom based low back disorders using motion measure models. *Spine* 20(23):2531–2546.

Marras, W., Ferguson, S., Gupta, P., Bose, J., Parnianpour, M., Kim, J., Crowell, R. (1999) The quantification of low back disorder using motion measures: methodology and validation. *Spine* 24(20):2091–2100.

Marras, W., Lewis, K., Ferguson, S., Parnianpour, M. (2000) Impairment magnification during dynamic trunk motion. *Spine* 25(5):587–595.

Marras, W., Davis, K., Ferguson, S., Lucas, B., Gupta, P., (2001) Spine loading characteristics of patients with low back pain compared with asymptomatic individuals. *Spine* 26:2566–2574.

Marras, W., Ferguson, S., Burr, D., Davis, K., Gupta, P. (2004) Spine loading in low back pain patients during asymmetric loading. *Spine J* 4:64–75.

Marras, W., Ferguson, S., Burr, D., Davis, K., Gupta, P. (2005) Functional impairment as a predictor of spine loading. *Spine* 30:729–737.

Mayer, T., Gatchel, R. (1998) *Functional Restoration for Spinal Disorders: The Sports Medicine Approach.* Lea and Febriger, Philadelphia.

Nachemson, A., Vingard, E. Assessment of patients with neck and back pain: a best-evidence synthesis. In: Nachemson, A., Johsson, E. (eds.) *Neck and Back Pain.* New York: Lippincott, Williams, and Wilkins, 2000, pp. 189–235.

National Research Council Institute of Medicine. (2001) *Musculoskeletal Disorder and the Workplace Low Back and Upper Extremity.* National Academy Press, Washington, DC.

Polatin, P., Mayer, T. Quantification of function in chronic low back pain. In: Turk, D., Melzack, R. (eds.) Handbook of Pain Assessment, 2nd edn. New York: The Guilford Press, 2001, pp. 191–203.

Simonsen, J. (1995) Coefficient of variation as a measure of subject effort. *Arch Phys Med Rehabil* 76:S16–S20.

Spitzer, W., LeBlanc, F., Dupruis, M. (1987) Scientific approach to the assessment and management of activity-related spinal disorders. A monograph for clinicians. Report of the Quebec Task Force on Spinal Disorders. *Spine* 12:S1–S59.

Waddell, G. Impairment. In: Waddell, G. (ed.) *The Back Pain Revolution.* New York: Churchill Livingston, 1998, pp. 119–134.

VIII

Ergonomic Industrial Interventions

38

Chairs and Furniture

Marvin J. Dainoff
Miami University

38.1 Introduction

This chapter consists of an overview of ergonomics appropriate to workplaces that utilize chairs and associated furniture. Such workplaces are typically described as "offices." However, in determining the scope of this material, it is necessary to consider a fundamental question: what is an office? In the modern electronic workplace, the answer is not straightforward. A traditional (*Oxford English Dictionary*) definition states that an office is "... a place for the transaction of private or public business" (Oxford University Press, 1971). With the profusion of portable computers and communications equipment, almost any location can fit that definition: an airport waiting room, a kitchen table, an automobile, even a park bench. For purposes of this chapter, we will focus on workplaces whose primary purpose is some aspect of information processing and transformation, where some sort of computer equipment is employed, and whose occupants are expected to remain in place for extended periods of time (i.e., several hours). These are the venues within which the bulk of the scientific research has been conducted. On the other hand, the general principles discussed here can be usefully applied to more temporary venues (e.g., setting up a temporary workspace with a laptop and modem in a hotel room).

38.2 Historical Overview

Attention to office ergonomics corresponded with the large-scale introduction of video display terminal (VDT) technology into the workplace in the late 1970s. This introduction was almost immediately accompanied by worldwide reports of associated health complaints. While initial concerns were focused on potentially harmful effects of electromagnetic radiation on pregnancy and vision, by the 1990s, ergonomic issues dominated the scientific and popular discussion (Dainoff, 2000; Dainoff and Dainoff, 1986).

In the United States, the Human Factors Society (now the Human Factors and Ergonomics Society) developed an ANSI Standard (ANSI/HFS 100-1989) on ergonomic aspects of computerized workplaces (Human Factors Society, 1988). The scope of this standard included ergonomic design specifications for monitor displays, input devices, furniture and chairs. A recent revision of this standard is now available as BRS/HFES 100 (Human Factors and Ergonomics Society, 2002). The International Standards Organization (ISO) is in the process of developing its own international standard: ISO 1492 — *Ergonomic standards for office work with visual display terminals*. This standard currently has 17 parts, each with its reference number and date of approval. (For information, see www.iso.org). In North America, guidelines which parallel BRS/HFES 100 are available from the Canadian Standards Association (Canadian Standards Association, 2000), and the Business and Institutional Furniture Manufacturers Association (Business and Institutional Furniture Manufacturers Association, 2002). A recent comprehensive overview of office ergonomics may be found in Kroemer and Kroemer (2001).

Why such attention to office ergonomics? The rationale underlying the design specifications contained in these standards could be summarized as follows: much office work with VDTs tends be characterized by prolonged periods of static posture coupled with high visual demands. In such postures, movement occurs in only a few muscle groups such as wrists and fingers. To the extent that poorly designed workstations require the operator to take up working postures that are inefficient ("awkward"), the onset of fatigue — with consequent discomfort and pain — will be relatively rapid. Conversely, the application of ergonomic principles to workplace design will delay the onset of fatigue, resulting in improved work efficiency and increased feelings of well-being. In the longer term, ergonomic improvements should reduce the incidence of those musculoskeletal disorders that might be linked to poor working postures.

This rationale appeared to be supported by published research evidence demonstrating effects of ergonomic interventions in office environments. See for example, Sauter, Dainoff, and Smith (1990).

However, toward the end of the century, office ergonomic issues became embedded in a more general concern about work-related musculoskeletal disorders (WRMSD) in all workplaces. In the United States, the federal authorities, through the Occupational Safety and Health Administration (OSHA), while initially focusing on the meat packing industry, moved towards publishing a general regulation applying to all workplaces, including offices. In the political reaction to this move, many opponents of regulation questioned the scientific basis underlying so called "ergonomic disorders." One outcome was that the U.S. Congress directed the National Research Council to conduct a thorough analysis of the issue, including an extensive review of the published literature (National Research Council, 2001). The cited literature in the National Research Council report was reanalyzed by Karsh, Moro & Smith (2001) for the specific purpose of evaluating efficacy of ergonomic interventions to control WRMSD. Earlier, in response to similar concerns about WRMDs in Europe, a parallel assessment of intervention effectiveness was published (Westgaard and Winkel, 1997).

38.3 Do Ergonomics Interventions in the Office Actually Work? Confusions and Contradictions

Let us start to answer this question by taking, as a point of departure, the three literature reviews just cited. The National Research Council's conclusion is forthright: (2001)

> . . .a clear and strong pattern of evidence emerges after considering the epidemiologic, biomechanical, basic science, and intervention literature collectively. We can conclude with confidence that

there is a relationship between exposure to many workplace factors and an increased risk of musculoskeletal disorders. (National Research Council, 2001, pp. 362–363)

This is a general conclusion, establishing a link between workplace risk factors and MSDs. Presumably, then, the purpose of an ergonomic intervention is to reduce such risk factors.

We first need to ask: what is an intervention? For purposes of this chapter, we can look at workplace interventions in two ways: *functional* and *structural*. The functional view classifies interventions in terms of the *goals* relative to reduction or control of musculoskeletal disorders in the workplace. The intervention can be *primary* (focused on preventing disorders before they occur), *secondary* (providing interventions in direct response to reported problems), or *tertiary* (providing interventions for the purpose of ameliorating already existing disorders.) The structural view classifies interventions in terms of *means* of accomplishing intervention goals. Office workplace interventions can consist of *engineering-based* approaches (providing "ergonomic" chairs, furniture, monitors, input devices which reduce biomechanical loads), *work methods-based* approaches (redesigning work procedures in order to change worker behaviors so as to reduce biomechanical loads and training workers to use equipment more effectively), and *administration-based* approaches (instituting organizational changes such as additional rest breaks and job rotation schemes). At a more macroscopic level, Westgaard and Winkel (1997) make an important distinction between interventions which are preplanned (i.e., providing a specific device to reduce load, such as a new chair), and those which include ongoing organizational involvement (i.e., employee contributions through participative ergonomics).

The conclusion from all three reviews is that there is some support for the general proposition that ergonomic interventions are effective in reducing risk factors for WRMSDs. However, with respect to the specific case of office workplaces, the evidence is less clear. The National Research Council (2001) report concludes:

> …there is some evidence that using ergonomic principles to modify chairs, workstations, and keyboards can be effective in reducing the prevalence of upper extremity symptoms; in the office setting, results concerning the effects of these interventions on physical findings are mixed (p. 359).

The analysis of Westgaard and Winkle (1997) is in some ways the most stringent. They argue that *none* of the intervention studies they review is definitive in the sense of meeting standards of scientific rigor associated with clinical trials (e.g., random assignment to experimental and control conditions, double-blind assessment). These authors recognize the difficulties of carrying out such studies in workplaces, and conclude that the body of research can support general conclusions about the effectiveness of ergonomic interventions. But, these conclusions are less clear when considering office work with VDTs (Westgaard and Winkel, 1997, p. 488). Karsh, Moro, and Smith (2001) make similar general arguments, but do not specifically address the question of office ergonomic interventions.

Few would deny that providing an ergonomic chair should be a key component of an office ergonomic intervention. However, Helander (2003) has recently published the startling assertion that we should (p. 1306): "Forget about ergonomics in chair design? Focus on aesthetics and comfort!" (This quote is the title of his paper.) His conclusions, which are based on empirical findings from several studies conducted with his colleagues (Helander, Little, and Drury, 2000; Helander and Zhang, 1997; Helander, Zhang, and Michel, 1995), is that so-called ergonomic features of chairs (e.g., adjustability of seat pan height and angle, backrest height and angle) are intended to address biomechanical risk factors associated with poor posture. Reduction of such risk factors should result in reduction in perceived discomfort. However, he finds that, in fact, users are unable to differentiate, on the basis of reduction in discomfort, ergonomic chairs with different ergonomic features. He attributes this failure to the demonstrated inability of users to distinguish small angles and pressure differences because of poor proprioceptive feedback from joints and ligaments. On the other hand, the same chairs could be differentiated on the basis of satisfaction and comfort. Thus, Helander concludes:

> Discomfort is based on poor biomechanics and fatigue. Comfort is based on aesthetics and plushness of chair design and a sense of relaxation and relief (Helander, 2003, p. 1315).

The implications for chair design are, according to this argument, that, while a minimal set of chair characteristics are necessary to avoid serious discomfort; the attention of designers should be focused on esthetics and comfort.

38.4   Implications for Best Practices: An Attempt at Clarification

How, then, are we to arrive at a set of best practices for office furniture? The approach we will follow is three-fold. First, we will present a critique of the existing literature on scientific grounds. Second, we will provide an alternative conceptual framework within which serious progress might be made. Finally, we will use this framework as an organizing device the purpose of which is to present ergonomic best practices in a usable format.

The overviews found in Westgaard and Winkel (1997) and Karsh et al. (2001) as well the more extensive discussion found in National Research Council (2001) provide thoughtful analyses of methodological issues in design of intervention studies and approaches to avoiding threats to internal and external validity (Cook and Campbell, 1979). The reviewers are well aware of the difficulties of conducting research in complex work settings. The literature reviewed by these authors had been filtered through different sets of criteria, so that is unlikely that any well-designed peer reviewed studies escaped their attention.

However, in examining the study summaries presented in these reviews, what appears as a problem with the entire literature is the level of specification of interventions. If we focus just on those interventions carried out in office environments, it seems that "ergonomic chairs" and "adjustable furniture" are typically treated as generic black boxes in the sense that they seem to be interchangeable linear components to be plugged in or not plugged in to the office environment. These studies often reflect a general awareness that chair and furniture adjustability should allow users to avoid awkward postures, but they lack a more detailed specification of the interrelationship between equipment functionality and actual user postural adjustment. A case in point is found in Westgard and Winkle's (1997) discussion of intervention impact. These authors indicate that the effectiveness of some intervention methods can be dependent on cooperation of employees or management; however, they go on to state, "If the intervention is considered to be inherently effective (e.g., introduction of new chairs without alternatives), this point is not considered relevant" (Westgaard and Winkel, 1997, p. 486). Presumably, the intention of this statement is to reflect that fact the employees in question had no choice in using their new chairs. Nevertheless, this statement is reflective of ergonomic chairs as interchangeable black boxes in the office system.

A different level of problem is in the underlying medical context of the above research. All three reviews assume a basic causative model in which physical attributes of the workplace (risk factors) present a possibility of biomechanical loading of soft tissues which are then expressed as discomfort and pain — later progressing toward impairment and disability. Social, organizational, and individual factors may act as moderators of this process. This underlying model is an extension of the more classic conception in which a single pathogen impacts on a target organ. Hence, the appropriate analytic tool is the use of random assignment to experimental and control groups to assess effectiveness of reduction of the pathogen and thus assess causality. In fact, the motivation of National Research Council (2001) was in support of ergonomic legislation, and the motivation of Karsh et al. (2001) was to evaluate why the effort failed.

However, as eloquently expressed by Moon (1966), the epidemiological concept of "level of exposure to risk factors" is problematic for office workplaces.

Although all occupational exposure occurs through an interface between the environment and the worker, the ergonomic interface differs significantly from a typical industrial hygiene model. Hazardous exposure to chemical substances is a function of quantitative levels in the work environment. Ergonomic exposure occurs across an interface largely *defined by human activity*. Physical exertion, contact, movement, or static positioning typically create the physical exposure . . . factors other than the physical work environment modify physical actions and may affect the resulting perception of sensory feedback relating to those actions . . .. These and other factors may influence the individual's capacity, opportunity, or motivation to alter the exposure interface in response to such feedback (Moon, 1966, p. 119).

Thus, part of the problem may be the inherent discontinuity between the traditional medical model and an ergonomic approach that seeks to optimize fit between characteristics of the individual and characteristics of the workplace. It is interesting that in his published dissent to the conclusions of the National Research Council, Szabo (2000, p. 452), while arguing there is no evidence that ergonomic interventions have reduced the incidence of medical conditions, goes on to state: "There is little doubt that that most ergonomic interventions increase comfort in the work environment, which is of great benefit to the worker."

Ironically, a recent study published in *Journal of the American Medical Association* (Stewart, Ricci, Chee, Morgenstein, and Lipton, 2003) estimated that common pain experienced by U.S. workers results in a productivity loss of $61.2 billion per year. This estimate is derived from analysis of an extensive telephone survey (the American Productivity Audit) of over 28,000 working adults. Most (76.6%) of the reported lost productive time was explained by reduced performance while at work, rather than absence from the workplace.

What can be concluded from these arguments? If the goal of ergonomics is to improve fit between the worker and the things with which he or she interacts (Dainoff and Dainoff, 1986) for the purpose of reducing discomfort and fatigue while increasing efficiency, even the critics (Szabo, 2000) might agree we are successful in achieving this goal. If, however, ergonomics is viewed as an adjunct of medical practice, the situation is not so clear. The core of the argument — the extent to which discomfort/pain is a precursor to a diagnosed medical disorder — is a medical issue, not an ergonomic one. In fact, we might argue that ergonomic practice should be more closely associated with industrial engineering than medicine, where criteria for effectiveness are somewhat different. We might imagine the reaction if we required the same level of evidence (randomized assignment to groups with double-blind assessment of outcomes) to justify purchase of a new computer system or software upgrade.

Given this argument, however, it is still the case that the scientific basis underlying application of ergonomic principles toward the office workplace is problematic. We will briefly speculate on why this has been the case and present a vision for how the situation could be improved.

Despite the appearance of a number of research studies on seating and seated posture in the recent several years, what has been lacking is a systematic assessment of the *integrated* relationships among the relevant "ergonomic" attributes of chairs and workstations, and the associated postural actions/reactions of workers using this equipment. In particular, chairs have too often been studied in isolation, rather than within the operational context of their use. We can posit a series of explanations for this situation.

With few exceptions, most the research studies reflect what might be called "brief encounters" rather than long-term programmatic research. Successful scientific attacks on serious problems are typically characterized by years of integrated research efforts by individuals or teams. This is rarely evident in ergonomic research on seated posture. This statement is not true in other areas of ergonomics. See, for example, Marras and Waters (2004) and McGill (2002).

One can attribute the lack of integrated research programs to the lack of sustained financial support for such research from either private or public sector. This is ironic in terms of the enormous potential cost savings from successful ergonomic implementation, but this is the current reality.

A second problem can be exemplified by Helander's (2003) work on the importance of esthetics in chair design. (The work of Helander and his colleague is one of the few examples of a sustained research effort over a number of years.) Their conclusion is that ergonomic chairs may be over designed. Users are not able to perceive differences among those features (e.g., seat pan and backrest angles) that are supposed to reduce discomfort through biomechanical means. The implications for chair design are clear-cut: ". . . the importance of ergonomics in design of chairs has been exaggerated.

38-6   *Interventions, Controls, and Applications in Occupational Ergonomics*

Unless there are ... obvious violations of biomechanics design rules, chair users will not complain about discomfort ..." (Helander, 2003, p. 1316). Minimal biomechanical requirements include: rounded front edge of the seat pan, cushioning, back angle adjustability to 120°, and support for the legs. Mechanisms which afford tilting and swiveling are considered "fun" but do not improve biomechanics (Helander, 2003, p. 1316).

There are several serious flaws in this approach. The key problem is in the fundamental principle underlying this research, namely, that it is appropriate to simply offer users a series of chairs for evaluation as consumer products using rating scales. While this approach may reflect certain realities as to current corporate purchasing practices, it is hardly a firm basis for systematic scientific conclusions about the effectiveness of chair design attributes. In particular, there is no indication that the users in these studies were given any instruction regarding *how* the various adjustment mechanisms worked or *why* they might be useful in helping the user accomplish his or her daily work tasks. See, for example, Dainoff (1998). Thus, it was simply up to the naïve user to perceive not only the chairs' *capabilities* for adjustment, but also the *rationale* for such features.

It is interesting that Helander himself had earlier provided a definitive analysis of the users' task in attempting to identify chair controls, and provided evidence that, properly instructed, users will utilize complex adjustment mechanism to find comfortable working postures (Helander et al., 1995). See also Dainoff and Mark (1987), and Gardner, Mark, Ward, and Edkins (2001). We will return to this issue later.

A second problem is that there was no serious attempt to assess the overall fit between each of the assessed chairs and the workstations within which these chairs were used. Some partial steps in the direction were attempted. "Smaller" chairs were assigned to female secretaries and "larger" chairs were assigned to male managers. Within these groups, participants were divided into subgroups by stature (Helander, 2003, p. 1311). However, stature is a poor surrogate for assessing the anthropometric dimensions of workstation fit. For example, the distance of the elbow above the seat pan is an important indicator of fit with respect to the work surface, but the correlation of seated elbow height with stature is typically close to zero ($r = 0.18$ for a large sample of U.S. Army females) (Gordon, Churchill, Clauser, Bradtmiller, McConville, Tebbetts, and Walker, 1989). As will be discussed later, Robinette, Nemeth, and Dainoff (1998) described how small differences in anthropometric dimensions can have major impacts on fit.

To summarize, the critique of the consumer product approach to ergonomic design of chairs as proposed by Helander (2003) is that it relies on the untrained user to understand and appreciate the functional articulation between chair features, and their capability for affording postural adjustments that enhance different kinds of work tasks. Since ergonomic professionals have not agreed upon these relationships (as reflected by a large literature with alternative perspectives) it is not surprising that users are not able to differentiate chairs with differing functional capabilities. However, rather than concluding that current ergonomic chairs are overdesigned and that chair designers should concentrate on esthetics, we propose an alternative approach. Instead of diverting our professional and scientific attention to a completely new arena "esthetics," we need to engage in a program of mapping out the functional capabilities required for seated posture under a variety of work constraints. If we can accomplish this goal, not forgetting the basic design precept "form follows function," the esthetics should come "for free." What follows is a strategy for accomplishing this goal.

38.5 Approaches to Understanding the Ergonomics of Chairs and Furniture

If there is a major theme in the critique and discussion up to this point, it might be the lack of a systematic approach to studying what is clearly a system problem: the integrated consideration of workplace furniture, person, and task. Two separate but related programmatic solutions will be proposed.

38.5.1   Need for Integrated Research Programs — The Jeffersonian Approach

Vicente (2000, p. 93) puts forth the ideal of Thomas Jefferson, who "...believed that scientific research could lead to a fuller understanding of nature, while simultaneously addressing a persistent social problem of national or global interest." Difficulties in attaining this ideal arise from inherent tensions between basic and applied research within a number of scientific communities. Vicente argues that these tensions can be understood, clarified, and ultimately resolved through the realization that basic and applied research efforts do not fall on a single continuum but in fact require a two dimensional representation of what might be called research problem space. This can be seen in Figure 38.1, which is adapted from Vicente (2000, p. 104).

Represented on the abscissa is the dimension of methodology. At the left of the diagram are those traditional experimental investigations which can be characterized as highly controlled but also quite unrepresentative of real world conditions. At the other extreme, on the right of the diagram, are field investigations that can be characterized as highly representative of real-world conditions, but lacking experimental control. Intermediate cases include more complex laboratory simulations in which some variables are deliberately confounded.

Orthogonal to the methodology dimension in research space is an intentionality or purpose dimension. Towards the bottom of ordinate fall those *knowledge-oriented* research investigations whose primary purpose is to gain theoretical understanding of basic underlying principles. The top region of the ordinate is devoted to *market-oriented* research efforts where the purpose is to solve practical problems leading to a production of useful product/tool. The bottom of the ordinate is anchored by *basic research*, while activities at the top can be called *development*. Intermediate cases include *strategic research* (investigations which, while still aiming at understanding basic principles, are constrained by market-oriented factors) and *applied research* (in which the primary purpose is now technology transfer to industry in service of product development.).

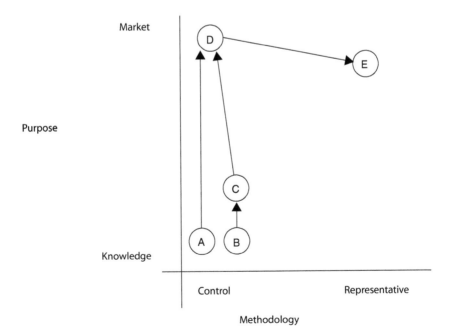

FIGURE 38.1   The Jeffersonian model of the relationship between basic and applied research. See text for details. Each circle represents a specific research study. (After Vicente, K.J. *Theoretical Issues in Ergonomics Science*, 1, 93–112, 2000. With permission.)

Vicente (2000) argues that the successful accomplishment of the Jeffersonian goals requires not single studies but programs of research that describe trajectories through the research space. He describes an example from his own laboratory of such a trajectory involving the specific problem of tool bars in three-dimensional (3D) software used in computer-aided design (CAD) applications, and how basic research in perception and cognition was used as basis for applied research leading to design innovations.

We argue that it is precisely this kind of integrated research program that is badly needed in the field of office ergonomics. Fortunately, we can identify a recent example of such a program which could be considered Jeffersonian, and which could serve as a model for future directions. This program resulted in the development and field assessment of the Steelcase Leap chair.

A digression is in order. It has been traditional in the ergonomic literature relating to chairs and office furniture that such products are not identified by brand name or manufacturer. Presumably the rationale is that such a practice leads to a more "scientific" or objective conclusion; avoiding the suspicion the researcher is "endorsing" a particular product. This is ironic since, as discussed earlier, the double-blind randomized assignment strategy characteristic of drug research has been held up as the "gold standard" for intervention research. However, there seems to be no problem with medical journals accepting published research funded by drug companies in which products are specifically identified by brand. In fact, a particular chair model offered by a particular company represents a specific design solution to a certain set of ergonomic constraints. If there is to be a serious scientific basis for ergonomic design, such design solutions must be specifically identified as, in a sense, hypotheses to be tested. The validity of the outcome of such studies can be assessed by traditional scientific criteria (Cook and Campbell, 1979), rather than by who funded the study. The bias of a researcher who crafts the best possible scientific case for a particular product is not different from the bias of a researcher who crafts the best scientific case for his or her theoretical perspective. Ultimately, it is the scientific community that must judge the adequacy of the evidence. (In the interest of full disclosure, the author had a small part in the development of the Leap chair as an outside consultant.)

38.5.2   Example of a Jeffersonian Research Program for Chair Design

The letters and arrows in Figure 38.1 represent the kind of trajectory suggested by Vicente (2000) connecting basic research to product development and evaluation for the case of the Leap chair.

Study A in the lower left corner, which is highly controlled and knowledge oriented, consists of the development of a methodology for postural analysis of seated posture (Bush, Hubbard, and Ekern, 1998). At an adjacent location on Figure 38.1, Study B, (Faiks and Reinecke, 1998a), involved a kinematic analysis of trunk movement while seated. Landmarks on thoracic and lumbar vertebrae were identified while seated subject change postures from forward-flexed to reclined positions. Results indicated different patterns of movement for the lumbar and thoracic spine. Study C is a follow-up to Study B (Faiks and Reinecke, 1998b). It is somewhat more product-oriented; utilizing a prototype chair with independent thoracic and lumbar supports. The forces required for comfortable static and dynamic seating for each support were measured for a sample of participants. Results indicated that larger forces were required for the thoracic support than for the lumbar support, and that the relative rate of required thoracic support was greater as the user reclined backwards. These data were later used in the design of the support mechanisms for the Leap chair.

Study D involves a controlled laboratory comparison of the now completed Leap chair with three competitive chairs available on the market (Bush, Hubbard, and Reinecke, 1999). (The three comparison chairs are not identified by brand name, but their pictures are available in the published article.) As indicated, study D is based on previous methodology and data. Participants were tested at a simulated keyboarding workstation while using each of the chairs. User postures relative to chair position, contact pressure, and lumbar curvature were assessed. Results indicated that the Leap chair performance exceed that of the comparison chairs along several dimensions. The Leap chair backrest supported users over a greater range of flexion and extension in both lumbar and thoracic regions. Pressure mappings were more consistent, and hand and head positions showed the least disturbances across the range of movements (Bush et al., 1999).

Study E is actually two separate investigations, an initial intervention study (Amick, Robertson, DeRango, Bazzani, Moore, Rooney, and Harrist, 2003b), and a replication at a different site (Amick, Robertson, DeRango, Bazzani, Moore, and Rooney, 2003a). Both studies are quasi-experimental designs with three separate groups — each assessed before and after the intervention. The Chair-plus-Training Group consisted of a group of employees who were provided Leap chairs along with a 45-min ergonomics workshop. The workshop not only provided instruction on chair adjustments, but also focused on the rationale for postural adjustments as part of an overall workplace system. The second group (Training Only), were given the training program using their existing chairs, and the third group (Control) received no intervention during the course of the study, but were provided the training program after completion of the study (for ethical reasons). For both study sites, the results indicated that musculoskeletal symptoms were reduced for Chair-plus-Training groups, but not the Training or Control groups. In addition, economic analyses indicated productivity gains associated with Chair-plus-Training group of $345 per work per day, resulting in 22:1 benefit–cost ratio. It might be argued that these studies should have included a fourth control group that was given a new chair but no training program. This conclusion would be correct on logical grounds. However, from practical considerations associated with the cost of yet another control group, it seems hardly necessary to scientifically validate the hypothesis that users provided with an advanced ergonomic chair with no training will not use the chair effectively. Anecdotal evidence and experience can easily confirm this hypothesis.

We have just described a program of research that started with calibration of method and apparatus and ended with a field demonstration of the effectiveness of a final product. This program required a serious commitment of time and effort. The published papers alone involve 13 separate researchers from 11 different organizations — university nonprofit, and corporate. Nevertheless, this program is presented as a model for advancement of the field in the hope of generating similar efforts.

38.5.3 Need for Conceptual Framework — The Ecological Approach

Galison (1997) has provided an intriguing study of the development of the field of microphysics, from the 19th century cloud chamber to the factory-like laboratories of the present day at places like CERN, Stanford, and Berkeley. His particular focus is on the way in which the development of laboratory apparatus transformed the *social/organizational* structure of microphysics from individual investigators working alone or in small groups with total control and understanding of their apparatus, to industrial-style organizations requiring collaboration among many professionals. In order for this collaboration to have occurred, Galison invokes the concept of "trading zones" (1997, p. 46). Derived from the field of linguistics, trading zones refer to simplified languages (creoles, pidgins) that arise when adjacent cultures require a mutually understandable means of communication in order to transact business.

It is our argument that Galison's insight has important implications for the field of ergonomics. Just in the area of ergonomics of chairs and furniture, the applicable research involves individuals from a number of academic specialties including biomechanics, epidemiology, economics, industrial engineering, industrial medicine, industrial design, muscle physiology, multivariate statistics, psychology of human performance, psychophysics, organizational design, orthopedics, optometry, etc. What is required is a trading zone; a conceptual framework within which specialists in one area can communicate with specialists in another. We argue that cognitive work analysis (Rasmussen, Pejtersen, and Goodstein, 1994; Vicente, 1999), with its underpinnings in ecological psychology (Gibson, 1979), provides such a framework.

38.5.4 An Ecological Approach to Ergonomics

The seminal research of James J. Gibson (1979) has formed the basis of the ecological movement within the field of perceptual psychology. This ecological approach has, in turn, provided an important set of conceptual tools for applied concerns within the field of human factors and ergonomics (Flach, Hancock, Caird, and Vicente, 1995). Rasmussen et al. (1994) and Vicente (1999), in particular, have utilized Gibson's work as a major theoretical underpinning of their approach to cognitive work analysis. These are the theoretical foundations upon which this chapter will be built. Previous versions of this structure have appeared in Dainoff (1998, 2000).

Ergonomics can be described as the fit between people and the elements of the physical environment with which they interact (Dainoff and Dainoff, 1986). As such, ergonomics is inherently relationship-oriented in that absolute dimensions and physical characteristics of objects in the work environment must be defined relative to the relevant characteristics of the user. The ecological perspective provides a principled approach to conceptualizing such relationships.

Core concepts in the ecological approach are *affordances* and *effectivities*. Affordances refer to the characteristics of the physical environment measured with respect to the individual or "actor." Affordances thus represent, in physical terms, the potential for action afforded by the environment for a particular individual. Effectivities are complementary in that they refer to action capabilities of individuals measured with respect to the physical environment. In the case of seating, for example, an ordinary office chair (physical object) affords sitting (action) for an adult but not a typical 2-yr-old child. The adult and child differ in their effectivities (action capabilities) because of their different sizes.

Goal–directed actions involve *perception–action cycles.* Any task can be broken down into a series of steps, in which perceived information results in action, which in turn reveals new information. For example, information is extracted from the home page of a website indicating the presence of a clickable button. Hand and finger muscles move the mouse to the location of the button and click. The effect of the click is to perform an action in the environment (a new web page appears). The new page has new information, some of which is extracted (perceived), and the cycle continues.

Perception–action cycles can be considered the fundamental unit of work analysis. Perception–action cycles are defined by certain classes of constraints, which can be classified into three groups. Task constraints reflect the functional requirements of a task. This includes individual task demands along with surrounding social and organizational factors. Workspace constraints reflect the layout of components within the workspace (e.g., display, input/output device, furniture), as well as relevant environmental constraints (lighting, air quality, temperature and humidity, gravity). Personal constraints reflect both physical (anthropometric and physiological) and psychological (cognitive, motivational, emotional) attributes of the actor.

The action components of perception–action cycles take place within a 3D postural envelope. Within the postural envelope, the operator must reach, lift, and manipulate, while parts of his or her body are or are not supported.

In order to carry out goal-directed perception–action cycles in the world, the individual actor must be able to perceive the affordances of that world. (A 250-lb adult will most likely not choose to sit in a chair made for a 2-yr-old!). However, most real-world work environments consist of collections of affordances, some of which may be nested within others. Characterizing such complex collections of affordances is the task of work domain analysis, a component of cognitive work analysis (Rasmussen et al., 1994; Vicente, 1999). Work domain analysis utilizes a conceptual tool called the *means–end abstraction hierarchy.* Central to this concept is the notion that groups of affordances can be hierarchically organized by function. At any point in the hierarchy, the affordances at one level act as the ends or goals with respect to the affordances (means) at the next level down.

The basic structure of the means–end abstraction hierarchy consists of five levels. (See Table 38.1.)

At the lowest end of the hierarchy, *physical form* refers to the representation of relevant components of the physical structure of the elements comprising the system. For example, consider a height-adjustable worktable. The actual range of adjustability in physical units (centimeters or inches) is represented at level 5. Moving upward, level 4 represents *physical function*, the ends for which level 5 is the means. In this case, physical function refers to the capability of, or requirement for, height adjustability. At level 3, a *purpose-related function* called *clearance*, which refers to the area under the worksurface, is an end for which height adjustability is a means. Clearance itself is subsumed under the broader

TABLE 38.1 Application of a Means–Ends Abstraction Hierarchy to the Office Workplace

Level		Fit				Perceptibility	
Level 1	Functional purpose	Fit				Perceptibility	
Level 2	Values and priorities	Posture–Workstation Fit		Flexibility			
Level 3	Purpose-related function	Reach	Clearance	Neutral Posture	Movement	Usability	Readability
Level 4	Physical function: interaction zones	Below the work surface		In front of the work surface		On top of the work surface	Surrounding the work surface
Level 5	Physical form: corresponding requirements	Furniture dimensions and ranges of adjustability		Chair characteristics, dimensions, and ranges of adjustability		Input device and monitor specifications; reach envelopes	Lighting and glare specifications

concept of *posture–workstation fit*, which is a level 2 *values and priorities* characterization. Finally, the overall system *functional purpose or goal* (optimal fit) is the found at the highest level of the hierarchy.

38.6   An Outline of Best Practices

The material that follows will consist of a review of best practices for an office workstation organized in terms of the ecological principles just described. In effect, this section will describe the work-domain of an office workplace in terms of the means–end abstraction hierarchy. In keeping with a best-practices approach, this section, particularly at the lower level of the hierarchy, will rely on the consensus-based BRS/HFES 100 document (Human Factors and Ergonomics Society, 2002).

38.6.1   Functional Purpose (Level 1)

The basic goal of ergonomics is to achieve an optimal *fit* between the individual, the task being accomplished, and the tools needed to accomplish the task. The purpose of achieving an optimal fit is to simultaneously enhance the individual's productivity, satisfaction, comfort, and health. To achieve *fit* means that the components of the workstation (chair, desk, keyboard, mouse, computer terminal, etc.) must be designed so that actions required to do the task are *efficient*, that is, they can be carried out with the least amount of effort necessary, while still achieving the desired level of performance. In ecological terms, we want to optimize the possibilities for perception–action systems in space and time by the organization and design of affordances (information and work objects).

38.6.2   Values and Priorities (Level 2)

The following abstract concepts serve as the means to accomplish the overall goal or ends of Level 1:

a. *Posture–workstation fit.* Do the components of the workstation allow users to get into those working postures needed to accomplish the task? Achieving posture–workstation fit requires that the interactions among task, workstation, and individual constraints are taken into account so as to allow perception–action cycles within acceptable postural envelopes, and acceptable levels of effort.

b. *Flexibility.* Do the components of the workstation allow users to change positions easily to meet new task requirements or relieve fatigue? This principle takes into account the fact that even optimum initial static postures are likely to be problematic over long periods of work. Moreover, different types of work tasks are likely to require different working postures.

c. *Perceptibility.* Do the components of the workstation allow users to read information on computer screen and paper copy easily or hear verbal information from communication devices or persons in the vicinity? Can the user perceive the affordances of the chairs and furniture in order to achieve optimum posture–workstation fit, and to adjust postures to meet changing task demands or relieve fatigue? (Although auditory speech and the acoustic environment are an important issue in the office environment, they are not part of the scope of this chapter and are mentioned here only for completeness.)

38.6.3   Purpose-Related Functions (Level 3)

The following functional characteristics listed are the means of achieving the previous values and priorities for an office workplace. Note that these are written as general requirements — independent of any particular piece of equipment.

a. *Reach.* Arms and hands should be in a position to easily reach the objects (keyboard, mouse, paper copy) that are used frequently; feet should be able to reach the floor or footrest.

b. *Clearance*. The user should be able to get in and out of a desired work posture without having lower limbs (legs, knees, thighs) come in contact with any portion of the workstation, including the underside of the worksurface and adjustment control mechanisms.

c. *Neutral posture*. The user should be able to avoid awkward working postures by achieving neutral posture as much as possible. Neutral posture is achieved when the neck, back, and trunk are aligned, the elbows fall naturally at the side of the body, the shoulders are relaxed, and the wrists and forearms are aligned.

d. *Movement*. The user should be able to change working postures easily to meet new task demands or to relieve fatigue.

e. *Usability*. Controls to adjust components of the workstation should be easy to understand, and easy to operate.

f. *Readability*. The computer display screen, paper copy, and any other source of text, should be easy to read.

38.6.4   Physical Function (Level 4) and Physical Form (Level 5)

The remaining two levels of the means end abstraction hierarchy will be collapsed, for purposes of discussion. In typical ergonomic guidelines and standards, the Physical Function level is where workplace specifications and requirements for individual components of furniture would appear. The specific numerical values for these specifications (which might differ according to situation or user population) would appear at the Physical Form level.

The BRS/ANSI 100 Standard section on component integration is organized in four specific workplace zones of user–equipment interaction (Human Factors and Ergonomics Society, 2002, pp. 8–28). These zones are described as: under, in front of, on top of, and surrounding the work surface. This chapter will utilize this organization as well.

38.6.4.1   Under the Work Surface

Clearance is the purpose-related function that is particularly relevant to the region under the worksurface. To achieve clearance, chair and worksurface affordances must be designed which allow the user's lower limbs to fit comfortably under the worksurface without contacting any portion of the worksurface (Human Factors and Ergonomics Society, 2002, p. 11). At the same time, the user's feet must be able to *reach* a support surface (footrest or floor) while maintaining the forearms and hands in appropriate *neutral posture* for keyboard work.

The relevant affordances in this situation are work surfaces and chair surfaces (seat pans) that are adjustable to match the appropriate effectivities of the users. Appropriate effectivities to be matched include: anthropometry (relevant body dimensions to achieve fit), strength and agility (needed to operate the adjustment mechanisms), and cognition (knowledge of *where* the control adjustments are located, *how* to operate them, and *why* they have been provided).

To understand the first set of effectivities, it is necessary to consider the anthropometric variation among the potential users in terms of how specific body dimensions relate to specific furniture dimensions. Next, it is necessary to determine which furniture dimensions must be adjustable and by how much in order to achieve a given amount of fit. Once this decision has been made, mechanisms for the control of furniture must be designed, and, in this process, assumptions must be made about strength and agility of the user population. Finally, there is the question of *usability* of the controls. Practically, this requires a shared responsibility between the designer (to consider usability principles), the manufacturer (to provide appropriate information about how and why of adjustment mechanisms), and the purchaser (if an organization — to provide training for end users; if an individual — to seek information). As can be seen, addressing these questions for just the basic question of clearance will necessitate consideration of a wide range of ergonomic issues.

To deal with the anthropometry issue requires that a user population be identified for which anthropometric data is available. This is straightforward in principle, but difficult in practice. The original ANSI/HFS 100 Standard defined the target population as U.S. civilians who were seated users of video display terminals. The Standard was to provide: "recommendations (that) specify the minimum parameters to accommodate the 5th percentile female dimensions through the 95th male dimensions. . ." (Human Factors Society, 1988, p. 41).

As an example, consider the clearance requirement for a small female. The data available at the time were from a 1954 anthropometric survey of a sample of the U.S. civilian population. The data were presented by Kroemer (1985). Examination of the actual body measurements obtained at the time indicates that two body dimensions were relevant; popliteal height (the vertical distance from the floor to the crease behind the knee, and seated thigh height (the vertical distance from the seat pan to the thickest part of the thigh). Thus, clearance could be achieved if the minimum height of the worksurface exceeded the combination of popliteal height and seated thigh height (with an additional allowance for shoe height).

Determination of clearance would have been straightforward if the sum of popliteal height and seated thigh height had been obtained for each person who was in the database. Unfortunately, the data presented by Kroemer — which was all that was publicly available — consisted only of summary tables containing 5th, 50th, and 95th percentile values for each anthropometric dimension (Kroemer, 1985). Hence, in this particular case, accommodating the 5th percentile female meant adding together 5th percentile female values of popliteal height and seated thigh height.

However, this approach is mathematically problematic unless the dimensions combined are highly correlated. In practice, the resulting outcome from the summary tables is an underestimation of measures for small females and an overestimation of measures for large males compared with what would have been obtained from the complete data set (e.g., adding together measures for each person in the sample.)

These errors of estimation can have profound practical implications. When the U.S. Air Force initially set out to design a training aircraft which was supposed to accommodate 1st to 99th percentile male and female pilots, the resulting cockpit dimensions failed to accommodated 30% of the white male pilots, 80% of the black male pilots, and 90% of the female pilots for whom they were designed (Robinette et al., 1998).

By the time the BRS/HFES 100 standard was being prepared, a major anthropometric survey of U.S. Army men and women had been conducted (Gordon et al., 1989). The Army survey was utilized by BRS/HFES 100 for two reasons: (1) it was judged that the ethnic composition of the U.S. Army survey was a better representation of the current U.S. civilian population than was the 1954 civilian survey; (2) the entire data base rather than just summary tables were available. Accordingly, 3D clearance envelopes could be constructed which would accommodate the upper 95% of the female population and lower 95% of the male population.

A complete description of how this was accomplished is contained in BRS/HFES 100 (Human Factors and Ergonomics Society, 2002, Appendix A); see Dainoff (2002) for an overview. In addition, a new source of anthropometric information is now available through the CAESAR project, which involved an extensive international civilian survey utilizing 3D laser scanning technology (Robinette, Blackwell, Daanen, Fleming, Brill, Hoeferlin, and Burnsides, 2002). These data, which include 3D computer graphic representations of each individual, will most likely require further revisions and rethinking of approaches to applications of anthropometry to meet ergonomic requirements. An extensive discussion of these issues may be found in a best-practices guideline (Human Factors and Ergonomics Society, 2004).

However, before we can fully discuss clearance envelopes, it is necessary to consider the affordance and effectivity implications of the interaction of worksurface height ranges of adjustability with chair adjustability. Thus, we now turn to consideration of the chair itself, which is discussed, in BRS/HFES 100, in the zone: "In Front of the Worksurface."

38.6.4.2 In Front of the Worksurface

In a seated office workplace, the chair necessarily serves as a primary focal point. Tenner (2003, pp. 104–160) provides some interesting reflections on the history and context of chairs as technologies. A more extensive discussion of some of these issues may be found in the previous edition of this handbook (Dainoff, 1998).

Each one of the Level 3 Purpose-Related Functions in Table 38.1 is a relevant higher level affordance for chair design; allowing for the possibilities of work-determined perception–action cycles. The chair design must afford *reaching* for the range of effectivities specified in the target user population. There are two aspects of reach: (1) the users' feet must be able to reach the floor or other support surface while in normal working position; (2) the user must be able to reach the primary work objects (e.g., keyboard and other input devices, telephone, documents) while avoiding awkward posture. Avoiding awkward posture is also part of the functional requirement for *neutral posture*. We have already discussed the necessity to consider interactions between seat pan height and worksurface height and their combined ranges of adjustment so as to satisfy *clearance* requirements. The chair design must allow for the possibility of *movement* both in allowing the user to deal with varying task demands (i.e., alternate between keyboard operation and answering the telephone), and in changing position in order to relieve fatigue. In order to accomplish all of the preceding functional requirements, the chair must have user-controlled adjustment mechanisms. However, unless these are *usable*, user will not operate the chair in the manner that was intended, and design functionality will be lost. Finally, the *readability* requirement entails that chair location and geometry place the users' head and eyes in a position where textual material which must be read can be read.

38.6.4.2.1 *Implementing Purpose-Related Functions for the Chair*
The most basic challenge in chair design is providing affordances for support (seat pans, backrests) which match users' effectivities (relevant anthropometric dimensions). This is straightforward in some cases; seat pan depths should be no longer than the distance from the buttocks to the back of the knee for a small (5th percentile) female; seat pan widths should be at least as wide as the distance between the hips of a large (95th percentile) female. This requirement would be located at Level 4: Physical Function. The actual values for these dimensions, which would be located at Level 5, Physical Form, are respectively, 43 and 46 cm according to BRS/HFES 100 (Human Factors and Ergonomics Society, 2002, pp. 81–82).

Specification of other components become more complicated. At the most basic level, the height of the seat pan is related to the user's popliteal height, as discussed during consideration of clearance. (The range of popliteal heights for 5th percentile females to 95th percentile males — including a 2.5 cm correction for shoes — is 37.6 to 50.1 cm.) However, the seat pans of modern ergonomic chairs will typically have variable tilt mechanisms that can either decline forward or incline rearwards. These angular corrections must be taken into account in specifying ranges of height adjustment. BRS/HFES 100 recommends a range of seat pan height adjustment of 38–56 cm (Human Factors and Ergonomics Society, 2002, p. 81). The upper limit of the adjustment range is higher than the popliteal height of the large male. This reflects the reality that adjustable chairs may be used with fixed height work surfaces, and that users will adjust the chair upwards to optimize posture with respect to that worksurface.

At the same time, the clearance specifications of BRS/HFES 100 require height adjustable worksurface. Ranges of 50–72 cm to the underside of the worksurface closest to the operator are specified for chairs with forward (declined) seat pan angles while the corresponding ranges are 50–69 cm for chairs with only rearward (inclined seat pan angles. See Human Factors and Ergonomics Society (2002, p. 76) for full details of clearance envelopes.

The issue of seat pan angle adjustability is part of broader question of the capabilities of chair affordances to support a variety of working postures. This is a complex question that is far from settled. What follows is a brief summary.

It has long been recognized that, in moving from a standing to a seated posture (trunk to thigh angle of 90°), that the pelvis only rotates about 60°. The remaining 30° is taken up by a reversal of the curvature of

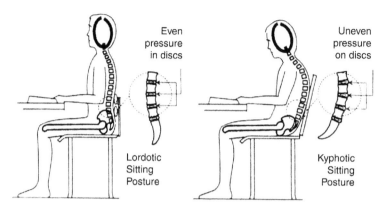

FIGURE 38.2   Illustrations of lordotic and kyphotic seating postures. Chair features are illustrative only and do not reflect necessary associations between features and postures.

lumbar spine from lordosis to kyphosis. This results in increased pressure on the spinal discs with the possibility of lower back disorders. For reviews of this material, see Chaffin and Andersson (1991), Dainoff (1998), Greico (1986), and Kroemer and Kroemer (2001). Figure 38.2 illustrates these relationships.

A solution to this problem is for the backrest to be inclined rearwards; thus removing pressure on the disks. This was proposed as a desirable solution by Grandjean, Hunting, and Pidermann (1983). At about the same time, Mandal (1981) argued that a forward-sloping seat pan would allow the pelvis to rock forward, thereby restoring the lumbar lordosis. Either solution allows approximation of trunk–thigh of 120°; a value which appears to be the neutral posture of these systems (Chaffin and Andersson, 1991). Dainoff and Mark (1987) argued that both solutions could be effective; depending on task constraints. The backward leaning posture is clearly effective in terms of spinal load and perceived comfort — approximating an "easy chair." However, as the backrest reclines, the head and eyes are moved further from the copy and display screen, thereby potentially impairing *readability*. This is less of a problem for work predominantly oriented towards the display screen since the size of display characters can usually be made as large as possible. Hence, display screen work — such as editing — can benefit from a backward posture. On the other hand, for tasks such as high-speed data entry, the user must read from paper text (which typically has font sizes smaller than on those on display screens), and the fingers must be in close proximity to the keyboard. Such task constraints benefit from a working posture in which the trunk is upright. In such conditions, a forward sloping seat pan can afford efficient work postures while alleviating disc pressure in the lumbar spine.

These findings may or may not have influenced chair designers directly, but, for the past 20 yr, most ergonomic chairs on the market have some degree of seat pan adjustability affording both forward and backward tilt as well as adjustable backrest angles. More recent chairs have armrests that adjust in two dimensions (up–down and in–out) and some have adjustable seat pan depths. BRS/HFES 100 requires an adjustable seat pan of at least 6°; at least 3° of which must be backward. The angle between backrest and seat pan must be adjustable between 90° and 105° with a recommendation of adjustability out to 120° (Human Factors and Ergonomics Society, 2002, p. 82).

Consequently, the modern ergonomic chair can be an intricate multifunctional piece of machinery. Backrest, seat pan height, and seat pan angle adjustability require coordination of multiple degrees of freedom. This coordination is solved in different ways depending on the designer and manufacturer. Helander et al. (1995) illustrate twenty different control mechanisms seen in chairs on the market at that time. For example, the seat pan angle and backrest angle may be independently adjustable or linked in some ratio (typically 1 : 2 or 1 : 3). Gardner, Mark, Dainoff, and Xu (1995) showed that independent control resulted in preferred user postures that would not have been attainable had the controls been linked. At the same time, if the controls are independent and the seat pan pivots directly over the backrest (a typical design), changes in seat pan angle will also raise or lower the knees — necessitating a compensatory height adjustment. Hence, moving from an upright trunk, forward seat pan angle posture to a backward trunk and seat pan angle posture will require three different control movements.

Other designs have dealt differently with this issue. As discussed earlier, the Steelcase Leap chair allows coordinated movement with a single control operation while maintaining differential support for lumbar and thoracic regions of the spine (Bush et al., 1999).

Let us summarize the situation with respect to chair affordances and effectivities. The basic ergonomic requirements for affordances that match *anthropometric* effectivities seem to be relatively well understood and implemented; this is not necessarily the case with respect to strength and agility and cognition. The multiple degrees of freedom of modern ergonomic chairs require that control mechanisms be provided that do not exceed the strength and agility of the users for whom they are intended, and that the purpose and rationale for these control actions are communicated to the user. Norman's (1988) distinctions between the designer's mental model of how the system, the translation of that model to the system interface (physical appearance and layout of controls) and the subsequent misinterpretation of the interface resulting in a user mental model discrepant from that of the designer, is particularly pertinent.

These control mechanisms are not simply aesthetic frills, as Helander (2003) would argue, but are critical in providing the required affordances for effective seated posture for a variety of work tasks. Moreover, recent work on spinal physiology has emphasized the critical importance of the affordance of *movement* (McGill, 2002; pp. 174–176). McGill argues that chairs should be adjusted every 10 min in order to avoid lower back pain from prolonged sitting. If this is to be accomplished, it is essential that users be effectively trained and motivated. Training was an important component of the successful field study presented by Amick et al. (2003a); see also Dainoff, Dainoff, and Cohen (2005).

38.6.4.2.2 *Input Device Support Surface*

BRS/HFES 100 considers the input device support surface as logically part of the zone in front of the worksurface. This organization allows for the consideration of attached keyboard trays as well workstations in which the keyboard and input devices are placed directly on a single or multilevel worksurface.

Relevant affordances are *reach, neutral posture, clearance*, and *movement*. The user should be able to reach the keyboard and mouse as well as other work-related objects. All of the issues and requirements regarding *clearance* discussed previous apply to the input device support surface, as well. With regard to *neutral posture*, BRS/HFES 100 defines neutral in terms of a requirement that the input device support surface afford forearm postures within a range of 20° above to 45° below the horizontal, and with no more than 20° of shoulder angle (abduction) (Human Factors and Ergonomics Society, 2002, p. 14). For a more systematic approach, see Kee and Karwowski (2001).

Alternative working postures for office work can also include standing. McGill (2002, pp. 174–176) emphasized the possibility of users getting up from chair. BRS/HFES 100 allows for the possibilities of users being able to alternate between sitting and standing postures by providing specifications for input device support surfaces that adjust between 56 and 118 cm (Human Factors and Ergonomics Society, 2002, p. 80) so as to accommodate the anthropometric effectivities of the target population. However, the same issues regarding *usability* of adjustment controls discussed previously are relevant here.

Discussion of alternative keyboard and input devices is beyond the scope of this chapter. However, utilization of such devices may change the relevant affordances necessitating changes in the chair and support surface configurations. For example, current evidence suggests that forward sloping keyboards will result in a closer approximation to neutral posture for the hands (Simoneau and Marklin, 2001). See also Aarås, Dainoff, Ro, and Thoresen (2001) for a field study demonstrating pain reduction from use of an alternative upright mouse.

38.6.4.3 On Top of the Work Surface

The workplace elements discussed by BRS/HFES 100 in the zone On Top of the Work Surface include the horizontal work envelope and the monitor support surface.

38.6.4.3.1 Horizontal Work Envelope

The horizontal work envelope defines the horizontal region on the work surface which affords *reach*. BRS/HFES 100 defines the primary work zone as: "...the shape swept out on the work surface by rotating the forearm horizontally at elbow height" (Human Factors and Ergonomics Society, 2002, p. 17). As such, this is a traditional definition of reach envelope (Barnes, 1958; Squires, 1959). However, an extensive series of studies have expanded and clarified our understanding of how reach envelopes should be defined.

By observing how people actually reach, (Gardner et al., 2001) categorized the different reach actions used as a function of the distance of the target from the prospective seated worker. At close distances people reach simply by extending their arm (arm-only reach). As reach distance increases, workers introduce other parts of their body, initially rotating their hips in order to extend their shoulder (arm-and-shoulder reach), then leaning forward (arm-and-torso reach), which is often combined with shoulder extension. With further increases in reach distance, workers have to stand, initially by raising their pelvis just off the seat, keeping the knees bent (partial standing reach), and later by standing upright and leaning forward (full standing reach).

An extensive series of experiments has consistently found that the transition between two reach actions does not occur at the maximum distance afforded by one of the actions, but at a closer distance (Choi, Mark, and Dainoff, 2003; Dainoff, Mark, and Gardner, 1999, 2003; Gardner et al., 2001; Mark, 2001; Mark and Dainoff, 2002; Mark, Nemeth, Gardner, Dainoff, Paasche, Duffy and Grandt, 1997). We have labeled these distances the *absolute critical boundary* and the *preferred critical boundary*, respectively. Results from these studies consistently show that the (preferred) transition between arm-only reaches and reaches involving other parts of the body occurred well before the absolute critical boundary for an arm-only reach. Other results show that these transitions are effected by task constraints (visual demand vs. motor demand). Finally, it has been demonstrated that when the angle of backrest is varied from upright to 20° backward, not only does the reach envelope shift with it, but the difference between preferred and absolute boundaries varies as well. In effect, the difference disappears as the backrest angle increases (Mark et al., 1997).

These findings are critical in considering the practical concerns of putting together the components of a workstation (system integration). There is not a single primary work area, but a dynamic region that varies with the position of the chair as well as with task constraints. The functional requirement of reach — together with that of neutral posture — is to determine the placement of important work objects so that such objects can be picked up or operated without excessive bending. If the reach envelopes to accomplish such requirements are set by strictly anthropometric means (e.g., 5th/95th percentile arm lengths), then these requirements will be defeated. Users will find themselves frequently bending or otherwise adopting awkward postures well before the anthropometric limits are reached.

This is particularly an issue for keyboard trays. Such devices are an inexpensive means of accomplishing height adjustability requirements, but, if extended too far in front of the worksurface, may have the unintended consequence of rendering the majority of the worksurface unreachable.

38.6.4.3.2 Monitor Support Surface

The location of the monitor support surface is relevant to affordances for *readability* and *neutral posture*. Issues of display quality — such as luminance contrast character geometry, etc., are beyond the scope of this chapter, but are discussed by BRS/HFES 100 (Human Factors and Ergonomics Society, 2002, pp. 45–71).

Conditions for optimal location of the display monitor on the display surface are still not well understood. It appears as if the optimal location for the visual system requires a more downward gaze angle whereas the optimal location for the muscles of the head and neck requires a gaze angle closer to the horizontal. See Somerich, Joines, and Psihogios (2001) and Burgess-Limerick, Mon-Williams, and Coppard (2000) for recent research attempting to elucidate these issues. BRS/HFES 100 indicates a reasonable compromise location places the center of the display region between 15° and 20° below the horizontal at distances of 75–93 cm (Human Factors and Ergonomics Society, 2002, p. 19). In addition, the gaze angle should be close to perpendicular (normal) to the center of the viewing area. Hence, lowering the monitor may require it to be tilted backward to maintain this orientation.

The presence of very large monitors, particularly in CAD/CAM work, presents particular problems of location. The size of such devices may place part of their viewing region above the horizontal, which is considered undesirable.

38.6.4.4　Surrounding the Worksurface

The primary concern in this workplace zone for this chapter is ambient illumination and its potential to effect *readability*. Readability can be impaired if light sources such as lighting fixtures and sunlight windows act as glare sources. Reflected glare on the display can reduce contrast, making reading more difficult. Or, glare from a source within the users' field of view can enter the eye directly — hampering visual performance. In such cases, if the glare cannot be avoided, it may require the user to adopt working postures that are no longer neutral.

Glare changes with monitor placement. As indicated in the previous section, if the monitor is lowered and the screen surface angled backwards so as to intersect the users' line of sight, it is more susceptible to reflected glare from overhead lighting fixtures. See BRS/HFES 100 (Human Factors and Ergonomics Society, 2002, pp. 19–20).

38.7　Conclusions

This chapter has attempted to describe the current situation with regard to ergonomic interventions related to chairs and furniture. With regard to the scientific evidence supporting the effectiveness of such intervention, we conclude that:

a. A more systematic approach is required to understand the functional interaction between task demands, postural adjustments, and specific chair and furniture design components.
b. Ecological psychology, in general, and cognitive work analysis in particular, can provide a useful framework for such investigations.
c. Current critiques of the scientific adequacy of ergonomic interventions in the office may be using an inappropriate medical model more suited to assessment of individual components (e.g., drugs) than the system-wide interactions commonly encountered in office environments.
d. In terms of the goals of reducing pain and discomfort and increasing work efficiency, the existing scientific data are more than sufficient to define a set of best practices for interventions. It is expected that these will be refined as better information becomes available.

References

Aarås, A., Dainoff, M., Ro, O., and Thoresen, M. (2001). Can a more neutral position of the forearm when operating a computer mouse reduce the pain level for visual display unit operators? A prospective epidemiological intervention study: Part II. *International Journal of Human–Computer Interaction*, 13(1), 13–40.

Amick, B.C., III, Robertson, M., DeRango, K., Bazzani, L., Moore, A., and Rooney, T. (2003a). The impact of a highly adjustable chair and office ergonomics training on musculoskeletal symptoms: two month post intervention findings. Paper presented at the *47th Annual Meeting of the Human Factors and Ergonomics Society*.

Amick, B.C., III, Robertson, M., DeRango, K., Bazzani, L., Moore, A., Rooney, T., and Harrist, R. (2003b). Effect of office ergonomics intervention on reducing musculoskeletal symptoms. *Spine*, 28(24), 2706.

Barnes, R. (1958). *Time and Motion Study* (4th ed.). New York: Wiley.

Burgess-Limerick, R., Mon-Williams, M., and Coppard, V.L. (2000). Visual Display Height. *Human Factors*, 42(1), 140.

Bush, T.R., Hubbard, R., and Ekern, D. (1998). Methodology for posture measurement in automotive seats: Experimental methods and computer simulations (ISATA Paper Number 9301110). Paper presented at the *International Symposium on Automotive Technology and Automation*.

Bush, T.R., Hubbard, R., and Reinecke, S. (1999). An evaluation of postural motions, chair motions, and contact in four office seats. Paper presented at the *43rd Annual Meeting of the Human Factors and Ergonomics Society*, Houston, TX.

Business and Institutional Furniture Manufacturers Association. (2002). *BIFMA G1–2002. Ergonomics Guideline for VDT Furniture Used in Office Work Spaces*. Grand Rapids, MI: Business and Institutional Furnature Manufacturers Association.

Canadian Standards Association. (2000). *CSA-Z412–00. Guideline on Office Ergonomics*. Toronto, Canada: Canadian Standards Association.

Chaffin, D.B. and Andersson, G.B.J. (1991). *Occupational Biomechanics* (2nd ed.). New York: Wiley Interscience.

Choi, H.J., Mark, L.S., and Dainoff, M. (2003). A performance-based model of normal working area. Paper presented at the *International Ergonomics Association XV Triennial Congress*, Seoul, Korea.

Cook, T.D. and Campbell, D.T. (1979). *Quasi-experimentation: Design and Analysis Issues for Field Settings*. Chicago, IL: Rand McNally.

Dainoff, M. (1998). Ergonomics of seating and chairs. In W. Karwowski and W.S. Marras (Eds.), *The Occupational Ergonomics Handbook*. Boca Raton, FL: CRC Press.

Dainoff, M. (2000). Safety and health effects of video display terminals. In R.L. Harris (Ed.), *Patty's Industrial Hygiene* (5th ed.). New York: Wiley.

Dainoff, M. (2002). The anthropometric basis for HFES 100 Computer Workstation Standard. Paper presented at the *Association of Canadian Ergonomists–Applied Ergonomics Joint Conference*, Banff, Canada.

Dainoff, M. and Mark, L.S. (Eds.). (1987). *Task and the Adjustment of Ergonomic Chairs*. Amsterdam: Elsevier (North-Holland).

Dainoff, M., Mark, L.S., and Gardner, D.G. (2003). Interactions of visual and motor demands on reaching actions at workstations. Paper presented at the *10th International conference on Human–Computer Interaction International*.

Dainoff, M.J. and Dainoff, M.H. (1986). People & Productivity: A Manager's Guide to Ergonomics in the Electronic Office. Toronto, London: Holt Rinehart and Winston of Canada.

Dainoff, M.J., Dainoff, M.H., and Cohen, B.G.F. (2005). The effect of an ergonomic intervention on musculoskeletal, psychosocial, and visual strain of VDT data entry work: The US part of the international MEPS study. *International Journal of Occupational Safety and Ergonomics*, 11(1), 49–63.

Dainoff, M.J., Mark, L.S., and Gardner, D.L. (1999). Scaling problems in the design of work spaces for human use. In P.A. Hancock (Ed.), *Human Performance and Ergonomics*. (pp. 265–290). San Diego, CA: Academic Press, Inc.

Faiks, F.S. and Reinecke, S. (1998a). Investigation of spinal curvature while changing one's posture during seating. In M.A. Hansen (Ed.), *Contemporary Ergonomics*. London: Taylor and Francis.

Faiks, F.S. and Reinecke, S. (1998b). Supporting the lumbar and thoracic regions of the back during sitting. Paper presented at the *North American Congress on Biomechanics*.

Flach, J., Hancock, P.A., Caird, J., and Vicente, K.J. (1995). *Global Perspectives on the Ecology of Human–Machine Systems*. Hillsdale, NJ: Lawrence Erlbaum Associates Publishers.

Galison, P. (1997). *Image and Logic*. Chicago, IL: University of Chicago Press.

Gardner, D.G., Mark, L.S., Dainoff, M., and Xu, W. (1995). Considerations for linking seatpan and backrest angles. *International Journal of Human–Computer Interaction*, 7, 153–165.

Gardner, D.G., Mark, L.S., Ward, J., and Edkins, H. (2001). How do task characters affect the transitions betwen seated and standing reaches? *Ecological Psychology*, 13, 245–274.

Gibson, J.J. (1979). *The Ecological Approach to Visual Perception*. Boston, MA: Houghton Mifflin.

Gordon, C.C., Churchill, T., Clauser, T.E., Bradtmiller, B., McConville, J.T., Tebbetts, I., and Walker, R.A. (1989). Anthropometric survey of U.S. Army personnel: Summary statistics interim report (Technical Report NATICK/TR-89–027). Natick, MA: U.S. Army Natick Research, Development, and Engineering Center.

Grandjean, E., Hunting, W., and Pidermann, M. (1983). VDT workstation design; preferred settings and their effects. *Human Factors*, 25, 161–173.

Greico, A. (1986). Sitting posture: an old problem and a new one. *Ergonomics*, 29, 345–362.

Helander, M.G. (2003). Forget about ergonomics in chair design? Focus on aesthetics and comfort! *Ergonomics, 46*(13/14), 1306–1307.

Helander, M.G., Little, S.E., and Drury, C.G. (2000). Adaptation and sensitivity to postural change in sitting. *Human Factors, 42*(4), 617–629.

Helander, M.G. and Zhang, L. (1997). Field studies of comfort and discomfort in sitting. *Ergonomics*, 40, 895–915.

Helander, M.G., Zhang, L., and Michel, D. (1995). Ergonomics of ergonomic chairs: a study of adjustability features. *Ergonomics*, 38, 2007–2029.

Human Factors and Ergonomics Society. (2002). *BRS/HFES 100 Human Factors Engineering of Computer Workstations*. Santa Monica, CA: Human Factors and Ergonomics Society.

Human Factors and Ergonomics Society. (2004). *Guidelines for Using Anthropometric Data in Product Design*. Santa Monica, CA: Human Factors and Ergonomics Society.

Human Factors Society. (1988). *ANSI/HFS 100–1988 American National Standard for Human Factors Engineering of Visual Display Terminal Workstations*. Santa Monica, CA: Human Factors Society.

Karsh, B.-T., Moro, F.B.P., and Smith, M.J. (2001). The efficacy of workplace ergonomic interventions to control muscluskeletal disorders: a critical analysis of the peer-reviewed literature. *Theoretical Issues in Ergonomics Science*, 2, 23–96.

Kee, D. and Karwowski, W. (2001). The boundaries for joint angles of isocomfort for sitting and standing males based on perceived comfort of static join postures. *Ergonomics*, 44, 614–648.

Kroemer, K.H.E. (1985). Office ergonomics: workplace dimensions. In D.C. Alexander and B.M. Pulat (Eds.), *Industrial Ergonomics*. Norcross, GA.: Institute of Industrial Engineering.

Kroemer, K.H.E. and Kroemer, A.D. (2001). *Office Ergonomics*. London: Taylor and Francis.

Mandal, A.C. (1981). The seated man (homo sedans). The seated work position. Theory and practice. *Applied Ergonomics*, 12.1, 19–26.

Mark, L.S. (2001). Two problems in the calibration of critical action boundaries. Paper presented at the *11th International Conference on Perception and Action*, Storrs, CT.

Mark, L.S. and Dainoff, M. (2002). Determination of safe and effective reach distances. Paper presented at the *Association of Canadian Ergonomists–Applied Ergonomics Joint Conference*.

Mark, L.S., Nemeth, K., Gardner, D., Dainoff, M.J., Paasche, J., Duffy, M., and Grandt, K. (1997). Postural dynamics and the preferred critical boundary for visually guided reaching. *Journal of Experimental Psychology: Human Perception & Performance, 23*(5), 1365–1379.

Marras, W.S. and Waters, T.R. (2004). State of the art research perspectives on musculoskeletal disorder causation and control. *Journal of Electromyography and Kinesiology*, 14, 1–5.

McGill, S. (2002). *Low Back Disorders: Evidence-based Prevention and Rehabilitation*. Champaign, IL: Human Kinetics.

Moon, S.D. (1966). A psychological view of cumulative trauma disorders: implication for occupational health and prevention. In S.D. Moon and S.L. Sauter (Eds.), *Beyond Biomechanics: Psychosocial Aspects of Musculoskeletal Disorders in Office Works*. London: Taylor and Francis.

National Research Council. (2001). *Musculoskeletal Disorders and the Workplace*. Washington, DC: National Academy Press.

Norman, D.A. (1988). *The psychology of Everyday Things*. New York: Basic Books,.

Oxford University Press. (1971). *The Compact Edition of the Oxford English Dictionary*. New York: Oxford University Press.

Rasmussen, J., Pejtersen, A.M., and Goodstein, L.P. (1994). *Cognitive Systems Engineering*. New York: Wiley.

Robinette, K., Blackwell, S., Daanen, H., Fleming, B.M., Brill, T., Hoeferlin, D., and Burnsides, D. (2002). Civilian American and European Surface Anthopometry Resource (CAESAR), Final Report, Volume 1, Summary (AFRL-HE-WP-TR-2002–0169). Wright Patterson AFB, Ohio: United States Air Force Research Laboratory, Human Effectiveness Directorate, Crew Systems Interface Division.

Robinette, K., Nemeth, K., and Dainoff, M. (1998). Percentiles: you don't have to take it anymore. Paper presented at the *Human Factors and Ergonomics Society 42nd Annual Meeting*.

Sauter, S.L., Dainoff, M.J., and Smith, M.J. (1990). *Promoting Health and Productivity in the Computerized Office: Models of Successful Ergonomic Interventions*. London: Taylor & Francis.

Simoneau, G.G. and Marklin, R.W. (2001). Effect of computer keyboard slope and height on wrist extension angle. *Human Factors*, 43, 287.

Somerich, C.M., Joines, S.M.B., and Psihogios, J.P. (2001). Effects of computer monitor viewing angle and related factors on strain, performance, and preference outcomes. *Human Factors*, 43(1), 39.

Squires, P.C. (1959). Proposed shape of normal work area. *Engineering & Industrial Psychology*, 12–17.

Stewart, W.F., Ricci, J.A., Chee, E., Morgenstein, D., and Lipton, R. (2003). Lost productive time and cost due to common pain conditions in the US workforce. *Journal of the American Medical Association*, 290, 2443–2454.

Szabo, R.B. (2000). Dissent. In N.R. Council (Ed.), *Musculoskeletal Disorders and the Workplace*. Washington, DC: National Academy Press.

Tenner, E. (2003). *Our Own Devices; The Past and Future of Body Technology*. New York: Alfred A. Knopf.

Vicente, K.J. (1999). *Cognitive Work Analysis: Toward Safe, Productive, and Healthy Computer based Work* Mahwah, NJ: Lawrence Erlbaum Associates, Inc.

Vicente, K.J. (2000). Towards Jeffersonian research programmes in ergonomics science. *Theoretical Issues in Ergonomics Science*, 1, 93–112.

Westgaard, R.H. and Winkel, J. (1997). Ergonomic intervention research for improved musculoskeletal health: A critical review. *International Journal of Industrial Ergonomics*, 20(6), 463–500.

39

The Computer Keyboard System Design

Jack Dennerlein
Harvard School of Public Health

39.1 Introduction

The goal of this chapter is to present specific ergonomic related research of the computer keyboard design and how it relates to interventions for musculoskeletal disorders (MSDs). As part of the many components of the computer workstation, the keyboard design imposes specific postural and biomechanical loads on the upper extremity that can be mitigated in part through the geometric design of the keyboard and the placement of the keyboard within the more global workstation. This chapter will present both laboratory and epidemiological research examining the relationship between keyboard design and exposure to biomechanical load and MSD health outcomes. Several laboratory and a couple small-scale epidemiological studies suggest that alternative split keyboards with specific keyswitch design reduce exposure and sometimes symptoms associated with MSDs disorders of the upper extremity.

39.2 Computer Use, Health, and Risk

Computer use has been associated with MSDs that affect many different components of the upper extremity (Faucett and Rempel, 1994; Bergqvist et al., 1995a,b; Matias et al., 1998; Gerr et al., 2002). MSDs associated with the neck and shoulders are relatively more prevalent than those associated with the hand and arm. In one prospective study of office workers, incidence rates for shoulder and neck complaints was 58 per 100 person-years and for the hand and arm the incidence was 39 cases per 100 person-years (Gerr et al., 2002).

While the injury mechanisms of chronic MSDS are not well understood, keyboarding includes many of the work-related physical risk factors identified through epidemiological studies. These include repetition, force, awkward posture, direct pressure, duration of exposure, and vibration (Silverstein et al., 1986, 1987; Armstrong and Silverstein, 1987; Rempel et al., 1992). The forces of the modern office environments are relatively low with peak forces ranging between 1 and 5 N per keystroke (Armstrong et al., 1994; Rempel et al., 1994); however, the number of repetitions are quite high. A modest typist typing 60 words per minute could type over 20,000 keystrokes per hour if typing continuously. The accelerations of the wrist during typing are also quite high and are similar to accelerations of high-risk industrial tasks (Marras and Schoenmarklin, 1993; Serina et al., 1999).

The largest identifiable risk factor for MSDs associated with computer use is duration of exposure (Faucett and Rempel, 1994; Bergqvist et al., 1995a,b; Matias et al., 1998; Anderson et al., 2003; Brandt et al., 2004; Gerr et al., 2004; Lassen et al., 2004). Anderson et al. (2003) reported that the prevalence of carpal tunnel was higher within individuals who self-reported using the mouse more than 20 h per week. However, duration is only one aspect of exposure, the others being intensity and frequency.

The most familiar factors are postural aspects and workstation configuration that affect posture (Gerr et al., 2004). Sauter et al. (1991) reported arm discomfort increased with increases in keyboard height above the elbow level. A large-scale, longitudinal epidemiological study demonstrated that important relationships exist between certain static postural aspects of an individual sitting at his or her workstation and the development of MSDs (Marcus et al., 2002).

In terms of specific keyboard design, cross-sectional studies have illustrated that keyboard and key-switch design can influence the pain and symptoms associated with carpal tunnel syndrome (Rempel, et al., 1999; Tittiranonda et al., 1999a). Furthermore, in one small cross-sectional laboratory study Feuerstein et al. (1997) observed that people with symptoms of MSDs of the upper extremity applied larger forces to the keyboard than people without symptoms. Hence, links between specific keyboard force — nominal and applied — and musculoskeletal symptoms do exist.

39.3   Reducing Awkward Postures through Keyboard Geometry

Typing on a typical computer keyboard is associated with wrist extension and ulnar deviation with the forearm fully pronated (Figure 39.1). Wrist extension and ulnar deviation have been reported in the

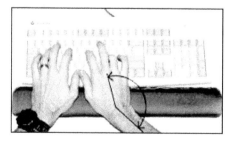

FIGURE 39.1 Typing on a conventional keyboard is associated with awkward extension and ulnar deviation postures of the wrist. (Modified from Occupational Safety and Health Administration (OSHA), 1997, *Working Safety with Video Display Terminals*, U.S. Department of Labor, Occupational Safety and Health Administration (OSHA), Report #: 3092, Washington, D.C.)

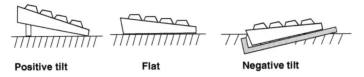

Positive tilt   Flat   Negative tilt

FIGURE 39.2 Side view of the keyboard illustrating keyboard tilt. Positive tilt can be achieved through the use of small legs on the back of the keyboard. Adjustable keyboard trays allow for negative tilt. As the tilt becomes negative the wrist extension decreases and becomes neutral with a negative keyboard. Traditionally flat keyboards still have a slope associated with the keys of approximately 8°. New keyboard designs have removed this baseline slope for the flat configurations to achieve more of a neutral wrist posture.

range of 15–25° and 11–22°, respectively, across several laboratory studies with the forearm fully pronated at 90° (Serina et al., 1999). Field studies have reported similar wrist extension values and smaller ulnar deviations with observations ranging between 3 and 9° (Huang, 1999; Gerr et al., 2000). The wrist extension angles are larger than the recommended guidelines of ANSI/HFES (1988) for keyboard designs, which are based on high internal pressures on the tissues of the wrist (Weiss et al., 1995; Diao et al., 2005). As a result, extensive design efforts have focused on the keyboard geometry with the goal of reducing the awkward wrist and forearm postures associated with typing.

The most prevalent and fundamental aspect of keyboard design is the tilt of the rows of the keys (Figure 39.2). Most keyboards are sloped upward with a positive angle such that the keys further away from the user are higher than the keys closer to the user. Most keyboards have tabs on the bottom toward the back that provide an option to increase the slope of the keys. These positive angles provide excellent visual access to the keys; however, the slope contributes to excessive wrist extension when the keyboard is at elbow height. When the keyboard is above elbow height, such as when it is put on a typical 29 in. desk, the positive slope reduces wrist flexion. By providing a negative keyboard slope support system, such as an adjustable keyboard tray, the top of the keys can be flattened. As a result, more neutral wrist postures can be achieved (Hedge and Powers, 1995; Simoneau and Marklin, 2001). Activities of the wrist muscles also decrease with negatively slopped keyboards suggesting that the biomechanical load on the muscles and the wrist are less (Kier and Wells, 2002; Simoneau et al., 2003). New keyboard designs have removed this baseline slope for flat configurations to achieve more of a neutral wrist posture.

The most common alternative design approach to achieve neutral wrist is the development of split keyboards. As the name implies split keyboards are divided into two halves with each half containing the keys typed by each hand of a touch typist. In addition to keyboard tilt, split keyboards can be configured (1) to change the distance between the two halves, (2) to rotate the two halves in the horizontal plane of the table surface (Figure 39.3), which reduces ulnar deviation, and (3) to rotate the two halves in the vertical

FIGURE 39.3 Top view of a flat split keyboard with the two halves rotated in the horizontal plane. As the angle between the two halves increases, the ulnar deviation of the wrist decreases.

FIGURE 39.4 The tenting action of a split keyboard. As the center of two halves are rotated up the required forearm pronation decreases. (From Occupational Safety and Health Administration (OSHA), 1997, *Working Safety with Video Display Terminals*, U.S. Department of Labor, Occupational Safety and Health Administration (OSHA), Report #: 3092, Washington, D.C.)

plane, tenting the keyboard such that the middle is higher than the rest of the keyboard (Figure 39.4), which reduces pronation.

Both the horizontal rotation and the distance between the two halves affect ulnar deviation (Tittiranonda et al., 1999b; Marklin and Simoneau, 2001). Using a commercially available split keyboard that only allowed for horizontal rotation of the two halves around a point located near the top of the keyboard, Tittiranonda et al. (1999b) reported that rotating the keyboard halves between 20 and 30° outward from a conventional keyboard design reduced ulnar deviation. On a prototype keyboard system Marklin and Simoneau (2001) reported decreasing ulnar deviation angles as the two halves were moved from a standard configuration to shoulder width apart. However, the shoulder joint externally rotates as the halves are separated adding strain to that joint.

Tenting can alleviate pronation of the forearm. Increasing the tent angles between 20 and 30° decreases pronation as much as 45° (Marklin and Simoneau, 2004). Some designs allow for the two halves to be rotated 90° to full vertical. As a result, compete visual access is denied (some designs offer rear-view mirrors to view the finger) and there are no temporary supports for the fingers, hands, and forearms, reducing performance and increasing error rates.

While these studies demonstrate that keyboard design can and does affect wrist and arm posture in both laboratory and field studies, only one has demonstrated relationships with health outcomes of MSDs. In a randomized, placebo-controlled 6-month trial Titteranonda et al. (1999a) reported a decline in self-reported musculoskeletal symptoms among computer users who used a fixed split keyboard incorporating tilt, horizontal rotation, and tenting of the keyboard compared to computer users who were given a flat keyboard. Smaller yet significant declines were observed in a group that used an adjustable split keyboard that provided some rotation of the two halves in the horizontal plane about a point in the top of the keyboard.

Alternative geometric keyboards come either as fixed or adjustable (Figure 39.5). In terms of reducing MSDs, usability, and productivity, adjustable keyboards fall short of fixed split keyboard designs and

FIGURE 39.5 Fixed and adjustable split keyboard designs that incorporate tilt, rotation, and tenting components. Studies have illustrated that fixed keyboards are easier to use and provide more neutral postures and better typing performance than the adjustable keyboards.

sometimes are no better than conventional flat keyboards. Titteranonda et al. (1999a) saw no difference in changes of self-reported symptoms between a completely adjustable keyboard and a flat fixed keyboard. Zecevic et al. (2000) reported less neutral wrist postures with an adjustable keyboard than compared to a fixed split keyboard. Zecevic also reported that productivity and usability were best with the fixed split keyboard. It should be noted that switching to an alternative keyboard is associated with loss of productivity; however, this loss of productivity often recovers within hours (Swanson et al., 1997; Tittiranonda et al., 1999a,b).

39.4 Changes in the Keyswitch Design

Another component for keyboard design is the force–displacement characteristics of the keyswitch design. The force–displacement curves for most keyswitches are not monotonic, but instead are characterized by rising to a peak force (local maximum), known as the activation or make force, then declining in force as the switch is further depressed, and finally increasing in force a second time as the keyswitch travel ends (Armstrong et al., 1994; Jindrich et al., 2004). The shape of the force–displacement curve characterizes the tactile feedback providing a click-feeling to the user as the key is pressed (Figure 39.6). Force–displacement relationships of computer keyswitches affect fingertip force, muscle activity, and fatigue while typing and have been associated with musculoskeletal symptoms (e.g., Rempel et al., 1999).

As the activation force of the keyswitch design increases so do the applied forces, forearm EMG, and musculoskeletal discomfort. As the keystroke force increases from 0.47 N (48 g or 1.6 ounces) to 0.89 N (90 g or 3.2 ounces), the forces applied to the keyboard by the fingers increase with peak forces some two to four times larger than the activation forces (Armstrong et al., 1994). Gerard et al. (1999) reported that forearm muscle activity is the highest with the highest activation force keyboard tested in their laboratory study of four keyboards, which was 0.83 N (85 g or 3 ounces). Self-reported discomfort was also significantly higher for the highest activation force keyboard.

Key travel also affects the endpoint forces associated with tapping (Radwin and Jeng, 1997; Radwin and Ruffalo, 1999). When travel increased from 0.5 to 4.5 mm average peak force decreased by 0.59 N. However, no differences were observed in EMG parameters associated with fatigue and in self-reported levels of fatigue.

Along with the activation force, the shape of the force–displacement curve also influences both the biomechanical exposures and the self-reported discomfort. Gerard illustrated that peak forces and

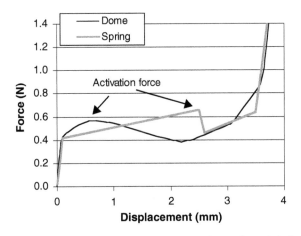

FIGURE 39.6 Typical key switch force–displacement characteristics for two keyswitch designs, a rubber dome and a buckling spring. The activation force is the maximum force necessary to fully press the key. The following decrease in force as the key is pressed provides tactile feedback during the keystriked.

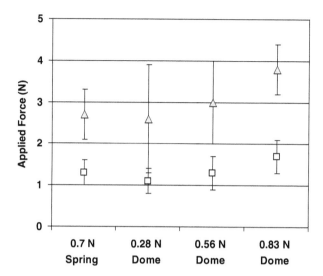

FIGURE 39.7 Distributions [50th (□) and 90th (△) percentiles] of applied forces for keyboards with different key switch designs. Muscle activity followed similar patterns. The spring design had applied forces that were not significantly different than the low and medium force rubber dome design even though its activation force was similar to the high force rubber dome. (Plotted data from Gerard MJ, Armstrong TJ, Franzblau A, Martin BJ, Rempel DM (1999) *Am Ind Hyg Assoc J* 60: 762–769.)

muscle EMG associated with a buckling spring key design are not as high as they are with a rubber dome keyboard with a similar activation force (Figure 39.7). The experimental subjects also preferred the buckling spring design over rubber dome switch designs that had smaller actuation forces. These two types of keyswitches have different shapes of force–displacement curves, as shown in Figure 39.6. Rempel et al. (1999) observed greater decreases in self-reported musculoskeletal symptoms among computer operators who used a keyboard that had a unique force–displacement characteristic (Figure 39.8) compared to a group that used a rubber dome keyboard.

Another factor is the orientation of the keyswitch movement relative to the finger and the hand. Balakrishnan et al. (In Press) reported reduction in finger joint torques and hence biomechanical

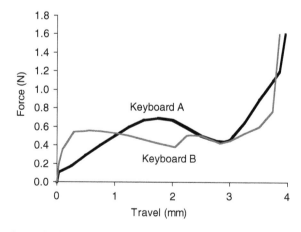

FIGURE 39.8 The two force–displacement curves implemented by Rempel et al. (1999). Hand pain complaints among symptomatic workers who used keyboard A decreased more than complaints from symptomatic workers who used keyboard B over a 12-week period. (Modified from Rempel D, Tittiranonda P, Burastero S, Hudes M, So Y (1999) *J Occup Environ Med* 41: 111–119.)

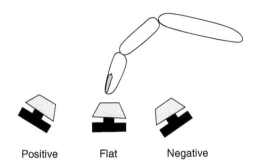

FIGURE 39.9 Changing the orientation of the key movement in the sagital plane changes the mechanical moment arm of the external applied force. As a result, joint torques are reduced for positive tilt conditions. (Modified from Balakrishnan AD, Jindrich DL, Dennerlein JT (In Press) *Hum Factors.*)

loading as the key orientation is rotated from a negative tilt to a positive tilt (Figure 39.9). The reduction in joint torque is achieved through a reduction of the effective moment arm of the fingertip force around the phalangeal joints of the finger. This is opposite to the direction of a negative keyboard slope that produces more neutral wrist postures. Therefore, to incorporate a positive slope keyswitch movement into a keyboard requires combining the movement direction with the overall keyboard geometry promoting less wrist extension.

39.5 Alternative Keyboard Layout and Speech Recognition

The preceeding discussion has focused on the mechanical design aspects of keyboards and their effects on the exposure to identifiable risk factors and their relationships to health outcome; however, there are two items that are often brought up in the discussion of keyboard designs — alternative key layout and speech recognition.

Extensive performance-related research has been completed on keyboard layouts and often discussed in the human performance and human computer interaction (HCI or CHI) literature. The most common alternative layout is the Dvorak keyboard, which remaps the most commonly used letters to the home row and to the center of the keyboard. Doing this improves typing speed and motor learning characteristics. With the advent of personal digital assistants and their touch screen capabilities other patterns of key layouts can be achievable and many performance-related design criteria can be applied (for more discussion, see MacKenzie and Zhang, 2001; Zhai et al., 2005). No data exist on how these alternative designs affect exposure to biomechanical factors and exposure to specific risk factors.

Speech recognition provides an opportunity to completely remove the keyboarding task altogether. Laboratory studies indicate that speech recognition systems reduce arm and shoulder muscle activity (Juul-Kristensen et al., 2004). For disability associated with MSD speech recognition provides an excellent alternative input device allowing those with extreme disorders to continue to interact with the computer. In terms of performance, speech recognition falls behind mouse and keyboard entry especially for multiple and varying tasks associated with computer work (Mitchard and Winkles, 2002). As a result, speech recognition is usually used to intervene when a disability has occurred, or when hands-free operations are necessary, such as multitask workstation design (Harisinghani et al., 2004). Speech recognition is often used for extensive text entry tasks such as medical transcriptions, and provides excellent performance and cost benefits for such specific applications (Schlossberg, 2005).

39.6 Integration with Workstation Design

The keyboard is only a small part of the computer workstation. Workplace interventions designed to reduce MSDs should include a holistic approach to computing in which the keyboard is an important aspect. With that in mind there are workstation characteristics that are directly related to the keyboard, mainly the placement of the keyboard within the workstation, that have been associated with MSD outcomes.

The vertical placement of the keyboard and its association with MSD has been discussed before in the literature (e.g., Sauter et al., 1991). Nonoptimal heights, where the keyboard height is different than elbow height (Sauter et al., 1991; Marcus et al., 2002), and nonneutral wrist postures (Simoneau and

Marklin, 2001) are associated with MSDs. In addition, Marcus et al. (2002) present new data linking the horizontal position of the keyboard with the incidence of hand and arm disorders. They found that having the keyboard more than 12 cm from the edge of the table was protective for hand and arm disorders. Such a factor may be associated with relieving awkward postures as well as providing an area for the forearm to rest; however, no existing studies exist examining the relationship between keyboard horizontal location and wrist and arm postures.

The use of wrist rests with computer keyboards is controversial; however, the scientific literature is limited with regard to the effect of wrist rests either on UEMSD morbidity or EMG activity. A greater risk of hand/wrist symptoms and disorders has been associated with the use of a wrist rest that is less than 7.5 cm in width (Marcus et al., 2002); however, more neutral wrist postures are associated with use of wrist rests (Gerr et al., 2000). Similarly, Visser et al. (2000) observed greater trapezius EMG activity with the use of wrist rests as compared with no arm or wrist support. Armrests, on the other hand, have had different results often reducing muscle activity and incidence of MSDs (Visser et al., 2000; Marcus et al., 2002).

Finally, the numeric keyboard integrated into the right-hand side of the keyboard affects the position of the mouse such that the right shoulder is externally rotated (Delisle et al., 2004). Removing the numeric pad may reduce the awkward posture associated with the mouse, but it is difficult to remove the numeric keys as they continue to be used by many data entry personnel and thus design efforts have focused on reducing the area of the board associated with the keypad (McLoone and Hinckley, 2003).

39.7 Conclusions

The studies highlighted here suggest that keyboards can make a difference as a tool to mitigate MSDs among keyboard workers. Fixed split keyboard provide alternative geometries that do affect wrist and forearm postures and hence reduce associated risk factors. Furthermore, keyswitch designs are directly related to the applied forces and forearm muscle activity. There is some small evidence that these changes can lead to decreased discomfort associated with computer keyboarding. Overall these alternative keyboards should be integrated within a complete systematic intervention including desk and chair designs.

References

Andersen JH, Thomsen JF, Overgaard E, Lassen CF, Brandt LP, Vilstrup I, Kryger AI, Mikkelsen S (2003) Computer use and carpal tunnel syndrome: a 1-year follow-up study. *JAMA* 289: 2963–2969.

ANSI (1988) *American National Standard for Human Factors Engineering of Visual Display Terminal Workstations*. Standard No. 100–1988. Human Factors Society, Santa Monica, CA.

Armstrong T, Silverstein B (1987) Upper-extremity pain in the workplace – role of usage in causality. In: NM Hadler (ed) *Clinical Concepts in Regional Musculoskeletal Illness*. Grune & Stratton, Orlando.

Armstrong TJ, Foulke JA, Martin BJ, Gerson J, Rempel, DM (1994). Investigation of applied forces in alphanumeric keyboard work. *Am Ind Hyg Assoc J* 55(1): 30–35.

Balakrishnan AD, Jindrich DL, Dennerlein JT (In Press) Keyswitch orientation can reduce finger joint torques during tapping on a computer keyswitch. *Hum Factors*.

Bergqvist U, Wolgast E, Nilsson B, Voss M (1995a) Musculoskeletal disorders among visual display terminal workers: individual, ergonomic, and work organizational factors. *Ergonomics* 38: 763–776.

Bergqvist U, Wolgast E, Nilsson B, Voss M (1995b) The influence of VDT work on musculoskeletal disorders. *Ergonomics* 38: 754–762.

Brandt LP, Andersen JH, Lassen CF, Kryger A, Overgaard E, Vilstrup I, Mikkelsen S (2004) Neck and shoulder symptoms and disorders among Danish computer workers. *Scand J Work Environ Health* 30: 399–409.

Delisle A, Imbeau D, Santos B, Plamondon A, Montpetit Y (2004) Left-handed versus right-handed computer mouse use: effect on upper-extremity posture. *Appl Ergon* 35: 21–28.

Diao E, Shao F, Liebenberg E, Rempel D, Lotz JC (2005) Carpal tunnel pressure alters median nerve function in a dose-dependent manner: a rabbit model for carpal tunnel syndrome. *J Orthop Res* 23: 218–223.

Faucett J, Rempel D (1994) VDT-related musculoskeletal symptoms: interactions between work posture and psychosocial work factors. *Am J Ind Med* 26: 597–612.

Feuerstein M, Armstrong T, Hickey P, Lincoln A (1997) Computer keyboard force and upper extremity symptoms. *J Occup Environ Med* 39: 1144–1153.

Gerard MJ, Armstrong TJ, Franzblau A, Martin BJ, Rempel DM (1999) The effects of keyswitch stiffness on typing force, finger electromyography, and subjective discomfort. *Am Ind Hyg Assoc J* 60: 762–769.

Gerr F, Marcus M, Ortiz D, White B, Jones W, Cohen S, Gentry E, Edwards A, Bauer E (2000) Computer users' postures and associations with workstation characteristics. *Am Ind Hyg Assoc J* 61: 223–230.

Gerr F, Marcus M, Ensor C, Kleinbaum D, Cohen S, Edwards A, Gentry E, Ortiz DJ, Monteilh C (2002) A prospective study of computer users: I. Study design and incidence of musculoskeletal symptoms and disorders. *Am J Ind Med* 41: 221–235.

Gerr F, Marcus M, Monteilh C (2004) Epidemiology of musculoskeletal disorders among computer users: lesson learned from the role of posture and keyboard use. *J Electromyogr Kinesiol* 14: 25–31.

Harisinghani MG, Blake MA, Saksena M, Hahn PF, Gervais D, Zalis M, da Silva Dias Fernandes L, Mueller PR (2004) Importance and effects of altered workplace ergonomics in modern radiology suites. *Radiographics* 24: 615–627.

Hedge A, Powers JR (1995) Wrist postures while keyboarding: effects of a negative slope keyboard system and full motion forearm supports. *Ergonomics* 38: 508–517.

Huang IW (1999) *Effects of Keyboards, Armrests, and Alternativing Keying Positions on Subjective Discomfort and Preferences Among Data Entry Operators*. Department of Environmental and Industrial Health, Univeristy of Michigan.

Jindrich DL, Balakrishnan AD, Dennerlein JT (2004) Effects of keyswitch design and finger posture on finger joint kinematics and dynamics during tapping on computer keyswitches. *Clin Biomech (Bristol, Avon)* 19: 600–608.

Juul-Kristensen B, Laursen B, Pilegaard M, Jensen BR (2004) Physical workload during use of speech recognition and traditional computer input devices. *Ergonomics* 47: 119–133.

Keir PJ, Wells RP (2002) The effect of typing posture on wrist extensor muscle loading. *Hum Factors* 44: 392–403.

Lassen CF, Mikkelsen S, Kryger AI, Brandt LP, Overgaard E, Thomsen JF, Vilstrup I, Andersen JH (2004) Elbow and wrist/hand symptoms among 6,943 computer operators: a 1-year follow-up study (the NUDATA study). *Am J Ind Med* 46: 521–533.

MacKenzie IS, Zhang SX (2001) An empirical investigation of the novice experience with soft keyboards. *Behav Inf Technol* 20: 411–418

Marcus M, Gerr F, Monteilh C, Ortiz DJ, Gentry E, Cohen S, Edwards A, Ensor C, Kleinbaum D (2002) A prospective study of computer users: II. Postural risk factors for musculoskeletal symptoms and disorders. *Am J Ind Med* 41: 236–249.

Marklin RW, Simoneau GC (2001) Effect of setup configurations of split computer keyboards on wrist angle. *Phys Ther* 81: 1038–1048.

Marklin RW, Simoneau GG (2004) Design features of alternative computer keyboards: a review of experimental data. *J Orthop Sports Phys Ther* 34: 638–649.

Marras WS, Schoenmarklin RW (1993) Wrist motions in industry. *Ergonomics* 36: 341–351.

Matias AC, Salvendy G, Kuczek T (1998) Predictive models of carpal tunnel syndrome causation among VDT operators. *Ergonomics* 41: 213–226.

McLoone H, Hinckley K (2003) Ergonomic principles applied to the design of the Microsoft® Office® computer keyboard. In: 15th Triennial Congress of the International Ergonomics Association (IEA 2003), Seoul, South Korea.

Mitchard H, Winkles J (2002) Experimental comparisons of data entry by automated speech recognition, keyboard, and mouse. *Hum Factors* 44: 198–209.

Radwin RG, Jeng OJ (1997) Activation force and travel effects on overexertion in repetitive key tapping. *Hum Factors* 39: 130–140.

Radwin RG, Ruffalo BA (1999) Computer key switch force–displacement characteristics and short-term effects on localized fatigue. *Ergonomics* 42: 160–170.

Rempel DM, Harrison RJ, Barnhart S (1992) Work-related cumulative trauma disorders of the upper extremity. *Jama* 267: 838–842.

Rempel D, Dennerlein J, Mote CD, Jr., Armstrong T (1994) A method of measuring fingertip loading during keyboard use. J Biomech 27: 1101–1104.

Rempel D, Tittiranonda P, Burastero S, Hudes M, So Y (1999) Effect of keyboard keyswitch design on hand pain. *J Occup Environ Med* 41: 111–119.

Sauter SL, Schleifer LM, Knutson SJ (1991) Work posture, workstation design, and musculoskeletal discomfort in a VDT data entry task. *Hum Factors* 33: 151–167.

Schlossberg EB (2005) Voice recognition software. Personal communication, January.

Serina ER, Tal R, Rempel D (1999) Wrist and forearm postures and motions during typing. *Ergonomics* 42: 938–951.

Silverstein BA, Fine LJ, Armstrong TJ (1986) Hand wrist cumulative trauma disorders in industry. *Br J Ind Med* 43: 779–784.

Silverstein BA, Fine LJ, Armstrong TJ (1987) Occupational factors and carpal tunnel syndrome. *Am J Ind Med* 11: 343–358.

Simoneau GG, Marklin RW (2001) Effect of computer keyboard slope and height on wrist extension angle. *Hum Factors* 43: 287–298.

Simoneau GG, Marklin RW, Berman JE (2003) Effect of computer keyboard slope on wrist position and forearm electromyography of typists without musculoskeletal disorders. *Phys Ther* 83: 816–830.

Swanson NG, Galinsky TL, Cole LL, Pan CS, Sauter SL (1997) The impact of keyboard design on comfort and productivity in a text-entry task. *Appl Ergon* 28: 9–16.

Tittiranonda P, Rempel D, Armstrong T, Burastero S (1999a) Effect of four computer keyboards in computer users with upper extremity musculoskeletal disorders. *Am J Ind Med* 35: 647–661.

Tittiranonda P, Rempel D, Armstrong T, Burastero S (1999b) Workplace use of an adjustable keyboard: adjustment preferences and effect on wrist posture. *Am Ind Hyg Assoc J* 60: 340–348.

Visser B, de Korte E, van der Kraan I, Kuijer P (2000) The effect of arm and wrist supports on the load of the upper extremity during VDU work. *Clin Biomech (Bristol, Avon)* 15 (Suppl 1): S34–S38.

Weiss ND, Gordon L, Bloom T, So Y, Rempel DM (1995) Position of the wrist associated with the lowest carpal-tunnel pressure: implications for splint design. *J Bone Joint Surg Am* 77: 1695–1699.

Zecevic A, Miller DI, Harburn K (2000) An evaluation of the ergonomics of three computer keyboards. *Ergonomics* 43: 55–72.

Zhai S, Kristensson P-O, Smith BA (2005) In search of effective text interfaces for off the desktop computing. *Interacting with Computers* 17(3): 229–250.

40

Intervention for Notebook Computers

Carolyn M. Sommerich
The Ohio State University

40.1 Introduction

In 1975, there were fewer than 200,000 computers in the United States.[1] By contrast, U.S. PC shipments for 2004 are expected to number 56 million units; worldwide shipments are expected to number around 165 million units.[2] A growing percentage of these computers are notebook units. In 1997, notebook PCs (NPCs) were already 40% of total computer output in Japan.[3] By 2007, notebooks are expected to constitute 40% of worldwide shipments and 47% of U.S. shipments; in 2004, they are expected to be 30 and 34% of those shipments, respectively.[4]

Though mobility is the primary feature of NPCs, increasingly NPCs are also the choice for use in place of desktop computers (DPCs) at small and large companies, and by those who work from home. Some companies provide peripheral devices for use with NPCs (monitor, keyboard, and pointing device), while others do not, because they feel unable to justify the extra expense. The trend of moving to NPCs is also seen for computers used for personal use. Students, from grade school through college, make up another growing segment of NPC users. The growing use of NPCs is seen in both private and public educational institutions. In some cases, the computer remains within the school, but in others, the student transports the computer to and from school, so that the computer can also be used at home.

Many studies have linked work-related use of DPCs with musculoskeletal discomfort in workers.[5-8] Neck, back, and upper extremity discomfort are prevalent in computer users, as are headache and eye discomfort. Recently, reports have linked school-related DPC use with musculoskeletal discomfort in younger people (students).[9-11] The growing use of NPCs raises concerns for development of discomfort in their users, as have some recent reports on student NPC users[12] and on adult workers who use

NPCs.[8,13] Harris and Straker[12] reported that 60% of a sample of 251 Australian schoolchildren reported some discomfort with using an NPC, with neck discomfort being most prevalent; 61% reported some discomfort with carrying the NPC, with shoulder and upper back discomfort most prevalent. Discomfort prevalence was greater for those who used a heavier NPC, and seemed to be related to the maximum amount of time students were likely to use the computer in one sitting. Heasman et al.[13] found similar reporting of discomfort frequencies between DPC and NPC users, with over 40% of respondents indicating discomfort occurred *sometimes* or *frequently* (versus *never* or *rarely*) in the back and neck, with similar frequencies of headache and irritated eyes. When specifically asked about discomfort associated with computer use, Sommerich[8] found the DPC and NPC users they surveyed reported similar rates of frequent discomfort* with computer use in the eyes (20% in both user groups) and the neck (21 and 17%, respectively). Among the NPC users, when moving/carrying the computer, 65% experienced discomfort in one or more body parts and 25% experienced discomfort in three or more body parts.

Musculoskeletal discomfort associated with computer work, in general, has been attributed to assumed postures, postural fixity, and inactivity.[14,15] While the level of muscle activity required to maintain typical computer work postures is generally low,[16] the static nature of the activation, the lack of variation in it, and, potentially most important, the length of time it is sustained, are thought to be reasons underlying the risk associated with computer work and other static tasks. In reviewing low-level static exertions, Sjøgaard[17] emphasized the importance of duration in elevating the risk of such exertions. Such exertions may result in problems with microcirculation within the muscle, muscle metabolism, or overworking of particular fibers ("Cinderella fibers"), as well as inadequate recovery between work periods. Visual discomfort has been associated with the location and quality of the display, as well as environmental factors (glare, dry air, etc.). There is much overlap in terms of factors that affect comfort of DPC and NPC users. However, some of the underlying reasons differ between the two. For example, it is easier to understand why someone would (have to) assume an awkward work posture when using an NPC in a temporary situation such as a waiting room, than when using a DPC or an NPC in a permanent office setting. Additionally, there are some concerns that are specific to NPC use and users, one of the most important of which is the transporting of the computer, sometimes over long distances or several times per day.

Figure 40.1 depicts a model of NPC design and usage factors and their potential effects on user comfort/health and performance. The model is based on epidemiological and laboratory DPC research, as well as limited NPC research. Each of the factors will be explored in this chapter, in order to inform the reader of the potential effects of NPC design and setup on a user's comfort and performance.

40.2 Notebook PC Design and Location Factors

40.2.1 Keyboard Design and Location

Some aspects of NPC keyboard design are shared with standard keyboards, such as layout of letter and number rows, but others are unique to the NPC, including alterations that may reduce keyboard size to comply with a notebook's footprint. Keyboard location decision is also important for DPCs and NPCs, but is more complex for notebooks.

40.2.1.1 Keyboard Layout

Problems with upper extremity posture during keyboard work have been recognized for more than three decades.[18] The root causes of postural problems include user stylistic methods, as well as adaptations in response to keyboard design or location. Pronounced ulnar wrist deviation during keying has consistently been linked to discomfort.[14,18,19] Nonneutral wrist postures have been associated with exposure of the median nerve to levels of pressure within the carpal tunnel that are known to adversely affect nerve physiology.[20] For some individuals, carpal tunnel pressure (CTP) can exceed physiologic levels

---

*"Frequent" discomfort meaning *quite often* or *almost always* rather than *never*, *rarely*, or *sometimes*.

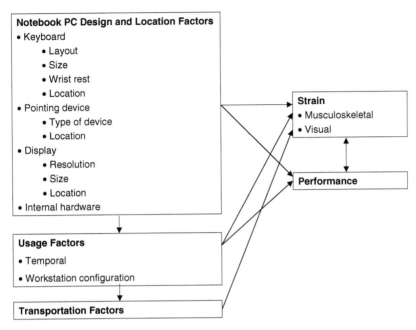

FIGURE 40.1   Model of design and other factors relevant to notebook computer use and potential effects on users.

during typing on a standard keyboard, due to the extent of ulnar deviation[21] or wrist extension.[22] These issues are common to DPC and NPC keyboards, but there are some design factors that may exacerbate problems with the NPC.

One layout situation that occurs primarily in NPCs is the lack of redundant keys. Standard DPC keyboards often have an integrated number pad to the right of the letter keys section, and in addition to the number row. They may also have two sets of function keys. These extra keys afford users opportunities, when typing numbers or using the function keys, to assume shoulder, elbow, and wrist postures that differ from text typing postures. Though some NPC letter keys can be converted to function as a number pad, their location does not afford an opportunity for posture changes. Table 40.1 shows upper extremity (UE) posture data from a study in which subjects performed a word typing task and a number typing task.[23] Subjects performed each task on a self-contained NPC and also on a standard

TABLE 40.1   Right Upper Extremity Joint Postures, during a Word Typing Task and a Number Typing Task, on a Standard Desktop Computer Keyboard with Number Pad (DPC) and a Built-In NPC Keyboard (NPC)

	Shoulder Lateral Rotation[a]	Shoulder Flexion[b]	Elbow Flexion[c]	Wrist Flexion[d]	Wrist Radial Deviation[e]
DPC — words/numbers	−23.5/1.0*	23.7/10.5*	106.5/98.1*	−20.1/−19.0	−12.1/−2.5*
NPC — words/numbers	−23.5/−22.6	24.0/24.8	112.1/105.6*	−11.7/−6.2*	−13.2/−15.3

Note: Values are means of ten subjects; units are degrees.
[a] Positive value indicates lateral rotation; negative value indicates medial rotation.
[b] Vertical humerus with elbow directly below the shoulder = 0° of shoulder flexion.
[c] Elbow fully extended = 0° of elbow flexion.
[d] Positive value indicates flexion; negative value indicates extension.
[e] Positive value indicates radial deviation (towards thumb); negative value indicates ulnar deviation.
*Indicates a statistically significant difference between postures during the two tasks.[23]
Source: From Sommerich, C. M., Starr, H., Smith, C. A., and Shivers, C., *IJIE* 30, 7–31, 2002. With permission.

full-size keyboard with a built-in number pad. Using the NPC keyboard, subjects remained in virtually the same posture for both the word typing and number typing tasks, indicating a high degree of postural fixity. However, using the number pad on the standard keyboard allowed subjects to vary UE posture, and assume a more neutral shoulder posture when typing numbers.

40.2.1.2 Keyboard Sizing

Sizing issues include key size, key spacing, and keyboard thickness. NPC keyboard dimensions can, but do not always, differ from those of a standard keyboard. In a study comparing text entry on a DPC with text entry on several NPCs, one NPC keyboard was only one third as long as the standard external keyboard, and another was 2.5 times as thick.[3] Yoshitake[24] determined that touch typists with large fingers (referring to finger tip breadth) were disadvantaged when working on keyboards with key spacing below 16.7 mm while typists with narrow fingers could type just as well with spacing as small as 15 mm. Spacing differences were achieved through changes in center-to-center spacing alone, or combined with changes in key top width. Loricchio and Lewis[25] tested numeric keypads of various key spacing/size configurations, and found no effect on error rate, but found effects on preference and input rate between a standard keypad and one with both narrower spacing and reduced width keys. Standard key spacing is 18 to 19 mm horizontally and 18 to 21 mm vertically; the horizontal strike surface is recommended to be at least 12 mm wide.[26]

Villanueva et al.[3] standardized the height of each keyboard relative to each subject, but still found differences in elbow flexion and wrist extension in comparing postures during text entry with a DPC and four progressively smaller NPCs. Keyboard sizes differed between the DPC and the three smaller NPCs. Medial shoulder (humeral) rotation increased as keyboard size decreased, though abduction was not affected. Keying difficulty was judged greater for the NPCs compared to the DPC, and was greatest for the two smaller keyboards. Keying performance was reduced for the two smaller NPCs.

40.2.1.3 Palm/Wrist Rests

Lack of UE support has been linked to UE discomfort in video display terminal (VDT) workers,[19] particularly in situations with limited rest break opportunities.[27] However, the means by which support is provided seems to be critical for achieving, simultaneously, the goals of reducing shoulder muscle loads, facilitating neutral wrist postures, not hampering productivity, and not introducing localized contact stress. Support provided only at the palm/wrist, as with support areas provided by keyboards built into NPCs or an external wrist pad, does not seem to achieve all four of these goals.[28–30] In contrast, Aarås and colleagues, using standard keyboards, have shown benefits to locating the keyboard back from the edge of the support surface, so as to afford provision of full support along the forearms.[31,32] It could be assumed that NPC users might benefit from such an arrangement, if thickness of the NPC were similar to a standard keyboard. Many are thicker, however, which could lead to pronounced wrist extension or imposition of contact stress on the forearm, depending on whether or not the computer had a built-in wrist rest. A study of novice users compared postures when using an NPC with a built-in wrist rest and one without.[33] For the two particular computers used in the study, similar upper body and upper arm postures were observed. There was a slight trend (not significant) toward greater wrist extension with the no-palm-rest NPC. There was a trend (not significant) toward greater ulnar deviation with the no-palm-rest NPC compared with the other computer when the computers were used on a desktop, and slightly less ulnar deviation (statistically significant) with the no-palm-rest NPC compared to the other NPC when the computers were used on the subjects' laps.

40.2.1.4 Peripheral Keyboards

With the exception of the wrist rest and thickness of the keyboard, there is little variation in keyboards built into today's NPCs. However, a wide range of options are available if the NPC user wishes to use a peripheral keyboard. Choices include the standard, tethered flat keyboard with or without built-in number pad, and a wide array of alternative keyboards.

40.2.1.4.1 *Alternative, Desktop-Style Keyboards*

Alternative desktop-style keyboards are those that vary from the standard flat keyboard, most often in terms of some degree of separation of the keys assigned to the left and right hands. The objective is usually to reduce the nonneutral postures of the wrist and the forearm. Ulnar deviation is reduced when the keys are "split," such that the two halves of the keyboard are rotated (in the horizontal plane) away from each other, with the pivot point located above the function row. Pronation is reduced when the two halves are "tented" (or laterally inclined), such that the two halves form a tent with the keys operated by the index fingers raised higher than the keys operated by the little fingers. Some alternative keyboards are user-adjustable, and others are not. Most studies that have been conducted on these keyboards have been short-term, laboratory-based studies. They have generally shown that postures are influenced by keyboard design, and in a predictable way, so that they can be effective in inducing postures that are less deviated from neutral than traditional flat keyboards.[34] Similar levels of productivity can be achieved with some alternative keyboards.[35] A 6-month-long interventional study of computer users with musculoskeletal disorders showed improving trends in pain severity and hand function for two of the alternative keyboards utilized in the study.[36]

In addition to alternative designs of configuration, there are alternative means of attachment: tethered or wireless. Both facilitate more open, less restricted postures than are typically associated with typing on an NPC in a stand-alone configuration (no peripheral display or keyboard).[23,37]

40.2.1.4.2 *Alternatives to Desktop-Style Keyboards*

Some keyboards, by virtue of their size, materials, or other design elements, might be considered portable enough to accompany a notebook computer. However, there is little published research that has been conducted on the use of such keyboards for this purpose. Therefore, these keyboards, which include mini-keyboards, gesture keyboards, virtual keyboards, soft keyboards, and folding keyboards, are only mentioned here to inform the reader of their availability in the marketplace. Note that these are specific terms that can be used to perform an internet search to learn more about the different keyboards.

40.2.1.5 Keyboard Location

The modular design of the traditional DPC allows for independent location of keyboard and screen. Straker et al.[37] state that "… independent adjustment of screen and keyboard is important to allow users to position the tactile and visual interaction components, in a way that encourages a good posture." While there is disagreement in the literature regarding appropriate monitor location,[38–40] most desired locations can be achieved as a result of this modular design. However, with most NPCs, screen and keyboard locations are not independent. The limited research on NPCs indicates that keyboard location has priority over display location.

Straker et al.[37] examined posture, discomfort, and performance in adults who regularly used computers. Their results indicated that subjects chose to position the keyboards of a DPC and a stand-alone NPC similarly, while the location of the NPC screen was, then as a consequence, lower than the location chosen for the DPC display. Price and Dowell[41] found significant differences in posture and discomfort associated with NPC placement. Comparing an NPC placed low (to optimize arm posture) to an NPC placed high (to optimize neck posture), discomfort in all body parts, including the neck, was greater with the higher placement. Results of these two studies, though only involving short-term use, appear to be consistent with epidemiological findings on DPC keyboard height. Elbow-level keyboard placement seems to offer a compromise between shoulder discomfort induced by higher locations and wrist discomfort induced by lower locations.[27]

Moffet et al.[33] found the effects of location (desktop versus laptop placement) to be greater than the effects of NPC design (keyboard with versus without a built-in wrist rest). Postures were generally further from neutral with laptop placement: head/neck flexion increased, upper arm extension was induced, and wrist extension increased, when compared with desktop placement. The further away from neutral that a joint is positioned, the harder the muscles have to work to maintain that position. This can lead to discomfort in the muscles if the position is maintained without rest breaks for an extended period of time. Additionally, nonneutral wrist postures can cause an elevation of the pressure in the space occupied by the median nerve as it passes through the wrist, from the forearm into the hand. Pressures great enough to interfere with nerve physiology and function can occur for some people with the wrist in just 30 to 40° of extension,[42] a range commonly observed in typing. Forearm pronation (palms down, the position assumed during typing) and ulnar deviation also increase CTP, and both also occur in typing.[21,43]

40.2.1.6 Keyboard Summary

It is important to recognize that the design and location of the keyboard directly affect the user's posture, comfort, and performance. User posture, either the arms/shoulders or head and neck, is compromised when using a notebook computer in a stand-alone configuration. Posture can be improved through use of an external keyboard, of which there are many designs from which to choose. Undersized keyboards, wherein key top size or key spacing are reduced, are generally not recommended for extended use by adults as they have been shown to comprise performance, thereby increasing the amount of time the computer must be used. Support of the forearms, with the keyboard at about elbow height, has been shown to have a beneficial effect on the shoulder discomfort in studies of desktop and notebook users.

40.2.2 Pointing Device Design and Location

Choices for external pointing device design and location are extensive. In contrast, the variety of devices that are built into today's NPCs is quite limited, and the types provided have been shown to adversely impact performance.

40.2.2.1 Type of Pointing Device

There are many types and designs of external pointing devices, and studies have shown performance differences between them. However, NPC users generally have a choice of only two types of integrated pointing devices: a touchpad or a pointing stick (nub positioned between the G, H, and B keys), both of which have been shown to hinder performance when compared to other types of devices. Sawin and Ark[44] demonstrated, however, that user performance with these internal pointing devices can be improved somewhat through attention to certain design factors, specifically a transfer function that incorporates negative inertia, roughening the texture of the pointing stick, employment of a palm rest, and use of active-matrix displays.

Several studies have conducted comparisons between internal and external pointing devices. For novice users, Card et al.[45] found positioning time and error rates for text selection were less for a mouse in comparison to a rate-controlled isometric joystick, step keys, or text keys. Epps[46] found a mouse was faster for target acquisition than a rate-controlled force joystick. Trackball performance was slightly better than mousing, and touchpad performance was better than joystick, but not as good as mousing. In contrast, MacKenzie et al.[47] found a trackball was worse than a mouse or a stylus and tablet, based on move time for dragging and pointing, and quantity of "dropping" errors. Comparing pointing devices in target acquisition and text selection tasks, Loricchio[48] also found mouse performance superior to that of a rate-controlled force pushbutton device. These studies are consistent in their findings of performance benefits of a mouse over touchpads, joysticks, and button-type pointing devices.

Results of several studies of children's performance using different types of pointing devices were reviewed by Barrero and Hedge.[49] Most relevant to this discussion was the finding that trackball performance was slower than mouse performance in grade school age children.

40.2.2.2 Pointing Device Location

Studies have documented effects of pointing device location on biomechanical and subjective outcomes. Cook and Kothiyal[50] compared muscle activity and posture when using a mouse located normally (to the side of a keyboard with built-in number pad), in a far right location, and in a compact location (to the side of a keyboard with no number pad). Muscle activity in the anterior and middle deltoid increased as mouse position became more lateral. Karlqvist et al.[51] found that shoulder posture and muscle activity were dependent upon medial–lateral and fore–aft mouse position. Subjects tended to prefer the mouse located directly in front of the shoulder, with taller subjects preferring the mouse further from the table edge and shorter subjects preferring it closer.

Kelaher et al.[52] identified effects of the location of an integrated touchpad on UE posture, discomfort (neck, upper back, and shoulders), perception of performance, and preference in a group of right-handed subjects. A bottom center location (most common in NPCs) and four alternatives (top center, top right, right side, and bottom right) were evaluated. Results suggested that the bottom center location may not be optimal. Rankings for it were generally in the middle for most outcome measures, including preference. Similar to findings from Karlqvist et al.,[51] the highest ranked location for the touchpad was directly to the right of the keyboard.

40.2.2.3 Integrated versus Peripheral Pointing Devices

When integrated and peripheral devices are compared, effects may be due to type of device, device location, or a combination of the two. Price and Dowell,[41] in studying various configurations of an NPC with and without peripheral devices, found greater discomfort in the right wrist/hand for two conditions requiring use of the integrated pointing device in comparison to four conditions requiring use of a mouse. This provides an interesting contrast to work by Harvey and Peper,[60] who reported that using a mouse to the right of an extended keyboard required more activity in the deltoid and trapezius muscles compared to a trackball centrally located below the space bar on an NPC. In considering the findings of these two studies, it may be that the benefits of intermittent postural change override the "cost" of intermittent increased muscle activity.

In a study designed to compare posture as a function of task (word typing versus pointing device use) and NPC configuration (with or without external input devices), the data indicated that use of the external mouse afforded subjects opportunities to vary shoulder medial/lateral rotation position between the tasks. In comparison, subjects assumed similar postures for both tasks when using the NPC alone.[23]

One other study worth mentioning here is one in which the effect of mouse size on user comfort was investigated.[53] This study is relevant to this discussion because of the proliferation of reduced size (mini) mice that are being marketed to today's mobile computer users. Holt and Sommerich[53] replaced the subjects' regular mice they used at work with three identical mice, which only differed in size (scale), one smaller than normal and one larger than normal. Each subject used each size mouse for 1 week, and then rated each mouse for comfort and performance. The researchers found a trend in the results based on subject's hand size. Subjects who ranked the smaller mouse as least preferred tended to have larger hands, while those who ranked the larger mouse as least preferred tended to have smaller hands.

40.2.2.4 Pointing Device Summary

It is important to recognize that the design and location of the pointing device directly affects the user's posture, comfort, and performance. Use of integrated pointing devices, touchpad or pointing stick, encourages postural fixity and increases performance time. Consistently, performance and comfort benefits have been shown to occur when an external mouse is used with an NPC. Given the modest differences in dimension and weight between a full-size mouse and a reduced-size mouse, most adult users would be advised to use a full size mouse. Children and other users with smaller hands might realize a benefit to using a reduced-size mouse.

40.2.3 Display Design and Location

The flat panel liquid crystal displays (LCD) provided in today's NPCs are being offered with increasingly higher resolution and with active-matrix (a.k.a. thin film transistor [TFT]) technology that provides a more responsive image and a wider viewing angle range than earlier passive matrix displays. The primary limitations are in the effects that the size, angle, and location of the display can have on users.

Work by Villanueva et al.[3] demonstrated that screen size is an important factor in terms of potential to affect user health and performance. In that study, smaller screen sizes were associated with visual discomfort, more nonneutral body postures (more flexion of the neck and trunk), and increased neck extensor muscle activity.

Screen size was directly related to subject-selected tilt angle — the smaller the display, the further back the display was reclined. In a more recent study that utilized a single NPC with an integrated TFT LCD, investigators studied the effects of different display tilt angles on the luminance levels of the LCD and on subjects' postures.[54] This was of interest to them, because display angle may be adjusted by NPC users not solely to achieve musculoskeletal comfort, but also in an effort to reduce glare on the display. The researchers found that luminance decreased as the display was reclined from 100 to 130°, which is in contrast to stable luminance levels of a similarly tested CRT display. Body posture and computer location relative to the table edge were all affected by the angle of the display. Viewing distance was not affected, however. The authors suggested that this was an important finding, which showed how the body will be used to try to achieve a preferred viewing situation. The decrement in luminance as the eye and display move relative to each other, combined with efforts to reduce glare, while simultaneously trying to maintain a preferred eye-display distance and angle may severely limit posture options and contribute to postural fixity while using an NPC.

Two recent studies have shown the benefits of utilizing an external keyboard so that the integrated NPC display can be raised to a more appropriate location. In their study of the effects of the use of peripheral input devices on NPC user posture, comfort, and performance, Sommerich et al.[23] found an average viewing angle of 35° when subjects used an NPC without any peripherals and positioned for appropriate keyboard height. This viewing angle was determined to be too low, based on a review of studies of preferred viewing angle, studies of neutral posture, and studies of monitor placement effects on user biomechanics. The authors of that review, Psihogios et al.,[40] concluded that mid-level monitor placement (10 to 20° for users without bifocals) achieves a balance between eye strain and body strain. Placing the NPC on a horizontal riser[23] or on a newer style of angled holder[55] improves head and neck posture, thereby reducing the mechanical load on the neck and neck extensor muscle activity as a consequence. Reduced discomfort and increased productivity were also demonstrated in these studies.

It is important, then, to recognize that the angle and location of the display directly affects the user's posture, comfort, and performance. When an NPC is used in mobile conditions, unless the user utilizes a portable keyboard, the location of the monitor will be dictated by the location of the work surface. It is essential, then, for the user to remain aware of how he or she is positioning his or her body in order to view the display and interact with the keyboard and pointing device, and to reduce any awkward posture by orienting the computer (rather than the body) to minimize glare on the display. In more permanent situations (office setting) strong consideration should be given to use of a support stand that permits the display to be raised to achieve an appropriate viewing angle.

40.2.4   Additional Hardware

Comments made in this subsection are only meant to address factors that affect the weight of the computer, if the NPC will be carried by the user. Type of processor and battery weight should be considered if the computer will often be used away from power sources. Certain types of processors are designed to draw less power than others, thereby potentially reducing the need to carry along an extra battery or a battery charger/power supply. The means by which files are backed-up or exchanged should also be a consideration when planning for mobile use of the computer. Computers that are designed to be highly mobile generally have limited built-in alternative storage abilities (no floppy drive, Zip drive, or CD-DVD). These devices, whether integrated or peripheral, can add considerable weight to the computer. The new flash drives, which are key chain size devices that are plugged into any USB port, practically eliminate the need to carry around other types of storage devices (either internal or external).

40.3   Usage Factors

40.3.1   Temporal Factors

Work patterns, specifically lack of rest breaks, have been associated with greater levels of discomfort in VDT workers.[27] In a cross-sectional study of professionals who used NPCs or DPCs for work, 34% of respondents reported only taking "very brief breaks" throughout the work day, and another 34% reported taking only very brief breaks plus a lunch break of more than 30 min. In that same group of workers, the reporting of frequent musculoskeletal discomfort was positively correlated with a score that was derived from each subject's propensity for working at the computer for various, uninterrupted extended periods of time. Yet, Galinsky et al.[56] demonstrated significant reductions in discomfort without reductions in productivity when hourly 5-min breaks were introduced into the schedules of data entry workers who normally only took only a 15-min break in the morning and also one in the afternoon, in addition to a lunch break. Regardless of the type of computer being used, taking hourly breaks is recommended. These breaks may be even more important for NPC users who work with few or no peripheral devices, as these configurations impose more sustained, awkward postures beyond those normally imposed by computer work.

40.3.2 Workstation Configuration

Previous sections of this chapter have already addressed, individually, the impact that the keyboard, display, and pointing device can have on the NPC user. The findings of Sommerich[8] provide further support for the importance of the use of peripheral devices when using an NPC. In a cross-sectional study of over 200 professionals who used NPCs for work, those users who used no peripheral devices with their computers had significantly higher scores on a scale that reflected frequency of discomfort associated with the use of the computer, when compared to those NPC users who used any peripheral device (2.1 [sd = 2.3] versus 0.9 [sd = 1.6]; on scale of 0 to 13), respectively. When subjects were regrouped, into those who used an NPC stand alone or with just a mouse or other external pointing device, there was still a significant difference between scores for those subjects compared with the others (1.6 [sd = 2.2] versus 0.8 [sd = 1.6]). The findings from this study, along with the results from studies cited earlier in this chapter, strongly support the use of some peripheral devices when using an NPC.

40.4 Transportation

Today's NPCs range in weight from "ultra mobile" units at approximately 1 kg to "desktop replacements" and "rugged" portables at close to 4 kg. It is important to remember, however, that users typically do not carry the computer by itself. Usually the computer is in its own carrying case or in a briefcase, and storage media, power supplies, and other devices might also be included in the package that is moved when the computer is carried.* Seventy percent of the notebook users in the study by Sommerich[8] reported moving the computer at least once per day, on average. Modes included holding in the hand (briefcase style), supported by one or both shoulders, or using a rolling cart. The percentage of subjects who used each mode appears in Table 40.2. Discomfort in one or more body parts was commonly associated with moving the computer, with the prevalence of shoulder discomfort being the greatest — reported by over half the NPC users (refer to Table 40.2). Associations were found between modes of moving the NPC and body part discomfort associated with moving the computer. Odds ratios describing the strength of those associations also appear in Table 40.2. The associations are not particularly surprising. Shoulder and neck pain were linked to carrying the computer on just one shoulder; two-shoulder support was associated with upper back discomfort; distal upper extremity pain was associated with carrying the computer in one hand. Only the rolling cart was not linked to any discomfort, yet with so few utilizing this method, it would be premature to make an unreserved recommendation of that method.

---

*The American Medical Association recommends children carry no more than 15% of their body weight in their school backpacks.[57] However, Fraser[57] found sixth-grade NPC users typically carried 8.2 to 10 kg packs, which included the NPC, books, and lunch kit.

TABLE 40.2 Prevalence of Methods of Moving NPCs; Prevalence of Discomfort Associated with Moving NPCs; Odds Ratios Linking Carrying Mode with Discomfort

		Modes of Carrying the NPC			
		In Hand	Supported by One Shoulder, Using a Shoulder Strap or Shoulder Bag	Supported by Both Shoulders, Backpack Style	Using a Rolling Cart
Body Part that Experiences Discomfort When Moving the NPC	Prevalence in Sample of 240 NPC Users, %	53	79	14	9
Hand	9	4.5			
Wrist	6	3.8			
Forearm	8	3.4			
Elbow	3	a			
Shoulder	57	0.5	3.5		
Neck	33	2.1			
Upper back	19		2.9		
Lower back	17				

a Indicates significant association, but odds ratio could not be calculated.

Source: From Sommerich, C. M., A Survey of desktop and notebook computer use by professionals, in *Proceedings of Human Factors and Ergonomics Society 46th Annual Meeting*, Human Factors and Ergonomics Society, Baltimore, MD, 2002.

40.5   Guidelines and Suggestions for NPC Selection and Use

While there are multiple standards and guidelines for the design and use of desktop-style computers, the extent of comparable information for NPCs is quite limited. Saito et al.[58] published a set of "Ergonomic Guidelines for Using Notebook Personal Computers," under the auspices of the Science and Technology Committee of the International Ergonomics Association (IEA), that are based on a set of Japanese guidelines published in 1998, through the Japan Ergonomics Society. The guidelines address the work environment and workstation layout, chair and desk, keyboards, display, nonkeyboard input devices, working posture, and other peripherals. Specific guidelines include: positioning the keyboard at elbow height; appropriate adjustment of the display (considering angle, brightness, contrast, and other settings); use of a mouse as the pointing device "if at all possible"; positioning the display 40 to 50 cm away from the eyes; looking away from the display for a few minutes every 30 min; consideration of use of an external keyboard if the built-in keyboard is undersized, more than 3.5 cm thick, or the task requires extensive numeric entry.

The U.S. Centers for Disease Control (CDC) provide suggestions for computer use that contain a short subsection on NPCs.[59] The CDC specifically does not recommend NPCs as a primary computer, but does recommend that a docking station be used with an NPC. When using an NPC, recommendations include mini-breaks every 20 to 30 min; a 46 to 76 cm viewing distance; keyboard at elbow height; use of an external mouse. The CDC also does not recommend resting the hands/wrists when typing; however, that recommendation is not consistent with the literature regarding the benefits of arm support while typing or mousing (support is best applied along the forearm and should not be concentrated at the wrist; further, use of the support should not result in the wrists being in pronounced flexion or extension or shoulders being elevated in order to make use of the support). With regards to carrying the NPC, the CDC recommends using a wheeled luggage cart when possible; if carrying on the shoulder, the bag should have a padded shoulder strap, and the CDC recommends shifting the load often to avoid overloading the shoulder. It should be noted, however, that repeated asymmetric lifting of the computer may impose some added risk to the back/spine, depending on how the load shifting is accomplished.

Other sources of recommendations for selection or use of NPCs include websites of manufacturers (such as the "healthy computing" section of IBM's website) and academic institutions. In general, reducing the load as much as possible, by choosing the appropriate computer at the outset, choosing to carry only essential items along with the computer, and using either a rolling cart or supporting the computer with both shoulders are factors that are likely to reduce the risk of developing discomfort while moving an NPC.

40.6 Summary

It is essential for computer users as well as computer providers (employers, parents, and teachers) to realize that use of a computer, the manner in which it is used, and its design can affect a user's physical comfort and productivity. This chapter provides information that all of these groups can use to help them make informed decisions regarding selection and use of NPCs.

Acknowledgments

Funding from the National Institute for Occupational Safety and Health (K01 OH00169), the Office Ergonomics Research Committee, and the North Carolina Ergonomics Resource Center supported some of the author's research cited in this chapter. Literature searches in support of this paper were conducted by Amy Asmus, Jeffrey Hoyle, and the author.

References

1. Juliussen, E. and Petska-Juliussen, K., *7th Annual, Computer Industry Almanac*, Computer Industry Almanac, Inc., Austin, 1994.
2. Spooner, J. G., IDC rethinks PC forecast, http://news.zdnet.co.uk/hardware/0,39020351,39116143,00.htm, Accessed: 2004, Last update: September 5, 2003.
3. Villanueva, M. B., Jonai, H., and Saito, S., Ergonomic aspects of portable personal computers with flat panel displays (PC-FPDs): evaluation of posture, muscle activities, discomfort and performance, *Ind Health* 36 (3), 282–289, 1998.
4. Spooner, J. G., Consumers keep notebook sales on a roll, http://www.cnet.com/4520-6018_1 105833.html, Accessed: 2004, Last update: January 9, 2004.
5. Bernard, B., *Los Angeles Times*, Department of Health and Human Services, Public Health Service, Centers for Disease Control, National Institute for Occupational Safety & Health, NIOSH Health Hazard Evaluation Report, Report No. HETA 90-013-2277, 1991.
6. Bergqvist, U., Wolgast, E., Nilsson, B., and Voss, M., The influence of VDT work on musculoskeletal disorders, *Ergonomics* 38 (4), 754–762, 1995.
7. Gerr, F., Marcus, M., Ensor, C., Kleinbaum, D., Cohen, S., Edwards, A., Gentry, E., Ortiz, D. J., and Monteilh, C., A prospective study of computer users: I. Study design and incidence of musculoskeletal symptoms and disorders, *Am J Ind Med* 41 (4), 221–235, 2002.
8. Sommerich, C. M., A survey of desktop and notebook computer use by professionals, in *Proceedings of Human Factors and Ergonomics Society 46th Annual Meeting*, Human Factors and Ergonomics Society, Baltimore MD, 2002.
9. Royster, L. and Yearout, R., A computer in every classroom — are schoolchildren at risk for repetitive stress injuries (RSIs)? In *Advances in Occupational Ergonomics and Safety*, Lee, G. C. H., ed., IOS Press, Amsterdam, 1999, pp. 407–412.
10. Katz, J. N., Amick, B. C., Carroll, B. B., Hollis, C., Fossel, A. H., and Coley, C. M., Prevalence of upper extremity musculoskeletal disorders in college students, *Am J Med* 109 (7), 586–588, 2000.
11. Noack, K. L., College student computer use and ergonomics, Unpublished master's thesis, North Carolina State University, 2003.

12. Harris, C. and Straker, L., Survey of physical ergonomics issues associated with school childrens' use of laptop computers, *IJIE* 26, 337–346, 2000.
13. Heasman, T., Brooks, A., and Stewart, T., Health and safety of portable display screen equipment, Health & Safety Executive, U.K., Contract Research Report, Report No. 304/2000, 2000.
14. Sauter, S. L., Schleifer, L. M., and Knutson, S. J., Work posture, workstation design, and musculoskeletal discomfort in a VDT data entry task, *Hum Factors* 33 (2), 151–167, 1991.
15. Grieco, A., Sitting posture: an old problem and a new one, *Ergonomics* 29 (3), 345–362, 1986.
16. Wærsted, M. and Westgaard, T., An experimental study of shoulder muscle activity and posture in a paper version versus a VDU version of a monotonous work task, *IJIE* 19, 175–185, 1997.
17. Sjøgaard, G., Low-level static exertions, in *The Occupational Ergonomics Handbook*, Karwowski, W. and Marras, W., eds., CRC Press, Boca Raton, 1999, pp. 247–259.
18. Duncan, J. and Ferguson, D., Keyboard operating posture and symptoms in operating, *Ergonomics* 17 (5), 651–662, 1974.
19. Hünting, W., Läubli, T., and Grandjean, E., Postural and visual loads at VDT workplaces. I. Constrained postures, *Ergonomics* 24 (12), 917–931, 1981.
20. Dahlin, L. B., Aspects on pathophysiology of nerve entrapments and nerve compression injuries, *Neurosurg Clin North Am* 2 (1), 21–29, 1991.
21. Sommerich, C. M., Marras, W. S., and Parnianpour, M., A method for developing biomechanical profiles of hand-intensive tasks, *Clin Biomech* 13, 261–271, 1998.
22. Rempel, D., Horie, S., and Tal, R., Carpal tunnel pressure during keying, in *Proceedings of Marconi Keyboard Research Conference*, UC Berkeley Ergonomics Program, Marshall, CA, 1994.
23. Sommerich, C. M., Starr, H., Smith, C. A., and Shivers, C., Effects of notebook computer configuration and task on user biomechanics, productivity, and comfort, *IJIE* 30, 7–31, 2002.
24. Yoshitake, R., Relationship between key space and user performance on reduced keyboards, *J Physiol Anthropol* 14 (6), 287–292, 1995.
25. Loricchio, D. F. and Lewis, J. R., User assessment of standard and reduced-size numeric keypads, in *Proceedings of Human Factors Society 35th Annual Meeting*, The Human Factors Society, Santa Monica, 1991, pp. 251–252.
26. HFES, *Human factors engineering of computer workstations*, The Human Factors and Ergonomics Society, Report No. BSR/HFES 100, 2002.
27. Bergqvist, U., Wolgast, E., Nilsson, B., and Voss, M., Musculoskeletal disorders among visual display terminal workers: individual, ergonomic, and work organizational factors, *Ergonomics* 38 (4), 763–776, 1995.
28. Horie, S., Hargens, A., and Rempel, D., Effect of keyboard wrist rest in preventing carpal tunnel syndrome, in *Proceedings of Marconi Keyboard Research Conference*, UC Berkeley Ergonomics Program, Marshall, CA, 1994.
29. Parsons, C. A., Use of wrist rests by data input VDU operators, in *Contemporary Ergonomics*, Lovesay, E. J., ed., Taylor & Francis, London, 1991, pp. 319–321.
30. Fernström, E., Ericson, M. O., and Malker, H., Electromyographic activity during typewriter and keyboard use, *Ergonomics* 37 (3), 477–484, 1994.
31. Aarås, A., Horgen, G., Bjorset, H. H., Ro, O., and Thoresen, M., Musculoskeletal, visual and psychosocial stress in VDU operators before and after multidisciplinary ergonomic interventions, *Appl Ergon* 29 (5), 335–354, 1998.
32. Aarås, A., Fostervold, K. I., Ro, O., Thoresen, M., and Larsen, S., Postural load during VDU work: a comparison between various work postures, *Ergonomics* 40 (11), 1255–1268, 1997.
33. Moffet, H., Hagberg, M., Hansson-Risberg, E., and Karlqvist, L., Influence of laptop computer design and working position on physical exposure variables, *Clin Biomech (Bristol, Avon)* 17 (5), 368–375, 2002.
34. Marklin, R. W., Simoneau, G. G., and Monroe, J. F., Wrist and forearm posture from typing on split and vertically inclined computer keyboards, *Hum Factors* 41 (4), 559–369, 1999.

35. Swanson, N. G., Galinsky, T. L., Cole, L. L., Pan, C. S., and Sauter, S. L., The impact of keyboard design on comfort and productivity in a text-entry task, *Appl Ergon* 28 (1), 9–16, 1997.
36. Tittiranonda, P., Rempel, D., Armstrong, T., and Burastero, S., Effect of four computer keyboards in computer users with upper extremity musculoskeletal disorders, *Am J Ind Med* 35 (6), 647–661, 1999.
37. Straker, L., Jones, K. J., and Miller, J., A comparison of the postures assumed when using laptop computers and desktop computers, *Appl Ergon* 28 (4), 263–268, 1997.
38. Bauer, W. and Wittig, T., Influence of screen and copy holder positions on head posture, muscle activity and user judgement, *Appl Ergon* 29 (3), 185–192, 1998.
39. Hill, S. G. and Kroemer, K. H. E., Preferred declination of the line of sight, *Hum Factors* 28 (2), 127–134, 1986.
40. Psihogios, J. P., Sommerich, C. M., Mirka, G. M., and Moon, S. D., A field evaluation of monitor placement effects in VDT users, *Appl Ergon* 32 (4), 313–325, 2001.
41. Price, J. A. and Dowell, W. R., Laptop configurations in offices: effects on posture and discomfort, in *Proceedings of the Human Factors and Ergonomics Society 42nd Annual Meeting*, Chicago, 1998, pp. 629–633.
42. Keir, P. J., Bach, J. M., and Rempel, D. M., Effects of finger posture on carpal tunnel pressure during wrist motion, *J Hand Surg Am* 23 (6), 1004–1009, 1998.
43. Werner, R., Armstrong, T., Bir, C., and Aylard, M., Intracarpal canal pressures: the role of finger, hand, wrist and forearm position, *Clin Biomech* 12 (1), 44–51, 1997.
44. Sawin, D. A. and Ark, W., Comparing in-keyboard pointing devices: considerations for the practitioner, in *Proceedings of the Human Factors and Ergonomics Society 43rd Annual Meeting*, The Human Factors and Ergonomics Society, Santa Monica, 1999, pp. 501–505.
45. Card, S. K., English, W. K., and Burr, B. J., Evaluation of mouse, rate-controlled isometric joystick, step keys, and text keys for text selection on a CRT, *Ergonomics* 21 (8), 601–613, 1978.
46. Epps, B. W., Comparison of six cursor control devices based on Fitts' Law model, in *Proceedings of the Human Factors Society 30th Annual Meeting*, The Human Factors Society, Santa Monica, 1986, pp. 327–331.
47. MacKenzie, I. S., Sellen, A., and Buxton, W., A comparison of input devices in elemental pointing and dragging tasks, in *Proceedings of the 19th Annual ACM Computer Science Conference*, Association for Computing Machinery, New York, 1991, pp. 161–166.
48. Loricchio, D. F., A comparison of three pointing devices: mouse, cursor keys, and a keyboard-integrated pushbutton, in *Proceedings of The Human Factors Society 36th Annual Meeting*, The Human Factors Society, Santa Monica, 1992, pp. 303–305.
49. Barrero, M. and Hedge, A., Computer environments for children: a review of design issues, *Work* 18, 227–237, 2002.
50. Cook, C. J. and Kothiyal, K., Influence of mouse position on muscular activity in the neck, shoulder and arm in computer users, *Appl Ergon* 29 (6), 439–443, 1998.
51. Karlqvist, L. K., Bernmark, E., Ekenvall, L., Hagberg, M., Isaksson, A., and Rosto, T., Computer mouse position as a determinant of posture, muscular load and perceived exertion, *Scand J Work Environ Health* 24 (1), 62–73, 1998.
52. Kelaher, D., Nay, T., Lawrence, B., Lamar, S., and Sommerich, C. M., An investigation of the effects of touchpad location within a notebook computer, *Appl Ergon* 32 (1), 101–110, 2001.
53. Holt, S. L. and Sommerich, C. M., Use of rapid prototyping in an ergonomic assessment of the relationship between hand anthropometry and mouse size preference, in *Proceedings of the 6th Industrial Engineering Research Conference*, Institute of Industrial Engineers (IIE), Miami, 1997, pp. 269–274.
54. Jonai, H., Villanueva, M. B. G., Takata, A., Sotoyama, M., and Saito, S., Effects of the liquid crystal display tilt angle of a notebook computer on posture, muscle activities and somatic complaints, *IJIE* 29, 219–229, 2002.
55. Berkhout, A. L., Hendriksson-Larsen, K., and Bongers, P., The effect of using a laptopstation compared to using a standard laptop PC on the cervical spine torque, perceived strain and productivity, *Appl Ergon* 35 (2), 147–152, 2004.

56. Galinsky, T. L., Swanson, N. G., Sauter, S. L., Hurrell, J. J., and Schleifer, L. M., A field study of supplementary rest breaks for data-entry operators, *Ergonomics* 43 (5), 622–638, 2000.
57. Fraser, M., Ergonomics for grade school students using laptop computers, in *Proceedings of the XVI Annual International Occupational Ergonomics and Safety Conference*, Toronto, 2002.
58. Saito, S., Piccoli, B., Smith, M. J., Sotoyama, M., Sweitzer, G., Villanueva, M. B., and Yoshitake, R., Ergonomic guidelines for using notebook personal computers, *Ind Health* 38, 421–434, 2000.
59. CDC. Computer workstation ergonomics, http://www.cdc.gov/od/ohs/Ergonomics/compergo.htm, Accessed: 2004, Last update: 11 August 2000.
60. Harvey, R. and Peper, E., Surface electromyography and mouse use position, *Ergonomics* 40(8), 781–789, 1997.

41

Meatpacking Operations

Steven L. Johnson
University of Arkansas

41.1 Distinctive Characteristics of Meat Processing Operations

The objective of this chapter is to address the ergonomics issues that are associated with production of a particular consumer product, meat (beef, pork, chicken, and turkey). The processing of these products has some specific characteristics that have ergonomics implications that are different than those experienced by other industrial environments. For example, in general manufacturing (i.e., fabrication and assembly), if there is a problem in the production operations, the process can be stopped with relatively few implications, other than a loss of production. The incoming materials may tend to pile up, but they do not tend to deteriorate over a relatively short time. In addition, the work-in-process can accumulate, which can cause a temporary problem with available space.

Production disruptions in meat processing, on the other hand, can have catastrophic affects. The incoming materials are live animals that can decline rapidly, particularly in the cold of February or the heat of August. The supply chain in meat processing is very integrated and not tolerant of disruption. In that product temperature and sanitation are critical, even a temporary buildup of product in the middle of the production system can have serious implications. Similarly, the handling of the output of the process is often critical in that there can be serious health consequences. Poor quality does not only result in customers being dissatisfied, it can also result in illness, or worse.

Other inherent characteristics of meat processing tasks also have important ergonomics implications. For example, in beef processing both the product handled and the tools required are often very heavy, awkward, and sometimes slippery. In poultry operations, the totes of product manually handled can be over 75 pounds. The birds can pass by an operator in a chicken evisceration line at a rate of 91 birds per minute. Another example of a high-speed task is when the final product is packed in cases as they exit a blast freezer.

During the last 20 yr, as consumer preferences have shifted from packages of hamburger or whole chickens to ready-to-microwave, cooked product, the nature of meat processing has changed significantly. Although much of the attention on to the industry has focused on the slaughter and cut-up activities, further processing and packaging operations also pose serious ergonomics issues.

The meat packing industry has been one of the most regulated industries in the United States. The federal regulatory organizations responsible for providing consumer protection associated with meat processing are the Department of Health and Human Service, Food and Drug Administration, U.S. Department of Agriculture (USDA), Food Safety and Inspection Service, the Animal and Plant Health Inspection Service, and the Environmental Protection Agency. The USDA regulates many aspects of the processing system, including ambient and product temperatures, line rates, personal protective equipment, staffing of certain lines, etc. The impact of these regulatory organizations, along with the need to comply with the Occupational Safety and Health Act (administered by OSHA) regulations, consume a much higher proportion of time and effort in meat processing operations than most other industries.

41.2 Indications of Ergonomics Issues in Meat Processing Operations

Both research and applications sources of information that accumulated during the 1980s indicated that the meat processing industry was experiencing disproportionately high levels of injuries and illnesses that had been associated with what became known as "ergonomics risk factors." In the early 1980s, it was noted that "repeated trauma" injuries for some employees in meat processing (butchers/meat cutters) were 75 times the rate of general industry. This, in combination with the attention directed toward the issue by organized labor resulted in a number of articles presented in the popular press that focused the public's attention on the industry.

Over time, the terms that have been used to characterize the task characteristics associated with these injuries have changed. They include: repetitive motion disorder, repetitive strain injury, cumulative trauma disorder. More recently, the terms musculoskeletal disorder (MSD) or work-related musculoskeletal disorder have been adopted. These terms have the advantage of focusing on a medical diagnosis and also tend to reduce the potential overemphasis on the single factor of repetition rate. Although the rapid repetition rate that occurs within beef, and particular poultry processing, can increase the impact of using poor biomechanics, the poor biomechanics (i.e., awkward wrist posture along with high grip forces) are generally the factors that need to be addressed. The risk factors for MSDs are discussed elsewhere in this handbook.

When the Occupational Safety and Health Act was initially passed in 1970, the meat products industry was designated by the Department of Labor to receive "priority attention" because it had the highest rate of occupational injuries. During the mid-1980s, there were a number of research results that supported the contention that meat processing operations involved tasks that could be stressful and have the potential to cause work-related injuries (Armstrong et al., 1982; Viikari, 1983; Pezaro, 1984; Roto and Kivi, 1984; Steib and Sun, 1984; Finkel, 1985).

The combination of public attention and research support led OSHA to focus on the industry and request the National Institute for Occupational Safety and Health (NIOSH) to conduct a number of

Health Hazard Evaluations in beef and poultry processing plants (NIOSH, 1987). The reports generated from these efforts were intended to aid the industry in addressing the safety and health issues; although they were also used in the context of enforcement when effective modifications were identified but not utilized. During this time period, both the government (U.S. Department of Health and Human Services, 1988) and industry (American Meat Institute (1988)) developed materials to address ergonomics issues in the meatpacking industry.

The U.S. Congress was also focusing on ergonomics in the House Subcommittee on Employment and Housing hearings of 1989. During this period, other countries were also focusing on work-related injuries and illnesses in meat processing operations that were associated with workplace design, work methods, tools, and equipment. For example, the Japan OSH Act was revised in 1990 to require safety officers in meat, poultry, and fish processing plants with more than 1000 employees to have university degrees in science plus a minimum of 5 yr of practical experience in occupational safety and health.

41.3 Early OSHA Enforcement Activities to Address Ergonomics Issues

During the 1980s, OSHA used the General Duty Clause (Section 5 (a)(1) of the OSH Act) as the enforcement mechanism to cite companies that had high incidences of work-related musculoskeletal disorders. Section 5(a)(1) of the Occupational Safety and Health Act requires an employer to:

> . . . furnish employment and a place of employment which are free from recognized hazards that are causing or are likely to cause death or serious physical harm to his employees . . .

The first use of the General Duty Clause to address this type of injury was a citation of the Kodak facility in Colorado in 1978 (Dickerson, 1978). That case was the first time that OSHA employed the terms "excessive ergonomics stress" and "excessive biomechanical stress" in a citation. Subsequently, it was used as the basis of fines for a number of companies (John Morrell and Company, ConAgra, IBP, Pepperidge Farms, and Tyson Foods). The fines ranged from 1 million up to 4.3 million dollars, although many of the fines were subsequently reduced when the companies agreed to institute an abatement program to address the ergonomics issues. The final fines were each still in the range of $200,000 to $400,000. The basis for these citations was the growing research literature that related certain task characteristics to occupational injuries (Eastman Kodak, 1978).

41.4 Cooperative Efforts to Provide Information to Reduce Ergonomics Hazards

In 1988, OSHA formed a task force to investigate "repetitive trauma disorders" in meatpacking (i.e., beef and pork processing). This effort included a coordinated effort by OSHA, industry representatives (American Meat Institute), and organized labor (United Food and Commercial Workers). The objective of this effort was to develop a "guideline" document that could be used by operational personnel to identify and document problematic tasks to implement and evaluate modifications intended to reduce the risk of injuries and illnesses.

The result of this effort was the *Ergonomics Program Management Guidelines for Meatpacking Plants* (OSHA, 1990). The rational for the guideline was stated as:

> In recent years, there has been a significant increase in the reporting of cumulative trauma disorders (CTDs) and other work-related disorders due to ergonomics hazards . . . CTDs represent nearly half of the occupational illnesses reported in the annual Bureau of Labor Statistics (BLS) . . . CTDs are particularly prevalent in the meatpacking industry.
>
> OSHA is therefore providing information and guidance on ergonomics program management to assist employers in meeting their responsibilities under the OSH Act.

Although the guidelines specifically addressed meatpacking operations (SIC code 2011), they have also been used effectively in industries with a wide variety of products.

According to what has become known as the "meatpacking guidelines," there are four requirements of an effective ergonomics program. The first is a *commitment by top management*. The primary aspect of this component is that the upper management must assign and communicate the responsibilities associated with the ergonomics program and also provide the resources necessary to ensure successful implementation of the program. The second requirement is that the program be detailed in a *written document* that can be communicated to all personnel within the organization. The documentation must include the implementation dates for each of the elements of the program. The third requirement is to include *employee involvement* in the program. The involvement of employees who perform the tasks and may experience the risk factors must be actively encouraged to participate in both the development and implementation of the program. The fourth requirement involves *regular program review and evaluation*. This review includes the continuous analyses of injury and illness trends, along with other evaluation methods, to establish the effectiveness of both the program, in general, and specific interventions, in particular. The guidelines emphasize that each of these requirements must be incorporated in an ergonomics management program for the effort to be effective at reducing the risk of work-related injuries and illnesses in the meatpacking industry.

41.5 Elements of the *Ergonomics Program Management Guidelines for Meatpacking Plants*

The first element of the guideline is a *Worksite Analysis* to identify and correct what OSHA referred to as "ergonomics hazards." In some ways, the term "ergonomics hazard" is unfortunate in that ergonomics is the solution to the problem, not an adjective that characterizes the problem. The phrase "biomechanical hazard" is probably more descriptive and less controversial. One source of information related to tasks that have a disproportionate number of injuries or illnesses in a workplace are the occupational safety and health records (medical claims, workers compensation files, OSHA Form 300, etc.). Ergonomics *checklists* are recommended to identify the "ergonomics risk factors." The guidelines focus on the factors that had been considered at that time to be associated with cumulative trauma disorders of the upper extremities (Putz-Anderson, 1988):

- Repetitive or prolonged activities
- Forceful exertions
- Prolonged static postures
- Awkward postures of the upper body
- Continued physical contact with work surfaces
- Excessive vibration from power tools
- Cold temperatures
- Inappropriate or inadequate hand tools

A *job hazard analysis* is recommended for the tasks that are determined to present risks to the operators as judged by the checklists or archival safety data. The tasks are documented with respect to each of the risk factors. Another tool that is suggested to be used in conjunction with the worksite analysis is a *symptom survey*. The survey is administered to assess whether the operators are experiencing any physical problems (aches, cramping, pain, stiffness, etc.) related to their jobs. Again, the term symptom survey is probably unfortunate from management's perspective. Many organizations have found that using the term "job improvement form" or "ergonomics survey" for the same content is less controversial and actually leads to more useful responses. By periodically administering the survey, problems can be detected and addressed before they become more serious and expensive for both the employee and the employer.

The second element of the guidelines is *Hazard Prevention and Control*. This phase of the program addresses modifications that have the potential to reduce or eliminate the risks that were identified by the previous analyses. The techniques recommended include engineering controls (e.g., equipment or tool redesign), work-practice controls (e.g., employee conditioning, use of correct biomechanics), personal protective equipment (e.g., correct fitting gloves), and administrative controls (e.g., rest breaks, job rotation).

Medical Management is the third element of the guideline. This phase of the program addresses the identification, evaluation, and treatment based on signs and symptoms of cumulative trauma disorders. The guidelines provide an "Upper Extremity Cumulative Trauma Disorder Algorithm" (i.e., flow chart) depicting the processes and decision points to be used by the health-care providers when dealing with the disorders. The medical management portion of the guideline also suggests the use of a symptom survey to allow the employees to report the location, frequency, and duration of discomfort associated with their job. The guide provides an example of a Symptom Survey Checklist that includes the question: "Have you had any pain or discomfort during the last year?" Although employee discomfort surveys can provide valuable information, the managements of many organizations were justifiably concerned that the administration of the survey would result in a large number of "false alarms." Experience has indicated that this can occur after the first administration of the survey; however, the transient "spike" does not persist. It is the trend information obtained as subsequent surveys are conducted and the results are compared that has been shown to be of most value.

The fourth element of the guidelines involve *Training and Education*. The recommended training involves somewhat different content for different stakeholders within the organization. It suggests that all affected employees be provided with training that was both job specific (i.e., correct methods of performing their jobs) and information that assists in early detection and identification of disorders. The training for engineers and maintenance personnel includes the characteristics of tasks that can increase the risk of disorders and illustrate how appropriately designed and maintained systems can reduce that risk. The guidelines also recommend that the supervisors and top management receive training on ergonomics and the ergonomics program for early detection and control of cumulative trauma disorders.

41.6 Effectiveness of the Meatpacking Guidelines

The guidelines became the basis for the implementation of ergonomics programs by many organizations within the meatpacking industry, as well as other very diverse industries (from auto parts to cosmetics). Although ergonomics had been an integral part of the occupational safety and health function within many companies, for the first time companies were using the term "ergonomics program" as a distinct entity. An important early observation was that the high costs that were anticipated to result from the implementation of the guidelines by its critics did not materialize. Although some companies found that the reported incidence of MSDs initially increased, the severity and total costs incurred were actually lower. The voluntary guidelines provide the flexibility that allows companies to use their existing safety policies and procedures, while adding those recommended by the guidelines when they are deemed appropriate.

The one area that can require an increase in resources is the documentation of the program. For example, many organizations have found that it is beneficial to maintain an "Ergonomics Handbook" that includes summaries of results from the application of the analysis tools (job-site analyses, employee ergonomics surveys, etc.), the modifications designed to address the issues, and a specific schedule indicating when the modifications would be completed. This "Handbook" significantly increases the communication among different functions within the organization (safety, medical, engineering, maintenance, etc.).

Although the guidelines are not a standard, they have been used by federal and state regulatory agencies with the result that a number of *Settlement Agreements* have been negotiated with companies to address ergonomics issues. The agreements generally have led to companies agreeing to initiate abatement programs that are based on the meatpacking guidelines. Many companies have also used the guidelines to set up ergonomics programs independent of regulatory agencies or as a preemptive measure.

41.7 Other Standards Developments That Affected the Meatpacking Industry

During the 1990s, there were also two other ergonomics standards development efforts. OSHA initiated the development of an Ergonomics Program Standard (OSHA, 2000). The promulgation of the standard was politically volatile with organized labor strongly supporting the development of the standard, while business groups actively opposing it (Scalia, 1994). Large sums of money were spent on lobbying efforts by both advocates and critics of the standard. The standard became effective in November 2001; however, Congress rescinded the regulatory rule in March 2002.

Beginning in 1990, the National Safety Council (2000) formed the "Standards Committee on Control of Cumulative Trauma Disorders (ASC Z-365)" that was accredited by the American National Standards Institute (ANSI). The objective of the effort was to develop a voluntary standard through a consensus among the stakeholders from business, labor, academia, and professional societies. To accommodate the change in terminology, the standard was subsequently retitled *Management of Work-Related Musculoskeletal Disorders* (National Safety Council, 2002). This title is more descriptive of the issues addressed than the OSHA standard. After 13 yr in development, the National Safety Council withdrew and the development was discontinued in 2003.

Although the general approach taken in the ANSI standard was similar to the meatpacking guidelines, a notable difference between the ANSI Z-365 conclusions and both the meatpacking guidelines and OSHA standards development was the categorization of "risk factors." The ANSI development divided the task related factors into two groups, risk factors and exposure properties:

Risk Factors	Exposure Properties
Force and contact stress	Magnitude
Posture and motions	Repetition
Vibration	Duration
Cold temperatures	Recovery time

The removal of repetition from the list of risk factors and discussing it as an exposure property is important to industries such as poultry processing due to the high rates experienced. The point of the change was to indicate that the repetition rate alone is not a risk factor; however, the repeated use of poor biomechanics in combination with grip forces can be problematic (depending upon the duration and recovery provided).

41.8 Effective Implementation of Ergonomics in Meatpacking

Many companies were independently implementing ergonomics programs that have been found to reduce occupational injuries (GAO, 1997). Rather than simply being concerned with being in regulatory compliance, many of these programs focused on the benefits of ergonomics programs for increasing productivity and product quality, as well as improving the quality of working life for their employees (Henderson and Cernohous, 1994; Jones, 1997; Moore and Garg, 1997a, b; Konz and Johnson, 2004).

Automation has also made significant advances in meat processing. For example, the "Advanced Meat Recovery" reduced the need for heavy, vibrating knives in beef and pork processing. This system is credited with a reduction of 38% in carpal tunnel syndrome, although during its implication other interventions were also occurring. In poultry processing, an extensive effort to automate the deboning tasks that are associated with MSD risk factors have met with mixed success. Although the automated equipment reduces the risk factors, the yield has often been significantly below manual deboning.

Industry organizations such as the American Meat Institute, National Chicken Council, and National Turkey Federation have developed templates for ergonomics programs that focus on:

- Education
- Early intervention of symptoms and diagnostic action
- Job rotation
- Engineering innovations/automation
- Employee participation and communication

State occupational safety and health agencies have also provided information. For example, in California, the *Ergonomics in Action: A Guide to Best Practices for the Food Processing Industry* (2003) was published by California Department of Industrial Relations/OSHA and Washington State (SHARP) published the *Healthy Workplaces: Successful Strategies in the Food Processing Industry* (Connon et al., 2001).

The federal OSHA has also formed "strategic partnerships" and "national ergonomics alliances" with individual companies and industries to work cooperatively with industry to address ergonomics issues. Part of this effort was the formation of the National Advisory Committee on Ergonomics to advise the Assistant Secretary of Labor for Occupational Safety and Health on ergonomics guidelines, research, and outreach assistance. The 15 members of the committee include the stakeholders (e.g., labor, industry, insurance companies, consultants, and academia).

41.9   Ergonomics Guidelines for Poultry Processing

Although there have been notable improvements with respect to reducing ergonomics hazards in the meatpacking industry, there are still opportunities. For example, in 2001, over 30% of the lost days in the poultry industry were related to MSDs. Poultry processing still experiences 1.5 times the incidence rate of manufacturing.

After the successful attempt to develop an ergonomics standard, the federal OSHA has taken the approach of developing targeted "guidelines" that are industry specific. In 2003, OSHA issued the "*Guidelines for Nursing Homes: Ergonomics for the Prevention of Musculoskeletal Disorders.*" They are currently developing an "*Ergonomics Guidelines for Poultry Processing.*" A draft of the guidelines was published in June 2003. The guidelines address muscle strains and back injuries that are considered to be work related if an event or exposure in the work environment either causes or contributes to the MSD, or significantly aggravated a preexisting MSD (specifically carpal tunnel syndrome, tendinitis, rotator cuff injuries, epicondylitis, trigger finger, muscle strains, and back injuries). The guidelines address four specific "ergonomics concerns": repetition, force, awkward or static postures, and vibration.

There are seven sections to the guidelines. The sections closely parallel the 1990 meatpacking guidelines. The first section addresses *Management Support* that is illustrated by providing adequate resources and establishing accountability of all individuals within the company. The second section discusses the *Employee Involvement* that results in prompt and accurate reporting of MSDs. It is encouraged that employee groups be formed to identify problems, analyze tasks, and recommend solutions. Employees at all levels should also be included in the design of the work practices, equipment, procedures, and training. *Training* is the third facet of the guidelines. As with the meatpacking guidelines, the training should involve the different functions (employees at risk, supervisors, engineers and maintenance, and healthcare providers). The content of the training for each function is consistent with the meatpacking guidelines and have previously been discussed in this chapter. The next section of the guidelines addresses the means of *Identifying Problems* that are associated with ergonomics risk factors. As with the meatpacking guidelines, this is accomplished through a combination of archival safety data, ergonomics surveys, and job hazard analysis tools. After the problem areas have been identified, the organization *Implements Solutions* such as modifying the workplace design, work methods, tools, and equipment. These solutions, along with a timetable for their accomplishment, are documented and their effectiveness evaluated. An important part of the guidelines is to establish a process for *Addressing Reports of Injuries*. In the jargon of the meatpacking guidelines, this is the medical management activity. The objective is to facilitate early reporting, diagnosis, and medical treatment to reduce the severity of injuries. The guidelines also require the facilities to *Evaluate the Ergonomics Efforts* by observing the injury and illness trends and reviewing the ergonomics surveys administered to the employees. The objective is to capitalize on the positive results and recognize when the solutions are not effective and must be modified.

About half of the ergonomics guidelines document involves examples of potential problems experienced in poultry operations and potential solutions that address those problems. The solutions are shown with illustrations (e.g., a cutout that allows the worker to get closer to the work), along with a discussion of when they have been found to be effective (when the horizontal reach is high or the operator must lean to reach an item). In addition, each example is accompanied with "Points to Remember" that address other associated solutions that might be beneficial.

As stated in the Executive Summary of the guideline, these guidelines provide recommendations for poultry processing facilities to reduce the number and severity of work-related MSDs. The guidelines are an attempt to illustrate both the effectiveness and the relative ease of implementing simple solutions to reduce operator effort, fatigue, and potential disorders.

41.10   Summary

Meatpacking was one of the first industries to receive extensive attention with respect to occupational safety and health issues related to ergonomics. There has been a significant amount of work applied to these issues from the industry, labor, and governmental agencies. In many ways, the inherent characteristics of many tasks in meatpacking operations (e.g., beef, pork, chicken, and turkey) illustrate how the design of workplaces, work methods, tools, and equipment can have important effects on productivity and product quality, as well as worker safety and quality of working life. Although there are still challenges to be resolved in the industry, there have been significant improvements that have been the result of the effective implementation of ergonomics programs.

References

American Meat Institute. *Management Strategies for Preventing Strains and Sprains: A Guide to Practical Ergonomics.* Washington, DC, 1988.

Armstrong, T.J., Foulke, J.A., Joseph, B.S., and Goldstein, S.A. (1982). Investigation of cumulative trauma disorders in a poultry processing plant. *American Industrial Hygiene Association Journal*, **43**: 103.

California Department of Industrial Relations. *Ergonomics in Action: A Guide to Best Practices for the Food-Processing Industries*, 2003.

Connon, C., Curwick, C., and Whittaker, S. *Healthy Workplaces: Successful Strategies in the Food Processing Industry.* Washington Department of Labor and Industries SHARP Program, June 2001.

Dickerson, O.B. *Occurrence and Cause of Tenosynovitis and Related Diseases.* Eastman Kodak Company, Windsor, CO, March 1978 (report of a NIOSH evaluation of the plant).

Eastman Kodak Company. Secretary of Labor v. Eastman Kodak Company OSHRC Docket No. 78-1518, 1978.

Finkel, M.L. (1985). The effects of repeated mechanical trauma in the meat industry. *American Journal of Industrial Medicine*, **8**: 375.

General Accounting Office. Worker Protection: Private Sector Ergonomics Programs Yield Positive Results. Publication No. GAO/HEHS-97-163. Washington, DC, 1997.

Henderson, C.J. and Cernohous, C. Ergonomics: a business approach. *Professional Safety*, January, 1994, 27–31.

Jones, R.J. (1997). Corporate Ergonomics programs of a large poultry processor. *American Industrial Hygiene Association Journal*, **58**(2): 132.

Konz, S. and Johnson, S. Managing an ergonomics program. Chapter 22 in *Work Design: Occupational Ergonomics*, HolComb-Hathaway, Scottsdale, AZ, 2004.

Moore, J.S. and Garg, A. (1997a). Participatory ergonomics in a red meat packing plant, Part I: evidence of long-term effectiveness. *AIHA Journal*, **58**: 127.

Moore, J.S. and Garg, A. (1997b). Participatory ergonomics in a red meat packing plant, Part II: Case studies. *AIHA Journal*, **58**: 498.

National Safety Council ASC Z-365. *Management of Work-related Muculoskeletal Disorders*. Accredited Standards Committee. National Safety Council, Secretariat, Itasca, IL, 2000.

National Safety Council. *Management of Work-Related Musculoskeletal Disorders Accredited Standards Committee Z-365* (Working Draft), August 2002.

National Institute for Occupational Safety and Health. *Health Hazard Evaluation Report for Longmont Turkey Processors*, HETA 86-505-1885, 1987.

Occupational Safety and Health Administration, U.S. Department of Labor. *Ergonomics Program Management Guidelines for Meatpacking Plants*, OSHA Report No. 3123, 1990.

Occupational Safety and Health Administration. *Final Ergonomics Program Standard*. 29 CFR Part 1910. November 2000.

Pezaro, A. *Critical Review Analysis for Injury Related Research in the Meatpacking Industry (SIC 2011)*. NIOSH Report 0014440, US Department of Health and Human Services, National Institute for Occupational Safety and Health, Cincinnati, 1984.

Putz-Anderson, V. (ed.). *Cumulative Trauma Disorders: A Manual for Musculoskeletal Diseases of the Upper Limbs*. Taylor & Francis, Bristol, PA, 1988.

Roto, P. and Kivi, P. (1984). Prevalence of epicondylitis and tenosynovitis among meat cutters. *Scandinavian Journal of Work, Environment and Health*, **10**: 203.

Scalia, E. *Ergonomics: OSHA's Strange Campaign to Run American Business*. National Legal Center for the Public Interest, Washington, DC, 1994.

Steib, E.W. and Sun, S.F. (1984). Distal ulnar neuropathy in meat packers: an occupational disease. *Journal of Occupational Medicine*, **26**: 842.

U.S. Department of Health and Human Services. *Safety and Health Guide for the Meatpacking Industry*. OSHA Publication No. 3108. Washington, DC, 1988.

Viikari, J.E. (1983). Neck and upper limb disorders among slaughterhouse workers. An epidemiologic and clinical study. *Scandinavian Journal of Work Environment & Health*, **9**: 283.

42

Poultry Processing

Carol Stuart-Buttle
Stuart-Buttle Ergonomics

42.1   Introduction

There are several Best Practice Guidelines that apply to the poultry industry (OSHA, 1990, 2004; National Broiler Council, 1990; Cal-OSHA, 2003). This chapter summarizes the main elements of best practices and expands on selective points based on this author's experience working with the poultry industry. The most effective use of this chapter is to consider it as a complement to current guidelines. The Occupational Safety and Health Administration (OSHA) issued the most recent guideline, entitled "Guidelines for Poultry Processing" (OSHA 2004). There is also a Poultry Processing Industry eTool at OSHA's website (OSHA, 2001).

Recently, there has been considerable press about the meat and poultry industries. In January 2005, the Human Rights Watch issued a report that expressed concern about safety and health in the meat and poultry industries, described the work as very hazardous and called for changes in Federal and State laws (HRW, 2005). In the same month the Government Accountability Office (GAO) issued its final report of an investigation into the safety of these industries as well. The conclusion of the GAO report was "safety in the meat and poultry industry while improving, could be further strengthened" (GAO, 2005).

Also in January 2005, the trade associations The National Chicken Council and The National Turkey Federation signed an alliance with OSHA (OSHA, 2005). They agreed to collaborate with developing training and education programs for the poultry industry on ergonomics techniques, program structure and applications, and to help communicate such information to employers and employees in the industry. Such materials will be another useful resource for those working in the poultry industry.

Both OSHA and the GAO discuss the great reduction in occupational injuries and illnesses in the poultry industry over the last 10 years (OSHA, 2004; GAO, 2005). In 1992, the total recordable cases for injuries and illnesses in the poultry industry was an incident rate of 23.2 (per 100 full-time workers). By 2001 the incident rate was 12.7, almost half that of 1992 (BLS, 2002). For the poultry industry in 2002, 30% of cases with days away from work were musculoskeletal disorders (OSHA, 2004).

Poultry processing tasks entail repetition, force, awkward and often sustained postures, contact stress, as well as exposure to vibration and cold temperatures. Lifting tasks are common and can be heavy when the birds are large, such as turkeys. Sharp tools are used and lacerations can be as big a problem as musculoskeletal disorders. Training that includes making sure employees pay attention when using knives and designing the workplace to reduce fatigue helps reduce laceration incidents.

Applying the science of ergonomics reduces injuries and illnesses and improves worker performance, which enhances productivity. In the poultry industry there is particular sensitivity to yield and yield is intimately connected to worker performance. A worker's performance is affected by the workstation design, equipment and tools, work methods, and the capability of the worker. If a worker is not given a sufficient window of time in which to perform the task then the outcome, often yield, will suffer and or the worker will get injured. The recent GAO report raises the issue of line speed and the need to study the relationship between speed and injuries (GAO, 2005). Currently, a maximum line speed is determined by the United States Department of Agriculture (USDA) and is related to the ability of USDA inspectors to inspect poultry during the process.

This chapter discusses the main elements of an ergonomics process with emphasis on the aspects that make a crucial difference to reducing injuries in poultry plants. The segment on implementing solutions includes administrative controls, engineering controls and the common hand tools used in the industry.

42.2 Ergonomics Process

Ergonomics principles alone cannot be effectively applied without the context of a process. An ergonomics process establishes a support structure by which to identify ergonomics problems, address them, and ensure effective improvements. These basic steps need to be accomplished concurrently with good medical management of medical cases. Training is a critical component of the process, not only in good task technique and skill but also for employees to understand how they can contribute to reduce injuries and improve the workplace. Part of effective training is gradual conditioning to perform the job. The elements of an ergonomics process are well known but they are not always given the attention that is recommended. In the poultry industry, success of ergonomics is especially dependent on a thorough and integrated process into safety and health and production processes.

There are many resources for establishing and managing an ergonomics process. Those in the poultry industry have long used the "Meat-Packing Guidelines" (as commonly called) although originally issued for the red-meat industry (OSHA, 1990). Also in the same year The National Broiler Council published The MET (Medical, Ergonomics and Training) Program (NBC, 1990). The National Institute for Occupational Safety and Health (NIOSH) issued a guidance document on elements of ergonomics programs that includes checklists and forms and other practical materials to meet a program's need (NIOSH, 1997). Ergonomics processes for large and small companies are discussed in other chapters of this handbook. The OSHA Guideline for Poultry Processing discusses program elements specifically as they pertain to the industry (OSHA, 2004).

There are three main areas in an ergonomics process, the first being, setting the stage for implementing ergonomics. This requires management support, involving the employees, which could be by establishing a team, and training everyone in basic ergonomics awareness. More in-depth training is necessary for those directly responsible and involved with the process. The second area needed in an ergonomics process is monitoring the workplace through a surveillance system. This may be connected to medical management as statistics are kept on those reporting injuries and illnesses. Risk management data such as turnover and absenteeism help to keep a barometer on the "health" of the workplace. Productivity data also indicate how well workers are performing and whether there may be some problems. The third area is an improvement process within the ergonomics process that looks at the tasks and includes, to use California OSHA terms, assessing, planning, doing, and verifying (Cal-OSHA, 2003).

The following elements of an ergonomics process are discussed in the following text: management; ergonomics team; training; monitoring the workplace; and the improvement process. Specifics relating to the poultry industry are given.

42.2.1 Management

When live product is being manufactured there is a strong production focus as the product cannot be left unprocessed. Such a focus has a tendency to override the balance that leads to the most efficient processing. Historically, a production mind-set has often gone hand in hand with not taking responsibility for safety and health. Safety and health responsibility has been laid at the feet of the safety and health personnel. The following suggestions are ways to integrate ergonomics into the production process:

- Integrate safety and health into production as this leads to the most effective ergonomics process. Managers should recognize how worker performance affects productivity.
- Build in accountability for safety and health into supervisors' and production managers' performance measures.
- Involve production managers in accident and near miss investigations.
- View safety and health personnel as a resource for production.

42.2.2 Ergonomics Team

Consider involving personnel with specific areas of expertise. Some suggestions follow on how to engage some of the different team members.

- *Employee involvement.* Employee involvement is essential and includes all production workers. They can contribute in many ways including helping to identify problems, making suggestions for improvement and providing feedback on interventions. The more opportunity for comments from the production workers the better. This can be during audits, through a suggestion system and when analyzing the task. To promote communication translate forms into any other languages used by personnel in the plant. In addition, if there is a predominance of another language used by employees, use bilingual personnel as much as possible when conducting surveys.
- *Maintenance and engineering department.* The department's work orders are usually dominated by production changes. Retrofit modifications are often viewed as irksome, especially when there are work volume pressures. A member of the maintenance department should be on the ergonomics task force or team. Consider involving the department more personally by addressing their specific ergonomics concerns as this can help them understand the production workers' perspectives. Ensure the maintenance and engineering personnel are versed in the tradeoffs that are necessary in design and the reiterative improvement process that sometimes occurs.
- *Medical provider.* The medical provider is an important team member. An in-house staff person can often shed light on the context of an injury or illness. Medical staff should provide close follow-up for workers returning to work and the staff should oversee the workers' progress through modified jobs as the workers improve and as medically permitted. Avoid keeping a worker on a single modified job. Attempt to rotate them through other suitable positions so that there is physical variety and as much retention of job skills as possible. Develop close relationships with external providers and encourage them to grow familiar with the jobs in the plant so that they can help with effective return-to-work.
- *The purchasing department.* Those responsible for purchasing can keep an eye on the market for improved products that better meet ergonomics criteria. This may include items such as gloves and tools. Those responsible for purchasing materials and equipment need to understand the

importance of specific features versus price. For example, a good edge on a knife may not be feasible with the cheapest model.

- *Quality personnel.* The quality department is responsible for testing product, for example, for contamination, as well as measuring yield. Those in the department can be on the alert for the relationship between changes in yield and human performance. In addition, quality personnel can conduct yield measures for pre- and postassessments of an ergonomics intervention.
- *Knife room employees.* Employees in the knife room should be included in the ergonomics process. They need to understand the importance of their role in maintaining and monitoring the sharpness and condition of the tools. Usually a knife room employee will be the most skilled to train production employees in knife-sharpening skills. The quality of the skill of knife sharpening is important as if it is not performed well, the knife can be made more blunt. Knife room personnel are often the best to keep an eye on the market for improved knives and related products.

42.2.3 Training

Everyone should have some level of ergonomics training. Each group should understand their role and responsibilities in the overall ergonomics process. An awareness level of training is appropriate for production workers. In addition to awareness level of information, supervisors should understand the importance of administrative controls such as rotation, and how to manage the return-to-work process. Managers need to be aware of the bigger role of ergonomics in the integration of safety and health and the relationship with performance, efficiency, and productivity. Those who are involved with analyzing jobs and developing improvements such as the ergonomics taskforce or team, require more specific training on methods of analysis and generating potential solutions. The engineering and maintenance personnel should have a more design-oriented ergonomics training that helps them understand how to improve the workplace and the design trade-offs that need to be made. Someone (or group) in the plant should have enough knowledge to be a relative expert for the plant, to know when outside help should be sought and where to find such assistance.

The combination of dedicated trainers providing technique training and ensuring employees are ramped up to a full production load has improved retention of employees as well as reduced injuries and illnesses of new employees. The plants that have invested in such training systems have kept them because of the overall cost effectiveness. Current employees also benefit as Trainers provide formal cross-training. Many injuries are averted when employees are prevented from performing jobs at which they are not qualified or on which they have not kept up their skill.

- *Trainers.* Some plants have relied upon a "buddy" system as the method of technique training. This does not ensure quality of training and often over time poor approaches are passed on. To maintain a standard and continually promote cross-training, a dedicated group of trainers is most effective. This ensures that there is always someone to conduct training and that they have not been pulled into the main production crew. Trainers can fulfill other functions that promote retention. They are on hand to answer questions, guide new employees, provide feedback and encouragement, and assist with ensuring that new employees take adequate breaks from the floor when they first start out so that they do not get injured. At times, a new hire may not be able to meet the skill level. A formal training program ensures such employees do not stay in tasks that they cannot manage and in which they are likely to get injured if they did remain, or else they may quit, neither outcome being desirable. If a plant has employees with a first language other than English, use trainers who are bilingual.
- *Technique training.* Many of the production tasks, especially those using knives, require skill. Adequate job-specific training is necessary to avoid poor technique that provokes increased forces on the body. Both new employees and those being cross-trained need to be given time to learn the skill before holding a full production load on that specific task. In addition to skill

development, the employees should be aware of appropriate techniques from an ergonomics viewpoint. Knife care and knife sharpening training is essential for all users of a sharp tool. Higher forces are transmitted through the hands as a blade grows dull so employees need to be competent at touch up sharpening of their tools or know when to change a blade. Good care and treatment of tools is important such as not stabbing the end of a knife into a table to momentarily let go of it.

- *Buddy system.* A buddy system is an important complement to trainers. Providing a mentor or buddy for a new employee helps welcome them into a plant and answer general questions including finding their way around. In addition, as employees ramp up to full production speed they will be working along side a buddy who has volunteered to help the new person along. The new employee can be given guidance on general safety and other protocols of the plant.
- *Conditioning.* New employees must be gradually ramped up to tolerate full production rates. They need to develop skills, condition their bodies, and train their muscles. Mixing orientation sessions in a training room with time on the production floor is a way to gradually expose employees to production. Training lines that go slower provide opportunity to learn the task steps. Performing on the line with a buddy so a full load is not held helps with the transition to full production. Trainers can oversee and document the progression and qualify an employee on specific tasks. Initial ramping in of a new employee on several tasks should take 3–4 weeks working up gradually from a 50% rate by the end of the first week to 100% production rate on a high skilled task at the end of training. Anyone cross-training should also ramp up for the same reasons although the length of time may be a little shorter as they have some experience. Those who have been absent for 2 weeks or more are likely to be slightly deconditioned and need to retune their muscles to the task by ramping in too. This is especially important for older workers.
- *Standard Operating Procedures.* Standard operating procedures (SOPs) or task methods are needed to ensure consistent training. Such descriptions can be developed to serve many purposes including, a job description that can be used by medical personnel, a basis for job analysis, and a description for hiring purposes. Such descriptions can be developed and maintained by a training department. Incorporate into the descriptions ergonomic principles pertinent to the task, such as specific task techniques that are beneficial from an ergonomics perspective.

42.2.4 Monitoring the Workplace

Monitoring the workplace involves gathering information to determine if there are problems or potential problems in the workplace. Analysis of information such as recordable injuries and investigation reports should be on-going. Records of near misses provide valuable information. Developing a culture in a company that encourages reporting of near misses is very positive as the message is not punitive but one of making collaborative improvements. Analyzing near misses and implementing interventions to ensure they do not happen again is truly proactive.

Conducting surveys, employee interviews, and having suggestion systems are well-known approaches to gathering employee input to the process. Employees' perspectives are important to understand so communication can be effective. By asking what production employees perceive is the problem, why and how they would improve it often raises systems issues that may otherwise be over looked. Good communication helps with acceptance of changes and improvements. Provide timely feedback to those who participate in surveys and suggestion systems. Consider posting results and planned action for all to see so that the process is more transparent and shared and the outcomes of participation are evident.

42.2.5 Improvement Process

The improvement process is well depicted by Cal-OSHA in their best practices guide for the food-processing industry (Cal-OSHA, 2003). Figure 42.1 shows the summary ergonomic improvement process.

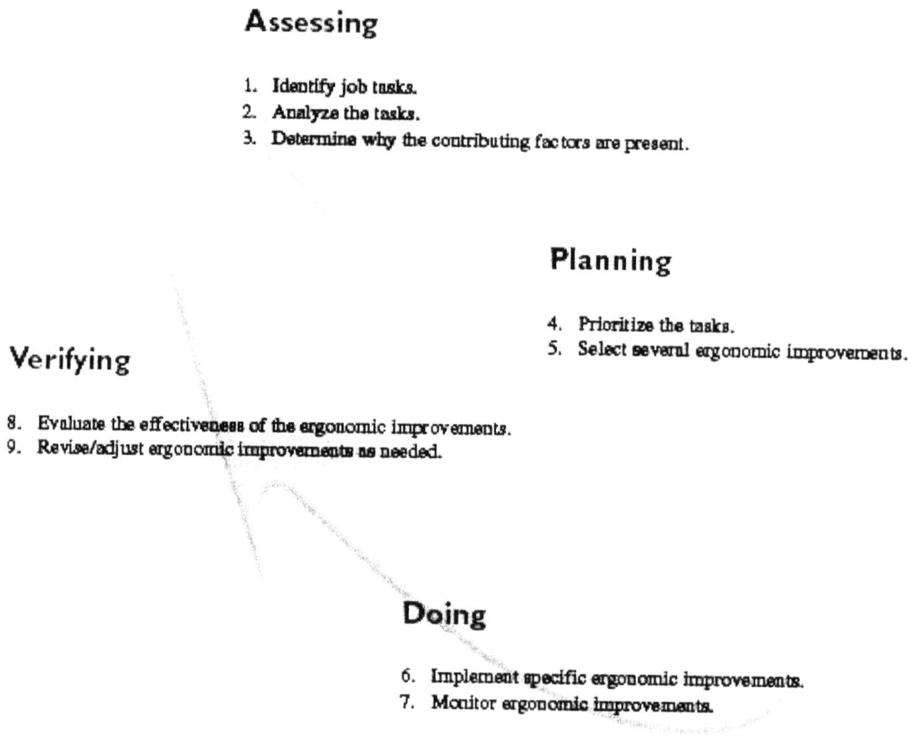

Assessing

1. Identify job tasks.
2. Analyze the tasks.
3. Determine why the contributing factors are present.

Planning

4. Prioritize the tasks.
5. Select several ergonomic improvements.

Verifying

8. Evaluate the effectiveness of the ergonomic improvements.
9. Revise/adjust ergonomic improvements as needed.

Doing

6. Implement specific ergonomic improvements.
7. Monitor ergonomic improvements.

FIGURE 42.1 Ergonomic improvement process. (From Cal-OSHA, *Ergonomics in Action: A Guide to Best Practices for the Food-Processing Industry*, 2003. With permission.)

Details of each of these steps are delineated and worksheets are also provided in the Cal-OSHA Best Practices. In addition to looking at the job steps, consider how the job is being done and whether it is physically awkward or fatiguing. Take into account the performance indicators such as the quality and error rates on the job, and system issues such as production bottlenecks. These are indicators of a problem job.

Evaluation of effectiveness of the ergonomic improvements (Step 8 under "Verifying" in Figure 42.1), should be reflected by an improvement in the measures that indicated a problem in the first place, such as improved quality or yield. A change in OSHA 300 logs of recordable injuries will take longer to manifest. However, a discomfort survey conducted before and after is a quick way to determine if there is improvement and also that no new problem has been created as the result of the change (Stuart-Buttle, 1994). Discomfort surveys can be combined with an opportunity for employees to comment on and rate the effectiveness of the changes.

42.3 Implementing Solutions

Interventions to eliminate or reduce a problem are often considered in two main categories: administrative and engineering solutions. Each group of solutions is discussed. As tools are important in the poultry industry, they are presented as a separate section.

42.3.1 Administrative Controls

The term administrative controls is used to refer to changes in the job design that can reduce risk exposure. For example, rotating employees between different jobs. Task technique controls (sometimes considered separate from administrative controls) are changing the way a job is performed so as to reduce risk. Many nonengineering factors can contribute to risk, such as poor fitting gloves, which are considered personal protective equipment (PPE). It is most convenient to collectively refer to nonengineering types of controls as administrative controls. The specific administrative interventions discussed below are: rotate employees; provide "floaters"; perform warm ups; fit gloves; and conduct preventive maintenance.

- *Rotate employees.* Rotating employees is an administrative practice that, if well managed, can provide job variety to offset fatigue and stress on one part of the body. In the poultry industry some departments find it difficult to find jobs that involve different muscle use and rotation can be limited by the extent an employee is cross-trained. However, there is benefit in rotating between similar jobs that have a different force profile or frequency, such as a cutting task rotating with an inspection task that only cuts occasionally. Such an exchange can still provide a break from the repetitive cutting task. Rotation from one side of a line to another is also important to ensure there is skill and the muscles are trained for working on either side. Rotation is important to maintain the skills that have been taught through cross-training. Injuries often occur when an employee is placed on a job that they are not used to doing, even if they were once trained for it. Cross-training helps to broaden the workforce and pool of workers that can perform a given job. When there are clusters of tasks that rotate the employees can control the rotation as long as a minimum amount of rotation is met. This encourages employee involvement.
- *Provide "Floaters".* A floater provides periodic breaks between scheduled breaks and must be fully cross-trained. Having a floater for this function ensures the line remains staffed and no one tries to hold the line for two people. Sometimes the floater can have a lead function as well, so long as this does not detract from the need to give people a break. Trainers should not be floaters as often they end up not training.
- *Perform warm ups.* Exercises for warming up the body by moving through a range of motion is common in many plants. Keeping the exercises viable and retaining the employees' enthusiasm is difficult. Often employee's skip out of doing them or do not perform the exercises in any beneficial way. Warming up the muscles and tuning into the skilled jobs that are performed at a demanding rate helps to get the body ready to respond to the demand on it and therefore reduces initial stress on the body. An alternative approach to traditional warm up exercises is to start the line slowly for the first 3 min. An advantage is everyone has to be involved and productive as they are on the line. The slower pace allows for tuning in the eye–hand coordination and muscles for the skills and bringing the blood to the muscles before there is a full demand on them. Encourage general range of motion exercises throughout the day to counter the effects of muscle fatigue. Overall fitness should always be encouraged.
- *Fit gloves.* Gloves are a primary PPE that should be improved as much as possible. For some tasks it is not unusual to wear three layers of gloves — one for warmth, one for protection against wet,

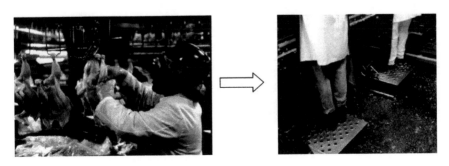

FIGURE 42.2 Work at appropriate height.

FIGURE 42.3   Raise the worker.

and a third as a metal mesh to protect against knife cuts. These layers must be as thin and light as possible, and a good fit to minimize strength deficits. Combine functions such as using a glove that is waterproof and warms the hand. The metal mesh style glove now comes in a lighter material that still gives cut protection. A heavy and too large a mesh glove can cause excess mesh at the fingertips which can get caught in machines as well as increasing the forces passing through the hand.

- *Conduct preventive maintenance.* Preventive maintenance (PM) is important on machinery and equipment. Apart from keeping the main machines running well, wheels on carts and dollies should not be forgotten. PM on wheels helps to keep pushing and pulling forces to a minimum. Tool maintenance is critical in the poultry industry where many cutting tools are used. Keeping a good edge on knives is also dependent on effective training of the employees on knife care and knife sharpening. See Section 42.3.3 for further discussion on types of tools used.

42.3.2   Engineering Controls

Engineering changes refer to improvements that are made to a workplace, equipment, or tool. However, it is very common that several solutions are implemented to resolve a problem and include both administrative and engineering changes. Several chapters in this handbook present general examples of

FIGURE 42.4   Raise the product.

FIGURE 42.5 Reduce reach.

administrative and engineering improvements to enhance ergonomics. Specific poultry industry examples of controlling musculoskeletal disorder risk and ergonomic improvements follow. Many tasks in poultry processing can be automated. The controls discussed below focus on improvements other than automation. Many design principles are reducing force on the body in some way. However, it can be useful to consider the principles separately rather than as one group of reducing force. Examples of the following principles are given: work at appropriate height; reduce reach; reduce sustained efforts; reduce repetition; provide space; minimize weight; control material handling; allow sit/stand; and control exposure to cold.

- *Work at appropriate height* (Figure 42.2). Working at the right height reduces awkward postures and reduces the forces that occur with such postures. The work height not only affects arm posture, but also the wrist angles and back position. The work level can be brought down to the right height or the person can be raised up to the work height.
  Raise the worker (Figure 42.3). The left picture shows the arms lifted too high. Platforms raise the worker to an appropriate work height so that the arms can be lowered. The platforms need to be light and easy to adjust.
  Raise the product (Figure 42.4). Product can be raised up to the worker so that the employees do not bend deeply, are less fatigued, and can work consistently over the shift.
  Reduce reach (Figure 42.5). There can be many incidents of extended reach including reaching over machines such as this gizzard harvester on the left. Collect product to be sent to a machine located away from the line as shown on the right, rather than reach over the machine. Note the convenient collection chutes.
  Divert product closer. Adjustable diverters can keep product close to the worker. Rails can be used to push shackles closer.
- *Reduce sustained efforts* (Figure 42.6). Muscular effort that is sustained is especially susceptible to fatigue. Avoid using the hands as a vise for holding and instead use supports. For example, balancers should be used for all air tools so the person does not have to hold the tool weight. Use

FIGURE 42.6 Reduce sustained efforts.

FIGURE 42.7   Use a frame to hold bags.

gravity as much as possible to reduce holding and use rails to help stabilize shackles rather using one hand while the other hand cuts. Automated bagging is now common but if it is still manual, sustained efforts can be minimized with using gravity and bag frames.

Reduce sustained efforts. The example on the left shows a sustained bent posture and holding of a bag with the left hand while filling with wings. The right hand example has a direct product feed to a small shelf for checking and the product is pushed into the bag that is held by the frame.

Use a frame to hold bags (Figure 42.7). This picture is an example of a multi-bag frame so the product can fall straight into the bag. Employees monitor the filling so that when one is full they rotate the frame to the next bag. An advantage of this system is the employees can perform another task during the filling.

• *Reduce repetition* (Figure 42.8). Repetition can be reduced by focusing on engineering out unnecessary handling or rehandling. For example, automatically orienting product eliminates the person turning and orienting the bird; sliding a bird rather than picking it up can reduce lifting frequency; bulk handling versus multiple box handling or shoveling reduces repetitions; and direct piping of product eliminates packing, transporting and unpacking product.

Orient the product (Figure 42.9). On the left the employees have to untangle and turn the birds to stuff giblets. On the right, the birds are separate and oriented for giblet stuffing. Good orientation also reduces awkward postures, especially of the wrist with giblet stuffing.

Slide product to reduce handling (Figure 42.10). Each packet was lifted and placed into the carton until the changes on the right were made where the product came down by gravity into the box and was guided for correct placement.

Eliminate repeated activity by bulk approach. Eliminate repeated shoveling of meat by using a hopper and controlling the dumping of the meat onto the line or into a machine. Another approach is to pipe product directly from a container or the line to the destination.

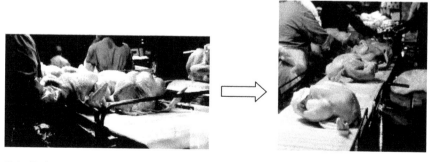

FIGURE 42.8   Reduce repetition.

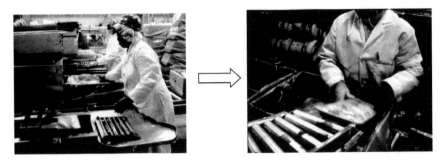

FIGURE 42.9   Orient the product.

- *Provide space.* Ensure there is an adequate work envelope to perform the task. If there is an insufficient window of time to make the cut the workers tend to travel along the line and become too close to one another and inadvertent lacerations result. This is not uncommon on very demanding cuts such as white ball and pop and score in deboning turkeys.
  Give a sufficient work envelope. Ensure there is room to perform the task and that the workers do not bunch up too closely together. In this turkey deboning example, the employees are turned to their left to start the cut up the line to give themselves enough time to complete the cut.
- *Minimize weight* (Figure 42.11). The weight of objects or product should be kept to a minimum or reduced. For example, control the volume of the product in a container to control weight; support the weight of a bird by a conveyor while it is handled, such as when hanging on a shackle; slide versus lifting; use hoists or lifts for containers; and use aids such as dollies under barrels.
  Control volume (Figure 42.12). Control the volume in a container by having lines marked inside or ensuring a certain weight on scales. This method requires compliance and cannot always be relied upon.
  Support the weight (Figure 42.13). Avoid lifting the whole weight of a turkey to hang it on a shackle as shown left. Hang the bird with the weight taken by the conveyor or table.
  Keep transitions smooth (Figure 42.14). Avoid lifting over uneven surfaces as shown on the left. Keep transitions smooth for reduced weight and effort and for quicker movement.
  Reduce forces with a dollie (Figure 42.15). Reduce lifting weight or dragging forces by using a dollie under barrels.
- *Control material handling.* There are many approaches to controlling manual material handling. Some of the principles are the same as above, such as working at the appropriate height. A chapter in this handbook addresses engineering controls for low back disorders in detail. To control lifting, an approach can be to eliminate manual handling altogether by automation.

FIGURE 42.10   Slide product to reduce handling.

FIGURE 42.11   Minimize weight.

When this is not possible, aids such as hoists can be used. To reduce the force on the back, slide rather than lift material, keep a lift at around waist height, and ensure the load is as close as possible. Transporting materials other than carrying them manually, includes

FIGURE 42.12   Control volume.

methods such as lift trucks, pallet jacks, carts and hand carts, conveyors, and carrying aids.

Use aids to reduce handling (Figure 42.16). Instead of manually handling product as shown on the left, use tippers to reduce handling. Note the lift table to keep the supply for the tipper at waist height.

Slide rather than lift (Figure 42.17). Keep scales in line so boxes do not have to be lifted on and off.

- *Allow sit/stand.* Standing to work is necessary for dynamic tasks as there is greater freedom of movement than when sitting. Consider a sit/stand choice for tasks that are less dynamic. Employees can rotate through these tasks so that many can experience a change from standing. There are stools that only allow leaning and others that allow full sitting. Backrests are preferable if there is full sitting. However, a backrest is not likely to be used if the person leans forward from the seat to do their job. Ensure the seat can be pulled in close enough to the job and that there is knee room.

Choice of sit/stand (Figure 42.18). A basic stool is shown in the left picture. Ensure there is knee room for any seated position. A lean stool is shown in the right picture. Ensure employees are not too far from their work when they are seated.

Provide footrails. (Figure 42.19) Footrails provide relief to the low back and should be available at standing positions.

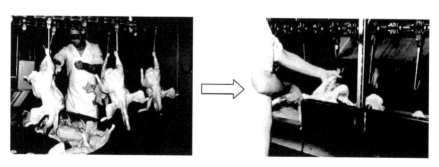

FIGURE 42.13   Support the weight.

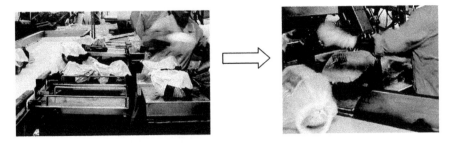

FIGURE 42.14 Keep transitions smooth.

FIGURE 42.15 Reduce forces with a dollie.

FIGURE 42.16 Use aids to reduce handling.

FIGURE 42.17　Slide rather than lift.

- *Control exposure to cold.* Cold temperatures restrict blood flow; however, good blood flow is needed for the working muscles. When working in a cold environment ensure workers understand the importance of dressing the core of the body warmly in layers so there is warmth spare for the arms and legs.

  Avoid unnecessary cold exposure such as working in ice. Dump the product onto a conveyor or hopper to free the product from the ice, rather than the workers digging for product. Employees should not manually process frozen product.

  Avoid working in ice (Figure 42.20). Prevent employees from digging manually through ice for product. Transfer the contents with a dumper into a hopper or on to a conveyor.

42.3.3　Hand Tools

Where possible, automate or use powered tools for cutting tasks. Most powered tools in the industry are air driven. Blunt tools, especially knives, have a large effect on the forces passing through the hands, wrists, and arms. As a knife dulls a higher force has to be exerted to make a cut. In addition, there is an increased possibility of making the cut twice. The knife can also slip if it is dull. Catching the bone or a metal surface dulls the knife as well.

FIGURE 42.18　Choice of sit/stand.

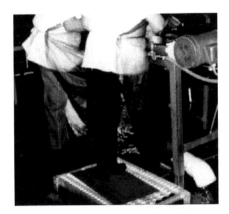

FIGURE 42.19   Provide footrails.

PM is necessary on all tools and cutting equipment so that the effort levels are kept to a minimum. Powered tools, such as circular blade knives, can have a vibration characteristic that should be monitored. PM on these tools helps to ensure the characteristics do not change due to wear.

FIGURE 42.20   Avoid working in ice.

At least two sets of stone-sharpened knives should be provided in a shift and effective sharpening skills taught for touch up on the line. Touch up sharpening should occur routinely after every few cuts before the body can feel that the knife is dull. Once there is awareness of a difference, the hand has already experienced more force through the tendons than necessary. Mousetraps are commonly provided on the line for touching up. Some employees in the industry argue that good steeling is best for touch up, while others say that if the skill of steeling is not mastered then a worse edge can be put on the knife. An emphasis on training employees in knife care and sharpening is critical. New employees should be watched closely for knife skill development and possible musculoskeletal stress as they are the most inclined to hit bone and metal until they are proficient.

Knives should be well cared for on the production floor and kept in plastic holders or holders with plastic liners so they are not dulled against metal as the tool is put away. The right knife should be provided for the task. For example, some fine cuts such as neck slit require a thin, short blade and the knife is held in a precision grip as the target is small. Other tasks need various width and lengths of blades. The

FIGURE 42.21   Keep tools in protective sheaths.

FIGURE 42.22   Ensure ample mousetraps or steels.

choice of tool affects the hand position. An angled handle on a tool may be suitable for some tasks and help to keep the wrist straight. The handles should have basic good ergonomic characteristics including providing traction, nonslip with poultry oils, at least 4 in. (10 cm) handle length and a maximum of 2 in. (5 cm) diameter. The quality of the steel in holding a good edge is important. Suspend or counterbalance heavy tools, such as powered scissors. The following points are illustrated next: protect the tools; keep an adequate work envelope; and typical hand tools.

- *Protect the tools*
  Keep tools in protective sheaths (Figure 42.21). Protect the tools from damage and people from inadvertent cuts. Scissors should be kept closed when not in use.
  Ensure ample mousetraps or steels (Figure 42.22). Provide plenty of mousetraps or steels so that they are in easy reach for quick and frequent sharpening.
- *Keep an adequate work envelope.* A tool should be used with a relaxed and natural movement as possible. Ensure there is adequate clearance between people and around the movement of the knife. An abrupt stop to a movement requires more force and control, especially if there is a consequence to not stopping, such as hitting a conveyor.

FIGURE 42.23   Clear obstructions to allow movement.

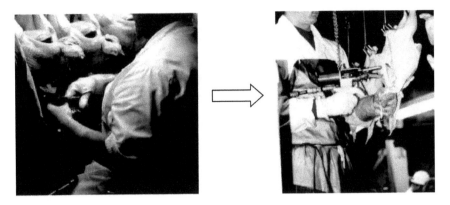

FIGURE 42.24   Scissors.

Clear obstructions to allow movement (Figure 42.23). The high conveyor causes the employee to abruptly stop the cut without following it through, which makes the task harder than it need be.
- *Typical hand tools*
Scissors (Figure 42.24). Steel scissors are still the tool of choice for tasks such as liver cutting. Cushioned handles help protect damage on the digital nerves. Spring assist scissors should be considered for tasks that are horizontal. Vertical tasks as shown here have the advantage of gravity to open the scissors. Use air scissors whenever possible to reduce the forces on the hand.

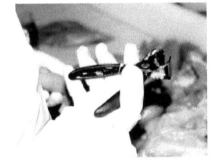

FIGURE 42.25   Tender pulling.

FIGURE 42.26   Lung gun.

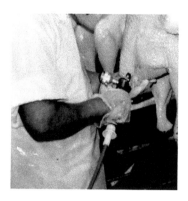

FIGURE 42.27   Circular bladed knives.

Tender pulling (Figure 42.25). The tool for pulling tenders has short handles, the ends of which land in the center of the hand, potentially on the nerve. However, this tool remains popular as it grips the tenders well. An alternative should be sought. Use a tender pulling machine and only perform this task manually for rework. Note the vise that holds the tender, against which the employee can pull with the tool.

Lung gun (Figure 42.26). Lung guns are long and awkward to handle. Suspend the guns from above by a balancer to reduce a constant hold. Ensure the hose does not act as a drag on the back of the gun. Note, if the task is performed vertically a balancer does not work so effectively.

Circular bladed knives (Figure 42.27). Pneumatic, circular bladed knives are used for many tasks including oil sack removal as shown on the left, and trim tasks as shown on the right. These knives should receive PM to minimize vibration.

Scoops (Figure 42.28). Choose a scoop that has the handle integrated close to the load as shown on the right. Biomechanically, this puts less stress on the wrist than the one on the left.

Shovels (Figure 42.29). Reduce "dead weight" that adds no value to the task. Choose a light plastic shovel over a heavy stainless steel shovel.

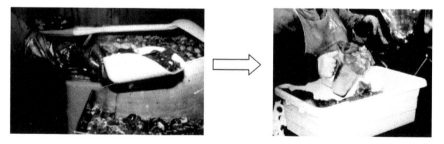

FIGURE 42.28   Scoops.

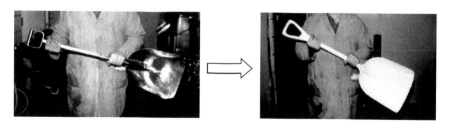

FIGURE 42.29   Shovels.

42.4 Summary

Effective ergonomics in the poultry industry involves implementing and integrating an ergonomics process into safety and health and production. Good training and gradual conditioning of employees is essential for reducing injuries. Although the industry is becoming more automated, there are common ergonomics principles that can be employed that make a difference. Both administrative and engineering control strategies are needed to make improvements to the workplace, reduce injuries and illnesses, and allow employees to perform at their best.

References

Bureau of Labor Statistics (BLS). 2002. *Detailed Occupational Injury and Illness Industry Data (1989–2001).* Series ID SHU30201531 and SHU30201532, poultry processing. http://data.bls.gov/cgi-bin/dsrv?sh.

California OSHA. 2003. *Ergonomics in Action: A Guide to Best Practices for the Food-Processing Industry.* 17 REU. Cal/OSHA Consultation Service, Dept. of Industrial Relations, CA.

Government Accountability Office (GAO). 2005. *Workplace Safety and Health: Safety in the Meat and Poultry Industry, While Improving Could Be Further Strengthened.* GAO-05-96. Government Accountability Office, Washington, DC.

Human Rights Watch (HRW). 2005. *Blood, Sweat and Fear: Workers' Rights in US Meat and Poultry Plants.* Human Rights Watch, Washington, DC.

National Broiler Council (NBC). 1990. *The MET Program for Supervisors.* National Broiler Council, Washington, DC.

National Institute of Occupational Safety and Health (NIOSH). 1997. *Elements of Ergonomics Programs: A Primer Based on Workplace Evaluations of Musculoskeletal Disorders.* DHHS (NIOSH) Publication No. 97-117. NIOSH, Cinncinnati, OH.

Occupational Safety and Health Administration (OSHA). 1990. *Ergonomics Program Management Guidelines for Meatpacking Plants.* OSHA 3123. Department of Labor/OSHA, Washington, DC.

Occupational Safety and Health Administration (OSHA). 2001. *E-Tool: Poultry Processing Industry.* Occupational Safety and Health Administration web page: http://www.osha.gov/SLTC/etools/poultry/index.html.

Occupational Safety and Health Administration (OSHA). 2004. *Guidelines for Poultry Processing: Ergonomics for the Prevention of Musculoskeletal Disorders.* OSHA 3213-00N Department of Labor/OSHA, Washington, DC.

Occupational Safety and Health Administration (OSHA). 2005. *National Alliances: National Chicken Council (NCC) and National Turkey Federation (NTF).* Occupational Safety and Health Administration web page: http://www.osha.gov/dcsp/alliances/ncc_ntf/ncc_ntf.html.

Stuart-Buttle, C. 1994. A discomfort survey in a poultry-processing plant. *Applied Ergonomics.* 25(1):47–52.

43

Working in Unusual or Restricted Postures

Sean Gallagher
*National Institute for
Occupational Safety and
Health*

43.1 Introduction

The human body is remarkably adaptable and capable of performance in a wide variety of environments and circumstances. It cannot be said, however, that the body is capable of performing equally well under all conditions. In fact, when faced with certain types of tasks or environmental demands, the body may have to adapt using methods that result in substantial performance limitations. Such a phenomenon is evident when workers must adopt unusual or restricted postures during performance of physically demanding work tasks. For the purposes of this discussion, the term "unusual posture" will be considered as any working posture other than typical standing or sitting positions. The term "restricted posture" designates postures that are forced upon workers due to restrictions in workspace.

The vast majority of ergonomics research has focused on establishing design criteria for work involving standing (e.g., Snook and Ciriello, 1991; Waters et al., 1993) or seated postures (e.g., Grandjean, 1988), and understandably so. However, it must be recognized that there are numerous jobs (e.g., underground miners, aircraft baggage handlers, plumbers, agricultural workers, mechanics, etc.) where workers must perform in less desirable postures such as kneeling, stooping, squatting, and lying down (Haselgrave et al., 1997). Unfortunately, experience has shown that many ergonomics techniques used to analyze or design standing or sitting workstations often do not adapt well to situations where a restricted posture is adopted (Gallagher and Hamrick, 1991). However, recent years have seen an increase in research examining the musculoskeletal risks and physical limitations associated with working in these postures. The purpose of this article is to summarize current knowledge in this area, and to establish principles for ergonomic design of jobs when working in unusual or restricted postures.

43.2   General Considerations

Workers typically enjoy the benefits of high strength capacity and mobility when they assume a normal standing position. This stance permits many powerful muscle groups to work in concert to accomplish occupational tasks. However, the muscular synergy present in the standing posture can be seriously disrupted when unusual or restricted postures are employed. One need only imagine a lift performed while lying down on one's side to understand that many powerful muscles (i.e., those of the legs, hips, and thighs) will be unable to fully participate in the lifting assignment. This example illustrates two important aspects of work in unusual or restricted postures. First, the number of muscle groups available to generate forces to accomplish a task is often reduced compared to standing. Second, the reduced number of participating muscles may lead to increased demands on those that can be recruited. It should be evident that each unique postural configuration will result in its own set of strength limits. The number and identity of the muscles that can be effectively recruited for the job will largely determine these limits.

Task performance in unusual or restricted work postures can also be affected by reduced mobility, stability, and balance. For example, when one is unable to stand on one's feet, mobility may be dramatically reduced. Reduced mobility can have a significant impact on the method of task performance. Consider an asymmetric lifting task performed in standing versus kneeling postures. When a worker is standing, it is reasonable to request that he or she avoid twisting the trunk simply by repositioning the feet when asymmetry is present. However, the task of repositioning is considerably more difficult when kneeling (especially when handling a load), and workers are not inclined to take the time nor the effort to do this. Instead, the worker will opt for the faster and more energy efficient twisting motion, at the expense of experiencing a sizable axial torque on the spine. Stability may also impact task performance in constrained postures. Workers may have to limit force application in certain postures to maintain balance.

As mentioned previously, these awkward work postures are often the consequence of restrictions in workspace, either vertically or laterally. For example, underground miners and aircraft baggage handlers often operate in workspaces where the available vertical space does not allow upright standing. Workspace restrictions of this sort put not only the worker in a bind, but also the ergonomist. The worker is affected by the limitations of the posture he or she must employ. The ergonomist may be deprived of favored techniques for reducing musculoskeletal disorder risk. For example, restricted space greatly limits the number and type of mechanical devices (cranes, hoists, forklifts, etc.) available to reduce the muscular demands on the worker. If mechanical assistance is to be provided, it frequently must be custom fabricated for the environment. Restrictions in workspace also limit opportunities to ease the strain arising from the worker's postural demands, often forcing the ergonomist to recommend working postures from a limited menu of unpalatable alternatives.

Restricted spaces may also result in more subtle effects. One is the tendency, as vertical space is reduced, to force workers into asymmetric motions. Lifting symmetrically (i.e., in the sagittal plane) is generally preferred in the standing posture, but becomes progressively more difficult if one is stooping in reduced vertical space. In fact, psychophysical lifting capacity in asymmetric lifts tends to be *higher* than in symmetric tasks under low ceilings (Gallagher, 1991). This represents a change from the unrestricted standing position, where asymmetry *reduces* lifting capacity (Garg and Badger, 1986). Finally, as Drury (1985) points out, space limitations tend to impose a single performance method on a worker. In unrestricted spaces, when a worker's preferred muscles fatigue, it is often possible for an individual to employ substitute motions which may shift part of the load off of fatigued muscles. Unfortunately, the opportunity to employ substitute motion patterns decreases as workspace becomes more limited. The result is intensified fatigue and a decrease in performance capabilities in restricted postures.

43.3 Epidemiologic Studies of Restricted Postures and Musculoskeletal Disorders

Unfortunately, the number of epidemiologic studies examining the association of restricted postures to the occurrence of musculoskeletal disorders remains sparse. However, studies that have investigated this relationship have tended to exhibit higher rates of musculoskeletal disorders in restricted as opposed to unrestricted postures.

Lawrence (1955) examined British coal miners to identify factors related to degenerative disc changes, and found that injury, duration of heavy lifting, duration of stooping, and exposure to wet mine conditions were the factors most associated with spinal changes. Another study investigating spinal changes in miners was reported by MacDonald et al. (1984). These investigators used ultrasound to measure the spinal canal diameter of 204 coal miners and found that those with the greatest morbidity had significantly narrower spinal canals. The study by Lawrence (1955) and other evidence suggests that the seam height of the mine has a marked influence on the incidence of low back disorders. In general, compensation claims appear to be highest in seam heights of 0.9–1.8 m (where stooping is prevalent). Claims are slightly lower in seams less than 0.9 m (where kneeling and crawling predominate), and are lowest when the seam height is greater than 1.8 m.

The finding of increased low back claims in conditions where stooping predominates is congruent with other evidence relating non-neutral trunk postures to low back disorders. For example, a case–control study by Punnett et al. (1991) examined the relationship between non-neutral trunk postures and risk of low back disorders. After adjusting for covariates such as age, gender, length of employment and medical history, time spent in non-neutral trunk postures (either mild or severe flexion) was strongly correlated with back disorders (OR 8.0, 95% CI 1.4–44) (OR, odds ratio; CI, confidence interval). In fact, this study disclosed a dose–response between the degree of torso flexion and the risk of low back disorder. Mild flexion was associated with an OR of 4.9 while severe flexion was associated with an OR of 5.7. Although it was difficult in this study to find subjects that were not exposed to non-neutral postures, the strong increase in risk observed with both intensity and duration of exposure were notable.

A study of 1773 randomly selected construction workers also examined the effects of awkward working postures on the prevalence rates of low back pain (Holmstrom et al., 1992). This study found that prevalence rate ratios for low back pain were increased for both stooping ($p < 0.01$) and kneeling ($p < 0.05$) when the duration of work in these postures were reported to be at least 1 h per day. Furthermore, a dose–response relationship was observed whereby longer durations of stooping and kneeling were associated with increased prevalence rate ratios for severe low back pain (Table 43.1). Thus, workers who adopt stooping or kneeling postures for longer periods of time appear to be at increased risk of experiencing severe low back pain.

TABLE 43.1 Age-Standardized Prevalence Rate Ratios with 95% Confidence Intervals for Low Back Pain and Severe Low Back Pain When Adopting Stooping and Kneeling Postures for Different Durations

	<1 h duration		1–4 h duration		>4 h duration	
	LBP	Severe LBP	LBP	Severe LBP	LBP	Severe LBP
Stooping	1.17	1.31	1.35	1.88	1.29	2.61
	(1.1–1.3)	(0.9–1.8)	(1.2–1.5)	(1.4–2.6)	(1.1–1.4)	(1.7–3.8)
Kneeling	1.13	2.4	1.23	2.6	1.24	3.5
	(1.0–1.3)	(1.7–3.3)	(1.1–1.4)	(1.9–3.5)	(1.1–1.4)	(2.4–4.9)

Source: From Holmstrom, E.B., Lindell, J. and Moritz, U. *Spine*, 17(16):663–671, 1992. With permission.

In addition to the effects on the back, working in certain unusual or restricted postures (particularly kneeling) has been shown to affect musculoskeletal disorders of the lower extremity (Lavender and Andersson, 1999). Sharrard (1963) reported on the results of examinations on 579 coal miners in a study examining the etiology of "beat knee." Forty percent of the miners reportedly were symptomatic or had previously experienced symptoms, characterized as acute or simple chronic bursitis. Incidence rates were found to be higher in seam heights lower than 4 ft and in workers required to kneel for prolonged periods at the mine face. The incidence of "beat knee" was found to be higher in younger mineworkers; however, this finding was thought to be due to a "healthy worker" effect. Specifically, it was thought that older workers with "beat knee" may have left the mining profession.

Studies have also indicated that other occupations where frequent kneeling is required experience higher rates of knee problems in relation to comparison occupational groups. Tanaka et al. (1982) found that occupational morbidity ratios for workers compensation claims involving knee-joint inflammation for carpet layers was over 13 times greater than that of carpenters, sheet metal workers, and tinsmiths. Knee inflammation among tile setters and floor layers were over six times greater than the same comparison groups. Workers in these occupations have been shown more likely to exhibit fluid accumulation in the superficial infrapatellar bursa, subcutaneous thickening of this bursa, and increased thickness in the prepatellar region (Myllymaki et al., 1993). The much higher incidence associated with carpet layers is probably also related to their use of a knee-kicker, a device used to stretch carpet during its installation. Knee impact forces during the use of this device have been shown to be as high as four times the body weight (Bhattacharya et al., 1985).

43.4   Performance Limitations in Restricted Postures

The past couple of decades have seen a number of studies that have examined the effects of working in unusual or restricted postures on a variety of performance measures. These measures have included psychophysical lifting capacity, muscular strength, metabolic cost, and electromyography. The following sections provide information regarding some of the effects of restricted postures on these performance measures.

43.4.1   Effects of Posture on Lifting Capacity

43.4.1.1   Lifting Capacity for a Single Lift

A comprehensive analysis of single lift psychophysical lifting capabilities in nontraditional working posture was performed by researchers at Texas Tech University under a contract from the U.S. Air Force (Gibbons, 1989). Under this contract, two lifting studies examined maximum psychophysical lifting capacities of both male and female subjects in standing, sitting, squatting, kneeling, and lying postures. The purpose was to simulate postures used during Air Force aircraft maintenance activities, which often involve use of unusual or restricted postures. Subjects were allowed to adjust the weight in lifting containers to the maximum they felt were acceptable for a single lift in each posture. It should be noted that the lifting tasks were standardized using percentages (35, 60, and 85%) of the vertical reach height of the subject in each posture. Thus, a lift to 35% vertical reach height in the standing posture will have a greater vertical load excursion than a lift to 35% vertical reach height in a kneeling posture.

Figure 43.1 and Figure 43.2 present data from male and female subjects, respectively, performing lifts in standing, kneeling (on one knee and on both knees), sitting and squatting postures. Inspection of these figures reveals several notable features. The first is that in all cases the standing posture resulted in the highest psychophysically acceptable loads compared to the restricted postures. One can also see from these figures that loads chosen in kneeling tasks result in the second highest estimates of lifting capacity (7 to 21% less than standing), and that one knee lifts did not differ from lifts on both knees in terms of load acceptability. The sitting posture resulted in acceptable lifting estimates just slightly below those achieved when kneeling (16 to 23% less than standing lifts), and squatting resulted in the lowest

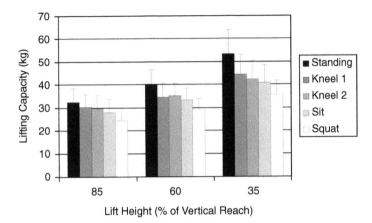

FIGURE 43.1 Acceptable loads selected by males for single lifts in several postures. Bars represent means, error bars represent standard deviations. (From Gibbons, L.E. Summary of Ergonomics Research for the Crew Chief Model Development: Interim Report for Period February 1984 to December 1989. Armstrong Aerospace Medical Research Laboratory Report No. AAMRL-TR-50-038. Wright-Patterson Air Force Bare, Dayton, OH, 1989. With permission.)

acceptable loads (20 to 33% less than standing). The squatting posture appears to be the least stable of the restricted postures, and it may be that the lower acceptable loads in this posture may be driven by the need to select a load that allows the subject to maintain his or her balance.

It is also apparent that the effects of posture on lifting capacity are more pronounced with lifts of 35% of vertical reach, and that the effect becomes progressively diminished (though still apparent) when lifts to 60% and 85% of vertical reach are performed. It may be that strength capbilities for lifts to higher heights may be controlled more limitations in shoulder and arm strength, and are thus not as dependent on body posture per se. Finally, comparison of male strength (Figure 43.1) versus female strength (Figure 43.2) indicates that posture effects are similar for both genders; however, the strength exhibited by females averaged about 50 to 60% of that achieved by their male counterparts.

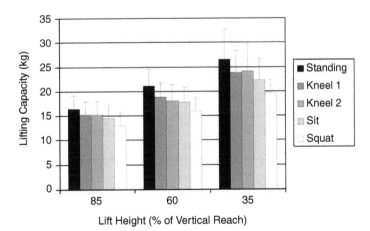

FIGURE 43.2 Acceptable loads selected by females for single lifts in several postures. Bars represent means, error bars represent standard deviations. (From Gibbons, L.E. Summary of Ergonomics Research for the Crew Chief Model Development: Interim Report for Period February 1984 to December 1989. Armstrong Aerospace Medical Research Laboratory Report No. AAMRL-TR-50-038. Wright-Patterson Air Force Bare, Dayton, OH, 1989. With permission.)

A separate study performed at Texas Tech looked at strength capacities in prone, supine, or side-lying positions (Gibbons, 1989). These postures exhibit drastic reductions in lifting capacity, with acceptable loads just 25 to 40% of standing values. The only exception was when the subject performed a two-handed lift in a face-up (supine) position, similar to a weightlifter's "bench press" exertion. In this instance, the average acceptable load actually exceeded the standing value by 20%. It appears that control of the load, and a balanced exertion of forces by both arms, play important roles in determining lifting capacity in the supine position.

43.4.1.2 Lifting Capacity for Longer Duration Tasks

It should be emphasized that the data discussed in the previous section represent *one-repetition maximum values*, and assume that workers would perform such tasks only occasionally, not for extended periods. However, periods of extended lifting in restricted posture are common in some industries. Examples include underground coal miners unloading supply items in a low-seam coal mine, or an aircraft baggage handler loading suitcases and packages inside the baggage compartment of a commercial airliner. Several recent studies have examined the lifting capacity of underground coal miners adopting restricted postures over more extended time frames (Gallagher et al., 1988; Gallagher and Unger, 1990; Gallagher, 1991; Gallagher and Hamrick, 1992). These studies also used the psychophysical approach, allowing subjects to adjust the weight in lifting boxes to acceptable loads during 20-min lifting periods. Most of these studies examined lifting capacities in kneeling and stooping postures, postures that predominate in underground coal mines having restricted vertical workspace.

In general, findings of these studies are congruent with limitations associated with these postures in the single lift studies described previously. Restricted postures (stooping and kneeling) were found to result in lower estimates of acceptable loads compared to the standing posture (Gallagher and Hamrick, 1992), and kneeling was found to have a significantly reduced estimate of acceptable load compared to stooping (Gallagher et al., 1988; Gallagher and Unger, 1990; Gallagher, 1991). Kneeling and stooping postures were examined under different vertical space constraints to see whether additional restrictions in space would further affect lifting capacity (i.e., is lifting capacity when kneeling different under a 1.2 versus 0.9 m ceiling? Is lifting capacity when stooping different under a 1.5 versus 1.2 m ceiling?). However, results indicated no additional decrements in lifting capacity were seen when comparing such conditions. The major determinant affecting lifting capacity in these studies was simply the posture adopted for the task (Gallagher and Unger, 1990). While posture was almost always an important determinant of lifting capacity in these studies, there were some factors, if present, that could reduce or eliminate the effect. In particular, it was found that if items had a poor hand–object coupling (no handholds), lifting capacity could be reduced to such an extent that posture effects were no longer evident (Gallagher and Hamrick, 1992).

A surprising (and somewhat unsettling) finding from these studies is that psychophysical lifting capacity in prolonged torso flexion (over a 20-min time frame) is not much different from unrestricted standing for the same lifting task (Gallagher and Hamrick, 1992). In one respect, this is not too surprising because the stooping posture is a position where considerable strength is available to lift a load. In fact, most workers prefer this position when initiating a lift off of the floor, probably due to the ability to employ the powerful hip extensor muscles in overcoming the inertia of the load. In his critique of the psychophysical method, Snook (1985) states that psychophysical method of establishing acceptable loads does not appear to be sensitive to bending and twisting motions that are often associated with the onset of low back pain, and the results reported above seem to support this limitation. Recent studies have indicated that prolonged stooping may be associated with ligament creep and an attendant reduction in the ability to recruit the back muscles (Bogduk, 1997; Solomonow et al., 1999). Furthermore, it has been suggested that potentially damaging shear forces may be present in this posture (McGill, 1999), and fatigue failure may occur more rapidly (Gallagher, 2003). Subjects may not get sufficient proprioceptive feedback regarding these matters; thus, they may not play into estimates of load acceptability. Nonetheless, these and other biomechanical factors may be important in development of low back disorders. It seems clear that development of lifting standards for a stooping posture must not rely solely on estimates of psychophysical lifting capacity, but should take into account biomechanical and physiological factors that may influence development of low back disorders in this posture.

43.4.2 Biomechanics of Unusual or Restricted Postures

As significant changes in whole-body posture are adopted, one would anticipate changes in both the magnitude and distribution of biomechanical stresses amongst the joints of the body, and available evidence appears to support this notion. The following sections describe results of studies examining various aspects of the biomechanics of working in restricted postures.

43.4.2.1 Effects of Restricted Postures on Strength

Studies examining static or dynamic strength capabilities in unusual or restricted postures are relatively rare. Isometric strength tests in kneeling versus standing postures have indicated that lateral exertions are weaker when kneeling; however, pushing forces are found to be equivalent or slightly higher when kneeling (Haselgrave et al., 1997). Static pulling and lifting forces in the kneeling posture exceeded those in the standing position, by 25 and 44%. Pushing upwards against a handle at eye height results in similar values in all postures (Gallagher, 1989).

Gallagher (1997) investigated isometric and isokinetic trunk extension strength and muscle activity in standing and kneeling postures. Findings of this study showed that trunk extension strength is reduced by 16% in the kneeling posture in comparison with standing, similar to decreases observed in psychophysical lifting capacity when kneeling. However, trunk muscle activity was virtually the same between the two postures. This indicates that the reduction in trunk extension strength when kneeling may be the result of a reduced capability to perform a strong rotation of the pelvis when the kneeling posture is adopted, as opposed to a change in function of the spinal muscles.

An intriguing set of strength data comparing isometric strengths of coal miners working in restricted postures to a comparison population of industrial workers, presented by Ayoub et al. (1981), is shown in Figure 43.3. Strength measures included back strength, shoulder strength, arm strength, sitting leg strength and standing leg strength. When compared with a sample of industrial workers (Ayoub et al., 1978), low-seam coal miners were found to have significantly lower back strength, but much higher leg strength. The authors ascribed the decrease in back strength to unspecified factors related to the postures imposed by the low-seam environment. Indeed, there is evidence to support this position. Low-seam coal miners may be obliged to work in a stooping posture for extended periods. In this posture, the spine is largely supported by ligaments and other passive tissues, "sparing" the use of the back

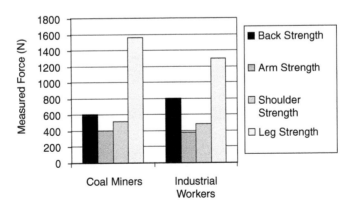

FIGURE 43.3 Comparison of strength measures for coal miners working in confined vertical space (Ayoub et al., 1981) to an industrial population (From Ayoub, M.M., Bethea, N.J., Deivanayagam, S., Asfour, S.S., Bakken, G.M., Liles, P., Mital, A., and Sherif, M. Determination and Modeling of Lighting Capacity, Final Report, Grant #5R010H-0054502, HEW, NIOSH. Texas Tech University, Lubbock TX, 1978. With permission.)

muscles. Studies of lifting in the stooping posture suggest that the gluteal muscles and hamstrings provide a large share of the forces in this position (Gallagher et al., 1988). The results of Ayoub et al. (1981) may reflect a relative deconditioning of back muscles when stooping (due to the flexion–relaxation phenomenon), perhaps the result of prolonged inhibition of muscular activity (e.g., Floyd and Silver, 1955) and damage associated with ligament creep (Solomonow et al., 2003). Furthermore, increased reliance on the leg and hip musculature may be necessary in situations where prolonged torso flexion is required (producing an increase in leg strength). Further research is needed to ascertain long-term adaptations in strength resulting from prolonged work in restricted postures.

43.4.2.2 Lumbar Spine Loads in Restricted Workspaces

Studies have suggested that one of the best predictors for low back pain is the external moment about the lumbar spine that results from the product of the force required to lift an object times the distance these force act away from the spine (Marras et al., 1993). As illustrated in Figure 43.4, recent evidence has shown that as vertical workspace is reduced, the moment experienced by the lumbar spine will be increased (Gallagher et al., 2001). Of course, such a response would be expected in the standing posture, where reduced ceiling heights would cause the trunk to bend forward increasing the moment on the lumbar spine. However, this study (which involved lifting heavy mining electrical cables) found no difference between stooping and kneeling postures in terms of the peak spinal moment experienced by the subject. The primary determinant of the lumbar moment was the ceiling height. The lower the ceiling was the higher the moment experienced (no matter which posture was chosen). The question raised by this study is why there was not a decreased moment when the kneeling posture is employed. Clearly, the trunk can maintain a more erect posture when kneeling. However, analysis of this position reveals that the knees create a barrier that prevents the worker from getting close to the load at the beginning (and most stressful part) of the lift. This creates a large horizontal distance between the spine and the load, resulting in a large moment, apparently offsetting the benefits of maintaining a more erect trunk position.

The point must be made, however, that though the spinal moments appear equivalent in these two postures, the same might not hold true for risk of experiencing a low back disorder. Biomechanical

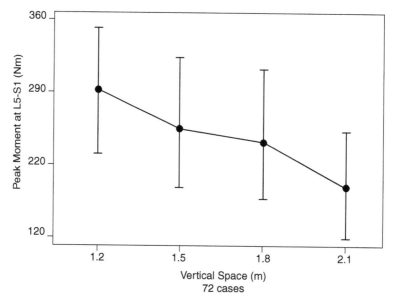

FIGURE 43.4 Lumbar moments, an indicator of strain experienced by the low back, are increased as vertical workspace becomes more confined (From Gallagher, S., Hamrick, C.A., Cornelian, K., and Redfern, M.S. *Occupational Ergonomics*, 2(4):201–213, 2001. With permission.)

analyses indicate that spinal shear forces are high when the spine is fully flexed. In addition, there are indications that the compression tolerance of the spine is decreased in this position. These factors would tend to favor the kneeling posture. However, one must also bear in mind the lower lifting capacity when kneeling. If the stooping posture is necessary due to strength demands, care should be taken to avoid the end range of spinal motion when performing the lift (McGill, 1999).

43.4.2.3 Trunk Muscle Activity in Restricted Postures

Changes in posture necessarily influence the roles and activation patterns of the muscles of the body. Studies examining the influence of posture on trunk electromyography (muscle electrical activity) have illustrated that restricted postures often result in significant changes in the manner in which muscles are recruited. One of the first studies of the muscle activity of the erector spinae muscles showed that when the trunk is placed in extreme flexion, these muscles become electrically silent (Floyd and Silver, 1955). It appears that the spinal ligaments and fascia assume responsibility for supporting the spinal column when it is fully flexed (either in standing or sitting postures). Biomechanical models suggest that this change results in an increased shear load on the lumbar spine compared to when muscles maintain control (Potvin et al., 1991). When lifting from a fully flexed posture, the back muscles remain silent during the initial stages of lifting weights of up to 28.5 kg (Floyd and Silver, 1955). Many authorities believe that the change from active muscle support to ligament support of the spine might entail increased risk of low back disorder (Basmajian and DeLuca, 1985; Bogduk 1997; Solomonow et al., 2003).

A recent study examined the influence of posture and load on the electromyographic activity of ten trunk muscles during a heavy cable-lifting task (Gallagher et al., 2002). Results of this study indicated that posture and load have quite different influences on trunk muscle recruitment (and thus loads experienced by the lumbar spine). No matter which posture was adopted, an increase in load resulted in increased muscle activity of all ten trunks muscles studied. However, changes in posture typically influenced the activity of trunk muscles in a more selective manner, usually involving only a small subset of the muscles (though the muscles affected by posture were often influential in terms of spine loading). Moreover, the effects of posture and load were found to be independent and additive (i.e., posture and load were found not to interact in terms of their influence on muscle activity).

43.4.2.4 Intra-Abdominal Pressure

Increased pressure within the abdominal cavity has been used by some researchers as a measure of stress on the spine, and has been used to assess restricted postures (Ridd, 1985). Analysis of intra-abdominal pressure (IAP) responses in standing and stooping postures reveal an almost linear decrement with progressively lower vertical workspace up to 90% of stature, whereupon the decrement levels off. In stooping positions ranging from 66 to 90% of full stature, the decrease in lifting capacity was a consistent 60%, according to the IAP criterion. The kneeling posture was found to incur only an 8% decrease in lifting capacity where the space restriction was equivalent to 75% of stature. There is some indication that lifting asymmetrically is less stressful than sagittal plane activities in restricted postures. Unfortunately, the assumption that IAP is a good indicator of spinal stress is still a contentious issue (McGill and Norman, 1987).

43.4.3 Physiologic Costs of Work in Unusual or Restricted Postures

The posture adopted in the performance of a work task has a decided influence on the metabolic demands incurred by an individual. Nowhere is this more evident than in the evaluation of metabolic demands of working in a restricted workspace. Several studies have indicated that restrictions in vertical space greatly increase the cost of locomotion. The most thorough experiment of the effects of stoop walking and crawling was reported by Morrissey et al. (1985). This study illustrated a progressive trend toward increasing metabolic cost as stooping becomes more severe (Table 43.2). Not only is the

TABLE 43.2   Physiological Cost of Erect Walking, Stoopwalking, and Crawling

Task	Sex	Heart Rate (beats/min)	Ventilation Volume (l/min)	Percent Work Capacity	Oxygen Uptake (ml kg^{-1} min^{-1})
Normal walk	Male	89.2 (5.4)	10.6 (0.4)	10.9 (0.9)	5.0 (0.9)
	Female	89.7 (3.6)	9.6 (0.7)	11.06 (2.2)	4.4 (0.6)
90%	Male	96.0 (9.3)	12.8 (0.9)	12.5 (2.0)	5.7 (1.4)
Stoopwalk	Female	107.5 (6.8)	12.4 (1.8)	15.3 (2.9)	5.8 (0.4)
80%	Male	86.8 (15.8)	13.9 (1.8)	14.7 (2.3)	6.8 (1.5)
Stoopwalk	Female	92.0 (12.7)	12.0 (0.6)	15.2 (2.2)	5.8 (0.2)
70%	Male	82.2 (7.2)	13.2 (1.7)	15.1 (4.1)	6.8 (1.5)
Stoopwalk	Female	89.9 (11.1)	11.0 (1.2)	15.7 (3.5)	6.0 (1.0)
60%	Male	88.5 (7.2)	17.0 (2.3)	18.1 (1.4)	8.3 (1.0)
Stoopwalk	Female	100.5 (21.6)	16.2 (5.3)	21.3 (5.0)	8.1 (1.8)
Crawling	Male	81.3 (11.3)	12.5 (1.3)	15.5 (2.3)	7.0 (0.5)
	Female	87.4 (7.8)	10.3 (1.0)	14.8 (2.7)	5.7 (1.8)

Note: Numbers in parentheses represent the standard deviation.
Source: From Morrissey, S.J., George, C.E., and Ayoub, M.M. *Applied Ergonomics*, 16, 99–102, 1985. With permission.

metabolic cost increased as stooping becomes more severe, the maximum speed attainable by subjects is reduced, particularly when stoopwalking at 60% normal stature and when crawling.

The metabolic cost of manual materials handling in restricted postures (stooping and kneeling) has also been studied. These studies suggest that the metabolic cost of manual materials handling is influenced by an interaction between the posture adopted and the task being performed. For example, the kneeling posture can be more costly than stooping when a lateral transfer of materials is done (Gallagher et al., 1988; Gallagher and Unger, 1990). However, other studies have illustrated that kneeling is more economical when the task requires increased vertical load displacement (Freivalds and Bise, 1991; Gallagher, 1991). A study of shoveling tasks found no difference in energy expenditure in standing, stooping and kneeling postures (Morrissey et al., 1983); however, only five subjects participated in this study and it may have suffered from a lack of sufficient statistical power to detect differences.

43.4.4   Recent Evidence on the Hazards of Torso Flexion

Torso flexion has long been considered on of the most hazardous positions in which to perform manual work. This belief has been reinforced by several recent studies that have uncovered some of the reasons why torso flexion may be so strongly related to the development of low back disorders. These studies have included an analysis of fatigue failure of the lumbar spine in flexed versus neutral postures (Gallagher, 2003), as well as studies that have investigated the neurological effects of creep of the posterior ligaments of the lumbar spine resulting from prolonged or repeated flexion (Solomonow et al., 2003). As will be seen, these studies suggest that deep torso flexion may be a significant pathway for the development of at least two different types of low back disorder.

Not only does deep flexion of the torso result in rapid fatigue failure of lumbar tissues when lifting, it also appears to be associated with neuromuscular dysfunction in the lumbar region. A series of studies summarized by Solomonow et al. (2003) using a cat model have shown that creep of lumbar ligaments can lead to a rapid and long lasting dysfunction in the lumbar musculature. In fact, these authors have shown that the creep developed in 20 min of static or cyclic flexion does not fully recover even after 7 h of rest (Solomonow et al., 2003). Flexion was also shown by these authors to elicit a large inflammatory response in the soft tissues of the lumbar spine, which may result from collagen micro-damage and which may explain the hyperexcitability observed in the multifidus muscle with ligament creep (Solomonow et al., 2003).

Fatigue failure of lumbar motion segments subjected to loads associated with lifting an object in different torso flexion postures was recently investigated by Gallagher (2003). This author simulated the spinal

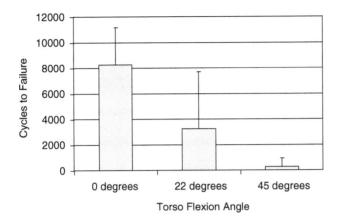

FIGURE 43.5 The number of cycles to failure for lumbar motion segments when exposed to spinal loads estimated when lifting a 9-kg box (From Gallagher, S., Ph.D. Dissertation, 2003).

loads associated with lifting a 9-kg weight in three torso flexion positions (neutral, partial, and full flexion), and subjected spinal motion segments to these loads repetitively until failure occurred. Results of this study are shown in Figure 43.5. As can be seen, the simulated loads associated with lifting 9 kg in the neutral posture could be tolerated for 8257 cycles on average; however, specimens in partial flexion lasted an average of 3257 cycles, while those at 45° lasted an average of only 263 cycles before failure. Results of this study suggest that lifting of loads in a flexed torso posture may result in rapid fatigue failure of tissues of the lumbar spine, and may be an important determinant in the development of low back disorders.

Epidemiologic studies have long revealed an association between torso flexion postures and low back disorders or pain. The etiology underlying the association has remained obscure, however. The recent studies described above suggest at least a couple of possible pathways by which low back disorders may develop during work in torso flexed postures. That is, the flexed trunk posture may lead to micro-damage of the ligaments of the spine, leading to a muscular dysfunction that, while recoverable, may affect the lumbar region for up to several days, or jobs such as lifting in trunk flexion may lead to more significant fatigue failure in motion segments of the lumbar spine, resulting in endplate fractures and disc degeneration which may lead to significant disability and pain, which may not be easily recoverable. It is hoped that continued research along these lines may further elucidate the etiology of low back disorders associated with torso flexion.

43.5 Intervention Principles for Unusual or Restricted Postures

The findings of recent studies that have examined the capabilities, limitations, and tolerances of unusual or restricted postures can assist in forming a basis for intervention principles designed to reduce the risk of musculoskeletal disorders to workers who must adopt them. The following sections discuss methods that may be useful in reducing injury risk for those who must work in restricted postures.

43.5.1 Avoid Full Flexion of the Torso

Perhaps the most important advice that can be given to reduce back injury risk is to avoid work in severe torso flexion. As discussed earlier, epidemiologic evidence indicates a clear association between flexion and low back disorders, and recent studies have highlighted several potential pathways associated with flexion that may lead to both short- or long-term low back disorders. If flexion cannot be avoided, it should be minimized, and frequent breaks should be allowed to assume a less stressful position on the back. Lifting in a flexed posture can lead to rapid fatigue failure of spinal tissues and should also be avoided entirely or, alternatively, minimized to the greatest extent possible. Any loads lifted in flexion should be as light as possible; however, it should be noted that even light loads may lead to fatigue failure over a short time frame.

Evidence of the adverse effects associated with the flexed torso posture continues to mount, and several potential pathways to the development of low back disorders have been recently identified. Of all the restricted postures discussed in this paper, the stooping posture seems most likely to lead to short- or long-term low back disorders. Eliminating or minimizing the amount of torso flexion workers must perform on the job may be the best single action to take to reduce the risk of low back disorders in the occupational working environment.

43.5.2   Design Loads in Accordance with Posture-Specific Strength Capacity

As detailed previously, many unusual or restricted postures are associated with a reduced strength capability. As a result, loads that are acceptable to lift in an upright standing posture may exceed those appropriate when workers adopt a restricted posture. In general, lifting capacity in the kneeling and sitting postures is reduced by up to 20% compared to standing; whereas, squatting lifting capabilities may be reduced by up to 33% of the standing value. Lifting capacity in lying postures is generally much lower, with acceptable loads just 25–40% those considered acceptable when standing. It should be apparent that if workers must adopt one of the postures listed above for lifting activities, loads need to be adjusted downward to reflect the reduced strength capabilities associated with specific postures. This may require working closely with suppliers or manufacturers of items that must be manually handled in specific work postures. Figure 43.6 shows an example of redesign of bags of rock dust, used to suppress coal dust in an underground mine. The traditional 50-lb (23 kg) bag is shown on the left. Based on a request from the mine, the manufacturer supplied rock dust in a 40-lb (18 kg) bag more acceptable to handle in the restricted postures workers used in the mine.

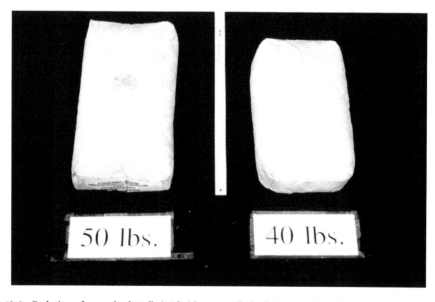

FIGURE 43.6   Redesign of a standard 50-lb (23 kg) bag to 40 lb (18 kg) was achieved by working with supplier, and creates a more acceptable load for workers operating in restricted postures.

43.5.3   Use of Mechanical-Assist Devices and Tools

Use of mechanical-assist devices and application-specific tools can often reduce the need to adopt awkward or restricted postures, or may reduce the stresses associated with operating in such postures. In unrestricted environments, examples of devices that can reduce the need to adopt awkward postures include lift tables and bin tilters. These devices may reduce the need for the worker to flex the trunk as would be needed to lift items off of the floor or to retrieve items from a large bin.

Often, it may be necessary to develop specialized devices or tools to reduce postural stress in restricted environments. While restrictions in workspace may limit the degree to which certain types of mechanical-assist devices can be employed, experience has shown that it is often possible to develop and fabricate specialized devices or tools that can reduce the risk of musculoskeletal disorders in restricted environments. An example from the coal mining industry is shown in Figure 43.7. This figure shows a specialized cart that rides on conveyor belt structure in a mine and can be used to move heavy supplies in restricted spaces. Various carts, jacks, and hoists can often be used quite effectively to assist with transport of materials in environments with restricted vertical space. Use of such equipment can significantly reduce the threat of musculoskeletal disorders when working in restricted space.

43.5.4   Rest Breaks/Job Rotation

As mentioned earlier in this chapter, restricted spaces tend to force workers into situations where the burden or work will be borne by specific muscle groups, with a limited ability to employ substitute motion patterns as these muscles fatigue. As a result, localized muscle fatigue is likely to develop more quickly in the stressed muscle groups, with an attendant reduction in strength capacity and an increase in the risk of cumulative soft tissue damage and the development of musculoskeletal disorders. As a result, it is important to provide workers with more frequent rest breaks or opportunities to perform alternative tasks that relieve the strain experienced by affected muscle groups. However, while rest breaks and job rotation may be an effective method for reducing fatigue and strain associated with work involving restricted space, use of these methods also serves as an indicator that redesign of the job should be considered.

FIGURE 43.7   Example of a specialized cart to eliminate manual transfer of supplies in a restricted environment.

43.5.5 Personal Protective Equipment

If workers are required to perform tasks in a kneeling posture for any significant period of time, a good pair of kneepads should be provided and worn by the worker so that the risk of inflammation and bursitis can be reduced. Kneepads should provide cushioning foam or gel to reduce contact stresses on the knee joint, especially the patella and the patellar ligament. Often, kneepads are designed with a stiff exterior of plastic or rubber to protect the knee against puncture wounds from sharp objects as might be encountered when kneeling in a rocky or debris-covered surface. Some kneepads are articulated so that they bend with the knee as workers adopt standing and kneeling postures.

43.6 Summary

Many workers adopt unusual or restricted postures during performance of their daily work. Recent research has shown that these postures can cause significant reductions in performance capabilities and are associated with an increase in musculoskeletal complaints. Performance limitations result from the combinations of increased biomechanical loads, higher physiological costs, reduced strength, decreased stability or balance, and by limiting the use of substitute motion patterns to relieve fatigued muscles. Special care needs to be taken in the design of jobs requiring the use of such positions, in order that reduced capabilities can be accommodated. Recommendations based on studies of lifting capabilities in the standing posture may far exceed what should be lifted in restricted postures. The data presented in this review article may provide a starting point for the development of ergonomics recommendations that apply to workers who must cope with work in restricted postures. Mechanical aids can reduce the risk of overexertion, but may need to be custom fabricated when restricted workspaces are present. In many cases, it may be possible to reduce object weights or strength requirements of a task, and increasing the frequency of rest breaks is advisable when awkward postures are used. Job rotation may be an effective strategy if the job to which the worker is rotated allows relief of the muscular fatigue or stress experienced in an unusual or restricted posture.

Though we have learned a substantial amount regarding such working postures in recent years, they remain a challenge to the ergonomics community. Continued development of models robust to changes in whole-body posture should do much to increase our insight into the structure and function of the musculoskeletal system.

References

Ayoub, M.M., Bethea, N.J., Deivanayagam, S., Asfour, S.S., Bakken, G.M., Liles, P., Mital, A., and Sherif, M. (1978). Determination and Modeling of Lifting Capacity. Final Report, Grant #5R010H-0054502, HEW, NIOSH. Texas Tech University, Lubbock, TX.

Ayoub, M.M., Bethea, N.J., Bobo, M., Burford, C.L., Caddel, K., Intaranont, K., Morrissey, S., and Selan, J. 1981. Mining in Low Coal. Volume 1: Biomechanics and Work Physiology. Final Report — U.S. Bureau of Mines Contract No. HO3087022. Texas Tech University, Lubbock, TX.

Basmajian, J.V. and DeLuca, C. (1985). *Muscles Alive: Their Functions Revealed by Electromyography.* 5th edn. Baltimore, MD: Williams and Wilkins.

Bhattacharya, A., Mueller, M., and Putz-Andersson, V. (1985). Traumatogenic factors affecting the knees of carpet installers. *Applied Ergonomics,* 16:243–250.

Bogduk, N. (1997). *Clinical Anatomy of the Lumbar Spine and Sacrum.* 3rd ed., New York: Churchill-Livingstone.

Drury, C.G. (1985). Influence of restricted space on manual materials handling. *Ergonomics,* 28: 167–175.

Floyd, W.F. and Silver, P.H.S. (1955). The function of the erectores spinae muscles in certain movements and postures in man. *Journal of Physiology.,* 129:184–203.

Freivalds, A. and Bise, C.J. (1991). Metabolic analysis of support personnel in low-seam coal mines. *International Journal of Industrial Ergonomics*, 8(2):147–155.

Gallagher, S. (1989). Isometric pushing, pulling, and lifting strengths in three postures. *Proceedings of the Human Factors Society 33rd Annual Meeting*, Human Factors Society, Santa Monica, CA, pp. 637–640.

Gallagher, S. (1991). Acceptable weights and physiological costs of performing combined manual handling tasks in restricted postures. *Ergonomics*, 34(7):939–952.

Gallagher, S., (1997), Trunk extension strength and trunk muscle activity in standing and kneeling postures. *Spine*, 22:1864–1872.

Gallagher, S., (2003), Effects of Torso Flexion on Fatigue Failure of the Human Lumbosacral Spine, Ph.D. Dissertation, The Ohio State University, Columbus, OH, 238 pp.

Gallagher, S. and Unger, R.L. (1990). Lifting in four restricted lifting conditions. *Applied Ergonomics*, 21(3):237–245.

Gallagher, S. and Hamrick, C.A. (1991). The kyphotic lumbar spine: issues in the analysis of the stresses in stooped lifting. *International Journal of Industrial Ergonomics*, 8:33–47.

Gallagher, S. and Hamrick, C.A. (1992). Acceptable workloads for three common mining materials. *Ergonomics*, 35(9):1013–1031.

Gallagher, S., Marras, W.S., and Bobick, T.G. (1988). Lifting in stooped and kneeling postures: Effects on lifting capacity, metabolic costs, and electromyography at eight trunk muscles. *International Journal of Industrial Ergonomics*, 3(1):65–76.

Gallagher S., Hamrick, CA., Cornelius, K., and Redfern, M.S. (2001). The effects of restricted workspace on lumbar spine loading. *Occupational Ergonomics*, 2(4):201–213.

Gallagher, S., Marras, W.S., Davis, K.G., and Kovacs, K. (2002). Effects of posture on dynamic back loading during a cable lifting task. *Ergonomics*, 45(5):380–398.

Garg, A. and Badger, D. (1986). Maximum acceptable weights and maximum voluntary strength for asymmetric lifting. *Ergonomics*, 29:879–892.

Gibbons, L.E. (1989). Summary of Ergonomics Research for the Crew Chief Model Development: Interim Report for Period February 1984 to December 1989. Armstrong Aerospace Medical Research Laboratory Report No. AAMRL-TR-90-038. Wright-Patterson Air Force Base, Dayton, OH, 390 pp.

Grandjean, E. (1988). *Fitting the Task to the Man*. 4th ed. London: Taylor and Francis.

Haselgrave, C.M., Tracy, M.F., and Corlett, E.N. (1997). Strength capability while kneeling. *Ergonomics*, 34(7):939–952.

Holmstrom, E.B., Lindell, J., and Moritz, U. (1992). Low back and neck/shoulder pain in construction workers: occupational workload and psychosocial risk factors. Part 1: Relationship to low back pain. *Spine*, 17(6):663–671.

Lavender, S.A. and Andersson, G.B.J. (1999). Ergonomic principles applied to prevention of injuries to the lower extremity. In: Karwowski and Marras, W.S., eds. *The Occupational Ergonomics Handbook*, Boca Raton, FL: CRC Press, pp. 883–893.

Lawrence, J.S. (1955). Rheumatism in coal miners. Part III. Occupational factors. *British Journal of Industrial Medicine*, 12:249–261.

MacDonald, E.B., Porter, R., Hibbert, C., and Hart, J. (1984). The relationship between spinal canal diameter and back pain in coal miners. *Journal of Occupational Medicine*, 26(1):23–28.

McGill, S.M. (1999). Dynamic low back models: theory and relevance in assisting the ergonomist to reduce the risk of low back injury. In: Karwowski, W., and Marras, W.S., eds. *The Occupational Ergonomics Handbook*. Boca Raton, FL: CRC Press, pp. 945–965.

McGill, S.M. and Norman, R.W. (1987). Reassessment of the role of intra-abdominal pressure in spinal compression. *Ergonomics*, 30(11):1565–1588.

Marras, W.S., Lavender, S.A., Leurgans, S.E., Rajulu, S.L., Allread, W.G., Fathallah, F.A., and Ferguson, S.A. (1993). The role of dynamic three-dimensional motion in occupationally-related low back disorders. The effects of workplace factors, trunk position, and trunk motion characteristics on risk of injury. *Spine*, 18(5):617–628.

Morrissey, S., Bethea, N.J., and Ayoub, M.M. (1983). Task demands for shoveling in non-erect postures. *Ergonomics*, 27:847–853.

Morrissey, S.J., George, C.E, and Ayoub, M.M. (1985). Metabolic costs of stoopwalking and crawling. *Applied Ergonomics*, 16: 99–102.

Myllymaki, T., Tikkakoski, T., Typpo, T., Kivimaki, J., and Suramo, I. (1993). Carpet layer's knee: an ultrasonogrphic study. *Acta Radiologica*, 34:496–499.

Potvin, J.R., McGill, S.M., and Norman, R.W. (1991). Trunk muscle and lumbar ligament contributions to dynamic lifts with varying degrees of trunk flexion. *Spine*, 16(9):1099–1107.

Punnett, L., Fine, L.J., Keyserling, W.M., Herrin, G.D., and Chaffin, D.B. (1991). Back disorders and non-neutral trunk postures of automobile assembly workers. *Scandinavian Journal of Work Environment and Health*, 17:337–346.

Ridd, J.E. (1985). Spatial restraints and intra-abdominal pressure. *Ergonomics*, 28:149–166.

Sharrard, W.J.W. (1963). Aetiology and pathology of beat knee. *British Journal of Industrial Medicine* 20:24–31.

Snook, S.H. (1985). Psychophysical considerations in permissible loads. *Ergonomics*, 28(1):327–330.

Snook, S.H. and Ciriello, V.M. (1991). The design of manual handling tasks: revised tables of maximum acceptable weights and forces. *Ergonomics*, 34(9):1197–1213.

Solomonow, M., Zhou, B.H., Baratta, R.V., Lu, Y., and Harris, M. (1999). Biomechanics of increased exposure to lumbar injury caused by cyclic loading. Part 1. Loss of reflexive muscular stabilization. *Spine*, 24(23):2426–2434.

Tanaka, S., Smith, A.B., Halperin, W., and Jensen, R. (1982). Carpet layer's knee. *The New England Journal of Medicine*, 307:1276–1277.

Waters, T.A., Putz-Anderson, V., Garg, A., and Fine, L.J. (1993). Revised NIOSH equation of the design and evaluation of manual lifting tasks. *Ergonomics*, 36(7):749–777.

44

Agriculture

Fadi A. Fathallah
James M. Meyers
University of California

Larry J. Chapman
Ben-Tzion Karsh
University of Wisconsin

44.1 Introduction

There is clear and consistent evidence that shows musculoskeletal disorders (MSDs) are the most prevalent and expensive of all work-related injuries. These injuries have been positively associated with occupational factors, and rank first in frequency among workers, with approximately half of the nation's workforce being affected (Bernard, 1997; National Research Council/Institute of Medicine, 2001; NIOSH, 2004). Consistent with other industries, MSDs are the most common of all occupational injuries and illnesses for farm workers (Villarejo, 1998; Villarejo and Baron, 1999; McCurdy and Carroll, 2000; McCurdy et al., 2003). MSDs in agriculture are commonly reported at rates near or above those of traumatic injury, respiratory illness, pesticide-related injury or illness, dermatological injury or illness, or other types of injuries and illnesses.

Production agriculture ranks among the top ten of industry subsectors for back pain (Clemmer et al., 1991; Guo et al., 1999). Strain and sprain are the most common disabling injuries in California, accounting between 31 and 43% of reported injuries (AgSafe, 1992; McCurdy et al., 2003), and it is the most common injury in the production of fruits and vegetables nationally (33%) (NIOSH, 2001).

The compelling evidence presented should place MSDs at the top of health and safety issues facing the agricultural community, in general, and the production agriculture industry in particular. However, unfortunately, preventing these disorders has not been a priority for most farm safety groups and organizations. Hence, the objectives of this chapter are to: (1) discuss the multitude of risk factors faced by the main subsectors of the production agriculture industry; (2) discuss, in general, the issue of ergonomics interventions in agriculture; and (3) present a sample of potential intervention strategies that could be implemented in various agricultural commodities. It is hoped that the intervention approaches and resources provided here will help in the process of reducing the prevalence and cost of MSDs in agriculture.

44.2   MSD Risk Factors in Agriculture

There is ample evidence of widespread exposure of those who work in agriculture to severe MSD risk factors on a daily basis. In many cases, MSD risk factor exposures can exceed those found in some of the nonagricultural industries now commonly cited as among the most hazardous for MSDs. In reviewing the work of the University of California Agricultural Ergonomics Research Center (UCAERC) for the past decade, three general risk factors were cited as both endemic and of highest riority throughout the agricultural industry (Meyers et al., 1998, 2000). They are lifting and carrying heavy loads (over 50 lb), sustained or repeated full body bending (stoop), and very highly repetitive hand work (clipping, cutting).

Each type of production agriculture has its own unique MSD hazards and musculoskeletal injury problems, although some hazards are similar throughout production agriculture in general. It should be noted that while many of the types of hazards reported can be said to be of general industrial concern and for which some generic approaches to reduction have been developed, each agricultural commodity imposes unique and specific demands and conditions on the worker. This means that most interventions, even where patterned on proven existing strategies, must be individually addressed. As a result, as compared to other industries, there simply are no ready, off-the-shelf tools and technologies for addressing most occupational MSD risk factors found in agricultural workplaces.

While by no means have all types of agricultural operations and workplaces been subject to risk exposure analysis, research in a number of differing types of crops and commodities clearly demonstrates the types of MSD risk factors present (Chapman et al., 2001). The following sections describe the sources of MSD risk factors in different sectors of the production agriculture industry.

44.2.1   Oil, Seed, and Grain Crops

This farm sector accounted for 24% of all U.S. farms in 1997 and includes soybean, oilseed, dry pea, bean, wheat, corn, rice, and other grain farming (USDA, 1990). Typical work activities with potential MSD hazards that have been studied include driving tractors to prepare, seed, and cultivate the soil; and driving grain combines and trucks to harvest the crops. These tasks involve exposure to various hazards including vibration, lengthy periods of sustained attention, and visual and postural confinement (Chapman et al., 2001).

44.2.2   Vegetable and Melon Crops

This sector accounted for 1.6% of U.S. farms in 1997, includes potatoes, and emphasizes vegetables for processing. Workers who prepare soil, plant, and cultivate are exposed to hazards such as extreme climates, vibration and noise from powered equipment, and poor lighting. Harvest workers are exposed to forceful exertions during cutting, prolonged static postures, and awkward postures (Gite, 1990; Meyers et al., 1995). Other risk factors include handling excessive or asymmetrical weights, repetitive lifting and carrying of heavy loads, highly repetitive handwork, and repetitive trunk flexion during cutting. All of these may increase the risk of having low back disorders. Cutting also entails continuous static gripping of the knife (Meyers et al., 1995, 2000) which may increase risk of hand or wrist disorders (Chapman et al., 2001).

44.2.3 Fruit and Tree Nut Crops

This industry sector includes citrus groves, apple orchards, and grape vineyards, and accounted for 3.5% of farms in the United States in 1997 (USDA, 1999). Workers are exposed to various hazards, particularly during harvesting, including (1) climbing ladders with heavy picking bags, which puts static loading on the shoulders, (2) long reaches, which can lead to instability on picking ladders, and (3) trunk flexion while setting the ladder, which may increase low back pain (Prussia, 1985; Miles and Steinke, 1996; Meyers et al., 2000). The use of shears or knives to cut the fruit also involves MSD risks such as whole body twisting, wrist twisting, high hand forces, and highly repetitive clipping, which contribute to risk of low back and hand or wrist disorders.

Vineyard workers experience multiple exposures to MSD risk factors (Duarj et al., 1999; Meyers et al., 2000; Fathallah et al., 2003). Workers must use high forces to lift, carry, and dump bins, exert heavy lateral forces by the legs to move bins around, and flex their trunk to pick up fallen grapes. These activities have been associated with low back pain. Vineyard workers experience neck extension, poor visibility through the leaves, static grips, and highly repetitive handwork during harvest and pruning. They experience heavy shoulder and arm forces during hand weeding and pruning. They also have to elevate harvest tubs overhead to empty them into field transport bins and they often have poor posture during the removal of excess leaves from bins (Chapman et al., 2001).

44.2.4 Greenhouse, Nursery, and Floriculture Crops

These operations comprised about 3% of U.S. farms in 1997 (USDA, 1999). Workers in this sector are exposed to risk factors such as forceful sustained pinch grips with both hands, severe contact stress on lateral surfaces of the fingers, cold, and repetitive severe trunk flexion (Stoffert and Wildt, 1987; Janowitz et al., 1999). The different tasks in greenhouse and nursery work have specific MSD risk factors (Miles et al., 1997; Meyers et al., 2000), including gripping stresses or highly repetitive clipping, which may increase the risk of hand disorders, and trunk flexion, which can increase the risk of low back disorders.

44.2.5 Fresh Market Vegetable Crops

Fresh market vegetable production made up about 1.6% of all U.S. farms (Chapman et al., 2000). Compared to the larger vegetable (largely for processing) and melons sector, fresh market operations are often smaller, less mechanized, and more labor intensive. Examples of MSD risk factors experienced during fieldwork are bending at the waist (trunk flexion), twisting at the waist (trunk twisting), complex trunk motion (combined forward and side-to-side bending with twisting), and shoulder/arm flexion/abduction. Other MSD risk factors include handling excessive or asymmetrical weights, repetitive lifting and carrying of heavy loads, highly repetitive handwork, and repetitive trunk flexion during cutting. Cutting also entails continuous static gripping of the knife (Meyers et al., 1995, 2000).

44.2.6 Dairy Cattle and Milk Livestock

The dairy sector represents 4.5% of farms in the United States (USDA, 1999). Workers are exposed to hazards such as poor working postures, slippery surfaces, poor lighting, hot and cold environments, poor tool design, prolonged trunk twisting and flexion, arms held above shoulder level, and squatting (Lundqvist et al., 1992; Nevala-Puranen et al., 1993). Milking typically requires the most labor hours. Large dairy farms often use parlors, exposing milkers to highly repetitive and specialized work involving repetitive grasping, reaching, and lifting. In one study, high discomfort rates were reported for the elbow, wrist, and hand and for the shoulder and the neck (Stal et al., 1996). On most U.S. dairy farms there are no specialized milking employees and no pit parlors. Instead, milking machinery is hand carried and attached to and from each cow. This entails repeated squatting, increasing risk of MSDs on lower backs, knees, hips, elbows, and shoulders. Other typical dairy tasks, such as handling hay bales, carrying or shoveling grain, and attaching equipment to tractors all may exceed back compressive force limits (Jorgensen et al., 1990).

44.2.7 Hog and Pig Livestock

Hog and pig farms were about 2.4% of all U.S. farms in 1997 (USDA, 1999). Workers in this sector are exposed to MSD risk factors such as heavy lifting and extended reaches (Hilton, 1997). In hog operations, moving dead animals and loads of feed materials inside short spaces within high hog density facilities may be strenuous and repetitive. Along with grasping and reaching to operate machinery to feed animals, work in swine confinement operations can lead to higher than normal rates of lower back and elbow and wrist discomfort (Chapman et al., 2001).

44.2.8 Poultry and Egg Production Livestock

The poultry and egg production sector totaled about 1.9% of farms in the United States in 1997 (USDA, 1999). Most poultry and egg production facilities involve animal confinement and use conveyors to gather eggs. In certain other types of loose poultry housing, egg collection can be the major repetitive task. Workers in the industry are exposed to hazards such as extended reaches into perches, reaches to collect eggs from the floor, climbing into perches, transporting cages, manually cleaning the nests, and bending to inspect nest boxes (Lundqvist et al., 1992; Scott and Lambe, 1996; Chapman et al., 2001).

44.2.9 General or Mixed Crop and Livestock Agriculture

In these sectors, a wide range of job activities are involved that may vary from one season to the next but that can be extremely repetitive during short periods like crop ripening and harvest. Much of general agricultural work is accomplished from the seat of a tractor. Most crop production agriculture in more industrially developed countries (e.g., corn, small grains like wheat, soybeans, and rice) involves work on mechanized platforms like tractors pulling machinery to plow, cultivate, and apply chemicals or the operation of self-propelled mechanical combines to harvest crops. Repetitive hand and arm movements and confined postures are often required to guide machinery, which may contribute to higher than normal elbow, wrist and hand, and neck and shoulder discomfort (Chapman et al., 2001).

44.3 Types and Examples of Interventions in Agriculture

To address the problem of MSDs in agriculture, interventions must be developed and implemented using a team approach. Management–union ergonomics teams have proved successful in other industries (e.g., Marklin and Wilzbacher, 1998). Such a team for agriculture should include:

- Farm workers
- Farmers
- Farm organizations
- Agriculture industry and other industries (e.g., manufacturers of agricultural equipment and machinery)
- Researchers
- The younger generation

At the top of this list are the farm workers themselves. Looking back at the history of technological innovation in agriculture, most major breakthroughs were initiated by farmers driven to improve their workplace. There is a need to increase worker participation in developing ergonomics interventions in order to provide the feedback on efficiency, comfort, and social and cultural issues that is necessary to improve worker acceptance and understand barriers to adoption. As an example of the importance of ethnocultural issues, prone workstations were rejected by East Indian workers in one California operation because it is not acceptable in their culture for women to be lying down in the company of men. Such ethnocultural barriers may be broken down only through participation.

As in other industries, one can classify the major types of potential interventions for controlling and preventing MSDs in agriculture into administrative and engineering controls.

44.3.1 Administrative Controls in Agriculture

Engineering controls are often preferred for ergonomics interventions, but they do not always exist for many work tasks due to lack of technology, affordability, or practicality, or they take considerable time to develop and implement. In these situations, risk can be mitigated by the use of administrative controls, which use workplace policy, procedures, and practices to change how the worker is exposed to risk. For much of the work in labor-intensive agriculture (e.g., harvesting of fresh produce), no feasible and economical engineering controls are currently available, especially for small farmers, so reducing the MSD risk must be accomplished using administrative controls.

There are several administrative controls available for manual work in agriculture. Pay structures can be examined to determine if switching from a piece-rate to an hourly wage system may reduce overexertion. Introducing programmed breaks, reducing the number of working hours, or hiring more workers may be effective in reducing the demands on the individual worker. Job rotation may be effective in reducing the cumulative trauma from any one particular work task. Moreover, if job rotation is not allowed or unavailable, then the potential of changing postures throughout the day should be explored so that a particular area of the body is not stressed all day. Finally, training managers and workers on reducing risks may be beneficial; however, this strategy has been questioned with regards to its efficacy in reducing MSDs in other industries.

The most potentially effective administrative control for reducing the risks of working in labor-intensive agriculture is organized rest breaks. Previous study by other researchers has evaluated rest breaks as intervention for MSDs in video display terminal usage, meat and poultry processing, and construction and have led to the following guidelines (Konz, 1998):

- Minimize the fatigue dose
- Use frequent short breaks
- Maximize the recovery rate
- Use work breaks
- Increase recovery/work ratio
- Have a work scheduling policy
- Optimize stimulation at work

In addition to these common administrative controls, several alternative ways of handling the problem of labor-intensive agriculture deserve consideration. For example, growers can eliminate the stooped work at harvest and its costs by having a "You-Pick" farm, where customers harvest their own fruits and vegetables for discounted prices. In France, this model has expanded to "tourist farming," where urban customers get the added benefit of experiencing the activities of rural life. Marketing the qualities of rural living and the traditions of farming may allow manually harvested commodities to be perceived as value-added and receive a premium, with the extra income being available to pay for workplace improvements. Promoting the benefits of rural life may also be used to improve the psychosocial environment for workers.

44.3.2 Engineering Controls for Agriculture

Engineering controls for labor-intensive agriculture can be classified into three classes of interventions: (1) approaches to alter workspace–worker interface; (2) mechanical worker protection or worker aids; and (3) fully mechanized operations.

44.3.2.1 Interventions that Alter Workspace–Worker Interface

One approach to eliminating some of the MSD risk factors in agriculture is through changing the physical work space. This can be done by changing the workplace or by giving the workers tools that allow them to interact with the workplace differently. There are several changes that can be made to the spatial workplace in agriculture. One is to raise the beds that crops are grown in. For example, strawberries are grown in raised beds and this reduces the degree of stooping compared to other crops. However, it is not known whether raised beds are feasible for other crops. Nursery and greenhouse work provide the best opportunities for changing the geometry of the workspace. Here crops can more feasibly be grown on raised beds or on tables to raise the working height. Employers can provide portable tables or carts and lifting aids and adjustable tables, as used in manufacturing, so that work operations can be performed at the proper working height. A further adaptation for greenhouses is the use of revolving carousel tables to increase the growing area while offering an adjustable working height. An alternative to changing the geometry of the growing environment is to change the geometry of the crop itself. This could consist of growing taller plant varieties or breeding or bioengineer plant varieties with the harvested commodity located at a more comfortable height for the worker.

Improved hand tool design may also reduce or eliminate the needs for stooped or squatting work. It is possible to redesign existing tools and methods to reduce risks, or introduce entirely new types of tools. An example is an improvement for rice paddy plowing in India. This job has high energy consumption and necessitates the use of a stooped posture to regulate forces applied to control plow depth. A research team determined an optimal height for the plow handle by studying discomfort, applied force, and oxygen uptake for different plow designs. Another class of hand tool development is developing tools with the functional equivalent of extending the arms. The UCAERC team was successful in developing extended handle carriers for potted plants that eliminate the needs to stoop or squat in order to move them. These examples exemplify that simple low-cost solutions exist for controlling the problem of stooped postures in agricultural work, but are often only realized through the focus of an ergonomics team.

44.3.2.2 Mechanical Worker Protection or Worker Aids

Mechanical worker protections or worker aids act to reduce the physical loading or contact stress on various joints and regions of the body during work. This approach is mostly applicable to the prevention of MSDs of the lower back and the lower limbs due to the prevalence of stooping and squatting postures in labor-intensive agriculture. There are three types of mechanical worker protections and worker aids that have the potential to control stooped or squat postures in agricultural production work: devices for kneeling, prone workstations, and load transfer devices.

44.3.2.2.1 Improving Kneeling Comfort

Kneeling postures dominate in work that is very low to the ground and tedious and requires a great deal of hand–eye coordination. In agriculture, such kneeling work is common in plant propagation for nursery and ornamental plant production. To reduce the discomfort of kneeling, workers can be provided knee pads. However, even with knee pads, kneeling and sliding oneself around on a rough soil surface can still be very discomforting. A much more stable and comfortable kneeling surface is provided by knee boards. They consist of a board covered with generous foam padding that workers can kneel entirely on. Knee boards offer protection for the shin and feet in addition to the knee. All-terrain wheels allow the boards to roll forward and backward as needed. Because pushing while kneeling is difficult, the UCAERC has developed a knee board with a hand-powered ratchet to drive the wheels. This knee board also has a storage area for work materials so they do not have to twist or bend as much.

Though it is questionable how much knee pads or knee boards prevent the MSD risks of working in a kneeling posture, favorable worker acceptance indicates they reduce some of the strain from the work.

44.3.2.2.2 *Prone Workstations*

The commonly observed stooped and squat postures can be eliminated in many agricultural tasks, such as harvesting or hand weeding, by using a workstation that supports the body in a more neutral spinal posture; for example, lying prone or sitting upright. Most development has focused on prone posture workstations because workers rarely have to deviate from a neutral spinal posture; in sitting posture workstations, the worker still has to frequently bend or twist the back at moderate angles to reach the plant material.

There are several issues to consider in optimizing the prone workstation for comfortable work. The first issue to consider is the prone posture itself, whether the body should be completely straight and level or whether the arms and legs should be bent and trunk inclined or declined. Research from underwater and zero-gravity studies has identified an optimal body posture of lying supine with the arms and legs bent at moderate angles, and this has led many to believe that inverting the zero-gravity posture is the optimal prone posture. The need for a head rest is a concern, as is whether the worker should face head-first or feet first. Related to the issue of comfort is whether the workstation should be tractor mounted, self-propelled, or human powered.

Additional considerations, besides comfort, should be also looked into when evaluating prone workstations. First, more research into human performance issues related to strength, endurance, and energy usage while working in the prone posture and attention to job rotation and worker scheduling are necessary in order to optimize worker efficiency and reduce the risks of prolonged static postures. Necessary ergonomics improvements for workstation design are enhancing motorized machines, by reducing whole body vibration and providing a feedback system for workers to set the proper pace, and developing guidelines for user-built prone workstations that minimize hazards and provide worker comfort. Finally, it is necessary to study the economic impact of adoption of prone workstations on larger operations for both management and laborers.

44.3.2.2.3 *Load Transfer Devices*

Load transfer devices are systems worn by workers and function to reduce the load on the lower back by transferring a portion of the external load to the hips and lower limbs. Examples of these devices are described in Section 44.4.

44.3.2.3 Mechanized Operations

Mechanization of agricultural tasks has the potential to eliminate the needs for stooped work in agriculture. Numerous past successes have reduced the need for stooping or hand labor in agricultural work, with some examples being the cotton picker, hay mower, mechanical planter, and the processing tomato harvester. Though these inventions have the desirable benefit of reducing stooped work, their successful implementation occurred primarily because they tremendously increased productivity. History shows that while such technological progress may reduce or eliminate one form of manual work, with it comes new types of ergonomics problems. Movement to mechanization may eliminate stooped work, but may increase exposure to risk factors for equipment operators such as noise and vibration, static work postures, and repetitive motion. Reducing the physical demands of work does not necessarily reduce the occupational health risks of the job. This is an important consideration in developing engineering controls for stooped and squatting postures in the workplace.

Despite the success of mechanization in many areas of agriculture, the production of most fruits and vegetables, even in industrialized nations, continues to be highly reliant on hand labor to perform many essential work tasks, especially harvesting and weeding. Current mechanical harvesting methods remain technically and economically unviable in many fruit and vegetable crops, especially for those intended for the fresh market, because they create excessive mechanical damage and cannot be used in crops with indeterminate maturity (e.g., strawberries).

Regardless of these limitations, several vegetable and fruit crops can currently be harvested mechanically. The UC Davis Postharvest Technology Group website (http://postharvest.ucdavis.edu) provides links to several manufacturers of fruit and vegetable harvesting equipment. For fresh market produce, possibilities exist to develop machines that redefine the ergonomics of the working environment. The stooping that is necessary to cut off vegetables at ground level for harvest can be eliminated by using a machine that cuts them and elevates them to waist height for sorting and packaging. Research work on elevating harvester machines has been conducted at UC Davis for cilantro, and there are commercial models for many leafy vegetable crops, such as spinach and lettuce varieties, that are for sale in North America and Europe. The use of such machines is annually increasing because of increased productivity and worker approval. Elevating harvesters are showing much promise in reducing the risk from stooped work in leafy vegetable harvest, and more options should be explored.

Complete automation of agricultural tasks is an emerging technology, which can eliminate the need for workers in high-injury risk tasks, especially those requiring heavy lifting and work in stooped postures. The first wide-scale use of automation to replace a traditionally manual labor task is automated (robotic) milking systems. First developed in Europe, they have become very common there and their usage is gaining in North America, particularly among smaller producers. The automated milking machines are shown to save time, eliminate a source of physical strain, and save money despite their high initial investment costs. Similar breakthroughs in automated agricultural systems are expected in the next 10 to 20 yr due to promising research currently being undertaken. The problem of stooped work while weeding may be tremendously reduced through a robotic weed control system that uses technologies being developed at the University of California at Davis (Lamm et al., 2002). Automated selective harvest of tree fruit and citrus is being researched in several industrial nations and such technology could lead to automated harvesters for fruit and vegetable crops. The outlook for the future suggests increased mechanization in agriculture, and, with it, a reduction in the demand for work tasks performed in stooped postures. However, these achievements are still several decades away and do not address the problem in nonindustrialized nations that cannot afford such technology.

44.4   Intervention Examples in Agriculture

In the past decade, a few small multidisciplinary teams of researchers and extension staff have undertaken efforts to develop organized intervention and prevention programs based on an ergonomics approach to the problems of specific tools and tasks encountered in agricultural workplaces. These programs have been largely successful in developing low-cost intervention strategies, which have proven acceptable to farmers and farmworkers and have also proven effective at significantly reducing specific risk exposures. A recent NIOSH publication entitled "Simple Solutions" highlights many of the most successful of these interventions (Baron et al., 2001). Some examples from various agricultural sectors follow.

44.4.1   Fruit and Tree Crops

To reduce the problems associated with citrus picking, a picking bag that fits as a vest over both shoulders instead of hanging from one shoulder has been used and picking from the top of the tree to the bottom helps to minimize load carried on the ladder (Wick, 1992). Changes can also be made to the picking ladder itself. Researchers have suggested redesigning ladders so that steps are more level, larger, and more slip resistant (Miles and Steinke, 1996). Miles and his colleagues in California demonstrated the impact of using smaller picking bags to reduce loads carried.

In the wine industry, awkward postures have been reduced using a waist-level hopper for dumping full tubs of grapes and a sorting conveyor for inspecting and removing poor-quality grapes (Duarj et al., 1999). Very significant reduction in back injury risk to winegrape harvesters were achieved by reducing the size of picking tubs for hand harvest (UCAERC, 2000). Harvesting aids have also been used to reduce postural problems associated with harvesting strawberries (Fitzpatrick, 1979; Theriault, 1988; Theriault et al., 1990). Harvesting aides for fruit trees that simplify fruit detachment such as extended poles to shake fruit loose or mechanical hand-held vibrators have been recommended as well (Prussia, 1985).

44.4.2 Plant Nurseries

New workstations to accommodate various worker heights were applied in plant propagation work (Meyers et al., 1995). To reduce risk while moving plants in cans, redesigned handles for comfortable power grips and reduced need for trunk flexion were applied (Janowitz et al., 1999). Transportable benches that can elevate for moving plants around the greenhouse have also been proposed (Kabala and Giacomelli, 1992), as have pipe-rail systems to allow for sitting or standing while working (Hendrix, 1986). Using lightweight handles for lifting and carrying cans, using hydraulic lifts to minimize lifting of cans, and using power clippers and cutters for pruning and cutting have been used to reduce repetitive gripping and lifting (Miles et al., 1997).

44.4.3 Fresh Market Vegetable Crops

One way to reduce the repetitive lifting and awkward postures that normally accompany the harvest and washing of leafy crops is use of mesh bags. By lining harvest containers with mesh bags before harvest, one can simply pick up a filled mesh bag from the container, dunk it in the wash tank, and then remove the entire bag after washing. This helps to eliminate repetitive bending and stooping during the wash process (Meyers et al., 1999). Standard containers that have sturdy handles make lifting and carrying easier on the body than bushel baskets, buckets, or wooden crates since they have good handles and a center of gravity close to the body (Newenhouse et al., 1999). Another means of reducing exposure to lifting and carrying hazards is to use hand pallet trucks with half-pallets to transport boxed produce (Meyers et al., 1999). Properly designed tools can also help. For example, in a study of three different shovel handle designs it was found that shovels with hollow fiberglass ribbed handles were associated with significantly lower arm muscle activity, grip force, and ratings of slipperiness and better efficiency than shovels with traditional wood handles or smooth fiberglass handles (Chang et al., 1999).

44.4.3.1 Prone Workstations Examples

Carts that allow fieldwork in seated or prone positions are less strenuous than working in stooped or crawling postures (Meyer et al., 1999; University of Wisconsin, 1999). Prone work platforms can take the form of human-powered individual carts, self-propelled individual carts, or work platforms with multiple prone workstations attached to a tractor or with their own integrated power unit. Most of the current development is occurring in Europe, although interest is increasing in North America. Prone workstations first appeared on American farms about 50 yr ago. Early development focused on mechanical design and improving productivity, while comfort seemed secondary. Studies from the 1970s describe use in strawberry and nursery operations and report productivity increases and preference for workers, though comfort was often a secondary concern. Subsequent development has centered on making prone workstations more comfortable. Researchers from the UCAERC have built and tested several human-powered and self-propelled carts for its agricultural ergonomics programs. The University of Wisconsin has provided guidelines for building human-powered seated and prone cars, and tested some of the currently manufactured self-propelled models (University of Wisconsin, 1999).

There are currently no commercial manufacturers of prone workstations in the United States. Almost all of the commercially available self-propelled prone workstations are designed and manufactured in Europe. There are at least two German manufacturers of motorized prone workstations (d'Heureuse Inc. and Kress Inc.). Another German company has produced a tractor-mounted, multistation prone work platform for cucumber harvesting. From Sweden, the Drängen is a single-person, track-driven motorized machine with a very adjustable and well-padded workstation and hydrostatic drive. The

Ryömijä is battery-powered, wheeled single-person machine produced in Finland with an adjustable workstation quite similar to the Drängen. Unfortunately, the European models are relatively expensive and either are not exported to the United States, or in case of the Drängen, lack local technical support. Given the low amount of technology, it is entirely possible that similar designs can be manufactured in the United States should prone workstations gain popularity.

44.3.4.2 Load Transfer Devices Examples

Load transfer devices function to reduce the load on the lower back by transferring a portion of the external load to the hips and lower limbs. A study concerned load transfer devices for use in stooped work in agriculture and construction was conducted (Barrett and Fathallah, 2001). The laboratory evaluation examined four load transfer devices for their ability to reduce back muscle activity. At the time of the study in 2000, there were three commercially available, with each using a different concept to support the upper body while working in a stooped posture. The Happyback was manufactured by Ergo-Ag of Aptos, CA, and utilized fiberglass rods and fabric to support a chest harness and transfer torso loads through a pad near the low back to thigh straps buckled above the knees. "Bending Non-Demand Return" has metal frame segments for the anterior torso and upper legs that are padded at the chest and thighs and connect to a resistive articulation and support belt at the hips. The Bendezy is an Australian design consisting of an aluminum frame with soft shoulder straps around the upper torso, low back pad, and abdominal straps to hold it in place. Resistance to torso load is provided by springs extending from a posterior counterweight lever to straps on the knees and feet. An additional fourth was tested that consisted of a modified Bendezy prototype having the springs and lower limb straps removed and a posterior counterweight added. The experiment consisted of nine subjects assuming stooped postures and lifting three weighs (0, 10, 20 lb) while wearing each device and for the control condition of a traditional stooped posture. During the lifts, muscle activities of the trunk and legs were measured by EMG. The EMG results showed reduced back erector muscle activity provided by all four load transfer devices. The devices all act to reduce the loading on the back, though the impact of this load reduction on low back disorder risk is uncertain. Additional evaluations by workers suggested that the concept of load transfer is worth exploring but that much improvement in comfort and adjustability is needed for them to be practical. Another laboratory study that explored the idea of using the weight picking tub to counterbalance the weight of the upper body weight during harvesting did not find a reduction in trunk muscle activities when compared to normal picking (Mirka and Shin, 2003).

Personal load transfer devices show the potential of providing significant reduction in loads imposed on the spine during stooped work. Further research is needed to evaluate their efficacy in reducing low back disorder risk, as well as risks arising from increased loading on the legs. More attention to anthropometric and comfort issues is needed in order to increase the feasibility and practicality of such devices. It is hoped that successful devices will become available to reduce the low back disorder risk of stooped work when other means of modifying the workplace are not possible.

44.4.4 Dairy and Cattle Production

A variety of techniques have been used to minimize ergonomics problems associated with dairy farming. For example, improving temperature and humidity, using better methods for confining animals, and improving lighting all improve the general work environment, while using a rail system during barn milking reduces repetitive squatting and stooping (Lundqvist et al., 1992; Nevala-Puranen et al., 1993; Josefsson et al., 1999a). Horizontal storage of silage instead of utilizing tower silos can reduce climbing and other hazards (Josefsson et al., 1999b). Training milkers in proper work methods, rest breaks, and proper tool use (Lundqvist et al., 1997) has been shown to significantly better work postures in dairy farmers (Nevala-Puranen, 1995). Ergonomics interventions have been recently introduced in the beef cow industry to reduce the risks of upper arm and low back disorders during manual calves weighing process (Southard et al., 2004).

44.4.5 Administrative Control Example — Rest Breaks

A study has been conducted looking at rest and recovery breaks as potential interventions for MSDs (Faucett et al., submitted). The motivation for testing organized rest breaks comes from symptom survey results for agricultural workers who frequently kneel or stoop that indicate the back and knees are the most affected body parts. Current rest break scheduling for agricultural workers does not have the frequency of rest breaks suggested by these guidelines. For example, in California, regulations require that employers allow employees to have a 10-min rest break for every 4 h worked and a 30 min meal period for any worker who works more than 5 h a day. Recent regulations in California require employers to provide a 5 min rest break for every hour of work during manual weeding operations. The Faucett et al. study conducted a controlled, randomized experiment to assess organized rest breaks as interventions for agriculture. Two jobs were analyzed: tree budding and grafting for nurseries and harvesting for strawberries. Measures were worker symptom surveys and productivity. Two rest break conditions were tested — the control case having scheduled breaks of 10 min mid-morning, and mid-afternoon and a 30-min lunch, and the intervention case having additional 5 min breaks each hour in which there was not already a scheduled break (approximately 20 additional minutes of rest per shift). The results of the study show that intermittent brief rest breaks appear to reduce the symptoms of fatigue and musculoskeletal discomfort, while productivity appears to be minimally affected. After cessation of the study, workers and managers in the tree nursery cooperatively agreed to continue the use of breaks during both hourly tasks and piece rate tasks. The UCAERC group has seen similar adoption with its engineering solutions for the wine grape and nursery industries. Workers continuing to use the intervention are often the best indicators of success.

The findings of this study emphasize the potential of rest breaks as an intervention for MSDs in the workplace, though much more research is needed. One such need is to develop and test optimal rest and work patterns for specific work tasks, like those requiring prolonged stooped or kneeling, repetitive cutting and clipping, sedentary postures, and mental strain. Another need is to understand the effects of rest break scheduling on musculoskeletal symptoms and on other outcome measures such as leg swelling, general fatigue, and heat stress. Lastly, there must be more research to evaluate the effects of rest breaks on health outcomes and on productivity over the course of the work day.

44.5 Useful Internet Resources

The following internet sites could be useful in identifying various ergonomics interventions in agriculture and for general health and safety issues in agriculture:

- The University of California Agricultural Ergonomics Research Center: http://ag-ergo.ucdavis.edu/
- The University of Wisconsin Healthy Farmers, Healthy Profits Project: http://bse.wisc.edu/hfhp/
- NIOSH document: "Simple solutions: ergonomics for farm workers": http://www.cdc.gov/niosh/01-111pd.html
- The National Institute for Occupational Safety and Health Agricultural Centers: http://www.cdc.gov/niosh/agctrhom.html
- The California Institute for Rural Studies: http://www.cirsinc.org/
- The National Center for Farmworker Health: http://www.ncfh.org/
- The National AgrAbility Project (interventions for people with disability): http://www.agrabilityproject.org/

44.6 Conclusions

Efforts to prevent MSDs must recognize that agricultural work is diverse and so agricultural ergonomics must bridge many specific problems with a flexible, generic approach. The rewards for careful attention to ergonomics include a more efficient production process, lower labor costs, reduced injury absences and turnover, and reduced expenditures for medical care and worker compensation as well as a reduced toll attributable to MSDs. With sufficient attention to the larger goals of whatever work is underway, investments in ergonomics can often pay for themselves many times over (Oxenburgh, 1997).

Acknowledgments

The authors would like to extend their thanks to investigators and supporting staff at the University of California Agricultural Ergonomics Research Center, including Dr John Miles, Dr Julia Faucett, Ira Janowitz, Victor Duraj, Diana Tejeda, and Mir Shafi, for their assistance. A special thank you goes to Brandon Miller for his valuable contribution to this chapter. We also would like to thank the National Institute for Occupational Safety and Health (NIOSH) for supporting the authors' research efforts.

References

AgSafe (1992) *Occupational Injuries in California Agriculture: 1981–1990.* AgSafe, University of California Division of Agricultural and Natural Resources, Oakland, CA.

Baron, S.L., Estill, C., Steege, A., and Lalich, N. (2001) *Simple Solutions: Ergonomics for Farm Workers.* Report 01–111. National Institute for Occupational Safety and Health, Cincinnati, OH.

Barrett, A. and Fathallah, F.A. (2001) Evaluation of four weight transfer devices for reducing loads on the lower back during agricultural stoop labor. Paper presented at the American Society of Agricultural Engineers Annual Meeting.

Bernard, B.P., ed. (1997) *Musculoskeletal Disorders and Workplace Factors: A Critical Review of Epidemiologic Evidence for Work-Related Musculoskeletal Disorders of the Neck, Upper Extremity, and Low Back.* U.S. Department of Health and Human Services, National Institute for Occupational Safety and Health, Cincinnati, OH.

Chang, S.R., Park, S., and Freivalds, A. (1999) Ergonomic evaluation of the effects of handle types on garden tools. *International Journal of Industrial Ergonomics* **24**, 99–105.

Chapman, L.J., Karsh, B., Meyers, J.M., Newenhouse, A.C., Meyer, R.H., Miles, J.A., and Janowitz, I., eds. (2001) *Ergonomics and Musculoskeletal Injuries in Agriculture.* American Conference of Governmental Industrial Hygienists, Cincinnati, OH.

Chapman, L.J., Joseffon, K.G., Meyer, R.H., Newenhouse, A.C., Karsh, B., and Miquelon, M. (2000) Intervention research to reduce injuries on dairy, berry, and fresh market vegetable farms. Paper presented at the NIOSH Meeting, Cooperstown, NY.

Clemmer, D.I., Mohr, D.L., and Mercer, D.J. (1991) Low-back injuries in a heavy industry I: worker and workplace factors. *Spine* **16**, 824–830.

Duarj, V., Miles, J.A., and Meyers, J.M. (1999) Development of a conveyor-based loading system for reducing ergonomic risks in manual harvest of wine grapes. Paper presented at the American Society of Agricultural Engineers Annual Meeting, Toronto, Canada.

Fathallah, F.A., Miles, J.A., Faucett, F., Meyers, J.M., Janowitz, I., Kato, A.E., Garcia, E., Reiter, D.A., Miller, B.J., and Tejeda, D.G. (2003) Ergonomic evaluation of pruning and harvesting tasks of winegrape trellis systems. Paper presented at the International Ergonomics Association 15th Triennial Congress, Seoul, Korea.

Faucett, F., Meyers, J.M., Miles, J.A., Janowitz, I., and Fathallah, F.A. (submitted) Rest break interventions in stoop or squat job tasks. *Applied Ergonomics.*

Fitzpatrick, J. (1979) Strawberry picking — an appropriate way. *Appropriate Technology* **6**, 4–5.

Gite, L.P. (1990) Quality of work life of farm workers and ergonomics. *Agricultural Engineering Today* **14**, 5–17.

Guo, H.R., Tanaka, S., Halperin, W.E., and Cameron, L.L. (1999) Back pain prevalence in US industry and estimates of lost workdays. *American Journal of Public Health* **89**, 1029–1035.

Hendrix, A.T. (1986) The pipe-rail system in glasshouse vegetable-cropping. *Acta Horticulturae* **187**, 137–143.

Hilton, W. (1997) Ergonomic job design in the hog industry. In *The Ergonomic Casebook*, Kohn, J., ed., Lewis, New York.

Janowitz, I., Faucett, J., Meyers, J.M., Tarter, M., Duraj, V., Miles, J.A., and Tejeda, D.G. (1999) *Reducing Work-Related Musculoskeletal Disorder Risk Factors and Symptoms in Plant Nurseries Through Material Handling Modifications.* University of California Agricultural Ergonomics Research Center, Davis, CA.

Jorgensen, M., Bishu, R.R., Chen, Y., and Reily, M.W., eds. (1990) *Handling Activities in Farming — An Ergonomic Analysis.* Taylor and Francis, London.

Josefsson, G., Miquelon, M., and Chapman, L. (1999a) *Long Day Lighting for Dairy Housing: A Work Efficiency Tip Sheet.* University of Wisconsin Biological Systems Engineering Department, Madison, WI.

Josefsson, G., Miquelon, M., and Chapman, L. (1999b) *Use Silage Bags: A Work Efficiency Tip Sheet.* University of Wisconsin Biological Systems Engineering Department, Madison, WI.

Kabala, W.P. and Giacomelli, G.A. (1992) Transportation and elevation system for greenhouse crops. *Applied Engineering in Agriculture* **8**, 133–139.

Konz, S. (1998) Work/rest: Part I — guidelines for the practitioner. *International Journal of Industrial Ergonomics* **22**, 67–71.

Lamm, R.D., Slaughter, D.C., and Giles, D.K. (2002) Precision weed control system for cotton. *Transactions of the ASAE* **45**, 231–238.

Lundqvist, P., Pinzke, S., and Gustafsson, B., eds. (1992) *Ergonomic Factors in the Work Situation for AI-Technicians on Animal Farms.* Taylor and Francis, London.

Lundqvist, P., Stal, M. and Pinzke, S. (1997) Ergonomics of cow milking in Sweden. *Journal of Agromedicine* **4**, 169–176.

Marklin, R.W. and Wilzbacher, J.R. (1998) Ergonomics intervention in warehousing. In *Advances in Occupational Ergonomics and Safety — Proceedings of the XIIIth Annual International Occupational Ergonomics and Safety Conference*, Kumar, S., ed., pp. 328–331. IOS Press, Amsterdam.

McCurdy, S.A. and Carroll, D.J. (2000) Agricultural injury. *American Journal of Industrial Medicine* **38**, 463–480.

McCurdy, S.A., Samuels, S.J., Carroll, D.J., Beaumont, J.J., and Morrin, L.A. (2003) Agricultural injury in California migrant Hispanic farm workers. *American Journal of Industrial Medicine* **44**, 225–235.

Meyer, R.H., Newenhouse, A.C., and Chapman, L.J. (1999) *Work Efficiency Tip Sheet on Specialized Field Carts.* University of Wisconsin Biological Systems Engineering Department, Madison, WI.

Meyers, J.M., Bloomberg, L., Faucett, J., Janowitz, I., and Miles, J.A. (1995) Using ergonomics in the prevention of musculoskeletal cumulative trauma injuries in agriculture: learning from the mistakes of others. *Journal of Agromedicine* **2**, 7–15.

Meyers, J.M., Miles, J.A., Faucett, J., Janowitz, I., Tejeda, D.G., Duraj, V., Kabashima, J., Weber, E., and Smith, R. (1998) *High Risk Tasks for Musculoskeletal Disorders in Agriculture Field Work.* University of California Agricultural Ergonomics Research Center, Davis, CA.

Meyers, J., Faucett, J., Miles, J., Janowitz, I., Tejeda, D., Duraj, V., Smith, R., and Weber, E. (1999) Effect of reduced load weights on musculoskeletal disorder symptoms in wine grape harvest work. Paper presented at the American Public Health Association.

Meyers, J.M., Miles, J.A., Faucett, J., Janowitz, I., Tejeda, D.G., Duraj, V., Kabashima, J., Smith, R., and Weber, E. (2000) *High risk tasks for musculoskeletal disorders in agricultural field work.* Paper presented at the IEA 2000/HFES 2000 Congress, San Diego, CA.

Miles, J.A. and Steinke, W.E. (1996) Citrus workers resist ergonomic modifications to picking ladder. *Journal of Agricultural Safety and Health* **2**, 7–15.

Miles, J.A., Meyers, J.M., Faucett, J., Janowitz, I., Kabashima, J., Tejeda, D.G., and Duraj, V. (1997) Tool design parameters for preventing musculoskeletal injury in nursery work. Paper presented at the American Society of Agricultural Engineers Annual Meeting, Minneapolis, Minnesota.

Mirka, G.A. and Shin, G. (2003) Harvesting ground-level crops: a biomechanical assessment and ergonomic intervention. Paper presented at the American Society of Agricultural Engineers Annual Meeting.

National Research Council/Institute of Medicine (2001) *Musculoskeletal Disorders and the Workplace: Low Back and Upper Extremities.* National Academy Press, Washington, DC.

Nevala-Puranen, N., Taatola, K., and Venaelaeinen-Juna, M. (1993) Rail system decreases physical strain in milking. *International Journal of Industrial Ergonomics* 12, 311–316.

Nevala-Puranen, N. (1995) Reduction of farmers' postural load during occupationally oriented medical rehabilitation. *Applied Ergonomics* 26, 411–415.

Newenhouse, A.C., Meyer, R.H., and Chapman, L.J. (1999) *Standard Containers: A Work Efficiency Tip Sheet.* University of Wisconsin Biological Systems Engineering Department, Madison, WI.

NIOSH (2001) *Injuries Among Farm Workers in the United States, 1995.* Report 2001-153. National Institute for Occupational Safety and Health (NIOSH), Cincinnati, OH.

NIOSH (2004) *Worker Health Chartbook, 2004.* Report 2004-146. National Institute for Occupational Safety and Health (NIOSH), Cincinnati, OH.

Oxenburgh, M.S. (1997) Cost–benefit analysis of ergonomics programs. *American Industrial Hygiene Association Journal* 58, 150–156.

Prussia, S.E. (1985) Ergonomics of manual harvesting. *Applied Ergonomics* 16, 209–215.

Scott, G.B. and Lambe, N.R. (1996) Working practices in a perchery system, using the OVAKO working posture analysing system (OWAS). *Applied Ergonomics* 27, 281–284.

Southard, S.A., Freeman, J.H., Drum, J.E., and Mirka, G.A. (2004) Ergonomic interventions designed for obtaining the birth weight of calves. Paper presented at the American Society of Agricultural Engineers Annual Meeting.

Stal, M., Moritz, U., Gustafsson, B. and Johnsson, B. (1996) Milking is a high-risk job for young females. *Scandinavian Journal of Rehabilitation Medicine* 28, 95–104.

Stoffert, G. and Wildt, U. (1987) Work physiological investigations of potplants handling on the ground. *Acta Horticulturae* 203, 95–100.

Theriault, R. (1988) Design of a harvesting-aid for strawberries. Paper presented at the American Society of Agricultural Engineers Annual Meeting.

Theriault, R., Mouceddeb, B., and Trigui, M. (1990) Power handling and self-steering for a strawberry harvesting-aid. Paper presented at the American Society of Agricultural Engineers Annual Meeting.

UCAERC (2000) *California Vineyard Ergonomics Partnership Project — Final Report.* U.C. Agricultural Ergonomics Research Center, NIOSH Community Partners for Healthy Farming Program, Davis, CA.

University of Wisconsin (1999) *University of Wisconsin Healthy Farmers, Healthy Profits Project.* Univeristy of Wisconsin, Madison, WI.

USDA (1990) *Characteristics of Agricultural Workforce Households.* U.S. Department of Agriculture (USDA), Washington, DC.

USDA (1999) *1997 Census of Agriculture.* U.S. Department of Agriculture (USDA), Washington, DC.

Villarejo, D. (1998) Occupational injury rates among hired farm workers. *Journal of Agricultural Safety and Health* (Special Issue), 39–46.

Villarejo, D. and Baron, S.L. (1999) The occupational health status of hired farm workers. *Occupational Medicine* 14, 613–635.

Wick, J.L., ed. (1992) *Ergonomic Factors in the Work Situation for AI-Technicians on Animal Farms.* Taylor and Francis, London.

45

Grocery Distribution Centers

Michael J. Jorgensen
Wichita State University

William S. Marras
W. Gary Allread
The Ohio State University

Carol Stuart-Buttle
Stuart-Buttle Ergonomics

45.1 Introduction

Musculoskeletal disorders (MSDs) such as strains and sprains to the low back and other body joints has historically occurred at a high rate in the food distribution centers compared with other industries. Published prospective epidemiological studies in retail warehouse settings with similar manual materials handling activities to grocery distribution centers have suggested that the intensity of the workload and lifting are positively associated with reporting of low back injuries (Kraus et al., 1997; Gardner et al., 1999). However, other research suggests that the workloads in grocery distribution centers may be greater than those found in the retail warehouse settings. Kuorinka et al. (1994) estimated that torso twisting occurred in 77% of box transfers observed in a grocery warehouse, whereas Drury et al. (1982) found that fewer than 20% of box handling tasks such as pallet loading and order picking were free from torso twisting. The National Institute for Occupational Safety and Health (NIOSH) reported that manual case handling rates in two grocery distribution centers they investigated exceeded 200 cases per hour (NIOSH, 1993a,b), back injuries among order selectors accounted for almost 60% of all lost workdays over a 5-yr period in one distribution center (NIOSH, 1993a), and the back injury incident rates ranged between 16 and 25 per 100 workers per year (NIOSH, 1993a,b). Furthermore, the National Association of Wholesale Grocers of America (NAWGA) and the International Foodservice Distribution Association (IFDA) reported that 30% of the injuries reported by food distribution warehouse workers were attributable to back sprains and strains (Waters, 1993). Thus, MSDs, and in particular those occurring to the low back, are a major source of lost time, and represent a significant occupational health concern to this industry. Manual material handling is a major activity in grocery distribution centers, where many aspects of the distribution centers contribute to the overall exposure to manual material handling, and, thus, contribute to the risk of MSDs. While it is recognized that it will be virtually impossible to eliminate all manual handling in distribution centers, strategies do exist that can reduce the frequency of manual handling, as well as the situations that increase the risk of MSDs due to manual handling. Thus, the objectives of this chapter are to identify and describe different grocery distribution centers currently in use, the ergonomic concerns related to each type, identify distribution center system attributes that increase the risk of MSD from material handling, and identify potential intervention strategies that can reduce the exposure to high-risk manual material handling tasks.

45.2 Grocery Distribution Center Systems

There are a number of different systems that are used to move product through grocery distribution centers. These include a traditional order pick system, a belt-pick, a cross-dock, and a flow-through system. Each of these different systems have their own ergonomics concerns; thus, many of the potential MSD intervention strategies will be dependent upon which system is in place at a particular location. Additionally, combinations of these four systems may be present in some facilities. The four different systems are described next.

45.2.1 Traditional Order Pick System

The traditional order pick system is the most common method of order picking in food distribution centers because it makes good use of vertical space in a warehouse. In this system, pallets of like product are typically stored above the actual pick slots and then are brought down as full pallets as needed and placed in slots. Employees who are referred to as "order selectors" or "selectors" typically drive powered pallet jacks through the aisles and pick cases from the slots and stack them onto pallets for shipment to stores. The selection or "picking order" is typically determined by a computer-generated list that identifies the location and quantity of the product to be picked. When the pallet is full, the selector then applies shrink wrap (either manually or applied at an automatic station) and transfers the pallet to the outgoing dock or appropriate outbound trailer for shipping. A typical traditional order pick system is shown in Figure 45.1.

Compared with other distribution systems, the traditional order pick system has several advantages. First, if full slots are used, as opposed to half slots or triple slots where pallets must be broken down (manually unloading cases) before being placed in the slots, each case is handled manually only once within this type of system, since cases are moved by entire pallets until they are removed from the pick slots by the selectors. Second, compared with systems such as a belt-pick, selectors are able to get small microbreaks while they are traveling from one slot to the next slot, or when driving the pallet onto the outbound trailer.

The main ergonomics concern of the traditional order pick system is that the nature of the selector's job involves exposure to a significant number of manual material handling factors associated with MSDs. These include repetitive lifting, sometimes of extremely heavy cases, elevated external moments about the shoulders and low back, and bending and twisting of the torso. With the traditional order pick system, each case must be handled a minimum of one time. However, if half slots or triple slots are used in a facility, cases would be handled more than once because pallets must be broken down for the smaller pallet loads to fit within the smaller vertical spaces. Specific MSD intervention strategies related to the traditional order pick system (e.g., case characteristics, slot issues, pallet and pallet jack issues) are discussed in later sections.

FIGURE 45.1   Example of a traditional order pick system where the selector transfers cases from a pick slot to the pallet jack.

45.2.2   Belt-Pick System

In a facility using this type of case handling system, the system performs the majority of the product transfer, where belts and chutes direct the cases automatically through most of the distribution center. Cases on pallets in storage are coded for specific stores and then loaded onto conveyor belts. Rather than having selectors transfer the cases from slots to pallet jacks, the product is routed to specific feed aisles, where it is manually palletized and shipped to stores. The palletizing portion of this system is shown in Figure 45.2.

Although the belt-pick system may have several advantages from the perspective of moving product through the facility efficiently, several ergonomics concerns exist. First, the employees are exposed to repeated manual handling of the cases. Belt-pick systems require each case to be handled a minimum of two times: (1) when depalletizing cases off the pallet and onto the conveyor belt, and (2) when palletizing cases from the feed slot to the out-bound pallet. This contrasts with the traditional order pick

FIGURE 45.2   A belt-pick system where the employee palletizes cases coming from a conveyor belt.

system, where, with the use of full slots, cases are only manually handled once, when selectors palletize the cases. Given the nature of this work, additional manual case handling (i.e., repetition) increases the risk of MSDs to selectors. Second, the belt-pick system increases the exposure to torso twisting when loading the cases onto the conveyor belt. Twisting of the torso has been associated with elevated risk of low back disorders (Kelsey et al., 1984; Marras et al., 1993). Third, this type of system promotes continuous repetitive manual handling of cases since the individuals performing the palletizing and depalletizing stay in the same general area of the facility for the duration of the day. This provides the selectors with less opportunity to take small breaks between handling of cases, as seen with selectors performing traditional order selecting. This could lead to more rapid muscular fatigue than in the traditional approach.

Although the belt-pick system has several ergonomics concerns, this system configuration does lend itself to the implementation of interventions to reduce the risk of MSDs due to material handling. Most pallets are lined up next to each other, and the belt itself is easily accessible. This lends itself to material handling aids that have been used to reduce manual material handing stressors in other production facilities. Lift assist devices that can run the length of the belt could reduce the weight of the product assumed by the selector, thus reducing the external moment, bring the product closer to the selector, and potentially reduce some of the torso twisting. Pallets can be placed on lift tables that raise each case layer to a more appropriate pick height and reduce torso bending (Marras et al., 2000). Similarly, conveyor belt heights can be adjusted to accommodate employees' working heights, whereas the use of lift and turn tables can reduce bending and reaching during palletizing. Administrative controls aimed at reducing the duration of exposure to the job stressors include rotating selectors to other jobs in addition to the current job, or adding more employees to the belt pick line to share the workload and reduce the exposure to these stressors.

45.2.3 Cross-Dock System

A Cross-Dock system is one where pallets are unloaded from the inbound trailers, broken down on the dock, and then transferred to outbound trailers on the dock. One such system is shown in Figure 45.3.

FIGURE 45.3 A cross-dock system.

The cross-dock system offers advantages over the traditional or belt-pick systems, namely in the reduction of storage space required for product.

Other advantages of using a cross-dock system include the potential to reduce repetitive manual case handling, the ability to use lifting aids, and a reduced order filling time. Reduced frequency of manual case handling can be achieved if full pallet quantities are shipped to the grocery stores. When partial pallets of goods are requested and pallets are broken down in the docking areas, they are not physically bounded by slots. This offers more latitude for engineering controls, such as lifting devices or lift tables to assist in case handling by reducing torso bending, twisting, and external moments. Finally, cross-dock systems have the potential to reduce the time that product is actually housed within the facility since the product can potentially go from the incoming dock to the out-bound dock, with little to no storage in slots.

Although cross-dock systems have the potential to reduced the frequency of manual case handling, there are some concerns as to the feasibility of this approach. First, many grocery stores do not need a full pallet of a specific item at any one time, so this system may not be practical for many distribution center operations. Second, because much of the transfer takes place between docks, the product is housed on the floor and not in vertical slots, as in traditional order pick systems. It can be expensive to provide this additional warehouse space. Third, with this system comes a high level of organization and structure, so that in-bound pallets are directed to the appropriate out-bound locations. This can create additional concerns and costs not found with the traditional order pick or belt-pick systems.

45.2.4 Flow-Through System

With a flow-through system, the products are unloaded from the inbound trailer and transferred directly to the outbound dock via conveyors. The product is coded after being unloaded and automatically routed to specific feed aisles, where it is manually loaded on the outbound trailer. A schematic figure of a generic flow-through system is shown in Figure 45.4.

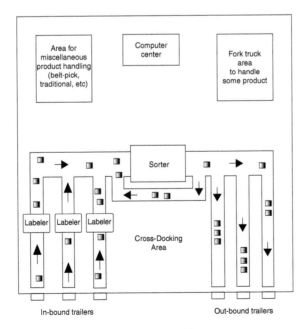

FIGURE 45.4 Schematic figure of a generic flow-through distribution system.

In comparison with the traditional or belt-pick systems, a flow-through system has the greatest potential to reduce or eliminate material handing activities. Selectors working in flow-through systems typically break down a pallet and may use a conveyor in some capacity to transport product from the receiving to the outgoing docks, reducing or eliminating the storage of cases in the distribution center. Repetitive manual case handling can be reduced if the quantities requested from the grocery stores are by-the-pallet, where fork trucks can transfer full pallets from in-bound to out-bound docks. When partial pallets of goods are requested and pallets are broken down in docking areas or are transferred to conveyors for transport to outgoing docks, they are not physically bounded by slots. This presents more latitude for engineering controls to reduce repetitive torso bending and twisting, such as lifting devices or lift tables to assist in case handling, especially for heavier items. These can be used at either the receiving or outgoing docks, or both, where feasible. Another potential advantage of the flow-through system compared to other systems is that the time that product is stored within the facility may be reduced, since much of the product can go directly from the receiving dock to the out-bound, store-specific dock with little or no storage in slots.

Although the flow-through system has the potential for reduced manual material handling, there are some concerns as to the feasibility of this approach. Because much of the transfer takes place between docks, some of the product is stored on the floor and not in vertical slots, like in traditional order pick systems, where it can be expensive to provide this additional warehouse space. Additionally, with this system comes a high level of organization and structure, so that in-bound pallets and individual product are directed to the appropriate out-bound locations. This can create additional concerns and costs not found with the tradition order pick or belt-pick systems. The complexity of routing the product on conveyors from in-bound to out-bound docks may require a large initial outlay of capital which may take a significant amount of time to recover. Thus, this may be more appropriate for large volume distributors.

45.2.5   Mixture of Different Systems

One additional type of distribution center system would include a mixture of several pick systems, with features from the different systems included in strategic locations that would increase the efficiency as well as reduce the risk of MSDs. For example, a cross-docking system approach could be used for the heaviest and most frequently picked products, whereas a traditional order pick system would be used for lighter or slower-moving product.

From an MSD prevention perspective, a distribution system hierarchy does exist. The flow-through system contains the greatest potential to reduce MSDs. This reduction potential stems from reductions of exposure to activities associated with repeated selecting of heavy cases from slots that require repetitive awkward trunk postures, as well as the added flexibility of being capable of using material handling devices and aids for raising and turning pallets. The systems with the next greatest potential include both the belt-pick and cross-dock systems. As with the flow-through system, material handling aids can be used to transfer product from pallets to the belt, or aids for raising and turning pallets can be used to reduce repetitive awkward postures during selecting. The traditional order pick system ranks behind the other three systems, as there are many case and slot features, among others, which increase the risk of MSDs. The feasibility of which distribution system to use in any particular distribution center, however, is dictated also by many other factors, such as space requirements and limitations, as well as product volume and other cost considerations.

45.3   Musculoskeletal Disorder Intervention Strategies for Specific Distribution Center Attributes

Each of the four distribution systems discussed earlier have their own set of advantages and disadvantages. They also have specific issues and potential remedies for improving these concerns. Engineering-based intervention strategies for prevention of MSDs related to more general distribution center issues are presented in this section. This includes issues related to the cases, the pick slots, pallets and pallet jacks, as well as shrink wrapping. For more detailed information on other engineering controls, as well as potential administrative controls, refer to Marras et al. (2005).

45.3.1   Case Features

The sizes, weights, and shapes of cases handled by selectors vary tremendously. Although this will always be the rule (e.g., no standard-case size will be available), there are features of these cases that can reduce stress on the joints of individuals.

45.3.1.1   Excessive Case Weights

Cases of product handled by selectors can be excessively heavy. For example, as identified by Marras et al. (2005), it was estimated that slightly more than 5% of all cases handled weighed more than 40 lb in some distribution centers. Case weights greater than 40 lb were shown to result in spinal loading levels reflective of high-risk jobs for similar case picking when lifting from the lower levels of the pallet (Marras et al., 1997). Thus, case weight is an issue that must be considered for selectors. In addition, loads handled by these individuals produce a bending moment about the spine that must be counterbalanced by the torso muscles. The further cases are held from the body or the higher the case weight, the greater the bending moment, which results in greater torso muscle forces to maintain the balance, which also results in greater loading on the spine. Moment generation about the spine during material handling has been identified as the single factor that best differentiates between jobs having either low or high risk for low back disorder (Marras et al., 1993). Intervention strategies to address excessive case weights include:

- *Work with suppliers to reduce case weights* — Work with suppliers to reduce the case weights, especially for the heaviest case weights. This would, of course, increase the number of cases handled by selectors (i.e., repetitive lifting). However, this additional material handling may be less stressful to the spine than the high case weights currently lifted.
- *Provide means to raise pallets from floor level* — Another intervention strategy to address excessive case weights, especially for the heaviest cases, would be to incorporate lift tables to raise the cases located at the lowest pallet layers. These lift assists may not be needed beneath every pallet in the facility and may be most useful and cost justified for the highest-moving products. Other options for raising cases from the floor include: installing risers in the slots for the pallets to be set on (Figure 45.5), putting pallets of product on top of empty pallets (Figure 45.6), and using full pallet flow racks (Figure 45.7).

FIGURE 45.5   Raising pallets off the floor by using risers in a full slot.

45.3.1.2 Lack of Handle Cut-Outs on Cases

Most cases in warehouses do not have handle cut-outs for selectors to use. As a result, cases often are more difficult to retrieve from slots, because the cardboard is difficult to grip and there may not be a location on the case that allows it to be pulled toward the selector. Intervention strategies with respect to handles include:

FIGURE 45.6 Raising a pallet in a full slot, by stacking empty pallets under the cases.

- *Work with suppliers to request handle cut-outs on cases* — Work with product suppliers to provide, especially for the heavier or more awkward-to-handle cases, handle cut-outs (Figure 45.8). Research on case weights between 40 and 60 lb has found that the use of handles reduces loading on the spine, thus reducing the risk of low back disorders (Davis et al., 1998; Marras et al., 1999). The handles give selectors an option of how to maneuver the cases as well. Additionally, this same research found that the largest reduction of risk of low back disorders and spinal loading as a result of adding handles occurred at the bottom layer of the pallets.

45.3.1.3 Handling Tray Packs (Plastic-Wrapped Cases)

Many cases are now wrapped with plastic rather than being completely packaged in cardboard cases (Figure 45.9). This can create an additional stressor for selectors, who, in an effort to reduce lifting requirements, often slide cases across one another on a pallet before lifting it from the slot. The force to slide tray packs across one another may be higher than the actual weight of the case. In summer months, tray packs likely will stick together even more than usual, further increasing the demands required of selectors.

FIGURE 45.7 Using full-pallet flow racks (on rollers) to raise pallets off the floor.

Several potential intervention strategies exist to reduce the physical stresses associated with the use of plastic wrapped tray packs.

- *Educate suppliers about ergonomics concerns with tray packs* — Discuss with and educate suppliers about the problems with the plastic on tray packs, in an effort to eliminate their usage.
- *Incorporate Lift Assists* — Under items in which tray packs are used, incorporate lift or turn tables, so that cases can be brought closer (both horizontally and vertically) to selectors.
- *Add slip sheets between layers* — Provide slip sheets or other materials between tray pack layers, to eliminate plastic laying on top of plastic, which would increase the ease of sliding the tray packs.
- *Treat tray packs as heavy cases* — Deal with tray pack items as if they were heavier cases, and consider the intervention strategies suggested above.
- *Use tray packs that are not wrapped completely* — Discuss with manufacturers the possibilities of wrapping the top and sides of the tray pack with plastic, leaving the bottom of the cardboard pack unwrapped. One such design is shown in Figure 45.10. This would eliminate the plastic-on-plastic sliding and reduce the force required to slide the case forward.

FIGURE 45.8 Handle cut-outs in cases, to ease lifting and carrying.

45.3.1.4 Layer-by-Layer Depalletizing of Cases

Some distribution centers train their selectors to depalletize cases layer by layer, as a way to maintain the integrity of the pallet (Marras et al., 2005). This differs from a pyramiding approach, in which cases are removed on a diagonal, from the top-front to the rear-back of the pallet. However, the layer-by-layer

FIGURE 45.9 Tray packs of product completely encased in plastic.

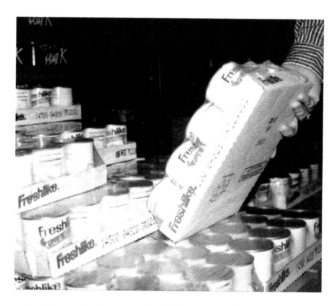

FIGURE 45.10   Tray packs without plastic wrapped on the bottoms of the cardboard trays.

method creates problems from a biomechanical standpoint. For most individuals, reaching to the rear of a pallet to handle a case is difficult and is likely to produce additional loading to the shoulders and back due to increases in the external moment. Several potential intervention strategies exist to reduce the physical stresses associated with the layer-by-layer depalletizing:

- *Allow pyramiding depalletizing, but with training* — The pyramiding approach will enable selectors to reach each case easier and likely will reduce the moment generated about the spine when lifting. However, it is necessary that training be given to selectors, so they only unload a case using this technique when it does not support the one above it. This will assure that the integrity of the entire pallet will be maintained. Additionally, care must be used when using the pyramiding method of depalletizing so that the pallets do not flip backwards when only cases to the rear of the pallet remain.
- *Rotate pallet after it has been half-way unloaded* — By instructing fork truck drivers to turn around pallet loads after selectors have removed about half the cases or by providing turn tables underneath these pallets, many of the cases will be easier to reach. This will reduce the physical requirements of the work. This is especially important for pallets that are placed on the upper tier when using half slots. Rotating the pallets in this way will reduce the reach requirements.

45.3.1.5   Unexpected Spinal Loading during Case Handling

Due to the large number of items that selectors must lift and new items continually being introduced, even experienced employees likely will not remember the general weights of all cases that are handled. This contributes to unexpected loading on the spine, that is, cases weighing more or less than anticipated by the selectors. Additionally, excessive glue on the cases may make them stick to each other, also contributing to unexpected loading of the spine when picking the cases. Finally, vendors' use of less than sturdy case material, or not enough glue used to construct the cases, may increase the risk of the case breaking during handling, thus contributing to unexpected loading of the spine when picking the case. Intervention strategies to reduce unexpected spinal loading during case handling include:

- *Tag slots to indicate case weights* — Provide feedback to selectors indicating the weight of the cases. This could be accomplished in two ways, especially for the heaviest cases: (1) mark each slot with the weight of the case, so that selectors can mentally and physically prepare for the lift, or

(2) color-code the floor in front of slots containing the heaviest cases with red markings, indicating a warning, so that selectors will be reminded to lift more carefully.

- *Request sturdy cases from suppliers* — Work with the suppliers of the cases or other distribution centers that may be having the same problems with cases breaking during selecting, to request sturdier cases. The same approach can be used to address the issue of excessive glue on the cases.

45.3.1.6 Slippery Cardboard Cases

Most cases have no handle cut-outs and often must be pushed and pulled along their sides. The cardboard is usually smooth and difficult for selectors to get a good grasp in this manner. This results in more difficulty accessing some cases (especially those housed in half slots or triple slots) and a longer pick time for the order. Interventions to address slippery cardboard cases include:

- *Use friction-increasing gloves* — Many types of gloves are commercially available (e.g., ones with rubber surfaces) that enable a better grip be placed on surfaces such as smooth cardboard. The use of gloves, however, has been shown for gripping tasks to actually increase the muscle force, which could lead to faster muscle fatigue. Therefore, if gloves are to be used, they should be selected very carefully and not impede other tasks performed by the selectors.

45.3.2 Slot Features

Major concerns for food distribution centers include the time taken for selectors to pick an order and the overall pick density, or the amount of picks able to be performed in a given area. These concerns translate to smaller warehouses that locate goods in as small an area as possible. In the short run, this approach may produce faster pick times. However, cumulatively, it may slow selectors down (e.g., due to physical fatigue) in warehouses that have attempted to condense the area too much. Several other slot features are at issue, as discussed next.

45.3.2.1 Posture of Individuals Working in Half Slots or Triple Slots

Given the nature of a selector's job, reaching cases is easiest when pallets are housed in full slots. In these types of slots, selectors are more able to lift cases closer to the body (which reduces the lifting moment) or while standing more upright (which reduces spinal loading). The use of half slots (Figure 45.11) or triple slots (Figure 45.12) creates several problems. First, for pallets brought to these slots that do not fit (vertically) into them, additional manual material handling is required to partially unload some of the cases. Second, access is limited in these slots, and selectors are less able to lift cases in a manner that can reduce loading to the back. Third, handling empty pallets is more difficult from these smaller slots. Intervention strategies to reduce the risk of MSDs when breaking down pallets and picking from slots include the following:

- *Use a mechanical load splitter* — When pallets must be broken down, either for placement into half slots or triple slots, or if being broken down in a cross-dock or flow-through system, the use of mechanical

FIGURE 45.11 Tight fit of a pallet in a half slot, making some cases difficult to retrieve and requiring selectors to bend or reach excessively.

equipment to remove several layers at a time eliminates repetitive manual case handling by the selectors. Figure 45.13 shows one piece of equipment called a load splitter, which uses a clamping device to move one or more layers of a pallet at a time. Other load splitters exist that has a flat bed that slides under a slip sheet and allow movement of one or more layers at a time.

- *Use full slots whenever possible* — The best practice found, from an ergonomics perspective, is the use of full slots to store cases. This reduces the potential for awkward postures for many of the cases located on a pallet and increases the chance that cases can be lifted more easily. However, this approach requires additional space when using full slots for all products. If the number of full slots is limited, place the heavier and faster-moving product in them.

FIGURE 45.12   Example of a triple slot, which requires more extreme body positions of selectors to reach cases.

- *Incorporate flow racks for slow-moving products* — One solution to the space problem is the use of gravity-fed flow racks (Figure 45.14). These flow racks can be placed above near-full slots, so accessing cases from them does not require awkward torso postures for selectors. This design increases the pick density with no increase in the travel time. The design in Figure 45.14 contrasts with another design (Figure 45.15) in which slower-moving items were gravity-fed from racks near floor level to above shoulder height. Full aisles of flow racks similar to that shown in Figure 45.15 could be used for lighter and slower-moving items. This type of system likely produces more stress on the back due to the need to lift some case from near the floor. Therefore, the lightest and slowest-moving of these items should be placed near the floor to reduce the repetition of picking at this level. This design also would increase the pick density for the items in these aisles and allow more space to have full slots for the heavier and faster-moving product.

FIGURE 45.13   Mechanical load splitter with a clamping device, used to break down pallets.

FIGURE 45.14 Gravity-fed flow racks used above a near-full slot.

Several other disadvantages exist with the use of flow racks. The flow racks must be filled from behind the racks, requiring another employee to perform material handling. Also, due to the design of the racks, selectors must lift the cases over the stop bar that keeps the cases from falling off the racks when not being picked. These lifting forces may become high if there are

FIGURE 45.15 Gravity-fed flow racks.

several cases behind the one being picked, essentially wedging it between the stop bar and the cases behind it.

45.3.2.2 Use of Half Slots

If half slots must be used due to space constraints, only store slower-moving and lighter products in them. Prioritize case movement through the facility and determine which items are slower moving. If many items fit within this category, then put the lighter-weight cases in half slots and keep the heavier cases in full slots. The use of half slots should be accompanied by a system that allows fork truck drivers to rotate pallets after they have been half picked, and that allow turn tables to be used under these pallets so that selectors can rotate the pallets themselves, or that enable pick sticks to be used to aid in case retrieval. Other MSD intervention strategies include:

- *Rotate pallets in half slots* — To reduce reach distances, especially for the upper slots, reconfigure the width of the slots such that the long side of the pallet is exposed to the aisle. This will reduce reach distances to cases at the back side of the pallet.
- *Maximize the flow rack/full slot ratio* — If half slots must be used, consider using a system as shown in Figure 45.14, where a gravity flow rack with slower moving product is housed above the half slot or near-full slot. This flow rack system will be useful only if slow-moving items are placed in the lower racks and if full slots house pallets in the rest of the facility. This will ensure that selectors will only need to access cases in these more awkward rack locations a minimum number of times.
- *Seldom use triple slots* — Abolish the use of triple slots in the distribution centers, as they require selectors to contort themselves in postures that substantially increase the risk of MSDs. Not only do triple slots require excessive bending to retrieve cases, but more of a pallet must be broken down for it to fit, producing additional times that each case must be handled within the facility.
- *Develop/revise the facility's slot-management system* — Review the optimization system used to manage the flow of goods through the warehouse, to ensure that pallets are placed in full slots according to case weight and movement speed of the product. This can greatly affect the physical requirements of selectors, as well as aid the selectors to help build stable bases for their orders.

45.3.2.3 Lifting Cases from the Lowest Pallet Layer

The highest loading to the spine and the greatest risk of low back injury to selectors arise from lifting cases off the lowest pallet layers, near the floor of the warehouse (Marras et al., 1997). It may be unrealistic to implement these interventions within every slot; therefore, priority should be given to the heaviest cases and the fastest-moving products. Strategies to elevate product on a pallet include:

- *Provide means to lift pallets from floor level* — Different methods are available, depending on the configuration of the distribution center (Figure 45.5 to Figure 45.7).
- *Implement lift tables* — These devices raise case heights as the pallet is being unloaded. They can be raised manually by selectors using hydraulics, or automatically using a spring-loaded system. These devices, however, can only be used in slots where there is sufficient vertical space to raise the product and allow selectors sufficient room for access. Therefore, priority should be given to the heavier cases used in full slots.
- *Stack pallets* — By placing a full pallet on top of one or more empty pallets, the bottom case layer will be further from the warehouse floor (Figure 45.6). This technique is best used when full slots (and, thus, more clearance) are being used.
- *Use roller conveyors* — Roller conveyors can be used in double-deep slot configurations, in which two pallets are placed end to end (Figure 45.7). These systems enable the rear pallet to be pulled out and accessed more easily after the front pallet is empty. This system results in the entire pallet being raised off the floor, which can ease the unloading of cases on lower pallet layers. The clearance of the top of the pallet in the slot must be considered, as should the potential for selectors tripping over these conveyors. Periodic maintenance of the rollers is imperative to reduce the force required to pull the pallets forward.

45.3.2.4 Difficulty in Reaching Cases Further Back in Slots

Regardless of the type of slot, spinal loading is increased when cases are removed from locations further back and lower on pallets (Marras et al., 1997). This is because selectors must bend over further to lift these cases, and they usually pick them up further from their bodies, which increases both the moment generated about the spine and risk of low back MSDs:

- *Provide turn tables* — In those slots housing especially the heaviest case weights, incorporate turn tables onto which pallets are placed. This will allow selectors to rotate the entire pallet after it is about half unloaded, for easier access to those cases previously furthest from the aisle. This would reduce the reach distance to the cases as well as the lifting moment generated about the spine. Note that these turn tables require additional floor space within a slot, so they may not be feasible for all locations.
- *Turn pallets around with fork trucks* — As an alternative to the use of turn tables in slots, have fork-lift drivers rotate the pallets after they have been half unloaded. This technique may also reduce the pick time of cases, since selectors will have easier access to the cases and can load them more quickly and with less effort. This approach would require close supervision, initially, to assure the proper timing of these pallet rotations.
- *Redesign racks to elevate the pallet* — One such design (Figure 45.16) allows selectors to access those cases located further back in the slot. This is appropriate for the heaviest cases handled in the facility. It should be assured, however, that the elevated height does not create space problems for the top cases on the pallet, and making them more difficult to retrieve. Additionally, designs as shown in Figure 45.16 may create tripping hazards and make it more difficult for selectors to walk between the pallets. Other designs may be possible that may reduce these hazards. One such design is shown in Figure 45.17, which elevates the pallets in the slot, but also allows increased access between the two pallets in the full slot, reducing reaching, and torso bending and twisting when accessing the rear product on the pallets. This design does, however, require the full slots to be widened, with curved risers placed on each side of the full slot.

45.3.2.5 Little Clearance between Pallets within the Same Slot

Often, fork lift drivers who move pallets into slots concentrate only on getting them out of the aisles. Pallets that are side by side in a slot do not enable selectors to easily reach those cases at the rear of the pallets (Figure 45.18), which can increase reach distances, external moments, and the risk of MSDs. Intervention strategies to reduce MSD risk include:

- *Provide fork lift driver and selector training to maintain clearance between pallets* — Instruct fork lift drivers to keep a clearance width of at least 16 in. or more between two pallets within a slot (Figure 45.19). Also, educate them as to the purpose of this action (that selectors can more easily reach cases and reduce their risk of MSD). In addition, train the selectors to use these clearances as well, to better reach those cases located further back on a pallet.
- *Increase slot width to increase clearance between pallets* — By increasing the slot width, this allows more clearance between the pallets. As shown in Figure 45.17, the slot width was increased, and risers

FIGURE 45.16 Risers used to elevate pallet from the floor and allow easier access to cases, but may create a tripping hazard.

FIGURE 45.17   Risers used to elevate pallets off floor and slot-widened to allow easier access to the rear of pallets.

were placed in the slot to raise the pallet off the floor as well as increase the ease of access between pallets.

45.3.3   Aisle Features

Depending on the type of pick system used in the distribution center, the width of some aisles may need to be considered. For example, along aisles where faster-moving product is stored, they may need to be wider, so that several selectors can pick at the same time. Congestion along narrower aisles will likely increase order pick times. However, the effects of these types of changes (i.e., more floor space required) need to be considered within a systematic optimization scheme for order picking within the distribution center.

45.3.4   Pallet Features

Many goods shipped to food distribution centers are placed on wooden pallets, which can weigh between 50 and 75 lb. Selectors often must move pallets of these weights in addition to their other tasks. This

FIGURE 45.18   Pallets side by side in a slot, resulting in no access between pallets.

FIGURE 45.19   Pallets set apart in a slot to increase the clearance between pallets and provide easier access to selectors.

creates more physical demands and loading on body joints. Intervention strategies to reduce MSD risk from handling pallets include:

- *Use plastic pallets when possible* — The substitution of plastic pallets, which weigh substantially less), for wooden ones, will reduce the cumulative weight handled on a daily basis by selectors. Within a distribution center, create a policy whereby all pallets used will be plastic.
- *Ask that suppliers use plastic pallets* — Food suppliers also must handle heavier wooden pallets within their facilities. By educating them as to the benefits of lighter, plastic pallets (reduced spinal loading, cost savings from fewer or lower-severity MSDs), suppliers may be more willing to have goods shipped on plastic pallets.
- *Use a pallet dispenser* — If wooden pallet must be used, one method of eliminating some or all of the handling of the heavy pallets by the selectors is to use a pallet dispenser (Figure 45.20). When selectors need an empty pallet to begin their order, they drive their pallet jack up to the pallet dispenser, and a pallet is dispensed right on the forks of the pallet jack.

45.3.5   Pallet Jack Features

Many of the intervention strategies addressed above were to enable cases be more easily accessed from slots. However, selectors in traditional order pick systems must place cases on pallet jacks, and pallet jacks are used in virtually every grocery distribution center. Therefore, the design of these devices with respect to risk factors for MSDs also must be considered.

45.3.5.1   Lack of Adjustability of Fork Vertical Heights

Even if cases are easily located in slots, pallet jacks usually require the first cases of an order to placed

FIGURE 45.20   Wooden pallet dispenser used to reduce the manual handling of pallets.

on a pallet near the floor level. Typically, the initial cases are used to build a sturdy base, and may consist of heavier, sturdier cases. As already indicated, working at this level with heavier product greatly increases the risk of MSDs to the low back. Intervention strategies for this issue include:

- *Use pallet jacks with risible forks* — Use pallet jacks that enables the selector to raise the pallet as needed, so that the initial cases can be placed on the lower layers more easily and without the extreme forward bending. This design uses the same concept as raising the pallets in slots higher off the floor to reduce the risk of MSD to the low back.
- *Use several pallets stacked on top of each other on the pallet jack* — To reduce the forward bending necessary for the first few layers on the pallet, raise the initial stacking height by placing more than one pallet on the pallet jack. This strategy may be most appropriate for less than full cube orders (Figure 45.21).

45.3.5.2   Added Physical Stress from Using Poorly Maintained Pallet Jacks

Pallet jacks develop much wear and tear from being used extensively by selectors. This can result in pallet jacks with reduced braking ability and more difficult steering and maneuverability. Increases of stress on the hands and wrists can result from difficulty in braking, and difficulty in steering and maneuverability of the pallet jack can increase the loading on the shoulders. Intervention strategies include:

- *Implement a system of routine, scheduled preventative maintenance on pallet jacks* — By periodically checking the working components of all pallet jacks, they can be kept in better condition and remain easier for selectors to use.
- *Provide pallet jack education* — Train selectors to report problems or difficulties with pallet jacks when they are first noticed, so that issues such as steering and braking are addressed before they create additional problems.

45.3.6   Plastic Wrapping Issues

Before pallets of product are sent to the stores, they are wrapped with plastic to maintain integrity to prevent movement during shipment. Some facilities have automatic pallet wrappers. However, others require selectors to wrap the plastic around cases on the pallet manually. Several issues exist for manually wrapping the pallets. First, to wrap the bottom layer of the pallet, selectors must bend forward extremely

FIGURE 45.21   Stacking of empty pallets on a pallet jack, to raise the initial stacking height of cases and reduce trunk bending.

far with the trunk. This places additional stress on the lower back that is eliminated when automatic wrappers are used. Second, although the weight of a full plastic roll may not be excessive as compared to some cases, the forward bending posture combined with the weight of the roll serves to increase the risk of MSD to the low back. Third, selectors have been observed putting their fingers in the plastic-wrap tube at each end, and walking around the pallet while rotating the tube around the fingers, which increases the likelihood of lacerations (Marras et al., 2005). Finally, some pallet jacks are not constructed to hold the plastic wrap in place when it is not in use. Selectors may injure themselves trying to catch a plastic roll that may fall off of the pallet jack due to inadequate storage space on the pallet jack. Intervention strategies related to plastic wrapping include:

- *Provide automatic wrappers* — Eliminating manual wrapping reduces one task of a selector's job and could provide additional rest time for selectors before a new order is begun. This rest time is important for the body to begin to recuperate from the physical demands of the job.
- *Provide handles for plastic wrap* — If manual wrapping must be done, develop a handle device onto which selectors can hold the wrap more properly. This not only will reduce the potential for lacerations to the hands, but it can reduce the time taken for the order to be picked by reducing the wrapping time.
- *Supply smaller rolls of plastic wrap* — For manual wrapping tasks, provide selectors with smaller rolls of plastic that are lighter weight and easier to handle.
- *Modify pallet jacks to securely hold plastic wrap* — On pallet jacks with no place for the plastic wrap to be held, selectors have been observed placing the wrap on the jack wherever it was believed it would not fall off (Marras et al., 2005). This often did not guarantee the wrap would not fall off the jack. By adapting all pallet jacks so that the plastic wrap can be securely held in place (Figure 45.22), the potential for accidents and waste will be reduced.

45.3.7 Pick Sticks

Pick sticks are tools used in distribution centers to reach and pull cases that are located out of easy reach of selectors, thereby reducing reach distances (Figure 45.23). Some distribution centers reportedly distribute pick sticks to all selectors, whereas other distribution centers make them available as needed (Marras et al., 2005).

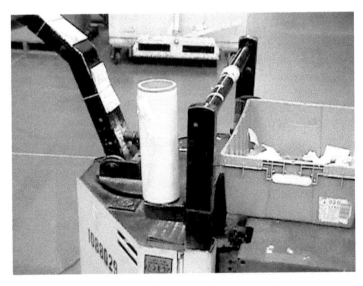

FIGURE 45.22 Example of holder on a pallet jack to secure plastic-wrap rolls.

FIGURE 45.23   A pick stick used by a selector to grab a hard-to-reach case.

45.3.7.1   Pick Sticks Not Used by Selectors or Not Available

Unfortunately, these assists come up missing or are seldom used by employees. As a result, difficult-to-reach cases (e.g., those in half slots or triple slots) must be retrieved by hand, adding to the physical requirements of these selectors. Interventions to increase proper use of pick sticks include:

- *Educate selectors and supervisors regarding pick sticks* — Selectors and supervisors need to be trained of the benefits to using pick sticks (i.e., less physical work), reminders should be posted in the facility, and the value of pick sticks should be addressed periodically, such as at department meetings.
- *Provide easier access to pick sticks* — If pick sticks are not being used in the distribution center, ask employees for the reasons why they are not used. Provide pallet jacks with holders for these sticks. Ensure that pick sticks are kept near slots that present the biggest problem, such as by half slots or triple slots housing the heavier or faster-moving products.

45.3.8   Coupling Issues

Different distribution centers use a variety of methods for coding cases before they go to the stores. Some place stickers on each case, whereas others use a small clipboard in which they check off items as cases are picked. Regardless of the method, selectors often hold these objects as they are handling cases. This reduces the coupling between the hands and the cases, that is, the ability for selectors to pick up and transfer the cases onto pallets. Poor coupling has been identified as a risk factor for low back disorders (Waters, 1993), and increases the risk of dropping loads, which can result in unexpected loading of the back and increase the risk of MSD. Intervention strategies to improve the coupling and reduce the risk of injury include:

- *Use a sticker dispenser* — If stickers must be applied to each case, provide a waist belt-held device that dispenses the stickers and eliminates the need for selectors to hold the pack while transferring cases (Figure 45.24).
- *Provide a clipboard hook* — If a clipboard system is used, provide a belt hook or other method whereby selectors can easily fasten the clipboard during material handling so that the hands are entirely freed up.
- *Implement headsets to replace the order sheets* — Headset technology allows the selector to identify the cases to be picked, without having to hold any orders (Figure 45.25). Both hands, therefore, would be free for the transfer of the cases.

FIGURE 45.24   A waist-held sticker dispenser, used to free up both hands for order selecting.

45.4   Conclusions

Musculoskeletal disorder rates in grocery distribution centers has been historically high compared with other industries, especially for the low back region. The process of transferring grocery products into and out of a distribution center results in physically demanding tasks, which arise mainly from the traditional need to move the product manually. Much of the elevated risk for MSDs in this work environment can be related to the physical nature of repetitively transferring the product, sometimes extremely heavy, to or from pallets, oftentimes performed in awkward postures of the torso or extended reaches.

While it is recognized that it may be impossible to eliminate all manual handling of product in distribution centers, the engineering-based MSD intervention strategies that were presented in this chapter were developed based on observations from several grocery distribution centers (Marras et al., 2005),

FIGURE 45.25   Headsets, used to direct selectors through their order, freeing both hands for selecting.

as well as from biomechanical studies on spinal loading and risk of low back disorders related to manual materials handling tasks similar to those found in distribution centers. Since many of the tasks that the selectors perform in their daily job results in cumulative loading throughout the day, consideration should be given to the feasible MSD intervention strategies aimed at reducing the cumulative physical loading, thereby reducing the overall risk of MSDs in grocery distribution centers.

References

Davis KG, Marras WS and Waters TR. Reduction of spinal loading through the use of handles. *Ergonomics* 1998; 41:1155–1168.

Drury CG, Law C-H and Pawenski CS. A survey of industrial box handling. *Hum Factors* 1982; 24: 553–565.

Gardner LI, Landsittel DP and Nelson NA. Risk factors for back injury in 31,076 retail merchandise store workers. *Am J Epidemiol* 1999; 150(8):825–833.

Kelsey JL, Githens PB, White AA, Holford TR, Walter SD, O'Connor T, Ostfeld AM, Weil U, Southwick WO and Calogero JA. An epidemiologic study of lifting and twisting on the job and risk for acute prolapsed lumbar intervertebral disc. *J Orthop Res* 1984; 2:61–66.

Kraus JF, Schaffer KB, McArthur DL and Peek-Asa C. Epidemiology of acute low back injury in employees of a large home improvement retail company. *Am J Epidemiol* 1997; 146(8):637–645.

Kuorinka I, Lortie M and Gautreau. Manual handling in warehouses: the illusion of correct working postures. *Ergonomics* 1994; 37(4):655–661.

Marras WS, Lavender SA, Leurgans SE, Rajulu SL, Allread WG, Fathallah FA and Ferguson SA. The role of dynamic three-dimensional trunk motion in occupationally-related low back disorders. *Spine* 1993; 18:617–628.

Marras WS, Granata KP, Davis KG, Allread WG and Jorgensen MJ. Spine loading and probability of low back disorder risk as a function of box location on a pallet. *Hum Factors Ergon Manuf* 1997; 7:323–336.

Marras WS, Granata KP, Davis KG, Allread WG and Jorgensen MJ. Effects of box features on spine loading during warehouse order selecting. *Ergonomics* 1999; 42:980–996.

Marras WS, Allread WG, Burr DL and Fathallah FA. Validation of a low-back disorder risk model: A prospective study of ergonomic interventions associated with manual materials handling jobs. *Ergonomics* 2000; 43:1866–1886.

Marras WS, Allread WG, Jorgensen MJ and Stuart-Buttle C. *A Best Practices Guide for the Reduction of Musculoskeletal Disorders in Food Distribution Centers.* The Institute for Ergonomics, Ohio State University, 2005.

NIOSH. Big Bear Grocery Warehouse, Columbus, OH. U.S. Department of Health and Human Services, Public Health Services, Centers for Disease Control, National Institute for Occupational Safety and Health, HETA Report No. 91-405-2340, 1993a.

NIOSH. Kroger Grocery Warehouse, Nashville, TN. U.S. Department of Health and Human Services, Public Health Services, Centers for Disease Control, National Institute for Occupational Safety and Health, HETA Report No. 93-0920-2548, 1993b.

Waters TR. Workplace factors and trunk motion in grocery selector tasks. Proceedings of the Human Factors and Ergonomics Society 37th Annual Meeting, pp. 654–658, 1993.

46

Patient Handling in Health Care

Audrey Nelson
James A. Haley VAMC

46.1 Introduction

While there has been a steady decline in the rates of most occupational injuries starting in 1992, work-related musculoskeletal disorders in nursing continue to rise.[1] Nursing personnel continue to be listed as among the top ten high-risk occupations, with injury rates reported as 8.8 per 100 in hospital settings and 13.5 per 100 in nursing home settings.[2] These incidence rates are considered to be low estimates, since underreporting of injuries in nursing is common.[3] Internationally, the prevalence of back injury in nursing, is estimated to be 17%, with an annual prevalence of 40 to 50%, and a lifetime prevalence of 35 to 80%.[4]

Many patient handling tasks have been identified as high risk, based on the physical demands of the tasks, workspace and working conditions, and unpredictable nature of the work. Unlike other industries, where the loads are modified to enhance safety, patients lack the convenience of handles, present with uneven distribution of weight, and have fixed weights, which cannot be altered. Additionally, patients present unique challenges, including variations in size, physical disabilities, level of cooperation, and fluctuations in the level of assistance they can provide during tasks. Further, many patient handling tasks are performed in sustained awkward positions such as bending or reaching over beds or chairs while the nurses' back is flexed.[5,6] Inadequate space and poorly designed work environments contribute to these awkward positions.

46.2 High-Risk Patient Handling Tasks

Patient handling is performed in diverse clinical settings, including acute care hospitals, long term care facilities, outpatient clinics, and in the patient's home. Each setting poses unique challenges related to occupational safety, eliminating the possibility of one universal ergonomic solution for safe patient handling. While physical environment, work practices, and safety culture influence which ergonomic interventions to implement, key to understanding the inherent risks associated with each clinical setting is the number, type, and frequency of each high-risk task performed.

High-risk patient handling tasks are defined as duties that impose significant biomechanical and postural stressors on the care provider. Factors such as the patient's weight, transfer distance, confined workspace, unpredictable patient behavior, and awkward positions such as stooping, bending, and reaching are threats to the nurse's safety. The frequency and duration of these tasks, varies from one clinical setting to another. It is important to identify and assess nursing staff perceptions of high-risk tasks. In addition to unique patient characteristics and nurses perceptions of risk, the variation between patient care units can also be explained by availability of equipment, physical layout, and work organization.

In 2003, Hignett and colleagues summarized research to create an evidence-base for patient handling and identified high-risk tasks.[7] (See Table 46.1.)

Few would argue that one of the highest risk patient handling tasks is a patient transfer. Patient transfers can start with the patient in a sitting position (vertical transfer) or when the patient is supine (lateral transfer).[21] However, high-risk tasks are not restricted to vertical and lateral transfers. Other high-risk patient handling tasks include repositioning a patient in bed, repositioning a patient in a chair, and transporting a patient in a bed or stretcher. Further, risk for injury extends beyond tasks that involve patient movement. Patient handling tasks can be designated as high risk if the tasks are performed in a forwardly bent position with the torso twisted, such as feeding, bathing, or dressing a patient. It is the combination of frequency and duration of these high-risk tasks that predispose a caregiver to musculoskeletal injuries.[21]

A few studies examined high-risks tasks in the operating room (OR).[22–24] Problem areas included: (1) standing for long periods of time, (2) lifting and holding patient's extremities, (3) holding retractors for extended periods of time, (4) transferring patients on and off OR beds, (5) reaching, lifting, and moving equipment, (6) repositioning patients in OR beds, (7) slippery shoe covers and floors, (8) tripping hazards, (9) sustained awkward positions, and (10) holding an arm or leg for extended periods during surgery.

Most of the research on high-risk tasks has been conducted in geriatric, long-term care settings.[8–14,25] The most stressful tasks included: (1) transferring patient from toilet to chair, (2) transferring patient from chair to toilet, (3) transferring patient from chair to bed, (4) transferring patient from bed to chair, (5) transferring patient from bathtub to chair, (6) transferring patient from chair lift to chair, (7) weighing a patient, (8) lifting a patient up in bed, (9) repositioning a patient in bed side to side, (10) repositioning a patient in a chair, (11) changing an absorbent pad, (12) making a bed with a patient in it, (13) dressing/undressing a patient, (14) tying supports, (15) feeding a bedridden

TABLE 46.1   Evidence for Identification of High-Risk Patient Handling Tasks

Task, Equipment, Intervention	Level of Evidence	Number of Studies	References
Hazardous tasks involve moving patients in bed; bed–chair transfers; toileting; bathing; and lifting from the floor	Moderate	7	Bell, 1979 (U.K.)[1] Garg et al., 1992 (U.S.A.)[2] Hui et al., 2001 (Hong Kong)[3] Owen, 1987 (U.S.A.)[4] Owen et al., 1992 (U.S.A.)[5] Schibye and Skotte, 2000 (Denmark)[6] Smedley et al., 1995 (U.K.)[7]
Ambulance work can result in harmful postures, with the highest risks involving the transportation of patients on equipment	Moderate	3	Massad et al., 2000 (Canada)[8] Furber et al., 1997 (Australia)[9] Doormaal et al., 1995 (Netherlands)[10]
A high-risk task for home care nursing is providing care in nonadjustable beds	Moderate	3	Bullard, 1994 (U.K.)[11] Knibbe and Friele, 1996 (Netherlands)[12] Skarplik, 1988 (U.K.)[13]

patient, (16) making a bed while the patient is not in it, (17) bathing patient in bed, (18) transferring from bed to dependency chair, and (19) picking a patient up off the floor post fall.

A few studies have examined high-risk tasks in acute care hospitals,[26] but due to differences in specialty units, it is difficult to generalize this information. Tasks identified included: (1) lifting a patient off the floor post fall, (2) transferring patients on and off stretchers, (3) transferring patients on and off cardiac chairs, (4) transfer of patient in and out of bed, (5) toileting, (6) and repositioning a patient in bed. There is limited evidence to date for clearly identifying high-risk patient handling tasks in critical care, medical/surgical units, psychiatry, emergency rooms, and outpatient departments.

Only one study has examined high-risk tasks in spinal cord injury units;[25] these tasks included: (1) transferring patient from wheelchair to bed, (2) transferring patient from bed to wheelchair, (3) lifting a patient up in bed, (4) repositioning a patient in bed side to side, (5) repositioning a patient in a wheelchair, (6) making an occupied bed, (7) dressing/undressing a patient, (8) bathing patient in bed, (9) making an occupied bed, (10) transferring a patient from bed to stretcher or prone cart, and (11) applying antiembolism stockings.

There are special challenges associated with safe patient handling and movement in home care. High-risk tasks in home care[18–20,27] included: (1) lifting a patient up in bed, (2) applying antiembolism stockings, (3) transferring a patient from chair to chair, (4) giving a tub bath, (5) repositioning a patient in a chair, (6) toileting a patient, and (7) providing care in non-height-adjustable beds.

Little work has been done related to the high-risk tasks in trauma and emergency setting. A moderate level of evidence is available to support that ambulance work can result in harmful postures, with the highest risks involving the transportation of patients on equipment.[15–17] Patient transfers in and out of personal vehicles are also considered a high risk task, particularly when the patient presents at the emergency room acutely ill. Unfortunately, this task has not been well studied.

Through job observation, questionnaires to employees, or brainstorming sessions with patient handlers, individual sites should determine what are the high-risk activities within their workplace. Figure 46.1 is a tool that can be used with nursing staff to identify and prioritize high-risk tasks. Keep in mind that there are likely to be variations of high risk tasks by unit as well as by shift.

46.3 Hazard Evaluation

In addition to identification of high-risk patient handling tasks, hazard evaluation of the physical work space and work processes needs to be completed.[24,28–30] Data can be collected through observation, survey, or expert panel reviews. Key assessment criteria include the following.

a. *Patient load.* The characteristics of the patients typically cared for in each setting needs to be considered. Patient characteristics include the medical diagnosis, comorbidities, level of dependence, cognitive and functional deficits that would interfere with the patient's ability to assist with care, height and weight, ease in lifting, stability, maneuverability, and any other inherent dangers.

b. *Nursing care provider.* Nursing personnel come in a variety of sizes, with variations in strength, physical fitness, experience, and training. Many tasks in health care are provided by two or more caregivers, so disparities in height and working capacity can impair synchronization of the task performance.

c. *Patient handling tasks.* The earlier discussion of high-risk task should be helpful in assessing this area. In addition to identifying high-risk tasks uniquely inherent in each clinical setting, one should ask:

- Is the task essential?
- Does it require loads to be held away from trunk?
- Does it involve excessive pulling or pushing?
- Is there risk of sudden, unexpected movement?
- Does it require excessive reaching?
- Are there insufficient rest or recovery periods between performance of high risk tasks?

Directions: Assign a rank (from 1 to 10) to the tasks you consider to be the highest risk tasks contributing to musculoskeletal injuries for persons providing direct patient care. A "1" should represent the highest risk, "2" for the second highest, etc. For each task, consider the frequency of the task (high, moderate, low) and musculoskeletal stress (high, moderate, low) of each task when assigning a rank. Delete tasks not typically performed on your unit. You can have each nursing staff member complete the form and summarize the data, or you can have staff work together by shift to develop the rank by consensus.

Frequency of Task H = High M = Moderate L = Low	Stress of Task H = High M = Moderate L = Low	Rank 1 = High-Risk 10 = Low-Risk	Patient Handling Tasks
			Applying anti-embolism stockings
			Bathing patient in bed
			Bathing a patient in the tub
			Changing an absorbent pad
			Dressing/undressing a patient
			Feeding a bedridden patient
			Holding an arm or leg for extended periods (e.g., surgery or dressing changes)
			Holding retractors for extended periods of time
			Making an occupied bed
			Patient transfers in and out of personal vehicles
			Picking a patient up off the floor post fall
			Providing care in non-height adjustable beds
			Repositioning a patient in a chair or wheelchair
			Repositioning a patient in bed side to side
			Toileting a patient
			Transferring a patient from bed to stretcher
			Transferring a patient from bed to chair
			Transporting patient off unit.
			Other Task:
			Other Task:
			Other Task:

FIGURE 46.1 Tool for prioritizing high-risk patient handling tasks. (Adapted from Owen, B.D. and Garg, A. *AAOHN J,* 39, 24, 1991. With permission.)

- Is the task too strenuous?
- Can the task only be performed by twisting of trunk?

d. *Physical environment.* The clinical setting needs to be evaluated for safety as well, as they environment may be contributing to risk. Key elements to evaluate include:

- Space constraints, which force the caregiver into awkward positions to perform a task. Confined spaces are problematic, particularly around bed or wheelchair, and in bathrooms. Consider the number of staff that need to perform the task, equipment, positioning needs, and other safety issues as tasks are performed.
- Floor surface, which is an issue when transporting patients in wheelchairs and stretchers; this is also an issue when there are uneven floor surfaces to transverse, such as the lip of a shower area or an elevator.
- Lighting
- Height of work surfaces
- Adequacy of storage areas
- Temperature, ventilation or humidity is unsuitable
- Selection and design of furniture

e. *Work practices.* The way caregivers organize and deliver nursing care needs to be evaluated as well. Some work practices may place personnel at higher level of risk. Issues associated with the level of complexity and responsibility can contribute to stress, predisposing staff to injuries unnecessarily. Other elements to be considered include:
- The process for defining workload and assigning work, including the number, duration, and timing of high risks assigned to specific caregivers
- Insufficient rest/recovery between high-risk tasks and between shifts
- Excessive lifting, lowering, or carrying of patient loads
- Rate of work that the worker cannot control
- Quality and consistency of communication between caregivers and between the patient and caregiver related to high-risk tasks and the way tasks are performed
- Assessment of patient handling needs at individual level (on admission, at intervals)
- Assessment of nonpatient manual handling tasks, such as moving furniture, pushing medication carts

f. *Patient handling devices and equipment.* The type, quantity, and location of existing patient handling equipment and devices needs to be inventoried. Consider also the process for routine maintenance of equipment, suitability and compatibility of equipment. In addition to an inventory, it is also important to assess which equipment is not being used and why. Research to evaluate nurses attitudes towards mechanical aids reveal there are complex reasons why nurses use or do not use equipment, including:
- The equipment is inconveniently located.[31–34]
- The equipment is poorly maintained.[31,32,35–37]
- The equipment takes too long to use, compared with manually performing the task.[11,31,32,34,36,38–45]
- The equipment takes too much effort or is burdensome to use[34,38,46] or staff feel the patient is light enough to do it manually.[47]
- The equipment cannot be readily used because there are not enough slings[31,45,46,48] or staff have difficulty attaching slings.[46] In one study, 59% said they did not know who was responsible for laundering slings and 57% said they had difficult attaching slings and would be unlikely to use the device for this reason.[48]
- There are insufficient number of patient handling devices, particularly since many staff need the devices simultaneously during peak work periods.[31,32,36,38,46,47,49]
- Staffing deficits hinder staff's ability to use equipment; often many tasks require two or more caregivers to perform safely and meeting these expectations is compromised when the staffing levels are low.[29]
- The time demands placed on busy staff to complete all assignments, combined with the organizational emphasis on productivity goals are in conflict with the staff's ability to integrate equipment use into their routines.
- Space restrictions for use of equipment (e.g., crowded bathrooms).[32,36,38]
- Patient fear or aversion to use of the equipment,[34,38,39,45] or patient reports of discomfort in using equipment.[46,50]
- Infection control issues, particularly when the patient is in isolation.[34,35,38,45]
- Low motivation, attitude[38] have staff to manually lift, do not need it;[45,47] lets aides use them.[45]
- Lack of training,[32,34,35,38,45–48,50] especially on units with high levels of turnover; annual training may be insufficient.[36] In one study, 21% nurses did not have training, 33% had formal training, and 61% had informal training by peers, with more nurses on night shift not being trained; more then half did not know weight limitations of equipment.[48] Use of the equipment is not intuitive.[37]
- Unsuitable equipment purchased (nurses not involved in selection)[32,39,47,50,51] (weight limitations).[46]

- Lack of space.[32,36,38,48] In one study, 49% nurses had difficulty using lifts due to space constraints, particularly when bathing and toileting patients. 28% said lack of space would result in not using a lift.[48] Storage issues/equipment is located in an inconvenient place.[31–34]
- Noisy at night, disrupting patient sleep.[32]
- Equipment (such as bed) not compatible with lifting device.[38]
- Arbitrary use of mechanical aides.[48] In one study, 62% nurses said there was a plan for handling each patient but that the specific type of lift was identified in less than 36% cases.[48] Where there was no specified plan, nurses were less likely to use the lift ($p = 0.0006$); whereas 67% said if the lift was specified they would use the lift.[48]
- Increased use of patient handling devices can be accomplished through making some improvements regarding design. For example, improvements in maneuverability of lifting devices can be accomplished with careful alignment of castors;[52] providing teaching packages for ongoing competency training;[52] improving patient dignity and stability of lifts;[53,54] matching equipment with physical environment and other equipment through universal design concepts;[55] improving spreader bar and sling attachment redesign to promote patient safety and ease in attaching slings;[53,56] providing instructions on selecting sling type and size;[56] improving design of brakes[56] and improvements in sling design.[57,58]
- Key to avoiding the problems with caregivers not using patient handling equipment include:[59] (1) provision of training in use of assistive devices, (2) purchase an appropriate number of devices on each unit, (3) involve end users in the section of devices they perceive to be less stressful and more comfortable, and (4) obtain support from management for use of devices. National Institute for Occupational Safety and Health (NIOSH) has established criteria for selecting the right equipment.[60] Key criteria include:
  - Device must be appropriate for the task
  - Device must be safe for resident and nurse
  - Device should be comfortable for resident
  - Device should be understood and used with relative ease
  - Device should be time efficient
  - Need for maintenance should be minimal
  - Device must be maneuverable in confined workspaces

g. *Organization.* The safety culture that emerges in a health care facility as well as policies, and other key organizational factors can also influence workplace safety. Aspects of the organization that need to be evaluated include:
  - Accident reporting system
  - Accident investigation system
  - Annual or quarterly safety audits/hazard evaluations
  - Identification and actions taken for high-risk problems
  - Education and training programs
  - External consultation
  - Administrative support

46.4 Ergonomic Solutions for High-Risk Patient Handling Tasks

Ergonomic solutions should address the specific high-risk tasks identified and needs to consider findings from the hazard evaluation. There is no "one size fits all" solution for safe patient handling and movement. To facilitate success in the ergonomic interventions, the solutions should match risks identified in the hazard evaluation. The key is to target interventions to mitigate the high-volume, high-risk tasks identify in each setting.

Further, solutions needs to be evidence-based, rather than based on tradition or habit.

Hignett[61] summarized 63 studies in a review article and graded the evidence related to patient handling interventions. She grouped the studies into three categories: multifactor interventions, single factor interventions, and training interventions; she concluded:

- There is moderate evidence that multifactor interventions based on risk assessment are successful.
- There is moderate evidence that multifactor interventions *Not* based on risk assessment are successful.
- There is moderate evidence that single factor intervention (patient handling equipment alone) is successful.
- There is moderate evidence that single factor intervention (lift teams) are successful.
- There is strong evidence that technique training interventions are not effective in improving work practices or reducing injury rates.
- There is moderate evidence that technique training interventions have mixed results in the short term.

In 1994, Lahad et al.[62] reviewed evidence for four interventions to prevent low back pain. Studies published between 1966 and 1993 were included ($N = 64$). There is limited evidence that exercises to strengthen back or abdominal muscles and improve fitness will decrease incidence and duration of low back pain episodes. There is minimal support for education strategies and insufficient evidence to recommend back belts. There were no published studies that examined risk factor modification (smoking cessation and weight loss), but it has merit for future research.

In 2003, Stetler et al.[63] summarized literature in the 1990's to develop a plan for a hospital that included external evidence (research findings) and internal evidence (systematic data collected on site to refine and improve program). They found:

- No simple solution or single intervention will be effective.
- Multifaceted interventions should include at least two of the following: elimination of risk factors, engineering controls, administrative controls (no lift policy), training/education.
- Successful evidence-based strategies for changing behavior include: opinion leaders, educational outreach, reminder systems, and audit/feedback, culture of safety.

Patient assessment/algorithms provide a standardized way to assess patients and make appropriate decisions about how to safely perform high-risk tasks. In Britain, Hayne[64] developed Hazard Movement Code; in Canada, a system was developed by the Health Care Occupational Health and Safety Association in Ontario.[65] In Australia, a system using logos was developed.[65] In the United States, Nelson and colleagues[30] developed a patient assessment form and series of algorithms for safe patient handling.

To support the selection of the right patient care ergonomic solution, research related to several proposal interventions will be described. Please note that in some cases the research supports use (e.g., patient handling equipment), while in other cases the research indicates that some commonly used strategies are ineffective.

a. *Back belts* are described as breathable, lightweight bands, with double-sided pulls, which allow for different levels of pressure and tautness. Although used in health care settings in the 1990s,[31] there is strong evidence that these belts are *Not* effective in reducing risk during patient handling tasks.[66–68]

b. *Patient handling equipment.* Several technological solutions have been found effective in addressing high-risk tasks. For example, the use of height-adjustable beds and electric beds is effective in reducing risks associated with bed-related patient handling tasks, such as bathing.[19,69,70] Patient lifting devices have been found to be effective in reducing injuries.[33,37,41,44,45] The technology of the traditional floor-based patient lifting device has evolved to ceiling mounted lifts, which provides all the ergonomic advantages of a mechanical lift, with added benefits further reducing caregiver strain, while reducing the time to complete the patient transfer by half.[71–73]

c. *No-lift policies.* Internationally, health care institutions individually, or as part of national legislation, have developed and implemented policies to address the work-related risk associated with patient handling and movement tasks. These policies vary in wording, but the intent is consistent — that care providers should avoid manual handling in virtually every patient care situation. Manual patient handling is broadly defined as the transporting or supporting of a patient by hand or bodily force, including pushing, pulling, carrying, holding and supporting of the patient or a body part.[74] Despite the high level of risk associated with *manual patient handling*, nearly every nurse in the United States was taught manual patient handling techniques as part of their basic nursing training.[28] National legislation to minimize manual patient lifting has been introduced in Europe, Canada, and Australia.[37,28] Simply mandating a policy, whether at the local facility level or through public policy is not likely to have an immediate impact on changing caregiver behavior. In countries where national policies have been enacted, it is still reported that while new nursing staff receive proper training, experienced nurses and nurse managers continue with manual handling approaches and seem to be unaware of new approaches and technologies.[7,28]

Key elements of a no lift policy include:
- Patients should be encouraged to assist in their own transfers and handling aids must be used whenever possible to help reduce risk if this is not contrary to a patient's needs.[34]
- Manual lifting may only be continued if it does not involve lifting most or all of a patient's weight.[34]
- A no-lift policy does not mean health care providers will never transfer or reposition any resident manually, but rather needs to be based on patients' physical and cognitive status as well as medical conditions.[75]
- Proper infrastructure must be in place before a no-lift policy is enforced. Infrastructure is defined as management commitment and support, availability of patient handling equipment, equipment maintenance, employee training, advanced training for resources, and a culture of safety. The culture of safety approach includes collective attitude of employees at all levels taking a shared responsibility for safety in the work environment and by doing so providing a safe environment for themselves and patients.[75]
- The effectiveness of no lift policies has been documented in research studies,[21,76] however the policy was included as part of a multifaceted program. In 2003, OSHA released ergonomics guidelines for nursing homes[77] to reduce the number and severity of work-related musculoskeletal disorders, which support alternatives to manual patient handling.

d. *Clinical tools (algorithms and patient assessment tools).* There is a lack of standardization regarding decisions about which patient handling technique and equipment is needed for a specific task. Key elements to be considered in these decisions are patient functional status, level of cognition, height and weight, and special patient conditions likely to affect the way the task is performed, such as a missing limb or surgical wound. Clinical tools are useful for applying research to practice and reducing unnecessary variation in practice.

Examples of such tools include patient assessment protocols and algorithms include:
- In Britain, a Hazard Movement Code was developed.[64]
- In Canada, a system was developed by the Health Care Occupational Health and Safety Association in Ontario.[65]
- In Australia, a system was developed using figures and pictorial notations.[65]
- In the United States, Nelson and colleagues[21] developed a patient assessment and series of algorithms for safe patient handling. Nelson's assessment and algorithms were included in the OSHA Ergonomic Guidelines for Nursing Homes.[77,78]

Each of these tools assist nurses in selecting the safest equipment, technique, and number of staff needed to perform safe patient handling tasks based on specific patient characteristics.[79] The use of assessment and algorithms ensure that patients receive assistance appropriate for their functional level, thus improving safety for patients as well as staff.

e. *Patient lift teams.* Several clinical trials have been conducted on patient lift teams and found this intervention to be effective in decreasing the lost days, restricted workdays and compensable injury costs.[80–82] A lifting team is defined as "two physically fit people, competent in lifting techniques, working together to accomplish high-risk patient transfers".[87] Lift teams seem to be most effective in small community hospitals, where the number of scheduled and unscheduled lifts is not so large that the logistics of providing these services is manageable. Further examination is needed to ascertain whether positive findings are due to the addition of more staff or the lift team itself.

f. *Classes in body mechanics and training in lifting techniques.* Although it is widely accepted that classes on body mechanics and training in lifting techniques prevent job-related injuries, 35 yr of research reveals that these efforts have consistently failed to reduce the job-related injuries in patient care settings.[4,43,88–103] While training may improve patient handling and lifting skills in the short term,[94] it has no impact in reducing injuries or musculoskeletal pain. Despite the fact that there is 35 yr of evidence that these interventions are not effective, classes in body mechanics and training in lifting techniques remain the primary solution used by health care facilities in the United States.

g. *Competency in use of patient handling equipment.* One effective educational strategy is the to educate and train nursing staff on the use of patient handling equipment. There is strong evidence that patient equipment is not used on units with high turnover due to a lack of training.[32,34,35,38,46–48,50] Several studies support the significance of training on equipment related to patient handling for a successful program in injury prevention.[21,34,104–106] Positive results indicate that there is a need for new education models to assure competency when using patient handling equipment. Ongoing training is key in health care settings for nurses to achieve proficiency and comfort on equipment use.[107,108]

h. *Peer leaders as new education model.* New education models are needed for assuring competency in use of patient handling technology and new ongoing training practices are needed in health care settings where nurses are employed.[107,108] One new model that shows promise is use of local peer leaders. A peer safety leader is defined as a nursing staff member who receives special training and then returns to the unit to share knowledge and skills with coworkers. Unit peer leaders foster knowledge transfer and forge a direct connection between staff and program goals (http://www.paraprofessional.org/publications/WorkforceStrategies2.pdf and http://www.frontlinepub.com/pdf/MentorSample.pdf.). Peer safety leaders have been called Back Injury Resource Nurses (BIRNs),[107] Ergo Rangers,[104] and Ergo Coaches. Further research is needed regarding this new educational model to promote nurse safety.

46.5 Summary

The purpose of this chapter was to describe a practical approach for reducing occupational injuries related to patient care. Work-related musculoskeletal injuries associated with patient care have been a problem for decades. The first step is to identify high risk tasks related to patient handling. There is significant variation across clinical settings in the frequency, duration, and level of risk associated with patient handling tasks. Additionally, patient characteristics, physical layout, availability of equipment, and work processes affect which high-risk tasks to target for mitigation. Given this variation across patient care settings, there is no one intervention, or even a set combinations of interventions likely to be universally effective in health care. The solutions need to match risks identified in a unit-based hazard evaluation.

Strategies to prevent or minimize injuries associated with patient handling are often based more on tradition and personal experience rather than scientific evidence. The most common patient handling approaches in the United States include manual patient lifting, classes in body mechanics, training in safe lifting techniques and back belts. Surprisingly there is strong evidence that each of these approaches is *Not* effective in reducing caregiver injuries.

Although research is needed to refine patient care ergonomic interventions and test their effectives across patient care settings, several evidence-based strategies have been identified, including (1) use of patient handling equipment/devices, (2) no-lift policies, and (3) patient lift teams. Promising new interventions, which are still being tested, include the use of unit-based peer leaders and clinical tools such as algorithms and patient assessment protocols.

References

1. Fragala, G. and Bailey, L.P., Addressing occupational strains and sprains: musculoskeletal injuries in hospitals, *AAOHN J*, 51, 252, 2003.
2. Bureau of Labor Statistics, Survey of occupational inquiries and illnesses, 2001, US Department of Labor, USDL December 19, 2002.
3. U.S. Department of Health and Human Services (HHS) Federal Register, Part II, Department of Labor, Occupational Safety and Health Administration, 29 CFR Part 1910; Ergonomics Program: Proposal Rule (Tuesday, 23 November 1999).
4. Hignett, S., Work-related back pain in nurses, *J Adv Nurs*, 23, 1238–1246, 1996.
5. Blue, C.L., Preventing back injury among nurses. *Orthop Nurs*, 15, 9, 1996.
6. Videman, T., et al., Low back pain in nurses and some loading factors of work. *Spine*, 9, 400, 1984.
7. Hignett, S., et al., *Evidence-based Patient Handling: Tasks, Equipment and Interventions*, Routledge, New York, 2003.
8. Bell, F., et al., Hospital ward patient-lifting tasks. *Ergonomics*, 22, 1257, 1979.
9. Garg, A., Owen, B., and Carlson, B., Ergonomic evaluation of nursing assistants' jobs in a nursing home, *Ergonomics*, 35, 979, 1992.
10. Hui, L., et al., Evaluation of physiological work demands and low back neuromuscular fatigue on nurses working in geriatric wards, *Appl Ergon*, 32, 479, 2001.
11. Owen, B., The need for application of ergonomic principles in nursing, in *Trends in Ergonomics: Human Factors IV*, 1987, 831.
12. Owen, B., Garg, A., and Jensen, R.C., Four methods for identification of most back-stressing tasks performed by nursing assistants in nursing homes, *Int J Ind Ergon*, 9, 213, 1992.
13. Schibye, B. and Skotte, J., The mechanical loads on the low back during different patient handling tasks. *Proceedings of the IEA2000/HFES 2000 Congress*, The Human Factors and Ergonomics Society, Santa Monica, CA, 5, 2000, 785.
14. Smedley, J., et al., Manual handling activities and risk of low back pain in nurses, *Occup Environ Med*, 52, 160, 1995.
15. Massad, R., Gambin, C., and Duval, L., The contribution of ergonomics to the prevention of musculoskeletal lesions among ambulance technicians. *Proceedings of the IEA2000/HFES 2000 Congress*, The Human Factors and Ergonomics Society, Santa Monica, CA, 4, 2000, 201.
16. Furber, S., et al., Injuries to ambulance officers caused by patient handling tasks, *J Occup Health Safety — Australia & New Zealand*, 13, 259, 1997.
17. Doormaal, M., et al., Physical workload of ambulance assistants, *Ergonomics*, 38, 361, 1995.
18. Ballard, J. (1994). District nurses—who's looking after them? *Occup Health Review*, Nov./Dec., 10–16.
19. Knibbe, J.J. and Friele, R.D., Prevalence of back pain and characteristics of the physical workload of community nurses, *Ergonomics*, 39, 186, 1996.
20. Skarplik, C., Patient handling in the community, *Nursing*, 3, 13, 1988.
21. Nelson, A.L. and Fragala, G., Equipment for safe patient handling and movement, in *Back Injury among Healthcare Workers*, Charney, W. and Hudson, A., Eds., Lewis Publishers, Washington, DC, 2004, 121–135.
22. Owen, B., Preventing injuries using an ergonomic approach, *AORN J*, 72, 1031, 2000.
23. Garb, J.R. and Dockery, C.A., Reducing employee back injuries in the perioperative setting, *AORN J*, 61, 1046, 1995.

24. Wicker, P., Manual handling in the perioperative environment, *Br J Perioper Nurs*, 10, 255, 2000.
25. Nelson, A., Unpublished research data from pilot study, James A. Haley VA Medical Center, Tampa, FL, 1996.
26. Owen, B.D., Keene, K., and Olson, S., Patient handling tasks perceived to be most stressful by hospital nursing personnel, *J Healthcare Safety, Compliance, Infect Control*, 5, 19, 2000.
27. Owen, B.D. and Staehler, K., Approaches to decreasing back stress in homecare, *Home Healthcare Nurs Manual*, 21, 180, 2003.
28. Corlett, E.N., et al., *The Guide to Handling Patients*, 3rd ed., National Back Pain Association and the Royal College of Nursing, London, 1993.
29. Collins, M., *Occupational Back Pain in Nursing: Development, Implementation and Evaluation of a Comprehensive Prevention Program*. Worksafe Australia, National Occupational Health and Safety Commission, Australia, 1990.
30. Nelson, A.L., et al., Safe patient handling and movement, *Am J Nurs*, 103(3), 32–43, 2003.
31. Nelson, A.L., Fragala, G., and Menzel, N., Myths and facts about back injuries in nursing, *Am J Nurs*, 103(2), 32–40, 2003.
32. McGuire, T. and Dewar, J., An assessment of moving and handling practices among Scottish nurses, *Nurs Stand*, 9, 35, 1995.
33. Yassi, A., et al., A randomized controlled trial to prevent patient lift and transfer injuries of health care workers, *Spine*, 26, 1739, 2001.
34. Retsas, A. and Pinikahana, J., Manual handling activities and injuries among nurses: An Australian hospital study, *J Adv Nurs*, 31, 875, 2000.
35. McGuire, T., Moody, J., and Hanson, M., Managers attitudes towards mechanical aids. *Nurs Stand*, 11, 33, 1997.
36. Green, C., Study of the moving and handling practices on two medical wards, *Br J Nurs*, 5, 303, 1996.
37. Garg, A., et al., A biomechanical and ergonomic evaluation of patient transferring tasks: bed to wheelchair and wheelchair to bed, *Ergonomics*, 34, 289, 1991.
38. Bell, F., Ergonomic aspects of equipment, *Int J Nurs Stud*, 24, 331, 1987.
39. McGuire, T., et al., A study into client's attitudes towards mechanical aids, *Nurs Stand*, 11, 35, 1996.
40. Takala, E.P. and Kukkonen, R., The handling of patients on geriatric wards, *Ergonomics*, 18, 17, 1987.
41. Daynard, D., et al., Biomechanical analysis of peak and cumulative spinal loads during patient handling activities: a substudy of a randomized controlled trial to prevent lift and transfer injury to healthcare workers, *Appl Ergon*, 32, 199, 2001.
42. Laflin, K. and Aja, D., Healthcare concerns related to lifting: an inside look at intervention strategies, *Am J Occup Ther*, 49, 63, 1994.
43. Owen, B., and Garg, A., Reducing risk for back pain in nursing personnel, *AAOHN J*, 39, 24, 1991.
44. Garg, A., et al., A biomechanical and ergonomic evaluation of patient transferring tasks: wheelchair to shower chair and shower chair to wheelchair, *Ergonomics*, 34, 407, 1991.
45. Evanoff, B., et al., Reduction in injury rates in nursing personnel through introduction of mechanical lifts in the workplace, *Am J Ind Med*, 44, 451, 2003.
46. Meyer, E., Patient lifter in a practical test. A spine-saving aid or bulk in the storage room? *Pflege Aktuell*, 49, 597, 1995.
47. Bewick, N. and Gardner, D., Manual handling injuries in health care workers, *Int J Occup Safety Ergon*, 6, 209, 2000.
48. Moody, J., McGuire, T., and Hanson, M., A study of nurses' attitudes towards mechanical aids, *Nurs Stand*, 11, 37, 1996.
49. Newman, S. and Callaghan, C., Work-related back pain. *Occup Health (Lond)*, 45, 201, 1993.
50. Switzer, S. and Porter, J.M., The lifting behavior of nurses — in their own words, in Darby, F. and Turner, P. Eds., *Proceedings of the 7th Conference of the New Zealand Ergonomics Society*, 2–3 August 1996 (pp. 33–43), New Zealand Ergonomics Society, Wellington, 1996.
51. Nelson, A., Ed., Patient Care Ergonomics Resource Guide. Veterans Administration Patient Safety Center, Tampa, FL, 2003. Also available at: www://patientsafetycenter.com.

52. Bell, F., *Patient Lifting Devices in Hospitals*, Croom Helm, London, 1984.

53. McGuire, T., Moody, J., and Hanson, M., An evaluation of mechanical aids used within NHS, *Nurs Stand*, 11, 33, 1996.

54. LeBon, C. and Forrester, C., An ergonomic evaluation of a patient handling device: the elevate and transfer vehicle, *Appl Ergon*, 28, 365, 1996.

55. Love, C., Ergonomic consideration when choosing a hoist and slings, *Br J Ther Rehabil*, 3, 189, 1996.

56. Olsson, G. and Brandt, A., An investigation of the use of ceiling mounted hoists for disabled people, Danish Centre for Technical Aids for Rehabilitation and Education, Denmark, 1992.

57. MDA. *Slings to Accompany Mobile Domestic Hoists*. HMSO, Norwich, A10, 1994.

58. Norton, L., An Ergonomic Evaluation into Fabric Slings Used during the Hoisting of Patients. Unpublished M.Sc. dissertation, University of Nottingham, 2000.

59. Owen, B.D., et al., An ergonomic approach to reducing back stress while carrying out patient handling tasks with a hospitalized patient, in Pheasant, S., Ed., *Ergonomics, Work and Health*, 298, 1991.

60. Collins, J.W. and Owen, B.D., NIOSH research initiatives to prevent back injuries to nursing assistants, and orderlies in nursing homes, *Am J Ind Med*, 29, 421, 1996.

61. Hignett, S., Intervention strategies to reduce musculoskeletal injuries associated with handling patients: a systematic review, *Occup Environ Med*, 60, e6, 2003. Retrieved February 8, 2005, from http://www.occenvmed.com/cgi/content/full/60/9/e6.

62. Lahad, A., et al., The effectiveness of four interventions for the prevention of low back pain, *J Am Med Assoc*, 272, 1286, 1994.

63. Stetler, C.B., et al., Use of evidence for prevention of work-related musculoskeletal injuries, *Orthop Nurs*, 22, 32, 2003.

64. Hayne, C., personal correspondence, cited in Collins, M., *Occupational Back Pain in Nursing: Development, Implementation and Evaluation of a Comprehensive Prevention Program*, Worksafe Australia, National Occupational Health and Safety Commission, Australia, 1990.

65. Health Care Occupational Health and Safety Association, *Transfer and Lifts for Caregivers*, HCOSHA Publications, Ontario, Canada, 1986.

66. Alexander, A., Woolley, S.M., Bisesi, M., The effectiveness of back belts on occupational back injuries and worker perception, *Professional Safety*, 10, 22, 1995.

67. NIOSH Back Belt Working Group, *Workplace Use of Back Belts: Review and Recommendations* (Publication #94-122). National Institute for Occupational Safety and Health, Rockville, MD, 1994. Retrieved July 8, 2004 from the NIOSH Web site http://www.cdc.gov/niosh/94-122.html.

68. Wassell, J.T., et al., A prospective study of back belts for prevention of back pain and injury, *J Ammed Assoc*, 284, 2727, 2000.

69. De Looze, M.P., et al., Effect of individually chosen bed-height adjustments on the low-back stress of nurses, *Scand J Work Environ Health*, 20, 427, 1994.

70. Knibbe, N. and Knibbe, J.J., Postural load of nurses during bathing and showering of patients, Internal Report, Locomotion Health Consultancy, The Netherlands, 1995.

71. Holliday, P.J., Fernie, G.R., and Plowman, S., The impact of new lifting technology in long-term care, *AAOHN J*, 42, 582, 1994.

72. Ronald, L.A., et al., Effectiveness of installing overhead ceiling lifts. *AAOHN J*, 50, 120–126, 2002.

73. Villeneuve, J., The ceiling lift: an efficient way to prevent injuries to nursing staff. *Healthcare Safety, Compliance Infect Control*, Jan, 19, 1988.

74. Manual Handling Operations Regulations (MHOR), *Manual Handling Operations Regulations 1992 Guidance for Regulations*, L23 (2nd ed.), HSE Books, London, 1992.

75. Interior Health Authority, *MSIP: A Practical Guide to Resident Handling* (Section 2.0 pp. 1–7). British Columbia: Author, Oct. 2004. Retrieved Feb. 2, 2002, from http://www.interiorhealth.ca/ NR/rdonlyres/4AF034BF-78B6-48BF-A18D-2AA6B39F99FE/1908/FullGuideforMSIPAPractical- Guide toResidentHandling.pdf.

76. Garg, A. (1999). Long-Term Effectiveness of "Zero-Lift Program" in Seven Nursing Homes and One Hospital, Contract No. U60/CCU512089-02, 1999. Milwaukee, WI: University of Wisconsin-Milwaukee.

77. U. S. Department of Labor, Occupational Safety and Health Administration (OSHA), 2002, Ergonomics Guidelines for Nursing Homes. Retrieved February 1, 2005, from http://www.osha.gov/ergonomics/guidelines/nursinghome/index.html

78. Nelson, A., Ed., *Patient Care Ergonomics Resource Guide*, US Dept. of Veterans Affairs Patient Safety Research Center, Tampa, FL, 2003. Retrieved February 1, 2005, from http://www.patientsafetycenter.com/Safe%20Pt%20Handling%20Div.htm

79. Nelson, A.L., et al., Preventing nursing back injuries: redesigning patient handling tasks, *AAOHN J*, 51, 126, 2003.

80. Caska, B.A., Patnode, R.E., and Clickner, D., Feasibility of nurse staffed lift team. *AAOHN J*, 46, 283, 1998.

81. Charney, W., Zimmerman, K., and Walara, E., The lifting team: a design method to reduce lost time back injury in nursing. *AAOHN J*, 39, 231, 1991.

82. Charney, W., The lifting team: second year data reported (News). *AAOHN J*, 40, 503, 1992.

83. Charney, W., The lifting team method for reducing back injuries: a 10 hospital study. *AAOHN J*, 45, 300, 1997.

84. Charney, W., Reducing back injury in nursing: a case study using mechanical equipment and a hospital transport team as a lift team. *J Healthcare Safety, Compliance, Infect Control*, 4, 117, 2000.

85. Davis, A., Birth of a lift team: Experience and statistical analysis, *J Healthcare Safety, Compliance Infect Control*, 5, 15, 2001.

86. Donaldson, A.W., Lift team intervention: a six year picture. *J Healthcare Safety, Compliance Infect Control*, 4, 65, 2000.

87. Meittunen, E.J., et al., The effect of focusing ergonomic risk factors on a patient transfer team to reduce incidents among nurses associated with patient care, *J Healthcare Safety, Compliance Infect Control*, 2, 306, 1999.

88. Anderson, J., Back pain and occupation. in Jayson, M.I.V., Ed., *The Lumbar Spine and Back Pain*, 2nd ed (pp. 57–82). Pitman Medical Ltd., London, 1980.

89. Brown, J., Manual lifting and related fields. An annotated bibliography. Labor Safety Council of Ontario, Toronto, Ontario, Canada, 1972.

90. Dehlin, O., Hedenrud, B., and Horal, J., Back symptoms in nursing assistants in a geriatric hospital, *Scan J Rehabil Med*, 8, 47, 1976.

91. Buckle, P., Epidemiological aspects of back within the nursing profession, *Int J Nurs Stud*, 24, 319, 1987.

92. Daltroy, L., A controlled trial of an educational program to prevent low back injuries, *N Engl J Med*, 337, 322, 1997.

93. Daws, J., Lifting and moving patients. A revision training programme, *Nurs Times*, 77, 2067, 1981.

94. Feldstein, A., et al., The back injury prevention project pilot study: assessing the effectiveness of back attack, an injury prevention program among nurses, aides, and orderlies, *J Occup Med*, 35, 114, 1993.

95. Hayne, C. Ergonomics and back pain, *Physiotherapy*, 70, 9–13, 1984.

96. Harber, P., et al., Personal history, training and worksite as predictors of back pain of nurses, *Am J Ind Med*, 25, 519, 1994.

97. Hollingdale, R. and Warin, J., Back pain in nursing and associated factors: a study, *Nurs Stand*, 11, 35, 1997.

98. Lagerstrom M. and Hagberg, M., Evaluation of a 3-year education and training program for nursing personnel at a Swedish hospital, *AAOHN J*, 45, 83, 1997.

99. Shaw, R., Creating back care awareness, *Dimens Health Serv*, 58, 32, 1981.

100. Snook, S., Campanelli, R., and Hart, J., A study of three preventative approaches to low back injury, *J Occup Med*, 20, 478, 1978.

101. Stubbs, D., et al., Back pain in the nursing profession. II. The effectiveness of training, *Ergonomics*, 26, 767, 1983.
102. Venning, P., Back injury prevention among nursing personnel, *AAOHN J*, 36, 327–333, 1988.
103. Wood, D., Design and evaluation of a back injury prevention program within a geriatric hospital, *Spine*, 12, 77, 1987.
104. Collins, J. W., Wolf, L., Bell, J., and Evanoff, B., 2004. An evaluation of a "best practices" musculoskeletal injury prevention program in nursing homes, *Injury Prev*, 10, 206–211, 2004.
105. Lynch, R.M. and Freund, A., Short-term efficacy of back injury intervention project for patient care providers at one hospital, *AIHA J*, 61, 290, 2000.
106. Owen, B.D., Keene, K., and Olson, S., An ergonomic approach to reducing back/shoulder stress in hospital nursing personnel: a five year follow up. *Int J Nurs Stud*, 39, 295, 2002.
107. Nelson, A.L., et al., Technology to promote safety mobility in elderly, *Nurs Clin North Am*, 39, 649, 2004.
108. NIOSH, *National Occupational Research Agenda for Musculoskeletal Disorders*. U.S. Department of Health and Human Services (Public Health Service, CDC, NIOSH), Washington, DC, 2001.

47

Prevention of Medical Errors

Marilyn Sue Bogner
*Institute for the Study of Human
 Error, LLC*

47.1 Medical Error

Ergonomics and its fraternal twin human factors are no strangers to the topic of medical error. In 1960, a nursing student published a two-part article coauthored by her mentor Alphonse Chapanis (Safren & Chapanis, 1960a, b) that reported medication mishap findings analogous to those of the Harvard Medical Practice Study (HMPS) published 36 yr later (Leape et al., 1991). A book *Human Factors in Health Care* (Pickett & Triggs) was published in 1975. In 1978, an article appeared that addressed human factors issues in anesthesia gas machines (Cooper, Newbower, Long, & McPeek, 1978). The book *Human Error in Medicine* (Bogner, 1994) presented a systems approach to medical error.

The 1978 article received considerable attention and lead to the redesign of anesthesia gas machines as well as the formation of the Anesthesia Patient Safety Foundation. The issues raised by the other writings received little interest. Even though the HMPS (Leape et al., 1991) reported injuries and deaths attributed to adverse events, medical error was not a topic of concern beyond a relatively small community. All that changed late in 1999 with the publication of the report on medical error by the Institute of Medicine (IOM) of the National Academy of Sciences (Kohn, Corrigan, & Donaldson, 1999) that stated 44,000 to 98,000 people die annually as result of medical care. The public was outraged, Congress reacted, and the research community responded.

47.2 Addressing Medical Error: The Quest for Patient Safety

Two major recommendations for addressing medical errors in the IOM report are to establish a Federal government agency to be the focus for and fund patient safety research, and a national error reporting database. Congress responded to the first recommendation by establishing the Agency for Healthcare Research and Quality (AHRQ) and providing $50 million a year to fund research with the directive that the products from that work would reduce the incidence of medical error by 50% by the end of 5 yr after publication of the report. The second recommendation for an error reporting database was elaborated in the report by the description of the two functions of an error-reporting activity: ". . . hold providers accountable for performance, or alternatively . . . provide information that can lead to improved safety" (Kohn et al., 1999, p. 74).

The emphasis in the report was the first of those functions — that of provider accountability, so it is not surprising that most of the research funded by AHRQ addresses some aspect of error reporting for provider accountability. This may seem to be the reasonable approach because the care provider commits the act that has the adverse outcome of death, serious injury, or need for extended treatment; however, underlying that apparent reasonableness is a potent presumption that appears in the title of the report, *To err is human*.

47.2.1 Potent Presumption

The presumption that appears in the title of the IOM report, that human have an innate propensity to commit error is pervasive in that document as it is in the literature and media reports and seemingly in everyone's thinking about medical error. Actually, rather than reflecting findings from empirical research, the phrase to err is human is a partial quotation from *An Essay on Criticism* by Alexander Pope (1711/2004) that compares humans with the Deity. The complete quotation is "To err is human, to forgive divine." A search of the literature across disciplines found no empirical support for an inherent human proneness to commit errors, so thinking and research on medical error apparently is influenced, if not determined, by a literary quote. Indeed, to some, "to err is human" might be interpreted that error is nearly uniquely human; however, that is far from accurate as illustrated by examples.

47.2.1.1 Shaky Basis for the Presumption

A tree that should grow in soil is found growing out of a rock — that can be considered an error. Sheepdogs as well as other herding dog breeds herd whatever entities are in their proximity such as two legged humans, poultry, even what has been considered impossible — they herd cats. The philosopher-physicist Ernst Mach in his discussion of error (1905/1976) relates the saga of the fly that lays her eggs on carrion so the larvae when they hatch have food; understandably, she is attracted by the odor of carrion. That attraction occurs whether the odor of carrion is emitted by a plant or rotting flesh. The fly with the best of intensions will lay her eggs on the carrion odor emitting plant and the larvae starve — an error with dire consequence. These examples of nonhuman error suggest two characteristics of error across species and across error activities — those of complexity and similarity.

When the complexity of a situation is not in harmony with the complexity of an acting entity, the act or behavior of the entity will be inappropriate or wrong — an error. The growing medium for the tree was complex — it included more than soil, so the tree grew inappropriately. Likewise, the sheepdog and fly were confronted with additional factors in their environments, did not differentiate among them, hence behaved in an unexpected, inappropriate manner; they committed errors. Similarity leads to misinterpretation — an item or entity that is similar to another may be substituted for what is intended. This is such a deep-seated tendency that similarity is one of Gestalt principles of perception (Kohler, 1958). The lessons provided by the tree, sheepdog, and fly by illustrating the power of complexity and similarity to influence activity have implications for medical error.

47.2.1.2 Presumption and Error Reporting

The presumption that humans are innately error-committing creatures is evident in the emphasis on the provider accountability function of error reporting recommended in the IOM report. When reporting error for accountability, the care provider typically reports his or her error in the form of who did what, "Dr. Kildare prescribed an inappropriate drug." The error committing entities in the examples, the tree, sheepdog, and fly when reporting for accountability would state "tree grew out of rock, sheepdog herded cats, fly laid eggs on a plant." How effective such error reports are in reducing the error is an empirical question. Fortunately, that information is at hand because November 2004 marked the fifth anniversary of the IOM report when the efforts to address medical error were to have reduced the incidence of error by 50%.

47.2.2 Research on Error Reporting for Accountability

A workshop was convened by the Commonwealth Fund to learn of the impact of 5 yr of research and other efforts on the goal of reducing medical error by 50%. Various presenters at that workshop stated in one way or another that patient safety is on the right track but the efforts to attain the 50% reduction in error not only did not meet that goal, the impact of those efforts on error is negligible (Commonwealth Fund, 2004).

The Reuters news service described the information presented at the workshop "... five years after the IOM report, little data exist showing progress and researchers are still debating not how to save lives, but what to measure" (Zwillich, 2004). One presenter stated that "a little bit of progress has been made in U.S. hospitals since 1999, but the difference is not striking" (Wachter, 2004). He went on to say that many states and private health systems require health workers to report medical errors or near misses in which a patient is put at potential risk, but researchers still have not figured out what to do with the reports once they have them.

Although the presenters at the workshop stated that more money, greater cooperation from the medical community, and legislation to protect data from being used in malpractice litigation would expedite progress toward the goal of 50% reduction in medical errors, the lack of progress from the efforts to date very likely indicate that the data gathered, the "who did what" for provider accountability, are not appropriate for reducing error. Another aspect of those data is that by focusing on the provider and his or her accountability for his or her performance, the blame culture is perpetuated. Only the care provider is identified as causing the error, hence only the care provider is blameworthy.

One needs to look no further than the functions of error reporting stated in the IOM report "... hold providers accountable for performance, or alternatively ... provide information that can lead to improved safety" (Kohn et al., 1999, p. 74) for an alternative to the who did what approach. The challenge is to report errors in a way that provides information that can lead to improved safety.

47.2.3 Information to Improve Safety

A very simple key exists to obtaining information that when addressed can enhance safety — a key so obvious that it is overlooked. To acquire information to reduce the incidence of error, that is, to identify the target(s) to address to prevent recurrence of the error, it is necessary to determine why the error occurred; to do that, it is necessary to understand what an error is, the nature of error. The simple key is the definition of error.

The literature is replete with what are purported to be definitions of error — diagnostic error, mistake, a planned procedure that is not completed, misadministration of medication. Those are not definitions; they do not express the nature of error. Rather, they are descriptions stating what occurred, what precipitated the adverse outcome — not the process by which such events occur. Those descriptions, however, have a common element that defines error; each describes the result of an action, a behavior. Because error is a behavior, to solve the problem of error, it is necessary to address the source of the problem — error as behavior. This simple definition has profound implications for the medical domain.

47.3 Error as Behavior

The topic of behavior has been addressed by centuries of research and theory in the social sciences and millennia of philosophical writings as well as research and theory in the physical sciences such as in the behavior of particles. In all of that work, behavior has not been considered as solely determined by the behaving entity; rather behavior reflects the interaction of the entity with factors in the context of the behavior. Behavior is expressed succinctly as $B = f\{P \times E\}$ (Lewin, 1946/1997); that is, behavior (B) is a function (f) of the person (P) (or for physical sciences, the entity) interacting with factors in the environment (E).

The profound impact on the study of medical error by the concept of error as behavior is that it introduces a category of factors to be considered, those of the environmental context in which the behavior occurs that actually affect the behaving individual. Those factors in the environmental context that do not affect the person do not contribute to error even though it may appear that way to an observer. If a factor does not affect the individual, then it is not relevant for the given behavior. Factors that affect the individual, the care provider, and impact his or her behavior exist in that person's life space.

47.3.1 Life Space

The life space of an individual contains all factors that affect the person at the point in time when the behavior occurs (Lewin, 1936/1966). Those factors can be distal from the person both in time and space yet interact with the person and as such affect his or her behavior. To reduce error, those factors that affect the person to commit an error must be identified and addressed. This is not the task of an expert who observes and identifies factors that influence behavior (Lewin, 1943/1997).

Only the behaving person knows the unobservable factors that induce error such as distracting pain of varicose veins, and the concern about an adverse outcome due to unanticipated anatomical abnormalities in the patient. It is that person's perception of the situation that has meaning for him or her (Davies, Ross, Wallace, & Wright, 2003). It is acknowledged that the person may not be aware of all life space factors that affect his or her behavior, but it can be assumed that the person is aware of the most salient, so the factors that contribute to error can be identified.

It may seem curious that this discussion of error as behavior draws heavily on work of Kurt Lewin originally published nearly three quarters of a century ago. Although the publication dates are old, Lewin's work reflecting his philosophy that "Nothing is as practical as a good theory" and his talent for effectively applying theory to real-world issues speak to contemporary issues. A relatively recent resurgence of interest in his work and the republication of two edited books of his work (Allport, 1948; Cartwright, 1951) in a single volume (Lewin, 1943/1997) testify to that. His insights on behavior, which have few if any rivals, are particularly relevant to preventing medical error.

47.3.1.1 The Systems of the Life Space

A framework for representing a person's life space was developed from findings from research on error by several industries (Bogner, 2000). These findings are sufficiently similar to be merged across industries to define eight systems of categories of factors that are represented as hierarchical nested systems comprising the system of a person's life space. They are the factors with which the person, the care provider interact; the factors that affect that person's behavior at any given point in time. Although the systems concept appears in the study of medical error or error in general (Perrow, 1984; Reason, 1990; Leape et al., 1995), only the error as behavior approach focuses on the person involved in an error and the affect of the system as factors that induce *that person's* behavior.

Typically, a system is considered as either the health care system, or a process involving patient care. Leape et al. (1995, p. 38) in their study of adverse drug events state that

> . . . the systems described are broadly defined. For example, "drug knowledge dissemination" is typically viewed a drug education whether in medical school, during residency, at hospital conferences, at a post-graduate course, or by continuing self-education, but it can also be usefully considered as

the sum of methods that are used to make information about drugs available to physicians, nurses, and pharmacists at the time of use.

Although "drug knowledge dissemination" may be a system and "failures" within that system may lead to errors, that system is so complex as to defy attempts to identify factors to be addressed to reduce error. Because of that and the goal of this discussion is to describe a means to identify factors that induce error, every reference to system in the discussion of error as behavior refers to the system of factors of the person's life space, those with which that person interacts and are manifest in behavior.

Three of the eight empirically defined categories of the life space of a person, a care provider, describe the basic care providing system of the point-of-care: the patient, the means of providing care such as a medical device, and the care provider who is affected by psychological phenomena as well as physiological factors. That basic care providing system is nested in the environmental context of care represented by the five systems of ambient conditions, physical environment, social environment, organization, and legal–regulatory–reimbursement–national culture factors (Figure 47.1). The eight systems of a person's life space are interrelated; perturbations in any system ripple through the other systems ultimately affecting the behavior of the care provider. See Bogner (2000, 2002, 2004a, b) for further discussion of this life space systems approach and its application to understanding medical error.

To aid in understanding and recalling the relatively complex representation of a care provider's life space, a characterization in keeping with those of others who study error is employed. As the representations of error as slices of Swiss cheese (Reason, 1990) and an onion (Moray, 1994) have been helpful in conveying their concepts of error, the life space of this systems model of behavior is represented as an artichoke.

47.4   Artichoke Systems Approach

The artichoke model of the life space of a care provider represents the interaction of factors in the environment and the person; the concentric circles of the leaves of the artichoke represent the systems of factors. The affect of those systems, the leaves, in defining the behavior of the care provider is illustrated by the artichoke life space with the care provider at the center, the heart of the artichoke in Figure 47.2. The effect of a system of factors, even those furthest removed from the care provider, the legal–regulatory–reimbursement–national culture factors, on the care provider can be conceptualized as squeezing the artichoke in Figure 47.2.

The particular factors in each circle of leaves can be conceptualized as the filaments of the leaves in that system. Squeezing not only compresses the outermost leaves, but when the pressure is sufficiently strong, all the layers of leaves are affected including the heart of the artichoke, the care provider. This underscores

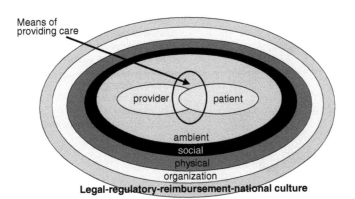

FIGURE 47.1   Systems of factors that influence behavior. (From Atlantic Health System 2005. With permission.)

FIGURE 47.2   Systems affecting the provider. (From Atlantic Health System 2005. With permission.)

the importance of considering all systems of factors in preventing error including the system rarely addressed in the literature — that of the legal–regulatory–reimbursement–national culture factors.

It should be noted that considering error as reflecting the affect of environmental factors on the individual care provider does not rule out the possibility of a care provider who for whatever reason consistently and in the light of favorable contextual factors does harm; however, rather than curse the many with the acts of a few, the focus in this discussion is on identifying factors that affect the care provider to commit an error.

47.4.1   Error Reporting for Safety

The life space framework of the eight layers of leaves of the Artichoke also is represented as a simple tool for reporting factors that induce error, the "incident worksheet" (Bogner, 2000). The Worksheet consists of the names of the eight systems in the order from the care provider to the outermost layer of leaves listed in a column on the left of a page (see Figure 47.3). The system names serves as prompt so that

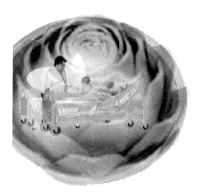

Patient
Means of providing care
Care provider
Are embedded in 5 systems of contextual factors:
   Ambient conditions
   Physical setting
   Social environement
   Organization
   Legal-regulatory-reimbursement-cultural factors

FIGURE 47.3   Conceptualization of tool. (From Atlantic Health System 2005. With permission.)

multiple factors might be recalled. The name of each system is followed by a line on which the care provider can identify those system factors that affected him or her to precipitate the error. These data indicate why the error occurred and define targets for change.

When the remediation is accomplished, its effectiveness in preventing error can be evaluated and the lessons learned applied to similar conditions as appropriate. Thus, the error reducing impact of the artichoke systems approach is multiplied beyond the specific error. This is in contrast to error reporting for accountability in which only the involved individual is addressed. Examples of medication misadministrations from the hypothetical hospital practice of Dr. Worthy illustrate these points.

47.4.2 Artichoke and Medication Misadventures

A systems analysis of adverse drug events (Leape et al., 1995) considered the drug delivery system in a hospital and reported the adverse event causes as physician ordering, transcription, pharmacy dispensing, and the nurse administration. This provides the who did what information characteristic of error reporting for provider accountability and attests to the difficulty in changing the focus from the individual. Even in this discussion of systems as causes of adverse events, examples are given in terms of individuals. With that focus, addressing those causes of error would involve exploring the activities of the providers in each of the drug delivery components — a formidable task indeed.

The artichoke problem solving orientation changes the focus from the care provider as cause of the adverse event to other systems of factors in the provider's life space. When obtaining data from each care provider that identifies a common factor, that factor can be identified as inducing error and efforts undertaken to address it. In this study (Leape et al., 1995), the common factor identified for each of the care providers is the drug. This transforms the findings to *drug* ordering, *drug* transcription, *drug* dispensing, and *drug* administration, which suggests that collecting data pertaining to characteristics of drugs involved in errors might provide insight into some aspect of medication error.

With this background, we join Dr. Worthy in her search for explanations for the misadministration of drugs to four of her patients — each time by a very competent, caring nurse. It is to be noted that although Dr. Worthy and the drug errors with four of her patients are hypothetical, the drugs involved and conditions surrounding them are not (U.S. Pharmacopeia, 2004).

47.4.2.1 Dr. Worthy's Exploration of Misadventures

Knowing that reporting errors put a nurse's license at risk, Dr. Worthy appreciated the nurses' reluctance to report what occurred (Weissman et al., 2004); however, she wanted to identify any problem that might have precipitated the errors so those conditions might be changed to prevent future errors. To do this, Dr. Worthy asked each nurse to describe the error scenario by completing the artichoke-based incident worksheet.

Wrong dose. The first case was a wrong dose error; the nurse administered a 60 mg dose of Geodon instead of the 20 mg dose that was ordered. This occurred despite the unit dose packaging to reduce the likelihood of dose error and the contextual factor that the drug was obtained from an automatic dispensing device (ADD) programmed to dispense the 20 mg dose for that patient.

Because the contents of the ADD are the jurisdiction of the pharmacy, Dr. Worthy sought the perspective of the pharmacist who examined the drawers of Geodon from which the ADD was loaded. The pharmacist found unit dose packages for both strengths in the same drawer. The technician who loaded the ADD responded to the incident worksheet by identifying factors that were in play when he loaded the ADD: he loaded it near the end of his shift and he was ill with the flu. Both the pharmacist and the technician believed that neither of those conditions would have caused the misloading of the drawer had it not been for the similar packaging (Figure 47.4).

Wrong drug. The second incident was a wrong drug error. Dr. Worthy used the incident worksheet to obtain the perspective of the second nurse regarding his administration of Timolol 0.5% instead of Levobunolol 0.5%. Although both are ophthalmic solutions used for the treatment of intraocular pressure, Dr. Worthy ordered Levobunolol 0.5% for the patient because of its known effectiveness

47-8   *Interventions, Controls, and Applications in Occupational Ergonomics*

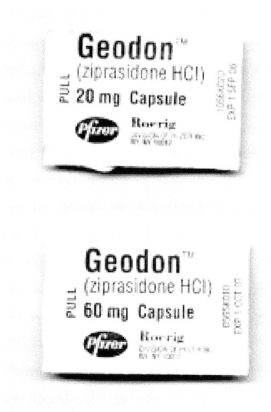

FIGURE 47.4   Geodon^{TM} 20 mg and 60 mg tablets. (From UPS Quality Review No. 78, 2004. With permission.)

with that particular patient. The nurse explored the context in which the mix-up occurred and found both drugs in the same location on the pharmacy supply shelf.

Dr. Worthy obtained the pharmacist's perspective. The pharmacist investigated why such a colocation occurred and discovered that both products were received from the wholesaler on the same day. Further investigation found that although the normal location for Levobunolol is on a different shelf, a technician assumed it and Timolol were the same product and placed them together on a shelf — an unacceptable although understandable assumption given the appearance of the packaging (Figure 47.5).

Wrong drug 2. The third case was another wrong drug error. Dr. Worthy asked the nurse to describe using the artichoke-based incident worksheet the situation when he administered Diphenhydramine (for allergic symptoms) when Lorazepam was ordered for the patient's anxiety disorder. In exploring the physical context, the nurse noted he obtained the Carpuject syringe with the pharmacy label stating it was Lorazepam from the drawer of drugs designated for the patient in the ADD.

After injecting the drug, he noticed Diphenhyrmine printed on the barrel of the syringe under the pharmacy label and promptly contracted Dr. Worthy to report what had occurred. Dr. Worthy obtained the pharmacist's perspective from the incident worksheet. The pharmacist following protocol spot-checks the loading of the drugs in the automatic dispenser; apparently the specific drawer in question was not checked. He spoke with the pharmacy technician who loaded the dispenser for her perspective on what occurred.

The technician in responding to the prompts of the incident worksheet recalled placing a Carpujet syringe with a green cap that was returned to the pharmacy in the Lorazepam bin with other syringes of the same size with green caps (Figure 47.6). She stocked a drawer of the ADD with a syringe from

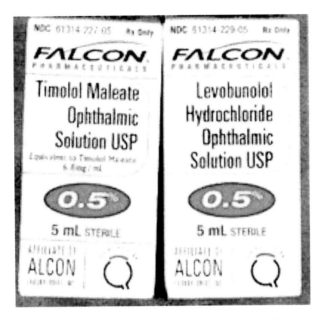

FIGURE 47.5 Timolol 0.5% and Levobunolol 0.5% ophthalmic solutions. (From UPS Quality Review No. 78, 2004. With permission.)

that bin. That technician typically is very conscientious, double checking all of her work, so the pharmacist further sought her perspective by asking her relate what might have affected her by responding to the system prompts on the incident worksheet.

In considering ambient condition, the technician recalled that a burned out light in the work area had not been replaced and the social system of factors prompted her memory that one of the other technicians had become ill and went home. There was no replacement so the workload was quite heavy, which necessitated her having to work quickly in a dimly lit area. Dr. Worthy pursued organization factors to determine why there was no replacement and was told that reimbursement policies had curtailed last minute hiring of temporary workers.

Wrong drug 3. The last case that Dr. Worthy explored was a wrong drug error involving tablets. The nurse reported to Dr. Worthy that she administered a tablet designated Diazepam 2 mg as Dr. Worthy prescribed. The patient's daughter was an active participant in her mother's care and inspected the tablet prior to the nurse administering it and approved it as the medication her mother receives. Later the patient complained of feeling dizzy. The nurse in seeking to identify what could have caused the dizziness asked the pharmacist to confirm that the tablet actually was Diazepam 2 mg.

The pharmacist checked the location of the bottle of Diazepam 2 mg to confirm that was the drug dispensed and found it in its usual location; however, the bottle of another drug to treat anxiety disorders, Alprazolam 2 mg atypically was next to it (Figure 47.7). Although both drugs are used to the same problem, they can have different side effects. The pharmacist examined the Alprazolam 2 mg tablets

FIGURE 47.6 Diphenhydramine and Lorazepam. (From UPS Quality Review No. 78, 2004. With permission.)

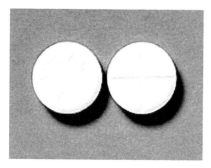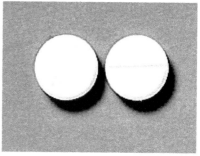

FIGURE 47.7   Diazepam 2 mg and Alprazolam 2 mg. (From UPS Quality Review No. 78, 2004. With permission.)

to determine if one if them might have been inadvertently dispensed as Diazepam 2 mg. The explanation was readily apparent — the appearances of the drugs in the adjacent bottles were nearly indistinguishable unless particular effort is made to note the embossed letters on each of the tablets, a task that is most readily accomplished using a magnifying glass.

Dr. Worthy's hypothetical pursuit of her curiosity about four instances of medication error provides a real-world, eye-opening window on a profound medical error problem. Addressing the misadministration using the artichoke systems approach incident worksheet paints a far different picture of accountability than the who did what of provider accountability.

47.4.3   Provider Accountability Versus the Artichoke

When investigating an incident with the focus on provider accountability as emphasized in the IOM report and the prevailing attitude about and research on medical error, the nurse in each of the four cases explored by Dr. Worthy would have been accountable for the misadministration. In each case, the information reported would be that the nurse administered the wrong drug or the wrong dose. The only interpretation of that information is that the nurse was solely responsible for the error. This focus on provider accountability in general and the attribution of error to the nurse reflect the application of a "stop rule" — a heuristic employed when assigning responsibility for an occurrence.

A stop rule is a tendency to cease searching for a cause of an act or event when a subjective, and pragmatic, and reasonable explanation is found (Rasmussen, 1990). Thus, because each of the nurses misadministered a drug, it is reasonable to conclude that he or she was solely responsible for the act, so no further enquiry about the cause of the error is necessary. The nurses are blamed for the misadministrations with the attendant negative professional and personal ramifications; they are reprimanded in some way, certainly admonished to read the drug label three times to confirm that the drug to be administered was the drug the physician ordered, and the misadministrations are considered as addressed. If the blame and reprimand are effective, then the likelihood of future misadministrations would be reduced for the specific nurses.

On the other hand, Dr. Worthy's explorations about why each of the four misadministrations occurred asked the nurses involved to describe the incidents following the prompts of the artichoke incident worksheet. It is to be noted that the prompts of the names of the systems elicit consideration of factors beyond the individual care provider and by so doing avoid the imposition of a stop rule. In all the cases Dr. Worthy pursued, a major contributory factor was similarity; in three cases the packaging of the drugs was very similar — in the fourth case, the actual appearance of tablets of different drugs essentially was the same. The ease with which the mix-up of drugs could occur is apparent when viewing Figure 47.4, Figure 47.5, Figure 47.6, and Figure 47.7.

Had industry, in developing the drug packaging, considered the hard-wired tendency that given similar objects, one object most likely will be misperceived as the similar one and the objects interchanged (Kohler, 1958), the misadministrations could have been anticipated and prevented by distinct packaging for each drug and obvious designations for the various strengths of the drug. In the absence of such actions by industry, the lessons of the tree, the sheepdog, and the fly that point to the inherent tendency to confuse similar items and the opportunities to do so afforded by the complexity of available pharmaceuticals could be heeded by all involved in administering drugs as well as the pharmacists. Special precautions could be instituted to make similar items distinct.

The prompts in the artichoke incident worksheet identified factors that contributed to the misadministrations. The tendency to misperceive is exacerbated when a care provider's cognitive functioning is compromised by factors in the systems of the context of care that cause fatigue and a sense of urgency due to the workload demands. The artichoke approach by identifying systems factors that induce error defines targets for change to decrease the likelihood of error not only for the specific nurses involved in the incidents, but for all care providers in comparable situations. This has profound implications for accountability.

47.5 Accountability

The targets for change identified by the artichoke systems approach actually point to sources for accountability. The accountability for the similar appearances in drug packaging and tablets lies with the pharmaceutical companies. They should be responsible for designing their products and their packaging to optimize the safety and ease of use from the perspective of the care provider. Given the life space of the individual, that perspective involves the context in which the products will be used. These are ergonomic issues that can and indeed should be addressed; those who monitor those companies should be accountable for seeing to it that those issues are addressed. Thus by identifying factors that induce error in the context of care, accountability for designing products to avoid the impact of those factors resides in those who have the capability to effectively address them, namely, industry and monitoring agencies.

Despite such sources of accountability the design of drug packaging places the burden of selecting the correct drug from among those that look alike on the care provider. It is not appropriate or fair from the ergonomic/human factors perspective of designing for the user, that the care provider be held accountable for the error provoking factors produced by pharmaceutical companies, but under the conditions of provider accountability, they are. Data gathered from incident worksheets and discussions of incidents guided by the systems of the artichoke should be given to those responsible for the similarities so they know the patient safety threat of their product. Knowing that, even if a pharmaceutical company was willing to change the packaging or appearance of a drug, it would take time — time during which other misadministrations could and very likely would occur.

During the time industry is making the changes or even if no changes will be made, occupational ergonomic efforts can address the factors identified as contributing to misadministrations to prevent errors by making stopgap changes locally. The artichoke findings that similarity in drug packaging and in tablet appearance induce errors can direct proactive steps to prevent similar errors by examining and identifying stock in the pharmacy that is similar in packaging or tablet appearance and differentiating those items by visual or tactile cues. Such cues could include as wrapping a cord around a bottle of pills, identifying bins by bright colors with tape configurations on the containers that are coded for strength of drug. Drug dispensing areas other than the pharmacy also should be checked for possible sources of confusion.

47.5.1 Preventing Errors That Do Not Happen

The findings from Dr. Worthy's information about why four drug misadministrations occurred can be used to anticipate and prevent incidents with other similar appearing packaging and drugs. Additional information could be gathered to prevent medical error using the artichoke systems approach and the incident worksheet to elicit factors that contribute to near misses or errors that almost happened but were averted. They are also useful in identifying factors that contribute to hazards or errors waiting to happen. The latter typically are not addressed but are legion in most if not all health care situations as evidenced by white tape on various devices and the preponderance of little sticky notes indicating that some adjustment in the typical use is necessary.

The importance of addressing hazards and near misses is illustrated by findings from a study of accidents in industry (Heinrich, 1931). Often represented as a pyramid, Heinrich reported that for every accident, there were 30 related reportable incidents or nears misses and 300 unreported occurrences or hazards that anticipated or contributed to the incidents and ultimately to the accident. Such hazards most likely are known and the care providers work around them. When circumstances change such as a contract nurse who is inexperienced in the specific work environment hence not aware of a hazard or a care provider who typically works around the hazard becomes so fatigued and harried that he or she forgets about the hazard, the care providers fall prey to hazards and an error ensues.

It is emphasized that the value of the artichoke systems approach relative to that of error reporting for provider accountability as illustrated by the four examples of medication misadministrations is that the identification and remediation of factors that induce error affects the behavior and reduces the likelihood of error for a number of care providers across situations. By acknowledging the error inducing quality of similar packaging and proactively making changes to differentiate such similarities, misadministrations very likely would have been prevented.

Blaming the nurses involved changes nothing, the error-inducing similar packaging continues to exist and misadministration by other nurses are likely to occur. The attribution of the error to a person stops the search for why the error occurred and the factors that contributed to the error remain to provoke another error. With the use of the artichoke systems approach, the pursuit of factors that contributed to the error stops only after the prompt for "why" for the eighth system of factors.

The use of why to elicit the causes of error is not unique to the artichoke approach although the prompts for it are. The use of why in understanding error was conceptualized as a search for the third order whys (Leape et al., 1995). If that concept were used, questions would be asked: "(1) Why did the incident occur? (2) Why did the error occur? (3) Why did the proximal cause occur?" (p. 40). It is to be noted that in this search for the third-order whys, the stop rule could be employed and each question answered by designating a person, thus staying firmly ensconced in the provider accountability mode although the orientation was for a systems approach. This could not occur using the artichoke approach — the prompts focus on factors.

47.6 Implications of the Artichoke

In addition to identifying error-provoking factors, the life space perspective of the care provider represented by the artichoke concept (Bogner, 2004b) can guide all aspects of health care from the conceptualization and development of medical devices and the packaging and instructions for the use of devices and drugs through the instructions given to patients to the design of health care facilities such as hospitals. A view towards the context-of-care as represented by the artichoke including the outermost leaves of the legal–regulatory–reimbursement–national culture systems should be applied throughout the life cycle of a health care related product; that is, in defining the requirements for its development, in delineating its design, in establishing operational procedures for its use, and when appropriate, in determining the means by which it will eventually be disposed of or salvaged.

47.6.1 Artichoke and Design

Infusion pump delivery problems identified in an analysis of adverse drug events (Leape et al., 1995) illustrate the value of the artichoke systems perspective in designing a medical device. That study attributed 4% of drug adverse events to device use; the proximal cause of some of the events was reported as error in setting up the pumps. Analyzing the process of setting up an infusion pump from the perspective of the care provider and applying the artichoke systems of factors, that is, considering the steps involved in setting up the pump with respect to each of the systems of the artichoke, identifies factors that contribute to if not provoke error.

Setting up an infusion pump involves programming the device with the dose and interval of the flow of the drug to the patient. This programming task is complex. If programming is interrupted, it must be resumed at the exact point of the interruption otherwise an inappropriate dose of the drug may be administered. Typically, no feedback of the programming is provided unlike a fax machine display of the phone number dialed. When setting up the pump is interrupted, the care provider has two options: start the process again which may involve more time than his or her workload allows, or hope that his or her recall of the point in the programming at which the interruption occurred is accurate and resume setting up the pump.

The study (Leape et al., 1995) collected data from hospital records, so the context in which the pumps are set up is a hospital. Interruptions are common in hospitals — ambient conditions can be noisy and social factors of other staff, patients, and their fiends and family cause distractions. Designing the device from the care providers' perspective to accommodate the affects of the artichoke systems of factors of the occupational conditions in which pumps are used could have prevented many errors and attendant adverse outcomes. This is a very different process from that of manufacturers who may test the device in the laboratory that is free of those occupational factors.

47.7 Paradigm Change: Error as Behavior Artichoke Systems Approach

The concept of error as behavior and the artichoke systems approach to assisting a care provider in identifying environmental factors that affect his or her behavior such that an error is committed, is a marked change from the typical if not traditional approach to medical error — that of provider accountability. Error reporting for provider accountability addresses the problem for one individual. By reporting who did what, there is no choice but to attribute the cause of error to the person.

Medical error is a problem; to prevent medical error it is necessary to employ a problem-solving approach. The paucity of information about the etiology of an error provided by the who did what of provider accountability does not provide information to reduce the incidence of error. From that approach to error, it is not surprising that the 5 yr of research funded by AHRQ did not reduce the incidence of error — the focus of that work solely on the person as the cause of error provides information that is incomplete and misleading.

Even when considering an aspect of health care as a system such as the drug delivery system, the affect of that system on the care provider becomes lost in the complexity and intricacies of the constituent processes. Thus, rather than identifying any factors, indeed any occupational ergonomic factors that might be addressed to prevent error, all that is stated is that errors are indicative of the system failures. It has been repeatedly proposed that the health care system is broken. If that is the case and the artichoke systems approach would argue to the contrary, then given the provider accountability orientation, far more than "all the kings horses and all the kings men" of the Humpty Dumpty nursery rhyme are necessary to make it right.

Rather than continue in the nonproductive provider accountability approach, it is time, indeed past time to acknowledge the need for another approach. This is not a new proposal; the IOM report states: "... Errors are due most often to the convergence of multiple contributing factors. Blaming an individual does not change these factors and the same error is likely to occur. Preventing errors and improving safety for patients require a systems approach in order to modify the conditions that contribute to errors" (Kohn et al., 1999, p. 42). The artichoke systems approach provides a structure and a tool for identifying those contributing factors.

That error as behavior systems approach not only considers those factors that affect the individual so that medical error problems can be solved, it is action oriented to solve problems effectively with the goal of preventing error across situations and care providers. Information about problems in the more complex aspects of health care, such as the drug delivery system, can be developed building upon the factors that are found to induce error in an individual.

Thus, a change in the paradigm of error is presaged. Rather than considering a complex series of health care activities such as the drug delivery system as the source of error — the complexity of which defies analysis for what actually causes an error so it can be addressed, the artichoke systems approach considers error for what it is — a behavior.

The benefits of a paradigm change to addressing error as behavior, identifying contextual factors that interact with the care provider to induce error, and changing those factors to prevent error are legion. Not only would patient safety be enhanced, care providers would become partners in preventing error rather than targets for blame. Change can be difficult; the need for a paradigm change demands that the difficulty be met and conquered.

References

Allport, G.W. (Ed.) (1948). *Resolving Social Conflicts: Selected Papers on Group Dynamics.* In K. Lewin (1997). *Resolving Social Conflicts & Field Theory in Social Science.* Washington, DC: American Psychological Association (pp. 1–154).

Bogner, M.S. (1994). *Human Error in Medicine.* Mahwah, NJ: Lawrence Erlbaum Associates.

Bogner, M.S. (2000). A systems approach to medical error. In C. Vincent & B. DeMol (Eds.), *Safety in Medicine.* Amsterdam: Pergamon (pp. 83–100).

Bogner, M.S. (2002). Stretching the search for the "why" of error: the systems approach. *Journal of Clinical Engineering*, 27, 110–115.

Bogner, M.S. (2004a). *Misadventures in Health Care: Inside Stories.* Mahwah, NJ: Lawrence Erlbaum Associates.

Bogner, M.S. (2004b). Understanding human error. In M.S. Bogner (Ed.), *Misadventures in Health Care: Inside Stories.* Mahwah, NJ: Lawrence Erlbaum (pp. 41–58).

Cartwright, D. (Ed.) (1951). *Field Theory in Social Science.* In K. Lewin (1997), *Resolving Social Conflicts & Field Theory in Social Science.* Washington, DC: American Psychological Association (pp. 155–410).

Commonwealth Fund (2004). See website Commonwealth Fund/Quality Improvement/Patient Safety/ *Quality Matters*: November Newsletter/more.

Cooper, J.B., Newbower, R.C., Long, C.D., & McPeek, B. (1978). Preventable anesthesia mishaps: a study of human factors. *Anesthesiology*, 49, 399–406.

Davies, J., Ross, A. Wallace, B., & Wright, L. (2001). *Safety Management: A Qualitative Systems Approach.* London: Taylor & Francis.

Heinrich, H.W. (1931). *Industrial Accident Prevention.* New York: McGraw-Hill.

Kohler, W. (1958). Relational determination in perception. In. D.C. Beardslee & M. Wertheimer (Eds.), *Readings in Perception.* Princeton, NJ: Van Nostrand Company, Inc. (pp. 353–358).

Kohn, L.T., Corrigan, J.M., & Donaldson, M.S. (Eds.), (1999). *To Err is Human: Building a Safer Health System.* Washington, DC: National Academy Press.

Leape, L.L., Brennan, T.A., Laird, N., Lawthers, A.G., Localio, A.R., Barnes, B.A., Hebert, L., Newhouse, J.P., Weiler, P.C., & Hiatt, H. (1991). The nature of adverse events in hospitalized patients. *New England Journal of Medicine*, 324, 377–384.

Leape, L.L., Bates, D.W., Culllen, D.J., Cooper, J., Demonaco, H.J., Gallivan, T., Hallisay, R., Ives, J., Laird, N., Laffel, G., Nemeski, R., Peterson, L.A., Porter, J, Serv, D., Shea, B.F., Small, S.D., Sweitzer, B.J., Thompson, T., & Vander Viet, M., (1995). Systems analysis of adverse drug events. *Journal of the American Medical Association*, 274, 35–43.

Lewin, K. (1966). *Principles of Topological Psychology.* New York: McGraw-Hill. (Original work published 1936).

Lewin, K. (1997). Defining the "field at a given time." In D. Cartwright (Ed.), *Field Theory in Social Science.* Washington, DC: American Psychological Association. (pp. 200–211). (Original work published 1943).

Lewin, K. (1997). Behavior and development as a function of the total situation. In D. Cartwright (Ed.), *Field Theory in Social Science.* Washington, DC: American Psychological Association. (pp. 337–383). (Original work published 1946).

Mach, E. (1976). *Knowledge and Error.* Dordrecht: D. Reidel Publishing Co. (Original work 1905).

Moray, N. (1994). Error reduction as a systems problem. In M.S. Bogner (Ed.), *Human Error in Medicine.* Hillsdale, NJ: Lawrence Erlbaum (pp. 67–92).

Perrow, C. (1984). *Normal Accidents: Living with High-Risk Technologies.* New York: Basic Books.

Pickett, R.M. & Triggs, T.J. (Eds.) (1975). *Human Factors in Healthcare.* Lexington: Lexington Books.

Pope, A. (2004). *An Essay on Criticism.* Whitefish, MT: Kessinger Publishing. (Original work published 1711).

Rasmussen, J. (1990). Human error and the problem of causality in analysis of accidents. *Philosophical Transactions of the Royal Society of London, 337,* 449–462.

Reason, J. (1990). *Human Error.* New York: Cambridge University Press.

Safren, M.A. & Chapanis, A. (1960a). A critical incident study of hospital medication errors — Part 1. *Hospitals, J.A.H.A.* 34, May 1, 32–66.

Safren, M.A. & Chapanis, A. (1960b). A critical incident study of hospital medication errors — Part 2. *Hospitals, J.A.H.A.* 34, May 16, 54–68.

U.S. Pharmacopeia (2004). *USP Quality Review, 78.* Rockville, MD: USP, February and www.usp.org/patientsafety.

Wachter, R. (2004). Patient Safety Five Years After *To Err Is Human.* Commonwealth Fund web site/Quality Improvement/Patient Safety/*Quality Matters*: November Newsletter/more/introduction/transcript.

Weissman, J.S., Annas, C.L., Epstein, A.M., Schneider, E.C., Clarridge. B., Kirle, L., Gatsonis, C., Feibelmann, S., & Ridley, N. (2004). Error reporting and disclosure systems: views from hospital leaders. *Journal of the American Medical Association,* 293(11):1359–1366.

Zwillich, T. (2004). Little Progress Seen in Patient Safety Measures. Washington: Reuters Health Information. November 04, 2004.

48

Bakery: Cake Decorating Process

Carolyn M. Sommerich

Steven A. Lavender

A. Asmus

A.-M. Chany

J. Parakkat

G. Yang

The Ohio State University

48.1 Introduction

Data compiled by the Bureau of Labor Statistics (BLS) show that musculoskeletal disorders, including back pain and carpal tunnel syndrome (CTS), are work-related illnesses that occur among bakery workers, including cake decorators. From 1997 to 2001, 403 cases of CTS were reported in this group[1] (R10 Tables). In 2002, almost half of all lost time injuries and illnesses in bakers were due to musculoskeletal disorders, which include such injuries as sprains, strains, back pain, and CTS[1] (Supplemental Table 12). The occurrence of musculoskeletal disorders in this population suggests that some issues may exist within the tasks performed as part of the cake decorating process. Work-related musculoskeletal disorders are associated with repetitive movements, forceful exertions, awkward postures, and extreme temperatures, among other factors.[2] Those mentioned are possible risk factors that cake decorators are exposed to on a regular basis. The aims of this chapter are to identify the main ergonomics risk factors that potentially contribute to the musculoskeletal symptoms experienced by cake decorators and to propose interventions that may reduce the exposure to the risk factors during cake decorating tasks.

The specialized job of cake decorating is included in the category "bakers" used by the BLS. In both small bakery operations and grocery store bakery departments, the cake decorator's jobs are similar in the tasks that the decorators perform. An outer layer of frosting is applied to cover the surface area of the cake. In this "icing" task several tools are used to develop the finished look of a frosted cake. Once the cake is iced it is ready for decorating. In preparation for decorating, pastry bags are filled with icing which is usually obtained from a 5-gallon bucket. Metal tips on the pastry bag control the volume and shape of the frosting material used in the decorating process. Pastry bags are used to apply a finished border, flower decor, and writing to the cake. The bag is squeezed by the worker who controls the amount of flow and the flow rate of the icing; hence, this job involves a set of skills requiring precisely controlled grip force and fine movement control. Cake decorating is a skill that is often learned in culinary institutions, and frequently requires full-time dedication of the workers.

Occupational Safety and Health Administration (OSHA) and the BLS have published *Ergonomics for the Prevention of Musculoskeletal Disorders: Draft Guidelines for Retail Grocery Stores*, supplying the grocery industry with suggestions for reducing the number and severity of injuries reported in their workplace.[3] The bakery is among one of the departments mentioned (also included are front end, stock, meat, and produce) due to the handling of heavy loads such as icing buckets, flour, and supplies, and the high force required when squeezing pastry bags. The guidelines also mention proper working height of tables and appropriate tool usage during the tasks. This guideline, although currently in draft form, demonstrates OSHA's awareness and concern with injuries in bakeries.

Research publications of ergonomics studies involving grocery stores are quite numerous regarding the "front-end" operations such as cashiering and bagging tasks. However, there is little information specifically targeting the cake decorating tasks. Osorio et al.[4] performed a study in one large grocery store in California, assessing a variety of jobs within the store where several cases of CTS were reported. Employees were tested for CTS via nerve conduction velocity testing. The jobs at the store were ranked into high-, medium-, and low-risk categories for cumulative trauma exposure. Cake decorators were ranked in the high-risk group due to the repetitive wrist motion. The main task contributing to this "high-risk" ranking was the icing task that entailed high grasping forces along with a flexed wrist.[4]

In a study focusing on bakery operations, Rigdon[5] reported that 630 employees participated in an effort to evaluate workstations in a commercial bakery operation through a union–management ergonomics evaluation committee. The results of the study included making workstation changes and tool modifications. The interventions were then monitored over a period of 4 yr. Cases of CTS reduced from 34 to 13, and the number of lost workdays in the same group reduced from 731 to 8 days.

The dissertation study by Chyuan[6] targeted a method of identifying ergonomics risk factors in food-service bakery operations, specifically examining the following three tasks: portioning pie fillings, dividing bread dough, and decorating cakes using a pastry bag. In addition to time studies and discomfort bodymaps, both the Postures Exposed Comparison (PEC) and Rapid Upper Limb Assessment (RULA) tools were used to identify risky postures assumed by the workers. Wrist positions were observed at or near the end of postural range with intermittent loads applied. The static nature of some tasks required neck and trunk postures with 20 to 60° of forward flexion. These postural conditions influenced the rating of cake decorating into the high-risk group.

Given that few studies have specifically focused on cake decorating tasks, and the occurrence of musculoskeletal symptoms and disorders among the bakery workers, we determined further investigation was warranted. In the following, we describe an observational study that was conducted to quantify the discomfort and the postures assumed by workers during common cake decorating tasks. The first step of the process was to determine the primary tasks involved in the cake decorating process, the tools used, and the postures assumed in the performance of the identified tasks. From the task, tool, and posture analysis, the ergonomics risk factors could be identified. Based on these findings several intervention methods are suggested.

48.2   Observational Study Approach

48.2.1   Participants

Six individuals with an average of 10.3 (5.9) yr of cake decorating experience participated in our observational study. All the participants were employed full time in bakery settings; three participants worked in small independent bakeries (ID# 1, 3, 5), three participants worked in grocery store chain bakeries

TABLE 48.1 Anthropometric Data from the Six Observational Study Participants

Participant ID#	1	2	3	4	5	6	Average (SD)
Age, years	50	44	40	51	35	25	41 (10)
Gender	M	F	M	F	F	F	—
Hand dominance	L	R	L	R	R	R	—
Standing height (cm)	189	161	178	157	165	160	168 (12)
Shoulder height (cm)	163	136	146	131	136	133	141 (12)
Elbow height (cm)	126	107	113	103	107	105	110 (8)
Waist height (cm)	116	102	110	97	100	102	104 (7)

(ID# 2, 4, 6); two participants were left-hand dominant (ID# 1, 3); and two participants were male (ID# 1, 3). The participants provided informed consent prior to their participation. Anthropometric data are provided in Table 48.1. Work experience data are summarized in Table 48.2.

On average, the participants spent 63.3% (range: 40 to 90%) of their time on "regular" days and 87.4% (range: 65 to 100%) on "busy" days on cake decorating tasks. On regular days, the participants decorated an average of 27 (range: 5 to 50) cakes per day or 129 (range: 42 to 275) cakes per week. On busy days, these numbers jump, on average, to 46 (range: 14 to 90) cakes per day and 228 (range: 65 to 500) cakes per week. The average time of decorating one cake was 13 min, depending upon its size and design. Apart from cake decorating, all the participants did other tasks, such as boxing cakes, cleaning tools, or decorating pastries.

48.2.2 Data Collection Procedures

Two digital cameras were used and synchronized to obtain front and side views of each task. Information collected from the side view included back and neck flexion, shoulder flexion, and wrist motions. The front view contained information on shoulder abduction and neck and back lateral bending (Figure 48.1). The video was then analyzed for postural and time information.

A hand-grip dynamometer was used to provide estimates of grip force requirements. Prior to observing the normal work activities, the participant provided maximum hand grip exertions with each hand in three wrist postures: neutral, flexed, and extended. The experimenter positioned the dynamometer at the participant's elbow height and held it for stability. Additionally, intermittently throughout the decorating process the participant was asked to squeeze the dynamometer with the same force used to squeeze the icing bag he or she was currently using. This provided an estimate of the force exerted during particular tasks.

A hand-held force gauge was used to measure the amount of force required to squeeze different types of icings and different types of bags. These measurements were obtained at the same time the participant was asked to squeeze the hand-grip dynamometer (the icing in the bag was in the same condition for both force assessments). The pastry bag was placed on the work surface. A plexiglass plate was placed

TABLE 48.2 Work Experience of the Six Observational Study Participants

Participant ID#	1	2	3	4	5	6	Average (SD)
Skills learning	School/job training	Self/job training	School	Self	School	Self	—
Years decorating	20	12	10	2	10	7.5	10.3 (5.9)
Days/week decorating	5.5	5.5	5	5	6	5.5	5.4 (0.4)
Hours/day decorating	8.5	9	7	8	11	8	8.6 (1.4)

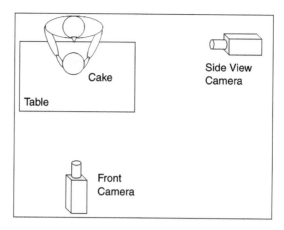

FIGURE 48.1   Camera setup for recording work postures.

on the center of the bag. The force gauge was applied to the plate for an equally distributed application of pressure on the bag. The pressure was applied until a continuous flow of frosting was observed from the end of the bag. The force required to obtain this flow was recorded.

A general information survey was administered that contained questions regarding daily work activities, production rates, and anthropometric information. In addition, a discomfort survey which was a modified Nordic questionnaire,[7] was administered. The survey asked questions concerning frequency and length of pain and discomfort of the upper extremity, the low back, the legs and the feet, as well as questions concerning lost time due to discomfort, work, and nonwork activities that aggravated the discomfort, and medical treatment sought due to discomfort.

48.2.3   Data Analysis Procedures

Postural data for each task were analyzed from the digital video images. The most extreme postures of these tasks were recorded from a video monitor using goniometric measurements. The postures of interest included shoulder abduction, flexion, and internal and external rotation, elbow flexion, wrist flexion/extension, wrist ulnar/radial deviation, wrist pronation/supination, neck flexion, neck forward and lateral flexion, upper back forward and lateral flexion, and lower back forward and lateral flexion. Using postural guidelines, including the RULA[8] guidelines, upper-extremity angle criteria were used to determine postural risk of the tasks (Table 48.3).

The data from the discomfort surveys, hand-grip strength, and applied bag pressure (force) were compiled and considered along with the postural cutoffs in the identification of risk factors and used in developing potential ergonomic interventions.

48.3   Observational Study Results

Through the videos 12 different cake decorating activities were identified as representative of a typical cake cycle common among all decorators. Table 48.4 summarizes the decorating tasks and where the postural criteria in Table 48.3 were exceeded.

48.3.1   Analysis by Task

Scooping icing from the bucket (Task 1) constituted, on average, 1.67% of the cake decorating cycle. Typically, a spatula was used to scoop the icing from the bucket. Postures observed when scooping from the bucket included shoulder abduction, neck flexion, and back flexion. In addition, wrist pronation/supination and wrist flexion/extension motions were observed in the dominant hand. All six

TABLE 48.3 Postural Limits Used to Determine Decorating Activities for Potential Intervention

Body Part	Movement	RULA Basis	Postural Limits Used in This Study
Shoulder	Abduction	Any abduction increased score by 1 for RULA	30°+ on dominant side 30°+ on nondominant side
Shoulder	Flexion	20°+ increased score by 2–4 for RULA	20°+ on dominant hand
Wrist	Flexion/extension	±15°+ increased score by 2–3 for RULA	If 3+ decorators had any F/E in addition to dynamic movement = risky If 3+ decorators had any F/E in addition to static postures = risky
Wrist	Radial/ulnar deviation	Any radial or ulnar deviation increased score by 1 for RULA	If 3+ decorators had any R/U in addition to dynamic movement = risky If 3+ decorators had any R/U in addition to static postures = risky
Wrist	Pronation/supination	Any pronation or supination increased score by 1–2 for RULA	If 3+ decorators had any P/S in addition to dynamic movement = risky If 3+ decorators had any P/S in addition to static postures = risky
Neck	Flexion	20°+ increased score by 3–4 for RULA	20°+
Neck	Lateral bending	Any lateral bending increased score by 1 for RULA	If 3+ decorators had lateral bending in addition to flexion = risky
Upper back	Flexion	N/A	20°+
Upper back	Lateral bending	Any lateral bending increased score by 1 for RULA	If 3+ decorators had lateral bending = risky
Lower back	Flexion	20°+ increased score by 3–4 for RULA	20°+
Lower back	Lateral bending	Any lateral bending increased score by 1 for RULA	If 3+ decorators had lateral bending = risky

participants exhibited shoulder abduction angles greater than 30° on the dominant side while scooping and two of the participants exhibited shoulder abduction angles greater than 80° (see Figure 48.2). Neck flexion angles exceeded 20° in four of the participants.

After scooping the icing from the bucket, the decorator would either fill the icing bag if he or she was speed icing the cake, or deposit the icing directly onto the cake. These activities were chosen for analysis because they were common to all decorators and also involved some kinematic issues. Filling the bag (Task 2) consumed 5% of the average cake cycle. In performing this task the participants flexed their upper backs, abducted their dominant shoulder, and moved their dominant arm and wrist in pronation/supination, flexion/extension, and ulnar/radial deviation. When depositing the icing onto the cake (Task 3) there was shoulder abduction, neck flexion, wrist pronation/supination, flexion, and ulnar deviation in the dominant hand. Depositing a large amount of icing from the spatula onto the cake could impose a large moment on the wrist given that this action typically requires a fast dynamic motion of the spatula that has a mass (the icing) located at the distal end of the spatula relative to the wrist.

In the cake decorating process some decorators used a speed icer. Speed icing involves the use of a large pastry bag with a rectangular, 2-in. wide but narrow, tip that allows the decorator to distribute the base layer of icing on the top and sides of the cake in long, wide ribbons that are ready to be smoothed, in contrast to "plopping" the icing onto the cake in mounds using a spatula, which then have to been distributed and smoothed. Use of the speed icer (Task 4) on average consumes 4.75% of a cake decorator's time, although in this study while it was used extensively in grocery stores, it was not used in the smaller

TABLE 48.4 Cake Decorating Tasks That Exceed Postural Criteria Identified in Table 48.3.

Task	Task Description	Posture											
		Shld abd	Shld flex	Neck flex	Neck latl	Up back flex	Up back latl bend	Low back flex	Low back latl bend	Wrist F/E	Wrist R/U	Wrist P/S	Grip force
1	Scooping from icing bucket	×		×		×				×		×	
2	Fill icing bag	×								×	×	×	
3	Plop icing onto cake from bucket using spatula	×		×						×	×	×	
4	Speed icing sides of cake			×	×	×	×		×	×			×
5	Applying icing ice with spatula	×		×	×	×				×		×	
6	Icing sides of cake with spatula	×	×	×	×	×	×		×	×	×	×	
7	Smoothing the icing on the cake	×			×					×	×		
8	Coming the icing on the sides of the cake			×	×	×	×		×	×			
9	Mix icing	×		×	×				×		×	×	
10	Applying decorative icing border			×		×	×						
11	Writing in icing on the cake			×	×	×				×			×
12	Cutting cake			×	×		×				×		×

FIGURE 48.2 Scooping from icing bucket. Icing buckets were often placed on the same work surface as the cake, forcing the decorator into severe shoulder abduction when scooping icing from the bucket.

bakeries. Use of the speed icer requires high grip forces as large volumes of icing are squeezed out of the bag. Grip forces, measured using the dynamometer force-matching approach ranged from 8.8 to 31.0 kg, with an average of 17.5 kg. For the individuals sampled, this average grip force represented 62.3% of the participant's maximum grip force. Postural concerns included neck flexion, neck lateral flexion, upper and lower back flexion, upper and lower back lateral flexion, and wrist extension. All four participants observed using the speed icer showed neck flexion angles above 30° as well as neck lateral bending. Three of the four participants also showed upper back lateral bending.

The spatula is the other primary method of icing application (Task 5), consuming an average of 12.8% of a cake decorator's time. Use of the spatula typically involved shoulder abduction, neck flexion, neck lateral bend, upper back flexion, wrist pronation/supination, and wrist extension with the dominant hand. All four participants measured for this task exhibited a neck flexion angle greater than 20° and three had back flexion angles greater than 20°. Use of the spatula, particularly on the side of the cake (Task 6), posed a higher risk with the combination of neck flexion and neck lateral bending, as well as upper and lower back flexion and lateral bending (see Figure 48.3). Wrist ulnar deviation was observed in all four participants.

Typically, if the decorator had deposited the icing on the cake using the speed icer, he or she would use a smoothing tool similar to that shown in Figure 48.4, to smooth the top and sides of the cake (Task 7). If the icing was deposited using a spatula, the spatula was used for this process. Four of the six decorators were observed performing the task, which consumed on average 7% of their time, with the smoothing tool. Smoothing resulted in shoulder abduction and neck lateral flexion. Similar to smoothing was the task of combing the sides of the cake with a comb tool for decorative effect (Task 8). Five out of the six decorators were observed performing this task which consumed on average 5.2% of the cake cycle time. Combing the sides of the cake involved neck flexion, upper and lower back flexion, and lateral bending. All participants showed wrist extension throughout the duration of the task.

The icing mixing task was observed in four participants (Task 9). This task consumed on average 1.67% of the cake cycle time. Although not measured by our methods, the icing required a high force to stir and was noted to be a difficult task for some of the decorators who may prepare several different color mixtures that will be used during the day. The stirring activity was performed with the neck and

FIGURE 48.3   Typical postures observed when using spatula to smooth icing on sides of cakes.

FIGURE 48.4   Smoothing tool being used to smooth icing on side of a cake.

upper back in flexion combined with neck lateral bending. Shoulder abduction above 40° was noted for three of the four participants observed performing this task.

After the base layer of icing was deposited on the cake, typically the decorator would put a border around the edge of the cake (Task 10). This would be done using a pastry bag filled with the same type of icing as the base layer. Four participants were observed putting borders on the cakes, which typically consumed 11% of the average cake cycle. The predominant postural concern here was the neck flexion, which exceeded 40° in three out of the four observed participants (Figure 48.5). Three out of four also had upper back flexion angles greater than 30° combined with upper back lateral bending. The wrists were in an extended posture and the grip forces ranged from 8.0 to 23.6 kg (average = 13.5 kg). When normalized, these grip forces used in border decoration averaged 50.6% of the maximum grip force.

Writing on the cake (Task 11) consumed 8% of the typical cake cycle. Similar to the border task, the participants showed neck flexion angles greater than 30°, with three showing neck flexion angles greater than 50°. Three participants also had neck lateral bending, as well as upper back flexion. Wrist extension and high grip forces were also a problem during the writing task. The force-matched grip forces while writing ranged from 6.7 to 22.3 kg with an average of 14.5 kg or 36% of maximum grip force.

FIGURE 48.5 Typical postures observed when applying a border.

The final task listed in Table 48.4 involved cutting the cake. This brief task, consuming on average just 3% of the cake cycle time, typically involved neck flexion combined with lateral bending, as well as upper back lateral bending. The dominant hand was in ulnar deviation in the four participants observed performing the task.

In addition to the 12 aforementioned tasks, a cake decorator's typical cake cycle time included decor other than writing and bordering (16.3% of cycle time), making roses as decor specifically (10.7% of cycle time), and performing other tasks such as preparation and cleanup in between the steps of cake decoration (38.3% of cycle time).

48.3.2 Discomfort Survey Results

All participants had discomfort in at least five body areas during the past 12 months. Five participants had discomfort in neck, shoulder, elbow/forearm, wrist/hand, and feet. Four participants had discomfort in the low back and leg and three participants had discomfort in the upper back. Most of the participants experienced discomfort either daily or at least weekly. The discomfort in one of the participants usually lasted less than 1 day. Two of the participants indicated that all their discomfort last longer than 3 months. The duration of the discomfort in the rest of the participants varied from less than 1 h to longer than 3 months. The intensity of most of the discomfort varied from moderate to severe.

Three of the six participants had seen a health-care provider for their discomfort within the last 12 months. One of the participants had his discomfort before starting the current job. For all the other participants their discomfort began after starting their cake decoration jobs. Two participants thought that the discomfort in their wrist and hand prevented them from doing normal activities (on or off the job). The other two participants thought the discomfort in their low back prevented them from doing normal activities.

Five participants considered that cake decorating made their discomfort worse. Four participants also thought that other activities at work made their discomfort worse. Activities off the job made upper extremity discomfort worse in two participants and low back discomfort worse in three participants. Two participants lost work time due to discomfort in their upper extremities and four participants lost work time due to their low back pain. All participants had discomfort in upper extremities, back, or legs during the past 7 days before the discomfort survey was taken.

48.4   Potential Interventions

48.4.1   Scooping from the Bucket

It was often observed that the icing bucket was located on the top of the work surface. Shoulder abduction while scooping can be minimized by relocating the bucket to a lower location. The OSHA guidelines for grocery bakeries[3] recommend placing buckets of icing "at the preferred work zone." We recommend placing the bucket on a stand at the height of about 64 cm (25 in.) above the floor. This height puts the bottom of the bucket at a level that corresponds to the average knuckle height (for U.S. population) minus the length of the spoon used for scooping. Likewise, this stand should allow most of the population to reach the bottom of the bucket without lateral bending. Tilting the bucket on this stand toward the worker would potentially eliminate shoulder abduction. To reduce the number of scoops necessary to fill the bag or cake with icing, a short-handled spoon is recommended as a replacement for the 15-cm (6-in.) spatula commonly used. The spoon should not increase the moment about the wrist, due to its short length (versus the longer spatula), even though the amount of icing moved at one time is increased.

48.4.2   Icing

Both the speed icer and the spatula were observed to be used by some decorators when initially icing the cakes. Even though there are tradeoffs between these two approaches, we tentatively recommend the use of the speed icer. While using the spatula to ice is a more dynamic task than using the speed icer to ice, more extreme awkward postures were observed while using the spatula (pronounced lateral neck bend, shoulder flexion, and shoulder abduction; moderate lateral thoracic bend). These are indicators that the speed icer may be the better tool for icing. Further evaluation is required to determine this, conclusively, however. To finish the task, use of a smoothing tool is recommended over a spatula, again due to the reduction of awkward postures observed with the smoothing tool (moderate reduction in thoracic lateral bend and large reductions in shoulder flexion and "abduction." The smoothers we saw in use were flat sheets of plastic, but Rigdon[5] mentioned "scrapers with handles," which might reduce the pinch-type exertion that the thumb must produce when using this tool. The smoother is guided, rather than strongly pinched, however, so without further study the specific benefits or disadvantages of a smoother with a handle cannot be predicted.

48.4.3 Bordering

Since bordering and decor tasks totaled 21.7% of the cake cycle and involved high sustained grip forces, alternative tooling should be considered. One commercial bakery developed an icing gun for employees to use, which required less squeezing force than traditional icing bags.[5] Ice press guns are also marketed to the public (see Figure 48.6 for an example). In addition to the need for less grip force to be exerted, the trigger mechanism may result in the task being more dynamic in nature than the use of

FIGURE 48.6 Commercially available ice press gun.

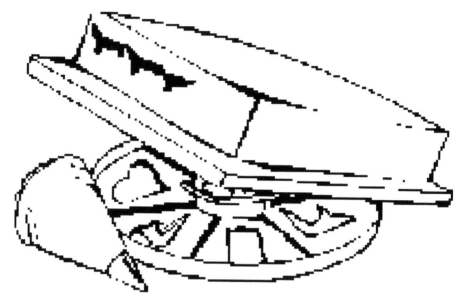

FIGURE 48.7   Tilting turntable.

a bag, thereby reducing localized muscle fatigue. However, given that none of the decorators in our study employed such a tool, it may be that there are disadvantages to use of such a device, as well. Therefore, this is another potential solution that requires further investigation before it can be recommended.

48.4.4   Bordering and Writing

The postural concerns associated with bordering and writing, namely neck flexion, upper back flexion, and shoulder flexion, suggest that the decorators may be trying to accommodate their visual system so that they can better see their work. In response to these postural concerns, we recommend tilting the cake toward the decorator. Tilting turntables are currently available (Figure 48.7). We believe that a moderate tilt (20° or so) combined with the ability to rotate the cake could minimize the postural issues associated with the boarding and writing tasks. OSHA[3] recommends adjusting the height of the turntable so that the cake is about elbow height. If the table is too low, this can be accomplished with use of an adjustable height turntable or with a simple riser placed under the turntable.

A number of other recommendations for modifications to bakery processes were provided by OSHA[3], and Rigdon,[5] which are listed in Table 48.5.

48.5   Summary

Twelve principal tasks were identified in this observational study of the cake decorating process. Some of these tasks require the decorator to use awkward postures or exert relatively high forces for extended periods of time. The goal of the current project was to identify which specific tasks were strongly associated with risk factors that may be responsible for the musculoskeletal discomfort experienced in cake decorators. Based on these analyses several alternative methods are proposed:

- Using a bucket stand to lower the height of the icing bucket
- Using a short-handle spoon rather than the spatula to obtain icing from the bucket
- Using a speed icer for applying the initial frosting layer as opposed to using the spatula

TABLE 48.5 Additional Interventions or Recommendations for Application to Cake Decorating and Other Bakery Operations

	Interventions/Recommendations	Source
Decorating	Use small decorating bags whenever possible to reduce the stress on the worker's hands. The larger the bag, the more force required to squeeze it	OSHA[3]
	Have an adequate number of mixing bowls available to reduce the need to transfer icing or batters that are mixed in the store to other containers	OSHA[3]
	Use powered mixers whenever possible to mix coloring into icing or purchase colored icing. This reduces the stress to workers' hands and arms from manually mixing colors into icing	OSHA[3]
Workstation modifications	Raise or lower tables and conveyor belts to appropriate heights for the task and worker	Rigdon[5]
	Conveyor belts were moved closer to workers who previously leaned forward	Rigdon[5]
	Pallets were stacked close to conveyor belts so it was not necessary to carry items between the two	Rigdon[5]
	Provide stools for workers who used to stoop	Rigdon[5]
Standing work	Provide foot rails for use while standing	Rigdon[5]
	Use footrests and antifatigue mats in areas where people stand	OSHA[3]
	Make sure that there is toe-clearance under counters and other work surfaces	OSHA[3]
Tools selection/ modifications	Customize tools to individual employees	Rigdon[5]
	Use spatulas, spoons, and other utensils that fit the worker's hand (not too wide or too narrow) and are not slippery	OSHA[3]
Alternative methods	Workers learned to carry large pans on their forearms instead of grasping the pans by their edges, which puts the entire weight on the wrists	Rigdon[5]
	Work from the long side of baking pans to reduce reaches when handling dough	OSHA[3]
	Ensure that the icing is of correct consistency. Icing that is too thick will be very difficult to squeeze through decorating bags. If icing is mixed in the bakery, add liquid to the recipe or warm the icing to obtain the correct consistency. If icing is purchased in buckets, store the buckets at room temperature or warm them before use—cold icing is very thick and hard to squeeze through decorating bags	OSHA[3]
Manual materials handling (MMH)	Use carts or rolling stands to move heavy items like tubs of dough or bags of flour	OSHA[3]
	When lifting keep large bags and containers of ingredients close to the body to reduce stress on the back	OSHA[3]
	Punch handles into plastic buckets that are currently picked up by jamming fingertips against narrow rims	Rigdon[5]
Administrative controls	Slow down "fast" workers	Rigdon[5]
	Job rotation, altering harder jobs with easier jobs	Rigdon[5]
	Whenever possible, break up continuous activities like cake decorating and dough handling with less strenuous tasks during the shift	OSHA[3]
	Add a worker	Rigdon[5]
	Medical management plan in writing, and in consultation with employees/ employee representatives	Rigdon[5]
	Ergonomics team — management and hourly	Rigdon[5]
	Employees watch out for each other and remind each other about methods	Rigdon[5]

- Using a smoother rather than a spatula for smoothing the icing
- Using a tilted turntable for bordering and writing tasks

We can hypothesize the biomechanical value of the proposed interventions, but a systematic evaluation would be required to validate those hypotheses. Several other recommendations were also provided, from OSHA's Draft Guidelines for Retail Grocery Stores[3] and from a commercial bakery based on that facility's own experiences.[5] When considering any intervention or workplace modification, even those that have been employed elsewhere or come highly recommended from an authoritative source, end users (cake decorators for the work discussed in this chapter) should be involved in the decision-making processes and evaluations of the interventions.

References

1. BLS, *Case and Demographic Characteristics for Work-related Injuries and Illnesses Involving Days Away from Work*, U.S. Department of Labor, Bureau of Labor Statistics, http://www.bls.gov/iif/oshcdnew.htm, accessed: 15 June 2004; last update: 25 March 2004.

2. Marras, W. S., Occupational biomechanics, in *The Occupational Ergonomics Handbook*, Karwowski, W. and Marras, W. S. eds., CRC Press, Boca Raton, 1999.

3. OSHA, *Ergonomics for the Prevention of Musculoskeletal Disorders: Draft Guidelines for Retail Grocery Stores*, U.S. Department of Labor, Occupational Safety and Health Administration, http://www.osha.gov/ergonomics/guidelines/grocerysolutions/, accessed: 14 June 2004; last update: 13 June 2003.

4. Osorio, A. M., Ames, R. G., Jones, J., Castorina, J., Rempel, D., Estrin, W., and Thompson, D., Carpal tunnel syndrome among grocery store workers, *Am J Ind Med* 25 (2), 229–45, 1994.

5. Rigdon, J. E., The Wrist Watch: How a Plant Handles Occupational Hazard With Common Sense — Many Small Changes Enable Sara Lee Bakery to Ease Carpal Tunnel Syndrome — Special Stools, Custom Tools, in *The Wall Street Journal*, Eastern ed. NY, 1992, pp. A1.

6. Chyuan, J. A., An ergonomic study for the control of upper extremity cumulative trauma disorders in foodservice bakery operations, Unpublished dissertation, Kansas State University, 1999.

7. Kuorinka, I., Jonsson, B., Kilbom, A., Vinterberg, H., Biering-Sørensen, F., Andersson, G., and Jørgensen, K., Standardised Nordic questionnaires for the analysis of musculoskeletal symptoms, *Appl Ergo* 18 (3), 233–237, 1987.

8. McAtamney, L. and Corlett, E. N., RULA: a survey method for the investigation of work-related upper limb disorders, *Appl Ergon* 24, 91–99, 1993.

49

Furniture Manufacturing Industry

Gary Mirka
North Carolina State University

49.1 Introduction

In 2002, U.S. Secretary of Labor Elaine Chao presented the four-pronged approach of the Occupational Safety and Health Administration (OSHA) to ergonomics. One of these components was an emphasis on enforcement — continued use of the General Duty Clause to issue citations relative to ergonomic hazards. The second was outreach and assistance to help business proactively address ergonomics in their workplaces. The third component of this approach was to develop a National Advisory Committee that was to be focused on the research and science of ergonomics, helping to identify gaps in our knowledge. The final component of this four-pronged approach was the development of ergonomics guidelines for specific industries. Initially, these guidelines were to be developed by OSHA in conjunction with the industry for which they were developed. To date, three guidelines have been developed with this model: one for the nursing home industry (OSHA, 2003), one for the poultry processing industry (OSHA, 2004a), and one for retail grocery stores (OSHA, 2004b). OSHA has also indicated that a draft guideline for the shipyard industry is in development. In addition to these OSHA-developed guidelines, OSHA also encouraged other industries to develop their own industry-specific guidelines and it is this opportunity that the furniture manufacturing industry has pursued.

49.2 The Development Team

The development committee for the Guideline was a mix of professionals from industry, government, and the research/consulting communities. The team included the eight active members of the Safety Committee of the American Furniture Manufacturers Association (AFMA), three individuals from the North Carolina Department of Labor, one researcher from academia, one person from the consultative services sector, one person actively supporting the AFMA in the development of training materials, and two individuals from the AFMA headquarters. The members of the AFMA Safety Committee that participated in the development of these guidelines represented some of the largest furniture manufacturers in the United States. These companies included Thomasville Furniture Industries, Broyhill Furniture Industries Inc., Bernhardt Furniture Company, Klaussner Furniture Industries Inc., Henredon Furniture Industries Inc., La-Z-Boy Incorporated, Pulaski Furniture Corporation and collectively had broad coverage of both the casegoods and upholstered furniture sectors of the industry. The primary role played by the industry members of the team was to provide work-proven solutions from their own experience as well as act as a conduit for work-proven solutions from their industrial colleagues who were not on the development team. The role of the members from the North Carolina Department of Labor were to provide their insight into the regulatory aspects of the project, maintain communications with the federal OSHA (keeping track of OSHA-developed guidelines, keeping OSHA apprised of our progress, etc.), and provide the day-to-day administrative function of the drafting and editing of the manuscript. The role of the researcher and consultant were to bring to the group a broader perspective of ergonomics including work-proven solutions (assessment tools, engineering solutions, program management techniques, etc.) from other industries as well as a more fundamental understanding of the broad issues in ergonomics. The individuals from AFMA headquarters and support staff provided legal/compliance implications and visual/graphical design perspectives, respectively.

49.3   The Process

Over a period of 12 months the development team met regularly to first identify the specific components to be included in the document, then to gather the necessary content, and, finally, to draft the document. In the early stages of this process, there was some discussion as to the form that the Guideline should take. It was decided that the final product was to be heavily weighted with the "best practices" currently employed by furniture manufacturers. Once this decision was made, a letter was drafted and sent to the AFMA membership asking for their active participation in the development process by providing effective solutions for problems that they had encountered in their own facilities. Accompanying this letter was a template that requested specific details of the intervention (cost of intervention, before and after description of work tasks, any challenges faced in the implementation of the intervention, any secondary benefits gained) so that the form and content of each of the ergonomics solutions page in the appendix was consistent across solutions. In addition to the request for effective engineering controls, the letter also asked for other materials such as surveillance tools, return to work policies, workplace assessment tools — anything that they have found to be effective and were willing to share with their furniture industry colleagues. When all of these materials were received, the committee began a screening process wherein a representative sample of engineering solutions from both the casegoods and upholstered sectors that spanned the full range of cost of implementation were identified and these were included as part of the extensive appendices of the Guideline (see Figure 49.1–Figure 49.5 for sample pages). Likewise all of the other submitted materials were screened and a representative sample of these other materials was included in the other appendices.

49.4   The Product

There are two formats for this guideline, a hardcopy version (AFMA, 2003) and an online version. The hardcopy version has two main components — the body of the document and the appendices. The body of the document contained the basic programmatic components that would be contained in any

AFMA Voluntary Ergonomics Guideline
Ergonomic Interventions

Name: **Height-Adjustable Upholstery Bucks**

Primary Task: **Upholstery**

Description: **The height-adjustable worktable can be adjusted using a foot pedal to present the furniture in the optimal position for the upholstery process.**

TASK DESCRIPTION	
BEFORE	AFTER
The operator maintains static awkward postures of the back (forward bending, side bending and twisting) while securing the fabric/leather to the frame. Operator will also often go into a deep knee bend position.	When upholstering the piece using the adjustable workstation the operator can raise the piece to the optimal position thereby reducing the awkward postures of the back and knees.

Ergonomic Impact: Reduces stress on the low back and knees.

Special Points of Interest: A model currently in development will also lower the work surface to a position 4 inches from the ground thereby eliminating the heavy lifting that often accompanies this task.

Estimated Cost to Purchase or Manufacture: Purchase for $600 - $800.

FIGURE 49.1 Sample page from the Engineering Controls appendix of the Guideline: height-adjustable upholstery bucks.

ergonomics guideline, regardless of industry (see Appendix for Table of Contents of the Guideline), while the appendices contained materials (solutions, processes, etc.) provided by the furniture manufacturing community. This approach of making use of work-proven solutions was taken in an effort to make the document more relevant to current problems faced by furniture manufacturers — as the solutions that would be submitted from industry would be those found to address injuries/illnesses of either high incidence levels or high severity, or both. The level of detail provided with each engineering solution allows the reader to more completely understand the problem that was being addressed (both before and after pictures as well as a brief verbal description of the problem addressed), the expected costs associated with the solutions as well as the documented benefits.

It was recognized that the hardcopy version of the Guideline was limited in the number of engineering controls that could be presented by both page limitations and the "cross-sectional" nature of the solution-gathering process. These limitations were overcome by the second component of the overall approach that the development team resolved to purse: the AFMA Engineering Controls webpage

AFMA Voluntary Ergonomics Guideline
Ergonomic Interventions

Name: **Spring-Loaded Fabric Buggy**

Primary Task: **Transport and acquisition of rolls of fabric**

Description: **Weight-calibrated springs are integrated into the cart
                     mechanism that raise the rolls of fabric for easy lifting.**

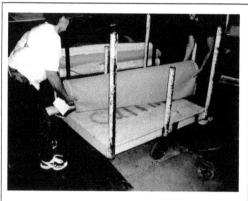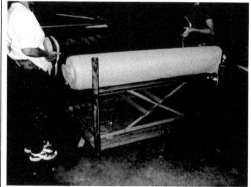

TASK DESCRIPTION

BEFORE	AFTER
The operator bent down into the fabric buggy (often to 6-8 inches from the ground) and lifted the roll of fabric to shoulder level.	The spring-loaded system raises the fabric to near waist level to reduce the awkward posture of the back.

Ergonomic Impact: Reduces the opportunity for back injury due to excessive bending
and lift heavy fabric rolls off of the buggy. Also reduces shoulder stress.

Special Points of Interest: These spring-loaded systems can be custom built and placed
in virtually any cart system.

Estimated Cost to Purchase or Manufacture: Purchased for $135.

FIGURE 49.2   Sample page from the Engineering Controls appendix of the Guideline: spring loaded fabric cart.

(www.ie.ncsu.edu/ergolab/engcontr/index.html). The basic idea behind this webpage is that AFMA members are continuously developing innovative engineering controls and a repository of these innovative solutions would facilitate the dissemination of ideas. This webpage has a hierarchical structure that quickly allows a user to identify any existing solutions to a problem that they might be experiencing in their own manufacturing facility. The wepages contain all of the same information regarding the engineering solutions that were included in the hardcopy version (description, cost, secondary benefits, etc.).

49.5   Discussion and Lessons Learned

One success story of this process was in the effective research — industry — government alliance that was utilized. Each of these groups brought important assets to the development process. The industry

AFMA Voluntary Ergonomics Guideline
Ergonomic Interventions

Name:	**Random Orbital Sander Interface**
Primary Task:	**Hand sanding**
Description:	**The operator wears a glove that has vibration-absorbing material in the palm. This glove is secured to the sander by means of a harness system that wraps around the circumference of the sander motor.**

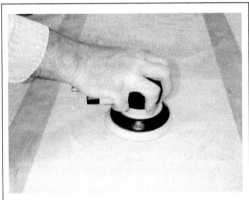 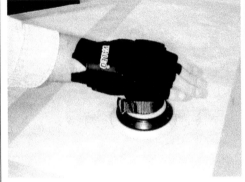

TASK DESCRIPTION	
BEFORE	AFTER
Using the random orbital sander requires a static gripping force on a vibrating handtool.	The glove reduces the amount of vibration reaching the operator and the harness system eliminates the need for the continuous gripping force.

Ergonomic Impact: Relieves stress and fatigue on the wrist and forearm.

Special Points of Interest: The nature of the harness system requires that an additional valve be placed in the hose for speed control. The harness system is also a bit cumbersome to take off and put on. Best results have been found for those individuals who spend >90% of their time hand sanding.

Estimated Cost to Purchase or Manufacture: Manufacture for $40.

FIGURE 49.3 Sample page from the Engineering Controls appendix of the Guideline: random orbital sander harness.

representatives brought a perspective on what the industry needs in terms of the document and supporting materials as well as representative samples of work-proven solutions for furniture industry problems. The government representatives were able to bring their perspective on the regulatory environment in which ergonomics is viewed as well as resources in terms of time and money for the development of the document itself. The research/consultant representatives brought a broader perspective on the state of ergonomics across the spectrum of industries, which included some solutions and analysis tools that were adopted for use in the Guideline. While each group had its own motivation for participation in this effort, the common goal of reducing the incidence and severity or work-related musculoskeletal disorders was the uniting factor.

There were also a number of challenges that the team met in this process. First, gaining the participation of the furniture industry at large in the process of developing the document was difficult. It was

AFMA Voluntary Ergonomics Guideline
Ergonomic Interventions

Name:	**Automatic Frame Spring Puller**
Primary Task:	**Attaching seat springs to the frames**
Description:	**The puller automatically stretches the springs and attaches them to the clips in the seat frame**

TASK DESCRIPTION	
BEFORE	AFTER
The employee had to grab each spring and pull against significant tension to the clip on the opposite end of the frame.	Employee attaches the spring to the mechanism and it pulls all springs simultaneously to the clip on the other end of the frame.

Ergonomic Impact: Relieved stress on the hands, wrists, elbows, shoulders and back because the employee doesn't have to stretch the springs by hand.

Special Points of Interest: The machine doesn't hinder the operation of the employee because the times between the hand springing and the machine springing are almost identical

Estimated Cost to Purchase or Manufacture: Manufacture for approximately $750.00 each.

FIGURE 49.4   Sample page from the Engineering Controls appendix of the Guideline: spring stretcher.

our intention to include the solutions of as many furniture manufacturers as possible in the best practices appendices. A letter that encouraged this participation was sent from the AFMA to its membership. Unfortunately, there was not much response from furniture companies outside those represented on the AFMA Safety Committee. Direct contact between committee members and various individuals in other companies was shown to be a more effective way of engaging these other companies, resulting in a wider level of participation. The second challenge was the timeframe for development. Most of the development team members had full-time duties at their regular appointments and typically fit the work on this guideline in when able. This led to a development process that took a full 12 months. A much shorter turnaround time would have been possible if a small group within the development team were to simply develop the document and then pass it around for approval. The consensus of the group, however, was that full participation in the writing of the document by all development team members was vital to establish ownership of the work, as the industry representatives on the

AFMA Voluntary Ergonomics Guideline
Ergonomic Interventions

Name:	**Suspended Table Rubbing Machine**
Primary Task:	**Rubbing dining table tops**
Description:	The suspended rub machine allows the operators to achieve a much higher quality and more consistent rub on dining table tops. It virtually eliminates all vibration associated with the constant vibration associated with the 40 pound machines. Also, operators do not have to ever lift the machines onto and off of the tables.

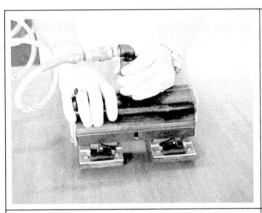 

TASK DESCRIPTION	
BEFORE	AFTER
Operator manually moves the heavy, vibrating rub machine across the work surface. Must lift the machine off of the finished piece.	The new system supports the mass of the rub machine and the framework structure isolates the rub machine vibration from the operator.

Ergonomic Impact: Loads on the low back, shoulder, elbow, and hand/wrist are reduced. Exposure to vibration is virtually eliminated.

Special Points of Interest: After installing the new machines, cycle time was reduced.

Estimated Cost to Purchase or Manufacture: Manufacture for $6,000 per station.

FIGURE 49.5 Sample page from the Engineering Controls appendix of the Guideline: suspended tabletop rubbing machine.

team were going to have to stand behind this document when completed and promote it among their peers.

Acknowledgments

This work was partially supported by a grant from the NCSU Furniture Manufacturing and Management Center (FMMC). The contents are solely the responsibility of the authors and do not necessarily reflect the views of the FMMC.

Appendix

Table of Contents

References

AFMA (2003) *American Furniture Manufacturers Association Voluntary Ergonomics Guideline for the Furniture Manufacturing Industry*, North Carolina Department of Labor.

OSHA (2003) *Guidelines for Nursing Homes: Ergonomics for the Prevention of Musculoskeletal Disorders*, U.S. Department of Labor, Occupational Safety and Health Administration.

OSHA (2004a) *Guidelines for Poultry Processing: Ergonomics for the Prevention of Musculoskeletal Disorders*, U.S. Department of Labor, Occupational Safety and Health Administration, OSHA 3213-09N.

OSHA (2004b) *Guidelines for Retail Grocery Stores: Ergonomics for the Prevention of Musculoskeletal Disorders*, U.S. Department of Labor, Occupational Safety and Health Administration, OSHA 3192-05N.

How to Obtain a copy of the Guideline

Electronic (pdf)

http://www.afma4u.org/resources/index.htm
http://www.nclabor.com/osha/ergoguideline.pdf

Hard Copy

AFMA, PO Box HP-7, High Point, NC 27261, U.S.A.
NC Department of Labor/ETTA, 1101 Mail Service Center, Raleigh NC 27699-1101, U.S.A.

50

General Construction

Steven F. Hecker
Jennifer Hess
Laurel Kincl
University of Oregon

Scott P. Schneider
*Laborers' Health and Safety Fund
of North America*

50.1   Introduction

The application of ergonomics in the construction industry is generally more recent and less developed than in manufacturing and other industries. However, the impact of ergonomics and human factors on safety and productivity in construction is now receiving significant attention, largely due to concerns with the incidence and costs of musculoskeletal disorders (MSDs), particularly of the lower back and upper extremities. In the United States, such injuries are more common in construction than any other industry sector except transportation (CPWR, 2002). Many construction tasks involve non-neutral and static body postures, high force demands, contact stress, and other risk factors such as vibration and adverse environmental conditions that contribute to acute and chronic musculoskeletal injuries.

It is ironic that modern ergonomics came lately to construction because bricklaying, one of the oldest trades and occupations, was the object of one of the earliest ergonomics studies. Frank Gilbreth was an apprentice bricklayer as well as a student of Frederick Winslow Taylor, the founder of scientific management and time–motion study. In the 1890s, Gilbreth made time–motion studies of bricklayers to improve their efficiency and introduced several innovations. He developed a two-tiered scaffold that positioned the materials on a higher level so that the worker did not have to bend down or stoop to get them. In 1891, he developed and patented the Gilbreth scaffold that could be jacked up incrementally to keep the work at waist height. He developed a "package" system for delivering bricks so they did not get broken. He also created work-rest schedules to reduce fatigue and improve efficiency. The result was an increase in productivity of more than 200%. His results and conclusions were detailed in his book *Bricklaying System* published in 1909 (Gilbreth, 1974).

Ergonomics concerns in construction are not limited to MSDs. Cognitive ergonomics can also have an important role in construction, particularly in addressing risks like working at heights or on uneven terrain, and operation of cranes and other heavy machinery. The constantly changing work environment means that as a rule construction workers must collect and process information more steadily and rapidly than those in fixed work environments (Helander, 1981). Failure to do so can result in impaired risk perception and avoidance.

50.2 Characteristics of the Construction Industry

The construction industry encompasses a wide variety of component industries and activities, but it can be classified in some major divisions. The two broadest categories are *building* and *civil engineering*. Building includes everything from apartments and houses to high-rise office buildings, factories, schools, government buildings, hospitals, etc. Civil projects are those related to transportation, including highways, bridges, railroads, tunnels, and other structures such as dams. This category is also often labeled *heavy/highway* construction. Building construction is divided between residential and nonresidential. The former includes single- and multi-family housing. Nonresidential can be divided into commercial and industrial. Commercial includes office buildings, retail establishments, and the like. Industrial construction is sometimes included as part of commercial, but has particular characteristics and describes the building and retrofitting of facilities such as large factories and oil refineries. Utility construction, such as power plants, is closely related to industrial construction. Residential construction tends to involve both smaller structures and smaller contractors, but in most countries the construction industry as a whole is characterized by small companies with fewer than 20 workers, and a significant percentage of self-employed workers (Hinksman, 1997).

50.2.1 Construction Workforce

We can also characterize construction work by the different trades or crafts that perform it. As many as 20 to 25 trades and subtrades might work on a single project (see Table 50.1). Construction projects of any size involve multiple parties and organizations coming together for each project. These include: owner (client), designer, general contractor or construction manager, specialty trade contractors (subcontractors), construction labor, and vendors and suppliers of materials. Supporting organizations include finance and insurance among others.

The construction industry has an increasingly diverse work population in the United States. In 2002, there was an average of 6.7 million construction workers, making up 5.2% of all workers in the United States (BLS, 2003). In 2000, women made up 9% of the construction workforce with 20% of those in production occupations and the rest working in clerical, administrative, and management positions. The average age of a construction worker was 39 yr, and the secular trend for a number of years has been toward an aging workforce. The percentage of minorities, most notably Hispanics, has been growing rapidly in the United States. A number of factors relevant to safety, health, and ergonomics

TABLE 50.1   Selected Construction Trades[a]

Trade/Subtrade	
Insulator	Laborer
Bricklayer	Concrete placement
Tile setter	Mason tender
Carpenter	Environmental abatement
Concrete form builder	Tunnel worker
Framer	Pipeline worker
Ceiling installer	Operating engineer
Interior/exterior finish	Cement mason
Scaffold erector	Plasterer
Drywall installer	Painter
Pile driver	Glazier
Floor coverer/carpet layer	Drywall finisher
Electrician	Plumber
Elevator constructor	Pipefitter
Floor coverer/carpet layer	Roofer
Iron worker	Sheet metal worker
Concrete reinforcement worker	HVAC
Structural steel erector	Architectural
Ornamental	Sprinkler fitter

[a]Based on North American trade taxonomy; trade jurisdictions may vary geographically.

may vary with age, gender (Messing, 2000; Welch et al., 2000), or race (Ringen et al., 1998; Dong et al., 2004), including:

- Anthropometric measurements, which affect tool design, work surface height, work practices, equipment dimensions, and personal protective equipment (Hsaio et al., 2003).
- The workplace social climate.
- Perception of risk, which may vary with age, experience, and cultural background.
- Access to health care for management of musculoskeletal symptoms, often more limited among immigrant and low-wage workers.
- The ability of an aging workforce to safely maintain work pace and intensity.

Finally, the U.S. construction industry is experiencing a skilled workforce shortage stemming from a variety of factors that have converged over the past 20 yr. These include the economic recessions of the 1980s and 1990s; a decline in construction training programs, both at the apprenticeship and secondary vocational level; the poor image of construction work; and competing occupational choices due to technological advances, particularly in computer-related jobs (Chini et al., 1999; CLRC, 1997). Koningsveld and van der Molen (1997), describing the situation in The Netherlands, suggest that at least some of these trends are not unique to the United States. Ergonomic improvements that lessen the physical workload and musculoskeletal injury risk for all construction workers could contribute to solving this shortage by improving working conditions, prolonging careers, and providing a wider population base from which to recruit new workers.

50.2.2   Challenges for Ergonomics

Certain characteristics of the construction industry present particular challenges to the application of ergonomic principles (Schneider, 1995):

- The work environment is constantly changing as construction progresses.
- The work force is generally mobile and most jobs are of short duration. This both limits employer incentives to invest in prevention of chronic conditions and makes timely intervention difficult.

- The multi-employer nature of most projects means that the work of one contractor may create hazards for others that the latter do not have direct control over.
- Communication, especially on large projects, is often complex.
- Many construction tasks are by definition at floor or ceiling levels.
- Commonly, construction workers supply their own hand tools, so tool interventions must take place through individual workers rather than through employers.
- Environmental conditions are not subject to control in an outdoor environment.
- To the extent that ergonomics is considered in the design of projects it is usually targeted at the end user of the facility rather than at those who construct it.

For these and other reasons MSD prevention in construction requires multiple approaches at multiple levels of the industry. Engineering and design controls are still a primary means of prevention, but they often need to be supplemented by administrative controls such as job rotation, rest breaks, working in teams, and improved planning and layout. Improved body mechanics and individual physical conditioning are often proposed as part of ergonomic intervention in construction, but data on their effectiveness is scarce, and they are not a substitute for higher-level controls (Hess and Hecker, 2003). Some interventions are within the control of individual workers, but others require participation and initiative by management, general contractors, project owners, equipment and material manufacturers and suppliers, and designers.

This chapter first characterizes the problems that ergonomics can address in construction, focusing on the incidence of MSDs and the risk factors that contribute to them. We describe a representative range of problem tasks, some of which are common to most construction trades and others that are trade-specific. We then discuss task analysis and exposure assessment methods specific to construction jobs. We examine ways in which the unusual structure of the construction industry affects the ergonomics intervention process and what research has taught us in that regard. Finally, we address some of the cognitive ergonomics issues related to construction safety.

50.3 Musculoskeletal Disorders in Construction Workers

A 2001 review of the injury data (Schneider, 2001) showed that construction workers experienced musculoskeletal injuries at a rate about 50% higher than manufacturing workers. This is in spite of a number of industry forces that could lead to understatement of MSDs of all types, including the tendency of construction workers to "work hurt" and not report symptoms because of the sporadic and insecure status of much construction work. In the United States, strains and sprains are the most frequent nonfatal injury type in construction, accounting for 36% of the total, while the low back is the most common region injured, accounting for 21% of all injuries (Schneider, 2001). Among construction occupations the highest rate of strains and sprains is found in masons, followed by painters and paperhangers (Figure 50.1). Construction ranks second only to transportation in levels of back injuries, and the rate of back injuries is 1.5 times the rate for all private sector industries (BLS, 2001; CPWR, 2002).

Overexertion is the most frequent cause of MSDs among U.S. construction workers, and again only the transportation sector exceeds construction in lost-time overexertion injuries (BLS, 2001). Overexertion from lifting accounts for 45% of injuries, while overexertion from all other sources, such as pushing, pulling, carrying, or throwing, accounts for another 32%. Roofing, siding, and sheet metal workers have the highest rate of overexertion injuries from lifting, while painters have the highest rate from overexertion other than lifting (Figure 50.2). Overexertion injuries related to lifting are 1.4 times the rate of all other industries (CPWR, 2002). Bending injuries account for 16.5%, while repetitive motion injuries account for about 7.0% of MSDs (BLS, 2001).

Carpal tunnel syndrome and tendonitis are likely underreported in this population for the reasons outlined above. Musculoskeletal injuries may have long-term impact on workers and productivity.

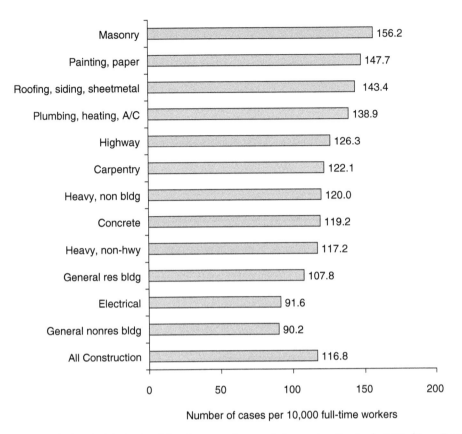

FIGURE 50.1 Strain/sprain injuries involving time away from work, by construction trade, 2001. (From Bureau of Labor Statistics (BLS), U.S. Department of Labor. 2001. Lost-worktime injuries and illnesses: characteristics and resulting time away from work, www.stats.bls.gov/iif/home.htm.)

According to a study by Welch et al. (1999), of those construction workers who visited a hospital emergency room with an acute musculoskeletal injury, 62% developed chronic symptoms. In those with chronic symptoms 24% reported that their symptoms adversely affected work assignments, productivity, and personal activities.

These statistics are generally mirrored in other industrial countries. Construction injury data for the Canadian province of Ontario from 1997 to 1999 show overexertion to be the largest single accident type at 25.3%; strains and sprains the largest injury type at 28.7%; and the back or spine the body part most often affected at 23.7% (CSAO, 2002). In The Netherlands more than 50% of sick leave among construction workers is due to musculoskeletal complaints (van der Molen and Delleman, 1997). A Swedish questionnaire study of 85,000 male construction workers (78% of the country's working population in construction) collected data on musculoskeletal symptoms over a 3-yr period, 1989 to 1992. MSD symptoms in nine body regions were significantly higher among all construction trades compared to male office workers and foremen in the construction industry, groups assumed to have lower on-the-job exposure (Holmström and Engholm, 2003). Table 50.2 shows MSD symptom prevalence in the five most affected body regions — low back, knee, shoulder, wrist/hand, and neck — for all construction workers and the trades with the highest risk for each region. Significantly elevated odds ratios were found for numerous trades in numerous body regions. For example, crane operators (4.5), painters (4.4), and insulators (4.1) had the highest odds ratios for neck pain. Low back odds ratios were highest in roofers (5.0) and floorers (3.5), while floorers (4.5) and plumbers (4.1) had the highest ratios for knee symptoms (Holmström and Engholm, 2003).

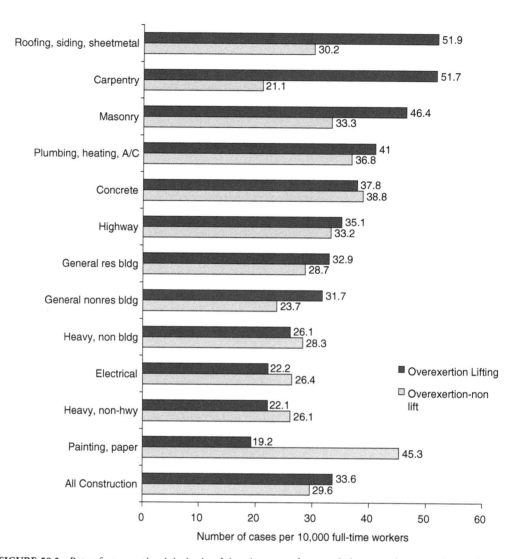

FIGURE 50.2   Rate of overexertion injuries involving time away from work, by type of construction work, 2001. (Adapted from the Center to Protect Worker's Rights (CPWR). 2002. *The Construction Chart Book, The U.S. Construction Industry and its Workers*, 3rd edition. Silver Springs, MD: CPWR and Bureau of Labor Statistics (BLS), U.S. Department of Labor. 2001. Lost-worktime injuries and illnesses: characteristics and resulting time away from work, www.stats.bls.gov/iif/home.htm.)

50.4   MSD Exposure and Risk Factors

As the Swedish and U.S. data indicate, there are some strong correlations between particular trades and musculoskeletal symptoms in specific body regions. The ergonomics literature further elucidates some of these relationships. The first general review of ergonomics hazards in construction was published 10 years ago (Schneider and Susi, 1994). Prior and subsequent studies have produced both detailed exposure assessments of specific tasks and broad taxonomies of construction tasks sorted by risk level. Risk factors for musculoskeletal injuries are usually categorized as follows.

- Repetitive motions
- Posture stresses (awkward postures)

TABLE 50.2 Prevalence and Odds Ratios for MSDs by Location with Foreman as the Reference Group; Swedish Construction Workers

Body Part	MSD Prevalence — All Construction Workers (%)	Occupations with Highest Odds Ratio	Odds Ratio (Foreman Reference Group)
Lower back	18.0	Roofers	5.01
		Floorers	3.53
		Scaffolders	3.37
		Insulators	3.07
Knee	16.2	Floorers	4.54
		Plumbers	4.13
		Roofers	3.49
		Sheet metal workers	3.41
Shoulder	8.8	Scaffolders	8.35
		Insulators	5.37
		Painters	4.51
		Crane operators	3.59
Wrist/hand	8.4	Scaffolders	9.10
		Insulators	4.47
		Sheet metal workers	4.43
		Floorers	3.94
Neck	7.0	Crane operators	4.50
		Painters	4.42
		Insulators	4.10
		Scaffolders	3.02

Source: Data from Holmström E; Engholm G. 2003. *American Journal of Industrial Medicine* 44(4): 377–384.

- Static postures
- Forceful exertions
- Localized contact stresses
- Temperature extremes
- Vibration

Everett (1997) developed a taxonomy of 65 major tasks of the 15 U.S. building trade unions. He then broke these tasks down into steps or stages and evaluated each step across the common ergonomic risk factors using a three-level measure: (1) insignificant risk, (2) moderate risk, and (3) high risk. This provides a useful matrix for screening and prioritizing tasks for ergonomics intervention. Some trades, bricklayers for instance, perform a limited repertoire of tasks, while other trades routinely perform a wider variety of tasks, encompassing multiple risk factors. This can make prioritizing ergonomic solutions a challenge.

50.4.1 Exposures Common to Multiple Trades

Construction work in general places workers at higher than average risk of injury (Schneider et al., 1998). Inherent in the nature of construction work are material handling, difficult access to the point of operation or installation, and work in uncontrolled outdoor environments. Frequent handling of heavy, bulky materials is an activity that virtually all trades will encounter. Common materials include mortar, cement, and concrete; lumber and plywood; roofing materials; steel reinforcement bar; doors and windows; electrical cable; drywall panels; and pipe and conduit. Manual materials handling places workers at risk due to high forces and frequent awkward postures required to manipulate the objects. Awkward postures and repetitive activities are also routine for many trades. Work tasks at awkward heights, from ground level and below to well above the shoulders and head, are common. Risk factors present in the general environment include crowded workspaces, uneven and slippery terrain, and excessively hot or cold temperatures that can exacerbate other ergonomic risk factors.

50.4.2 Trade Specific Exposures and Controls

Table 50.1 lists many of the specialized trades performing construction work. There are consistent patterns between work tasks and specific MSD risks in a number of trades. This section describes some of these patterns and presents examples of ergonomics improvements that have demonstrated success in reducing exposure. Such improvements commonly fall into the following categories:

- Hand or power tools
- Work practices
- Work organization, including scheduling, sequencing, site layout, etc.
- Substitution of materials
- Changes in material handling
- Design, including access
- Personal protective equipment

The trades described here are not an exhaustive list, and it is not our intention to suggest that other trades not mentioned are free of significant risks.

50.4.2.1 Carpet/Floor Coverers

Carpet and floor installation has long been recognized as an ergonomic problem. National Institute for Occupational Safety and Health (NIOSH) studies of carpet installation in the 1980s showed carpet installers had a high risk of knee injury associated with use of a knee kicker to stretch carpeting (Tanaka et al., 1982, 1989; Thun et al., 1987; Tanaka and Habes, 1990). A 1990 NIOSH Alert recommended use of power stretchers whenever possible and use of knee protection (Tanaka and Habes, 1990). A Swedish study compared carpet and floor layers with painters and found significantly higher knee morbidity among carpet and floor layers associated with kneeling work (Kivimäki et al., 1992). Danish studies of floor layers and carpenters also showed higher prevalence of knee disorders related to kneeling work among carpenters and floor layers (Jensen et al., 2000). Intervention strategies for carpet installers were outlined in a biomechanical study of carpet stretching devices in 1993 (Village et al., 1993). Jensen and Kofoed (2002) reviewed prevention strategies for floor layers through the use of mechanical aids, tools to allow standing height work, better job planning, and personal protection (knee pads). They found that prevention efforts were possible but were complicated because of the culture of construction that makes change difficult.

50.4.2.2 Concrete Work

Trades and tasks associated with concrete work involve multiple MSD risk factors. Concrete work involves constructing forms, setting the concrete reinforcement rods (or rebar), and pouring and finishing the concrete floors and support beams. Concrete reinforcement work has long been recognized as hazardous. Wickström (1978) and Wickström et al. (1985) reported high rates of back disorders among concrete reinforcement workers because their work requires them to bend over for long periods of time tying the rebar together. Buchholz et al. (2003) performed an ergonomic analysis of concrete reinforcement workers during highway construction and found they spent a significant portion of their time (over 30%) in nonneutral postures, particularly during ground level and ventilation work. Manual handling of rebar also presents a significant risk due to its length and weight. As much as possible, rebar should be moved by crane to minimize this risk.

Once the rebar is placed and tied, concrete is poured into the forms and must be spread, leveled, and finished. The weight and viscosity of the concrete can make this work difficult. The increasing use of concrete pumping has reduced some of the heavy physical work involved, but the handling of the hoses on the "slick line" itself can involve high forces and awkward postures. Use of steel skid plates under the

FIGURE 50.3 Steel skid plates under a concrete hose junction. (From Hess et al. Elsevier, 2003, With permission.)

junctions between hose sections (Figure 50.3) was found to reduce low back risk in laborers pulling concrete hoses during concrete placement (Hess et al., 2004). Leveling the concrete is done by a process called screeding which requires workers to work in a bent-over posture gripping a screeding device (often a piece of lumber), usually with a pinch grip, and sliding it back and forth while pulling the concrete (Figure 50.4). Recently, several variations of laser-guided machines called laser screeds have been developed to do this work mechanically (Figure 50.5). The StewartPrezant Ergonomics Group (2003) completed an ergonomic analysis of the various screeding techniques and machines that showed the ergonomic advantages of mechanical screeding. The viscosity of the concrete affects the force workers must use to level it. Viscosity, measured by a slump test, is determined by engineering and architectural specifications and may be set at a level greater than is structurally necessary. The viscosity can be modified or adjusted by chemical additives called plasticizers to reduce the ergonomic risk.

Concrete must be vibrated as it is poured into forms to eliminate voids. The worker operating the vibrator is exposed to hand–arm vibration as well as risk of back/shoulder strain, especially while vibrating wall forms where the vibrator must be pulled up repeatedly from the base of the wall through the concrete.

Building and stripping concrete forms is also ergonomically challenging work because of the size, weight, and position of the forms. Once set, concrete sticks to the forms, frequently

FIGURE 50.4 Leveling concrete manual screeding.

FIGURE 50.5 Leveling concrete by laser screeding.

necessitating use of high forces to pull them apart. Spielholz et al. (1998) published a review of the ergonomic risks of this work showing a high percentage of awkward body postures and a high prevalence of symptoms.

50.4.2.3 Mechanical and Electrical Trades

These trades, including plumbers, pipefitters, sheet metal workers, and electricians, have a number of exposures in common. Neck and shoulder injuries appear to be connected with occupations requiring significant overhead work, such as sheet metal workers and electricians. Sheet metal workers manufacture and install the air handling systems for heating and cooling a building. Much of their work is at ceiling level, where they install ductwork and the hangers that support it. Anton et al. (2001) found greater forces in the shoulder muscles during overhead drilling when the worker reached outward and upward, compared to keeping the drill close to the body. Drill bit extensions help to reduce awkward shoulder postures (Figure 50.6 and Figure 50.7). Plumbers and pipefitters do similar overhead tasks with pipe and hangers, while electricians face overhead exposures when routing cable and installing overhead fixtures.

All of these trades make frequent use of hand tools, which can expose workers to excessive gripping and awkward hand and wrist postures. Sheet metal workers use metal shears both in the shop and in the field. Plumbers and pipefitters use pipe cutters and reamers as well as wrenches. Electricians use a range of hand tools including wire strippers and cutters, screwdrivers, and pliers. Welch et al. (1995) did a survey of musculoskeletal symptoms among sheet metal workers and found high rates of symptoms, including hand cumulative trauma disorders, from use of cutting tools and shoulder problems from overhead work in installation.

Rosecrance et al. (1996) reported on a survey of musculoskeletal symptoms among over 500 members of the plumbers and pipefitters union. High rates of back, neck, and knee problems were reported and related to awkward and static postures, including overhead work and kneeling. Rosecrance et al. (2002) tested apprentices from the electrical, plumbing/pipefitting, sheet metal, and operating engineer trades for carpal tunnel syndrome (CTS) and found the following prevalence rates of CTS: electricians, 6.1%; sheet metal workers, 9.6%; plumbers/pipefitters, 8.5%; operating engineers, 8.2%. These compare with general population statistics of about 3%. A recent NIOSH report presents findings from a workshop on ergonomics risks and interventions in the mechanical and electrical trades (Albers et al., 2004).

FIGURE 50.6 Awkward pasture when drilling overhead.

FIGURE 50.7 Drill extension to modify posture.

50.4.2.4 Drywall Installation

Drywall installation is fast-paced work involving lots of material handling. Drywall workers are the highest occupational risk classification for back disorder compensation claims in Washington State (Lipscomb et al., 2000). U.S. Bureau of Labor statistics data also show drywall installers to be at high risk of overexertion, back, and sprain and strain injuries (Chiou et al., 2000). The risk is associated with the weight of the sheets of drywall, their awkward size, and the distances they have to be carried by hand. Sheets can weigh between 60 and 130 pounds and are normally 4-ft wide. Many times they can be moved by carts, but manual handling is common (Figure 50.8 and Figure 50.9). Extension handles are available to make manual handling easier, but they are not commonly used. Switching to 3-ft (90 cm) wide drywall has been proposed in Scandinavia (Björklund et al., 1991), and pilot studies have shown it to be ergonomically easier to carry, in part due to the lighter weight and also the increased visibility for the carrier (Lappalainen et al., 1998). Such a change, though, would require redesign of the stud system to which the boards are attached. Lifts are also available for raising and mounting drywall on the ceiling in residential applications. Screwing the drywall to the studs also involves ergonomics challenges from manual screwdriving. This risk can be reduced by micropauses (Andersson, 1991).

FIGURE 50.8   Worker manually handling drywall sheet.

FIGURE 50.9 Drywall transport cart.

Most drywall installers have now switched to battery-powered screw guns, which reduce the postural and repetitive motion risks but also speed up the work. Many drywall workers in the United States are paid on a piece-rate system, which exacerbates the risk of musculoskeletal injury by increasing the pace of work.

50.4.2.5 Carpentry

Concrete form building and drywall are particularly high exposure tasks for carpenters, but a variety of other carpentry activities also pose risks. Among these are framing of both residential and commercial buildings, ceiling and floor installation, finish work and cabinet/fixture installation, and pile driving. Workers' compensation claims for work-related musculoskeletal injuries among union carpenters have shown the highest risks of injury to those doing light commercial work and drywall installation (Lipscomb et al., 1996, 1997). Body discomfort surveys of carpenters pointed to drywall, ceiling work, and concrete formwork as the high-risk activities, particularly for back and knee injuries (Dimov et al., 2000). These findings correlate with ergonomics exposure surveys of carpentry work (Bhattacharya et al., 1997). Mirka et al. (2003) studied residential carpentry tasks and introduced several assistive devices for framers that significantly reduced low back stresses. In 1994, the Carpenters Union developed an "Ergonomics for Carpenters" curriculum that has been used in apprentice- and journey-level training classes to provide members with information on proper work techniques to minimize the risk of injury (UBC, 1995).

50.4.2.6 Scaffold Erection

This job is carried out by a variety of trades, often carpenters, and is characterized by heavy material handling demands. Frames are large, heavy, and awkward to carry. They are often assembled overhead and disassembled from unstable positions. The frames often get stuck together requiring a lot of force to disassemble. NIOSH has studied the biomechanical problems of scaffold tasks (Hsaio and Stanevich, 1996) and the best postures for disassembly (Cutlip et al., 2000). Researchers in The Netherlands used a participatory approach to develop interventions for scaffold erectors. They developed new carts for transport of the pieces, worked with suppliers to make sure parts were arranged when delivered to minimize handling and assembly, and developed hoists for moving parts up to the top of the scaffold during assembly. These interventions were evaluated and found to decrease manual handling, heart rate, and awkward postures (Koningsveld et al., 1998).

50.4.2.7 Masonry

Gilbreth's early work on bricklaying ergonomics was described in the introduction. Bricklaying is one of the oldest trades and the best studied. The work itself is fairly repetitive and more amenable to ergonomics analysis than other more varied trades. Since Gilbreth's time, numerous researchers have studied bricklaying. Ontario data show that overexertion is responsible for a higher percentage of bricklayer lost-time injuries than construction workers as a whole, and that the acts of lifting and laying are by far the most common activities associated with the injuries (CSAO, 2002). The primary musculoskeletal exposures are to the low back from bending, lifting, and twisting, and to the shoulders, arms, and hands from gripping and lifting bricks and blocks.

In the 1980s, researchers at Bygghälsan in Sweden developed recommendations for the weight of masonry blocks (Hammarskjöld, 1987). Block weight was limited to 3 kg if lifted one-handed, 12 to 20 kg blocks were to be lifted only between knees and shoulders, and blocks over 20 kg had to be lifted mechanically. These recommendations were adopted as Swedish standards in 1987. In the same period, German researchers studied the lumbar load of bricklayers and concluded that keeping the wall at waist height produced the lowest load on the muscles and resulted in the highest productivity (Jäger et al., 1991; Luttmann et al., 1991). Research on how to improve the work of bricklayers led to the development of new blocks with handholds to improve the grip and reduce bending (as the block is lifted from the top). Dutch researchers developed new systems and carts/dollies for transporting bricks and blocks, reduced the size and weight of blocks, and proposed task enlargement for bricklayers (Vink and Koningsveld, 1990; TNO, 1993; van der Molen et al., 1998). Researchers in the United States and Canada have also been studying the effects of lower weight masonry block on risk of injury and productivity with mixed results thus far (Zellers and Simonton, 1997; Vi, 2000/2001).

More recently, setting bricklayers' materials at a height of 31 cm has been shown to reduce the frequency and duration of deep trunk flexion compared to working with the materials on the ground (van der Molen et al., 2003a). However, the researchers note that bricklayers actually have two workstations, one for the storage of materials and one at the wall. An evaluation of the adoption of work height and other interventions showed that demonstrating improvements in productivity and cost–benefit advantages were key to adoption by companies and workers (de Jong et al., 2003). American researchers have also been studying bricklayers and developing guidelines for improving their work (Cook et al., 1996). Adjustable height scaffolding has become widespread in the United States over the past few years allowing bricklayers to keep their work at waist height. As early as 1990 this was shown to cause significant improvements in productivity (20–35%) (Suprenant, 1990).

50.4.2.8 Painting

Painting ceilings and high walls results in awkward neck and upper extremity postures. Painters also must frequently move heavy equipment and buckets of paint. The painting trades do the finishing work on drywall after it has been installed, sometimes using stilts to access high walls and ceilings. Various interventions have been evaluated for painters' work. For painting ceilings, a small flange (painter's disk) on the pole helps the painter support the weight with his or her entire hand and not just his or her grip, thereby reducing muscle demand and fatigue (Cederqvist and Örtengren, 1985). Easy to handle packages with handles were developed for painter's putty by the Swedish Packaging Research Institute (Erneling and Kling, 1983). More recently, a new spring-powered flat box to mechanically apply taping compound (mud) to a ceiling drywall joint was evaluated and found to reduce exposure risk of musculoskeletal injury (Shaw, 2003).

50.4.2.9   Roofing

U.S. data indicate that roofers have higher rates of lost-time overexertion and back injuries than any other construction occupation (CPWR, 2002). Roofers work at floor level spending extended amounts of time in crouched or bent-over postures. They move heavy rolls of material and asphalt buckets, while standing on unlevel surfaces, and they remove old roofing (tear off). Vink (1992) looked at an intervention for fastening roofing materials to the roof deck from a standing height and found the tool presented an improvement in work methods. There are numerous other mechanical devices available to tear off old roofing, move materials around on the roof, spread gravel on a new roof, and pick up old gravel before a tear off.

50.4.2.10   Construction Laborers

Laborers do some of the heaviest and hardest jobs in construction, including heavy manual materials handling and "tender" or assistant work to many trades like masons and plasterers supplying them with the materials they need. They also do a substantial amount of environmental cleanup work (asbestos and lead abatement, hazardous waste work). The wide variety of tasks makes their exposures harder to characterize. One study of Danish "semiskilled" workers found many tasks that created risk of low back injuries particularly from lifting and pushing/pulling (Damlund et al., 1986). The operation of jack hammers and other chipping hammers exposes laborers to potentially hazardous levels of whole-body and hand–arm vibration and noise as well as dust. Several manufacturers have developed reduced-vibration jack hammers, and antivibration gloves meeting the ISO 10819 standard can provide some protection (Wasserman and Wasserman, 2002). Another problematic task is shoveling, which has been extensively studied (Freivalds, 1986a,b; Degani et al., 1993). New shovel designs with a bent handle or an additional D handle on the shaft can be helpful in reducing the risk of injury for some shoveling tasks (Degani et al., 1993). Many shoveling tasks are being mechanized and done with backhoes or other construction equipment. Mechanization of materials delivery among "bricklayer assistants" or mason tenders has also been shown to reduce the risk of injury (van der Molen et al., 2003b). Several recent studies have identified high-risk tasks for subsets of construction laborers (mason tenders, concrete workers, and highway workers) and developed training materials for them as an ergonomics intervention (OIOC, 2000, 2003).

50.4.2.11   Equipment Operators

This group of workers, also called operating engineers, have a particular set of exposures different from most other trades. While they do not do the manual handling that many other trades do, they operate heavy equipment all day and are exposed to whole-body vibration, awkward seating postures related in some cases to viewing their work, and risks of injury while dismounting and maintaining the equipment. An ergonomic assessment has been published of an excavator operator showing repetitive hand–arm movements to be a problem (Buchholz et al., 1997). A study of whole-body vibration among equipment operators tested 14 different equipment types and found eight types that exceeded ISO vibration standards (Cann et al., 2003). Zimmermann et al. (1997) note that work environment interventions are potentially more feasible and effective for operating engineers because they spend nearly all of their time in an environment designed by the equipment manufacturer. Their recommendations include improved cab design with better seating, reduced vibration, better designed controls, and regular breaks. One study of newly designed equipment showed significantly reduced stress to the operators (Andersson et al., 1995). A checklist for evaluating cab design has been published recently (Kittusamy, 2003). One evaluation showed that improvements in the design of the equipment reduced workers' compensation premiums significantly and produced significant savings to the contractor (Zimmermann and Cook, 1999).

50.5 Exposure Assessment Methods

Three dimensions of exposure — frequency, duration, and magnitude of job duties or tasks — can increase a worker's risk of sustaining a musculoskeletal injury. Assessment of these variables is more challenging in construction than in most fixed industries. Construction jobs and tasks tend to be less repetitive and have longer cycle times than those typical of manufacturing and assembly work. Whereas a highly repetitive factory job might be adequately analyzed from a brief observation or video recording of a few job cycles, a construction worker's day might consist of a variety of tasks with some cycle times that are hours long (Buchholz et al., 1996). The longer cycle times and irregular pattern of many construction tasks have led researchers to develop other methods of exposure assessment.

Exposure assessment tools range from the simple to the complex. A construction site safety coordinator might use a relatively short checklist to observe tasks that are thought to include typical MSD risk factors (Hollman et al., 1999). The observer could estimate the percentage of time the worker spends in particular awkward postures or how many times per minute he or she performs a specific operation. Perception surveys can provide self-reported data on symptoms, perceived exertion, or specific tasks, tools, or equipment that cause discomfort or pain (Wiktorin et al., 1993). A good starting point for an ergonomic evaluation of construction tasks is the risk factor analysis done by Everett (1997) of 65 tasks of 15 different trades. These ratings provide an initial screening tool to prioritize where the greatest risks might be.

Beyond a checklist, several observational tools have been developed to provide more detailed, accurate, and reliable measures of exposure in the field. These allow sampling of task variables such as postures, tool use, loads handled, grasp type, and activity frequencies. Observational methods entail systematic recording of actual work activities, at specific intervals over a representative time period. Some methods employ trained observers recording data on forms during real time, while others rely on video-computer analysis of work postures and activities. One such observational method is the Ovako working posture-analyzing system (OWAS), a Finnish postural analysis tool (Matilla et al., 1993). ARBAN, a systematic method that married OWAS with video for analysis of construction tasks, was developed in Sweden (Wagenheim et al, 1986). The Posture, Activities, Tools and Handling (PATH) instrument is another sampling-based job analysis method based on OWAS and developed specifically for evaluating construction jobs (Buchholz et al., 1996). Observational methods are particularly effective for assessing awkward postures and manual material handling simultaneously in a dynamic work situation, and they are relatively inexpensive, though labor intensive. Reliability is always an issue when evaluating job tasks with high variability, and especially so when using observational methods. For example, PATH has been shown to have high inter- and intra-observer agreement for the trunk and lower extremity but not for evaluations of the neck (Buchholz et al., 1996).

Most of the rigorous attempts to evaluate risk factors through collection of quantitative biomechanical data have been conducted in laboratories with subjects doing simulated construction tasks. Researchers have looked at repetitive wrist motions during hand tool use, awkward shoulder postures during overhead work, whole-body vibration experienced by operating engineers, and the low back force experienced by carpenters and laborers lifting and carrying heavy materials and equipment (Zimmermann et al., 1997; Cook et al., 1998; Spielholz et al., 1998; Chaffin et al., 1999; Anton, et al., 2001; Mirka et al., 2000). These methods provide more detailed, accurate information about risk, but they are expensive to perform, and the applicability of laboratory simulations to actual construction activities has not been documented.

Very recently, ergonomists and researchers have been experimenting with biomechanical assessment tools to provide direct measurement and more precise quantitative data in the field on construction workers doing their actual tasks at the work site. Tools such as surface electromyography (EMG), electrogoniometers, dynamometers, and nerve conduction equipment have been used for many years to conduct laboratory studies. Surface EMG provides information on muscle activity and effort. Electrogoniometers can collect position data over time, which can then be used to calculate velocity and acceleration in three dimensions. Dynamometers can be used to collect force information. Recent advances in technology have made these tools portable, making it possible to collect quantitative measurements on construction workers during actual work activities. For example, electrogoniometers were placed on the wrists of workers using different types of orbital sanders to evaluate changes in wrist position (Spielholz et al., 2001). The lumbar motion monitor (LMM) (Marras et al., 1993), a specialized electrogoniometer, has been used to evaluate dynamic low back activity in laborers moving concrete hoses (Hess et al., 2004). Direct measurement equipment is frequently expensive to purchase, and it can be difficult to use in some work situations where worker activities may preclude wearing the instrument.

Each technique for evaluating construction worker exposure has tradeoffs. Self-reported surveys are easy and inexpensive to administer but they yield the least precise and objective information. Observational measurements are superior to self-reported worker perceptions but they still provide only moderate precision and accuracy, and there is no mechanism for assessing exposure duration. They are less expensive than direct measurements, but they are labor intensive and time consuming because of the large number of measurements needed to describe jobs with high variability. Direct measurement tools provide the most accurate quantitative information but they are expensive in terms of capital investment. Burdorf et al. (1997) provide a useful guide for choosing the appropriate risk factor evaluation tool.

50.6 Organizational, Management, and Logistical Considerations in Construction Ergonomics

When examining the tasks of construction workers from an ergonomics perspective, we quickly recognize that many of the determinants of the conditions under which people work are found upstream from the task itself. While some of these conditions are not controllable, many others are. The following are some categories of upstream factors that bear examination:

- *Material logistics*: The delivery, storage, handling, and transport of construction materials at the project site can have significant bearing on the physical workload of construction employees. The distance from delivery location to point of installation depends on the site layout, planning for a storage or laydown area, and timing of deliveries. Just-in-time delivery can minimize the need for storage space and eliminate or minimize the distance that materials need to be carried, whether manually or mechanically. Wegelius-Lehtonen and Pahkala (1998) analyzed material logistics and delivery on a series of Scandinavian construction sites and developed best practices for use by site superintendents. In addition to horizontal transport considerations, vertical transport must also be planned. Options include loading platforms, material hoists, elevators, stairs, cranes, and forklifts. In some cases these devices may eventually be part of the permanent structure being built, so that installation early in the construction project may benefit ergonomic conditions for workers. Many of these logistics concerns need to be addressed in the planning and design phases of a project, as site layout and scheduling can severely limit the possibilities once the construction work is underway.

    Overexertion in material transfer is a common accident type in construction, accounting for 32% of injuries and 36% of absence days in one large company (Perttula et al., 2003). However, the same study noted that in a database of serious construction accidents, 25% of the serious material handling injuries occurred during mechanical vertical lifts. Thus, the replacement of manual transfers by mechanical transfers can be effective in reducing overexertion injuries, but can introduce its own set of hazards, including struck-by-object and fall incidents.

- *Scheduling and sequencing*: The time allotted for construction and the order in which various scopes of work are undertaken also have significant implications for the ergonomics of construction work. Trade stacking, or workers of multiple crafts working next to or on top of each other, is one consequence of a compressed schedule and poor sequencing. The order in which certain installations are made can affect the access later contractors and trades have to do their own work. For example, electricians and pipefitters may compete for space to route their installations if insufficient space has been allocated and one trade gets in before the other. If sheet metal workers install large-diameter overhead ducts before the electricians get in to route their cable and fixtures, the electricians may be forced into awkward overhead postures by having to work around the ducts.

*   *Housekeeping*: This covers a wide range of items including new materials; waste from demolition, packaging, or excess materials; and the use and storage of power cables, ladders, and other construction equipment. Orderliness on the construction site affects safety, efficiency, and quality. Congestion caused by poorly strung power cords and awkwardly placed ladders can cause traumatic accidents and create access problems making for poor ergonomics. Poorly planned housekeeping can also lead to materials being moved around multiple times, thereby increasing the exposure of employees to material handling related injuries.

*   *Management*: Most aspects of the preceding three areas are the domain of construction management and require planning well before construction and consistent monitoring and feedback during construction. The general contractor or construction manager must play a coordinating role among the multiple subcontractors who work on projects of any size. Potential problems with materials logistics, housekeeping, and scheduling and sequencing are multiplied in the multi-employer environment and can be minimized only through a strong planning, coordination, and communication effort led by management.

50.7   Construction Design and Ergonomics

Architects and engineers commonly design facilities, whether buildings, roads, or bridges, with the end user in mind rather than the people who will construct the facility. Safety, and less frequently human factors, are major considerations in just about any design because of building and fire codes, environmental or safety and health regulations, contract specifications, and other government or owner directives. There is now broad acceptance that design, whether of buildings, machinery, or vehicles, should consider ergonomics and human factors in order to optimize the safety and comfort of the user (Christensen and Manuele, 1999). Unfortunately, studies of how well safety through design is practiced suggest that this ideal is more easily stated than followed. Wulff et al. (1999a,b) investigated the implementation of ergonomic requirements in a large engineering design firm designing offshore installations for the oil industry. The researchers found that many designers had limited knowledge of human factors, accepted little responsibility for addressing human factors in design, and as a result did not implement human factors considerations. Persons in lead positions tended to have greater knowledge and appreciation of inclusion of human factors in the design, but numerous significant barriers were identified in the design field as a whole. Designers complained of time pressures, lack of specificity of human factors requirements, their own unfamiliarity with human factors requirements, and cost constraints. Among the researchers' conclusions are that at least three measures are needed for facilitating the inclusion of human factors in design: (1) external pressure, for example, from the customer or a government body; (2) formal procedural requirements incorporated into design documents; and (3) an active human factors resource person integrated in the design organization (Wulff et al., 1999a).

If design for ergonomics/human factors is this challenging when designing for end users, it has traditionally received even less attention for construction workers themselves. Some of the same barriers listed earlier apply to the construction phase with the addition of designers' concerns about assuming liability for their designs if a worker is injured on the job. Still there is growing recognition that design decisions impact construction ergonomics and safety and cannot be ignored (Hislop, 1999; Hecker and Gambatese, 2003). A study of ergonomics training and task interventions on an industrial construction site identified numerous instances in which design choices and material specifications created material handling hazards that were difficult to remedy at the worker or contractor level (Hecker et al., 2001). Two European studies suggest that from 47% (Gibb et al., 2003) to 60% (European Foundation for the Improvement of Living and Working Conditions, 1991) of construction accidents could have been eliminated, reduced, or avoided if different choices had been made prior to the construction process itself. While it is methodologically challenging to identify specific cause and effect relationships, the weight of evidence suggests that design and other preconstruction activities do significantly influence accidents.

In the case of musculoskeletal injuries, material handling and storage, access, and specification of materials require particular attention. Since these factors can affect ergonomics and safety during maintenance, operations, and retrofitting of a facility, as well as during construction, it is advisable to take a life-cycle approach in looking at the design. Examples of design elements that can address some of these risks include:

- Designing abundant fall protection tie-off points in structural steel and concrete components so that they can easily be accessed by construction and maintenance workers.
- Specifying coverings for roof openings that will withstand sufficient force to minimize the danger of workers falling through.
- Where fixtures and alarms must be located in high or remote locations specifying low-maintenance units so that need for replacement or servicing is infrequent.
- Designing components that can be prefabricated at ground level or off-site to reduce installation work that would otherwise be done in awkward postures or at elevation.
- Placing fans, chillers, and other components where workers have easy access for installation and maintenance.
- Considering the weight of materials that are specified to insure that where items are too heavy to move manually, the design allows for access by mechanical lifts.

There are clearly ties between design, construction methods, and planning and scheduling that will impact project ergonomics, so it is important that there be communication and knowledge transfer between designers and those experienced in construction management and methods. This is more likely to be found in design-build projects (DBIA, 2003) but, with proper planning and commitment, is possible in traditional design-bid-build project delivery methods as well (Hecker and Gambatese, 2003). New software tools that enable enhanced visualization of building components and processes during the design phase should improve the integration of design and construction (Slaughter, 2003). Hecker et al. (2004) provide a review of the state of the art surrounding the impact of design on construction safety and ergonomics.

50.8 Cognitive Factors in Construction Safety and Ergonomics

Many of the same cognitive factors addressed in human factors analysis in fixed industry are relevant to the construction industry. Among these are psychomotor skills, sensory perceptive skills, affective responses and motivation, attention, learning and memory, language and communication, problem solving, and decision making (Grandjean, 1988). For example, the visual, vestibular, and somatosensory afferent sensory inputs provide information continuously to the central nervous system. This allows workers to maintain upright postural balance by providing information about the body's position and the environment (Shumway-Cook and Woolacott, 1995). Postural balance is related both to slips, trips, and falls, a major concern in the construction industry, and to maintaining proper body mechanics to complete the construction task.

In the United States, the incident rates for falls per 10,000 full-time construction workers are 40.0 for falls to a lower level, 24.0 for same-level falls, and 9.2 for slips or trips without a fall. Among construction workers slips, trips, and falls together are the second event leading to injury behind contact with objects, constituting 23% of the nonfatal injuries involving days away from work (BLS, 2000). Construction workers encounter many situations that demand a combination of balance, attention, and exertion. Hsiao and Simeonov (2001) describe the work of roofers, which includes handling heavy or bulky equipment and materials; temporary work environments with incomplete support surfaces; inclined, compliant, and slippery surfaces; walking on narrow beams; and working close to edges and around openings. Other conditions and factors that can impact the information provided to the worker's sensory systems or the way information is processed in the central nervous system include: personal protective equipment, the attentional demands of the job (Woollacott and Shumway-Cook, 2002), work experience, mental and physical fatigue, age (Gary, 1991), and vision. Kincl et al. (2004) found that job experience increases the accuracy of visual spatial perception, even though age has the opposite effect, while fatigue and inclined work surfaces decrease the accuracy of visual spatial responses. Visual spatial perception along with depth perception, visual ambiguity, and moving visual scenes will affect a worker's postural balance. Overall, some cognitive ergonomic aspects of construction work must be addressed to reduce the risk of both traumatic and chronic musculoskeletal injury.

Advances in education and training technology may play an important role in better preparing construction workers in the areas of cognitive ergonomics. Such advanced training technologies are a major area of development in occupational safety and health training (Hudock, 1994). Interactive methods of training, such as virtual reality, have been used in the medical fields and are currently being used in mining safety and health training (Filigenzi, 2000). A European Union project, Virtual Environments for Construction Workers' Instruction and Training (VECWIT), has developed and tested interactive 3D graphics tools for training construction workers on the hazards of scaffolding and working at heights (Arcangeli et al., 1999a, b). Virtual reality also can be used for occupational research purposes. It has been demonstrated that virtual reality simulations are adequate for studying the risk factors of occupational slips and falls (Simeonov et al., 2002). NIOSH researchers are using virtual reality technologies to investigate the safety of workers on scaffolding and roofers on compliant or sloped surfaces (Dotson and Hsiao, 2000). Further research and improved interactive training using advanced technologies such as virtual reality may address some cognitive ergonomic issues in construction.

50.9   Challenges of Ergonomic Interventions in Construction

Intervention to improve ergonomics in construction takes a variety of forms. The intervention matrix displayed in Figure 50.10 is one representation (Hecker et al., 2001). It reflects the complexity of the industry in several ways. Construction projects bring together multiple parties for a limited time, and some of those parties do not even interact concurrently. Therefore, the project is one level of intervention while the individual organizations, that is, contractors, owners, suppliers, unions, designers, etc., constitute another locus or level of intervention. The matrix demonstrates that some interventions are within the control of individual construction workers. For instance, workers often can select and purchase their own hand tools, so individuals can make choices for improved ergonomic design of hammers, screwdrivers, metal snips, etc. Power tools and heavier equipment, however, are normally purchased by the contractor. Workers may influence the contractor's decisions, but they do not have extensive control over these choices.

Earlier we discussed the importance of planning and sequencing to avoid conflicts between different trades working in the same space. While workers and foremen of different crews can resolve some conflicts over limited amounts of space at the time of the work, their degrees of freedom are often severely constrained by the time the work has commenced. Everyone from architects and engineers to the general contractor and project owner may have responsibilities for decisions that determine these workers' conditions of work. Intervention may be required at all of these levels to change the ergonomics of the situation.

Similarly, the materials that workers work with influence risks. Four-foot wide drywall panels come in lengths from 8 to 20 ft. Thicker panels of 20-ft length can weigh as much as 200 pounds each. Clearly, the

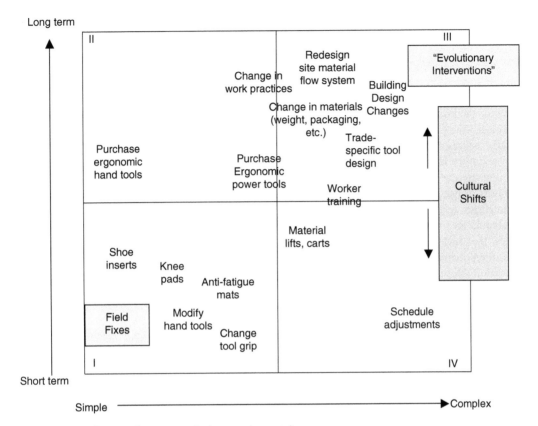

FIGURE 50.10 Construction ergonomics intervention matrix.

physical stresses of heavier and bulkier panels are greater than shorter and lighter ones for the workers installing the panels. However, there is a tradeoff in this case because the workers who finish the walls by taping the joints will have more joints to tape with shorter panels. Similarly, there are differences in exposure to masons depending on the size and weight of block that they lay.

Packaging and availability of materials also can affect the ergonomics of construction work. The packaging of mortar or cement in 90-pound bags poses greater physical demands than 45- or 60-pound bags, but that is a manufacturer or supplier decision, perhaps influenced by the market. A light-weight concrete building block has been developed that is easier to handle but has structural and appearance properties equal to a traditional heavier block. Such a block may cost more but productivity gains could offset the cost. However, the common aggregates used to manufacture light-weight blocks are not readily available in all parts of the country, making the cost of lightweight block noncompetitive in those regions.

Sometimes exposures can be better controlled by changing the location of the work through prefabrication. On one project a system of large steel ceiling trusses was prefabricated off-site and delivered on large flatbeds so that less welding and connecting work needed to be done at heights when the trusses were put in place. In the construction of the first tunnel of Boston's "Big Dig" project workers had to drill 17,000 holes in the concrete ceiling to install supports for the ceiling ventilation system. This exposed workers both to silica dust and to postural stresses from overhead work. The tunnels were formed from prefabricated "tube tunnel sections." When theses hazards were recognized, the sections for subsequent tunnels were fabricated with embedded supports to eliminate the vast majority of the drilling. Both ergonomic and respiratory risks were dramatically reduced, and the redesign led to a greater than 60% reduction in the cost per square foot of the ventilation system (Moir, 1999).

Though not, strictly speaking, an ergonomic intervention, increasing numbers of construction employers are introducing prework stretching programs in an effort to control the incidence of MSDs in their work forces (Simonson and Iannello, 1994). Systematic data are lacking, but it appears likely from anecdotal evidence and the trade press that stretching programs are more common than true ergonomics programs in the U.S. construction industry. These programs vary but usually include 10 min of generalized whole-body stretching or a combination of warm-ups, aerobics, and stretching. A recent review of the literature found no studies that justify the use of stretching alone for the prevention of workplace injury (Hess and Hecker, 2003). Moreover, existing studies fail to establish that stretching actually reduces injuries, and some authors go so far as to suggest that being more flexible may actually increase injury risk (Biering-Sorensen, 1984). A recent Swedish study of a construction site 10-min morning exercise program, which included warm-ups, aerobics, and leg stretches, found that after 3 months of exercise the experimental group had greater trunk, hip flexor, and hamstring flexibility (Holmstrom and Ahlborg, 2005). However, no association between levels of flexibility and injury rates could be drawn from this study. Self-report survey data and articles in trade publications suggest that such programs are popular with workers, but more rigorous studies are needed to truly assess the potential benefits of stretching and other workplace-based exercise programs in MSD injury prevention (Hecker and Gibbons, 1997; Smith, 1990).

50.9.1 Intervention Process

Participatory ergonomics, defined in various ways, has become popular in many industries (Noro and Imada, 1991). The complex and fragmented structure of the construction industry necessitates such a process, though at the same time it also presents obstacles (Moir and Buchholz, 1996). The multiple parties to the construction process, the various levels at which intervention may be needed, the craft nature and tradition of much of construction labor, and the rapidity with which the environment may change all argue for the participation of workers, contractors, tool and material manufacturers and suppliers, and designers in identifying and implementing ergonomic improvements.

The process of changing work methods in construction is an interaction of technology, work organization, management, and the workers themselves. Many craft workers develop their work practices through a socialization process in which apprentices absorb from more experienced members a craft identity and attitude toward problem solving as well as the use of tools and work methods (Jensen and Kofoed, 2002). For a new tool or piece of equipment to be readily adopted it must be easy to use and easy to learn to use. Otherwise, potential long-term benefits may never be realized because the innovation may not be given a fair chance due to the time pressures of the job and time required for familiarization (Cedarqvist and Lindberg, 1993; Jensen and Kofoed, 2002). Evidence from construction and other industries suggests that the involvement of end users is a key to successful implementation of ergonomic changes (Noro and Imada, 1991; Moir and Buchholz, 1996; de Jong et al., 2003).

Moir and Buchholz (1996) provide additional compelling reasons why a participatory approach may be essential in construction. The dynamic nature of the workplace, in which workstations are regularly constructed and deconstructed, requires that those doing the work be intimately involved in decisions about and implementation of ergonomic changes. Construction workers have greater autonomy than most other groups of employees and because their workplace changes so constantly, solving problems is an integral part of their job. While this talent is most often applied to solving production problems, it is equally applicable to ergonomic problems. Finally, it is important to include the economic and cultural values of both workers and contractors in designing interventions to improve their chances of acceptance (Moir and Buchholz, 1996).

There is ample support in the community health intervention literature for the importance of participation in implementing ergonomic innovations with construction workers. Rogers (1995) posits five characteristics of innovations that will affect their rate of acceptance and adoption. Empirical data as to the *relative advantage* of the innovation is one factor. *Compatibility* is the extent to which the innovation is compatible with the norms and practices of the system to which it is introduced. Where innovators might adopt an innovation even in the light of uncertain information about comparative advantage, low levels of compatibility may be a sufficient deterrent to innovation adoption. The craft structure of many segments of the construction industry makes this particularly critical, in that new tools or processes that are seen to deskill, automate, or otherwise interfere with the practice of the craft will likely provoke resistance from the skilled trades (Schneider, 1995). The other three factors, *complexity, trialability,* and *observability,* all have relevance to the construction work force and industry. The fast pace and time constraints of construction projects mean that typically there is insufficient time for a complex innovation to be learned or tried. This, of course, reduces the observability of the benefit of the innovation.

While greater participation is usually seen as enhancing the likelihood of ergonomic changes being appropriate to the task and thus more likely to be adopted, gaining this participation is a challenge. More and more construction projects are now built on a fast-track schedule. Time away from production for craft workers is not easily granted so that ergonomics practitioners interested in using participatory methods need to be creative and specific in their approach. In a recent intervention study with concrete laborers, the ergonomic innovation being considered was skid plates, 2-ft-diameter steel disks that were placed under junctions in the concrete hose through which concrete was pumped (Hess et al., 2004). The skid plates reduced the frequency of the quick-release latches holding the hose sections together becoming caught on the rebar mat, and generally reduced friction in pulling the hose across the surface. The research team worked with both the project superintendent and the labor crew to introduce the skid plates. Researchers presented a 1-h introduction to ergonomic principles and discussed potential innovations that could reduce the physical strain of the laborers' job. This took place immediately after a work shift, and workers were paid 1 h of overtime and provided with food and beverages. Assessment of the effects of the skid plates on various low back biomechanical variables took place during normal work tasks, and researchers collected perceptual feedback from workers informally during breaks and lunch. A closing 1.5-h session was held after the intervention took place, again after the work shift. Here the researchers used videotape of the crew members themselves to demonstrate good body mechanics and to present the quantitative results of the assessment. All of this took place within a 6-week period and required 2.5-h of overtime pay for approximately nine crew members.

Contractors and safety professionals can do this type of intervention development at company safety meetings or other gatherings. Alternatively, new tools or work practices can be introduced through training programs for apprentices and journeymen at apprenticeship training centers or other construction training venues. These, however, are more difficult to coordinate with introduction of the innovation on actual worksites if cooperation from contractors is not forthcoming.

50.10 Sources for Ergonomic Solutions in Construction

In recent years as more attention has been focused on developing solutions and evaluating them in the field, a number of databases of ergonomic innovations have emerged. There is a wide range in terms of how rigorously these innovations have been tested, but that is simply the current state of the art. Table 50.3 lists the website addresses that are discussed in this section. Washington State has developed a solutions database on their website and has concluded numerous demonstration projects, including many in construction, as a part of the implementation of their ergonomics standard prior to its repeal in 2003 by a voter referendum. Reports of all these demonstration projects are available on their website. NIOSH has been developing a best practice guide for construction ergonomics with "tip sheets" that is expected to be completed in 2006 and will be available on the agency's website.

The Laborers' Health and Safety of North America has numerous tip sheets on its website on potential interventions for many construction laborer subtrades such as asbestos abatement, lead abatement, plasterer tender, and hazardous waste workers. Other factsheets on potential interventions in construction can be found at the websites of the following agencies and centers: Worksafe Western Australia, the

TABLE 50.3   Construction Ergonomics Websites

Washington State Department of Labor and Industries
   www.lni.wa.gov/Safety/Topics?HazardInfo/Ergonomics/default.asp
The Laborers' Health and Safety of North America
   www.lhsfna.org
Worksafe Western Australia
   www.safetyline.wa.gov.au/default.htm
Canadian Center for Occupational Safety and Health
   www.ccohs.org/oshanswers/ergonomics/
Ohio Bureau of Workers Compensation Construction Best Practices Guide
   http://www.ohiobwc.com/downloads/brochureware/publications/ConstSafeGrant.pdf/
British Columbia Workers Compensation Board
   www.worksafebc.com/publications/health_and_safety_information/bulletins
   /constructive_ideas/default.asp/
University of Massachusetts at Lowell
   www.uml.edu/Dept/WE/COHP/Documents/bridea.htm
Arbouw
   www.arbouw.nl
State of Oregon
   www.cbs.state.or.us/external/osha/consult/ergonomic/const_ergo.html

---

Canadian Center for Occupational Safety and Health, the Ohio Bureau of Workers' Compensation Construction Best Practices Guide, and the British Columbia Workers' Compensation Board's "Constructive Ideas." The University of Massachusetts at Lowell has developed several best-practice worker-initiated solution tip sheets called Bright Ideas, available on their website. Arbouw has also produced numerous best-practice guides for various construction sectors (in Dutch) called "A Blads," which include recommendations for ergonomics interventions. Finally, the State of Oregon also has developed a comprehensive construction ergonomics website with solutions information.

50.11   Conclusions

Ergonomics offers valuable tools to improve the safety and health of construction workers and to improve construction efficiency. These improvements range from the use of ergonomically improved hand tools to system-level innovations to manage the logistics of construction materials. The fragmented and highly variable nature of the construction industry undeniably poses challenges to the application of ergonomic analysis and redesign, but the last decade has seen a significant increase in attention to ergonomic issues in construction. The personal and financial costs of musculoskeletal injuries, especially in an aging workforce, continue to drive the need for improvements, and advances in ergonomic assessment tools are providing better means to target and prioritize risky trades and operations. A key element of continuing progress in this area is the participation and collaboration of workers, contractors, owners, unions, safety and health professionals, designers, and all other parties to the construction process.

References

Albers J; Estill C; MacDonald L. 2004. *Mechanical and Electrical Construction Trades — WMSD Risk Factors and Interventions*. Cincinnati: National Institute for Occupational Safety and Health.
Anderson P. 1991. Manual screw tightening with and without micropauses. Bhyggalson Bulletin 91-09-16, 39–40.
Andersson B; Norlander S; Wos H. 1995. Evaluation of muscular stress in construction machine operators: an EMG study. *Applied Occupational and Environmental Hygiene* 10(3):161–169.
Anton D; Shibley LD; Fethke NB; Hess J; Cook TM; Rosecrance J. 2001. Effect of overhead drilling position on shoulder moment and electromyography. *Ergonomics* 44(5):489–501.

Arcangeli G; Assfalg J; Bagnara S; Del Bimbo A; Mariani M; Parlangeli O; Tartaglia R; Vicario E. 1999a. Use of virtual reality for instruction and training of young construction workers (European Project "SAFE"). *Proceedings of the International Conference on Computer-Aided Ergonomics and Safety*, Barcelona.

Arcangeli G; Assfalg J; Tartaglia R; Vicario E. 1999b. A virtual environment for construction workers instruction and training. *Proceedings of the IEEE International on Conference on Multimedia Computing and Systems*, Florence, Italy, June 7–11.

Bhattacharya A; Greathouse L; Warren J; Li Y; Dimov M; Applegate H; Stinson R; Lemasters G. 1997. An ergonomic walkthrough observation of carpentry tasks: a pilot study. *Applied Occupational and Environmental Hygiene* 12(4):1204–1278.

Biering-Sorensen F. 1984. Physical measurements as risk indicators for low-back trouble over a one-year period. *Spine* 9:106–119.

Bjorklund M; Helmerskog P; Nordberg M; Saderman U; Holmquist L; Lindblad B; Makynen J; Ahrman S. 1991. 90-sheets for the 90s, *Bygghalsan Bulletin*, 91-09-16.

Buchholz B; Paquet V; Punnett L; Lee D; Moir S. 1996. PATH: a work sampling-based approach to ergonomic job analysis for construction and other non-repetitive work. *Applied Ergonomics* 27:177–187.

Buchholz B; Moir S; Virji MA. 1997. An ergonomic assessment of an operating engineer: a pilot study of excavator use. *Applied Occupational and Environmental Hygiene* 12(1):23–27.

Buchholz B; Paquet V; Wellman H; Forde M. 2003. Quantification of ergonomic hazards for ironworkers performing concrete reinforcement tasks during heavy highway construction. *American Industrial Hygiene Association Journal* 64:243–250.

Burdorf A; Rossignol M; Fathallah FA; Snook SH; Herrick RF. 1997. Challenges in assessing risk factors in epidemiologic studies on back disorders. *American Journal of Industrial Medicine* 32:142–152.

Bureau of Labor Statistics (BLS), U.S. Department of Labor. 2000. Tables R4, R8. www.bls.gov.

Bureau of Labor Statistics (BLS), U.S. Department of Labor. 2001. Lost-worktime injuries and illnesses: characteristics and resulting time away from work, www.stats.bls.gov/iif/home.htm.

Bureau of Labor Statistics (BLS), U.S. Department of Labor. 2003. Industry at a glance: construction, www.bls.gov/iag/construction.htm, December, 2003.

Cann AP; Salmoni AW; Vi P; Eger TR. 2003. An exploratory study of whole-body vibration exposure and dose while operating heavy equipment in the construction industry. *Applied Occupational and Environmental Hygiene* 18:999–1005.

Cederqvist T; Lindberg M. 1993. Screwdrivers and their use from a Swedish construction industry perspective. *Applied Ergonomics* 24:148–157.

Cederqvist T; Örtengren R. 1985. An electromyographic study of the effect of a new tool for ceiling painting on hand grip force, Applications of Biomechanics, *Proceedings of the Precongress Meeting of the X International Congress of Biomechanics*, June 12–15, Linköping, Sweden, pp. 125–126.

Center to Protect Worker's Rights (CPWR). 2002. *The Construction Chart Book, The U.S. Construction Industry and its Workers*, 3rd edition. Silver Springs, MD: CPWR.

Chaffin DB; Andersson GBJ; Martin BJ. 1999. *Occupational Biomechanics*, 3rd edition. New York, NY: Wiley-Interscience.

Chini AR; Brown BH; Drummond EG. 1999. Causes of the construction skilled labor shortage and proposed solutions. *ASC Proceedings of the 35th Annual Conference*, pp. 187–196.

Christensen WC; Manuele FA. 1999. *Safety Through Design*. Chicago, IL: NSC Press.

Chiou SS; Pan CS; Keane P. 2000. Traumatic injury among drywall installers, 1992 to 1995. *Journal of Occupational and Environmental Medicine* 42(11):1101–1108.

Construction Labor Research Council (CLRC). 1997. *Craft Labor Supply Outlook, 2001–2010. Confronting the Skilled Construction Work Force Shortage, A Blueprint for the Future*. Washington, DC: The Business Roundtable.

Construction Safety Association of Ontario (CSAO). 2002. *Injury Atlas: Ontario Construction*. Toronto: Construction Safety Association of Ontario.

Cook TM; Rosecrance JC; Zimmermann CL. 1996. Work-related musculoskeletal disorders in bricklaying: a symptom and job factors survey and guidelines for improvements. *Applied Occupational and Environmental Hygiene* 11(11):1335–1339.

Cook T; Rosecrance J; Zimmermannn C; Gerleman D; Ludewig P. 1998. Electromyographic analysis of a repetitive hand gripping task. *International Journal of Occupational Safety & Ergonomics* 4(2): 185–198.

Cutlip R; Hsiao H; Garcia R; Becker E; Mayeux B. 2000. A comparison of different postures for scaffold end-frame disassembly. *Applied Ergonomics* 31:507–513.

Damlund M; Goth S; Hasle P.; Munk K. 1986. Low back strain in Danish semi-skilled construction work. *Applied Ergonomics* 17(1):31–39.

Degani A; Asfour SS; Waly SM; Koshy JG. 1993. A comparative study of two shovel designs. *Applied Ergonomics* 24(5):306–312.

de Jong AM; Vink P; De Kroon, JCA. 2003. Reasons for adopting technological innovations reducing physical workload in bricklaying. *Ergonomics* 46(11):1091–1108.

Design-Build Institute of America (DBIA). 2003. *Safety Opportunities Under Design-Build.* Washington, DC: Design-Build Institute of America, Publication #704.

Dimov M; Bhattacharya A; LeMasters G; Atterbury M; Greathouse L; Ollila-Glenn N. 2000. Exertion and body discomfort perceived symptoms associated with carpentry tasks: an on-site evaluation. *American Industrial Hygiene Association Journal* 61(5):685–691.

Dong X; Platner JW. 2004. Occupational fatalities of Hispanic workers from 1992-2000. *American Journal of Industrial Medicine* 45:45–54.

Dotson BW; Hsiao H. 2000. Safe work at elevation through virtual reality simulation. NOIRS 2000 — *Abstracts of the National Occupational Injury Research Symposium 2000*, Pittsburgh, PA, October 17–19. National Institute for Occupational Safety and Health.

Erneling L; Kling J. 1983. Easy-to-handle packagings at building sites. *Bygghalsan Bulletin* 1983–05–01.

European Foundation for the Improvement of Living and Working Conditions. 1991. *From Drawing Board to Building Site.* London: HMSO Books.

Everett JG. 1997. *Ergonomic Analysis of Construction Tasks for Risk Factors for Overexertion Injuries.* Ann Arbor, MI: Center for Construction Engineering and Management.

Filigenzi MT; Orr TJ; Ruff TM. 2000. Virtual reality for mine safety training. *Applied Occupational and Environmental Hygiene* 15(6):465–469.

Freivalds A. 1986a. The ergonomics of shoveling and shovel design — an experimental study. *Ergonomics* 29(1):19–30.

Freivalds A. 1986b. The ergonomics of shoveling and shovel design — a review of the literature. *Ergonomics* 29(1):3–18.

Gary A. 1991. Ergonomics of the older worker: an overview. *Experimental Aging Research* 17(3):143–155.

Gibb A; Haslam R; Hide S; Gyl D. 2003. The role of design in accident causality. Paper presented at the *Designing for Safety and Health in Construction Symposium*, Portland, Oregon, September 15–16.

Gilbreth FB. 1974. *Bricklaying System.* Easton: Hive Publishing Company.

Grandjean E. 1988. *Fitting the Task to the Man: A Textbook of Occupational Ergonomics*, 4th edition. London, Taylor & Francis.

Hammarskjöld E. 1987. Blocks and bricks of masonry — weight recommendations — a new Swedish standard. *Bygghalsan Bulletin*, 1987-05-01.

Hecker S; Gambatese J. 2003. Safety in design: a proactive approach to construction safety and health. *Applied Occupational and Environmental Hygiene* 18(5):339–342.

Hecker S; Gibbons W. 1997. Evaluation of a prework stretching program in the construction industry, *Proceedings of the 13th Triennial Congress of the International Ergonomics Association*, Vol. 6, pp. 115–117. Helsinki: Finnish Institute of Occupational Health.

Hecker S; Gibbons B; Barsotti A. 2001. Making ergonomic changes in construction: worksite training and task interventions. In Alexander D; Rabourn R, eds. *Applied Ergonomics*, pp. 1162–1189. New York: Taylor & Francis.

Hecker S; Gambatese J; Weinstein M. 2004. *Designing for Safety and Health in Construction*. Eugene, OR: University of Oregon Press.

Helander M. 1981. *Human Factors/Ergonomics for Building and Construction*. New York: John Wiley.

Hess JA; Hecker S. 2003. Stretching at work for injury prevention: issues, evidence and recommendations. *Applied Occupational and Environmental Hygiene* 18(5):331–338.

Hess JA; Hecker S; Weinstein M; Lunger M. 2004. A participatory ergonomics intervention to reduce risk factors for low-back disorders in concrete laborers. *Appl Ergon* 35(5):427–41.

Hinksman J. 1997. Construction: major sectors. Chapter 93: 93.1–93.52, in Stellman J, ed., *ILO Encyclopedia of Occupational Health and Safety*. Geneva: International Labour Office.

Hislop R. 1999. *Construction Site Safety: A Guide for Managing Contractors*. Boca Raton, FL: Lewis Publishers.

Hollmann S; Klimmer F; Schmidt KH; Kylian H. 1999. Validation of a questionnaire for assessing physical work load. *Scandinavian Journal of Work, Environment & Health* 25:105–114.

Holmström E; Ahlborg B. 2005. Morning warming-up exercise-effect on musculoskeletal fitness in construction workers. *Applied Ergonomics* 36(4):513–519.

Holmström E; Engholm G. 2003. Musculoskeletal disorders in relation to age and occupation in Swedish construction workers. *American Journal of Industrial Medicine* 44(4):377–384.

Hsiao H; Simeonov P. 2001. Preventing falls from roofs: a critical review. *Ergonomics* 44(5): 537–561.

Hsiao H; Stanevich RL. 1996. Biomechanical evaluation of scaffolding tasks. *International Journal of Industrial Ergonomics* 18:407–415.

Hsiao H; Bradtmiller B; Whitestone J. 2003. Sizing and fit of fall-protection harnesses. *Ergonomics* 46(12):1233–1258.

Hudock SD. 1994. The application of educational technology to occupational safety and health training. *Occupational Medicine: State of the Art Reviews* 9(2):201–210.

Jäger M; Luttmann A; Laurig W. 1991. Lumbar load during one-handed bricklaying. *International Journal of Industrial Ergonomics* 8:261–277.

Jensen L; Mikkelsen S; Loft I. 2000. Work-related knee disorders in floor layers and carpenters. *Journal of Occupational and Environmental Medicine* 42: 835–842.

Jensen K; Kofoed LB. 2002. Musculoskeletal disorders among floor layers: is prevention possible? *Applied Occupational and Environmental Hygiene* 17:797–806.

Kincl L; Bhattacharya A; Bagchee A; Succop P. 2002/2003. The effect of workload, work experience and inclined standing surface on visual spatial perception: fall potential/prevention implications. *Occupational Ergonomics* 3(4):251–259.

Kittusamy NK. 2003. A checklist for evaluating cab design of construction equipment. *Applied Occupational and Environmental Hygiene* 18:721–723.

Kivimäki J. 1992. Occupationally related ultrasonic findings in carpet and floor layers' knees. *Scandinavian Journal of Work, Environment & Health* 8:400–402.

Koningsveld EAP; van der Molen HF. 1997. History and future of ergonomics in building and construction. *Ergonomics* 40(10):1025–1034.

Koningsveld EAP; Vink P; Urlings IJM; de Jong AM. 1998. *Reducing Sprains and Strains in Construction through Worker Participation: A Manual for Managers and Workers with Examples from Scaffold Erection*. Washington, DC: The Center to Protect Workers' Rights.

Lappalainnen J; Kaukianen A; Sillenpaa J; Vijanen M; Roto P. 1998. Effects of Gyproc ERGO plasterboard on the health and safety of workers — pilot study. *Applied Occupational and Environmental Hygiene* 13(10):698–703.

Lipscomb HJ; Kalat J; Dement JM. 1996. Workers' compensation claims of Union Carpenters 1989–1992: Washington state. *Applied Occupational and Environmental Hygiene* 11(1):56–63.

Lipscomb HJ; Dement JM; Loomis DP; Silverstein B; Kalat J. 1997. Surveillance of work-related musculoskeletal injuries among union carpenters, *American Journal of Industrial Medicine* 32(6):629–640.

Libscomb HJ; Dement JM; Gaal JS; Cameron W; McGougall V. 2000. Work-related injuries in drywall installation. *Applied Occupational and Environmental Hygiene* 15(10):794–802.

Luttmann A; Jager M; Laurig W. 1991. Task analysis and electromyography for bricklaying at different wall heights. *International Journal of Industrial Ergonomics* 8:247–260.

Marras WS; Lavender SA; Leurgans SE; Rajulu SL; Allread WG; Fathallah FA; Ferguson SA. 1993. The role of dynamic three-dimensional trunk motion in occupationally related low back disorders. *Spine* 18:617–628.

Matilla M; Karwowski W; Vilkki M. 1993. Analysis of working postures in hammering tasks on building construction sites using the computerized OWAS method. *Applied Ergonomics* 24: 405–412.

Messing K. 2000, Ergonomic studies provide information about occupational exposure differences between women and men. *Journal of the American Medical Women's Association* 55(2):72–75.

Mirka GA; Kelaher DP; Nay T; Lawrence BM. 2000. Continuous assessment of back stress (CABS): a new method to quantify low-back stress in jobs with variable biomechanical demands. *Human Factors* 42:209–225.

Mirka GA; Monroe M; Nay T; Lipscomb H; Kelaher D. 2003. Ergonomic interventions for the reduction of low back stress in framing carpenters in the home building industry. *International Journal of Industrial Ergonomics* 31:397–409.

Moir S. 1999. Big Dig. Electronic Library of Construction Occupational Safety and Health (eLCOSH), http://www.cdc.gov/elcosh/docs/d0100/d000067/d000067.html.

Moir S; Buchholz B. 1996. Emerging participatory approaches to ergonomic interventions in the construction industry. *American Journal of Industrial Medicine* 29:425–430.

Noro K; Imada AS (eds.). 1991. *Participatory Ergonomics*. London: Taylor & Francis.

Occupational and Industrial Orthopaedic Center (OIOC). 2000. *Ergonomics Manual, Mason Tenders*. New York: OIOC.

Occupational and Industrial Orthopedic Center (OIOC). 2003. *Ergonomics Working for Cement and Concrete Construction Laborers*. New York: OIOC.

Perttula P; Merjama J; Kiurula M; Laitinen H. 2003. Accident in material handling at construction sites. *Construction Management and Economics* 21:729–736.

Ringen K; Seegal J; Weeks J. 1998. Construction: health, prevention and management, in Stellman J, ed., *ILO Encyclopedia of Occupational Health and Safety*, Chapter 93. Geneva: International Labour Office.

Rogers EM, 1995. *Diffusion of Innovations*, 4th edition. New York, NY: The Free Press.

Rosecrance JC; Cook TM; Zimmermannn CL. 1996. Work-related musculoskeletal symptoms among construction workers in the pipe trades. *Work* 7:13–20.

Rosecrance JC; Cook TM; Anton DC; Merlino LA. 2002. Carpal tunnel syndrome among apprentice construction workers. *American Journal of Industrial Medicine* 42(2):107–116.

Schneider SP. 1995. Implement ergonomic interventions in construction. *Applied Occupational and Environmental Hygiene* 10(10):822–824.

Schneider SP. 2001. Musculoskeletal injuries in construction: a review of the literature. *Applied Occupational and Environmental Hygiene* 16(11):1056–1064.

Schneider SP; Susi P. 1994. Ergonomics and construction: a review of potential hazards in new construction. *American Industrial Hygiene Association Journal* 55(7):635–649.

Schneider SP; Griffin M; Chowdhury R. 1998. Ergonomic exposures of construction workers: an analysis of the U.S. Department of Labor Employment and Training Administration database on job demands. *Applied Occupational and Environment Hygiene* 13(4):238–241.

Shaw G. 2003. New tools for drywall finishers may reduce injury risk. Mid-State Central Labor Council (NY). Electronic Library of Construction Occupational Safety and Health (eLCOSH), http://www.cdc.gov/elcosh/docs/d0100/d000067/d000067.html.

Shumway-Cook A; Woollacott M (eds.). 1995. Control of posture and balance, in *Motor Control: Theory and Practical Applications*, Chapter 6, pp. 119–142. Baltimore, MD: Williams & Wilkins.

Simeonov P; Hsiao H; Dotson B; Ammons D. 2002. Comparing standing balance at real and virtual elevated environments. *Proceedings of the Human Factors and Ergonomics Society 46th Annual Meeting*, September–October, pp. 2169–2173.

Simonson BW; Iannello P. 1994. Company's exercise program mobilizes its industrial athletes before work. *Occupational Health and Safety*, September. 44–45.

Slaughter ES. 2003. The link between design and process: dynamic process simulation models of construction activities, in Raja RA; Issa IF; O'Brien WJ; eds., *4D CAD and Visualization in Construction: Developments and Applications*. Exton, PA. Swets & Zeitlinger/A.A. Balkema.

Smith RB. 1990. Work-place stretching programs reduce costly accidents, injuries. *Occupational Health and Safety* March:24–25.

Spielholz P; Wiker SF; Silverstein B. 1998. An ergonomic characterization of work in concrete form construction. *American Industrial Hygiene Association Journal* 59:629–635.

Spielholz P; Bao S; Ninica H. 2001. A practical method for ergonomic and usability evaluation of hand tools: a comparison of three random orbital sander configurations. *Applied Occupational and Environmental Hygiene* 16(11):1043–1048.

StewartPrezant Ergonomics Group. 2003. *Ergonomic Analysis of Concrete Screeding and Finishing Techniques*. Washington, DC: NIOSH.

Suprenant BA. 1990. Tower scaffolding increases productivity by 20 percent. *The Magazine of Masonry Construction*, January.

Tanaka S; Habes D. 1990. *Request for Assistance in Preventing Knee Injuries and Disorders in Carpet Layers.* NIOSH Alert #90–104.

Tanaka S; Smith AB; Halperin W; Jensen R. 1982. Carpetlayer's knee [letter to the editor]. *New England Journal of Medicine* 307:1275–1276.

Tanaka S; Lee ST; Halperin WE; Thun M; Smith AB. 1989. Reducing knee morbidity among carpetlayers. *American Journal of Public Health* 79(3):334–335.

Thun M; Tanaka S; Smith AB; Halperin WE; Luggen ME; Hess EV. 1987. Morbidity from repetitive knee trauma in carpet and floorlayers. *British Journal of Industrial Medicine* 44:611–620.

TNO-Netherlands Organization for Applied Scientific Research. 1993. Laying the foundations for better brickwork. *Applied Research*, number 52.

United Brotherhood of Carpenters (UBC). 1995. *Ergonomics for Carpenters*. Washington, DC: United Brotherhood of Carpenters Health and Safety Fund of North America.

van der Molen HF; Delleman NJ. 1997. Arbouw guidelines on physical workload for the construction industry. Abstracts from the *1st International Symposium on Ergonomics in Building and Construction*, Tampere, Finland.

van der Molen HF; Bulthuis BM; van Duivenbooden JC. 1998. A prevention strategy for reducing gypsum bricklayers' physical workload and increasing productivity. *International Journal of Industrial Ergonomics* 21:9–68.

van der Molen HF, Grouwstra R, Kuijer P, Paul FM, Sluiter JK, Frings-Dresen MHW. 2003a. Effect of working height adjustment on the physical work demands during construction work. *Proceedings of the 15th World Congress of the International Ergonomics Association*, Seoul, August.

van der Molen HF; Grouwstra R; Kuijer P; Paul FM; Sluiter JK; Frings-Dresen MHW. 2003b. A Systematic review on the effectiveness of interventions to reduce physical work demands. *Proceedings of the 15th World Congress of the International Ergonomics Association*, Seoul, August.

Vi P. 2000/2001. Getting a grip: better design of concrete blocks can reduce risk of injury. *Construction Safety Magazine* 11(4). Construction Safety Association of Ontario.

Village J; Morrison JB; Layland A. 1993. Biomechanical comparison of carpet-stretching devices. *Ergonomics* 36(8):899–909.

Vink P. 1992. Application problems of a biomechanical model in improving roofwork. *Applied Ergonomics* 23(3):177–180.

Vink P; Koningsveld EAP. 1990. Bricklaying: a step by step approach to better work. *Ergonomics* 33(3):349–352.

Wagenheim M; Samuelson B; Wos H. 1986. ARBAN — a force ergonomic analysis method, in Corlett EN; Wilson JW; Manenica I, eds., *The Ergonomics of Working Postures*, pp. 243–255. London: Taylor and Francis.

Wasserman DE; Wasserman JF. 2002. The nuts and bolts of human exposure to vibration. *Sound and Vibration* 36(1):41–42.

Wegelius-Lehtonen T; Pahkala S. 1998. Developing material delivery processes in cooperation: an application example in the construction industry. *International Journal of Productivity and Economics* 56(1):689–699.

Welch L; Hunting KL; Kellogg J. 1995. Work-related musculoskeletal symptoms among sheet metal workers. *American Journal of Industrial Medicine* 27(6):783–791.

Welch LS; Hunting KL; Nessel-Stephen L. 1999. Chronic symptoms in construction workers treated for musculoskeletal injuries. *American Journal of Industrial Medicine* 36(5):489–501.

Welch L; Goldenhar L; Hunting K. 2000. Women in construction: occupational health and working conditions. *Journal of the American Medical Women's Association* 55(2):89–92.

Wickström G. 1978. Symptoms and signs of degenerative back disease in concrete reinforcement workers. *Scandinavian Journal of Work, Environment & Health* 4(suppl 1):54–58.

Wickström G; Niskanen T; Riihimaki H. 1985. Strain on the back in concrete reinforcement work. *British Journal of Industrial Medicine* 42:233–239.

Wiktorin C; Karlqvust L; Winkel J. 1993. Validity of self-reported exposures to work postures and manual materials handling. *Scandinavian Journal of Work, Environment & Health* 19:208–214.

Woollacott M; Shumway-Cook A. 2002. Attention and the control of posture and gait: a review of an emerging area of research. *Gait and Posture* 16:1–14.

Wulff IA; Westgaard RH; Rasmussen B. 1999a. Ergonomic criteria in large-scale engineering design — I. Management by documentation only? Formal organization vs. designers' perceptions. *Applied Ergonomics* 30:191–205.

Wulff IA; Westgaard RH; Rasmussen B. 1999b. Ergonomic criteria in large-scale engineering design — II. Evaluating and applying requirements in the real world of design. *Applied Ergonomics* 30:207–221.

Zellers KK; Simonton KJ. 1997. An optimized lighter-weight concrete masonry unit: biomechanical and physiological effects on masons. *Safety & Health Assessment & Research for Prevention*. Technical Report Number 45–1-1997.

Zimmermann CL; Cook TM. 1999. Ergonomics, economics and operating engineers, in Singh A; Hinza J; Coble R, eds. *Implementation of Safety and Health on Construction Sites*. Rotterdam: AA Balkema.

Zimmermann CL; Cook TM; Rosecrance JC. 1997. Operating engineers: work-related musculoskeletal disorders and the trade. *Applied Occupational and Environmental Hygiene* 12(10):670–680.

51

Introducing Ergonomics to Developing Countries

David Caple
David Caple & Associates Pty Ltd

51.1 Introduction

This chapter reviews the difficulties associated with the transfer of research findings from industrially advanced countries (IACs) into industrially developing countries (IDCs). The significant differences in culture, religion, environment, health, working conditions, and economics, confound the application of many of the ergonomics principles directly into IDCs. The majority of workplaces remain in IDCs agricultural industry with 71% estimated to be living a subsistence lifestyle (O'Neill, 2000). Ergonomic challenges associated with the transfer of technology into IDCs reflects the requirement to address macroergonomics issues in understanding the needs within the recipient community, rather than just focusing on the microergonomic aspects relating to human–interface design issues.

The future development of ergonomics within IDCs will rest within ongoing collaboration and cooperation between governments, nongovernment organizations (NGOs) such as the United Nations, World Health Organization (WHO), International Labour Organization (ILO), and professional associations such as the International Ergonomics Association (IEA). The establishment of effective communication and development of trust between local communities and those willing to provide assistance, will form for foundation of the holistic approach. There are numerous examples where this has already been undertaken and the opportunities are evident for extensions of this development with goodwill by the individuals from within these organizations.

O'Neill (2000) discusses the extremes of the working environment with 71% of people in IDCs surviving on less than US$1.00 per day. They primarily rely on subsistence agriculture and face the poverty cycle involving low economics, poor health and exposure to disease. The spread of communicable diseases such as AIDS, has contributed towards the challenges of survival within many of the poorest developing countries.

The United Nations together with agencies such as the WHO and the ILO are working with NGO in developing a range of practical initiatives to assist the survival and development of people within these countries. The professional association representing the science of ergonomics, the IEA have promoted a range of initiatives to assist in the promotion and application of ergonomics within developing countries.

51.2   Understanding Developing Countries' Needs

Ahasan (2003) outlines many of the inhibitors to transferring ergonomics knowledge and learnings from industrially advanced countries (IACs) to IDCs. These include:

Resources. These countries have limited resources available in their education sector to enable ergonomics teaching. There are limited numbers of qualified teachers, limited access to books and many of the teaching institutes and governments do not have access to computers, and audio visual equipment to utilize the ergonomics training materials. Further, many of these publications are in the major languages of the world and are not translated into languages or dialects which are able to be used within many of the IDC communities.

Limited research. Despite the interest in ergonomics within developing countries, there has actually been very limited research on the specific needs for ergonomics and the successful application of ergonomics in any major studies. Kogi and Sen (1987) provided early illustrations on how well-designed ergonomics workplaces can be achieved within these countries. However, these are generally viewed in the literature as successful case studies, rather than systematic and sustainable changes.

Reporting of injuries. Researchers such as Mwaura (1992) have indicated that the reporting of work-related injuries in developing countries is poor. Consequently, there are problems of reliability from the available data as to the incidents and severity of work-related injuries. Without this data, it is difficult to develop baseline measures from which improvements can subsequently be compared.

51.3   Occupational Health and Safety Legislation

In many developing countries, legislation which would support the introduction of ergonomics is either nonexistent, or poorly enforced. Ahasan (2003) describes how developing countries that do not have government backed legislation which is enforceable results in opportunities for industries to continue in neglecting the safety requirements of the workers. Ahasan (2003) commented that workers have almost no power or influence. This results in ergonomics potentially being seen as an expense, and impact on production, rather than a positive influence in protecting the workers. Gurr et al. (1998) and others have described work in these areas as more hazardous, strenuous, and involving a non-ergonomically designed environment placing the workers at greater risk of illness and injury. This is further compounded by living conditions involving low nutrition, as well as the incidents of infection or disease.

51.4   Education in Ergonomics

Various approaches have been trialled in the education of "change agents" to introduce ergonomics to developing countries.

51.4.1 Studying Abroad

There are some individuals from developing countries that have the opportunity to study ergonomics abroad in a developed country. Ahasan (2003) indicated the difficulties related to the sustainability of this approach. More specifically such problems include:

- Obtaining official permission to leave the IDC and study abroad is often difficult.
- The economics associated with studying abroad is generally outside the scope of the majority of workers.
- Due to the political involvement in funding and permission, often persons of senior level are granted permission to study abroad. However, these are not necessarily those individuals who would subsequently be in a position to implement their learnings in a practical environment when they return.
- Many of those who study abroad, stay aboard and do not return to their country of origin.
- Feedback on external programs have indicated that they do not necessarily translate easily back to the local work environment. For example, the NIVA (1993) Stockholm project targeting ergonomics for developing countries, indicated that 82% of the IDC participants agreed that the program, whilst practical in influencing specific work requirements in developing countries, they were not in themselves, an appropriate vehicle to help solve the real problems. This is because the course is conducted away from the local context where many of the social, political, and environmental factors restrict the application of generic learnings.

51.5 Ergonomics Education in IDCs

With the cooperation of external countries and agencies, ergonomics short courses and training programs have been conducted within developing countries. For example, Shahnavaz (2000) has conducted numerous ergonomics courses within developing countries through the University of Lulea, Sweden, during the 1980s and 1990s. These programs have helped to bring the application of ergonomics into these countries and to train people within their own culture and workplace environments.

Shahnavaz (2000) concludes that there is a need to have scientific and technology infrastructure and training facilities to enable ongoing operation, maintenance and development of technology within these countries.

One limitation of this approach, is the availability of qualified educators who have the cultural sensitivities, language, and resources to support and undertake these programs.

Within many of the larger developing countries, these programs tend to be only practical within the major cities thus alienating rural workers where transportation and living expenses restrict their ability to participate.

51.6 Government Inspectors and Union Representatives

Education of government inspectors and union representatives provides the opportunity for communicating ergonomics principles through a wide cross section of industry areas. These groups have the opportunity through their work to visit multiple work places and to have the communication opportunities to explain and implement ergonomics principles at a practical level. Within Asia, Kogi (2004) has described a range of programs where ergonomics had been targeted to this sector. This would be consistent with the targeted "agents of change" and consistent with a more contemporary approach to facilitate wide penetration of the ergonomics knowledge across industry sectors. Within the IEA programs for IDC, Prof. Pat Scott (Rhodes University, South Africa) has been convening and conducting a range of ergonomics workshops within the southern African countries. This has involved expertise from within South Africa, and also using visiting ergonomists from abroad. These programs have assisted in raising the profile of ergonomics within these communities and assisted also in developing case studies and encouragement for local participation.

51.7   Cultural Influences

There are many aspects of culture within developing countries that severely limits the direct transfer of ergonomics research and data from developed countries. For example, the role of women within agricultural communities presents a wide range of ergonomic challenges derived from their cultural functions. O'Neill (2000) described manual handling in the rural informal work sector as primarily being undertaken by women. This involves the manual transfer of goods such as the carrying of crops from the field to the village and to the market. It was estimated that 25 kg of crop could be carried on the head of the women for up to 15 km, two to three times per week. This is primarily due to the condition of tracks and paths that limit the use of load carrying equipment.

Additionally, the women are also involved in the fetching of water and fire wood. Bryceson and Howe (1993) estimated that a typical household could annually carry 220 tonnes per kilometre. This could take up to 4800 h per year involved in the load bearing activities. The use of children within the work force is a primary cultural issue in many developing countries with need of their skill and strength in the crop growing activities. The ILO has a major project concerned with the welfare of children within the work force. From an ergonomics perspective, the majority of the research relating to work place safety is focussed on the needs of adults. There is limited research applicable to children within the work force, particularly undertaking those tasks found in developing countries. There are religious and cultural sensitivities also within many developing countries which direct the expectation and behaviours of population groups. This may range from the traditional methods of undertaking tasks which are communicated and expected to continue through the generations, as well as the expectation that children would necessarily learn from and maintain the lifestyle of their forbearers. The introduction of change within these environments due primarily to ergonomics research, is needing to be sensitive to the implications for these cultural and religious requirements.

51.8   Technology Transfer

Ahasan et al. (1999) studied metal workers in Bangladesh. He found stresses from thermal factors (heat and humidity), poor air quality (dust and fumes), awkward body postures, and noise amongst the ergonomics risk factors. In assessing the control of these work factors, constraints in this country included:

- Negative attitudes and a general acceptance of unsafe work practices
- Poor access to training and information
- The workers in a poverty cycle involving low wages and poor nutrition resulting in acceptance of the working conditions
- Weak trade unions and corruption
- Lack of support to change the working conditions by the international community

Asogwa (1987) looked at working conditions in Nigeria and the seriousness of workplace injuries. He suggested the method of resolving these as follows:

- Development of a positive attitude towards health and safety
- National effort by the governments to improve legislation and enforcement
- Greater international cooperation

Shahnavaz (1992) discussed the processes of technology development in developing countries and indicated that technology should not be regarded as the objective of development, but the principal means of its attainment.

Three basic needs for this technology to be successful in developing countries were identified. These are:

- Economy (to meet the basic needs and to provide equality of work opportunities)
- Ecology (environment and protection)
- Energy (involving use of materials and human resources)

Shahnavaz (1992) advocates a "holistic process approach." This involves a much broader approach than focusing on the transfer of engineering capabilities to the developing countries. This takes into consideration the environment such as the impact on economic, physical, cultural, political, and social factors.

This is referred to as part of the macroergonomic approach towards technology change. This must take into consideration the following factors:

- Organization design
- Management systems
- Communication networks
- Worker participation to optimize the technology systems

In balance, the traditional ergonomics approach with the "microergonomic systems" still needs to be considered including anthropometry, work capacity, functional ability, work environment conditions, as well as cognitive and cultural characteristics of the workforce.

The transfer of technology is as much the responsibility of the supplier, as it is of the recipient. Without appropriate legislation and inspection policies of the government, inappropriate selection and installation of technology may occur resulting in ergonomics hazards for the workers. Shahnavaz (2000) encourages a greater consideration of the cultural values and behaviour patterns between suppliers from developed countries to the IDC recipients to ensure that the successful technology transfer is based on a holistic approach.

O'Neill (2000) discussed the factories and mines that are developing in IDCs as part of the movement from agriculture industries, to secondary industries. That many of the industries involve heavy manual work such as steel manufacturing, mining, textiles, chemicals, as well as building and construction.

Kogi et al. (1999) make the point that the solution to many of the physical ergonomics problems must be simple and reached through a participatory approach. Low-cost solutions are important, but the returns to the employer need to be obvious and quick. This appreciation of "cost benefiting" is a critical element towards the acceptance of ergonomics change within the technology systems. Within agricultural systems, many of the same principles apply. Interventions needs to assist low capacity to apply strength, skills, and ability. For example, the improved design of a agricultural hoe was discussed by O'Neill (2000) as a simple example of applying ergonomics principles to the agricultural industry.

51.9 Future Ergonomic Challenges

Kogi (1995) continues to provide simple practical guidance in advancing ergonomics interventions in developing countries. Amongst the guidelines include:

- Building on local practice to enhance its adaptation based on ergonomics principles
- Looking at multiple needs within a community together, rather than focussing on individual factors
- Focussing on practical solutions that can be quickly and easily implemented and return a "cost benefit"
- Procedures which are adopted must be action orientated
- Participation of the users within the identification of the solution is critical to its sustainability
- Low-cost solutions should be the primary focus to embrace the ergonomics principles

In the area of training and education in developing countries, Kogi (1995) continues to advocate:

- Methods that lead people to immediate action to address the ergonomics risk factors in multiple work aspects
- Programs that are not driven by technical expertise and advise, but primarily focused around the participation of the workers themselves, in identifying the issues and developing the appropriate actions
- Flexible use of support tools, techniques, and training to enable broad application across a variety of work and lifestyle situations

Shahnavaz (1983) commented that working and non working conditions in IDCs are so interrelated that they may be considered as an undivided totality. Consequently, interventions of ergonomics need to embrace the holistic needs of the community and not just those focused primarily around the tasks defined as work related.

51.10 Discussion

It is evident from this review of introducing ergonomics into developing countries that a holistic approach is required that is sensitive to the cultural, religious, social, environment, as well as human-related needs. The role of participation as a primary methodology in identifying the ergonomics issues and developing the resolution requires the approach to training and development to be targeted towards the needs of the primary "change agents." These individuals who have the opportunity to visit and interface with workers within their local work environments. They include people such as government inspectors, union officials, NGO agents, as well as ergonomics consultants and researchers with sensitivities towards the specific needs of the IDC areas.

The development and promotion of simple resources such as the ILO publication "Ergonomic Checkpoints" and associated publications on other targeted areas of applied ergonomics relevant to developing countries form a useful basis for practical based training and education with the "change agents." The ongoing programs of the IEA, including the translation of distance learning programs into languages relevant in Africa and the extension of the basic training materials and case studies, will assist in providing opportunities for further promotion of ergonomics in these countries. The longer-term development of ergonomics within developing countries will continue to depend on collaboration and cooperation between the major stakeholders such as governments, aid agencies, peak organizations such as the United Nations, WHO, ILO, as well as the professional associations, such as the IEA.

It is evident from this review that there are limited resources available within IDCs for the development and application of ergonomics to address the disparate working condition standards compared with IACs. Resourcing of ergonomics programs will require collaborating and cooperation between governments, NGOs, professional associations, and international agencies. The leadership provided by the United Nations, WHO, and the ILO indicates the importance placed on working conditions within developing countries. Whilst the problems appear to be immense, there are significant opportunities for Federated Ergonomics Societies, Universities and Governments in IACs to assist in the developing countries. The Ergonomics Society from The Netherlands has taken an initiative through the IEA of "twinning" to support specific programs within Indonesia. This particularly was targeted to Bali Island after the terrorist bombings in 2003. Direct relationship and support will ensure that ongoing effective communication will develop programs targeted to local needs.

The IEA has introduced the membership of "networks." These primarily are countries or Federated Societies wanting to work together to achieve common goals. Whilst in Europe, the FEES (Federation of European Ergonomic Societies) provides an opportunity for joint communications on European Union related issues. Within Latin America, ULAERGO (Union of Latin American Ergonomic Societies) has provided a vitality to countries within this region to work in a supportive nature together.

These examples of collaboration and cooperation provide the foundation for supporting regions or local areas within developing countries by ergonomics societies from other regions of the world. For example, the

Portuguese Ergonomics Society is translating distance learning materials from English to support the opportunities for learning ergonomics in Portuguese speaking countries in Africa. This IEA initiative further provides evidence that cooperation and collaboration will foster programs of mutual benefit.

The lack of research and legislation within developing countries are areas where ongoing support can be provided. For example, within East Timor, occupational health and safety specialists have had an opportunity to meet with the new Inspectors of the Department of Labour in 2003. Caple (2003) concluded through this meeting that there is limited resources available for industry wide programs and there was no legislation in place at this time to support enforcement of safe working conditions. Tangible support can be provided by developed countries in assisting in the drafting of template legislation and the development of supportive programs of training and resource materials that can be translated into local languages. Caple (2003) noted that expatriate companies wishing to develop industry within these poorer countries were amongst the worst employers from an ergonomics perspective. They were not adopting the same standards of injury prevention within their local businesses, as they would be expected to apply within developed countries. It is therefore important that international agencies working within these areas have the ability to identify such discrepancies and encourage local governments to enforce more stringent standards.

The leaders of ergonomics education within developing countries may come from those individuals who have the opportunity to study aboard. However, the major focus for ergonomics training needs to be provided from within the country itself.

With recognition of the culture, social, religious, and environmental sensitivities, these programs need to be developed in close cooperation between external ergonomists and researchers with the local communities. It is therefore not recommended that the concept of a "traveling roadshow" approach across developing countries would necessarily have any sustainable impact. The presentation of generic ergonomics training is likely to have little relevance to local issues and workplace conditions.

It would be recommended, in turn, that relationships be established between federated members and local communities, through government agencies, NGOs, and IEA federated ergonomics societies to more accurately identify their local needs before the development of assistance programs are implemented. As part of the IDC program within the IEA, it is becoming more relevant to identify programs that can be evaluated from a qualitative and quantitative perspective. Without appropriate evaluation, it is unclear whether the investment of time and resources necessarily achieves the desired outcomes. It will therefore be appropriate that a methodology be developed to include processes of clear goals and objectives, as well as measurable outcomes for future IDC programs.

The adoption of the "holistic process approach" proposed by Shahnavaz (2000) should be seen as an approach of general application. This would enable the assessment of technology transfer programs into IDCs to be planned and evaluated on a broader level than merely the engineering and maintenance ergonomics design issues. This is consistent with the "macroergonomic" area of research that should be adopted as part of the IDC programs. It should also include broader issues such as the effects on the environment and ecology, as well as the economic benefits to the local community. To facilitate this holistic approach, it is necessary for ergonomists to better understand and work more closely with fellow professional groups. These include business leaders, economists, health planners, union groups, as well as other NGO agents. The opportunities are becoming evident that ergonomics as a science is incapable of significant impact in developing countries if working in isolation of other groups who are also endeavouring to address the "poverty cycle" in many of these areas.

Through ongoing collaboration and cooperation between the IEA federated societies and other agencies, the integration of ergonomics within a holistic model can be achieved.

51.11 Conclusions

It is evident from the limited research within IDCs that there are significant problems in directly transferring ergonomics research findings and programs to address the significant workplace issues within their work environments. The confounding factors relating to cultural, social, religious, environment, health, and human expectations within these areas confound the direct application of ergonomics findings from western cultures. The simple example relating to ergonomics requirements of seating at work arising from research in developed countries has little relevance to the general postures of prolonged squatting and sitting on the floor more commonly seen in many of the developing countries.

Whilst many ergonomists have an ethical and moral drive to assist in improving working conditions in developing countries, there is still much to be learnt on the methodologies to achieve this goal. The requirement for development of dialogue and trust between the local communities and stakeholders with the members of the federated societies of the IEA needs to be recognized as an essential step in identifying the range of contributions that can be developed. Through the ongoing cooperation and collaboration that is now becoming evident between the IEA and many government and nongovernment agencies, there are new emerging possibilities of contributing towards the development of ergonomics within IDCs.

References

Asogwa, S.E. (1987). Prevention of accidents and injuries in developing countries. *Ergonomics* 30(2):379–386.

Ahasan, M.R., Mohiuddin, G., Vayrynen, S., Ironkannas, H., and Quddus, R. (1999). Work-related problems in metal handling tasks in Bangladesh: obstacles to the development of safety and health measures. *Ergonomics* 42(2):385–396.

Ahasan, R. (2003). Work-related research, education and training in developing countries. *Work Study* 52(6/7):290–297.

Bryceson, D.F. and Howe, J. (1993). Reducing household transport in Africa: reducing the burden on women? *World Development* 21(11):1715–1728.

Caple, D. (2003). Report on East Timor study tour. Victorian Workcover Authority. Melbourne, Australia.

Gurr, K., Straker, L., and Moore, P. (1998). Cultural hazards in the transfer of ergonomics technology. *International Journal of Industrial Ergonomics* 22(45):397–404.

Kogi, K. (1995). Ergonomics and technology transfer into small and medium-sized enterprises. *Ergonomics* 38(8):1691–1707.

Kogi, K. and Sen, R.N. (1987). Third world ergonomics. *International Reviews of Ergonomics* London: Taylor & Francis.

Kogi, K. (2004). Report to International Ergonomics Association, International Development Committee.

Kogi, K., Kawakami, T., Itani, T., and Batino, J.M. (1999). Impact on productivity of low-cost work improvement in small enterprises in a developing country. In: *Global Ergonomics* Oxford: Elsevier.

Mwaura, W. (1992). Shortfalls in occupational injury and disease data collection in Kenya. *African Newsletter on Occupational Health and Safety* 2(11):21–27.

NIVA — Nordic Institute for Advanced Training in Occupational Health and Safety (1993). Course Report — OHS Research in Developing Countries. Finish Institute of Occupational Health, 1–20.

O'Neill, D.H. (2000). Ergonomics in industrially developing countries: does its application differ from that in industrially advanced countries? *Applied Ergonomics* 31(6):631–640.

Scott, P. (2004). Personal communication, University of Rhodes, South Africa.

Shahnavaz, H. (1983). Ergonomics in developing countries: a conceptual approach. In: *Proceedings of the First International Conference on Ergonomics of Developing Countries, Centre for Ergonomics of Developing Countries*. Lulea, Sweden.

Shahnavaz, H. (1992). Ergonomics and Industrial Development, Impacts of Science and Society, No. 165, Taylor & Francis, London.

Shahnavaz, H. (2000). Role of ergonomics in the transfer of technology to industrially developing countries. *Ergonomics* 43(7):903.

52

The Future of Work and an Ergonomics Response

52.1 Introduction

The world we live in is changing so much and so rapidly that "there is nothing permanent except change," although that statement is attributed to Heraclitus writing about 500 BCE. This applies across society and not just to work, as detailed, for example, by Fukuyama (1999) and Kennedy (2000). The predictions of how this future will affect work have a wide spectrum, from the cautiously optimistic (Friedman, 1999) to the thoroughly pessimistic (Rifkin, 1995; Mander and Goldsmith, 1996). Note that these sources were from the more optimistic 1990s (at least in the United States) where the Cold War had ended, the Internet was expanding explosively and the influence of international terrorism was relatively minor.

This chapter develops the trends noted in the preceding paragraph and uses them to derive predictions about how work will change from an ergonomics/human factors viewpoint. It then examines ways in which ergonomics can (perhaps *Must*!) change to meet new challenges. The genesis of this chapter was the author's work on the relationship between ergonomics and the quality movement (e.g., Drury, 1997), the human factors implications of service quality (Drury, 2003) and specifically the author's commission to examine the future of work as part of the Institute of Medicine/National Research Council (IOM/NRC) report on Musculoskeletal Disorders and the Workplace (NRC, 2001).

52.2 Societal and Workplace Changes

52.2.1 The Societal Background

Fukuyama (1999) argues that the past 30 yr have seen "The Great Disruption" in our world. He primarily identifies this disruption as the information society, which will naturally increase freedom and equality while forcing outmoded bureaucracies, such as the Soviet Union or AT&T to either disappear or reinvent themselves as smaller, more decentralized entities. These changes, in turn, affect the "social capital" as organizations that once claimed our loyalty (the state, the church, the large corporation) begin to lose, and hence cause individuals to lose a sense of kinship. He sees this loss of social capital as the cause of long-term worsening of social indicators such as crime rates, divorce rates, births to single mothers, and measures of interindividual trust. However, more recent trends were positive through the 1990s with research cited by Fukuyama to support the hypothesis that social capital is now mainly generated by much smaller, more decentralized grouping. These include neighborhood societies activist groups, smaller religious groups and even internet groups. Set against this societal backdrop, the (perhaps) decentralized workplace becomes a vital part of the generation of social capital. Ergonomists ignore the social side of work at their peril.

Workforce demographics are the second largest societal change. Figures will be given for the United States, but similar changes are happening around the world. For an excellent view of the future of work from New Zealand, which covers their workforce demographic changes in some detail, see http://www.dol.gov.nz/PDFs/fow-stocktake.pdf. This White Paper is from 2004 but remains unfortunately anonymous.

In the United States the population is rising, probably by 25% from 1990 to 2010 (Stenburg and Coleman, 1993), leading to an increase in the overall labor force. The labor force participation rate was 65.8% at the end of 2003 (Bureau of Labor Statistics, 2004; www.bls.gov/cps/cescps-trends.pdf). Of those employed, the age group 55–64 is the fastest growing, with the median age increasing by 3 yr over the past decade (NRC, 2001). Day (1996) estimates that by 2050, the fraction of the U.S. population classified as "white" will have decreased from about three-quarters to just under one-half. Two thirds of the expected increase in population by 2050 will have come from immigration. Participation rates of women were about 60% in 2003, compared to about 76% for men (Bureau of Labor Statistics, 2004). For African Americans, the figures were considerably higher for women (65%) but slightly lower for men (72%). Finally, participation rates for people with disabilities is low (30% in 1994) but increasing.

From these data, we see that the workforce is getting older, particularly as retirement pension age increases in the United States. The workforce will have more women, more minorities, and more immigrants as well as increasing numbers with disabilities. The trend is towards increased diversity, as in many other aspects of society. Those once excluded because of gender roles, race, country of origin or disability, have struggled to participate and have to some extent succeeded. There are still major discrepancies in type of work, and hence in income, between different groups as documented by Barbara Ehrenreich in her (2001) book, *Nickel and Dimed*. For the ergonomist, the challenge is to design for a more diverse population, not only physically but cognitively and socially.

The third societal change impacting work is the development of technology. Writers for many years have been documenting not only the changes wrought by technology but the increasing rate of technological change. The well-known Moore's law (Moor, 1965) describes a doubling of computing power every 15 months or so, suggesting an increasing rate of technological change. Thus, Peeters' and Pot (1979) plotted time between major discoveries against year of discovery to predict a major disruption early in the current century. The idea of a singularity or disruption in the first half of the 21st century, caused by great increases in computing power and connectedness, has also been proposed by Vinge (1993).

The technological changes mentioned earlier are primarily changes in information technology. Computers and communications technology are general purpose machines (like steam engines, aircraft, and electric motors) that can be used in a wide variety of systems. What Friedman (1999) calls the "democratization of technology" has, through lowered costs and better distribution, put these tools in the hands of much of the world. This information technology explosion over the final decades of the 20th century has both enabled and forced enterprises to change, in much the same way as steam power and the assembly line forced manufacturing to change in earlier times. We now are in a position to examine the impact of these changes on work.

52.2.2 Changes to Work

As we have seen the world has been changing, so it is hardly surprising that work has changed. This is at times driven by changes in the world, and at times an integral part of that change: an expected result in a highly connected system. Examples of mutual effects are:

- The forces of technology that help change the world are the product of work and enterprises.
- Profound changes from globalization are at least partly the result of inexpensive communications and computing power.
- Population migrations are partly driven by the desire for work, leading to high immigration levels in developed countries.
- As work becomes more productive less people provide more output, driving the historic changes from agriculture through manufacturing to service industries.

In this section we examine a number of specific trends that impact the nature of work, and thus provide possible predictions for the future of ergonomics.

52.2.2.1 The Move to Service

Currently, only about a quarter of the workforce is involved in agriculture or manufacturing (Silvestri, 1997). Since the 19th century, the agricultural sector has been losing ground to manufacturing as less people are needed in modern farming methods to produce greatly increased foodstocks. In the 20th century, this change has been superceded by the move from manufacturing to service, and again the total output of manufacturing has increased. In developed countries such as the United States, the total workforce in manufacturing has remained relatively constant while the fraction in this sector has been decreasing. Note that even in manufacturing and agriculture, many of the jobs are in fact service jobs: engineers, finance specialists, information technology (IT) workers, and cleaners (Rust and Metters, 1996).

Service industries have a unique set of characteristics, based on their offering (extended from Grönroos, 1990):

1. Services are more or less intangible
2. Services are often produced and consumed simultaneously
3. Services cannot typically be inventoried
4. Services are activities rather than things
5. To some extent, the customer participates in the production process
6. Service industries are relatively easier to start up than manufacturing and agriculture

The implications for the workplace can be quite profound. Drury (2003) takes up the idea of the customer being highly involved in service production:

> A customer interacts personally, or via communications media, with an agent (human or machine) of the service company, rather than being at the end of a long chain of operations. Thus, many service employees see customers continually, rather than only occasionally. In effect, this interaction between customers and employees is often the essence of the service. The interaction may be needed so that the customer and service agent can jointly define the correct offering.

This customer involvement provides one argument for a human factors approach to service. The customer can be viewed as the operator of a system attempting to achieve a desired result, a typical description of human factors characterization (Chen and Drury, 1999).

52.2.2.2 Technology and the Workplace

Since the writings of E. R. W. F. Crossman and P. M. Fitts in the 1960s, human factors professionals have been predicting that work is moving from the physical to the cognitive, implying that human factors/ergonomics must make similar changes. Malone (2004) makes this a major point of his book on the future of work. He predicts that computer and communication technologies have profoundly altered the nature of work, allowing large but decentralized enterprises to remain agile. He cites the success of the Linux operating system as an example of a product developed by a "loosely-coordinated hierarchy of thousands of volunteer programmers all over the world" (p. 6).

We have seen false claims of the new age of information technology before. In 1994, the NRC in the United States found almost no evidence of productivity improvement arising from the huge IT invest-ments of the previous decades (NRC, 1994). Recently, however, IT is seen as having been the driver of the large productivity improvements of the past decade. In fact, the advent of ubiquitous computing has changed most jobs profoundly. Most office tasks have been taken over by the person requiring the work to be done, using increasingly powerful computers. Thus, typing, formatting, form filling, filing, and budget tracking are all normal functions of most jobs rather than being the province of the "office staff." This has freed such workers to take on roles of administrative assistant rather than typist or file clerk. Similar changes have taken place in manufacturing, where production workers use computer and communications tools for machine setup, quality monitoring, and even production scheduling.

The ergonomic implications are that the whole workflow must be able to understand and use these new tools effectively, and indeed to be able to suggest improvements. Some of these IT tools are quite poorly designed and implemented. For example, Kleiner et al. (1999) showed how most implementations of large Enterprise Resource Planning Systems fell far short of user needs. Eason (1988) gives many examples of the dangers of poorly implemented IT systems. The New Zealand White Paper referenced in Section 52.2.1 noted that these IT changes produced a skill-bias at work, demanding changes to the education and training systems.

A second human factors implication is that, to quote the New Zealand White Paper,

"By making information sharing cheap and easy, IT allows for the decentralization of decision making and flatter management structures, potentially enhancing flexibility and process innovation."

As has been noted by socio-technical systems (STS) researchers for many years (e.g., Taylor and Felten, 1993), the technical parts of the STS profoundly change the overall STS design. Malone (2004) sees new organizational structures emerging with "the participation of people in making the decisions that matter to them."

This sounds encouraging, but we need to ask whether our workforce is willing to spend much effort in such decision-making when other forces on work are leading to a new Taylorism. For example, globaliza-tion of manufacturing has produced endless goods for consumers, but these need to be delivered from their origins (often half a world away) to the consumer. There has been a rise in warehousing and dis-tribution centers as well as an increase in home delivery services. Such services are highly labor intensive, with much physical labor (e.g., Schulte et al., 1994; Krost, 2000) and high potential for musculoskeletal injury. This brings up another ergonomics issue: despite the moves to service industry and the introduc-tion of pervasive computing, the physical aspects of work do not seem to go away. New jobs contain old elements, and even old jobs remain, often as craft industries. Indeed, Samuel R. Delany has noted "today's technology is tomorrow's handicraft," so we should expect nothing less.

52.2.2.3 Globalization

The movement known as globalization is not a single entity or force but rather a resultant of a set of forces acting on society and enterprise. It is enabled by cheap, ubiquitous computing and communi-cations, direct demands from customers and owners of capital for better and faster gratification, and a worldwide demand for work. It is manifested in worldwide supply chains, international capital markets and movement of work to places where it can be performed at least cost. (Note this often implies moving manufacturing to low-wage countries, but can also mean taking advantage of education, skills, and infrastructure of high-wage countries.)

The impact of globalization on work has been discussed in the NRC (2001) study of musculoskeletal disorders and by Drury (2000a,b, 2003). The following discussion is summarized from all three sources, with some additions.

Drury (2003) summarizes the forces of globalization as:

"The globalization of customers, finance and the production of goods and services has been driven by the forces of deregulation, inexpensive transportation and the rapid diffusion of distributed computing (Friedman, 1999). Industry is spreading increasingly across more regions of the world and is shifting towards communications and services. Furthermore, the global capital markets are forcing "creative destruction," that is, the often brutal flow of capital away from enterprises with low shareholder value towards enterprises where the capital will generate the greatest return. Investments are moving rapidly, forcing industries to respond quickly to changing customer demands. We have moved from managerial capitalism of the first part of this century to investor capitalism with more demanding shareholders (e.g., large pension funds) and more information available instantly (Whitman, 1999). Even national governments are being forced by the investment community to reduce their costs and become more open and transparent, which Friedman (1999) calls the "golden straightjacket." Kanter (1995) characterizes these changes as a greater mobility (of capital, people and ideas), a greater simultaneity (of technology or investment information), a greater bypass (of other choices besides large corporations) and a greater pluralism (with smaller headquarters and decentralized decision-making). She advocates that industry "think like a customer.""

Implications for work are numerous. First, there will be continued pressure on all the workforce to be effective and efficient to produce high-quality, low-cost, and on-schedule delivery. These demands may be mutually incompatible, for example, any speed–accuracy trade-off at work will tend to make quality and speed antagonistic. This has already been found for NASA's "Faster, Better, Cheaper" missions where a retrospective study compared successful to failed missions, finding the failures underfunded and on too tight a schedule (Scott, 1999).

A second implication is that the globalization of customers, and their position as drivers of globalization, will see the rise of "mass customization" (Pine, 1993) where customers participate in the design of their product. Jeans manufacturers in the United States will custom-produce jeans based on a customer's measurements. Dell computers and Saturn cars are "built to order" rather than "built for stock." Even the traditional Leica range-find camera can now be built with custom combinations of features known as "Leica *a la carte*." Such mass customization means that assembly of each individual unit of production is potentially unique, although manufacturing organizations will attempt to use mass production methods to achieve economics of scale (e.g., Dell computers) rather and bespoke hand assembly (e.g., Leica cameras). For the workforce, this means having a system (line or cellular manufacturing) where each item can be different, with increased information-processing demands on the operator as well as increased opportunity for error. Conversely, the uniqueness of each item adds at least some variety to monotonous jobs, and could be used to further "product identification" by the operator.

The most prominent aspect of globalization in popular culture is the movement of jobs. Factories that have formed the mainstay of communities close down when the jobs are moved to low wage areas. Whole industries, such as travel agencies, have seen extreme consolidation with the migration to internet booking systems (and the elimination of commissions by airlines seeking shareholder value!). New industries do develop, for example, the United States created more jobs than it lost in the 1990s. However, the new jobs are rarely created in the areas of greatest job loss, and they often require quite different skills from those of the suddenly unemployed. What may be good for the enterprise, and even the world (debatable) may be devastating for the individual and the community. There are large ergonomic implications for training and retraining, as well as design of newer, IT-based jobs to perhaps conform to older stereotypes. The re-employment of displaced workers may merely add to the other sources of workforce diversity noted earlier.

52.2.2.4 Working Times

There have been historical changes in the total work week, and even work year. These are still continuing with some evidence (Schor, 1991) that hours of work per year are in fact increasing in the United States. However, the Bureau of Labor Statistics data (Rones et al., 1997) found little change in mean hours of work from 1976 to 1993, but an increase in the share of persons who are working very long workweeks (48 h plus). This study did find a steady increase in the hours worked *per year*, from 1805 to 1905 hours for men and from 1293 to 1526 for women. Harrington (2001) reviews evidence for deleterious effects of both long working hours and shift work, conducting shift work in particular, has effects on performance, sleep patterns, accident rate, mental health, and cardiovascular mortality. He notes that in Europe one in five workers is engaged in night work while one in 20 work more than 48 h per week. A union background paper (ACTU, 2003, based on Harcourt and Kenna, 1997) suggests that in Australia "almost one third of full-time employees work more than 48 h per week, and a third of these work more than 60 h per week." There has also been an increase in the number of people working less than full time: the unemployed, the underemployed and those in nontraditional employment such as subcontracted work. Jacobs and Gerson (1998, 2004) refer to this, coupled with the increase in the number of people working very long hours as a "bifurcation of working time." People's perceptions of their working hours tend toward the nominal 40-h week: those working less desire more while those working more desire less. Goldenhar et al. (2003) call this the "Goldilocks Hypothesis." We should also note that in many parts of the world the unemployment rate is not the 5–10% typical of developed economics but can be 50–80%, especially where war or famine have disrupted the economy. These conditions show that for some countries, the future of work does not include much of their population.

The overall picture is one of more hours worked per year and more shift work, particularly as worldwide financial transactions and e-commerce happen around the clock. For ergonomists, the implications are that a smaller fraction of the workforce work a traditional week, with larger numbers of part time workers (who may not maintain the skills of full timers) and of those with excessive work weeks (who may suffer from increased error rates, injury and illness due to accumulated fatigue). Again, the diversity of the workforce is seen to be increasing, with implications for job design. For example, an aviation maintenance technician (AMT) may start a shift at 23:00, when aircraft finish passenger service, work 8 h and be asked to work another 8 h as more maintenance problems than expected were discovered. This AMT, already fatigued, may have to interact with several part-time personnel during this single work period, giving increased opportunity for communications error (e.g., Taylor and Christiansen, 1998).

52.3 Managing Change in Ergonomics/Human Factors

So far we have seen a number of rather large changes to the world and to work. Along the way we have noted potential implications for human factors/ergonomics, but now we need to consider our total response. This must necessarily be at two levels: the profession and the enterprise.

52.3.1 Response at the Profession Level

In the NRC (2001) report on musculoskeletal disorders, an earlier review of the future of work led to implications for the ergonomics profession brought about by potential changes to musculoskeletal disorder incidence. There the approach taken was to examine each of several trends, for example, the rise of distribution centers based on globalized production, and determine whether it would increase or decrease the potential for injury. This approach was extended in Drury (2000a,b) to consider broader implications beyond musculoskeletal disorders, for example, quality, STS.

The chapter on Future of Work in NRC (2001) had as its main conclusion an increase in diversity of both jobs and workers. This will arise from nontraditional work (shifts, long hours, part time) and nontraditional workers (minorities, immigrants, disabilities) so the ergonomist's task of fitting the job and worker together will increase in complexity. There is a greater a priori probability of task demands exceeding worker capabilities as the variance of both increases. Work may change qualitatively as more workers are aided by computer systems, and more workers are exposed directly to customers, via service industries and mass customization. As work is outsourced, then more of the enterprise effort will be devoted to managing long supply chains, creating potential for time urgency of coordination.

These changes can only be responded to at the level of the profession, for example the International Ergonomics Association, national societies and professional certification organizations. Other, more local, responses are also needed.

52.3.2 Response at the Enterprise Level

We have seen how changes in society and the workplace drive changes in ergonomics, but so far only concentrated on changes at the level of the whole profession: where should ergonomics/human factors emphasis go in the future? As ergonomists we do not always make our major contribution at the level of the whole profession, rather we manage our own human factors/ergonomics group in an enterprise, an educational institution, or a government department. The challenge at this level is to make changes at a strategic level within our own organizations so as to position our activities sensibly to meet (anticipated) future demands. I will concentrate on the enterprise ergonomics group to illustrate a way of planning strategically for the future.

An enterprise exists to produce an "offering" (i.e., product or service) for a customer (we will ignore its needs for self-preservation and generation of shareholder value in this section). The maturity of the offering will place the enterprise at a specific maturity stage:

Introduction stage: innovation matters
Maturity stage: quality and features matter
Commodity stage: price and delivery matter

The stage of maturity drives the emphasis on the product from innovation, through design and quality to commodity manufacturing concerns. Ergonomists need to provide different product support at each stage, from product design through quality/error analysis to production and delivery efficiency. To be able to plan, and then provide, this support, the human factors function needs to be aware of the strategic business plans of the enterprise. As an ergonomist, have you read the annual report, strategic plan, and business plan of your organization? At a more advance level, were you involved in writing the plans? Ergonomics is not just a reactive function.

These business plans are typically manifested in projects at the ergonomics level, either actual projects tasked by the management (e.g., installing a new manufacturing line or developing a new version of a product) or potential projects seen by the ergonomist as necessary even though not yet planned by management. Both these "top down" and "bottom up" projects need to be addressed with the resources of the ergonomics function. They are the task demands typical of any ergonomic analysis, and clearly imply that we need to compare them with human capabilities.

Capabilities of a human factors function comprise those of ergonomists within that function (e.g., human factors department) but may also include others in the organization. For example, design staff may have knowledge of anthropometry or human–computer interaction (HCI), while industrial engineers may have capabilities in manual materials handling analysis, simulation, or flowcharting. Similarly, the human resources function may hold expertise in self-directed teams, or (less likely) sociotechnical systems analysis. An inventory of human factors capabilities across the organization is well worth making and keeping updated.

After listing these demands and capabilities, the obvious matching comes from a matrix where we have the demands (projects) as rows and the capabilities as columns. Table 52.1 shows an example of such a table where some projects are likely to come from management (e.g., NIOSH lifting analysis for heavier products) while others are generated bottom up (e.g., sociotechnical systems design for new assembly system). With such a matrix it is possible to plan ahead for the activities of the ergonomics function, matching future workloads, or even hiring new ergonomists to fit any missing capabilities required to meet future demands. The table also shows underused capabilities as empty columns. These can be seen as opportunities for intervention or as an indication that retraining will be needed to match capabilities and demands in the future.

Ergonomists typically perform projects and, like other workers, appear to be seeing increased workload. What if a matrix such as Table 52.1 reveals that the total workload to meet all demands exceeds

TABLE 52.1   Demand/Capability Matrix for Human Factors/Ergonomics Function

Enterprise Demands	HCI Design	Physical Workplace Design	MMH Injury Reduction	Decision Analysis	...	Sociotechnical Systems
New assembly system		Anthropo-metry project				Design for self-regulating teams
Heavier new products			NIOSH lifting analysis			
Software to be interagted with existing hardware	Integrate HCI with hardware					
New process control system for product A				Analyze decisions required		
⋮						

the human factors resources available? We have a choice of increasing the available ergonomic workforce or choosing less projects than are available. For reasons of enterprise agility noted earlier, it may not be possible to increase staff size: also staff size increases take time. There is always the possibility of outsourcing, for example to a consulting organization, or of taking on temporary help, from the same source or from student interns. The problem with these alternatives is that even the best ergonomists need time to learn an organization: its products, its process, its people. Often the only available option is to choose among the projects.

As Ashby (1956) noted many years ago, design and choice are the same concept. In designing a system we make choices from available alternatives: the choices we make determine the final design and its properties. Thus, making choices among projects means that we are "designing" the human factors function and to some extent the organization. We can adopt a passive approach to design, essentially taking on all of the projects given by management until our total capacity is reached, after which all projects are refused. We have plenty of excuses in an organization for adopting a passive approach. It is certainly easier bureaucratically; "we are just doing our job," and so on. However, such an approach denies leadership roles to ergonomists, even when they have legitimate technical (and even interpersonal) capabilities to help lead an organization in an increasingly competitive world. Active choice means both seeing new project possibilities (e.g., from Table 52.1) and deliberately choosing which projects to perform and which to (politely) decline. Such a course will mean a constant designing or redesigning of the ergonomics function, but will help contribute the long term agility of the enterprise. To borrow another concept from Ashby, if we are to help control the enterprise's vital parameters (e.g., by keeping profits and injuries within desired bounds) then our control system must obey the law of requisite variety. The ergonomics function must have at least as much variety, that is, different responses, as the changing organizational demands. Using a third concept from Ashby (1956), looking to the future by anticipating enterprise needs provides an anticipatory behavior that can help stabilize a system with natural time lags. Any function within an organization needs to do this, but it is rarely found in the job objectives or mission statements of the ergonomics function.

References

ACTU (2003). Working Hours and Work Intensification Background Paper, http://www.actu.asn.au/congress2003/papers/workinghoursbp.html.

Ashby, W.R. (1956). *An Introduction to Cybernetics*, Chapman and Hall, London.

Bureau of Labor Statistics (2004). www.bls.gov/cps/cescps-trends.pdf.

Chen, A.-J. and Drury, C.G. (1999). Human errors and customer service quality. In: *Proceedings of the International Conference on TQM and Human Factors — Towards Successful Integration*, Linkoping, Sweden: Linkoping University and Institute of Technology, 76–81.

Day, J.C. (1996). *Population Projections of the United States by Age, Sex, Race, and Hispanic Origin: 1995 to 2050*. U.S. Bureau of the Census, Current Population Reports, P25-1130, U.S. Government Printing Office, Washington, DC. February 1996.

Drury, C.G. (1997). Ergonomics and the quality movement (The Ergonomics Society 1996 Lecture). *Ergonomics*, 40(3), 249–264.

Drury, C.G. (2000a). External forces on the work system: a framework for human factors implications. In: *Proceedings of the IEA 2000/HFES2000 Congress*, San Diego, CA, August 2000, 2-758–761.

Drury, C.G. (2000b). Quality, globalization and the future of work. In: *Proceedings of the IEA 2000/HFES2000 Congress*, San Diego, CA, August 2000, 2-459–462.

Drury, C.G. (2003). Service quality and human factors. *AI and Society*, 17(2), 78–96.

Eason, K. (1988). *Information Technology and Organizational Change*. London: Taylor & Francis.

Ehrenreich, B. (2001). *Nickel and Dimed*. New York: Henry Holt and Company.

Friedman, T.L. (1999). *The Lexus and the Olive Tree: Understanding Globalization*. New York: Farrar, Straus and Giroux.

Fukuyama, F. (1999). *The Great Disruption: Human Nature and the Reconstitution of Social Order*. New York: The Free Press.

Goldenhar, L,M., Hecker, S., Moir, S., and Rosencrarce, J. (2003). "The Goldilocks Model" of overtime in construction: not too much, not too little, but just right. *Journal of Safety Research*, 34(2), 215–226.

Grönroos, C. (1990). *Service Management and Marketing*. MA: Lexington Books.

Harcourt, T. and Kenna, S. (1997). Discussion Paper on Working Time, http://www.actu.asn.au/public/papers/wtimedpd39of97.html.

Harrington, J.M. (2001). Health effects of shift work and extended hours of work. *Occupational and Environmental Medicine*, 58, 68–72.

Jacobs, J.A. and Gerson, K. (1998). Who are the overworked Americans? *Review of Social Economy*, 56(4), 442–459.

Jacobs, J.A. and Gerson, K. (2004). The *Time Divide: Work, Family, and Gender Inequality*, Cambridge, MA: Harvard University Press.

Kanter, E.M. (1995). *World Class: Thriving Locally in the Global Economy*. New York: Simon & Schuster.

Karasek, R. and Theorell, T. (1991). *Healthy Work*. New York: Basic Books.

Kennedy, A.A., (2000). *The End of Shareholder Value: Corporations at the Crossroads*. Cambridge, MA: Perseus Publishing.

Kleiner, B.M., Bishu, R.R., Drury, C.G., Nair, M., Getty, R., and Muralidhar, A. (1999). Human factors implications of enterprise resource planning (ERP) solutions. In: *Proceedings of the Human Factors and Ergonomics Society 43rd Annual Meeting*, Houston, TX, 675–677.

Krost, M. (2000). An aggressive lift training program at a grocery distribution center yields positive results. In: *Proceedings of the Human Factors and Ergonomics Society/International Ergonomics Association Congress*, San Diego, CA.

Malone, T.W. (2004). *The Future of Work: How the New Order of Business Will Shape Your Organization, Your Management Style and Your Life*. Boston, MA: Harvard Business School Press.

Mander, J. and Goldsmith, E. (1996). *The Case Against the Global Economy*. San Francisco, CA: Sierra Club Books.

Moore, G.E. (1965). Cramming more components onto integrated circuits. *Electronics*, 38.8.

NRC (1994). *Organizational Linkages: Understanding the Productivity Paradox*. D.H. Harris (ed.), Panel on Organizational Linkages, Committee on Human Factors. Commission on Behavioral and Social Sciences and Education, Washington, DC: National Academy Press.

NRC (2001). *Musculoskeletal Disorders and the Workplace*, Washington, DC: National Academy Press.

Peeters, M.H.H. and Pot, F.D., (1993). Integral organizational innovation in the Dutch clothing industry: the myth of new production systems. *International Journal of Human Factors in Manufacturing*, 3, 275–292.

Pine, B J. (1993). *Mass Customization — The New Frontier in Business Competition*. Boston, MA: Harvard Business School Press.

Rifkin, J. (1995). *The End of Work*. New York: Putnam.

Rones, P.L., Llg, R.E., and Gardner, J.M. (1997). Trends in hours of work since the mid-1970s. *Monthly Labor Review*, 3–14.

Rust, R.T. and Metters, R. (1996). Mathematical models of service. *European Journal of Operational Research*, 91, 427–439.

Schor, J.B. (1991). *The Overworked American*. New York: Basic Books.

Schulte, M., Ge, S., and Hertting-Thomasius, R. (1994). An analysis of the subjective stain perceived by parcel delivery. *International Journal of Industrial Ergonomics*, 13(1), 15–23.

Scott, W.B. (1999). Mars probes: did 'cheaper' and 'faster' preclude 'better'? *Aviation Week and Space Technology*, December 20/27, 11–12.

Silvestri, G.T. (1997). Occupational employment projections to 2006. *Monthly Labor Review*. November, 58–83.

Stenberg, C.W. III and Colman, W.G. (1993). *American's Future Work Force: A Health and Education Policy Issues Handbook*. Westport, CT: Greenwood Press.

Taylor, J.C. and Christensen, T.D. (1998). *Airline Resource Management: Improving Communication*. Warrendale, PA: Society of Automotive Engineers, Inc.

Taylor, J.C. and Felten, D.F. (1993). *Performance by Design*. Prentice Hall. Englewood Cliffs, NJ.

Vinge, V. (1993). Technological singularity. *Whole Earth Review*, 18.

Whitman, M. (1999). Global competition and the changing role of the American Corporation. *The Washington Quarterly*, 22.2, 59–82.

Index

Q